Advanced Transport Phenomena

An integrated, modern approach to transport phenomena for graduate students, featuring traditional and contemporary examples to demonstrate the diverse practical applications of the theory. Written in an easy-to-follow style, the basic principles of transport phenomena and model building are recapped in Chapters 1 and 2 before progressing logically through more advanced topics including physicochemical principles behind transport models. Treatments of numerical, analytical, and computational solutions are presented side-by-side, often with sample code in MATLAB, to aid students' understanding and develop their confidence in using computational skills to solve real-world problems.

Learning objectives and mathematical prerequisites at the beginning of chapters orient students to what is required in the chapter, and summaries and over 400 end-of-chapter problems help them retain the key points and check their understanding. Online supplementary material including solutions to problems for instructors, supplementary reading material, sample computer codes, and case studies completes the package (available at www.cambridge.org/ramachandran).

P. A. Ramachandran is a Professor in the Department of Energy, Environment, and Chemical Engineering at Washington University, St. Louis. He has extensive teaching experience, mainly in transport phenomena, mathematical methods, and chemical reaction engineering, and he has also held many visiting appointments at various international institutions. He has written or co-written two previous books, as well as over 200 journal articles in which he has pioneered many new concepts and computational tools for the modeling of chemical reactors. He is a recipient of the Moulton Medal from the Institution of Chemical Engineers, UK, the NASA certificate of recognition, USA, and the NEERI award from the Institution of Chemical Engineers, India.

"Anyone who teaches transport phenomena will treasure this book because it provides an integrated approach to help students better understand the core theories through both traditional and contemporary examples of transport phenomena problems, along with side-by-side presentations of both analytical and numerical methods and sample MATLAB codes – the long-awaited, all-in-one solution."

Roger Lo
California State University

Advanced Transport Phenomena

ANALYSIS, MODELING, AND COMPUTATIONS

P. A. RAMACHANDRAN

Washington University, St. Louis

CAMBRIDGE
UNIVERSITY PRESS

CAMBRIDGE
UNIVERSITY PRESS

University Printing House, Cambridge CB2 8BS, United Kingdom

Cambridge University Press is part of the University of Cambridge.

It furthers the University's mission by disseminating knowledge in the pursuit of
education, learning, and research at the highest international levels of excellence.

www.cambridge.org
Information on this title: www.cambridge.org/9780521762618

© P. A. Ramachandran 2014

First published 2014

Printed in the United Kingdom by TJ International Ltd. Padstow Cornwall

A catalog record for this publication is available from the British Library

Library of Congress Cataloging in Publication data

Ramachandran, P. A., author.
Transport phenomena : analysis, modeling and computations / P.A.
Ramachandran.
pages   cm
ISBN 978-0-521-76261-8 (Hardback)
1. Transport theory–Textbooks.   I. Title.
TP156.T7R36 2014
530.13′8–dc23   2014014317

ISBN 978-0-521-76261-8 Hardback

Additional resources for this publication at www.cambridge.org/ramachandran

Cover illustration: close-up of ink spreading in water mixed with droplets of paint. This image was created by
dropping a small amount of oil- and xylene-based gold paint onto the surface of colored water. Different inks
were then dropped onto the gold paint until the weight of the ink caused the gold paint to dip and allowed the
inks to burst into the water. The complex colors, shapes, and patterns are a result of varying levels of flow rate,
ink density, and surface tension. This creates eddies and vortices that affect the way light is reflected from the
surface. Image © Perry Burge/Science Photo Library.

CONTENTS

PREFACE

The analysis, modeling, and computation of processes involving the transport of heat, mass, and momentum (transport phenomena) play a central role in engineering education and practice. The study of this subject originated in the field of chemical engineering but is now an integral part of most engineering curricula, for example, in biological, biomedical, chemical, environmental, mechanical, and metallurgical engineering both at undergraduate and at graduate level. There are many textbooks in this area, with varying levels of treatment from introductory to advanced, all of which are useful to students at various levels. However, my teaching experience over thirty years has convinced me that there is a need for a book that develops the subject of transport phenomena in an integrated manner with an easy-to-follow style of presentation. A book of this nature should ideally combine theory and problem formulation with mathematical and computational tools. It should illustrate the usefulness of the field with regard to practical problems and model development. This is the primary motivation for writing this book. This comprehensive textbook is intended mainly as a graduate-level text in a modern engineering curriculum, but parts of it are also useful for an advanced senior undergraduate class. Students studying this book will understand the methodology of modeling transport processes, along with the fundamentals and governing differential equations. They will develop an ability to think through a given physical problem and cast an appropriate model for the system. They will also become aware of the common analytical and numerical methods to solve these models, and develop a feel for the diverse technological areas where these concepts can be used.

Goals and outcome

The book is written with the objective that students finishing a first-year-level graduate course in this field should acquire the following skills and knowledge.

- **Fundamentals and basic understanding** of the phenomena and the governing differential equations. They should develop an ability to analyze a given physical problem and cast an appropriate model for the system. They should be exposed to the philosophy of the modeling process and appreciate the various levels at which models can be developed, and the interconnection and parameter requirements of various models.

- **Analytical and numerical skills** to solve these problems. They should develop the capability to solve some of the transport problems in a purely analytical setting and also expand their capability using numerical methods with some common software or programming tools. Often solving the same problem by both methods reinforces the physics and speeds up the learning process.

- **An understanding of technological areas** where transport models are useful. Students should develop an understanding of the diverse range of applications of this subject and understand how the basic theory, models, and computations can be used in practical applications.

To achieve these goals the book focuses on analysis and model development of transport process in detail, starting from the very basics. It illustrates the solution methods by using the classical analytical tools as well as some common computational tools. The application of the theory is demonstrated with numerous illustrative problems; some sample numerical codes are provided for some problems to facilitate learning and the development of problem-solving abilities. References to many areas of application are provided, and some case-study problems are included.

Intended audience

The level and the sequence of presentation are such that the book is suitable for a first-level graduate course or a comprehensive advanced undergraduate course. In a modern graduate engineering curriculum, the entering students often have diverse backgrounds, and some graduate students might not have taken introductory undergraduate courses in transport phenomena. The introductory part of the book presented in the first two chapters is expected to bring these students up to speed.

Style and scope

The style of presentation is informal, and has more of a "classroom" conversational tone rather than being heavy scholarly writing. Each chapter starts off with clearly defined learning objectives and ends with a summary of "must-know" things that should have been mastered from that chapter. Computer simulations are also illustrated, together with analytical solutions. Often solutions to the same problem obtained by both analytical and numerical methods are shown. This helps the students to validate and benchmark their solutions, and to develop confidence in their computational skills. Also sample packages are included to accelerate the application of computer-aided problem solving in the classroom. These sample codes are presented in separate subsections or are boxed off for easier reading of the main text. Key equations are shown in boxes for easy reference. Case studies are given in several chapters, although the space limitation prohibits an extensive discussion of these applications. Additional material and computer codes will be posted on the accompanying website, which is being developed as supplementary material. This web-based material can be viewed as a living and evolving component of the book.

For instructors

Instructors will find the presentations novel and interesting and will be able to motivate the students to appreciate the beauty in the integrated structure of the field. They will also find the worked examples and exercise problems useful to amplify the class lectures and illustrate the theory. Also the mathematical prerequisites listed at the beginning of each chapter will help the instructor to adjust the lecture content according to the students' mathematical preparedness. Additional web-based material that will aid the teaching of these necessary mathematical tools in a concise manner is being planned.

The book has more material than can be covered in one semester, and it can be used in the following manner in teaching.

- For an integrated course for students entering a modern graduate program with **diverse** undergraduate background, Chapters 1–13 can be covered at a reasonable pace in a one-semester course with some reading materials assigned from the other chapters.

- For a course focused mainly on flow problems Chapters 3–6 followed by Chapters 14–17 will provide a nice one-semester textbook.

- For a course focused mainly on heat and mass transfer the course can start with Chapters 7–13 and end with Chapters 18–22.

Distinguishing features

The book provides an integrated approach to the field. Theory is illustrated with many worked examples and case-study problems are indicated. The book also discusses many important and practically relevant topics that are not adequately covered in many earlier books. Some novel topics and features of the current book are indicated below.

- Discussion on multiscale modeling, model reduction by averaging and "information" flow.
- Solution of illustrative problems by both numerical and analytical methods.
- Sample codes in MATLAB for help in the development of numerical problem-solving skills.
- Detailed analysis of coupled transport problems.
- Introduction to non-Newtonian flow, microfluid analysis, and magnetohydrodynamics.
- Introduction to perturbation, bifurcation, and stability analysis.
- Detailed discussion on analysis of transport with chemical reaction.
- Detailed analysis of multicomponent diffusion with many worked examples.
- A full chapter on electrochemical systems and ionic transport.
- Application examples drawn from a wide range of areas and some suggested case-study problems.

Acknowledgement

Washington University, St. Louis, provided me with an academic home, and I wish to express my gratitude. Many summers of being visiting professor at Kasetsart University,

Bangkok, helped me to teach and fine-tune many topics. I would like to mention my appreciation of my *alma mater*, ICT, Mumbai, formerly known as UDCT. In a significant manner, I have been beneficiary of the rigorous and often disciplinary system of education in India, starting from my elementary school and continuing all the way to UDCT. I would like to acknowledge my many mentors and colleagues, too numerous to thank individually, from whom I have benefited throughout my career. Most of all I would like to thank all my students. My real education started with them, and still continues.

I would like to express my appreciation of my immediate family in the USA, Nima, Josh, Gabe, and Maya, and my brothers, sisters, and sisters-in-law in India for all their support and encouragement. I would like to express my appreciation of my friends in University City, Missouri, and to thank Dawn, who stressed the importance of diet and nutrition when training for a marathon.

On the editorial side, many thanks are due to Cambridge editors and especially to Claire Eudall, who provided valuable advice on the style and structure of various chapters. Also I appreciate the help of Ramesh Prajapati for the preparation of many figures in the text.

TOPICAL OUTLINE

The topical organization of this book is as follows.

Chapter 1 is the basic introductory material which illustrates the richness of the subject, spanning applications to a wide range of problems in science and engineering. This chapter also provides the introduction to the basics of model building and shows the relationships among models of various levels of hierarchy. The basic vocabulary is introduced, and the physical properties needed in transport problems are discussed. The link between continuum and molecular models is indicated. The chapter concludes with a brief note on the historical development of the subject.

Chapter 2 illustrates the formulation of model equations for many common transport problems using a basic control-volume-balance type of approach. All three modes of transport are illustrated so that the student can grasp the similarities. Some "standard" problems are illustrated. This chapter is written assuming no significant earlier background knowledge in this field, and is therefore useful to bring such students up to speed.

The next few chapters, Chapters 3–6, provide the detailed framework for the analysis of momentum transport problems. The kinematics of flow are reviewed in Chapter 3, while the kinetics of flow are discussed in Chapter 4, leading to the derivation of the differential equations for the stress field and the velocity field in Chapter 5. Solutions to illustrative flow problems are then reviewed in Chapter 6, and here some "standard" flow problems shown in Chapters 1 and 2 are revisited in a more general setting, and solutions to some additional complex problems are reviewed. Flows involving non-Newtonian fluids and magnetohydrodynamics are also treated briefly, since they find extensive applications in practice and it is necessary to expose the student to these topics.

Chapters 7 and 8 deal with the differential equations for energy transport and the temperature field, with many illustrative heat-transfer problems in Chapter 8. Similarly, Chapters 9 and 10 deal with differential equations for mass transport and illustrative applications. These chapters bring out the close analogy and common problem-solving strategies for these two transport processes. In the heat-transfer context entropy balance is introduced in a simple manner and the relation to the second law is pointed out in a succinct manner. In the mass-transfer context several important topics such as gas–liquid reactions, membrane transport, and dispersion are presented in detail. Numerical methods involving MATLAB for both ODE and PDE are presented. Sample codes are provided as examples, and side-by-side comparisons with analytical solutions are provided for many problems, so that the students can benchmark their results. The transient problems for both heat and mass are then analysed in Chapter 11 in a unified setting, while some convective transport problems are studied in Chapter 12.

Chapter 13 provides an analysis of a number of coupled problems, for example natural convection, simultaneous heat and mass transfer, condensation, fog formation, and temperature effects in porous catalysts.

Chapter 14 develops some tools to analyze transport problems in further detail. The dimensionless analysis is revisited using novel matrix-algebra-based methods. The concept of scaling and pertubation methods is introduced together with many applications. The scaling tools also provide the necessary background to the boundary-layer flows discussed in Chapter 15. Chapter 15 also discusses additional topics in fluid mechanics such as low-Reynolds-number flow and irrotational flows. Chapter 16 deals with bifurcation and stability analysis. Chapter 17 provides an introductory treatment of turbulent flows.

Chapters 18 and 19 deal with additional topics in heat transfer, including convection in turbulent flow, boiling, condensation, and radiation heat transfer (Chapter 19). The final three chapters (Chapters 20–22) discuss some topics in mass transfer, including more discussion on convective transport and axial dispersion (Chapter 20), multicomponent systems (Chapter 21), and transport of charged species (Chapter 22).

NOTATION

a_w	activity of water or solute indicated in the subscript in Section 10.7
A	area of cross-section for flow
A	Arrhenius pre-exponential factor in Section 8.3.2
A	amplitude of surface temperature oscillation in Section 11.10, K
A_1, A_2	usually integration constants
A_p	projected area of solid in the direction of flow
Ar	Archimedes number
B	dimensionless parameter defined as $(L/R)Pe$ in Section 12.4
Bi_G	Biot number in gas–liquid mass transfer, $k_G H_A/k_L$
Bi_h	Biot number for heat transfer, hL_{ref}/k_{solid}
Bi_m	Biot number for mass transfer, $k_m L_{ref}/D$
Bo	Bond number
Br	Brinkman number for viscous production of heat, Eq. (13.23)
C	total molar concentration of a multicomponent mixture, mol/m^3
Ca	capillary number, $\mu v_{ref}/\sigma$
C_A	local concentration of species indicated in the subscript (A here), mol/m^3
C_A^*	concentration of A in liquid if in equilibrium with the bulk gas (Section 10.5)
$\langle C_A \rangle$	cross-sectionally averaged concentration
C_{Ab}	concentration of species A indicated in the bulk phase, mol/m^3
C_{Ab}	cup mixed average concentration of species A, Section 12.4
C_{Ai}	concentration of species A at the interface, mol/m^3
$C_{A,i}$	inlet concentration of species A for flow reactor, Chapter 2, mol/m^3
C_{As}	concentration of species A at a solid surface, mol/m^3
C_{AG}	Concentration of species indicated in the subscript in the bulk gas, mol/m^3
C_{AL}	Concentration of species indicated in the subscript in the bulk liquid, mol/m^3
C_{AL}^*	hypothetical concentration of A if in equilibrium with the bulk gas, mol/m^3
$C_{A,e}$	concentration of species indicated in the subscript exit
C_b	cup mixed (flow) average concentration of species A, Section 20.5
C_{BL}	concentration of liquid-phase reactant in bulk liquid in Section 10.5
c	molecular speed in Chapter 1 (kinetic theory)
$\bar{c}$	average molecular speed in Chapter 1 (kinetic theory)
$\bar{c^2}$	average of the squares of the molecular speed in Chapter 1 (kinetic theory)
c	speed of sound in Chapter 2
c	speed of light in radiation heat transfer in Chapter 19
c_A	dimensionless concentration of species indicated in the subscript (A here), C_A/C_{ref}
$\bar{c}$	average speed of molecules in Section 1.8.1
C_D	drag coefficient
C_L	lift coefficient
c_p	specific heat of a species, mass basis, under constant-pressure conditions, J/kg · K

C_p	specific heat of a species, mole basis, J/mol · K
c_v	specific heat of a species, mass basis, under constant-volume conditions, J/kg · K
d	diameter of the molecules treated as rigid spheres in Section 1.8.1
d, d_t	diameter of a tube or pipe
d_I	impeller or pump diameter, Sections 14.1.5 and 14.1.6
d_P	particle or solid diameter
D_e	effective diffusivity of a species in a heterogeneous medium
D_i	molecular diffusivity of species i
D_K	Knudsen diffusion coefficient for small pores
D_t	turbulent mass diffusivity, m^2/s
Da	Damköhler number Vk/Q
e	charge on an electron in Chapter 22
e	pipe roughness parameter in Sections 5.5 and 17.6.1
e	total energy content per unit mass
e_x	unit vector in the x-direction
$\boldsymbol{E}$	electric field
E^2	operator defined by Eq. (3.53) or Eq. (3.55)
E^4	Stokes operator defined as $E^2 E^2$
E	emissive power of a gray body
E_b	emissive power of a black body, W/m^2
E_{bk}	emissive power of a black body from surface k, W/m^2
$E_{b\lambda}$	spectral emissive power, W/m^2 nm
$\tilde{E}$	rate-of-strain tensor
f	dimensionless streamfunction in boundary-layer flow
f	Fanning friction factor
F_{ik}	radiation view factor, surface i to k
F	Faraday constant = 96 485 C/mol
$\boldsymbol{F}$	force acting on a control volume
$\mathcal{F}, \mathcal{F}_m$	correction factor for unidirectional mass transfer, Sections 10.1 and 20.2.1
$\mathcal{F}_h$	augmentation factor for heat transfer due to blowing
g	acceleration due to gravity
g_s	rate of production of entropy per unit volume, $W/K \cdot m^3$
G	pressure-drop parameter defined as $-dP/dx$
$\dot{G}$	superficial gas velocity, $kg/m^2 \cdot s$
Gr	Grashof number
$\hat{h}$	enthalpy per unit mass
h	heat transfer coefficient (usually from solid to fluid), $W/m^2 \cdot k$
h	elevation or height from a datum plane for flow problems
h	Planck's constant in radiation chapter, 6.6208×10^{-34} J · s
h_f	head loss due to friction
h_G	heat transfer coefficient in the gas film
h_L	heat transfer coefficient in the liquid film
$\hat{h}_{gl}$	heat released on condensation of a species, J/kg
$\hat{h}_{lg}$	heat of vaporization, J/kg
$\hat{h}_{sl}$	heat needed for melting a solid, J/kg
H_A	Henry's-law constant for solubility of A defined by $P_A = H_A C_A$, $Pa\,m^3/mol$

Ha	Hartmann number
Ha	Hatta number for gas–liquid reactions
i	current density in Chapter 22, A/m^2
i	square root of -1 in Section 11.11
I	intensity of radiation, W/m^2
j_A	mass diffusion flux of A (mass reference), $kg/m^2 \cdot s$
J_A	molar diffusion flux of A (mole reference), $mol/m^2 \cdot s$
J_k	radiosity of a surface in radiation, W/m^2
k_G	mass transfer coefficient from gas to interface (partial pressure driving force), $mol/Pa \cdot m^2 \cdot s$
k_L	mass transfer coefficient from an interface to bulk liquid (concentration driving force), m/s
k	thermal conductivity of a species, subscript l for liquid, g for gas, s for solid, $W/m \cdot K$
k	turbulent kinetic energy per unit mass, m^2/s^2
k	rate constant for reaction, general
k_B	Boltzmann constant, 1.38×10^{-23} J/K
k_0	rate constant for a zeroth-order reaction, $mol/m^3 \cdot s$
k_1	rate constant for a first-order reaction, 1/s
k_2	rate constant for a second-order reaction, $m^3/mol \cdot s$
k_m	mass transfer coefficient from a solid to fluid (concentration driving force), m/s
k_m°	mass transfer coefficient under low-mass-flux conditions, m/s
$\tilde{K}$	diffusivity matrix in Section 21.4
$\tilde{K}$	matrix of multicomponent diffusion coefficient in Section 21.4
K_G	overall mass transfer coefficient from a bulk gas to a bulk liquid (gas phase partial pressure driving force), $mol/m^2 \cdot s \cdot Pa$
K_L	overall mass transfer coefficient from a bulk gas to a bulk liquid (liquid concentration driving force), m/s
L	length of the plate or tube or catalyst slab, m
M	local Mach number, v/c
m	mass of a molecule in Section 1.8.1
$\dot{m}$	mass flow rate, kg/s
$m_{A,tot}$	total mass of A in an unit or control volue, kg
$\dot{m}_{Ai}$	mass flow rate of A entering a unit, kg/s
$\dot{m}_{Ae}$	mass flow rate of A exiting a unit, kg/s
$\dot{m}_{AW,tot}$	total mass of A transferred to walls from an unit or procss, kg/s
$\bar{M}$	average molecular weight of a mixture, kg/g.mol
M	momentum flow rate vector, N
M_A	molecular weight of species indicated in the subscript, kg/g.mol
M_w	molecular weight in general
$\mathcal{M}$	total moles present in a control volume, g-mol
$\dot{\mathcal{M}}$	moles per second entering/leaving the unit, i = inlet, e = exit
$\mathcal{M}_A$	moles of A in the system or control volume
Nu	Nusselt number, usually defined as hd_t/k or hx/k
N_{Av}	Avogadro number = 6.23×10^{23} molecules/g-mol
n	number density of molecules in Section 1.8.1

n	normal vector outward from a control surface
n_A	mass flux vector of species A, stationary frame, kg-A/m$^2 \cdot$ s
n_{Ax}	component of mass flux vector of A in the x-direction, kg-A/m$^2 \cdot$ s
N_{tu}	number of transfer of unit parameter
p	fluid pressure; equal to the average normal stress, Pa
p_{vap}	vapor pressure of a species, Pa
P	thermodynamic pressure used in equation of state, Pa
p	the concentration gradient or temperature gradient in the p-substitution method
p^*	dimensionless pressure, $p/\rho v_{ref}^2$
p^{**}	dimensionless pressure, $p^* Re$
P	power input for agitated vessels, W
P_c	critical pressure of a species, Pa
$\mathcal{P}$	modified pressure defined as $p + \rho gh$
p	temperature gradient in Example 8.3 and concentration gradient in Section 10.4.6
Pe	Péclet number, $d_t \langle v \rangle / \alpha$
Pe_R	Péclet number based on pipe radius, $d_t \langle v \rangle / D$
Pe^*	dispersion Péclet number in Section 12.5, $\langle v \rangle L / D_E$
Po	power number as $p/(\rho N_i^3 d_i^5)$ in Section 14.15
Pr	Prandtl number, $c_p \mu / k$
q	dimensionless stoichiometric ratio defined by Eq. (10.44) in Section 10.5
Q	volumetric flow rate in a pipe, m^3/s
$(\dot{Q})_V$	internal heat generation rate, W/m^3
q	heat flux vector (molecular) W/m^2
$q^{(m)}$	heat flux vector (molecular), same as q, W/m^2
q_s	heat flux from a surface or wall to a flowing fluid
$q^{(t)}$	heat flux vector due to turbulence, W/m^2
q_x	component of the heat flux vector in the x-direction
q_y	component of the heat flux vector in the y-direction
q_w	heat flux to the wall of a pipe from a fluid
$\dot{Q}$	rate at which heat is added to the control volume; unit volume basis, W/m^3
$\dot{Q}_V$	rate at which heat is generated within control volume per unit volume, W/m^3
q_z	component of the heat flux vector in the z-direction
r	radial coordinate in cylindrical and spherical system
R	radius of cylinder or catalyst particle
r_A	local rate of mass production of A by reaction per unit volume, mass units, kg/m$^3 \cdot$ s
R_A	local rate of mole production of A by reaction per unit volume, mole units, mol/m$^3 \cdot$ s
R^*	gas constant defined as R_G / M_w
R_A	rate of production of a species A by reaction
Re	Reynolds number, $L_{ref} v_{ref} \rho / \mu$
R_G	gas constant, 8.314 Pa m^3/mol $\cdot$ K
$\hat{s}$	entropy energy per unit mass of fluid, J/K $\cdot$ kg
s	entropy flux vector, W/K $\cdot$ m^2
s	shape parameter for conduction or diffusion, 1 for slab, 2 for long cylinder, 3 for sphere

Sc	Schmidt number, $\mu/(\rho D)$
Sh	Sherwood number, $k_m x/D$
St	Stanton number, $Nu/(RePr)$ or $Sh/(ReSc)$
t	time variable
t_E	exposure time for a gas–liquid interface
T	local temperature in the medium
T_a	temperature of the surroundings
$\langle T \rangle$	cross-sectionally averaged temperature
T_b	cup mixing (flow-averaged) temperature
T_c	critical temperature of a species
T_f	temperature of the surrounding fluid in contact with a solid
T_i	temperature of a gas–liquid interface
T_w	temperature of a wall or tube
T_∞	temperature of the approaching fluid
$\hat{u}$	internal energy unit mass of fluid, J/kg
$\hat{U}$	internal energy per unit mole of fluid, J/mol
U	overall heat transfer coefficient from hot fluid to cold fluid, $W/m^2 \cdot K$
$\hat{v}$	specific volume, $1/\rho$, m^3/kg.
$\boldsymbol{v}$	velocity vector; also mass-fraction-averaged velocity in a multicomponent mixture, m/s
$\boldsymbol{v}'$	fluctuating velocity vector in turbulent flow
$\bar{\boldsymbol{v}}$	time-averaged velocity vector in turbulent flow
$\boldsymbol{v}^*$	mole-fraction-averaged velocity in a multicomponent mixture, m/s
v_x	x-component of the velocity; v_y and v_z defined similarly
v_z	axial (z-) component of velocity in cylindrical coordinates
v_θ	velocity component in the tangential (θ) direction
$\boldsymbol{v}_A$	velocity of species A in a multicomponent mixture, stationary frame, m/s
v_e	velocity component in the fluid outside the boundary layer, m/s
V	total control volume
$\hat{V}$	molar volume, m^3/mol
V	speed of a moving solid in shear flow in flow direction, m/s
v_b	molecular volume at boiling point of solvent
V_f	friction velocity defined as $\sqrt{\tau_f/\rho}$ used in turbulent flow, m/s
$\dot{W}$	rate at which work is done on the control volume, W/m^3
$\dot{W}_s$	rate at which work is done by a moving part on the control volume, W/m^3
$\dot{W}_f$	rate at which heat energy is produced by friction, W/m^3
x	distance variable in the x-direction, y and x defined similarly.
x_i	mole fraction of species indicated by the subscript (usually in the liquid phase)
y	distance variable in the y-direction
y_i	mole fraction of species indicated by the subscript (usually in the gas phase)
y^+	dimensionless length used in turbulence analysis near a wall
$y_B(l.m)$	log-mean mole fraction of the non-diffusing component
z	axial distance variable in cylindrical coordinates
z^*	dimensionless axial distance variable in cylindrical coordinates, z/R
z_i	number of charges on an ionic species
Z	frequency of molecular collisions in Section 1.8.1

Greek letters and other symbols

α	thermal diffusivity of a solid, m^2/s
α	absorptivity of a surface in radiation
α_t	turbulent heat diffusivity, m^2/s
ϵ_H	turbulent heat diffusivity, m^2/s
β	bulk modulus of elasticity, N/m^2
β	angular velocity vector
γ	dimensionless activation energy in Section 13.7 and Example 16.1
γ	ratio of specific-heat values, c_p/c_v
∇	gradient operator
∇_*	gradient operator in dimensionless coordinates
∇^2	Laplacian operator defined by Eqs. (1.56)–(1.58) for scalars
∇^2	Laplacian operator defined in Sections 5.3.1 and 5.3.2 for vectors
∇^4	biharmonic operator defined by Eq. (5.31)
Δ	difference operator, out – in,
Δ	ratio of boundary-layer thickness, heat/mass to momentum
ΔH	heat of reaction, J/mol
ΔH_v	heat of vaporization, mole basis, J/mol
$\Delta\pi$	osmotic pressure diffference in Section 10.7, Pa
δ	parameter in Frank-Kamenetskii model
δ	thickness of momentum boundary layer in general
δ_f	film thickness for mass transfer, abbreviated as δ in Chapter 10
δ_m	thickness of mass-transfer boundary layer
δ_t	thickness of thermal boundary layer
ϵ	dielectric permittivity of a medium in Chapter 22
ϵ	emissivity of the medium
ϵ	energy dissipation rate in turbulent flow analysis
ϵ	a parameter in Lennard-Jones model in Chapter 1
η	effectiveness factor of a porous catalyst in Chapter 10
ζ	dimensionless axial distance, z^*/Pe
η	similarity variable defined by Eq. (11.30) in Chapter 11 for heat conduction
η	similarity variable defined in Chapter 12.2 for convective heat transfer
κ	circulation (line integration of tangential velocity) in Section 15.4.3
κ	conductivity of an ionized liquid in Section 22.1.3
κ	ratio of radius values, R_c/R_o, in Chapter 6
κ	Boltzmann constant, also denoted as k_B
λ	Debye length in Sections 22.5 and 22.6
λ	mean free path in Section 1.8.1
Λ	consistency index parameter for power law fluids
θ	angular direction in polar coordinates

θ	latitude direction in spherical coordinates
θ	dimensionless temperature in heat transfer examples
μ	coefficient of viscosity, Pa $\cdot$ s
μ_i	mobility of charged species i in Chapter 22
μ_w	chemical potential of water in Section 10.7
ν	coefficient of kinematic viscosity, μ/ρ, m^2/s
ν_t	turbulent kinematic viscosity, μ_t/ρ, m^2/s
ν_T^+	dimensionless total (molecular + turbulent) kinematic viscosity
ρ	density of the medium or the fluid, kg/m^3
ρ_c	electric charge density in Chapter 22
ρ_A	density of A in a multicomponent mixture, kg/m^3
σ	surface tension, N/m
σ_{xx}	total stress (viscous and pressure) in the x-direction
σ	Staverman constant in Section 10.7.1
σ_{yx}	same as τ_{yx} since shear stress has no pressure contribution
σ	Stefan–Boltzmann constant
τ	dimensionless time in Chapter 11, t/t_{ref}
$|\tau_w|$	stress exerted by the wall opposite to the flow direction in response to $-\tau_w$
τ_w	stress exerted by the solid on the fluid in pipe flow, $\mu\, dv_z/dr$ at $r = R$, usually negative in the flow direction
τ_f	stress exerted by the fluid on the solid, $\mu\, dv_x/dy$ at $y = 0$
τ_0	yield stress for Bingham flow
τ_{xx}	viscous stress in the x-direction on a plane whose unit normal is in the x-direction
τ_{yx}	viscous stress in the x-direction on a plane whose unit normal is in the y-direction; other components are defined similarly
ϕ	blowing parameter in Section 13.6.1
ϕ	electric potential in Chapter 22
ϕ	longitude in the spherical coordinate system
ϕ	velocity potential defined by Eq. (3.49) in Section 3.10
ϕ	Thiele parameter for a first-order reaction
ϕ_0	Thiele parameter for a zeroth-order reaction defined by Eq. (10.34)
Φ_v	rate of heat production by viscosity per unit volume, Eq. (7.12), W/m^3
ψ	streamfunction defined by Eq. (3.39) or Eq. (3.40)
ω	frequency of oscillation in periodic flow, s^{-1}
ω^*	dimensionless frequency of oscillation in periodic flow, ωt_{ref}
ω	vorticity for a plane flow defined as ω_z
ω	vorticity vector for a general 3D flow, $\nabla \times V$
ω	specific energy-dissipation rate in turbulent flow
ω_A	mass fraction of species indicated by the subscript, kg-A/kg-total
ξ	dimensionless radial position, r/R or x/L
Ω	angular velocity, rotational speed
Ω_i	speed of rotation or agitation in Section 14.1.5 and 14.1.6, r.p.s.

Common subscripts

b   bulk conditions
g, G gas-phase properties
e   exit values (Chapter 2)
i   inlet values (Chapter 2)
i   interface conditions (Chapters 9 and 10)
l, L liquid-phase properties
s   conditions at a surface of a solid or catalyst

1 Introduction

Learning objectives

You will learn from this chapter

- the basic concepts and the framework for analysis of transport problems;
- the notion of conservation laws and the transport laws, including their use in setting up models for some simple problems in transport;
- the various hierarchical levels (micro, meso, and macro) at which the models for transport processes can be developed;
- the interrelation between models of different levels;
- the need for and the definition of parameters such as friction factors and heat and mass transfer coefficients for macro- and meso-scale models;
- the range of application areas where the transport phenomena and the models have been and can be used.

Mathematical prerequisites

- The notion of scalars and vectors.
- Common coordinate systems (rectangular, cylindrical, and spherical) and the notion of components of vectors in various coordinates.
- The gradient of a scalar field and its physical significance.
- The dot product of two vectors.
- The definition and meaning of the directional derivative.
- The expression for gradient in common coordinate systems.

The goal of this chapter is to introduce the scope of the subject matter of transport phenomena, which essentially deals with flow of fluids, flow of heat, and flow of mass. This subject provides us with the basic tools to build models for systems or processes. These tools find application in a wide range of areas drawn from different disciplines. The analysis of transport processes which you will learn from this book can be effectively used in these areas. At the end of this chapter we will provide a run-down of some of these important areas of applications.

The chapter starts off with an introduction of the basic methodology of analysis and modeling of transport processes. Some important definitions and terminology that will reappear throughout the text are presented. Since transport phenomena are closely related to model development, we then provide a general discussion on the (philosophy of) modeling of processes and systems. The need for modeling at different levels and scales is emphasized and "information flow" from one scale to another is discussed. Simple examples are then given to show how models of different levels can be developed and what information is needed to complete the model formulation. These examples are important in order to get a feel for the problems which we will study in detail in later chapters and hence it is sufficient to understand only the basic ideas from this chapter. The idea is to start speaking the language. The problems at the end of the chapter will help you to practice the words. If you have already taken similar undergraduate courses, you can use these problems as a refresher.

Transport models are developed on the basis of a continuum hypothesis, assuming that matter is distributed continuously in space. Hence the molecular structure of matter is not reflected in the analysis and is captured by using transport properties. Some common transport properties are the viscosity and thermal conductivity of the fluid/material and diffusion coefficients. These properties are, however, related to the molecular structures of the species under consideration. Hence an understanding of the molecular-level models and how they can interact with transport-phenomena models is important, and a brief introduction is provided in this chapter. The detailed study of molecular models is outside the scope of this book.

The chapter concludes with a brief indication of the historical developments in transport phenomena and some thoughts on the future.

1.1   What, why, and how?

1.1.1   What?

The subject matter of transport phenomena is the unified study of the heat, mass, and momentum transport and applications to systems of importance in engineering analysis. Traditionally this subject has been a core course in chemical engineering education but it is now an integral part of most branches of engineering.

Most chemical transformation and biological processes involve one or all of the above-mentioned transport processes: the momentum transport (fluid flow), transport of heat, and mass diffusion and convection. These processes are often accompanied by chemical reactions as well. Hence a unified study of the principles governing these transport processes provides an underpinning for the development of process models and system analysis. Such a study is the focus of this book. Some application areas are indicated in Section 1.9 to indicate the range of problems amenable to this type of analysis.

Transport phenomena are encountered in our everyday life. In the morning we boil a pot of water to make the coffee, which is a two-phase heat-transfer process with a phase change.

What is the rate of boiling as a function of the rate of heat input?

We then add a lump of sugar to the coffee. If we do not stir the cup, the sugar dissolves by a process of diffusion. If we stir it, the dissolution takes place by the combined action of diffusion and convection.

How much time is needed to dissolve the sugar?

The car we drive to work is fitted with a catalytic converter, which is a set of flow channels coated with an active layer of catalyst such as Pt. This is an example of a flow system accompanied by mass transfer and chemical reaction to reduce the pollutants such as CO, NO_x, etc. The extent of pollutant removal depends on the flow rate of the gas, the temperature, the rate of mass transfer to the catalytic surface, and the rate of reaction at the surface itself.

How does the extent of pollutant removal change with the gas flow rate in the channel? How does it vary in the start-up period when the engine is cold?

The heart of the computer you use is made of a silicon chip, but the electronic activity arises due to the fact that the chip has undergone a diffusion process to incorporate phosphorus, boron, or other dopants.

If phosphorus diffusion is done for 2 hours, to what depth does phosphorus penetrate into the silicon substrate?

Relation to system analysis

Transport phenomena can be viewed as a subset of the general field of mathematical modeling and system analysis. Mathematical modeling involves the quantification of a process and developing an understanding of it. System analysis involves relating the effects of the input variables to the system outputs. Transport phenomena are closely related to these subjects; of course one deals primarily with chemical and biological systems. A system is an interconnected entity intended to achieve a desired objective. The human body can be thought of as a very complex and well-designed biological system. A simple example of a chemical system is a packed-bed reactor, which is simply a tube packed with solid catalyst particles. A knowledge of transport-phenomena principles is an essential prerequisite to build such system models. This connection of transport with modeling and system analysis is not often clearly stated and emphasized in some textbooks. You should always keep the perspective of system analysis when you study the various transport processes in this book. Thus one of the goals of this study would be to understand the basic principles so that you can build a mathematical model for a given system. You can then use this model for various purposes as discussed later in Section 1.6.

1.1.2   Why?

Three distinct subjects, *viz.* fluid dynamics, heat transport, and mass diffusion, are studied together in transport phenomena. Why should they be studied together rather than in three separate courses? The answer is the basic premise that all three modes of transport can be analyzed using the same (or similar) methodology. In fact, a close analogy exists between the model equations which describe the transport of heat and mass, and these models are often used interchangeably with only changes in the definitions of the parameters. The analogy between momentum transport and heat/mass transfer is not so close, which will become evident as we go along. In spite of this and some other differences, the method of problem formulation and solution techniques for all three modes of transport are closely related and therefore provide motivation for a unified study.

A second reason for a unified study is that systems of practical importance rarely have only one mode of transport. Thus a basic problem in momentum transfer is the flow of a fluid in a pipe under isothermal conditions. But in process industries the flow is often accompanied by heat transfer or mass transfer, and the system may also undergo a chemical transformation. Hence all three modes of transport often need to be analyzed together. A unified study is therefore very important and provides the framework to analyze such coupled effects in a systematic manner.

Examples of coupled problems

As an example consider the hot-water radiator which is used in many old homes in St. Louis. This can be a blessing on a chilly wintry night but certainly puts a burden on the utility payments in the winter and indirectly contributes to global warming. But here we are more interested in the mechanism of heat transfer from hot water to air. The phenomenon is referred to as natural convection heat transfer. Here the heat is transferred from the water to the surface of the radiator. The heat is then transferred from the surface to air by conduction, and the air gets heated. The hot air rises due to buoyancy, creating a natural circulation flow in the system. The flow in turns causes additional heat to be transferred from the radiator to air by convection, and hence the process has a feedback effect due to the coupling between the momentum transfer and heat transfer. The rate of heat transport cannot be calculated in isolation, and one needs information on the velocity profile, which in turn depends on the rate of heat transfer. An illustrative plot of the upward fluid velocity and temperature distribution near a hot vertical plate is shown in Fig. 1.1 and serves as a prototype problem in natural convection heat transfer, which is a coupled problem in flow and heat transfer.

As a second example, consider a continuous-flow tubular reactor. This is simply a tube into which the reactants are fed at one end and from which the products are withdrawn at the other end. Now, in a fluid, the different fluid elements have different (axial) velocities, with the fluid at the center of the tube moving a lot faster than the fluid near the walls. The extent of reaction depends on the residence time of the fluid elements, as discussed later, and hence is different at different radial positions in the tube. The difference in the extent of reaction at various locations sets up a concentration difference, which causes diffusion to occur. Hence it is obvious that the performance of a tubular reactor is strongly affected by the fluid flow pattern in the tube and by mass diffusion. Chemical kinetics (the rate at which a species reacts) alone is not sufficient to predict the reactor's performance. Reactions are also either exothermic or endothermic; therefore, the extent of heat generation, the rate of heat transport, and the temperature distribution in the reactor are also of importance. It may be noted that the rate of reaction is usually an exponential function of temperature, and hence this coupling can be quite strong. Thus we encounter here a coupled problem involving flow and heat and mass transfer accompanied by a chemical reaction, which can

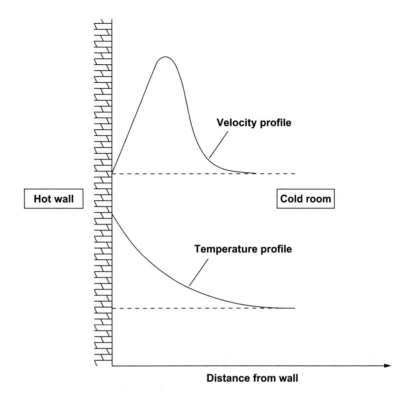

Figure 1.1 A simple example of a coupled momentum and heat-transfer problem, namely natural convection near a radiator. Velocity and temperature profiles near a hot vertical plate are shown schematically, and later we shall calculate these.

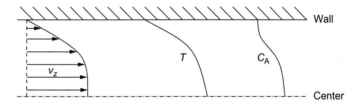

Figure 1.2 Schematic velocity, temperature, and concentration profiles in a tubular reactor. An example would be a polymerization reactor where the molecular weight (more formally, relative molecular mass) changes as the reaction proceeds, causing a dramatic change in fluid viscosity. This significantly alters the velocity profile in the reactor, which in turn affects the temperature and concentration profiles.

be quite challenging to compute. It can be said without exaggeration that the central theme in reactor design and scaleup is the question of how to evaluate the coupling between transport and chemical reaction.

Illustrative velocity, temperature, and concentration profiles in a tubular reactor with an exothermic reaction are shown in Fig. 1.2.

How do we calculate these? What are the governing equations?

1.1.3   How?

This is the main focus of the book. It will be shown that the analysis can be done on many levels and at many scales as well. But the basic methodology is to define a control volume and then apply the basic conservation laws of physics to this control volume. The conservation laws are the energy balance (heat transport), the mass balance for total mass and for each species constituting the mixture (mass transport), and Newton's second law (momentum transport).

A system or control volume can be defined as any part of the equipment, the entirety of the equipment, or even a large-scale system as a whole. The basic conservation laws apply to the system. Thus the total mass and energy content of the system are conserved. The change in a species' mass depends on the extent of chemical reaction that the species has undergone. The rate of change of momentum in the system has to be balanced by the forces acting on the system in accordance with Newton's second laws. These are universal principles of physics. The application of conservation laws is, therefore, the starting point of each analysis of transport processes. Translation of these principles into mathematical equations leads to a basic model for the system.

The system model based on the application of conservation laws alone (which we call the basic model) is, however, not complete. Additional equations governing the rate of transport are needed, which are referred to as constitutive equations or, more generally, as transport models. These equations are specific to the chemical composition of the system and are not universal, unlike the conservation laws. Basic models supplemented with transport laws then provide the system model. The methodology can therefore be shown in the simple-minded diagram illustrated in Fig. 1.3. We first elaborate on the use of conservation laws to start building the model.

1.1.4   Conservation statement

The conservation principle is rather simple. For example, the total mass is conserved in the system. Mass can enter and leave through the system boundaries, which are also called control surfaces. Mass can also accumulate in the system if the control volume size is changing in time. The conservation law is simply a balance of these quantities as discussed below.

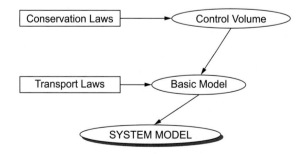

Figure 1.3 A schematic flowchart of the basic methodology for development of system models.

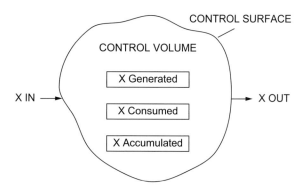

Figure 1.4 A schematic sketch of the conservation principle applied to a control volume.

Consider any quantity X of interest. Then the law of conservation of this quantity can be stated in words as follows:

$$Input + Generation = Output + Depletion + Accumulation$$

The simple-minded statement written above is of fundamental importance and will be used extensively in this book. The statement is shown pictorially in Fig. 1.4.

The input and output terms represent the interaction of the control volume and the surroundings. Thus heat, mass, or momentum can flow in or out of the control volume or "diffuse" across the control surface boundary. Transport due to flow across a control surface boundary is called convective transport, and the rate of transport is readily calculated if the velocity at that point is known. This will be elaborated on later. Transport can also take place at a molecular level, and this is referred to as diffusive transport. Additional equations are needed in order to calculate the rate of transport by this mechanism. These laws are called constitutive laws. Commonly encountered transport laws are now enunciated, and the need for such laws is indicated.

1.1.5 The need for constitutive models

If there is a temperature difference between the two points on either side of a control surface, it is a common observation that heat moves in or out of the control surface depending on the direction of the temperature gradient. In addition, if there are concentration differences in the system then it is observed that the matter tends to move from regions of higher concentration to those of lower concentration. Finally, if two fluid layers are moving at different velocities, the faster-moving layer tends to drag the slower-moving layer; in other words, there is a drag force on the slower-moving layer (and *vice versa* on the faster-moving layer due to Newton's third law). This force can be viewed as an exchange of momentum between the two points where a velocity difference exists.

Equations describing the rate of transport indicated in the preceding paragraph are called the constitutive laws or simply transport laws. These laws have to be included in the basic model of the system to describe the rate of heat/mass/momentum crossing the control surface. Hence a system model can be developed by combining the constitutive equations and the conservation laws as indicated earlier in Fig. 1.3.

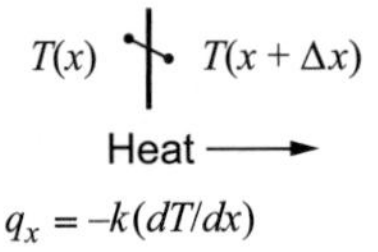

$$T(x) \uparrow \quad T(x + \Delta x)$$

Heat $\longrightarrow$

$$q_x = -k(dT/dx)$$

Figure 1.5 An illustration of Fourier's law of heat conduction: a temperature gradient drives heat transport.

1.1.6   Common constitutive models

Fourier's law

Consider the situation shown in Fig. 1.5, where there exists a temperature difference between two points separated by a distance Δx (on the adjacent side of a control surface for instance). Then there will be a heat flow across surfaces, which can be modeled by Fourier's law of heat conduction. This law states that the heat flux is proportional to the negative of the temperature gradient:

$$q_x = -k\frac{dT}{dx} \tag{1.1}$$

Here q_x is the heat flux, which is defined as the heat transported per unit time per unit transfer area in the x-direction. The units would then be $\text{J/s} \cdot \text{m}^2$ or W/m^2 in the SI system. The parameter k is a constant of proportionality, which relates the heat flux to the temperature gradient and is called the thermal conductivity. By matching the units on both sides we find that k has the units of $\text{W/K} \cdot \text{m}$.

The magnitude of k varies widely for different materials and is dependent on the molecular structure of the system. This shows that the transport properties are closely linked to the molecular nature of the matter.

Later we will extend Fourier's law to the case of heat flow in all three coordinate directions, and this will require the definition of the gradient of a scalar field and the notion of a directional derivative.

> **Historical vignette.** The law was proposed by Fourier in 1807 in a landmark paper entitled "On the propagation of heat in solid bodies". His name is included on the list of 72 names inscribed on the Eiffel tower, and students of transport phenomena will recognize many of these luminaries.

Fick's law

In 1862 Fick developed a similar law for mass transport. The mass flux of a species A is the mass of A transported per unit time per unit transfer area in a specified direction. This can be measured in either mass units or mole units. If J_{Ax} is then the moles of species A transported in the x-direction (per unit area of transport per unit time) then this is found to be proportional to the concentration gradient of A. Fick's law then states that

$$J_{Ax} = -D_A\frac{dC_A}{dx} \tag{1.2}$$

Fick's law holds, strictly speaking, for binary systems, and more refined models are needed for multicomponent systems. Also Fick's law needs a definition of the system velocity and is applied to a system moving with a mixture velocity. These details are not important at this stage. Later we will study these in detail. At this point we simply note the similarity between Fourier's law and Fick's law, and also note that the diffusion coefficient D_A has units of m^2/s. The magnitude of D_A is vastly different for mass transport in gases, liquids, and solids, and depends on the molecular structure of matter. As an abbreviation we will often use D for diffusivity instead of D_A if we are focusing on the mass transport of just one species.

Newton's law of viscosity

In a similar manner, if there is a velocity difference between two points, i.e., if two fluid layers are moving at different velocities, then there is a force acting tangential to the surface. The magnitude of this force per unit area is called the shear stress. Newton proposed a law that related the stress to the velocity gradient:

$$\tau = \frac{F_x}{A} = \mu \frac{dv_x}{dy} \qquad (1.3)$$

This is known as Newton's law of viscosity. Here τ is the shear stress. For the purpose of Chapters 1 and 2 the above definition of shear stress is sufficient. However, stress is a tensor type of quantity, as will be addressed in more detail in Chapter 4.

The schematic drawing in Fig. 1.6 makes the concept somewhat clearer. Let two fluid layers on adjacent sides of a control surface have different velocities. This velocity gradient causes a tangential force on the control surface as shown in Fig. 1.6. The force divided by area is the shear stress.

On comparing the units in Eq. (1.3). we find that viscosity has the units of Pa · s or, in expanded terms, kg/m · s.

A remark on the sign convention is in order. The stress or the corresponding force is the force exerted by the $y+$ layer (the top layer in Fig. 1.6) on the $y-$ (bottom) layer. This will have a positive numerical value if the top layer is moving somewhat faster than the bottom layer and *vice versa*. It may also be noted that some books use the opposite sign convention to define the stress.

What is the force exerted by the $y-$ layer on the $y+$ layer?

Note that the extension to a case in which the velocity varies in all directions is deferred to a later chapter. This requires the notion of the gradient of a vector, which is a tensor

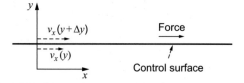

Figure 1.6 A schematic diagram to explain Newton's law of viscosity; the force on the surface divided by the surface area is the shear stress, denoted as τ.

quantity. Also many fluids do not obey such a simple linear behavior, and are called non-Newtonian fluids. Common fluids such as air, water, etc., with simple molecular structures do indeed obey Newton's law of viscosity.

The three modes of transport described above arise due to molecular-level interaction, and are therefore called molecular transport.

1.2 Typical transport property values

From the above discussion we find that the basic properties needed are therefore μ, k, and D. The molecular transport arises due to interaction of the molecules comprising the system and hence we would expect that molecular models would be useful to predict these properties. This link to molecular properties of matter is discussed briefly in Section 1.8

From a modeling point of view on a continuum basis (discussed in Section 1.3), we need only typical property values, and how the property changes as a function of temperature and pressure. Such information is briefly summarized below. It is also important to know how the property can be estimated for mixtures if no experimental data are available. Some information is provided in this section. The book by Reid, Prausnitz, and Poling (1987) provides further information on physical property values and prediction methods. The website http://www.engineeringtoolbox.com is a useful resource as well.

1.2.1 Viscosity: pure gases and vapors

Values of viscosity for air are given in the Table 1.1.

Note that the values in Table 1.1 must be multiplied by 10^{-5} to get the μ values. Thus the viscosity at 300 K is 1.846×10^{-5} Pa·s. Students often make such mistakes when looking at tabulated values of the physical properties from books or web sources. It is also important to have an appreciation of the range of values for typical compounds.

Viscosity values for steam are given in Table 1.2. It may be noted that the viscosity values for steam are about five times larger than those for air.

Table 1.1. Viscosity values for air at 1 atm pressure

	Temperature (K)					
	280	300	320	340	360	400
μ (10^5 Pa·s)	1.7503	1.846	1.9391	2.0300	2.1175	2.2857

Table 1.2. Viscosity values for steam at 1 atm pressure

	Temperature (K)					
	380	400	450	500	600	800
μ (10^5 Pa·s)	12.75	13.42	15.23	17.03	20.64	27.86

The effect of temperature is to increase the viscosity, as can be seen from the tables above. The effect of pressure (not shown above) is mild for moderate changes in pressure. For vapor species the viscosity increases with pressure mildly, and at or near the critical pressure the viscosity becomes comparable to that for a condensed liquid at that temperature.

In practical design calculations, the viscosity of a mixture of gases is often needed, and there is no simple relationship. For example, for a binary mixture the viscosity is not a linear function of the mole fraction, and one may even see a maximum in viscosity at an intermediate mole fraction, especially for a polar–non-polar mixture.

1.2.2   Viscosity: liquids

Values of viscosity for water as a function of temperature are shown in Table 1.3. It may be noted that the effect of temperature is to decrease the viscosity, a trend opposite to that for gases.

Data for many other liquids at 25 °C can be found from the engineeringtoolbox website. Typical values abstracted from this source for two common liquids are 80.28 Pa · s for glycerol and 0.0008 Pa · s for water at $T = 300$ K. On comparing these two values, we find a large difference in the viscosity values for these liquids, as can be verified by observation.

The viscosity of a liquid below the normal boiling point is not affected by moderate changes in pressure. Under a large increase in pressure an order-of-magnitude change (increase) in viscosity is observed. For mixtures, the effect of composition is not well established. Some recommenced equations are available in the book by Reid *et al.* (1987).

1.2.3   Thermal conductivity

The values of the thermal conductivity for some common materials are shown in Table 1.4.

For gases the thermal conductivity increases with temperature. Over a small temperature range the relation is nearly linear. The thermal conductivity increases with pressure, although the effect is relatively small at low and moderate pressures. Thermal conductivity for mixtures is not a linear function of mole fraction. For polar–non-polar mixtures a maximum in conductivity is observed at an intermediate mole fraction.

Data for thermal conductivity of various liquids at temperature of 25 °C can be found from the engineeringtoolbox website. In general the thermal conductivity decreases with an increase in temperature, while the effect of pressure is modest. For mixtures the values are somewhat smaller than those predicted using linear interpolation.

Table 1.3. Viscosity values for water

	Temperature (K)						
	280	300	320	340	360	380	400
μ (10^4 Pa · s)	14.50	8.67	5.84	4.31	3.29	2.67	2.25

Table 1.4. Thermal conductivity at 25 °C for common materials; note the wide range of values

Material	k (W/m · K)
Air	0.0246
Hydrogen	0.168
Water	0.58
Benzene	0.16
Insulators	0.03–0.08
Iron	80
Copper	401

Table 1.5. Illustrative values of the diffusion coefficient for common gas mixtures at temperature 298 K and pressure 1 atm

Gas A	Gas B	$D = D_{AB} = D_{BA}\,(m^2/s)$
Hydrogen	Air	7.12×10^{-5}
CO_2	Air	1.64×10^{-5}
Benzene	Air	0.88×10^{-5}

For solids a wide variation in thermal conductivity is observed, with some materials with very low values being insulators, while others (metals) exhibit high values, acting as good conductors of heat.

1.2.4 Diffusivity

Binary gas mixtures

Illustrative values of diffusion-coefficient values for a binary gas mixture (gas in air) are shown in Table 1.5. Note that there is one value of diffusivity, i.e., $D_A = D_B$, which will be shown later. Air is normally treated as a single species, although it is essentially a mixture of nitrogen and oxygen. Also note from Table 1.5 the effect of molecular weight. The lighter gases such as hydrogen have a higher value of D.

The effect of temperature is to increase the diffusion coefficient (usually proportionally to $T^{1.75}$). The diffusivity is inversely proportional to pressure.

Diffusion in liquids

The values for diffusion of solutes in water are shown in Table 1.6. These are for dilute concentrations and the effect of concentration on the diffusivity will be discussed later. Also the thermodynamic non-ideality plays a role, as will be discussed later. Note that the values for D for liquids are an order of magnitude lower than those for gases. The effect of temperature is often predicted using the condition that $D\mu/T$ is a constant, the constant being a function of the solute size only. The relation is called the Stokes–Einstein relation.

Table 1.6. Illustrative values of the diffusion coefficient for solutes in water at dilute concentrations, for temperature 298 K and pressure 1 atm

Solute	$D\ (\text{m}^2/\text{s})$
Hydrogen	5.13×10^{-9}
CO_2	1.77×10^{-9}
Phenol	0.84×10^{-9}

Table 1.7. Illustrative values of solid-state diffusion coefficients in the material indicated at temperature 293 K (Cussler, 2009)

Solute	Material	$D\ (\text{m}^2/\text{s})$
He	Pyrex	4.5×10^{-15}
Sb	Ag	3.5×10^{-25}
Al	Cu	1.3×10^{-34}

Illustrative values of the diffusivity of a solid in a second solid are given in Table 1.7. Note the extremely small values for these in comparison with those for gases and liquids. The diffusion in solids is therefore remarkably slow, but increases with temperature since the diffusion coefficient increases with temperature.

It should be noted here that the constitutive laws are not general, but are specific to a class of materials and require careful consideration. For example, a marked difference among the thermal conductivities of various materials may be noted from Table 1.4. Similarly, the diffusion of a material in air is different from that in water. The study of the various models that are useful to describe the constitutive laws of transport is another important area of research, which is closely related to transport phenomena.

Further complications arise in transport in heterogeneous media and composite materials. For example, they say that blood is thicker than water. This is true, and the rheological (flow) behavior of blood, as characterized by viscosity values, is different than that of water. But even more important is the fact that blood is a heterogeneous medium consisting of plasma, red cells, and white cells. Thus the apparent viscosity of blood is a function of composition, and no unique value can be assigned. How these properties are related to the molecular structure of matter is a challenging area for research.

1.3   The continuum assumption and the field variables

1.3.1   Continuum and pointwise representation

Differential models

With transport models one starts by defining a control volume. Now, depending on the type of control volume, transport models at three levels can be developed within the context of the continuum assumption. This leads to models of various hierarchical levels. These levels and more details are discussed in the next section. Here we discuss the model at the topmost level, *viz.*, the differential models and the subtle assumption involved in transport modeling. In the differential models (also called microscopic models or pointwise models) the control volume is a differential element of space with dimensions, for example, of Δx, Δy, and Δz in a Cartesian coordinate system. These dimensions are then made to tend to zero, thus reducing the control volume to a point in space. The resulting model is a **differential model**, which provides a point-to-point or detailed description of the quantities, for example, the temperature distribution in the system as a function of spatial location and time.

The continuum assumption

The fact that the control volume can tend to zero is not pleasing because, if the size of the control volume becomes so small, it may approach atomic dimensions. This is resolved by making the **continuum** assumption, which says that the matter is distributed continuously in space. The continuum assumption is central to transport analysis and leads to the concept of the field and field variables. The important field variables modeled in analysis of transport phenomena are presented in the following subsection.

The basic simplification from the continuum view of the world is that point values for all the properties such as density, temperature, pressure, etc. can be assumed. This is the central assumption of continuum models. Obviously this breaks down as we approach the molecular dimensions, but this is ignored in the continuum models. This does not mean that the molecular properties are not important. These properties will be indirectly reflected in the constitutive models as described later.

The continuum assumption is justified if there is a statistically representative number of molecules even in a microscopic control volume (whose size is defined relative to the total system volume being modeled). Let us take a simple example to illustrate the continuum principle.

Example 1.1. Calculate the number of gas molecules in a cube of size 1 μm in each dimension at 1 atm pressure and temperature 298 K, and examine the validity of the continuum assumption. Also calculate the density of the gas, if the gas is air, based on an equation of state for the gas.

Solution.
The problem is rather simple, but serves to illustrate the units used and some of the recurring calculations for many problems.

Pressure units

The following quantities should be noted:

$$1 \text{ std atm} = 1.013\,25 \times 10^5 \text{ Pa}$$

Note that this (std atm or simply atm) is only a unit of pressure and does not represent the actual value of pressure on a given day.

We use the units of pascals in most cases, since the pascal is the unit of pressure in the SI system. Also the pressure is often expressed as bars:

$$1 \text{ bar} = 1 \times 10^5 \text{ Pa} = 0.986\,923 \text{ atm}$$

The units for all the quantities indicated above should be noted. For calculations, pressure in pascals (Pa) should be used, while pressure expressed in bars is sometimes more convenient.

Note that, in many old literature sources and databases for mass transfer coefficients, atm is still used as the unit of pressure.

The gas constant

The next recurring parameter is the gas constant:

$$R_G = 8.314 \, \frac{\text{Pa} \cdot \text{m}^3}{\text{g} \cdot \text{mol} \cdot \text{K}} = 8.314 \, \frac{\text{J}}{\text{g-mol} \cdot \text{K}}$$

This is closely related to Avogadro's number, N_{Av}, and the Boltzmann constant. Avogadro's number is the number of molecules in 1 mol (g-mol) of any substance, $N_{Av} = 6.023 \times 10^{23}$ molecules per mole.

How is the Boltzmann constant related to the gas constant?

The ideal-gas law

Now to find the number of moles in a given volume, we use the ideal-gas law. The number of moles of gas in volume V is $\mathcal{M} = PV/(R_G T)$, which computes to 4.086×10^{-17} mol for our data.

The concentration of the gas is in turn calculated as the number of moles divided by the volume or as $P/(R_G T)$ and has a value of $40.68 \, \text{g-mol/m}^3$. Note the units for concentration in the above example. This is expressed in g-mol/m^3 or simply written as mol/m^3. Also note that, although the kg is used as the unit of mass in SI units, the moles are expressed in gm-moles. This can also be a source of confusion to students.

The number of molecules in a volume can be obtained by multiplying the number of moles by the Avogadro number. Hence we have $= \mathcal{M} N_{Av} = 4.086 \times 10^{-17} \times 6.0233 \times 10^{23} = 2.46 \times 10^7$ molecules in our small differential volume.

Since we are dealing with such a large number of molecules, a continuum approach seems valid even if the system size is of the order of $1 \, \mu\text{m}^3$. This dimension should be compared with the size of the system to be modeled, which may be a tubular reactor of diameter 10 mm or larger. This therefore justifies the continuum assumption.

The density of a gas

The mass of an oxygen molecule is 5.32×10^{-26} kg. Hence the density of oxygen under the given conditions is $1.3104 \, \text{kg/m}^3$.

Alternatively, the density can be computed using the ideal-gas law as

$$\rho = \frac{M_w P}{R_G T} \tag{1.4}$$

where M_w is the molecular weight. A word of caution on units is called for here. The molecular weight in the SI units has the units of kg/g-mol, a somewhat clumsy unit. But this keeps the system of units consistent.

For example, 1 mol of $O_2 = 32.04 \times 10^{-3}$ kg. Students often inadvertently substitute 32.04 for the quantity. Not using the proper value for the molecular weight is a common error in using the SI units. On substituting we find that the density of oxygen is 1.3105 kg/m^3.

Note that the ideal-gas law is used to calculate the density which is valid for low-pressure systems. For higher pressures, modified equations are needed. Gas laws that relate density to temperature and pressure are known as equations of state. The *van der Waals* equation of state is commonly used of conditions for high pressures.

1.3.2 Continuum vs. molecular

Continuum properties are manifestations of the molecular nature of matter. For example, the pressure exerted by a gas can be viewed as a force due to molecular bombardment on the walls of the container. Water flows differently from glycerine because the intermolecular forces in the two systems are different. These differences in intermolecular forces lead to different values of viscosity for water and glycerine. Polymer solutions have even more complex flow properties. Hence continuum models will need additional information relating to the properties of the matter. Molecular information is needed in the form of some equation, empirical data, or detailed molecular-level model. However, a continuum description is most useful for engineering practice and leads to the concept of the field variables.

What is a field? Mathematically it can be defined as a one-to-one mapping between a point in space and a corresponding value for the variable. The variable can be a scalar variable (scalar field, temperature, for example), a vector variable (velocity, for example), or a tensor variable (stress field, for example).

A continuum description of the model at the differential level of modeling involves the following information for the *field* variables.

We now provide a summary of the common field variables.

1.3.3 Primary field variables

The primary field variables are shown in Box 1. The transport phenomena can now be redefined as the methodology to compute these field variables.

1.3.4 Auxiliary variables

The calculation of the primary field variable requires the calculation of many secondary field variables, which are defined below in Box 2.

Box 1 Primary field variables in continuum models

Primary variables
Fluid velocity (vector) $v = v(x, y, z, t)$
Pressure $p = p(x, y, z, t)$
Temperature $T = T(x, y, z, t)$
Species (A) concentrations $C_A = C_A(x, y, z, t)$

Box 2 Auxiliary field variables in continuum models

Density ρ; calculated from pressure and temperature by an equation of state, e.g., the ideal-gas law

Stress $\tilde{\tau}$; a tensor field

Heat flux q; a vector field

Mass flux of species (A) n_A; a vector field, or equivalently its molar flux N_A

Vorticity ω; a vector field (curl of velocity vector), a measure of local fluid rotation

Box 3 Common governing equations for transport variables

Momentum
- Equations of motion
- Navier–Stokes (N–S) equations (Newtonian fluids)
- Vorticity transport equation and the Biot–Savart law (flow computations and turbulence-generation models)
- Biharmonic equation (slow flow of a viscous liquid)
- Time-averaged N–S equations (turbulent flow systems)
- Volume-averaged multiphase flow models (study of flow in dispersed systems, e.g., gas–solid suspensions, gas–liquid systems with small bubbles)
- Multiboundary two-phase-flow models (study of two-phase segregated flow, e.g., flow in gas–liquid systems with large bubbles)

Heat
- Equations of energy
- Enthalpy balance equation
- Equation for kinetic energy
- Equation for temperature field
- Poisson/Laplace equation for temperature (conduction in solids)
- Graetz equation (heat/mass transfer in a laminar flow in ducts)
- Equation for entropy transport (prelude to second-law analysis of the system and exergy balance)

Mass
- Equations of mass transfer; species mass balance
- Concentration equation for binary or pseudo-binary systems
- Stefan–Maxwell equations for mass transport in a multicomponent mixture
- Diffusion–reaction equation (e.g., transport in a catalyst particle)
- Nernst–Planck equation (transport in ionic systems)

The field variables are described by a set of differential equations that can in principle be used to compute the field at each and every point in space. Common differential equations are shown in Box 3, and we will have a chance to derive and look at most of these equations as we progress through the text. The primary equations are the equations of motion, overall mass balance, equations of energy, and the species mass balance. The other equations listed below are special cases of these primary equations.

1.4 Coordinate systems and representation of vectors

Some field variables, such as temperature, pressure, and concentration, are scalars, while some, such as velocity, are vectors. Hence it is useful to have a brief review on representation of vectors and coordinate systems that are useful for such representation. A quick reading is recommended, since you would have been introduced to these ideas in earlier studies.

1.4.1 Cartesian coordinates

The representation of any vectorial quantity is simplified if one defines the component of the vector with reference to a specified coordinate system. The rectangular Cartesian coordinate system can be used for this purpose. It may be noted here that we are dealing with a Euclidian space here. This is defined as a space that is curvature-free (no Einsteinian relativity for us!), and a set of rectangular Cartesian coordinates can always be constructed in a curvature-free space. Three mutually perpendicular lines then define a set of Cartesian coordinates. The coordinate directions are represented by either the traditional x, y, z notation or the index notation x_1, x_2, and x_3 with $x_1 = x$ etc., and both notations are used in this book. The index notation is more useful to represent the results in a general manner. Likewise the components of the velocity vectors are designated as v_x, v_y, and v_z or as v_1, v_2, and v_3 respectively.

Unit vectors along the three coordinate directions are denoted by e_1, e_2, and e_3. This is more convenient than the more simpler i, j, and k notation. Thus any general vector can be represented in the index form as

$$v = v_1 e_1 + v_2 e_2 + v_3 e_3 \tag{1.5}$$

where e_1 etc. are unit vectors in each coordinate direction. The unit vectors are also denoted as e_x, e_y, and e_z in this book.

Note that the components are specific to the coordinate system under consideration. If a different coordinate system is chosen the components of the vector change but the vector itself remains the same, i.e., it still represents the given physical entity. This is similar to selecting a set of units for a scalar quantity. Thus the freezing point of water can be 0, 32, 273, 460 etc. depending on the scale chosen to express it, but it has the same physical meaning. Similarly, a vector always represents a given physical entity but its components are different in different coordinate systems.

A velocity field can then be represented as a vector function,

$$v = v(x_1, x_2, x_3, t) \tag{1.6}$$

or in terms of the components

$$v_i = v_i(x_1, x_2, x_3, t); \qquad i = 1, 2, 3 \tag{1.7}$$

This representation is also known as the Eulerian representation (viewed from a coordinate system fixed in space) as defined in Chapter 4. A compact notation for Eq. (1.6) is

$$v = v(\mathbf{x}, t) \tag{1.8}$$

where $\mathbf{x}$ is the triplet x_1, x_2, x_3 or commonly (x, y, z).

1.4.2 Cylindrical coordinates

Cartesian coordinates are not suitable for every problem and, depending on the geometry, other coordinate systems are useful. Two commonly used coordinate systems are the cylindrical coordinates and spherical coordinates as discussed below. In the cylindrical coordinates a position vector is indicated by its r, θ, and z location. The cylindrical coordinate system is shown schematically in Fig. 1.7. The Eulerian representation of the velocity in cylindrical coordinates is

$$v = v_r e_r + v_\theta e_\theta + v_z e_z \tag{1.9}$$

where e_r, e_θ, and e_z are unit vectors pointing in the r, θ, and z directions, respectively.

The relation between the components of a vector in cylindrical coordinates and in Cartesian coordinates is also shown in Fig. 1.7.

Flow fields that are not dependent on θ are referred to as axisymmetric. Flow fields that are confined to the (x, y) plane are referred to as plane flow or 2D flows. Such flows can be represented either using (x, y) coordinates or the polar coordinates $(r$ and $\theta)$.

It should be noted that the unit vectors do not point in the same direction. This will create problems when we are performing differential or integral operations on vector quantities. The unit vectors do not change in magnitude, but the direction may change, as θ changes, as noted in the following equations:

$$\frac{\partial}{\partial \theta}(e_\theta) = -e_r \qquad \text{and} \qquad \frac{\partial}{\partial \theta}(e_r) = e_\theta \tag{1.10}$$

It is easy to visualize this graphically by drawing unit vectors at different points in polar coordinates. The fact that unit vectors change direction leads to concepts such as centrifugal

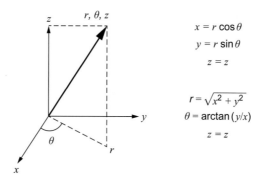

Figure 1.7 An illustration of the cylindrical coordinate system.

forces when dealing with angular motion. These arise naturally as a result of the coordinate system being used! The terms representing these forces naturally occur as a part of coordinate transformation and need not be explicitly modeled!

1.4.3 Spherical coordinates

A third common coordinate system is that constituted by spherical coordinates. Here again unit vectors in the three coordinate directions (r the distance from the center, θ the polar angle, and ϕ the azimuthal angle) are defined, and the vector is composed from the components in these three directions. A schematic representation of the spherical coordinate system is shown in Fig. 1.8. The velocity is represented as

$$v = v(r, \theta, \phi, t) = v_r e_r + v_\theta e_\theta + v_\phi e_\phi \tag{1.11}$$

Note that some unit vectors do not point in the same unique direction and have derivatives with respect to the coordinates θ and ϕ.

1.4.4 Gradient of a scalar field

Gradient: Cartesian

The gradient of a field variable is an important quantity in transport analysis, and in this section we define this and show the physical meaning of the gradient. The concept is an extension of the simple slope or gradient concept dT/dx in one dimension, where T varies as a function of x only. If T varies as a function of all three coordinates, one can think of slopes in the x, y, and z directions and together these three components constitute a vector, the gradient vector. This is defined as

$$\nabla T = e_x \frac{\partial T}{\partial x} + e_y \frac{\partial T}{\partial y} + e_z \frac{\partial T}{\partial z} \tag{1.12}$$

Each component of the gradient vector should be viewed as the contribution to the change in temperature (or any other property) along that direction. Hence, if we move a distance δx along the x-axis the temperature will change by $[\partial T/\partial x]\delta x$. Generalization leads to the notion of a directional derivative. The temperature change per unit length along a direction pointing with a unit vector n is therefore

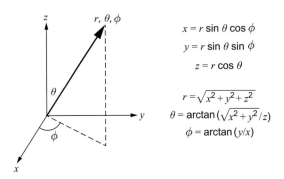

Figure 1.8 An illustration of the spherical coordinate system.

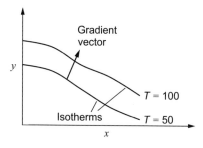

Figure 1.9 An isotherm contour plot and the notion of a gradient vector. Note that the gradient vector is perpendicular to the isolines.

$$\nabla T \cdot \boldsymbol{n} = \boldsymbol{n} \cdot \nabla T$$

which is called the directional derivative.

Temperature will not change along an isotherm, a line of constant temperature. The directional derivative must be zero along this direction. The dot product of two vectors is zero only if the vectors are perpendicular to each other. Hence it is obvious that the gradient vector at any given point is a vector perpendicular to an isotherm passing through that point. This is illustrated in Fig. 1.9.

The maximum change in temperature is therefore in a direction perpendicular to the isotherm. This is nothing but the direction of the gradient vector. Hence the gradient points in the direction of maximum change in the temperature and its magnitude is the rate of change per unit length along this direction. This provides a physical meaning of the gradient.

Generalization of Fourier and Fick

The gradient concept permits extension of Fourier's law of heat conduction:

$$\boldsymbol{q} = -k\,\nabla T \tag{1.13}$$

The quantity of heat crossing per unit time per unit area perpendicular to a vector $\boldsymbol{n}$ can therefore be expressed in terms of the directional derivative $\boldsymbol{n} \cdot \boldsymbol{q}$.

Similarly, Fick's law can be expressed as

$$\boldsymbol{J}_A = -D_A\,\nabla C_A \tag{1.14}$$

Note that this is applicable to ideal binary systems with a suitably defined frame of reference. These subtleties will be discussed later.

The extension of Newton's law of viscosity is, however, not that straightforward. The gradient of velocity is not a vector but a tensor quantity with nine components. Hence the generalization of Newton's law of viscosity is deferred to a later chapter.

Gradient: cylindrical and spherical coordinates

The forms of the gradient vector in cylindrical and spherical coordinates are also of interest, and these are defined below. The gradient operator ∇ is involved here, and is a vector operator. This is an important operator and is used to derive expressions for many vector operations in several places in the text.

How do we get these expressions?

Gradient operator

In Cartesian coordinates,

$$\nabla = e_x \frac{\partial}{\partial x} + e_y \frac{\partial}{\partial y} + e_z \frac{\partial}{\partial z} \qquad (1.15)$$

In cylindrical coordinates,

$$\nabla = e_r \frac{\partial}{\partial r} + e_\theta \frac{1}{r} \frac{\partial}{\partial \theta} + e_z \frac{\partial}{\partial z} \qquad (1.16)$$

In spherical coordinates,

$$\nabla = e_r \frac{\partial}{\partial r} + e_\theta \frac{1}{r} \frac{\partial}{\partial \theta} + e_\phi \frac{1}{r \sin \theta} \frac{\partial}{\partial \phi} \qquad (1.17)$$

1.5 Modeling at various levels

The fundamental problem in transport can be described as computation of the field variables for a given situation. The detailed description of the field variables as a function of position $T(x, y, z)$ and time is called a detailed model, 3D transient model, or distributed parameter model. If there is no time dependence the model is referred to as a steady-state model. All the field variables are interrelated in many transport problems and hence a complete description will often be difficult. Computational fluid dynamics and detailed computational calculations are of great help, but often burdened with many problems. (A further comment here is that the initial conditions and some of the boundary conditions needed for the model solution are neither clearly known nor can be defined precisely.) Models at some simplified levels are often used in engineering design and analysis; these are described in this section.

1.5.1 Levels based on control-volume size

We now discuss the various levels of models that can be constructed depending on the size of the control volume. Also an overall process or item of equipment contains many subsystems, and models of various levels can be built for each of these subsystems. These subsystems cover various length scales and processes that occur at various time scales. Thus one can build multiscale models for the processes, and these concepts are explained towards the end of this section.

Differential level

Here a differential control volume is used to develop the transport models. Detailed or pointwise balances are used to represent the system, resulting in a set of differential equations for the field variables. These equations represent models at the highest level of hierarchy.

Macroscopic scale

A large control volume or the whole equipment is used. The conservation laws are applied to this region. The volume element is not taken to tend to zero, and the balances based on the selected larger volume are used directly. Such models are useful to get the relations between engineering quantities in an approximate way. Point-to-point information (details) will be lost, and the effects of these variations have to be incorporated in some way (model averaging) or other (empirical fitting). This leads to our information-loss principle. *Information on length (or time) scales lower than that at which analysis is done is lost and has to be supplemented in some suitable manner.*

Appreciation and awareness of this principle will be important when modeling systems. An illustration of the loss of information and an explanation of the use of multiscale modeling to overcome the loss will be given in Chapter 2 and in various other places in the text.

Mesoscopic or 1D models

Such models are useful when there is a principal direction over which the flow takes place, e.g., pipe flow. The control volume is differential in this direction but spans the entire cross-section in the other two directions. Some approximation of the information lost due to area or radial averaging needs to be made.

Nota bene. In some cases 1D models are obtained as a result of some assumptions made in the differential model (such as the assumption of fully developed flow in uniaxial flow), and these are then an exact representation of the prevailing situation. No supplementary information is needed here.

Relations between the models at various levels are shown in Fig. 1.10. Differential models contain the detailed information, and mesoscopic models can be derived by integration of these models over a flow cross-section. The process is known as area averaging. Similarly macroscopic models can be derived by integrating the differential models over the entire control volume. The procedure is called volume averaging.

Compartmental models

Several macroscopic models are interconnected to form a system analysis model for a complicated process. For example, the transport of pollutants in the environment is modeled in this manner; here it is common to use four compartments as shown in Fig. 1.11 to simplify an otherwise complex situation.

Compartmental models are also widely used in biomedical systems modeling. A simple two-compartment model to represent the body is shown in Fig. 1.12. A simulation example based on this will be shown in Chapter 2.

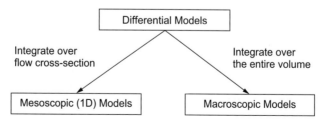

Figure 1.10 The relationship between the models at various levels; also indicating the concepts of area and volume averaging.

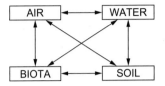

Figure 1.11 Modeling at a megascopic level: an illustration of a compartmental model for pollutant distribution in the environment. Species can be transferred across compartments as indicated by the arrows in the figure.

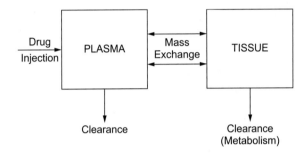

Figure 1.12 A schematic sketch of a simple two-compartment model to study drug uptake and metabolism.

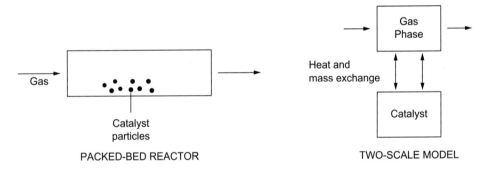

Figure 1.13 An example of modeling at two scales: a packed-bed catalytic reactor.

1.5.2 Multiscale models

Models at different levels often need to be combined to develop a process or reactor model. An example of such a model is the model for a packed-bed catalytic reactor shown in Fig. 1.13. The model is developed at two levels: (i) the reactor-scale or the gas-phase models and (ii) particle-scale models. At each scale, models can be at various levels. For example, the gas phase can be modeled at a differential or mesoscopic level. Particles can be modeled by differential models or by some lumped model. What type of model should one use? The answer is not obvious and will depend on how the model will be used and how much experimental information there is to support the level at which modeling is done. It may be a good idea at this stage for the student to ponder over these issues. A common theme can be summarized as follows: *combination of a larger-scale reactor or equipment model with a sub-model at a local scale is a common way of setting up process models.*

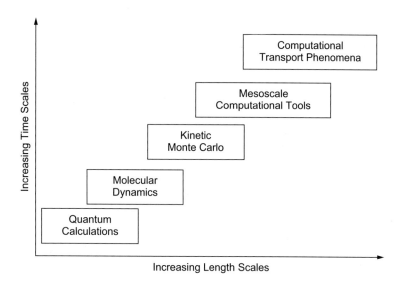

Figure 1.14 Various tools suitable for various scales and times.

1.5.3   Multiscale modeling below the continuum level

In a general sense, molecular modeling refers to the use of computational methods to describe the behavior of matter at the atomistic or molecular level (Maginn, 2009). However, there is a hierarchy of length and time scale bringing the span from near continuum to the atomistic electron level, and multiscale modeling refers to the process of analyzing the data and integrating them suitably to the continuum level (de Pablo, 2005).

For the continuum context, multiscale modeling provides useful information that can be used for the parameters appearing in the continuum model or for improving some approximations, especially at phase interfaces. It is useful to provide a (partial) list of important types of information that can be garnered by the multiscale or molecular approach:

- physical property estimation
- improved constitutive models
- interfacial models to update boundary conditions at the interface.

The common tools useful for multiscale modeling are (i) quantum chemistry calculations, (ii) molecular dynamics, (iii) kinetic Monte Carlo, and (iv) mesoscale computations such as lattice-Boltzmann calculations. A schematic representation of these tools is shown in Fig. 1.14. and the length and time scales that can be captured by these models are also indicated.

1.6   Model building: general guidelines

Analysis, modeling, and computation

The formulation of the system models is done by making use of the conservation equations and combining them with the material-specific constitutive laws from the analysis phase of the transport modeling. The analysis phase provides a conceptual working frame for the

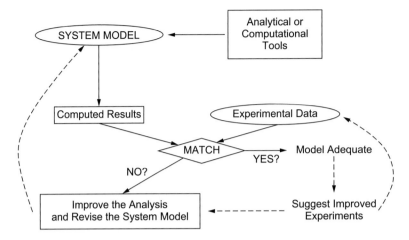

Figure 1.15 Iterative process for model development of engineering systems.

problem under investigation. This is then translated into a set of appropriate mathematical equations, which is the modeling phase. Model equations have then to be solved, which constitutes the computational phase of modeling. (It may be noted that the term computation is used in a general sense here and includes both analytic and numerical solutions.) Computational results are then validated experimentally to improve the models. Hence modeling is always an iterative process requiring the three phase of analysis, computation, and experimentation. The cycle is aptly described in Fig. 1.15. In this book you will mainly learn the analysis, modeling, and computations. Occasional references to experimental tools to study transport processes will be presented, but the study of experimental techniques, *per se*, is not the central theme of this book.

Level of complexity

Models can be developed at various levels of complexity and sophistication. A simple rule is the Einstein principle: a mathematical model should be as simple as possible but not simpler. Thus the model has to capture the necessary information sought but must not be overly complicated. The level of analysis is usually kept compatible with the end use of the model and consistent with the level of accuracy of the measurable properties of the system and the requirements. The following quote from Professor Villadsen is very apt:

It is easy to overload the model with physical phenomena which are not really relevant but lead to exhaustive numerical calculations. It is equally important to recognize and include in the model all the relevant phenomena of the particular system; otherwise the model has no predictive power.

Thus we have models of the following types, depending on what we are looking for.

• Preliminary model: here a quick reasonable approximate solution (e.g., input/output relations) is needed, for instance, for the purpose of feasibility studies. A macroscopic scale analysis may be sufficient.
• Design model: here the model is used to design or scale up a process, and the level of complexity is higher than that of a preliminary model. Also, a high level of confidence

in the model predictions is needed at this stage. The effects of all the important design variables need to be assessed here. A mesoscopic or differential model is called for.

- Control model: the model is essentially used to control an existing process. The model has to be transient in nature and should predict the output of selected variables as a function of time for specified input disturbances. Since the transient models are computationally more difficult, some approximations are usually made here. For example, the non-linear terms may be linearized using a Taylor series around the steady state, and the concept of a transition state matrix could be used. A simplified linear model may be sufficient, since often we are interested in controlling a process operating in a steady state.
- Learning model: here the goal is more academic, and the model is developed to obtain a detailed understanding of an existing process. Such models will lead to process improvements, revamping, and retrofitting, or to an entirely new and more efficient process. The level of complexity is usually very high here. It can be as high as the professor wants it to be for a Ph.D. thesis! A multiscale model is ordered from the menu.

The above philosophical discussion should stimulate some thought when looking at models presented in various scholarly journals and when developing your own models.

The analysis phase discussed in the preceding paragraph constitutes outlining the conceptual framework of the model. The skills needed for this are a combination of intuition and experience. The questions to be addressed at this stage can be outlined as follows.

- What information is being sought?
- How will this information be used?
- What is the appropriate modeling level consistent with the objectives above?
- What type of transport laws or closures may be needed?
- What approximations may be made in this context?
- Are these approximations good enough, or do we need to bring in information from other scales?

1.7 An example application: pipe flow and tubular reactor

In this section, we take the simple example of flow in a pipe to illustrate various features of transport analysis. The objective of the discussion is mainly to help the student gain an overall perspective of the methodology which will be studied in detail in later chapters. Some useful quantitative results are derived as well.

Flow in a pipe is a very simple example of a transport problem but is of great importance. Pipes are used to transport a variety of fluids, including drinking water. If the pipe is used to remove heat, we have an example of a simple pipe heat exchanger. If the pipe walls were made of a porous membrane, we could use it as a dialyzer or reverse osmosis device, and these are generally referred to as mass exchangers. If a chemical reaction takes place, we have an example of a tubular chemical reactor. Thus a simple pipe provides an elegant example of all three modes of transport. A number of questions may be of importance to an engineer dealing with such a system. Depending on the information sought, models at various levels may be needed, as discussed in Section 1.3. Here we provide some illustration of these levels and also provide some key useful formulas that are commonly used.

1.7.1   Pipe flow: momentum transport

In this section we discuss simply the flow in a pipe. There is no heat or mass transport or chemical reaction.

An engineer may need answers to the following types of questions.

- What is the relation between the mass flow rate and the imposed pressure drop in the system?
- Suppose experiments are done with a pipe of small diameter and the relation is measured. Can we use this information to design a larger and longer pipe? This is called the scaleup problem.
- Suppose we know the flow rate in a pipe of a reasonable diameter. If the size of the capillary were halved, by what factor would the flow be reduced? This is referred to as a scaledown problem, a problem of great importance in biomedical engineering. This might be a problem of life and death if the pipe is a blood vessel or artery and the size of the artery is reduced by clogging.

Let us now look at some answers to these questions. In these problems we need to find a relation between the volumetric flow rate and the pressure drop across the capillary or pipe. Even this simple problem does not have a unique answer! It depends on the flow regime prevailing for the specified flow conditions. The flow in a pipe can be laminar (at low flow rates) or turbulent (at high flow rates). The flow rate vs. pressure drop relations are different for the two cases.

1.7.2   Laminar or turbulent?

The existence of two distinct flow patterns is well established and commonly observed as well. The smoke from a lighted cigarette is seen to flow smoothly and uniformly for a short distance and then turns to a puffy irregular pattern. Similarly the fluid emanating from a faucet looks like a smooth jet near the entrance and becomes wavy and irregular later. The key observation was made by Reynolds (1883), who introduced a dye into water flowing in a pipe. At low flow rates the flow was regular and formed a single line of color, indicating that the flow was laminar. At higher flow rates a fluctuating dispersive pattern was observed. A schematic diagram of the track of a dye in laminar vs. turbulent flow is shown in Fig. 1.16.

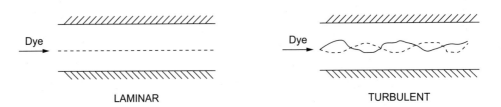

Figure 1.16 Laminar and turbulent flow in a pipe.

Flow transition

The laminar to turbulent transition in circular pipes normally occurs when a dimensionless parameter Re is less than 2100. The Reynolds number for a pipe of diameter d is defined as

$$Re = \frac{d\langle v\rangle \rho}{\mu} \tag{1.18}$$

Here we use an average velocity $\langle v\rangle$ in the pipe in the axial direction. This is a cross-sectional average of the local velocity, and calculation of the average velocity given the local velocity profile will be illustrated later in Example 2.4. It is given by

$$\langle v\rangle = \frac{\text{Volumetric flow rate}}{\text{Cross-sectional area}} = \frac{Q}{\pi R^2} = \frac{\dot{m}}{\pi R^2 \rho} \tag{1.19}$$

where $\dot{m}$ is the mass flow rate. Note that $\langle v\rangle$ can also be written as $4Q/(\pi d^2)$, where d is the pipe diameter.

The classical experiments done by Reynolds showed that the flow was laminar if $Re < 2100$.

For $2100 < Re < 4000$ the flow is not fully characterized and is classified as transitional flow. For $Re > 4000$ the flow is usually turbulent. It may be noted that these values are only indicative, not absolute. The transition is a slow process, not a catastrophic event, and the transition depends on the magnitude of the disturbance as well.

The transition from laminar flow can be analyzed by a mathematical tool called linear stability analysis discussed in Chapter 16, and some of these issues will be revisited at that point.

Flow relation: laminar flow

For laminar flow in straight horizontal pipes, the flow rate Q vs. pressure drop $(p_0 - p_l)$ relationship is found to be linear:

$$Q = \frac{\pi R^4}{8\mu L}(p_0 - p_l) \tag{1.20}$$

Here R is the pipe radius, L is the length, and μ is the fluid viscosity. The equation is often referred to as the Poiseuille law or the Hagen–Poiseuille equation. Its general form was published in 1841 by Poiseuille as an empirical relation for flow of water through glass capillary tubes. No theory was involved in this at that stage. The formal theoretical derivation of the equation was done later on the basis of momentum balance, which will be studied in a later chapter. The history of the development of the Poiseuille equation and subsequent applications in many fields has been provided by Sutera and Skalak (1993), and makes interesting reading.

The most significant aspect of this equation is the dependence on the radius to the power of four. Thus a 10% reduction in diameter can lead to a 46% decrease in flow rate at a fixed pressure difference across the capillary. The student can figure out the biomedical significance of this phenomenon.

Several other points may be noted. The volumetric flow rate is not a function of the density of the fluid. The mass flow rate, however, is.

The linearity of the equation indicates that a resistance concept similar to Ohm's law is useful, and this is examined in an exercise problem.

The flow-rate equation for turbulent flow

For large flow rates the flow becomes turbulent. The pressure drop, using an equation modified from the book of Bird, Stewart, and Lightfoot (2007) (henceforth referred to as BSL) is related to the flow rate as

$$Q = 14.79 R^{19/7} \rho^{-3/7} \mu^{-1/7} [(p_0 - p_L)/L]^{4/7} \qquad (1.21)$$

The equation is normally recommended for the range $10^4 < Re < 10^5$ and is valid only for smooth pipes. For rough pipes additionally a relative roughness parameter appears, as indicated later. This shows that the turbulent flow cannot be modeled in a unique manner, unlike laminar flow. Also note that the linearity between pressure drop and flow does not hold. The concept of resistance is no longer appealing.

These equations show that the flow rate is affected by some key parameters such as the density, viscosity, pressure drop, and the length and radius of the pipe. The parametric relations among the variables in transport phenomena are correlated in dimensionless groups rather than in terms of the primitive variables. Here we show the compact representation of the pressure drop in terms of dimensionless groups.

1.7.3   Use of dimensionless numbers

For pipe flow the following have been found to be useful.

The friction factor f is defined as a dimensionless wall shear stress for pipe flow:

$$f = \frac{|\tau_w|}{\rho \langle v \rangle^2 / 2}$$

It turns out that $\rho \langle v \rangle^2 / 2$ has the same units as the stress and is a convenient scaling factor for the friction factor. More commonly it is defined in terms of the pressure drop as

$$f = \frac{1}{4} \frac{p_0 - p_L}{\rho \langle v \rangle^2 / 2} \frac{d}{L} \qquad (1.22)$$

and correlated with the Reynolds number defined earlier by Eq. (1.18). Both definitions of friction factor are the same since the shear stress and the pressure drop can be related as shown in Example 1.2.

The laminar flow equation, Eq. (1.20), can be compactly represented as

$$\text{laminar } f = \frac{16}{Re} \qquad (1.23)$$

Likewise the equation for turbulent flow, Eq. (1.21), can be represented as

$$f = \frac{0.0791}{Re^{1/4}} \qquad (1.24)$$

for turbulent, smooth pipes. Thus we find that f and Re alone are sufficient to correlate a large amount of pipe-flow data. This is a first example of the use of dimensionless groups, and we will have many more examples of these as we progress. We can clearly see the advantage offered by dimensionless groups.

We now illustrate what information can be gained by construction of a macroscopic model for pipe flow.

Example 1.2. Macroscopic model for pipe flow

Set up a macroscopic model for a pipe flow (force balance) and relate the pressure change to the frictional forces at the walls.

Solution.

The whole pipe is treated as the control volume. Also we assume that the inlet and outlet have the same velocity profiles. This is referred to as the assumption of a fully developed flow. Hence there is no change in the momentum of the fluid, and the sum of the forces acting on the system is zero as per Newton's second law.

The forces acting on the system are shown in Fig. 1.17. These are the force due to the inlet pressure p_0, the force due to the exit pressure p_L, and frictional forces due to the viscous force exerted by the wall. Note that the pressure forces are compressive and act inwards on the control volume. The force due to shear stress is shown opposing the flow as expected, and is denoted as $|\tau_w|$.

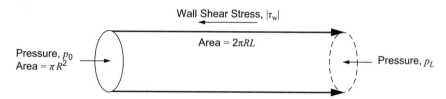

Figure 1.17 A schematic diagram of the control volume and the forces acting on it for a macroscopic description of flow in a pipe.

Balancing these leads to

$$p_0(\pi R^2) - p_L(\pi R^2) - |\tau_w|(2\pi RL) = 0$$

On rearranging we have a relation for the pressure drop and the wall friction:

$$\frac{p_0 - p_L}{L} = \frac{2}{R}|\tau_w|$$

This is a useful and important relation showing how the forces are balanced in a fully developed pipe flow. But such a macroscopic model cannot give any additional information (the information-loss principle) and therefore cannot provide us with the value of $|\tau_w|$. This information has to come either from a detailed microscopic model or by simply fitting the experimental data by an empirical model.

The friction factor was defined earlier as

$$f = \frac{|\tau_w|}{\rho \langle v \rangle^2 / 2} \tag{1.25}$$

Since the pressure drop is directly related to τ_w this justifies the correlation of pressure data using the friction factor shown in Section 1.7.3. Thus one conclusion that can be drawn from the macroscopic model is that the friction factor is the proper dimensionless number to express the pressure loss due to friction. This parameter enables us to compact a large amount of data for different fluids such as air, water, etc.

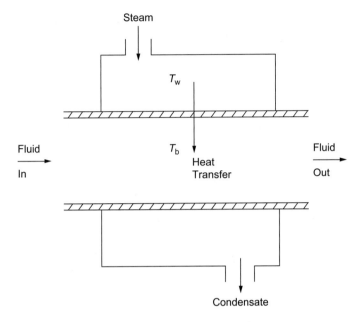

Figure 1.18 A schematic diagram of a single-stream heat exchanger.

1.7.4   Pipe flow: heat transport

A simple example of a pipe flow with heat transport is found in an item of equipment called a single-stream heat exchanger. This is shown in Fig. 1.18.

If the pipe is heated by steam we have a case where the wall temperature (T_w) can be assumed to be a constant. There are actually two streams in this case (water to be heated and steam), but the situation is referred to as a single-stream heat exchanger since only the water temperature is changing in the process.

If the walls are heated by an electrical means we have a case where the wall heat flux (heat input to the system) is constant. The wall temperature will then change as a function of position as well. This is truly a single-stream case since there is only a single flowing stream.

The designer may want to know simply the average temperature of the fluid in the system. This then calls only for a mesoscopic (1D) model.

Example 1.3. Mesoscopic model for heat transport in a pipe
Develop a mesocopic model for heat transport in a pipe.

Solution.
We use a control volume of thickness Δz in the pipe. The control volume covers the whole cross-section of the pipe. The heat in and out terms are identified in Fig. 1.19.

We propose in the context of the mesoscopic models that a flow-averaged temperature, also called the bulk temperature, T_b, can be defined such that the enthalpy crossing any plane is equal to $\dot{m}c_p(T)_{b,z}$. The precise expression for T_b as an integral is discussed later, but the concept of bulk temperature is central to mesoscopic modeling.

The various terms needed for the energy valance in Fig. 1.19. can be specified as follows.

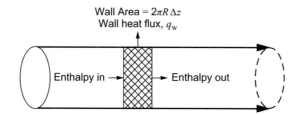

Figure 1.19 A schematic diagram of a mesoscopic (1D) control volume for analysis of a single-stream heat exchanger.

Enthalpy in by flow at z, $\dot{m}c_p(T)_{b,z}$.

Enthalpy out by flow at $z + \Delta z$, $\dot{m}c_p(T)_{b,z+\Delta z}$.

Heat transferred to the walls = area $\times$ flux = $(2\pi R\,\Delta z)q_w$, where q_w, is the heat flux at the wall.

Balancing all the terms: in by flow $-$ out by flow $-$ out by transfer to walls $= 0$, or, expressed mathematically,

$$\dot{m}c_p(T)_{b,z} - \dot{m}c_p(T)_{b,z+\Delta z} - (2\pi R\,\Delta z)q_w = 0$$

On dividing by Δz and taking the limit as $\Delta z \to 0$ we obtain the differential equation for the system:

$$\boxed{\dot{m}c_p\frac{dT_b}{dz} = -(2\pi R)q_w} \qquad (1.26)$$

This is our basic model obtained purely from conservation principles. Additional closure for the heat flux at the wall is needed if this is not specified, for example if the wall temperature is maintained constant.

The heat flux at the walls can in principle be calculated if dT/dr at the walls is known, by using Fourier's law at the wall. Here T is the local temperature, not T_b. This information is available only in a microscopic model. Hence some model or transport law has to be used for this quantity in the absence of this detailed information. The heat transfer coefficient to the wall is defined as

$$q_w = h(T_b - T_w) \qquad (1.27)$$

This is only a definition of the heat transfer coefficient, not a fundamental law. The coefficient provides a means of closing the basic model given by Eq. (1.26). Using this we have

$$\dot{m}c_p\frac{dT_b}{dz} = -(2\pi R)h(T_b - T_w) \qquad (1.28)$$

which is our (mesoscopic) system model for the single-pass heat exchanger. The model requires knowledge of the heat transfer coefficient in the system. This information can come from fitting a large amount of experimental data (the empirical method) or from a detailed microscopic model (the theoretical method).

Note that the wall heat flux is equal to $-k(dT/dr)$ evaluated at $r = R$. Hence the heat transfer coefficient is given as

$$h(T_b - T_w) = -k\left(\frac{dT}{dr}\right)_{r=R} \tag{1.29}$$

which formally ties the microscopic and the mesoscopic models. More discussion on this is deferred to later chapters.

An example to conclude the discussion of the single-stream heat exchanger is now provided.

Example 1.4.
Assume that the heat transfer coefficient is constant along the tube length. Derive an expression for the temperature in the system as a function of axial position.

Solution.
Equation (1.28) can be rearranged and integrated to give

$$\int_{T_i}^{T} \frac{dT_b}{T_b - T_w} = -\int_0^z [2\pi Rh/(\dot{m}c_p)]dz \tag{1.30}$$

The temperature at any axial location is therefore related to the inlet temperature by the following equation:

$$\ln\left(\frac{T_w - T_b}{T_w - T_i}\right) = -\frac{2\pi Rh}{\dot{m}c_p}z \tag{1.31}$$

The exit temperature is given by substituting $z = L$, where L is the length of the pipe or the heat exchanger. Again it is convenient to express the results in terms of a dimensionless form. A dimensionless temperature can be defined as

$$\theta = \frac{T_w - T_b}{T_w - T_i} \tag{1.32}$$

With this definition the range of temperature is one (inlet) and tends to zero (for a large enough exchanger). Similarly the LHS of Eq. (1.31) calls for a dimensionless number N_{tu}, which can be defined as

$$N_{tu} = \frac{2\pi RhL}{\dot{m}c_p} \tag{1.33}$$

which is known as the transfer-units number.

The exit temperature is then represented in a nice compact form as

$$\theta = \exp(-N_{tu}) \tag{1.34}$$

For larger values of N_{tu} the exit temperature (dimensionless) becomes zero, i.e., the actual exit temperature becomes equal to the wall temperature, as could be anticipated from physical considerations.

1.7.5 Pipe flow: mass exchanger

Here we present a simple model for transport of a solute across a porous membrane. The situation is common in dialysis devices. A mesoscopic model is quite useful to correlate the data and to predict the extent of purification that is achievable in the device. Again a bulk average concentration C_{Ab} is used, such that the number of moles of A crossing any cross-sectional area is equal to QC_{Ab}, where Q is the volumetric flow rate. The application of the mass balance for the solute A then leads to

$$Q\frac{dC_{Ab}}{dz} = -(2\pi R)N_{Aw} \tag{1.35}$$

where N_{Aw} is the number of moles of species A transferred across the membrane per unit area of the walls. This is the basic model, which is similar to that for the heat exchanger.

A model for mass transfer across the membrane is needed. This is the transport law. Use of the transport law will then complete the system model. A simple law may be

$$N_{Aw} = K_m C_{Ab}$$

where K_m is an overall permeability parameter across the membrane. Details of the calculation of this parameter are presented in a later chapter, but here one should appreciate the close similarity between the mass exchanger and the heat exchanger. One should also understand the steps involved in setting up and closing a process model.

1.7.6 Pipe flow: chemical reactor

We now show how modeling can be done for a pipe where a chemical reaction is taking place. It is important to note that the axial velocity profiles affect the performance of the reactor. Now, depending on the flow regime, various velocity profiles can exist as shown in Fig. 1.20. Laminar flow of a Newtonian fluid has a parabolic profile, whereas the profile is very steep in turbulent flow. The plug flow shown in this figure is an idealization that is often used to simplify the models. Here the velocity is assumed to be the same at each local radial position and is also equal to the average velocity $\langle v \rangle$.

The extent of reaction of each of the various fluid elements depends on the residence time of that element in the reactor. The local residence time is defined as

$$\text{Local residence time} = \frac{\text{Length traveled}}{\text{Local velocity}}$$

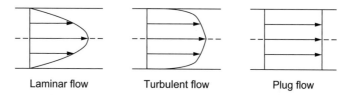

| Laminar flow | Turbulent flow | Plug flow |

Figure 1.20 A schematic diagram of axial velocity profiles for laminar and turbulent flow in a pipe as a function of radial position. For turbulent flow the velocity should be interpreted in a time-averaged sense, which will be discussed in a later chapter. The plug flow is an idealization that assumes a flat velocity profile.

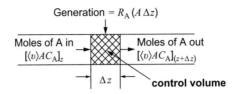

Figure 1.21 The control volume for plug-flow analysis of a tubular reactor.

The residence time is different at different radial positions, except for the idealized case of plug flow. Hence a detailed understanding of the flow pattern is of importance in reactor modeling.

As a first approximation, tubular reactors are often modeled as plug-flow reactors. Since the concentration is not varying in the radial direction, the control volume can span the entire cross-section. Thus the plug-flow analysis and model can be developed using a mesocopic control volume. An illustrative control volume is shown in Fig. 1.21.

The mole balance of A, in − out + generation equals zero, then leads to

$$\langle v \rangle \frac{dC_A}{dz} = R_A$$

R_A is the rate of production by a chemical reactor which has to come from a kinetic model for the reaction. For a simple first-order reaction

$$R_A = -k_1 C_A$$

where k_1 is the reaction rate constant (note the negative sign since A is being consumed). Hence the plug-flow model is

$$\langle v \rangle \frac{dC_A}{dz} = -k_1 C_A$$

Here we assume that the velocity at any axial position is constant, which is true for constant-density systems, e.g., isothermal systems where equal numbers of moles are produced for a gas-phase reaction.

The integrated form can be represented as

$$C_A = C_{Ai} \, \exp(-k_1 z / \langle v \rangle) \tag{1.36}$$

again for a first-order reaction, where C_{Ai} is the inlet concentration.

1.8   The link between transport properties and molecular models

Continuum mechanics plays a foundation for analysis of transport processes. However, as noted earlier, the information on the molecular scale is missing and has to be supplemented. This is often done using experimental data for transport properties such as (μ, k, and D for instance) but molecular-level modeling can provide a good platform for predicting these properties. This section provides a brief review of the kinetic theory of gases, simple molecular models for transport in liquids and diffusion in solids, and how these models can be used to predict the transport properties.

A detailed study of this section might not be needed in order to follow the other chapters in the book, but it is important to get an overview of the material presented in this chapter, both from an educational perspective and to gain an understanding of the underpinning behind the continuum models.

We now provide a brief overview of the information that can be gained from the kinetic theory of gases.

1.8.1   Kinetic theory concepts

The kinetic theory of gases is a classical example of a molecular-dynamics approach. The essential features are reviewed here, and the prediction of physical properties using this theory is illustrated. A very thorough study of kinetic theory is not required for the study of transport phenomena, but some basic concepts are indispensable. The kinetic theory provides a link between molecular-level models and continuum model for gases, as explained in the following section. Useful equations from the kinetic theory of gases are summarized in this subsection.

Postulates and parameters

The basic postulate is that the gas molecules are in a state of random motion and collide with each other. The motion is random in all directions and there is no preferred direction. There is a distribution of molecular velocities, and correspondingly a distribution of molecular speeds can be formulated (see Problem 22). The distribution function can be averaged and a mean molecular speed can be assigned to the gas at equilibrium.

This average speed, according to the kinetic theory of gases, is given by

$$\bar{c} = \sqrt{\frac{8\kappa T}{\pi m}} \tag{1.37}$$

where κ is the Boltzmann constant $1.380\,66 \times 10^{-23}$ J/K. Note that the Boltzmann constant is also equal to the gas constant R_G divided by Avogadro's number, $\kappa = R_G/N_{Av}$.

Hence the above equation can also be written as

$$\bar{c} = \sqrt{\frac{8R_G T}{\pi M_g}} \tag{1.38}$$

where M_g is the molecular weight of the gas expressed in kg/g mol.

The **average of the square** of the velocity is also important and is given by the following equation (also see Problem 23):

$$\bar{c^2} = \frac{3\kappa T}{m} = \frac{3R_G T}{M_g} \tag{1.39}$$

Note that the mean of the square of the velocity is not the same as the mean velocity squared. The square root of this is called the root-mean-square (r.m.s) velocity. Additional formulas that are basic to the kinetic theory are summarized below.

The frequency of collisions per unit area of any stationary surface is given as

$$Z = \frac{1}{4}n\bar{c} \tag{1.40}$$

where n is the number density of molecules, i.e., the number of molecules per unit volume. The average distance a gas molecule travels before colliding with another molecule is called the mean free path. The expression is given as

$$\lambda = \frac{1}{\sqrt{2}\pi d^2 n} \tag{1.41}$$

In this section d is defined as the diameter of the molecule treated as a rigid sphere and should not be confused with the same notation for pipe diameter used earlier.

The average distance a molecule can travel before colliding with another molecule is equal to 2/3 of λ.

The continuum properties of a gas can be viewed as a result of the random motion of the molecules and the resulting gas–gas collision. One requirement for the continuum model to be valid is that gas–gas collision should be the dominant momentum-transfer mechanism on a molecular level, rather than collisions of gas molecules with solid walls. Hence the continuum representation is valid if the mean free path is much smaller than the characteristic length of the containing vessel. Note that the mean free path increases as the gas pressure decreases due to the corresponding decrease in n and hence the continuum approximation may be less accurate at low pressures. The problem of reentry of a space craft into the Earth's atmosphere is an example for which the continuum model might.

Thermodynamic parameters

The pressure exerted by a gas can be viewed as the momentum change of molecules hitting a wall and bouncing back. The kinetic theory of gases provides a description of molecular motion and the following expression for momentum exchange can be derived from this theory:

$$P = \frac{1}{3}mn\bar{c^2} \tag{1.42}$$

On substituting for $\bar{c^2}$ from Eq. (1.39) and rearranging we find

$$P = n\kappa T \tag{1.43}$$

which is the ideal-gas law. This is the same as $PV = NR_GT$, where N is the number of moles present in a volume V of gas since κ is equal to R_G/N_{Av}.

Likewise temperature can be viewed as a consequence of the kinetic energy of the gases and the following result can be verified:

$$\frac{1}{2}m\bar{c^2} = \frac{3}{2}\kappa T \tag{1.44}$$

This can be viewed as the defining equation for temperature on the basis of the kinetic theory of gases.

The only way the energy can be transferred between rigid spheres is as translational kinetic energy. Hence Eq. (1.44) can be used to define the internal energy of an ideal gas based on the kinetic theory model:

$$U \text{ per molecule } = \frac{3}{2}\kappa T \tag{1.45}$$

The change in internal energy with respect to temperature is called the specific heat at constant volume C_v. It follows that

$$C_v = \tfrac{3}{2}\kappa \text{ per molecule or } C_v = \tfrac{3}{2}R_G \text{ per mole.}$$

The heat capacity at constant pressure, C_p, is equal to $C_v + R_G$ and is therefore equal to $\tfrac{5}{2}R_G$ in units of J/(mol) · K.

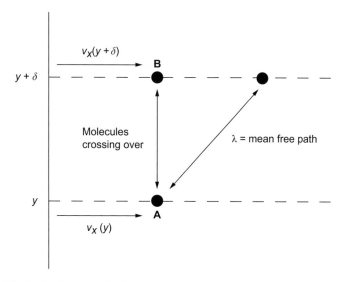

Figure 1.22 The lattice jump model for transport properties.

Transport parameters

The transport properties can be derived on the basis of the kinetic theory from the following considerations. A simple model is the lattice model shown in Fig. 1.22. Here we assume that molecules can occupy discrete locations in a lattice and jump to any of the six adjacent sides in the lattice. The lattice spacing is assumed to be equal to the mean free path.

The mass crossing over from A to B is equal to $nm\bar{c}$ per unit area. The momentum exchange from A to B is mass times velocity $= nm\bar{c}[v_x(y + \delta) - v_x(y)]$. This is equal to

$$-nm\bar{c}(\delta y)\frac{dv_x}{dy}$$

The force on the layer is the negative of the momentum exchange.

On comparing this with Newton's law of viscosity we find

$$\mu = nm\bar{c}\delta y$$

An estimate of δy is 1/3 of the mean free path or $\lambda/3$. Also $nm = \rho$. Hence

$$\mu = \frac{1}{3}\bar{c}\rho\lambda$$

where $\bar{c}$ is given by Eq. (1.38). Thus we are able to relate the viscosity, a continuum property, to the mean speed of molecular motion $\bar{c}$, albeit based on a simple molecular model. On substituting for $\bar{c}$ and λ, with some minor rearrangement, we obtain

$$\mu = \frac{2}{3\pi}\frac{\sqrt{\pi m\kappa T}}{\pi d^2} \tag{1.46}$$

a result derived by Maxwell in 1860. The term πd^2 in the denominator is known as the collision cross-section. The key prediction that viscosity is independent of pressure holds up to a pressure of 10 atm, while the prediction of a square-root dependence on temperature is somewhat less satisfactory.

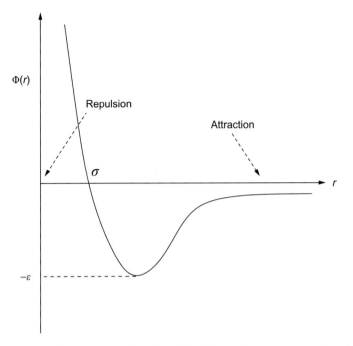

Figure 1.23 The potential energy as a function of the distance between two molecules.

The model of Chapman and Enskog

The basic kinetic theory was modified by Chapman and Enskog. The key modification was that the potential energy of the interacting molecules between the collisions was included in the model. The intermolecular potential energy has a functionality shown in Fig 1.23 as a function of the separation between the molecules.

The potential-energy distribution is not exactly known, but can be approximated by the 6–12 function for non-polar molecules.

$$\phi(r) = 4\epsilon \left[\left(\frac{\sigma}{r} \right)^{12} - \left(\frac{\sigma}{r} \right)^{6} \right] \tag{1.47}$$

introduced by Lennard–Jones. Here σ is the effective diameter of the molecules, which is close to d in the earlier section on the kinetic theory but not the same; the second parameter, ϵ, is the minimum potential energy.

Note that as the distance decreases the repulsive forces dominate, while as the distance increases the attractive forces dominate. The values of σ and ϵ have been tabulated for many substances, but, as a quick approximation, they can be related to the properties of the fluid at the critical point as follows:

$$\epsilon/\kappa = 0.77T_{\mathrm{c}}$$

and

$$\sigma = 2.44(T_{\mathrm{c}}/P_{\mathrm{c}})^{1/3}$$

The units here are K for ϵ/κ and σ is in Å. The critical pressure is in atm.

The viscosity of a pure monatomic gas can be computed in terms of the Lennard-Jones model as

$$\mu = 2.6693 \times 10^{-6} \frac{\sqrt{MT}}{\sigma^2 \Omega_\mu}$$

where Ω_μ is a slowly varying function of temperature known as the collision integral for viscosity. Its value is one at 293 K if the gas were indeed assumed to be rigid spheres. The molecular weight is taken as g/mol here, while the calculated viscosity is in Pa · s (SI units).

Tabulated values for Ω_μ are available, but it is easier to use the following fitted correlation for Ω_μ as a function of the dimensionless temperature, Θ, defined as $\kappa T/\epsilon$:

$$\Omega_\mu = \frac{1.161\,45}{\Theta^{0.14874}} + \frac{0.524\,87}{\exp(0.773\,20\Theta)} + \frac{2.161\,78}{\exp(2.437\,87\Theta)} \qquad (1.48)$$

The predictions based on this model are close to experimental values, and also the model predicts the correct dependence on temperature and pressure.

Thermal conductivity

A similar picture holds for heat transport. The lattice jump model indicates that there is a transfer of heat from the regions of high temperature to that of lower temperature, and the thermal conductivity can be related to the molecular model as

$$k = \frac{1}{3}\bar{c}\rho\lambda C_v \qquad (1.49)$$

On substituting for the mean molecular speed the following equation can be derived:

$$k = \frac{2}{3\pi}\frac{\sqrt{\pi m \kappa T}}{\pi d^2}C_v \qquad (1.50)$$

The equation is useful if the molecule is approximated as a rigid sphere with an effective diameter equal to d. The equation predicts that the thermal conductivity is independent of pressure, which is also observed experimentally up to a pressure of 10 atm or so. The equation predicts that the thermal conductivity is proportional to the square root of temperature. This dependence is weak compared with experimental results, and can be corrected by using the Chapman–Enskog approach as was done for viscosity. The corrected model is

$$k = \frac{25}{32}\frac{\sqrt{\pi m \kappa T}}{\pi \sigma^2 \Omega_k}C_v \qquad (1.51)$$

where Ω_k is the collison integral for thermal conductivity. The value is taken to be the same as Ω_μ, the collisional integral for viscosity.

Diffusivity

A similar model holds for diffusivity and the self-diffusivity (the diffusivity of a gas A in a mixture of a second gas with identical molecular properties) is

$$D_{AA} = \frac{1}{3}\lambda\bar{c} \qquad (1.52)$$

The development of a formula for D_{AB} based on rigid-sphere theory for two gases A and B with unequal molecular diameters and unequal molecular weights is more complex, and we simply state the result from other sources as

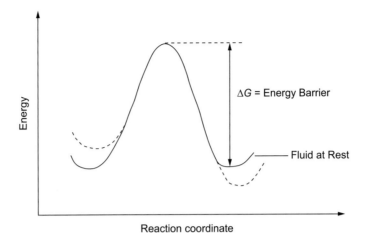

Figure 1.24 A model for viscosity of liquids. The dotted line is the energy for a fluid under stress.

$$D_{AB} = 1.8583 \times 10^{-7} \sqrt{T^3 \left(\frac{1}{M_A} + \frac{1}{M_B} \right)} \frac{1}{P\sigma_{AB}^2 \Omega_{D,AB}} \tag{1.53}$$

where P is in atm and D_{AB} is in m^2/s. The parameter σ_{AB} is taken as $(\sigma_A + \sigma_B)/2$. The parameter $\Omega_{D,AB}$ is the collision integral for diffusivity. For rigid spheres the values are unity, but the values for many gas pairs have been tabulated in the literature (see, for instance, Bird, Stewart, and Lightfoot (2007)).

A useful interactive program is `http://demonstrations.wolfram.com/BinaryDiffusionCoefficientsForGases/`.

1.8.2 Liquids

The Eyring model is often used to describe liquid behavior. Here the liquid molecules are assumed to be trapped in cages formed by a potential barrier and an energy barrier for viscous flow, ΔG_v, is needed in order to overcome this barrier. The jump frequency in either direction is given by the kinetic theory of gases as

$$\omega = \frac{\kappa T}{h} \exp[-\Delta G/(R_G T)]$$

In the presence of an applied stress the energy-barrier height is altered, and it becomes directional. The height in the direction of the stress is

$$\Delta G = \Delta G_v \pm [l/(2\delta)](\tau_{xy}\tilde{V})$$

where l is the jump length and δ is a length scale over which the stress changes significantly.

The net velocity of particles slipping past one another can be related to the average jump length times the jump frequency. The viscosity of the liquid can then be related to this energy barrier and the frequency as

$$\mu = \frac{N_{Av}h}{\tilde{V}} \exp[\Delta G_v/(R_G T)]$$

where $\tilde{V}$ is the molar volume of the liquid. The energy barrier can be predicted by molecular-dynamics calculation, but the following formula is used as a rough guide:

$$\Delta G_v = 3.8 R_G T_B$$

where T_B is the boiling point of the liquid. This then leads to the following equation:

$$\mu = \frac{N_{Av}h}{\tilde{V}} \exp(3.8 T_B/T)$$

Group-contribution-based methods are widely used to predict the viscosity of liquids. Common methods are described in the book by Reid et al. (1987).

Thermal conductivity

The thermal conductivity of liquids is predicted more on an empirical basis than on molecular considerations since the nature of thermal conduction in liquids is less well understood than is that in gases. The range of values is somewhat narrow for organic non-polar liquids, usually between 0.1 and 0.3 W/m · K. This makes the empirical correlation somewhat easier.

A reasonable correlation based on a kinetic-theory type of model is

$$k = \frac{8\pi\rho^2\kappa^3 N_{Av}^2}{M_w^2}$$

The model is based on the assumption that the molecules are arranged in a cubic layer and energy is transferred from one lattice plane to the next at the speed of sound for the given fluid. This leads to the following equation of Bridgman

$$k = \kappa \left(\frac{N_{Av}}{\tilde{V}}\right)^{2/3} \left(\frac{8c_v}{\pi c_p}\right)^{1/2} v_s \tag{1.54}$$

where κ is the Boltzmann constant and v_s is the speed of sound in the liquid. The ratio of specific heats c_p/c_v is generally taken to be one for liquids. The speed of sound in turn is equal to $\sqrt{c_p/c_v}/\sqrt{(\rho\beta_c)}$, where β_c is the compressibility factor defined as

$$\beta_c = \frac{1}{\rho}\left(\frac{\partial\rho}{\partial P}\right)_T$$

and is a measure of density change with pressure. The value is usually very small since liquids are essentially incompressible. (The value for water is 4.6×10^{-10} Pa^{-1}). With these simplifications the equation for thermal conductivity reduces to

$$k = 3(N_{Av}/\hat{V})^{2/3}\kappa v_s$$

which is a simplified form of the Bridgman equation.

For liquid metals, conduction takes place predominantly by the movement of free electrons and hence the Bridgman model based purely on lattice vibrations is inadequate and predicts much lower values for the conductivity.

Diffusion coefficients: liquids

For liquid diffusivities, the Stokes–Einstein equation is widely used:

$$D = \frac{\kappa T}{6\pi \mu r_A} \tag{1.55}$$

where r_A is the solute radius.

The prediction of the above equation is fairly accurate if the ratio of the solute radius to the solvent radius is larger than five. The reason for this is that the equation was developed for a rigid sphere (solute) moving in a continuum (solvent). For molecules with complex structure a correction factor f is often applied and a modified equation is used.

An alternative equation that is also widely used is the Wilke–Chang equation:

$$D = 7.4 \times 10^{-12} \frac{(\phi M_S)^{1/2}}{\mu \hat{V}_s^{0.6}} T$$

where $\hat{V}_s$ is the molar volume of the solute (in cm^3/mol) and M_S is the molecular weight of the solvent (in g/mol). Also μ is in g/cm s, which is equal to 0.1 Pa s, the SI unit for viscosity. The units of D are m^2/s here.

The empirical parameter, ϕ, in the equation above is known as the association parameter and has a value of 1 for most organic solvents, 1.5 for alcohols, and 2.6 for water. Note that the equation has a similar form to the Stokes–Einstein equation ($D\mu/T$ is a constant) and the dependences of the temperature and the liquid viscosity are identical. The equation is not very accurate for concentrated systems and here the diffusion coefficient varies strongly with concentration. An activity correction is usually applied, as discussed in Chapter 9.

More information on prediction of transport properties of various gases and liquids is given in the book by Reid, Prausnitz, and Poling (1987), which is a standard textbook on this subject.

1.8.3   Transport properties of solids

Thermal conductivity

Again, due to the wide variation in the arrangement of molecules in solids the *a priori* prediction of this property is a difficult task. Solids can be classified into dielectrics and metals in general, and there is a clear distinction between the transport mechanisms in these two classes of materials.

Dielectrics

For dielectrics the vibrational mode of energy transport is the dominant mechanism and hence the conductivity can in principle be related to the speed of sound. The suggested formula obtained from the analogy with the kinetic theory of gases is

$$k = \frac{1}{3}\rho c_v v_s \lambda_f$$

where λ_f is a parameter analogous to the mean free path. This can be interpreted as the smallest linear dimension of the crystal, but in general is affected by lattice defects and the presence of other impurities. Impurities act as scattering sites for the sound waves and in general cause a decrease in conductivity.

Metals

For metals the movement of free electrons induced by the thermal gradients is the main mechanism for energy transport. Thus we expect the thermal conductivity to be directly proportional to the electrical conductivity. This is the basis of the Weidmann–Franz equation stated as

$$k = Lo\,\sigma\,T$$

where σ here is the electrical conductivity and Lo is the Lorenz constant,

$$Lo = \frac{\pi^2\kappa^2}{3e^2} = 2.45 \times 10^{-8}\ \text{W}\,\Omega\,\text{K}^{-2}$$

which is a measure of the ratio of the thermal conductivity to the electrical conductivity. Here e is the charge on an electron.

Both mechanisms (phonon transport and electron transport) can operate together, and the overall thermal conductivity is then taken as the sum of the two contributions.

Diffusion in solids

Diffusion involves a jump of the diffusing molecule to a vacancy in the solid. A theoretical model based on this leads to the following equation for crystalline materials:

$$D = R_0^2 f_v \omega$$

where R_0 is the spacing between atoms, f_v is the fraction of vacant sites in the crystal, and ω is the jump frequency.

1.9 Six decades of transport phenomena

The subject of transport phenomena is very old, with some topics dating back several centuries. Some important landmarks are shown in Box 4. Thus the three aspects studied in transport phenomena were being developed independently a long time ago as a part of classical physics. The main landmark in this field was the publication of the book by Bird, Stewart, and Lightfoot (1962) entitled simply *Transport Phenomena*, which has been and still is a widely used book and is a classic in this field. The book is commonly referred to as BSL. The book unified the field and provided an important pedagogical tool for a systematic study of the subject. The field then found applications in many disciplines and many branches of engineering as noted below in this section. The article entitled "Five decades of transport phenomena" written by Professor Bird (2008) is a useful summary of the developments in this field and is an important reading assignment for students.

It is also interesting to note how Bird choose engineering as a career, and I quote from another article by him (Bird, 2010).

When it was time for college, I wanted to study foreign languages but my father promptly vetoed the idea and said that I would study engineering "because you can always get a good job if you study that". He further said that ChE [chemical engineering] was the newest and most difficult branch of engineering.

Chemical engineering is no longer the newest branch of engineering, but is often still considered the most difficult. It is to be hoped that the present book will expose this myth. Incidentally, the chemical engineers are paid the second-highest starting salary currently, second only to petroleum engineers, as revealed in recent surveys, and this could be a motivation for pursuing this field for some students. Whatever the reason for you to study this subject may be, I hope you will be able to appreciate the logic, majestic structure, and beauty of the subject and share my strong enthusiasm for this field.

Box 4 Some early landmarks in the development of transport phenomena

Buoyancy	Archimedes	260 B.C.
Viscosity	Newton	1672
Conductivity	Fourier	1807
Diffusion	Fick	1851
Mechanical energy balance	Bernoulli	1738
Inviscid fluid flow	Euler	1755
Pipe flow	Hagen	1841
Pipe flow	Poiseuille	1843
Flow analysis	Navier	1822
Slow viscous flow	Stokes	1845
Flow in porous media	Darcy	1856
Turbulent flow	Reynolds	1876
Boundary-layer theory	Prandtl	1925
Integral analysis	von Kármán	1921
Mass transfer (film model)	Whitman	1923
Mass transfer (penetration model)	Higbie	1935
Micromixing	Danckwerts	1952
Axial dispersion	Taylor	1953
The Book	BSL	1962

We now give an overview of the growth of applications of transport phenomena in various branches. Some key references are cited. Also some illustrative problems in different branches are mentioned to motivate your interest.

Environmental applications

Transport phenomena (principles and modeling) have found extensive applications in environmental engineering, providing a modern perspective and new approach to this field. Typical problems that may be addressed using the transport modeling methodology are as follows.

- Fate and transport of contaminants in atmosphere. The problem is usually solved by dividing the system into four or more compartments and considering transport and reaction in each of the compartments (see Fig. 1.11).
- Groundwater transport is another example. Contaminant leakage from nuclear waste tanks into rivers could be a major problem, and some of these issues can be analyzed by transport models. What is the rate of leakage and what measures can be employed to alleviate the problem?
- Transport of excess nutrients to water bodies leads to growth of algae and destruction of other organisms, a process known as eutrophication. What are the rates of transport in such systems?

The book by Clark (1996) provides a nice introduction to this field.

Biomedical applications

The focus in this field is to bring together fundamentals of transport models and life-science principles. Key areas where the transport-phenomena approach can be utilized are as follows.

- Pharmacokinetics analysis, concering the distribution and metabolism of drugs in the body. See Fig. 1.12 for a simple-minded compartmental model for this.
- Tissue engineering and development of artificial organs and assistance devices such as dialysis units.
- Quantitative characterization of blood flow and circulation.
- Bioheat transport models.

An early book by Lightfoot (1974) and recent books by Truskey, Yuan, and Katz (2004) and by Fournier (2011) are illustrative of this approach. A recent book edited by Becker and Kuznetsov (2013) is another valuable addition to the field of transport in biological systems.

The book by Lightfoot provides the following bizarre (scaleup) problem: if 2 micrograms of LSD is hallucinatory but safe for a mouse, how large a dose is safe for an elephant? You may wish to ponder this problem and see what the controlling steps in the metabolism are. What transport effects could be rate controlling?

Transport analysis of semiconductor materials processing

The book by Middleman and Hochberg (1993) is a classic in this area. Illustrative problems are as follows.

Thin-film solar cells. Silicon is deposited on a surface by the chemical reaction of pyrolysis of silane (SiH_4). What conditions would lead to a deposit of uniform thickness?

Crystals from melt. Silicon crystals are being formed from a molten melt of silicon. What is the growth rate (pulling rate) of the crystals? The pulling rate cannot exceed a certain value if one is to form single crystals. What changes in the operating conditions would you suggest to obtain single crystals?

Application to metallurgy and metal winning

Transport-phenomena analysis and models are widely used in metallurgy and metal winning. The books by Szekely and Themelis (1991), Sindo (1996), and Geiger and Poirier (1998) are examples in this field.

Ore smelting in blast furnace, gas–solid reactions, steel and copper making, and alloy formation by melt-drop solidification are examples where transport principles are needed. Since these are high-temperature processes, energy optimization plays a role as well, and system analysis based on transport models is useful.

Product development and product engineering

Transport phenomena are increasingly used in product development. Properties of common household products such toothpastes, creams, lotions, etc. can be improved by these techniques since modeling provides a strong handle on the relationship between the processing conditions and the quality of the final product. The book by Cussler and Moggridge (2001) is a useful resource for further reading in this field.

Electrochemical processes

Electrochemical processes have a wide range of applications, including batteries, solar cells, electro-deposition, thin films, microfluidic devices, etc. Transport-phenomena principles are increasingly being used to design and improve these devices. The book by Newman and Thomas-Alyea (2004) is a good treatise on this subject.

Multiphase transport models

Boilers, condensation equipment, chemical reactors, etc., are examples of systems where there are multiple phases. The book by Faghri and Zhang (2006) is a good introduction to this subject. Chemical reactors such as trickle-bed reactors, where gas and liquid flow over a packed bed, and bubble-column reactors, where a gas is sparged into a pool of liquid, are other examples of industrial processes involving multiphase flow. The book by Ramachandran and Chaudhari (1983) is a good starting source on this subject. Computational flow modeling of these reactors is dealt with, with a lot of examples, in the book by Ranade (2002).

Computational transport phenomena

Advances in computational power permit us to solve numerically many complex problems, which was not possible five decades ago. Some illustrative methods and examples will be presented in later chapters. The books by Ramachandran (1993), Schiesser (1994), White and Subramanian (2010), etc. are illustrative of the development in this field. An excellent monograph on computational fluid flow is the book by Peyret and Taylor (1983). The book by Farmer *et al.* (2009) covers various aspects of this field including multiphase flow and also provides FORTRAN source codes. The code does not require a lease and can run on a PC or a supercomputer.

1.10   Closure

As a closure to this chapter we note that the subject of transport phenomena has shown tremendous growth, having now grown into a full tree and yielded many fruits. The field is quite mature. But maturity does not mean that the field is not exciting or has no potential for growth. On the contrary, the field will see tremendous development and continues to provide excitement to researchers all over the world. Some thoughts on future trends are presented below. The student will be able to come up with even more innovative ideas.

- Seamless integration of modeling at various levels and scales.
- Improved computational tools and system-specific computations.
- Integration of computations with the physics and software tools to apply these novel methodologies.
- Improved models for transport in membrane and nanostructured materials.
- Improved understanding of fundamentals of multi-phase flows and closure laws for such systems.
- Applications to many novel areas related to energy and environment. Energy generation and resource management are key areas for the future, and you will be able to find innovative solutions using the tools in this book.

Summary

- The development of a mathematical model to describe a process is central to the analysis and design of that process. Transport phenomena provide the basic tools to develop such models.

- Transport phenomena are based on the continuum approximation. The general model is the pointwise or differential model to calculate essentially the primary field variables, which are the velocity, pressure, temperature, and concentration or mole fraction of the species in a multicomponent mixture.

- The differential models are formulated using the general conservation laws coupled with some constitutive models for transport of heat and mass and the relations for the viscous stresses in a fluid. The constitutive laws are system-specific, not general, unlike the conservation laws. However, for many fluids the laws of Fourier, Fick, and Newton are commonly used. These laws lead to the definition of basic material properties, *viz.*, thermal conductivity, diffusivity, and viscosity.

- Models can be developed at various levels depending on the end use of the model. A model is only a crude imitation of the reality, and hence the level of the model should be compatible with the end use. One person's reality is another person's illusion. Hence modeling can be viewed as an art of finding the right amount of illusion. In general some *a priori* knowledge of the system is needed in order to determine at what level the model should be developed.

- In transport-phenomena analysis one usually develops a model at three levels. (i) Macroscopic models provide overall information on the system. In general these models require many assumptions and some closure parameters. (ii) Mesoscopic models provide information on the variation of an averaged property as a function of the main flow direction. The main closures are the transport coefficients and the friction factor or the drag coefficient. (iii) Differential models provide complete point-to-point information on the field variables. Physical properties such as viscosity, thermal conductivity, and diffusivity are needed in these models. If the flow is turbulent, additional closure parameters are needed to account for the effect of velocity fluctuations since we usually deal with time-averaged values of the field variables here.

- Macro and meso are the volume-averaged and cross-sectionally averaged descriptions of the differential models, respectively.

- The student should be able to apply the conservation principles to a simple problem. He/she should have an appreciation of the level at which these laws are being applied and the meaning of the variables depending on the context or level of modeling.

- Simple examples of the application of model formulation to transport in a pipe were presented and should be looked at closely to understand the methodology. More such examples will follow in Chapter 2.

- The basic transport properties are tied to molecular models for the species. A simple example using the kinetic theory was shown to indicate how the primary properties such as viscosity, thermal conductivity, and diffusivity arise from molecular motion. Some additional discussion of the molecular basis for transport parameters in liquids and solids was also presented.

ADDITIONAL READING

A useful reference on vector algebra and calculus is the book by Arfken, Weber, and Harris (2013), and it may be useful to review/study Chapter 3 of their book.

Another useful book on vectors is the one by Aris (1962), which focuses especially on fluid-dynamic applications of the vector calculus and hence is particulary useful for applications of the vector calculus to transport phenomena.

Mathematical methods as applied to chemical engineering problems are discussed in the book by Morbidelli and Varma (1997), and this will be a good reference source for you to refresh the mathematical prerequisites needed for the study of the various chapters in this book.

The classic text on the molecular theory of gases and liquids is the book by Hirshfelder *et al.* (1954). This is not a book for the faint of heart when it comes to the mathematics, but it does provide a comprehensive treatment of the physical properties and interactions that govern the behavior of gaseous and liquid-phase molecules, and shows the link between the molecular and the continuum model in complete detail.

The estimation of physical properties and transport properties from molecular structure using group contributions and other, similar, methods is discussed in the books by Reid *et al.* (1987) and Poling *et al.* (2000).

Problems

1. Explain briefly the following terminologies.
   - Continuum models
   - Control volume and control surface
   - Conservation laws
   - Constitutive models or laws
   - Differential or microscopic models
   - 1D or mesoscopic models
   - Macroscopic models
   - Gradient vector and directional derivative
   - Laminar and turbulent flow
   - Friction factor
   - Heat transfer coefficient
   - Concept of plug flow

2. Nuclear reactors are accompanied by conversion of mass into energy according to the famous equation $E = mc^2$. Hence the mass balance does not hold in such systems. Calculate the mass loss in a typical nuclear reactor and calculate the percentage mass loss in such systems. Can we still use mass balance as an approximation for such systems?

3. Indicate some situations where the continuum models are unlikely to apply.

4. How many molecules are there in a cube of size 1 nm in each direction for (a) air under standard conditions and (b) water under standard conditions? If you are modeling transport in a nanostructured material, would you expect continuum behaviour to hold? Under what conditions?

5. The viscosity data for water as a function of temperature are given in Table 1.3. Fit an equation of the type

$$\mu = A \exp(B/T)$$

to represent the data. This equation is called the Andrade equation or Andrade correlation, and is widely used for correlating the effect of temperature on viscosity of liquids. Equations fitted to data are useful for computer simulation.

6. Viscosity data for air as a function of temperature are shown in Table 1.1. Fit a power-law model of the type

$$\mu = AT^n$$

to represent the data. Note that the viscosity of gases increases with increasing temperature, whereas that of liquids decreases.

7. A tank is being filled by having a flow at the rate of $\dot{m}$ into the system. Unfortunately there is a tiny hole at the bottom and the water is leaking out of the system. Develop a basic model for the variation of the tank level as a function of time. Identify the additional transport law needed to complete the model. The transport law commonly used here is the Torricelli equation as discussed in Chapter 2.

8. Verify the following relations by (i) graphical construction of vectors and the parallelogram law and (ii) by transforming the vectors to Cartesian coordinates and then taking the derivatives:

$$\frac{\partial}{\partial \theta}(e_\theta) = -e_r$$

$$\frac{\partial}{\partial \theta}(e_r) = e_\theta$$

What would the corresponding relations in spherical coordinates be?

9. How are the unit vectors in cylindrical coordinates related to unit vectors in Cartesian coordinates? Write this in a vector–matrix form as

$$E = LE^*$$

where E are the components of the unit vectors in one coordinate while E^* are the components of the unit vector in the other coordinate system. L is a matrix that is often called the coordinate transformation matrix. Verify that the matrix L is symmetric. What is its inverse?

Use the above properties of L or other ways to express a given vector in Cartesian coordinates related to the same vector in cylindrical coordinates. Show that the vector components transfer by the same rule as the coordinate components. This is referred to as an invariance property of a vector.

State the coordinate transformation matrix in spherical coordinates.

10. State or derive the relation between partial derivatives such as $\partial/\partial x$ and those in cylindrical coordinates. Use this or some other method to derive a formula for ∇T in cylindrical coordinates. Repeat for spherical coordinates to find the gradient operator.

11. The Laplacian of a scalar is important in mass and heat transfer applications, and is obtained by applying the operator $\nabla \cdot$ the gradient of a scalar, e.g., to ∇T. By applying this to Eq. (1.15) verify that the Laplacian in Cartesian coordinates is given by

$$\nabla \cdot \nabla T = \nabla^2 T = \frac{\partial^2 T}{\partial x^2} + \frac{\partial^2 T}{\partial y^2} + \frac{\partial^2 T}{\partial z^2} \qquad (1.56)$$

Derive the following corresponding expression for the cylindrical coordinates

$$\nabla^2 T = \frac{1}{r} \frac{\partial}{\partial r} \left(r \frac{\partial T}{\partial r} \right) + \frac{1}{r^2} \frac{\partial^2 T}{\partial \theta^2} + \frac{\partial^2 T}{\partial z^2} \qquad (1.57)$$

What is the form of these expressions if the temperature is a function of r only?

12. Derive the following corresponding expression for the Laplacian in spherical coordinates:

$$\nabla^2 T = \frac{1}{r^2} \frac{\partial}{\partial r} \left(r^2 \frac{\partial T}{\partial r} \right) + \frac{1}{r^2 \sin \theta} \frac{\partial}{\partial \theta} \left(\sin \theta \frac{\partial T}{\partial \theta} \right) + \frac{1}{r^2 \sin^2 \theta} \frac{\partial^2 T}{\partial \phi^2} \qquad (1.58)$$

13. Flow experiments are done with water in a pipe of diameter 40 cm . Find the flow rate at which the flow ceases to be laminar.

14. Note that every equation has to be dimensionally consistent. The units on the RHS of any equation have to be the same as the units on the LHS. Verify that the Hagen–Poiseuille equation Eq. (1.20) is dimensionally consistent.

    The history of the Hagen–Poiseuille equation has been summarized in a paper by Sutera and Skalak (1993). What are the main points discussed in this paper?

15. Glycerine is flowing in a small capillary of diameter 0.25 cm and length 30 cm. For a pressure difference of 3 bars applied across the tube a flow rate of 0.001 m^3/s was obtained. Find the viscosity of the liquid using the Hagen–Poiseuille equation.

    If the density of glycerine is 1260 kg/m^3, calculate the Reynolds number, Re, and check whether the Hagen–Poiseuille equation is valid under the experimental conditions.

16. The linear relation given by the Hagen–Poiseuille equation, Eq. (1.20), is reminiscent of Ohm's law for current as a function of voltage in a current-carrying wire,

$$I = \frac{V}{\Omega}$$

where I is the current, V is the voltage difference, and Ω is the resistance. The analogy is complete here, with Q, the volumetric flow rate, being similar to the current, while the pressure difference is analogous to the voltage; thus the pressure difference drives the flow just as the voltage drives the electric current. Thus

$$Q = \frac{p_0 - p_L}{\Omega_p}$$

where Ω_p is the resistance to flow. What is the expression for resistance for laminar flow problems?

17. A 32 km-long pipeline delivers petroleum at the rate of 2000 m^3 per day. The pressure drop of the pipe across the system is 34.5 MPa. Find the resistance of the pipe to the flow. Express the resistance per unit length of the pipe.

    Now, if an additional parallel line of the same size is laid along the last 18 km of the pipe, what will the new flow rate be? Assume laminar flow holds in both cases. **Hint:** treat this as a resistance in series and two resistances in parallel, and find the overall resistance.

18. Repeat the analysis for the above problem if the flow is turbulent. Note that the resistance concept is not useful here. Why?

19. An Alaska pipeline is 48 in in diameter and 800 miles long. It carries crude oil from Prudhoe bay to Valdez and is designed to deliver about 2.1 million barrels per day. Use viscosity data from the literature. Find the total pressure drop in the system and the power needed to maintain the flow.

    Note that power is the rate at which work is done (the familiar PV term in thermodynamics). Hence

    $$\text{Power} = \text{Pressure drop} \times \text{volumetric flow rate}$$

20. A heat exchanger is operated at a flow rate of 30 kg/s and water enters at 20 °C and leaves at 80 °C. Find the length of exchanger needed if the (average) heat transfer coefficient is 200 W/K·m².

    Solve the above problem if a constant inward wall flux of 2 W/m² is introduced instead of a constant wall temperature.

21. Derive the integrated equations for the concentrations in a plug-flow reactor if the reaction is of (i) second order and (ii) zeroth order.

22. The speed of molecules according to kinetic theory is given by the Boltzmann distribution function f. Thus $f(c)dc$ represents the probability that c lies between c and $c + dc$ and the distribution function is

    $$f(c) = \frac{4}{\sqrt{\pi}} \alpha^{3/2} c^2 \exp(-\alpha c^2)$$

    where $\alpha = m/(2\kappa T)$.

    Show that the area under the distribution function is unity as expected for any probability distribution function.

    Find the mean speed defined as

    $$\bar{c} = \int_0^\infty c f(c) dc$$

    Find the mean kinetic energy defined as

    $$\text{KE} = \frac{1}{2}m \int_0^\infty c^2 f(c) dc$$

    Verify that the expressions given in the text are valid by doing these integrations. Use of a table of integrals or other software such as Mathematica or MAPLE is useful to find the integrals.

23. What is the relation between the square of the mean velocity and the mean of the velocity squared in the context of the Boltzmann distribution?

24. The distribution of the energy of the molecules is also of importance in the kinetics of chemical reactions. The fraction of molecules with energy in the range between E and $E + dE$ is given $F(E)dE$, where $F(E)$ is an energy distribution function. Show that

    $$F(E) = \frac{2}{\sqrt{\pi}} \left(\frac{1}{k_B T}\right)^{3/2} E^{1/2} \exp(-E/(k_B T))$$

    Hint: the kinetic energy is $E = mc^2/2$ and hence the substitution $dE = mc\,dc$ in the expression for the KE distribution can be used.

Note that an equation of this type is used for the study of the effect of temperature on the reaction rate constant.

25. Find the following properties for oxygen at 298 K and 1 atm based on the kinetic theory of gases.

    (a) The average speed of the molecules.
    (b) The r.m.s. value of the speed.
    (c) The mean free path assuming a molecular diameter of 3 Å.
    (d) The viscosity of oxygen based on the simple model.
    (e) The viscosity of oxygen at 298 K using the Lennard-Jones modification to the simple kinetic-theory model for transport properties. Compare your answer with the experimental value and the value from part (d) above.
    (f) The thermal conductivity of oxygen.

26. The simple formula $C_p = (5/2)R$ (molar units) is valid only for monatomic gases. A simple extension that has been suggested is

$$C_p = (5 + N_r)\frac{1}{2}R$$

where N_r is the rotational degree of freedom, which is equal to 2 for linear molecules and 3 for a non-linear molecule.

Using this, compute the specific heat of methane at 350 K and 1 atm and compare your result with experimental values. Report your answer as both $J/mol \cdot K$ and $J/kg \cdot K$.

Similarly, for the thermal conductivity a relation of the following form is often used:

$$k = k_{mono} + 1.32 \left(c_p - \frac{5}{2}\frac{R}{M} \right) \mu$$

which is known as the Eucken correction to account for the internal degrees of freedom. Here k_{mono} is the value calculated for a monatomic gases, and c_p is in mass units.

Calculate k for methane under the same conditions based on this model and compare with the experimental value.

The Prandtl number, Pr, is an important dimensionless group in heat transfer and is defined as $c_p\mu/k$. Estimate its value for methane.

27. Show the details leading to the equation (1.52) in the text,

$$D_{AA} = \frac{1}{3}\bar{c}\lambda$$

where D_{AA} is the self-diffusion coefficient.

28. Find the binary pair diffusivity for the system methane (A)–ethane (B) at 293 K and 1 atm by using the Lennard-Jones method given by Eq. (1.53) by the following methods.

    (a) Use the following parameters (from the BSL book): $\sigma_A = 3.780$ Å and $\sigma_B = 4.388$ Å, $\epsilon_A/\kappa = 154 K$, and $\epsilon_B/\kappa = 232$ K.

        The collision integral is approximately 1.168 for these conditions as interpolated from the tables in BSL.

    (b) Use the pure-component parameters calculated from the critical properties using the relations on page 40.

    (c) Find the diffusivity using the MATHEMATICA tool box cited on page 42. To use this tool you will need to download and install in your computer the Wolfram computable download format (CDF). The Wolfram tool box has many other tools that you may find useful for computation of transport phenomena. Compare the results.

29. Estimate the thermal conductivity of water if the speed of sound in water is 1498 m/s. Compare your answer with the experimental values.

30. For most liquids other than water and substances with polyhydroxy groups, the thermal conductivity decreases with an increase in temperature. Water shows an anomalous effect. The conductivity, k, first increases with temperature (up to 420 K) and then decreases with further increases in temperature. Provide some arguments to justify these observations.

    Note. The effect of temperature is small and often ignored, or a linear relation with temperature is used.

2 Examples of transport and system models

Learning objectives

After a careful study of this chapter and working through a number of problems at the end of the chapter, you will be able to

- appreciate the different levels at which models can be formulated and the utility and limitations of each level,
- formulate the model equations starting from the basic conservation principles stated in Chapter 1,
- identify the missing link in the models derived from conservation laws, and understand and use the common ways of supplying this link,
- solve the models by simple analytic or numerical methods for the type of problems considered in this chapter, and
- extract useful engineering information from the solutions.

Mathematical prerequisites

- Solution of first-order linear differential equations of the type

$$\frac{dy}{dt} + a(t)y = b(t) \tag{2.1}$$

- Solution of second-order differential equations of the type

$$\frac{d^2y}{dx^2} - \phi^2 y = 0 \tag{2.2}$$

and similar equations.
- The concept of matrix notation and the matrix representation of a system of linear differential equations:

$$\frac{d\boldsymbol{y}}{dt} = \tilde{A}\boldsymbol{y} + \boldsymbol{R} \tag{2.3}$$

where $\boldsymbol{y}$ is the solution vector for example, y_1, y_2 etc. Here $\tilde{A}$ is the coefficient matrix (assumed to be constant for simplicity), not a function of t or y, and $\boldsymbol{R}$ is the vector of RHS constant (non-homogeneous) terms.

- Numerical solution of the above equations using toolbox kits in MATLAB, such as the ODE45 solver, or similar programs in other software. Note that MATLAB is used only for convenience and popularity; students may use other solvers or write their solvers in FORTRAN with IMSL or, if brave enough, in C++. A program similar to MATLAB is OCTAVE, which runs on the LINUX platform and can be used in lieu of MATLAB.
- The concept of matrix eigenvalues and eigenvectors will also be useful, although it is not an essential requisite for the moment. The student can come back to the relevant sections of this chapter dealing with this topic at a later time.
- The concept of an exponential matrix and an analytic solution to Eq. (2.3). Again this is optional at this stage.

Conservation principle

The conservation principle which forms the basis of the model formulation in this chapter is repeated here for completeness and ease of reading.

$$\boxed{Input + Generation = Output + Depletion + Accumulation}$$

If the system is at steady state the accumulation term is set as zero:

$$Steady\ State \Longrightarrow Accumulation = Zero$$

The goal of this chapter is to start simple and to illustrate the general methodology of formulation of transport problems for many standard and commonly encountered problems. Model formulation at all three levels, micro, meso, and macro, will be illustrated. For microscopic models, the molecular information needed will be assumed to be contained in the three basic transport properties, the *viscosity, thermal conductivity,* and *diffusivity,* and the three basic transport laws, *Newton's law of viscosity, Fourier's law of conduction,* and *Fick's law of diffusion,* are assumed to hold. Also, only simple examples, where the field varies in only one coordinate direction, are considered, and problems amenable to analytical solutions are discussed. The emphasis is therefore on analysis and model formulation using the methodology described in the previous chapter. In particular we illustrate the methodology described in Fig. 1.3 of Chapter 1 by deriving the models for some cases. It may be noted here that for more general analysis and complex problems it is easier to work with the general differential equations of transport which are derived and used in later chapters. Thus we can go from general to particular cases. This chapter, however, illustrates a case-by-case approach to get a feel for the method of analysis.

Although the microscopic models provide detailed point-to-point information and are therefore useful in detailed design calculations or models of a process, it may be useful to construct a macroscopic model first. For macroscopic analysis, the control volume is sufficiently large and hence the information from the micro-scale is lost, so simple empirical laws or assumptions are needed in order to "close" the problem. The *transport coefficients* are used for heat transfer problems, and the *friction* or *drag factor* is commonly used to do the closure for momentum applications. Likewise some assumptions on the mixing pattern in the control volume are needed in conjunction with macroscopic mass balance analysis. Macroscopic models provide an overall description of the system and are useful for initial design or a quick estimation of the effect of various parameters on the system's performance.

Mesoscopic balances can be thought to be in between the above two models. These again need some empirical information (e.g., transport coefficient values). These models are used in the design of many types of engineering equipment such as heat exchangers, mass exchangers, dialyser, membrane transport processes, etc., and provide a strong backbone for engineering models. Hence it is important for the student to appreciate the underlying methodology in setting up such process models. Some applications are illustrated in this chapter.

Compartmental models can be viewed as interconnected macroscopic models for a system consisting of several subunits or compartments. The modeling of such systems is therefore a basic extension of macroscopic analysis. They usually require additional *inter-compartmental exchange parameters*. These types of model find applications, for instance, in biomedical engineering (e.g., drug distribution in a complex system such as the human body) and in environmental engineering. Computations of compartmental systems are illustrated using MATLAB using a simple two-compartment model as an example.

2.1 Macroscopic mass balance

We start off with macroscopic balances for no particular reason, perhaps because they are widely used in engineering analysis and preliminary models. These should be the first models to be developed in order to get a feel for the numbers. Mass balance and applications are shown first. This is followed by macroscopic momentum and then energy balance. Applications where all three balances are needed are also common, and a few examples will be used to illustrate such problems.

2.1.1 Species balance equation

General mathematical form

Consider a system as depicted in Fig. 2.1, which has one inlet boundary and one exit boundary, and is enclosed by walls (interpreted here in a general sense). Mass can enter the system from the inlet and leave via the outlet. The main mode of transport here is convection rather than diffusion. Mass can also enter/leave through the walls of the system if the walls are porous (e.g., a membrane). The mechanism of transport here can be either convection or mostly diffusion. Mass can also enter/leave the system through the "walls" if the walls

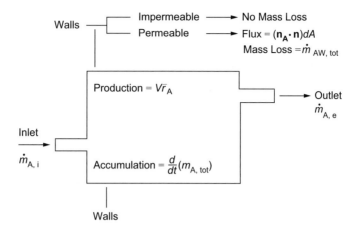

Figure 2.1 The control volume for illustration of the macroscopic species balance.

are made of a catalytic surface or made of a dissolving solid, or the wall is a gas–liquid interface. The main mechanism is diffusion through the walls here.

Species A can be produced by chemical reaction in the volume, and this represents a mass production term. Note that, if A is consumed by reaction, there is depletion of A, but we treat this as a generation term with a negative value. Finally, species A can accumulate within this system if things are changing with time. Putting all these processes together, we can write

$$\text{In} - \text{Out} - \text{Transferred to the walls} + \text{Generation} = \text{Accumulation}$$

Mathematically this is represented as

$$\dot{m}_{A,i} - \dot{m}_{A,e} - \dot{m}_{AW,tot} + V\bar{r}_A = \frac{d}{dt}(m_{A,tot}) \tag{2.4}$$

Each term in the above expression has the units of kg-A/s, or mass of A per unit time. Note that the superscript dot symbol is generally used to indicate quantities per unit time. Also note that the equation can also be written in mole units. We simply divide by the molecular weight M_A throughout.

The first term is the mass of A entering the system per unit time at the inlet boundary. The subscript i is used for the inlet stream.

The second term is the mass of A leaving the system, with e denoting exit.

The third term, $\dot{m}_{AW,tot}$, is the total rate at which mass of A is transferred from the system to the walls. This is defined exactly as the integral of the normal component of the mass flux vector.

$$\dot{m}_{AW,tot} = \int_{A_w} [n \cdot n_A] dA$$

where n is the unit outward normal to the walls. The term n_A is the local mass flux vector at any position in the system (a local or microscopic quantity). Note that $n \cdot n_A$ is the directional derivative of the mass flux vector and hence it represents the mass crossing a unit area perpendicular to the control surface. The total mass leaving the wall is therefore the integral of this term as indicated above.

For an impermeable wall the third term in Eq. (2.4) is zero, provided that no heterogeneous catalytic reaction is occurring at the wall. A catalytic wall is treated as though it were a porous wall, and there is a mass loss/gain of A due to the reaction at the walls. A membrane wall is also treated in a similar manner. Note that in the context of macroscopic models this term has to be closed by defining an overall permeability or mass transfer coefficient, or by other means.

The fourth term is the generation of A in the system due to the chemical reaction. An average rate of production of A per unit volume denoted as $\bar{r}_A$ is used to calculate the rate. The rate can be modeled locally using the laws of chemical kinetics and depends on the local concentration at that point. The detailed variation of this concentration is not available in the macroscopic model and hence an average rate is used as a representative value. Mathematically the average rate can be defined easily by the following equation but in the macroscopic analysis there is no way to calculate this (this is the information-loss principle):

$$\bar{r}_A = \frac{1}{V} \int_V r_A \, dV$$

where r_A is the local rate (in mass units) at any point in the system corresponding to the local concentration. V is the volume of the system.

Finally the term on the RHS is the accumulation term. This is the time rate of change of the total mass of A in the system. The total mass is given as the integral of the local density of A at any point:

$$m_{A,\text{tot}} = \int_V \rho_A \, dV$$

Mole basis

Equation (2.4) can also be expressed in terms of molar units by dividing throughout by the molecular weight of species A. We use $\mathcal{M}$ to denote the number of moles here, and $\dot{\mathcal{M}}$ for the number of moles per unit time:

$$\dot{\mathcal{M}}_{A,i} - \dot{\mathcal{M}}_{A,e} - \dot{\mathcal{M}}_{A,W,\text{tot}} + V\bar{R}_A = \frac{d}{dt}(\mathcal{M}_{A,\text{tot}}) \tag{2.5}$$

Molar units are more convenient for chemically reacting systems since the various species react or form according to their stoichiometric proportions. R_A is the molar rate of production, which is equal to r_A/M_A.

This defines all the terms in the species balance equation in a formal way. Since local information is not available, the various integrals shown above cannot be evaluated in an exact manner and have to be approximated in some suitable manner with some reasonable assumption on the density (or equivalently the concentration) variation in the system.

Use in engineering analysis

With so many unknowns it looks like the macroscopic species mass balance is not very useful. This is not so. The species mass balance equation (2.4) (or on a mole basis as in Eq. (2.5)) can be used in various ways in engineering analysis. If some assumptions regarding the mixing conditions in the unit are made, then some average value for the concentration can be assigned and the term $\bar{r}_A$ can be calculated and used to find other quantities such as the exit flow rate of A in the system. If the exit flow rate of A (or the total flow rate and mass fraction of A) is measured experimentally, the equation can be used to find a value for $\bar{r}_A$, the average rate of production of A in the system. If the mole fractions of all the species in the exit are measured, the equation can be used to check the overall mass balance and therefore the consistency of the experimental data. A few examples will demonstrate these applications.

In our first example the problem of finding the average rate of reaction is closed by assuming that the system is well mixed, which is often used as an assumption in reactor modeling.

Example 2.1. A well-mixed reactor model for waste treatment.

A waste stream is to be treated by holding it for some time in a surge tank prior to discharging it into a river. The waste stream has a dissolved toxic species A, which is unstable and undergoes a first-order chemical reaction. The purpose of the surge tank is to allow enough residence time for some decomposition of A to occur in order to reduce the concentration of the pollutant. Derive an expression for the exit concentration as a function of time and the final steady-state concentration in the system.

Assume that the tank is well stirred and use the backmixed assumption. The problem is sketched in Fig. 2.2.

Solution.

Moles of A in $= QC_{Ai}$

Moles of A out $= QC_{Ae}$

The backmixed assumption implies that the tank is well mixed, meaning that the concentration is the same everywhere and also equal to the exit concentration:

$$\text{Backmixed tank model: } C_A(\text{tank}) = C_{Ae} \tag{2.6}$$

since the tank contents have the same composition, which is also the same as the exit concentration. Hence the generation rate due to reaction is calculated as $-kC_A(\text{tank})$ $V = -VkC_{Ae}$. Here k is the rate constant of the first-order reaction.

Note that the generation rate is assigned a negative sign since A is consumed by reaction. Also the backmixed assumption permits us to write the rate in terms of the exit concentration which we want to calculate. Finally, we have Accumulation $= V \, dC_{Ae}/dt$. Again the

Figure 2.2 A schematic representation of a continuous-flow well-mixed reactor, the continuously stirred-tank reactor (CSTR).

accumulation is calculated based on the following assumptions: (i) the concentration in the tank is assumed to be uniform and equal to the exit concentration (the backmixed assumption); and (ii) the tank volume V is assumed to be maintained at a constant value throughout the operation. This assumption is not valid during the start-up period when an empty tank is being filled.

Putting all the terms together, we have the following differential equation for the change in concentration in the tank as a function of time:

$$V\frac{dC_{Ae}}{dt} = QC_{Ai} - QC_{Ae} - VkC_{Ae} \qquad (2.7)$$

The equation is a first-order differential equation and needs an initial condition.

The initial condition is $C_{Ae} = C_{A0}$ at time $t = 0$. This depends on what was in the tank at the beginning of the process. Let us assume a value for this for now, since the start-up conditions are not specified, and proceed further. Note that for a complete simulation the initial period during which the tank fills up needs to be included as well. The value of C_{A0} can be taken as zero if we assume that the tank has already been filled up with, say, pure water prior to switching to the waste stream.

Also it is convenient to work with dimensionless quantities.

Let $t^* = tQ/V$ be a measure of time.

Let $c_A = C_A/C_{Ai}$ be a measure of concentration.

Let $Da = Vk/Q$ be a dimensionless rate constant; this is usually referred to as the Damköhler number in the reaction engineering parlance. It is the ratio of the holding time (or the residence time) to the reaction time. The differential equation is now

$$\frac{dc_{Ae}}{dt^*} = 1 - (1 + Da)c_{Ae}$$

where c_{Ae} is the dimensionless exit concentraion. The equation is of the form of Eq. (2.1) with $a(t) = 1 + Da$ and $b(t) = 1$. The solution is the sum of a homogeneous solution and a particular solution:

$$c_{Ae} = \alpha \exp[-(1 + Da)t^*] + 1/(1 + Da)$$

with α as the constant of integration. The equation also needs an initial condition (to evaluate α). The initial concentration in dimensionless form is

$$c_{A0} = C_{A0}/C_{Ai} \text{ at } t^* = 0$$

The use of the above initial condition gives the concentration profile in the system as a function of time:

$$c_{Ae} = [c_{A0} - 1/(1 + Da)]\exp[-(1 + Da)t^*] + 1/(1 + Da) \qquad (2.8)$$

The solution has two parts: (i) a time-decaying function and (ii) a steady-state part. The initial transients die out in a time of approximately $3/(1 + Da)$, and the value of the steady-state exit concentration is equal to $1/(1 + Da)$. Note that the time needed to reach steady-state is a function of the Damköhler number.

We now examine the same problem with no chemical reaction taking place and illustrate how the model may be used to ascertain whether the assumption of backmixing is reasonable or not.

2.1.2   Transient balance: tracer studies

Example 2.2. Is it backmixed? Step and pulse tracer response.

The simplified case of Eq. (2.8) with no reaction ($Da = 0$) is important and provides a very useful experimental method to determine whether and to what extent a reactor is backmixed. The method is known as tracer analysis, also known as stimulus–response experiments. The stimulus represents a change in the inlet conditions, and the change in concentration in the outlet as a function of time is measured and represents the response curve of the system. The exit response can be predicted from the above analysis and matched with the experimental data to test some of the model assumptions, for example, whether the assumption that the system is well mixed holds or not.

Two types of tracer injection are common: (i) step injection and (ii) pulse or bolus injection. The outlet responses for these two cases are presented below. This section is also useful for reaction engineering studies, where stimulus–response experiments are common.

Step injection response.

Equation (2.8) is applicable with $Da = 0$ and $C_{A0} = 0$ (assuming that there is no tracer at the start of the process). The inlet concentration is taken as one since we are dealing with dimensionless quantities. The response is then an exponential function of time,

$$c_{Ae} = 1 - \exp(-t^*) \tag{2.9}$$

Hence an experimental result that is close to an exponential rise in concentration of the tracer indicates that the system can be treated as a well-mixed reactor.

Bolus injection response.

Here the tracer is introduced all at once at or around time zero, and no further tracer is added after time zero. Equation (2.8) can still be used, with a slightly different connotation to the variables. However, it is easier to start with the system equation and solve it directly for better clarity. The starting mass balance is Eq. (2.7), which is reproduced below for ease of reference:

$$V\frac{dC_{Ae}}{dt} = QC_{Ai} - QC_{Ae} - VkC_{Ae}$$

The simplifications applicable for this case are (i) inlet concentration $C_{Ai} = 0$, since we are dealing with $t > 0$, i.e., after the bolus injection; and (ii) that there is no reaction of the tracer 0. This leads to

$$V\frac{dC_{Ae}}{dt} = -QC_{Ae}$$

and the solution is

$$C_{Ae} = C_0 \exp(-Qt/V) = C_0 \exp(-t^*)$$

where C_0 is the integration constant, which has the physical meaning of the initial concentration in the tank at $t = 0^+$, i.e., immediately after the tracer injection. The initial concentration is not known *a priori* and depends on the quantity of the tracer injected at time zero. To determine it one needs an overall mass balance of A for the total duration of the experiment. Let the quantity of tracer injected be M. The amount of tracer leaving the system over an interval of time Δt is $QC_{Ae}\,\Delta t$. The total tracer leaving the system is the integral of this from time zero to infinity. By virtue of there being an overall mass balance for A this must be also equal to the quantity of tracer injected. Hence

$$M = \int_0^\infty QC_{Ae}\, dt = \int_0^\infty QC_0 \exp(-Qt/V)dt$$

On performing the integration and rearranging we have $C_0 = M/V$.
   Hence the response to a bolus injection is

$$C_{Ae} = (M/V)\exp(-Qt/V)$$

The bolus injection response is therefore an exponentially decreasing function of time, which can be expressed in dimensionless form as

$$c_{Ae} = \exp(-t^*)$$

where c_{Ae} is equal to $C_{Ae}/(M/V)$.

Tanks in series approximation

If these experiments show that the system is not well mixed, a model can be developed assuming that the system is made up of a number of tanks or compartments connected in series. The resulting model is called the tanks-in-series model and is commonly used in chemical-reactor analysis and in environmental applications. An illustration is shown in Fig. 2.3, where we represent the system as two tanks in series. The system is modeled by writing the mass balance equations for each tank, with the output from tank 1 serving as the input to tank 2. Some exercise problems will give you practice in deriving and solving the tanks-in-series model.

   What type of results would you expect if you were to model the system assuming a large number of tiny tanks connected in series?

   An illustrative response to a system modeled as two tanks in series is shown in Fig. 2.4. An important equation is the overall mass balance, which is taken up now.

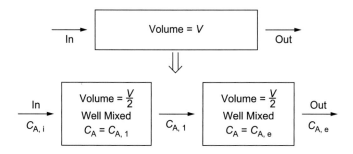

Figure 2.3 A reactor and an equivalent two-tanks-in-series approximation to the reactor.

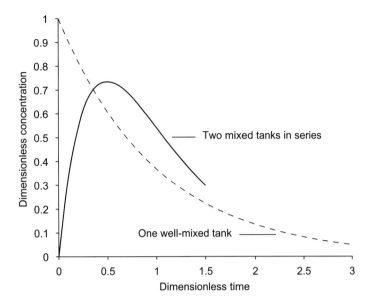

Figure 2.4 Transient concentration profiles in response to a bolus input. Two cases are shown: (i) a system modeled as one backmixed tank and (ii) a system modeled as two tanks in series.

2.1.3   Overall mass balance

On adding Eq. (2.4) for all of the N species (denoted by the subscript j) and noting that

$$\sum_{j=1}^{N} r_j = 0$$

we get the overall mass balance equation

$$\dot{m}_i - \dot{m}_e - \sum_{j} \dot{m}_{j\,W,\text{tot}} = \frac{dm}{dt}$$

where m is the total mass in the system. For the case in which there is no wall mass transfer

$$\dot{m}_i - \dot{m}_e = \frac{dm}{dt}$$

which is a simple statement that the mass in minus the mass out equals the accumulation.

When Eq. (2.5) is summed for all the species we get an equation for the change in the total number of moles in the system:

$$\dot{M}_i - \dot{M}_e + V \sum R_{A,\text{tot}} - \sum \dot{M}_{AW,\text{tot}} = \frac{d}{dt} M_{\text{tot}}$$

Note that $\sum_{j}^{N} R_j$ is not equal to zero in general, unlike $\sum r_j$; only when the number of moles of products is equal to the number of moles of reactants is this term zero.

Overall mass balance

A simple illustration of overall balance is shown by the following example.

Example 2.3. A fluid enters a vessel through an inlet pipe of cross-sectional area A_1 and leaves via a pipe of area A_2. Find a relation between the average velocities in the inlet and exit pipes. Assume steady state.

Mass balance: In − Out = 0 leads to

$$\rho_1 \langle v \rangle_1 A_1 = \rho_2 \langle v \rangle_2 A_2$$

where $\langle v \rangle$ represents a cross-sectional average value for the velocity. If the densities are the same then

$$\langle v \rangle_2 = \langle v \rangle_1 \frac{A_1}{A_2} \tag{2.10}$$

This relation is useful for many problems in macroscopic flow analysis.

The concept of average velocity denoted by $\langle v \rangle$ is quite useful, and the following example demonstrates the calculation of this quantity.

Example 2.4. Average velocity across a plane.

A fluid in a pipe or channel has an axial velocity profile described by a function of a radial position or the cross-flow direction. Let $v_z = v_z(r)$, where v_z is the local axial velocity and r is the cross-flow direction. Derive an expression for the average velocity at that plane.

Solution.

By definition of the average velocity the mass crossing the plane is equal to

$$\dot{m} = \rho A \langle v \rangle$$

Locally over a differential area the mass crossing is equal to $\rho v_z(r)dA$. Hence the total mass crossing the plane is equal to the integral of this over the total cross-sectional area

$$\dot{m} = \int_A \rho v_z \, dA$$

Equating the two gives

$$\langle v \rangle = \frac{1}{A} \int_A v_z \, dA \tag{2.11}$$

which is the formal definition of the cross-sectional average.

If the velocity variation is known in the cross-flow direction, this expression can be used to find the average velocity.

For a circular pipe the equation can be simplified using the equation for the differential area in polar coordinates, which is $dA = r \, dr \, d\theta$. Hence

$$\langle v \rangle = \frac{1}{A} \int_{r=0}^{R} \int_{\theta=0}^{2\pi} v_z r \, dr \, d\theta$$

or

$$\langle v \rangle = \frac{1}{\pi R^2} \int_{0}^{R} 2\pi r v_z \, dr$$

If the form of v_z is known as a function of r, the average velocity can be calculated.

An example for the time needed to drain a tank is now shown. This is a simple example, but it shows that two components (the conservation law and the transport law) go into building a mathematical model.

Example 2.5. Draining of a tank.
A tank has a volume V of liquid at time zero and is draining from the bottom. An expression for the height change with time is needed.

Solution.
The overall mass balance should include accumulation now, since this is a transient problem

$$\text{In} - \text{Out} = \text{Accumulation}$$

The whole tank is taken as the control volume and we have In $= 0$ and Out $= \rho v_o A_o$, where v_o is an average exit speed and A_o is the exit area.

Finally, accumulation is the time rate of change of the total mass of liquid in the tank $= (d/dt)(\rho A h)$, where h is the instantaneous height and A is the cross-sectional area of the tank.

On putting all these together, the overall mass balance gives

$$\text{Accumulation} = -\text{Mass Out}$$

or

$$A \frac{dh}{dt} = -A_0 v_0$$

The use of the conservation law is completed now, but we cannot proceed further because information on v_o is needed. This is the missing information which has to come from a detailed flow model or empirical observations.

The Torricelli law is commonly used here. It states that the speed with which the fluid exits is the same as if a fluid parcel had been dropped from a height h in a gravitational force field. The kinetic energy gained is then equal to the potential energy lost. Hence

$$v_o = \sqrt{2gh}$$

This law can be derived from the Bernoulli equation for energy and completes our model. This is the second component of the model, *viz.* a transport law. The model formulation is now completed, and reads

$$A \frac{dh}{dt} = -A_0 \sqrt{2gh}$$

This, together with the initial condition that, at time $t = 0$, $h = h_0$, the initial height, completes the problem, and the differential equation can then be integrated to find the variation of the level of the liquid in the tank as a function of time. The integration and final solution can be readily worked out, but what you should focus on here is the methodology employed to construct system and transport models. The example shows how the conservation laws are used first. Then some law for transport has to be used as a supplement to complete the problem. This is a common approach for many situations.

The macroscopic mass balance for each species is also the backbone for setting up compartmental models for complex systems, and this procedure will be discussed now.

2.2   Compartmental models

2.2.1   Model equations

Complex systems such as the human body and the environment are often modeled using compartmental models. Conceptually this is similar to the tanks-in-series model discussed earlier; the only difference is that the interconnectivity of compartments can be quite complex. The complex system is therefore treated as a network of interconnected compartments and macroscopic balances (usually) are applied to each compartment. Thus a complete system is viewed as a network of perfectly mixed cells connected in some suitable manner. The compartments can exchange mass with each other. The schematic basis of the compartment model is shown in Fig. 2.5. The mass balance for each species takes the following form:

$$V_j \frac{d(C_{s,j})}{dt} = Q_{j,i} C_{s,j,i} - Q_{j,e} C_{s,j} + \sum_{m=1}^{N} K_{m,j}(C_{s,m} - C_{s,j}) + V_j R_{s,j} \tag{2.12}$$

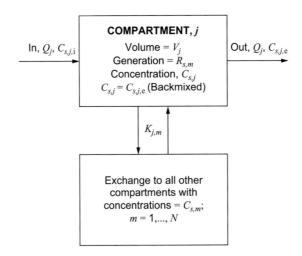

Figure 2.5 A schematic diagram of a compartmental model.

The notations are pertinent to those shown in Fig. 2.5. Here the subscript s refers to a species, while j refers to a tank. The total number of tanks is N. All of the tanks are assumed to be interconnected, with each tank exchanging mass with other tanks. The summation term on the RHS accounts for this interaction. The rate of exchange is $K_{m,j}$, which represents the volume of fluid exchanged per unit time from compartment m to compartment j. Various species may have different exchange coefficients. Hence $K_{m,j}$ should ideally be tagged as $K_{m,j}^{(s)}$ rather than simply as $K_{m,j}$.

Examples of compartmental models are widespread in biomedical systems, pharmacokinetic analysis, environmental systems, and chemical-reactor modeling. We now discuss some computational aspects of the compartmental model with one application. Additional problems are provided at the end of the chapter.

2.2.2   Matrix representation

Compartmental models can be generalized using matrix–vector representation:

$$\frac{dy}{dt} = \tilde{A}y + R$$

You should work out what the coefficient matrix $\tilde{A}$ and the vector R are for the case represented by Eq. (2.12). We discuss the solution here for the case where $\tilde{A}$ and R are constants. The formal solution to such problems can be represented using the concepts of eigenvectors and eigenvalues of a matrix and the concept of a matrix exponential. Recall that if y is a scalar variable then the solution is

$$y = B_1 \exp(At) - R/A$$

where B_1 is an integration constant. The solution consists of two parts: (i) a homogeneous solution and (ii) a particular solution.

The constant of integration B_1 can be evaluated from the initial condition. If $y(t=0)=y_0$ then the solution can be written as

$$y = [y_0 + R/(A)]\exp(At) - R/A$$

The solution for multiple systems of initial-value problems (IVPs) is represented in exactly the same manner as that for a single equation. The solution is

$$y = \exp(\tilde{A}t)[y_0 + \tilde{A}^{-1}R] - \tilde{A}^{-1}R \tag{2.13}$$

where exp is the exponential of a matrix, which can readily be computed using the function *expm* in MATLAB. Here y_0 is the vector of initial values. Using this format the MATLAB implementation of the compartmental model is relatively simple and useful.

Note that the exponential of a matrix is defined similarly to an exponential function. This is simply a power series in matrix form:

$$\exp(\tilde{A}) = \tilde{I} + \tilde{A} + \frac{\tilde{A}^2}{2!} + \frac{\tilde{A}^3}{3!} + \cdots$$

but the numerical implementation is different and uses the Padé approximations.

The use of MAPLE provides the exponential matrix in a symbolic form rather than the numerical values at a given instant of time. Some sample codes are given in the book by White and Subramanian (2010).

Table 2.1. MATLAB code for solution of an initial-value problem

```
function odemain
global param1 param2 % parameters to be given to function dydt.
param1 = 2.0; param2 = 0.0;;
y0 = [1] % vector of initial valus
tspan = linspace (0, 2, 11) ; % time intervals of solution
[t y] = ode45 (@fun1, tspan, y0) % Calling routine
plot (t,y) % solutions; A plot is generated

% function subroutine follows; This could be a separate file.
function dydt = fun1 (t, y)
global param1 param2
dydt = -param1*y; % A simple first-order equation shown.
```

2.2.3   A numerical IVP solver in MATLAB

Matrix methods are extremely elegant but useful only for linear systems. For non-linear systems of equations, numerical integration is the preferred method. The Runge–Kutta method with fourth- and fifth-order discretization is the workhorse here. The function ODE45 does the work for you, and a driver that will be useful for many run-of-the-mill problems is presented below. The driver for a single equation is shown in Table 2.1, but the extension to multiple equations is straightforward. The detailed mathematics behind the Runge–Kutta method is not shown here, so other books should be consulted for this. The emphasis here is more on application of the method in transport-model computations.

An alternative implementation using CHEBFUN is shown as an exercise in Problems 8 and 9. The use of CHEBFUN provides a symbolic "feel" for the results with the efficiency of numerical computations. The final results can be manipulated as though they were ordinary functions rather than numerical data.

Example 2.6. A two-compartment model for pharmacokinetic analysis.
Set up the model equations for a two-compartment model shown in Fig. 1.12 for analysis of the distribution of a drug in a body.

Solution.
Apply the equation given by (2.12) for each compartment. The following equations should result.

For compartment 1 the blood compartment:

$$V_1 \frac{dC_1}{dt} = Q_{1,\text{in}} C_{1,\text{in}} - Q_{1,\text{Out}} C_1 + K_{\text{ex}}(C_2 - C_1) - V_1 k_1 C_1 \qquad (2.14)$$

For compartment 2, the tissue compartment:

$$V_2 \frac{dC_2}{dt} = Q_{2,\text{in}} C_{2,\text{in}} - Q_{2,\text{out}} C_2 + K_{\text{ex}}(C_1 - C_2) - V_2 k_2 C_2 \qquad (2.15)$$

We obtain a system of two differential equations, which can be expressed in the compact matrix–vector form shown in Eq. (2.3). Note that C_1 is the concentration in tank 1 and at the exit of tank 1 as well due to the backmixed assumption. A similar condition holds for tank 2 as well. The solution can then be computed directly using the expm function or can be implemented numerically.

An illustrative result is shown in Fig. 2.6.

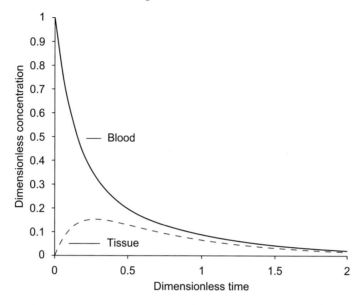

Figure 2.6 Transient concentration profiles in response to a bolus input for a two-compartment model. The parameters used were $V_1 = 1$, $V_2 = 3$, $k_1 = 0.2$, $k_2 = 2.0$, and $K_{ex} = 5.0$ in arbitrary units. The flow rates are taken as zero, which is representative of the drug-distribution case.

Time constants

A note on the time constant is in order. The system time constants for the two-compartment model can be approximated as the reciprocals of the eigenvalues. For $K_{ex} = 5$ (the case shown in Fig. 2.6) the eigenvalues are $\lambda_1 = -7.402$ and $\lambda_2 = -1.4465$.

The results for $K_{ex} = 50$ (not shown here) are interesting in the sense that the system shows only one decay. The eigenvalues are $\lambda_1 = -67.3$ and $\lambda_2 = -1.54$. There is an initial fast decay where the tracer exchanges mass with the tissue corresponding to the first eigenvalue. Blood and tissue acquire the same concentration quickly. This is followed by metabolism in both compartments, leading to a slow decay in the system. In some cases it may be difficult to experimentally detect the initial fast decay; Problem 7 deals with some of these issues.

The stiffness factor

A related concept is that of stiffness of the differential equation. The ratio of the largest to smallest eigenvalues is defined as a measure of the stiffness:

$$\text{Stiffness parameter} = \frac{\max|\lambda|}{\min|\lambda|}$$

In our case above for $K_{ex} = 50$, the stiffness parameter is 40. If the equations are stiff the solver ODE15s should be used in lieu of ODE45.

We now turn our attention to macroscopic momentum balances.

2.3   Macroscopic momentum balance

2.3.1   Linear momentum

General form: the momentum theorem
In the conservation law for momentum the generation of momentum is due to the forces and hence the generation is put as the sum of the forces acting on the system. The input and output are combined as the net outflow term. Hence the starting point is the macroscopic momentum balance, which can be stated in words as

> Sum of the forces = Net rate of momentum outflow from the control surface + Rate of accumulation of momentum within the control volume

The sum of the forces is simply represented as $\sum F$ here. This can be split into the contributions of the individual forces if needed. Likewise the momentum contained in the control volume, c.v., is simply the volume integral of mass (i.e., $\rho \, dV$) times the velocity v and the accumulation term (the last term above) is the time derivative of this term.

Finally the momentum outflow term from a control surface with area A can be expressed as follows. On noting that the momentum efflux is equal to the mass efflux times the velocity, we can calculate this as the integral of $\rho(v \cdot n)v \, dA$ over the control surface, c.s.

Hence the macroscopic momentum balance can be stated mathematically as

$$\sum F = \int_{c.s.} \rho(v \cdot n)v \, dA + \frac{\partial}{\partial t}\left(\int_{c.v.} \rho v \, dV\right) \qquad (2.16)$$

which is referred to in fluid mechanics as the momentum theorem. Note that all quantities are vectors.

Simplified form: single in and single out flow
For many systems, the fluid enters at only one part of the control surface and leaves at another part. Thus there is only a single entry point and a single exit point. The situation is shown schematically in Fig. 2.7, and a simplified form of the momentum theorem can be used for such cases.

Let the normal vector in the outflow direction e_2 be chosen to be the same as the direction of the local velocity vector at the outflow boundary. Then n in the momentum theorem at the exit is equal to e_2. Hence the term $(v \cdot n) = v_2$, where v_2 is the magnitude of the velocity vector.

Then the momentum leaving with the exit fluid can be calculated as

$$\dot{M} = \int_A \rho(v \cdot n)v \, dA = e_2 \int_A \rho v^2 \, dA \qquad (2.17)$$

2.3 Macroscopic momentum balance

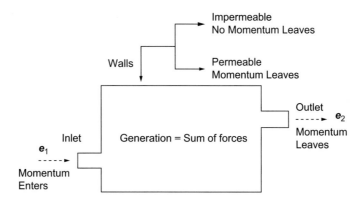

Figure 2.7 The control volume for illustration of the macroscopic momentum balance with single inflow and single outflow. e_1 is a unit vector pointing in the inlet flow direction and e_2 is a unit vector in the outlet direction.

This simplifies further if the velocity over the cross-section is constant (plug flow). Here v_2 is a constant across the entire exit area and can be pulled out of the integral. Then

$$\dot{M} \approx (\rho v_2^2 A)n = (\rho v_2^2 A)e_2 \tag{2.18}$$

Using the mass flow rate, $\dot{m} = \rho v_2 A_2$, this becomes

$$\dot{M} = (\dot{m}v_2)e_2 \tag{2.19}$$

If the velocity profile exists over the cross-section A then it is convenient to introduce a correction factor

$$\dot{M} = \beta_M \dot{m}\langle v_2 \rangle e_2 \tag{2.20}$$

and use this expression for the momentum leaving the system.

On further comparing Eqs. (2.19) and (2.20) the momentum correction factor β_M can be formally defined as

$$\beta_M = \frac{\langle v^2 \rangle}{\langle v \rangle^2}$$

and is the ratio of the cross-sectional average of the velocity squared to the average velocity squared. Note that the square of the average is not always equal to the average of squared values. But the value is close to one in turbulent flow and is often neglected for further simplification. This is because the velocity profile here is nearly uniform except near the walls. For laminar flow with a parabolic profile, the correction factor is seen to be 4/3, and you are asked to verify the result in Problem 10.

Expression Eq. (2.20) holds for the momentum leaving the system in the exit stream. A similar expression holds for the momentum entering the system, since the mass flow rate is the same (unless the walls are porous) we can write

$$\dot{M}_{in} = \beta_M \dot{m}\langle v_1 \rangle e_1 \tag{2.21}$$

where β_M is the momentum correction factor evaluated for the inlet velocity profile.

For a system with one inlet and one outlet the momentum balance can be therefore written in a simpler form as

$$\sum F = \rho \left\langle v_2^2 \right\rangle A_2 e_2 - \rho \left\langle v_1^2 \right\rangle A_1 e_1 + \frac{\partial}{\partial t} \left(\int_{\text{c.v.}} \rho v \, dv \right) \tag{2.22}$$

or, if the mass flow rate and momentum correction are introduced and the steady-state assumption is used,

$$\dot{m} \left(\beta_{M,2} \langle v_2 \rangle e_2 - \beta_{M,1} \langle v_1 \rangle e_1 \right) = \sum F \tag{2.23}$$

Using the Δ notation which represents the difference between the exit value and the inlet value the equation can be expressed in a compact form as

$$\dot{m} \Delta (\beta_M \langle v \rangle e) = \sum F \tag{2.24}$$

where the Δ term is defined as

$$\Delta = \left(\beta_{M,2} \langle v_2 \rangle e_2 - \beta_{M,1} \langle v_1 \rangle e_1 \right)$$

which is another convenient form for problem-solving purposes. Note that vector subtraction is implied on the RHS.

Force components
The force term is usually decomposed into pressure, gravity, and contact forces exerted by the "walls". Hence

$$\sum F = p_1 A_1 e_1 - p_2 A_2 e_2 + \rho V g + F_{\text{sf}}$$

where F_{sf} is the total force acting on the control volume from the walls of the container, the total solid-to-fluid contact force. This force includes the viscous drag and any additional force due to uneven pressure variation on the walls. This information is an integral of the microscopic values and will not be available in the macroscopic model. Some empiricism is needed to fit the data, usually in terms of a drag coefficient or lift coefficient.

A simple application
A simple but very pedagogically useful example is shown in Fig. 2.8. The key point to note is that the input and output terms are vector quantities and if the flow direction changes the vectors e_1 and e_2 change. Hence care has to be used in these calculations. The illustrative sketch in Fig. 2.8 is quite useful to understand this concept.

In case (a) neither the cross-sectional area nor the flow direction changes. The net momentum efflux is zero.

In case (b) the cross-sectional area changes but the flow direction is not changing. The net momentum efflux is non-zero due to velocity changes but has a component only in the x-direction.

In case (c) the cross-sectional area does not change but the flow direction is changing. The net momentum efflux is non-zero and will have components both in the x- and in the y-direction now.

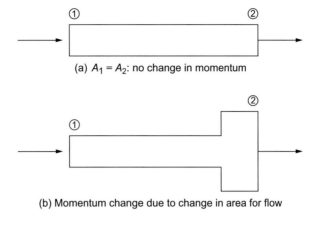

(a) $A_1 = A_2$: no change in momentum

(b) Momentum change due to change in area for flow

(c) $A_1 = A_2$, but momentum change due to change in flow direction

Figure 2.8 A simple example to show the rate of change of momentum.

Finally, both the cross-sectional area and the flow direction can change (this situation is not shown in Fig. 2.8) and the rate of change in momentum can be calculated by accounting for both these effects. Students should work out the expressions for the net momentum efflux for all these cases. Always express the momentum flow rate as a vector quantity in terms of the unit vectors in the chosen coordinate directions. Then perform vector addition/subtraction componentwise. For applications of the macroscopic momentum balance, approximations are usually needed. Some common simplifications are as follows.

(1) Frictionless fluid: the viscous force contribution to F_{sf} is neglected.
(2) Open systems: pressure terms are the same everywhere.

Example 2.7. Fluid impinging on a turbine blade.
A jet of fluid is impinging on a blade. The fluid comes in the x-direction and leaves in the y-direction. See Fig. 2.9. It is required to calculate the force on the blade. Problems of this type are of importance in design of turbines.

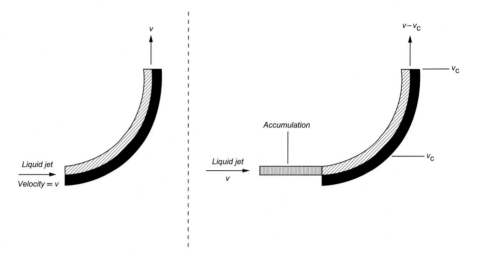

Figure 2.9 Schematic representations of fluid flowing past a curved vane: (a) stationary vane and (b) moving vane seen in a fixed coordinate system.

An overall macroscopic momentum balance is sufficient if detailed information is not needed. For instance we may wish to know the force exerted on the vane and how much power is generated if the vane is moving. Macroscopic momentum balance provides this information, as shown below.

Stationary vane

Momentum in per unit time = mass flow rate $\times$ velocity = $(\rho v A)v e_x$.

Momentum out per unit time = mass flow rate $\times$ velocity = $(\rho v A)v e_y$.

Note that these terms have the same magnitude, but they are not equal since they are vectors. Also one key assumption used here is that the magnitude of the exit velocity is the same as that of the inlet velocity. This is true for a frictionless fluid, as dictated by the Bernoulli equation, which will be discussed later.

The force acting on the control volume equals the rate of change of momentum and hence

$$F_{sf} = \rho A v^2 (e_y - e_x)$$

or, in terms of the components of F_{sf} in the x- and y-directions,

$$F_x = -\rho A v^2$$

and

$$F_y = \rho A v^2$$

These are the force components exerted by the surface on the fluid. By Newton's third law the force exerted by the fluid on the surface is minus the above quantity.

Moving vane

A modification of the problem is when the vane or the blade is moving, say in the x-direction with a velocity v_c. This can be addressed by using a frame of reference moving with the velocity of the vane. The corresponding results are shown without details here and parallel the stationary case:

$$F_x = -\rho A (v - v_c)^2$$

and

$$F_y = \rho A (v - v_c)^2$$

Note that the relative velocity $v - v_c$ is used as the representative velocity.

A moving vane can do work, and the work done by the fluid is equal to

$$\text{Rate of Work} = -F_x v_c = \rho A (v - v_c)^2 v_c$$

Note that the analysis is simplified by using a coordinate system moving with the vane. If a fixed control volume is used, then the accumulation of momentum has to be included, and the calculations are somewhat more involved. But the final result will be the same.

2.3.2   Angular momentum

This section may be omitted without loss of continuity.

The macroscopic balance for angular momentum is very useful in many problems, especially in connection with rotating machinery. The concept is also needed for differential models in fluid–particle or gas–liquid systems and other cases where the particles or the dispersed phase can have a rotational component of motion as well as a translational component. The balance is simply stated as

Rate of change of angular momentum = Net torque acting on the control volume

To describe angular momentum one has to select the origin (usually the location of the axis of rotation) and designate the distance vector of any point from the origin as r. See Fig. 2.10.

The change in angular momentum can be due to (i) the accumulation within the control volume if the velocity is changing with time and (ii) inflow and outflow of angular

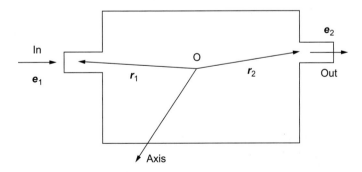

Figure 2.10 A diagram to illustrate angular-momentum calculations. O is the origin or the axis, while r_1 and r_2 are the distance vectors from the origin to the inlet and outlet, respectively. The direction of the axis vector is perpendicular to the plane of the paper.

momentum from the inlet and outlet sections of the controlling surface. These two terms can be represented as follows:

$$\text{Accumulation} = \frac{dL}{dt}$$

where L is the total angular momentum within the system, which can be formally defined as

$$L = \int_V \rho[r \times v]dV$$

where $\times$ represents the vector cross product.

The angular momentum outflow can be represented as

$$\text{Outflow} = \int_A \rho(v \cdot n)[r \times v]dA \approx \rho v_2^2 A[r_2 \times e_2]$$

where e_2 is the unit vector in the outflow direction. In the approximate form above, we have neglected the momentum correction factor and assume a constant velocity across the outflow surface.

This can be further simplified to $\dot{m}v_2[r \times e_2]$, using the mass flow rate $\dot{m}$.

Similarly, the inflow of angular momentum is given by

$$\text{Inflow} = \int_A \rho(v \cdot n)[r \times v]dA \approx \rho v_1^2 A[r_1 \times e_1]$$

which can be expressed in terms of the mass flow rate as $\dot{m}v_1[r_1 \times e_1]$. Hence a useful workhorse equation for the angular momentum is

$$\frac{dL}{dt} + \dot{m}v_2[r_2 \times e_2] - \dot{m}v_1[r_1 \times e_1] = T \qquad (2.25)$$

This is a convenient working formula to use for many simple applications.

What are the units of these quantities? In what direction does the torque vector point?

The net torque depends on the forces acting on the system, which are sometimes difficult to calculate in the context of macroscopic balances. Hence, for problem-solving purposes, the torque can be calculated if other quantities are specified, and *vice versa* other quantities such as rotational speed can be calculated if the torque acting on the system is specified. Problem 12 and the following example are illustrative simple examples to show the usefulness of this for engineering calculations.

Example 2.8.

A centrifugal pump has a diameter of 20 cm and delivers 0.4 m^3/s of water at a tangential velocity of 9 m/s. The pump operates at 1200 r.p.m. The fluid enters axially near the center and is discharged in a direction tangential to the impeller.

Determine the torque on the pump shaft and the power needed.

Solution.

The axis for torque calculation is taken in the axial or z-direction. Since the fluid enters along the axis, the angular momentum of the entering fluid is zero. The angular momentum of the exit stream is calculated as

$$\dot{m}v_2 r_2 \times e_2 = (\dot{m}v_2 R)e_z$$

The mass flow rate is

$$\dot{m} = 1000 \, \frac{\text{kg}}{\text{m}^3} \times 0.4 \, \frac{\text{m}^3}{\text{s}} = 400 \text{ kg/s}.$$

Hence the exit angular momentum is 360 N-m. The torque is therefore the change in angular momentum, which is the same as the exit value 360 N-m since the inlet is at the center and therefore does not contribute to the angular momentum.

The power needed to pump the fluid is calculated as the torque multiplied by the angular velocity, which comes out to be 45 kW. Note that the angular velocity has to be in radians per second and in our case is 125 rad/s.

2.4 Macroscopic energy balances

2.4.1 Single inlet and outlet

The analysis is presented for systems with a single inlet and a single outlet for simplicity. Extension to multiple inlets/outlet can be done by adding the corresponding energy-influx/efflux terms. Also we consider steady state, since this is the most common situation where the macroscopic energy balance is applied. The extension to transient systems by including an energy-accumulation term in the control volume is straightforward.

The energy balance stated in words is Energy leaving per unit time − Energy entering per unit time = Rate of addition of heat to the system + Working rate of the surroundings on the system. The macroscopic control volume used for energy balance together with the terms being balanced are shown schematically in Fig. 2.11.

Entering and leaving terms are calculated by multiplying the mass flow rate by the energy content per unit mass, which is represented as $\dot{m}e$, where e is the total energy per unit mass. This can be written as

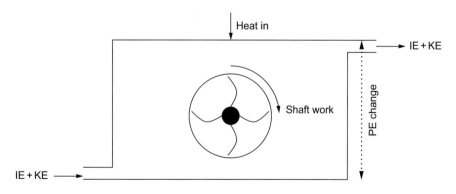

Figure 2.11 The control volume for illustration of the macroscopic energy balance with single inflow and single outflow. IE, internal energy; KE, kinetic energy; and pE, potential energy.

$$e = \hat{u} + \frac{1}{2}v^2 + gy$$

where $\hat{u}$ is the internal energy per unit mass, $v^2/2$ is the kinetic energy per unit mass (assuming a uniform velocity profile or plug flow), and gh is the potential energy per unit mass. Here y is the elevation from a datum plane. Note that the potential energy is actually the contribution of the work done by gravity but it is easier to include it as an energy term.

Hence the balance equation can be written as

$$\dot{m}\left[\hat{u} + \frac{1}{2}v^2 + gy\right]_{\text{out}} - \dot{m}\left[\hat{u} + \frac{1}{2}v^2 + gy\right]_{\text{in}} = \dot{Q}_{\text{tot}} + \dot{W}_{\text{tot}} \tag{2.26}$$

The working-rate term is usually split into two parts: (i) shaft work due to moving parts such as pumps and turbines within the control volume, denoted by $\dot{W}_s$, and (ii) flow work. The flow work represents the work done by the pressure forces in the inlet and outlet boundaries. Then

$$\dot{W}_{\text{tot}} = \dot{W}_s + \dot{W}_{\text{flow}}$$

The flow work can be calculated as follows by noting that the work equals the pressure times the volumetric flow rate. At the inlet the work rate done is $P_{\text{in}}Q$ on the fluid, which is equal to $P_{\text{in}}\dot{m}/\rho$. At the outlet the working rate is $P_{\text{out}}\dot{m}/\rho$ by the fluid. The net flow work is equal to $\dot{m}[P_{\text{in}}/\rho - P_{\text{out}}/\rho]$, and this term is usually moved to the LHS of the energy balance given by Eq. (2.26). Hence

$$\dot{m}\left[\hat{u} + \frac{p}{\rho} + \frac{1}{2}v^2 + gy\right]_{\text{out}} - \dot{m}\left[\hat{u} + \frac{p}{\rho} + \frac{1}{2}v^2 + gy\right]_{\text{in}} = \dot{Q}_{\text{tot}} + \dot{W}_s \tag{2.27}$$

This is the general form of the macroscopic energy balance. Note that ρ at the inlet may be different from that at the outlet if the fluid is compressible. Also note that the kinetic-energy term is based on a uniform velocity profile; otherwise a kinetic energy correction factor needs to be applied (see Problems 10 and 11). This factor is nearly unity for turbulent flow conditions, whereas it is equal to two for a parabolic velocity profile encountered in flow of Newtonian fluids under laminar conditions.

The macroscopic energy equation can also be written as

$$\Delta\left[\hat{u} + \frac{p}{\rho} + \frac{1}{2}v^2 + gy\right] = \frac{\dot{Q}_{\text{tot}}}{\dot{m}} + \frac{\dot{W}_s}{\dot{m}} \tag{2.28}$$

where Δ represents the difference in values at the outlet and the inlet.

The enthalpy form

The pressure term on the LHS of Eq. (2.27) can be combined with the internal-energy term if one uses the enthalpy per unit mass $\hat{h}$:

$$\hat{h} = \hat{u} + \frac{p}{\rho} \tag{2.29}$$

Hence the energy balance is expressed as

$$\dot{m}\left[\left(\hat{h} + \frac{1}{2}v^2 + gy\right)_{\text{out}} - \left(\hat{h} + \frac{1}{2}v^2 + gy\right)_{\text{in}}\right] = \dot{Q}_{\text{tot}} + \dot{W}_s \tag{2.30}$$

which is the enthalpy form of the macroscopic energy balance.

The engineering Bernoulli equation

In many fluid-flow problems the system operates under nearly isothermal conditions and there is no external supply of heat. For such problems the energy balance is usually represented as

$$\Delta\left[\frac{p}{\rho} + \frac{1}{2}v^2 + gy\right] = \frac{\dot{W}_s}{\dot{m}} - \frac{\dot{W}_\mu}{\dot{m}} \tag{2.31}$$

where we add a viscous loss term corresponding to the energy lost by friction. This term represents the internal energy change due to friction and is simply added as a work loss term. A more detailed derivation can be done by volume averaging of the complete energy equation, which is shown in a later chapter (in Section 7.8), and this will show how the loss term arises naturally by a volume averaging of the differential energy balance equation.

The head form of this equation is more commonly used. This is obtained by dividing by g,

$$\Delta\left[\frac{p}{\rho g} + \frac{v^2}{2g} + y\right] = h_{pump} - h_{turbine} - h_{friction} \tag{2.32}$$

where the work term is expressed as a "head" generated by dividing the work term by g. Also this term is split into the pump and turbine terms. The pump term h_{pump} is positive and therefore is a head gain (work done on fluid) due to any pump present in the system, while $h_{turbine}$ is the head lost by the fluid in doing work to drive a turbine.

Friction loss calculation

The friction loss due to viscous forces is usually calculated from the pressure drop for a horizontal pipe. For a straight pipe in fully developed flow, $h_{friction}$ is equal to $-\Delta[p/(\rho g)]$ as per Eq. (2.32) above (since there is no change in velocity and no change in elevation).

The pressure drop in turn depends on whether the flow is laminar or turbulent, and in both cases it can be calculated using the friction factor. Using these two relations, the following result can be generated for the calculation of the head loss due to friction:

$$h_{friction} = 2f\frac{\langle v \rangle^2}{g}\frac{L}{d_t} \tag{2.33}$$

for a circular pipe.

This is very useful equation in pipe-flow analysis. For a network of pipes of different diameters connected in series, each part of the straight pipe contributes its own friction loss as per the above equation. Additional losses arise due to bends, expansion, obstacles to flow such as valves, fittings, etc., and these are correlated by empirical equations. The total friction loss is the sum of all such losses.

2.4.2   The Bernoulli equation

A special case of Eq. (2.31) is the classical Bernoulli equation which applies to a frictionless fluid:

$$\Delta\left[\frac{p}{\rho} + \frac{1}{2}v^2 + gy\right] = 0$$

This equation should be used only when the friction losses are expected to be small compared with the other terms in the equation. It should not be used inadvertently. More discussion is provided later, in Section 15.3.2, where we derive this starting from a differential balance and show the range of its applicability. Furthermore, although we show this as a simplified limiting case arising from the energy balance, it can also be derived purely on the basis of momentum balance considerations for a frictionless fluid.

Some examples of the use of the Bernoulli equation are now demonstrated.

Example 2.9.

A viscous liquid leaves a pipe and emerges as a jet as shown in Fig. 2.12. The diameter of the jet is smaller than that of the pipe from which the jet is emanating. Calculate this diameter from the energy balance. Assume that the flow in the pipe is occurring under laminar conditions.

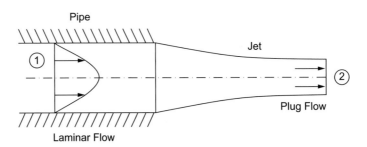

Figure 2.12 A schematic diagram of fluid leaving a die, showing the contraction of a jet.

Solution.

The kinetic energy of the fluid in the pipe is $2 \times \langle v_1^2 \rangle /2$ for laminar flow. Note that we use the kinetic-energy correction factor here, assuming a parabolic velocity profile. Here v_1 is the average velocity in section 1.

The kinetic energy of the fluid in the jet is $\langle v_2^2 \rangle /2$. Here we claim that the velocity profile in the jet is to be plug flow since there is no constraining wall here to reduce the velocity. Hence the kinetic-energy correction factor is taken as one. Here v_2 is the average velocity in section 2. Assuming that there is no friction loss between cross-section 1 and cross-section 2, the kinetic-energy values can be equated:

$$2 \times \langle v_1^2 \rangle /2 = \langle v_2^2 \rangle /2$$

From mass balance for an incompressible fluid $\langle v_1 \rangle A_1 = \langle v_2 \rangle A_2$. On using this in the previous equation, we find $A_2/A_1 = 1/\sqrt{2}$. Hence the diameter ratio of the jet to the pipe is $d_2/d_1 = \sqrt{A_2/A_1} = 0.84$.

Note that some fluids exhibit elastic effects. Such fluids are called viscoelastic fluids (they will be discussed briefly in Section 6.11). Jet swelling effects are observed in such fluids, with d_2/d_1 being larger than one.

Example 2.10.

A Venturi type of pipe as shown in Fig. 2.13 is used for flow measurement. In this problem we examine the pressure profile and how the pressure measurements can be used to find the flow rate in the system. The analysis is done by writing a Bernoulli equation between section 1 and section 2 assuming no friction losses:

$$\frac{p_1}{\rho} + \frac{1}{2}v_1^2 = \frac{p_2}{\rho} + \frac{1}{2}v_2^2$$

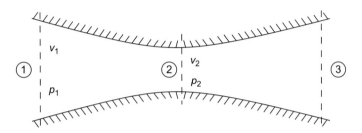

Figure 2.13 A schematic diagram of a Venturi pipe and its pressure profile.

The velocities v_1 and v_2 are related by mass balance:

$$v_1 = v_2 \left(\frac{d_2}{d_1}\right)^2 = \beta^2 v_2$$

where β is the diameter ratio d_2/d_1, which is less than one. We note that the velocity increases as the area decreases. Correspondingly, the kinetic energy at the throat is greater, and to balance this the pressure has to be less, i.e., $p_2 < p_1$.

Using the expression for v_1 in the energy balance, we have

$$v_2 = \frac{1}{\sqrt{1 - \beta^4}} \sqrt{\frac{2(p_1 - p_2)}{\rho}}$$

This expression does not account for the friction losses. A correction factor is applied for this, but the correction factor is close to 0.98 (nearly one) for a well-designed Venturi pipe. The pressure measurements thus provide a means of velocity and flow-rate measurement, which is the principle behind a Venturi meter. The mass flow rate can be calculated from v_2 if the area of the throat (and the fluid density) are known.

Example 2.11. A simple model for a windmill.

In this section we illustrate a simple macroscopic model to analyze a wind turbine. The schematic diagram is shown in Fig. 2.14. Fluid approaches the wind turbine at a velocity of v_1 and pressure of p_∞. It leaves at a velocity of v_2 as shown in the figure. We assume frictionless flow and hence the fluid downstream is at the same pressure p_∞ as that at the inlet. The rate of working and the efficiency are to be calculated by doing a macroscopic analysis.

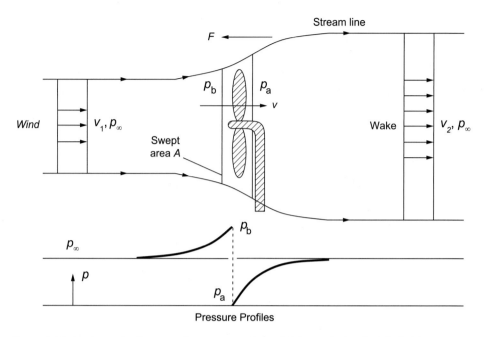

Figure 2.14 A schematic diagram of a macroscopic model for analysis of a windmill. Top: control volume; bottom: pressure profile.

Solution.
There has to be a force on the system as shown in the figure in order to keep the propeller in place. An overall momentum balance (between planes 1 and 2) gives this force, which is equal to the rate of change of momentum. Note that F is the force, exerted by the fluid on the solid:

$$F = \dot{m}(v_1 - v_2)$$

Force balance across the section between a and b gives

$$F = A(p_b - p_a)$$

where A is the swept area.
On equating the two, we have

$$\dot{m}(v_1 - v_2) = A(p_b - p_a) \tag{2.34}$$

Now, applying Bernoulli between 1 and b we get

$$p_\infty - p_b = (1/2)\rho(v^2 - v_1^2)$$

and similarly applying Bernoulli between a and 2 gives

$$p_a - p_\infty = (1/2)\rho(v_2^2 - v^2)$$

Hence

$$p_b - p_a = (1/2)\rho(v_1^2 - v_2^2)$$

Using this in Eq. (2.34), we find

$$\dot{m} = (1/2)\rho A(v_1 + v_2)$$

The overall energy balance gives the working rate of the shaft. The work done per unit time is equated to the change in the kinetic energy:

$$\dot{W}_s = \dot{m}(v_1^2 - v_2^2)/2 = (1/4)\rho A(v_1 + v_2)(v_1^2 - v_2^2)$$

We now discuss the use of the macroscopic balance in some problems where the compressibility of the fluid is of importance. Such problems arise in converging–diverging nozzles and in supersonic flows.

2.4.3   Sonic and subsonic flows

Flow in a converging–diverging nozzle is an example of a problem where all three macroscopic balances are needed. We illustrate this here as a simple case study. The basic analysis starts with a geometry of varying cross-section as shown in Fig. 2.15.
    We write all three balances, starting with the mass balance first.

The mass balance
The mass crossing any section is constant and therefore we have

$$\rho A v = \text{constant} = \dot{m} \qquad (2.35)$$

at any cross-section.
    Note that here we use the simplified notation v to represent $\langle v \rangle$, the cross-sectionally averaged velocity.

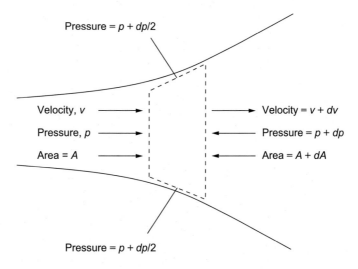

Figure 2.15 A schematic diagram for illustration of the macroscopic momentum balance leading to the Euler equation.

Equation (2.35) can be represented in a differential form (also called the d-form) as

$$d(\rho A v) = 0 \tag{2.36}$$

All of the three variables ρ, A, and v are functions of distance along the nozzle axis.

The equation of state
The density is a function of the local pressure and temperature, and is given by an equation of state. For example, the ideal-gas law is often used:

$$\rho = \frac{M_\mathrm{w} P}{R_\mathrm{G} T} \tag{2.37}$$

We need two more equations for the pressure variation and temperature variation in the system. These are given by momentum and energy balances (simplified with some assumptions).

Momentum balance
The macroscopic x-momentum balance is developed by balancing the change in momentum of the x-component of all of the pressure forces acting on the system. The pressure forces are labeled in Fig. 2.15, and it should be noted that the pressure force from the curved wall labeled as $p + dp/2$ has a component in the x-direction. This component should be included in the x-momentum consideration. The net pressure force can be shown to be $-A\,dp$. The resulting final equation is

$$d(\rho A v^2) = -A\,dp \tag{2.38}$$

Upon expanding the LHS of (2.38) and using the overall mass balance, the x-momentum balance reduces to

$$\frac{dp}{\rho} + v\,dv = 0 \tag{2.39}$$

which is referred to as the Euler equation of motion. It may be noted here that the above equation can also be shown to be a limiting case of the more general Navier–Stokes equation when applied to 1D flow.

Energy balance
Locally we can use the energy balance in terms of the internal energy in the following form:

$$d\left[\hat{u} + \frac{p}{\rho} + \frac{1}{2}v^2\right] = 0 \tag{2.40}$$

where we assume that no heat is added and that we have a frictionless fluid.

Note that an alternative form for quick calculation is the enthalpy form:

$$d\left[\hat{h} + \frac{1}{2}v^2\right] = 0 \tag{2.41}$$

which is the enthalpy form which will be used in some example problems later. Here $\hat{h}$ is the enthalpy per unit mass, which is defined as $\hat{u} + p/\rho$. Also note that $d\hat{h} = c_p\,dT$.

The equation in the internal energy form (2.40) is rearranged by expanding the terms as

$$d\hat{u} + pd\left(\frac{1}{\rho}\right) + \frac{dp}{\rho} + v\,dv = 0$$

Now, using the momentum balance, we find that the last two terms cancel out, leaving us with

$$d\hat{u} + pd\left(\frac{1}{\rho}\right) = 0 \tag{2.42}$$

For further analysis we can use the thermodynamic relation that

$$T\,d\hat{s} = d\hat{u} + pd\left(\frac{1}{\rho}\right) \tag{2.43}$$

On comparing Eqs. (2.42) and (2.43) we find that

$$d\hat{s} = 0$$

This is a remarkable result, which can be stated as follows: *frictionless flow under adiabatic conditions is an isentropic flow.*

This is often stated as a basic postulate in thermodynamic books, but it is interesting to note that it arises directly using the basic transport considerations.

Properties of isentropic flow

The following expressions, which are valid for an adiabatic and isentropic process, are known as adiabatic compression laws. It may be noted that these are not separate laws but merely a combination of the ideal-gas law and isentropic conditions:

$$\frac{T_2}{T_1} = \left(\frac{p_2}{p_1}\right)^{(\gamma-1)/\gamma}$$

where $\gamma = c_p/c_v$, and

$$\frac{T_2}{T_1} = \left(\frac{\rho_2}{\rho_1}\right)^{\gamma-1}$$

$$p\hat{v}^\gamma = \text{constant}$$

Since the flow in the nozzle is isentropic, we can use any one of these relations in lieu of the energy balance.

Summary of equations

We now summarize the equations needed to model a frictionless flow under adiabatic conditions: (i) an energy equation in enthalpy or temperature form, (ii) mass balance, (iii) the ideal-gas law, and (iv) the adiabatic compression law.

We have four equations for four variables, v, ρ, T, and p, and this completes our analysis and modeling of this system. Thus, if we know how the area changes in the flow direction, we can then calculate all these quantities. For many applications the area is calculated as a variable for a given change in velocity together with the other three variables, *viz.* density, pressure, and temperature. Thus there can be many variations to the problem, depending on what is specified and what needs to be calculated, but there are only four equations and hence only four quantities are independently fixed.

Table 2.2. Key thermodynamic relations. Note that in transport calculations these are used locally. This is called the microscopic equilibrium assumption.

Caloric equations of state: a postulate or a constitutive law
$$d\hat{u} = T\,d\hat{s} - p\,d\hat{v}$$

Internal energy variation
$$d\hat{u} = c_v\,dT \text{ at constant volume conditions}$$

Enthalpy definition
$$\hat{h} = \hat{u} + p\hat{v}$$

Enthalpy variation
$$dh = c_p\,dT \text{ at constant pressure}$$

Relation between c_p and c_v (ideal gas)
$$c_p - c_v = R^* = R_{\mathrm{G}}/M_{\mathrm{w}}$$

or
$$C_p - C_v = R_{\mathrm{G}}$$

Ideal-gas law
$$p\hat{v} = R^*T = R_{\mathrm{G}}T/M_{\mathrm{w}}$$

Adiabatic compression law (holds if $d\hat{s} = 0$)
$$p(\hat{v})^\gamma = \text{constant if isentropic}$$

We will show only some key results here, leaving additional problems as exercises. Compressible flow has many complications that are not evident in incompressible flows. See, for instance, Problem 22.

The key thermodynamic relations often needed for transport analysis are summarized in Table 2.2.

The enthalpy form of the energy balance

The enthalpy form of the energy balance is often simpler to use in many applications. It is of the following form:

$$c_p\,dT + d(v^2/2) = 0 \text{ with no external heat transfer}$$

The enthalpy balance states that the enthalpy change and the kinetic-energy change are equal to zero, i.e., the system energy is conserved. The temperature and velocity are therefore related by the above equation.

The value of the speed of sound in a medium is often used as the scaling parameter for velocity, and we discuss the formula for the calculation of this quantity first.

The speed of sound

For an adiabatic flow it can be shown from thermodynamics that

$$\left(\frac{\partial \rho}{\partial p}\right)_{\hat{s}} = \frac{M_{\mathrm{w}}}{\gamma R_{\mathrm{G}}T} = \frac{1}{\gamma R^*T}$$

The pressure change with density at constant entropy is a measure of the speed of propagation of sound waves in the medium and is denoted as c^2. Hence the speed of sound is given as

$$c = \sqrt{\gamma R^* T}$$

Historical vignette. It turns out that Newton derived the equations first but without the γ term. The error was that the partial derivative was evaluated under constant-temperature conditions rather than under constant-entropy conditions. The predicted value was not in agreement with the experimental value. He attributed the disparity to experimental error due to unclean air. The thermodynamics was not developed at that time, and the correct expression was derived later by Laplace.

We also note for our interest that the expression can also be rewritten in terms of bulk modulus of elasticity K:

$$c = \sqrt{\frac{K}{\rho}}$$

which is then applicable to both liquids and gases.

Now we use the equations developed for converging–diverging geometry to show the area–velocity relations in a nozzle of duct of variable cross-section under compressible flow conditions.

Area–velocity relations in a nozzle

The mass balance equation given by (2.36) is written as

$$\frac{dA}{A} + \frac{d\rho}{\rho} + \frac{dv}{v} = 0$$

The term $d\rho/\rho$ is written as $dp/(c^2 \rho)$ using the velocity of sound as a parameter. From momentum analysis $d\rho/\rho$ is equal to $-v\,dv/c^2$, and hence we have

$$\frac{d\rho}{\rho} = -v\,dv/c^2 = -(v/c)^2 \frac{dv}{v}$$

Using this in the mass balance, the following equation for velocity change with cross-sectional area holds:

$$\frac{dv}{v}\left[(v/c)^2 - 1\right] = \frac{dA}{A}$$

where v/c is the local Mach number M. This equation is very useful in understanding some properties of compressible flow in converging–diverging nozzles.

The above equation is often written with v as the independent variable and the cross-sectional area as the dependent variable:

$$\frac{dA}{dv} = \frac{A}{v}\left(M^2 - 1\right)$$

This equation shows that, if $M < 1$ (subsonic flow), the area must decrease to accommodate an increased velocity. The velocity increase is accompanied by a decrease in pressure in accordance with the momentum balance. This behavior is the same as that for an incompressible fluid.

If the flow is supersonic, $M > 1$, the area must increase to accommodate an increased velocity. This is because the density decreases faster than the velocity increase if $M > 1$. Thus the area must increase in order for $\rho A v$ to remain constant. The dependence is now opposite to that for an incompressible fluid. Also, to achieve supersonic flow, the area must first decrease and then increase further. The fluid therefore has to pass through a converging–diverging nozzle.

The sonic velocity can therefore be reached at the throat and nowhere else. This follows from the equation that dv can be non-zero only if $M = 1$.

An example is presented to illustrate these calculations.

Example 2.12.

A converging–diverging nozzle produces a velocity of Mach three at the exit at a flow rate of 1 kg/s. The reservoir conditions are pressure 90 kPa and temperature 298 K.

Find the exit conditions.

Find the conditions at the throat.

Solution.
The inlet density is computed from the ideal-gas law as 1.0523 kg/m^3.

We use the energy equation in temperature form:

$$c_p \, dT + \frac{1}{2} d(v^2) = 0$$

We also note that

$$c_p = \frac{\gamma R^*}{\gamma - 1}$$

Hence

$$v^2/2 = \frac{\gamma R^*}{\gamma - 1}(T_0 - T)$$

The speed of sound at the exit is $\sqrt{\gamma R^* T}$. Hence, on dividing both sides of the above equation by c^2 and rearranging, we have

$$M^2 = \frac{2}{\gamma - 1} \frac{T_0 - T}{T}$$

Since the exit Mach number is three we can now solve for the exit temperature. We find that the exit temperature is equal to 106 K for $\gamma = 1.4$. The student should verify the answer. Considerable cooling has taken place in the nozzle.

The exit pressure is now calculated by using the adiabatic-compression laws using the above value for exit temperature. Exit pressure is equal to 2.45 kPa.

Finally the exit density is calculated from the ideal-gas law as 0.08 kg/m^3.

The area at the exit is therefore equal to $\dot{m}/ > (sv) = 0.0197$ m^2.

Using the exit conditions, the conditions at the throat can be calculated. We use a Mach number of one at the throat since the sonic velocity is reached here if supersonic flow occurs in the diverging section.

This problem illustrates a widely used concept in compressible flow, namely that all of the properties ρ, p, T, and the A ratio can be computed if the Mach number is known at any point.

Supersonic flows can develop shock. Shock means a sudden change in flow properties. There is an entropy gain across the shock, and the isentropic relations no longer apply at the shock front. The following example shows how the property changes across a shock can be calculated using macroscopic balances.

Example 2.13. Momentum and energy balance across a shock.

A normal shock wave occurs in the flow of helium. The conditions at the point of shock are pressure 1/14.7 atm; temperature 5 °C, and velocity 1250 m/s. $\gamma = 1.66$ for helium.
   Find the conditions at the point after the shock wave.
   Determine the entropy and enthalpy gain in the system.

Solution.
The inlet density is computed from the ideal-gas law.
   All conditions are known at the inlet.
   We need four equations for the exit. These are

(1) continuity, $\rho_1 v_1 = \rho_2 v_2$;
(2) momentum $p_1 - p_2 = \rho_2 v_2^2 - \rho_1 v_1^2$;
(3) energy $c_p(T_1 - T_2) = v_2^2/2 - v_1^2/2$, where $c_p = \gamma R^*/(\gamma - 1)$; and
(4) the ideal-gas law $\rho = p/(R^* T)$.

   These equations can be set up and solved. It may be easier to solve them iteratively by assuming an exit velocity. The iterations then proceed as follows. The exit density is calculated from the mass balance. The exit temperature is calculated by enthalpy balance. The exit pressure is calculated by momentum balance. The exit density can now be calculated independently from the ideal-gas law. This should match the exit density from mass balance. Otherwise a new velocity is assumed and the calculations are repeated.
   The entropy change across the shock can be calculated as

$$\Delta_{\hat{s}} = c_v \ln(T_2/T_1) + R^* \ln(\rho_1/\rho_2)$$

The value is 0.8017 J/kg·K The results are summarized in Fig. 2.16.

We now move on to some applications of the macroscopic energy balance to heat transfer problems.

2.4.4   Cooling of a solid: a lumped model

We develop here a macroscopic model for cooling of a sphere in a large body of a cooling liquid held at a constant temperature. We also interpret the meaning of the temperature appearing in the model as a volume-averaged value rather than a point value.

Transient model equation

For many but not all heat transfer problems the contributions of kinetic and potential energy can be neglected and the general energy balance can be simplified to

Accumulation of internal energy = Heat in − Heat out + Heat added from surroundings

+ Internal rate of generation of heat

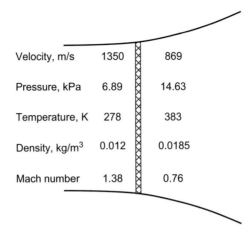

Velocity, m/s	1350	869
Pressure, kPa	6.89	14.63
Temperature, K	278	383
Density, kg/m³	0.012	0.0185
Mach number	1.38	0.76

Figure 2.16 Conditions across a shock for Example 2.13.

For the present problem there is no flow in or out of the system and hence the in and out terms are zero. The accumulation term is evaluated as

$$\text{Accumulation} = \frac{d}{dt}\left[\int_V \rho c_v T \, dV\right]$$

where T_{ref} is taken as zero. (Note that this could be any arbitrary constant quantity, since it appears under the differential sign.) Hence the heat balance is

$$\frac{d}{dt}\left[\int_V \rho c_v T \, dV\right] = \dot{Q}$$

where $\dot{Q}$ is the heat transferred from the fluid to the solid. The heat transferred from the solid to the fluid needs to be modeled by a transport law in terms of a heat transport coefficient. Hence this is equal to $hA_p(T_s - T_f)$. Again this is an approximation and assumes that the heat transfer coefficient is the same at every point on the surface of the solid and that the surface temperature of the solid is also uniform at every point on the surface of the solid. We define a volume-averaged temperature for the solid as

$$\bar{T}_p = \frac{1}{V}\int_V T \, dV$$

Hence the accumulation is represented as

$$\text{Accumulation} = V\rho c_v \frac{d\bar{T}_p}{dt}$$

Using this in the heat balance, the final expression is:

$$\boxed{V\rho c_v \frac{d\bar{T}_p}{dt} = -hA_p(T_s - T_f)} \qquad (2.44)$$

which is the transient macroscopic heat balance for the system. Note that the temperature appearing on the LHS is the volume-averaged temperature of the solid, while that appearing

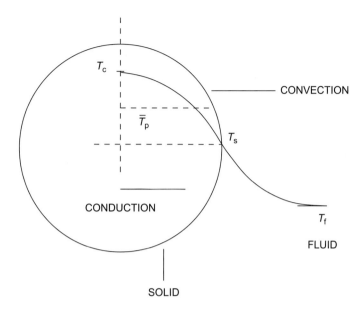

Figure 2.17 The lumped model for the cooling of a solid. The diagram shows the various temperatures in the system.

on the RHS is the surface temperature of the solid. This is an exact representation of the system. The various temperatures are shown in Fig. 2.17. The center temperature is higher than that at the surface, and $\bar{T}_p$ is some average temperature in the solid. Since the internal profiles are not being solved in the macroscopic model, the model is not closed and the equation cannot be solved unless some assumptions are made. One useful approximation is the lumped-model analysis, which will now be discussed.

The lumped-model approximation

The temperatures $\bar{T}_p$ and T_s are assumed to be nearly the same, which is called the lumping approximation. The conditions for the validity of the lumping assumption will be examined later. To cut the story short, the lumping analysis (a first-level model) applies if the volume-averaged temperature of the material is nearly the same as the center temperature of the solid. This means that the heat conduction within the solid has to be quite fast. The conduction resistance must be smaller than the convection resistance. The representative distance for conduction is taken as V_s/A_p, where V_s is the solid volume and A_p is its external surface area. Hence a measure of the conduction resistance is equal to $(V_s/A_p)/k$. The convection resistance is equal to $1/h$. The ratio of two is a dimensionless group known as the Biot number,

$$Bi = \frac{\text{Conduction resistance}}{\text{Convection resistance}} = \frac{h(V_s/A_p)}{k}$$

If $Bi < 0.2$ the lumped assumption may be assumed to be valid. On the other hand, for large-Biot-number conditions, internal temperature gradients within the solid are important, and a differential model has to be used for the particle. The corresponding model is a PDE, as will be shown later.

The time of cooling is then obtained by direct integration for the lumped model, which takes the form

$$\rho c_p V \frac{d\bar{T}_p}{dt} = -h A_p (\bar{T}_p - T_f) \tag{2.45}$$

This equation is often referred to as Newton's law of cooling; the rate of cooling is proportional to the temperature difference.

A dimensionless average temperature can be defined as

$$\bar{\theta} = \frac{\bar{T}_p - T_f}{T_0 - T_f}$$

where T_0 is the initial temperature. A time constant for convective heat loss t_c of $\rho c_p V / (h A_p)$ suggests itself. Hence the dimensionless time is defined as

$$t^* = t/t_c = \frac{h A_p t}{\rho c_p V}$$

The solution of the lumped model for cooling in dimensionless terms is

$$\bar{\theta} = \exp(-t^*)$$

In time equal to t_c cooling by 63% would have taken place.

We now show a case where the solid is cooling in a liquid of finite volume. Two differential equations, one for the solid and one for the liquid, are needed.

Example 2.14. Cooling of a sphere in a liquid of finite volume.

The heat balance for the solid remains the same, except that now the liquid temperature also varies with time. Hence a heat balance for the liquid must also be established. The balance equation is

$$V_f \rho_f c_{vf} \frac{dT_f}{dt} = h A_p (T_s - T_f)$$

The assumption is that the tank is well mixed that the temperature is uniform in the tank.

This has to be solved together with the equation of the solid (which remains the same but is reproduced below for completeness) from the simple lumped-parameter model:

$$\rho c_p V \frac{dT_s}{dt} = -h A_p (T_s - T_f) \tag{2.46}$$

Note again that the lumping assumption is valid only for small values of the Biot number, and hence T_s is taken as the representative temperature of the solid in the accumulation term.

The dimensionless version of the equation is obtained by defining the following dimensionless quantities:

$$\bar{\theta}_s = \frac{T_s - T_{f0}}{T_{s0} - T_{f0}}$$

and

$$\theta_f = \frac{T_f - T_{f0}}{T_{s0} - T_{f0}}$$

where T_{f0} and T_{s0} are the initial temperature of the fluid and that of the solid, respectively. This choice makes the dimensionless temperature go from zero to one. Also the initial temperature of the solid is one and the initial temperature of the fluid is zero.

The time is defined in the same way as before. An additional parameter, C is needed. This is the capacity ratio:

$$C = \frac{\rho c_p V}{V_f \rho_f c_{vf}}$$

With these definitions, the equations in dimensionless form are

$$\frac{d\bar{\theta}_s}{dt^*} = -[\bar{\theta}_s - \theta_f] \tag{2.47}$$

and

$$\frac{d\theta_f}{dt^*} = C[\bar{\theta}_s - \theta_f] \tag{2.48}$$

The details of the solution are not shown here and should be worked out. The final results are as follows:

$$\bar{\theta}_s = \frac{C}{C+1} + \frac{1}{C+1} \exp(-(C+1)t^*)$$

and

$$\theta_f = C(1 - \theta_p)$$

The steady-state values of the solid and fluid temperatures are equal, and the numerical value is $C/(C+1)$.

An additional example of heating of fluid in a stirred tank is now illustrated.

Example 2.15. Develop a dynamic model for a heating of a fluid in tank. Heat is supplied from a jacketed vessel with steam heating. The transfer rate across the system may be assumed to be $UA(T_s - T)$, where U is the overall heat transfer coefficient. Here T_s is the steam temperature and T is the process fluid temperature in the tank. The tank is assumed to be well mixed in this case.

Solution.
In this case the heat in and out terms due to flow in and out are to be included:

$$\text{Accumulation} = \text{Heat in} - \text{Heat out} + \text{Heat added from surroundings}$$

The mathematical representation is then given as

$$\rho c_p V \frac{d\bar{T}}{dt} = \dot{m} c_p (T_i - T_e) + UA(T_s - \bar{T})$$

where $\bar{T}$ is the average temperature in the tank. If the tank is well mixed then $\bar{T} = T_e$, which we simply denote as T, the temperature of the fluid in the tank. The governing equation then is

$$\rho c_p V \frac{dT}{dt} = \dot{m} c_p (T_i - T) + UA(T_s - T)$$

A dimensionless temperature can now be defined as

$$\bar{\theta} = \frac{T - T_i}{T_s - T_i}$$

Defined this way, the inlet temperature is equal to zero and the steam temperature is one.

The dimensionless parameters needed can be shown to be

$$U^* = \frac{UA}{\dot{m}c_p}$$

which is a measure of the heat transfer coefficient in dimensionless units, and

$$t^* = t/t_c = \frac{\dot{m}c_p}{\rho c_p V}t = \frac{\dot{m}}{\rho V}t$$

which is the dimensionless time.

The dimensionless form of the differential is then given, after a minor algebraic rearrangement, as the following:

$$\frac{d\theta}{dt^*} = -(1 + U^*)\theta + U^* \tag{2.49}$$

Thus the parametric representation of the problem is

$$\theta = \theta(t^*; U^*)$$

Note that the representation is compact, being in terms of just two parameters. This is an advantage of writing the problem in dimensionless form. The solution is obtained by integration of the model with the given initial condition:

$$\theta = \frac{U^*}{1 + U^*}\left\{1 - \exp[-(1 + U^*)t^*]\right\} \tag{2.50}$$

and the final steady-state temperature in dimensionless form can be shown to be

$$\theta_\infty = \frac{U^*}{1 + U^*}$$

This expression for the steady-state value can also be independently verified by setting the time derivative in Eq. (2.49) to zero and directly solving for the tank temperature. You should verify such details and examine the limiting-case situation for all transport problems. This also acts as a check on your solution.

Problems of this type are important in process control, where the tank temperature has to be controlled to remain within some narrow and precise range of temperature. A dynamic simulation is a prerequisite for control, since it sets the transient response to any process disturbance. The extension to a case where the jacket temperature in the system varies is shown as Problem 32. Reacting systems can also be analyzed by adding the mass balance equation and then adding an additional heat-generation term in the heat balance.

Now we move to simple differential balances. Some simple examples are provided to illustrate the problem formulation methodology. The examples considered are 1D problems where the variable of interest changes in one spatial dimension only.

2.5 Examples of differential balances: Cartesian

2.5.1 Heat transfer with nuclear fission in a slab

Consider a slab of material containing a fissionable material. The nuclear reaction in the material produces heat at a rate of $\dot{Q}_v$ per unit volume of the slab. Assume that the top, bottom, front, and back sides are insulated, and the heat flow is therefore solely in the x-direction. We develop here a differential model for this system and the differential equation for the temperature in the system.

The control volume is a differential element of thickness Δx as shown in Fig. 2.18. A balance leads to

Heat in at x − Heat leaving at $x + \Delta x$ + Generation (Heat produced in the volume) = 0

On translating this into mathematics term by term, we have the following.

The heat entering surface x by conduction is $(q_x)_x A$.

The heat leaving surface $x + \Delta x$ by conduction is $(q_x)_{x+\Delta x} A$.

The rate at which heat is produced in the volume is $\dot{Q}_v A\,\Delta x$.

On putting all this together, we obtain the following heat balance:

$$(q_x)_x A - (q_x)_{x+\Delta x} A + \dot{Q}_v A\,\Delta x = 0$$

Taking the limit as $\Delta x \to 0$, we obtain a differential equation for the variation of the heat flux in the system:

$$\frac{dq_x}{dx} = \dot{Q}_v$$

which is the basic model obtained by use of the conservation principle alone. The model is then closed by applying a constitutive equation. Fourier's law is now used to complete the problem:

$$q_x = -k\frac{dT}{dx}$$

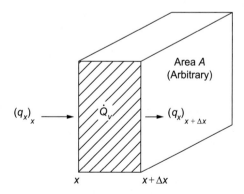

Figure 2.18 The control volume for 1D differential analysis in Cartesian coordinates. The case of heat conduction with generation is shown here.

Hence

$$\frac{d}{dx}\left(k\frac{dT}{dx}\right) = -\dot{Q}_v \qquad (2.51)$$

This is a second-order differential equation. The nature of the equation depends on the form of the function for k and $\dot{Q}_v$. In general, the thermal conductivity is a function of the local temperature in the slab, i.e., $k = k(T)$. This leads to a non-linear differential equation. The non-linearity is referred to as material non-linearity since it arises from the material-property variation. However, to simplify the problem, the thermal conductivity is often assigned a constant value evaluated at some average temperature in the slab.

The model then simplifies to

$$k\frac{d^2T}{dx^2} = -\dot{Q}_v \qquad (2.52)$$

Now, referring to the RHS, the rate of heat generation depends on the temperature and the quantity of active fissionable material present at that point. Hence $\dot{Q}_v$ is in general a function of x and T, i.e., $\dot{Q}_v(x, T)$. The differential equation becomes non-linear if $\dot{Q}_v$ is any non-linear function of T. This is often referred to as source/rate non-linearity. The case where $\dot{Q}_v$ is a constant is the most simple and a general expression for the temperature profile is derived in the following example.

Constant $\dot{Q}_v$ case
For the case where $\dot{Q}_v$ is a constant a general expression for the temperature profile in the system can be derived by integration of Eq. (2.52) twice:

$$T(x) = -\frac{\dot{Q}_v}{2k}x^2 + A_1x + A_2 \qquad (2.53)$$

Here A_1 and A_2 are integration constants that are determined by the problem-specific boundary conditions. The following example shows this for a specific problem. But in general we note here that the temperature has to be a quadratic function of x. In passing we note that the temperature profile across the slab is a linear function of x in the absence of heat generation.

A simple example to illustrate the calculation of the integration constants is shown below. These cases of variable thermal conductivity and a non-constant rate of heat generation are analyzed in a later chapter.

Example 2.16.
A slab of thickness $2L$ is maintained at a surface temperature of T_s at both ends ($x = \pm L$). There is a constant internal heat generation at a rate of $\dot{Q}_v$. Find the temperature profile, the maximum temperature, and the heat being transferred to the surroundings.

Solution.
In view of the symmetry we set the center of the slab at $x = 0$. We state that dT/dx must be zero here due to symmetry. Using this in Eq. (2.53), we can show that $A_1 = 0$.

At the surface $x = L$ the temperature takes a value of T_s. Hence A_2 can be calculated and substituted back into Eq. (2.53). The final solution for the temperature profile is

$$T(x) = T_s + \frac{\dot{Q}_v}{2k}(L^2 - x^2)$$

The maximum temperature occurs at $x = 0$ and is equal to $T_s + \dot{Q}_v L^2/(2k)$.

By applying Fourier's law at the surface the heat leaving the solid from $x = L$ can be calculated:

$$q_s = -k\frac{dT}{dx} = \dot{Q}_v L$$

which also agrees with the overall heat balance. The heat leaving the surface is seen to be equal to the total heat produced in half of the slab.

2.5.2 Mass transfer with reaction in a porous catalyst

Consider a porous catalyst in the form of a long slab with a chemical reaction taking place at a rate of R_A per unit volume of the slab. Assume that the mass flow of species A is therefore only in the x-direction. Develop a differential model for this system and the differential equation for the concentration of A in the system.

We proceed in a similar manner to the heat transfer problem.

Moles of A entering surface x by diffusion = $(J_{A,x})_x A$.

Moles of A leaving surface $x + \Delta x$ by diffusion = $(J_{A,x})_{x+\Delta x} A$.

Moles of A produced in the volume = $R_A A \, \Delta x$.

A balance leads to

Moles produced in the volume + Moles in at x − Moles leaving at $x + dx = 0$

On translating this into mathematics, we have

$$R_A A \, \Delta x + (J_{A,x})_x A - (J_{A,x})_{x+\Delta x} A = 0$$

On taking the limit as $\Delta x \to 0$ we obtain a differential equation for the variation of the heat flux in the system:

$$\frac{dJ_{A,x}}{dx} = R_A$$

This is the basic model for the system. The model can be completed if a constitutive (transport) law for diffusion of species A in a porous solid is available or can be developed. Fick's law is now used here to complete the problem:

$$J_{A,x} = -D_{eA}\frac{dC_A}{dx}$$

where D_{eA} is the (effective) diffusion coefficient of A in the porous medium. Using Fick's law and the mass balance Eq. (2.5.2), we get

$$\boxed{D_{eA}\frac{d^2 C_A}{dx^2} = -R_A} \qquad (2.54)$$

This equation is known as the diffusion–reaction equation and is widely applied in the analysis of transport in catalysts.

Further information on the kinetics of the reaction is needed. In general R_A is a function of the concentration of species A and of the temperature. It is also often a function of the concentrations of other species in the system (e.g., bimolecular reactions, reactions with strong product inhibition, etc.). These dependences are given by a kinetic model for the process. In the following examples, we examine the solution to the problem for simple forms for R_A, viz. a zeroth-order reaction and a first-order reaction.

Example 2.17. Diffusion with zeroth-order reaction.

Here we set

$$R_A = -k_0$$

where k_0 is the volumetric rate of consumption of reactant A. Equation (2.54) can be integrated twice to obtain

$$C_A(x) = \frac{k_0}{2D_{eA}}x^2 + A_1 x + A_2$$

which is the general solution for the concentration profile. The constants of integration A_1 and A_2 can be found if the boundary conditions are specified. Note the analogy with the corresponding equation for the temperature profile in the earlier problem.

A common application of this model is in trickling-bed filters used in waste-water treatment.

Example 2.18. Diffusion with first-order reaction.

In this case

$$R_A = -k_1 C_A$$

where k_1 is the rate constant for a first-order reaction. Equation (2.54) now takes the following form:

$$D_{eA}\frac{d^2 C_A}{dx^2} - k_1 C_A = 0 \tag{2.55}$$

and can be solved by treating it as a linear second-order differential equation. The general solution can be expressed in two equivalent forms:

$$C_A(x) = A_1 \cosh\left(x\sqrt{k_1/D_{eA}}\right) + A_2 \sinh\left(x\sqrt{k_1/D_{eA}}\right)$$

with A_1 and A_2 being integration constants, or

$$C_A(x) = B_1 \exp\left(x\sqrt{k_1/D_{eA}}\right) + B_2 \exp\left(-x\sqrt{k_1/D_{eA}}\right)$$

where B_1 and B_2 now represent the integration constants.

How are B_1 and B_2 related to A_1 and A_2?

A common application is in diffusion models for a catalyst pore. Further details are discussed later in Section 10.4. At this stage the model development and the nature of the resulting differential equation should be noted. The general solution with two integration constants is also noteworthy, since it can be applied to many different boundary conditions.

Non-linear reactions and other complexities associated with non-linear diffusion are treated in a later chapter. In many cases numerical solutions are needed, and MATLAB-based programs are also discussed later.

2.5.3   Momentum transfer: unidirectional flow in a channel

A fluid is contained between two parallel plates. The motion of the fluid can be caused by either (i) applying a pressure gradient or (ii) moving one of the plates. The first case is referred to as pressure-driven flow. The second case is called shear-driven flow. One can also have a combination of pressure-driven and shear-driven flow. The problem examined here is how the differential models can be used to find an expression for the velocity profile in this system. The analysis is based on the assumption that the flow is laminar.

The momentum balance is based on a differential control volume as shown in Fig. 2.19. The forces shown in this figure are balanced as shown below.

The shear force on the top surface is $\tau_{y+\Delta y}\,\Delta x\,W$. Here W is an arbitrary width in the plane of the paper.

The shear force on the bottom surface is $-\tau_y\,\Delta x\,W$.

The pressure force at the fluid entrance is $p_x\,\Delta y\,W$.

The pressure force at the fluid exit is $-p_{x+\Delta x}\,\Delta y\,W$.

In a fully developed unidirectional flow, the velocity is only in the x-direction and is not changing as a function of x. Hence there is no net convective flow of momentum. Therefore, by Newton's second law, the sum of the forces is equal to zero. Hence

$$\tau_{y+\Delta y}\,\Delta x\,W - \tau_y\,\Delta x\,W + p_x\,\Delta y\,W - p_{x+\Delta x}\,\Delta y\,W = 0$$

On dividing throughout by $\Delta x\,\Delta y$ and taking the limit we have.

$$\frac{\partial \tau}{\partial y} - \frac{\partial p}{\partial x} = 0 \qquad (2.56)$$

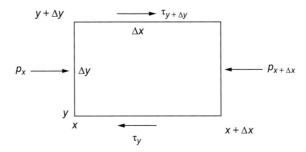

Figure 2.19 The control volume for flow analysis in a rectangular channel and the forces acting on the control volume.

The model is the basic model in the stress form. The model can be completed by application of the Newton's law of viscosity, which applies to common fluids such as air and water, giving

$$\tau = \mu \frac{dv_x}{dy}$$

The following differential equation is then obtained for the velocity profile in the system:

$$\mu \frac{\partial^2 v_x}{\partial y^2} = \frac{\partial p}{\partial x} \qquad (2.57)$$

The velocity profiles can be computed if the pressure-gradient term is specified. Often the pressure changes linearly in the flow direction, which is again a consequence of unidirectional flow. Hence the pressure gradient is treated as a constant for such cases. Let

$$\frac{\partial p}{\partial x} = -G$$

which is a constant. Then the differential equation for the velocity takes the form

$$\mu \frac{d^2 v_x}{dy^2} = -G \qquad (2.58)$$

The equation can be integrated to get the following general expression for the velocity profile:

$$v_x = -\frac{G}{2\mu} y^2 + A_1 y + A_2$$

The constants can be evaluated from the boundary conditions.

For a channel of height d we use the no-slip boundary condition. The velocity is zero at $y = 0$ and at $y = d$. Using this, we can find A_1 and A_2 and back substitute into the solution. The final solution for the velocity profile then is

$$v_x = \frac{G}{2\mu} y(d - y)$$

which is known as the plane Poiseuille flow equation.

Note that the equations for heat, mass, and momentum for this problem have similar forms, as can be noted from Eqs. (2.52), (2.54), and (2.58). The solutions too have similar forms.

2.6   Examples of differential models: cylindrical coordinates

Now we repeat the analysis for the case of a cylindrical geometry to show some nuances that arise primarily due to the nature of the coordinate system. We primarily look at problems where the primary variable is a function of radial position. The point to bear in mind is that the area over which conduction/diffusion takes place is a function of the radial position.

2.6.1   Heat transfer with generation

The problem is similar to that in Section 2.5.1, except that the generating solid is a cylinder or concentric annular cylinder. These two geometries can be handled together. The heat flow is now in a radial direction.

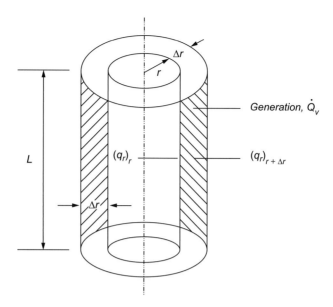

Figure 2.20 The control volume for 1D differential analysis in a long cylinder (with axisymmetry).

The heat conduction analysis is done in the following manner.

The heat entering surface r by conduction is $(q_r)_r A(r)$, where $A(r)$ is the area for heat transfer at the position r.

The heat leaving surface $r + \Delta r$ by conduction is $(q_r)_{r+\Delta r} A(r + \Delta r)$.

The heat produced in the volume is $\dot{Q}_v(2\pi r\, \Delta r\, L)$ since the volume element contained in Δr can be approximated as $(2\pi r\, \Delta r\, L)$.

A balance leads to

$$\text{Heat produced in the volume} + \text{Heat entering by conduction at } r$$

$$-\text{ Heat leaving at } r + dr = 0$$

Hence

$$\dot{Q}_v(2\pi r\, \Delta r\, L) + (q_r)_r A(r) - (q_r)_{r+\Delta r} A(r + \Delta r) = 0$$

Taking the limit as $\Delta r \to 0$, we have

$$\dot{Q}_v(2\pi rL) - \frac{d}{dr}[(q_r)A(r)]$$

Note that $A(r)$ remains inside the derivative since it is a function of r and cannot be pulled out. Further simplification is done by assigning

$$A(r) = 2\pi rL$$

leading to the following differential equation for the radial heat flux:

$$\frac{d}{dr}[rq_r] = \dot{Q}_v r$$

If Fourier's law is now used for q_r ($q_r = -k\,dT/dr$), the following differential equation is obtained for the temperature:

$$\frac{k}{r}\frac{d}{dr}\left(r\frac{dT}{dr}\right) = -\dot{Q}_v \tag{2.59}$$

Here we have used a (constant) mean thermal conductivity for the system and have also moved the r-term to the LHS.

The general solution depends on the nature of the source term, $-\dot{Q}_v$. If this is a non-linear function of temperature then analytic solutions are often difficult or impossible to find. Numerical solutions may be needed in such cases. If this term is a constant, a general solution for the temperature profile is obtained by integrating (2.59) twice:

$$T(r) = -\frac{\dot{Q}_v r^2}{4k} + A_1 \ln r + A_2$$

The solution depends on the boundary conditions and the overall geometry, i.e., whether the nuclear material is contained in a solid cylinder or a concentric cylinder. These specific cases are taken as exercises and also discussed in a later chapter, since the focus of this chapter is mainly on the analysis and the model formulation. Further details are taken up in Section 8.2.1.

2.6.2   Mass transfer with reaction

Consider the analogous problem of mass transfer in a porous catalyst, with a chemical reaction of a species A taking place at the rate R_A. The shape of the porous catalyst is assumed to be a long cylinder (or an annular cylinder) so that diffusion is occurring in the radial direction only. The governing equation can be readily written down by inspection, with the corresponding heat transfer problem as

$$\frac{D_{eA}}{r}\frac{d}{dr}\left(r\frac{dC_A}{dr}\right) = -R_A \tag{2.60}$$

The details leading to the derivation of the above equation have been omitted, but it is rather obvious.

The solution can be obtained for simple cases for R_A, but in general a numerical solution is needed. The solution for a zeroth order reaction is obtained easily by integration twice and has the same form as the temperature equation.

The solution for a first-order case,

$$R_A = -k_1 C_A$$

is somewhat complicated. Use of this rate form in Eq. (2.60) leads to

$$\frac{D_{eA}}{r}\frac{d}{dr}\left(r\frac{dC_A}{dr}\right) = k_1 C_A \tag{2.61}$$

and the solution of this is in terms of the Bessel functions. The general solution is

$$C_A(r) = A_1 I_0\left(r\sqrt{k_1/D_{eA}}\right) + A_2 K_0\left(r\sqrt{k_1/D_{eA}}\right)$$

where I_0 and K_0 are the modified Bessel functions of the first and second kind, respectively.

For a solid cylinder the boundary conditions to be used are generally as follows.

At $r = 0$ the concentration is finite but the K_0 function becomes equal to $-\infty$. Hence A_2 is set as zero.

The concentration at $r = R$ is set as the surface concentration C_{As} as the second boundary condition and the following expression can be derived for the concentration in a solid cylinder with a first-order chemical reaction:

$$C_A(r) = C_{As} \frac{I_0 \left(r\sqrt{k_1/D_{eA}}\right)}{I_0 \left(R\sqrt{k_1/D_{eA}}\right)}$$

2.6.3   Flow in a pipe

We now analyze flow in a pipe subject to some assumptions. The flow is assumed to be fully developed and the velocity is assumed to have only an axial component. Also the flow is assumed to be laminar. The control volume used in the analysis is an annular ring element as shown in Fig. 2.21. Since the flow is unidirectional there is no contribution due to momentum change. Hence the force balance is done here, and the sum of the forces acting on the ring element is equated to zero.

The following expressions can be written for the forces.

The shear force on the bottom surface is the shear stress times the area of the bottom surface.

The force at r is $-\tau_r A(r)$, where $A(r) = 2\pi r \, \Delta z$, a function of the local radial position. Also note that a negative sign is used here (why is it needed?).

Similarly the shear force on the top surface is $(\tau)_{r+\Delta r} A(r + \Delta r)$.

The pressure force at the fluid entrance is $p_z[2\pi r \, \Delta r]$.

Note that the area over which the pressure acts is equal to $2\pi r \, \Delta r$.

Finally the pressure force at the fluid exit surface is $-p_{z+\Delta z}[2\pi r \, \Delta r]$.

Note the negative sign on this term (why is it needed?).

By Newton's second law the sum of the forces is equal to zero. Hence

$$-\tau_r A(r) + \tau_{r+\delta r} A(r + \Delta r) + p_z[2\pi r \, \Delta r] - p_{z+\Delta z}[2\pi r \, \Delta r] = 0$$

On dividing throughout by $\Delta r \, \Delta z$ and taking the limit, we have

$$\frac{1}{r}\frac{\partial (r\tau)}{\partial r} - \frac{\partial P}{\partial z} = 0 \qquad (2.62)$$

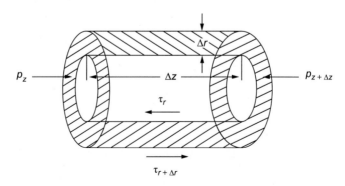

Figure 2.21 The control volume for flow analysis in a pipe and the forces acting on the control volume.

The model is the basic model in the stress form. The model can be completed by application of Newton's law of viscosity:

$$\tau = \mu \frac{dv_z}{dr}$$

leading to (after some minor simplification)

$$\boxed{\frac{\mu}{r} \frac{d}{dr} \left(r \frac{dv_z}{dr} \right) = -G}$$

(2.63)

where $G = -dp/dz$, which is equal to the inlet pressure minus the exit pressure divided by the pipe length. This term is therefore used as a source term driving the flow. The integrated form of Eq. (2.63) provides a general solution to the velocity profile:

$$v_z = -\frac{Gr^2}{4\mu} + A_1 \ln r + A_2$$

(2.64)

This general form is suitable both for pipe flow and for flow in a concentric cylinder. The solution particular to pipe flow is presented below.

Example 2.19.
Apply Eq. (2.64) and develop an expression for the velocity profile for flow in a circular pipe.

Solution.
The A_1 term in Eq. (2.64) should be dropped since the velocity at the center will become infinite otherwise. Next we apply a boundary condition at the pipe wall. We claim that the layer next to the wall is at rest (since the wall itself is not moving). The fluid velocity at $r = R$, the pipe wall, is then set as zero. This is called the no-slip condition. Using this we can find A_2 and backsubstitute into Eq. (2.64). The final result is

$$v_z = \frac{G}{4\mu} (R^2 - r^2)$$

Thus we find that the velocity profile is parabolic, as mentioned earlier in Chapter 1. The volumetric flow rate is given by integrating the local velocity (v_z) times the local area for flow ($2\pi r \, dr$). Thus the expression for the volumetric flow rate is

$$Q = \int_0^R v_z(r) 2\pi r \, dr$$

The resulting expression upon integration is the Hagen–Poiseuille equation discussed in the last chapter, which is repeated here for convenience:

$$Q = \frac{\pi R^4}{8\mu L} (P_0 - P_L)$$

(2.65)

2.7   Spherical coordinates

Some examples of problems posed in spherical coordinates are now presented. An illustrative control volume is a spherical shell contained between r and $r + \Delta r$ as shown in Fig. 2.22. The key point to note is that the area for diffusion is $4\pi r^2$ at any location r.

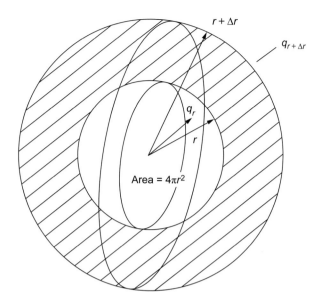

Figure 2.22 The control volume for 1D differential analysis in a sphere.

A shell balance approach leads to the following equation for diffusion with reaction in a spherical geometry:

$$\frac{D_{\text{eA}}}{r^2}\frac{d}{dr}\left(r^2\frac{dC_{\text{A}}}{dr}\right) = -R_{\text{A}} \tag{2.66}$$

The details leading to the derivation of the above equation have been omitted, but it is rather obvious.

A similar equation holds for heat transfer with generation:

$$\frac{k}{r^2}\frac{d}{dr}\left(r^2\frac{dT}{dr}\right) = -\dot{Q}_v \tag{2.67}$$

The solution can be obtained for simple cases for R_{A}, but in general a numerical solution is needed. The solution for a zeroth-order reaction (or for the case of constant generation for the analogous heat problem) is obtained easily by integration twice. The solution can be shown to be a quadratic function of r.

First-order reaction
The solution for a first-order case,

$$R_{\text{A}} = -k_1 C_{\text{A}}$$

is somewhat complicated, and the general solution involves what are called spherical Bessel functions. For a solid sphere the solution can be obtained as follows:

$$\frac{C_{\text{A}}(r)}{C_{\text{A,s}}} = \frac{R}{r}\frac{\sinh\left(r\sqrt{k_1/D_{\text{eA}}}\right)}{\sinh\left(R\sqrt{k_1/D_{\text{eA}}}\right)}$$

What happens to the concentration at $r = 0$?

The transient diffusion problem

Consider the cooling of a sphere exposed to a fluid. A shell balance approach leads to the following equation for transient heat conduction with no generation in a spherical geometry:

$$\frac{k}{r^2}\frac{\partial}{\partial r}\left(r^2\frac{\partial T}{\partial r}\right) = \rho c_p \frac{\partial T}{\partial t} \tag{2.68}$$

The boundary conditions are rather interesting and of general importance. If the surface temperature of the sphere is known, then this can be used as the boundary condition at $r = R$. This is known as the boundary condition of the first kind or the Dirichlet boundary condition.

Often the surface temperature of the sphere is not known and only the external fluid temperature is known. In this case a heat balance at the surface leads to:

$$-k\left(\frac{\partial T}{\partial r}\right)_{r=R} = h(T_{r=R} - T_f)$$

where T_f is the external fluid temperature and h is the heat transfer coefficient from the solid to the fluid. This is known as the boundary condition of the third kind or the Robin condition. Further discussion is provided in a later chapter.

The solution can be found by various methods including separation of variables, using the Laplace transform, and numerical methods, and these are shown in a later chapter. Our goal is to show how the models can be developed in this chapter and how the boundary conditions can be formulated.

The problem presented in this section is the more general model for cooling of a sphere and should be used in lieu of the lumped model if the Biot numbers are large. Also note that the model is now governed by a partial differential equation rather than the simple first-order differential equation of the lumped model.

2.8   Examples of mesoscopic models

. . . .

Some examples of formulation of mesoscopic models are presented in this section.

2.8.1   Tubular reactor with heat transfer

Here we extend the mesoscopic model for a plug-flow type of tubular reactor including significant heat effects.

Since the concentration is not varying in the radial direction, the control volume can span the entire cross-section. Thus the plug-flow analysis and model can be developed using a mesoscopic control volume. An illustrative control volume was shown in Fig. 1.21, and species mass balance led to the following equation:

$$\langle v \rangle \frac{dC_A}{dz} = -k_1 C_A$$

In the presence of heat transfer k_1 is a strong function of temperature and is given by the Arrhenius equation:

$$k_1 = A \exp[-E/(R_G T)]$$

where E is the activation energy and A is the pre-exponential factor.

Now a heat balance to relate the temperature is needed. Let us write the heat balance in words first:

Enthalpy in at z – enthalpy out at $z + \Delta z$ + Heat produced by chemical reaction

– Heat transferred from the control volume to the coolant = 0

Putting the various terms together leads to the following equation for the temperature profile as a function of axial length:

$$\rho c_p \langle v \rangle \frac{dT}{dz} + (-\Delta H)k_1 C_A + UA(T - T_C) = 0$$

where U is the overall heat transfer coefficient for the tube to the coolant side and T_C is the coolant temperature. It may be also noted here that the value of c_p is the local specific heat at any position z and should be calculated from the local mixture composition. Thus it can be a function of the position.

The species balance and temperature equation are now solved simultaneously using numerical methods. For moderately exothermic reactions the ODE45 code is useful but, for systems with large heats of reaction, the differential equations become stiff. ODE15s may be useful, but you may wish to note that the system can have many complexities such as runaway leading to very steep temperature profiles in the reactor. These problems are not addressed here. The book by Varma *et al.* (1999) is a good source for further study of non-isothermal reactor modeling.

2.8.2 Heat transfer in a pin fin

Consider the heat lost from an extended surface attached to a primary surface as shown in Fig. 2.23. The surface loses heat to the surroundings at a temperature of T_a with a heat transfer coefficient of h. The temperature distribution in the fin and the total heat lost from the surface are to be computed. If the fin is thin enough in cross-section a lumped model can be used, leading to a mesoscopic model. The goal of this section is to develop such a model.

The heat balance in words is

Heat in by conduction – Heat out by conduction

– Heat out by transfer to surroundings = 0

The various terms can be put as follows.

The heat in by conduction at x is $-k(A \, d\langle T \rangle / dx)_x$, where A is the local cross-sectional area, which can be a function of x. The temperature $\langle T \rangle$ is the cross-sectional average of the pointwise values.

The heat out by conduction at $x + \Delta x$ is $-k(A \, d\langle T \rangle / dx)_{x+\Delta x}$,

The heat transferred from the surface to the surroundings is $(P \, \Delta x) h (T_s - T_a)$, where P is the local perimeter, which can be a function of x.

On putting all the terms together and taking the limit as $\Delta x \to 0$, we obtain the differential equation for the macroscopic model:

$$\frac{d}{dx}\left(A(x)k(T) \frac{d\langle T \rangle}{dx} \right) = h(x)P(x)(T_s - T_a)$$

It may be noted that this is an exact model; some students may infer that this is only an approximate representation since we are not constructing a point-to-point differential

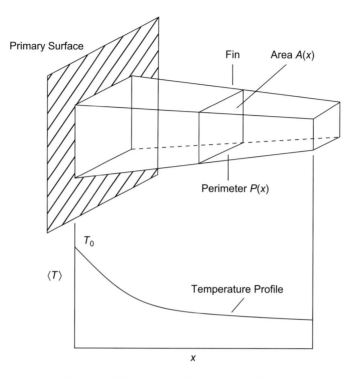

Figure 2.23 A schematic diagram of heat transfer from an extending surface attached to a primary surface.

model. The model has an unknown parameter T_s, which can be approximated as $\langle T \rangle$ in order to close the system; this is a lumping assumption. This is where the approximation, which is valid only for low values of the Biot number, is introduced.

The resulting model for a constant-area and constant-perimeter fin is

$$\frac{d^2 \langle T \rangle}{dx^2} = \frac{h}{k} \frac{P}{A}(\langle T \rangle - T_a) \tag{2.69}$$

Finally we need to specify two boundary conditions to complete the model formulation.

At $x = 0$, $T = T_0$ is the base-plate temperature.

At $x = L$ we have a few choices; the simplest is to neglect the heat loss from the edge and assume it to be insulated. This leads to $dT/dx = 0$ at $x = L$. This is justified since the heat loss from the edge is small compared with that from the exposed surface of area PL.

Equation (2.69) is a standard second-order differential equation and the solution can be written as the combination of sinh and cosh functions in terms of two integration constants. The constants are in turn evaluated by using the appropriate boundary conditions. The details of the solution are left as Problems 30 and 31.

2.8.3   Countercurrent heat exchanger

A schematic diagram of a counterflow heat exchanger is shown in Fig. 2.24. Here we have two streams flowing in opposite directions and exchanging heat across the tube walls.

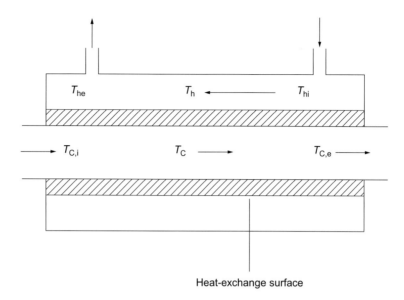

Heat-exchange surface

Figure 2.24 A schematic diagram of a countercurrent heat exchanger.

The case with only one flowing fluid was illustrated in Chapter 1. Here we extend the analysis to a countercurrent heat exchanger. The key goals are to illustrate the mesoscopic modeling and to show the mathematical features of the resulting problem. The system equations will be shown to be two first-order ordinary differential equations. But the notable point here is that the problem becomes a boundary-value problem rather than an initial-value problem. Hence this example is also a good starting point to understand boundary-value problems and the solution methods for such problems.

The balance equations are now needed both for the hot stream and for the cold stream.

For hot fluid designated by subscript h, the balance equation is

$$\dot{m}_h c_{ph} \frac{dT_h}{dz} = -(2\pi R)q_r = Pq_r \tag{2.70}$$

where q_r is the radial heat flux across the tube wall from the hot fluid to the cold fluid. P is the perimeter, equal to $2\pi R$. The student should verify the heat balance before proceeding further. Also note that the temperature T_h should be interpreted as a cup mixing temperature of the hot fluid and does not represent the temperature at any point in the system. The cup mixing average is taken in a plane perpendicular to the z-direction.

For the cold fluid (subscript c),

$$\dot{m}_c c_{pc} \frac{dT_c}{dz} = \pm(2\pi R)q_r \tag{2.71}$$

The plus sign is used if the hot fluid also flows in the same direction as the cold fluid, i.e., cocurrent flow, whereas the minus sign is used if the cold fluid is flowing in the opposite direction.

An overall heat transfer coefficient defined as

$$q_r = U(T_h - T_c)$$

is needed in order to calculate the rate of heat flow from the hot liquid to the cold liquid.

Using this definition in the heat balance equation, the following equations are obtained:

$$\dot{m}_h c_{ph} \frac{dT_h}{dz} = -UP(T_h - T_c) \tag{2.72}$$

$$\dot{m}_c c_{pc} \frac{dT_c}{dz} = \pm UP(T_h - T_c) \tag{2.73}$$

Note that the area of the heat exchanger is PL. This area could be based on the area of the hot-fluid side or the cold-fluid side. For either case we can define an appropriate U such as $U_h A_H$ or $U_c A_C$. The value of the heat transfer coefficient is adjusted such that $U_h A_h = U_c A_C$, depending on what A is used.

The equations can be made dimensionless by using the dimensionless parameters listed in Table 2.3. The dimensionless form is directly presented below and the student should verify the details:

$$\frac{d\theta_h}{d\zeta} = -U^*(\theta_h - \theta_c) \tag{2.74}$$

$$\frac{1}{R_C} \frac{d\theta_c}{d\zeta} = \pm U^*(\theta_h - \theta_c) \tag{2.75}$$

The various dimensionless quantities needed are summarized in Table 2.1.

The matrix representation of the differential equations is useful at this stage:

$$\frac{d\theta}{d\zeta} = \tilde{A}\theta$$

where the coefficient matrix is

$$\tilde{A} = \begin{pmatrix} -U^* & U^* \\ \pm R_c U^* & \mp R_c U^* \end{pmatrix}$$

The plus sign is used in the first column of the second row for cocurrent flow.

Cocurrent flow

Let us consider the cocurrent flow first.

From the definitions for the dimensionless temperatures we note that the hot fluid enters at a "temperature" of one, while the cold fluid enters at a temperature of zero. Hence we use the initial condition that $\theta_h = 1$ and $\theta_c = 0$ at $\zeta = 0$, leading to an initial-value problem in two variables.

The matrix solution is convenient for a quick and simple representation of the problem.

Table 2.3. Dimensionless parameters used in the simulation of the heat exchanger. The subscript i refers to the inlet values.

Parameter	Name	Definition
Temperature of hot fluid	θ_h	$(T_h - T_{ci})/(T_{hi} - T_{ci})$
Temperature of cold fluid	θ_c	$(T_c - T_{ci})/(T_{hi} - T_{ci})$
Axial length	ζ	z/L
NTU parameter	U^*	$UA/(\dot{m}_h c_{ph})$
Capacity ratio parameter	R_C	$\dot{m}_h c_{ph}/(\dot{m}_c c_{pc})$

The initial condition vector is $\theta_0 = [1; 0]$. Hence the solution is simply obtained as

$$\theta = \text{expm}(\tilde{A}\zeta)\theta_0$$

The exit temperatures are of more interest and can be obtained by setting $\zeta = 1$ in the above equation. Thus the matrix method permits a direct calculation using simple MATLAB commands.

System invariants

Let us examine the eigenvalues of $\tilde{A}$ more closely. The eigenvalues are 0 and $-U^*(1 + R_C)$. These can be calculated manually from the characteristic quadratic, and the student should verify the results.

Whenever one of the eigenvalues is zero, the number of system equations can be reduced by one, and the number of differential equations to be solved can be reduced by one.

This is a general rule and has many applications in transport problems in general. For the current problem this means that the two differential equations can be combined to produce

$$R_C \frac{d\theta_h}{d\zeta} = -\frac{d\theta_C}{d\zeta} \tag{2.76}$$

This equation can be integrated to give

$$\theta_h = -R_C\theta_c + C$$

which relates the hot- and cold-fluid temperatures at any point in the exchanger. From the inlet conditions the constant of integration is equal to R_C. Hence

$$\boxed{\theta_c = R_C(1 - \theta_h)} \quad \text{Cocurrent flow} \tag{2.77}$$

This is an invariant of the problem. In other words the cold- and hot-fluid temperatures are always related by the above relation. The invariant expression is a mathematical representation of the overall heat balance which states that the heat gained by the cold fluid must match the heat lost by the hot fluid.

If the exchanger were infinitely long then thermodynamic equilibrium would be achieved with the hot- and cold-fluid temperatures being the same. This final value can also be found from the invariant, and is equal to $R_C/(1 + R_C)$. We would expect any solution to reach this value if the parameter U^* were assigned a large value.

On substituting the invariant property into the differential equation (say for hot fluid) we have

$$\frac{d\theta_h}{d\zeta} = -U^*(1 + R_C)\theta_h + U^*R_C \tag{2.78}$$

This can be integrated to find the temperature profile of the hot fluid as well as the exit conditions at the end of the heat exchanger. The integrated solution is

$$\theta_h = \frac{R_C}{1 + R_C} + \frac{1}{1 + R_C}\exp[-U^*(1 + R_C)\zeta]$$

The exit temperature for the hot fluid is then obtained by using $\zeta = 1$:

$$\theta_{he} = \frac{R_C}{1 + R_C} + \frac{1}{1 + R_C}\exp[-U^*(1 + R_C)]$$

The exit temperature for the cold fluid is calculated from Eq. (2.77) as

$$\theta_{ce} = \frac{R_C}{1 + R_C} \left\{ 1 - \exp[-U^*(1 + R_C)] \right\} \tag{2.79}$$

We find that the temperature of both the cold and the hot fluid in the limiting case of $U^* \rightarrow \infty$ is equal to $R_C/(1 + R_C)$, as is also indicated by the invariant analysis.

These equations can be used to find the exit temperature if U^* is specified (the simulation problem); alternatively, if the exit temperature of one of the fluids is specified then the required size of the heat exchanger can be calculated, which is a design problem. Note that in the design problem both exit temperatures cannot be specified since they are tied together by the invariant equation. Only one exit temperature can be chosen. The system is characterized by two equations, and hence only two quantities can be calculated.

This completes the cocurrent flow analysis. Let us now move to the countercurrent flow.

Countercurrent flow

Now we note that the hot fluid enters at a "temperature" of one. Hence we can use the following condition: at $\zeta = 0$, $\theta_h = 1$.

However, the cold fluid enters at the other end and hence the condition to be satisfied is at $\zeta = 1$, $\theta_c = 0$, leading to a boundary-value problem since the initial conditions for both temperatures cannot be specified at the same position. In general boundary-value problems are more difficult to solve than initial-value problems. For this case, however, the solution can be obtained analytically, as shown below.

The equations can be combined to yield

$$R_C \frac{d\theta_h}{d\zeta} = \frac{d\theta_C}{d\zeta} \tag{2.80}$$

This equation can be integrated to give

$$\theta_c = R_C \theta_h - R_C + \theta_{ce}$$

where the inlet condition for the hot fluid has been used. Now, using the exit condition, we have

$$0 = R_C \theta_{he} - R_C + \theta_{ce}$$

or

$$\theta_{ce} = R_C(1 - \theta_{he}) \tag{2.81}$$

This is an invariant for the countercurrent exchanger.

Hence the equation for the cold fluid at any location is

$$\boxed{\theta_c = R_C(\theta_h - \theta_{he}) \text{ for countercurrent}} \tag{2.82}$$

This can now be used in the differential equation for θ_h to give

$$\frac{d\theta_h}{d\zeta} = -U^*(1 - R_C)\theta_h - U^* R_C \theta_{he} \tag{2.83}$$

Note that an undetermined parameter, namely θ_{he}, appears in this equation as a coefficient.

The solution is obtained keeping this parameter unknown as

$$\theta_h = \frac{1 - R_C + R_C\theta_{he}}{1 - R_C} \exp[-U^*(1 - R_C)\zeta] - \frac{R_C\theta_{he}}{1 - R_C}$$

The exit temperature is then

$$\theta_{he} = \frac{1 - R_C + R_C\theta_{he}}{1 - R_C} \exp[-U^*(1 - R_C)] - \frac{R_C\theta_{he}}{1 - R_C}$$

which can be rearranged to give the following explicit expression for the exit temperature:

$$\theta_{he} = \frac{(1 - R_C)\exp\left[-U^*(1 - R_C)\right]}{1 - R_C \exp[-U^*(1 - R_C)]} \tag{2.84}$$

The exit temperature of the cold fluid can be calculated using this value and the invariant of the system.

An alternative method is to use the matrix exponential function. This is demonstrated next.

2.8.4 Counterflow: matrix method

The matrix method can still be used for counterflow, although the inlet conditions are not completely known, since this is now a boundary-value problem.

In this case we assume θ_c at the exit such that θ_c at the inlet must be zero.

Let the initial condition vector be $\theta_0 = (1; \theta_{c0})$, where θ_{c0} is some unknown number in the range of 0 to 1 representing the exit temperature of the cold fluid.

Hence the exit temperature vector should be

$$\theta(\zeta = 1) = \mathrm{expm}(\tilde{A})\theta_0$$

The second row corresponds to the inlet temperature of the cold fluid, which is zero. We can write the second row as

$$E_{21}(1) + E_{22}\theta_{c0} = 0$$

where E represents the exponential of matrix $\tilde{A}$: $E = \mathrm{expm}(\tilde{A})$. The above relation gives the exit temperature of the cold fluid directly in terms of the coefficient of the E matrix:

$$\theta_{c0} = -E_{21}/E_{22}$$

This is an interesting application since it shows the power of linear algebra combined with MATLAB matrix manipulations to solve such problems in a concise manner.

Example 2.20.

Calculate the performance of the heat exchanger specified below for both cocurrent and countercurrent flows.

The hot fluid flow rate is 3 kg/s, and the fluid is to be cooled from 400 K.

The cold stream is cooling water at the rate of 4 kg/s available at 300 K.

The overall heat transfer coefficient is 800 W/m² · K.

The heat transfer area is 30 m².

The specific heat for both fluids is taken as that of water, 4180 J/kg · K.

Also find what the temperatures of the fluids leaving the system would be if the heat exchanger were infinitely long.

Solution.

The dimensionless parameters needed are

$$U^* = \frac{800 \times 30}{4 \times 4180} = 1.4354$$

The parameter U^* is also known as the NTU parameter in the heat-exchanger design literature.

The second parameter is the ratio of the flow rates times the heat capacity:

$$R_C = \frac{3 \times 4180}{4 \times 4180} = 0.75$$

The coefficient matrix is then set up and solved. The simulation model for cocurrent flow gives 0.47 for the exit temperature of the hot fluid. The model gives a temperature of 0.39 for the cold fluid. The actual temperature is therefore $400 - 0.47(400 - 300)$.

For countercurrent flow the results are temperatures of 0.37 for hot fluid and 0.47 for cold fluid.

If the exchanger were infinitely long the fluid would leave at a temperature 0.43 for cocurrent flow. For countercurrent flow the hot fluid will leave at a temperature of 0 while the cold fluid will leave at a temperature of 0.75.

Note that all temperatures are stated in dimensionless units, and it is an easy task to backcalculate the actual values.

Summary

- The combination of conservative laws and transport laws leads to system models. System models can be developed at various levels. The interpretation of the variables such as temperature differs for different levels of models, and this can be a source of confusion when setting up the models. For example, in the macroscopic model the temperature should be viewed as a volume average of the local temperature rather than as point values. Also the basic transport laws (e.g., Fourier's law) have to be replaced by empirical laws or appoximations when doing, for example, macroscopic balances.

- The differential models are the most general and require very few assumptions. However, they can be complex to compute, and hence macroscopic models are quite popular for engineering analysis. These (macroscopic) models require a number of assumptions and simplifications. Often these assumptions are part of what is called engineering judgement and there are no general rules as such. Hence the construction of macroscopic models often presents the most difficulty to students.

- The macroscopic species mass balance is given by Eq. (2.4). In order to apply this, for example, to a chemical reactor, some assumptions on the conditions and the extent of mixing in the control volume must be invoked. A benchmark simplifying assumption is that the system is well mixed. This is known as the backmixed assumption. This closes the rate term and permits us to evaluate the performance of a reactor. The actual reactor performance is usually better than this, and hence the backmixed assumption leads to a conservative design.

- The transient mass balance equation provides the response to a system due to changes, for instance, in the inlet conditions. These can be used as probing tools to understand the system behavior. The technique is known as stimulus–response studies or the bolus-response studies in pharmacokinetics. For example, we can determine whether a system is well mixed or not by use of these tools. We can study how an injected drug is distributed and eventually removed from a human body.

- The response curve to a bolus injection in a well-mixed system is an exponentially decreasing function. If this response is not seen then the assumption of backmixng is not suitable. Often this is modified by considering the reactor to be two or three zones connected in series, with each zone being assumed to well mixed on its own.

- Compartmental models are developed by connecting many separate macroscopic units together with interconnections. These interconnections are modeled by an exchange parameter between two compartments. Such models find wide use in biomedical systems. Megascale modeling of pollutant distribution in the environment is another important application of the compartmental models.

- If a linear kinetics is assumed in the compartmental models the system can be represented as a set of differential equations in a compact matrix–vector form. Solution methods based on the eigenvalues and eigenvectors of the coefficient matrix provide a powerful tool for such problems. A MATLAB implementation of this was illustrated in the text.

- For non-linear systems, compartmental models lead to initial-value problems involving a set of first-order differential equations. A solution procedure based on ODE45 is a well-established workhorse for such problems. A sample code was provided in the text to get you started.

- The overall macroscopic mass balance is an important equation, which is also needed for momentum and energy balance models. An important result is that $\rho \langle v \rangle A$ is constant across a section in a flow system. Thus, for an incompressible fluid, if the flow area decreases the velocity increases and *vice versa*. (Note that this is not valid for compressible fluids under certain conditions due to density variations.)

- The macroscopic momentum balance is given by Eq. (2.16) in the text. The vector nature of this equation is worth reemphasizing. Thus momentum can change if there is a change in flow area, flow direction, or both. The best way to solve these problems is write each term as a vector represented in terms of the components referred to a set of coordinates. Componentwise addition can then be used. Examples provided in the text are useful to understand this.

- Momentum balance can be done in a stationary frame or in a moving frame. The interpretation of the terms such as the rate of change in momentum, fluid acceleration, forces, etc. differs for these cases, but the final answer is the same. This reinforces the material-frame-indifference concept, which is well established in physics.

- The change in angular momentum is balanced by the torque imposed on the control volume, and Eq. (2.25) provides the necessary relationship. The angular-momentum balance has important applications in the design of pumps, turbines, rotating machinery, etc.

- The macroscopic energy balance given by Eq. (2.26) is an important relation in engineering analysis. When applied to a frictionless fluid it leads to the classical Bernoulli equation, which has many applications. The friction loss is added usually in an empirical way, leading to the engineering Bernoulli equation (2.31).

- For many problems all of the three balances are needed, with each providing a different item of needed information. Flow in a converging–diverging nozzle is an example of such a problem. In addition, the flow is often compressible and an additional relation for the density variation in terms of an equation of state is needed. A key result is that the sonic velocity can be reached, if at all, only at the throat, and supersonic speeds can be achieved only if the flow geometry has a converging section followed by a diverging section.

- Energy balance applied to heat transfer problems involves mainly the balance of system enthalpy. Kinetic energy and other terms are usually neglected. For many applications the system is assumed to be well mixed or at nearly uniform temperature, which is also called the lumped-parameter model in heat transfer. The cooling of a highly conducting solid is an example where a lumped model is suitable, while the cooling of a poorly conducting solid is not. For the latter case differential (or microscopic) models to calculate the internal temperature gradients are needed. Lumped models find many applications in many other problems, such as in jacketed-reactor heating and cooling, and temperature control of such systems.

- Differential models can be constructed using a differential control volume or the so-called shell balance. Examples for heat, mass, and momentum when applied to a simple 1D case show a commonality that it is useful to appreciate.

- Differential balance in 1D takes a different form in cylindrical (and spherical) coordinates (due to the change in diffusion area with r-direction) from that in Cartesian coordinates. It is well worth noting the difference, which we simply call the geometry effect.

- Differential momentum balance when applied to a laminar flow in a pipe leads to the classical Hagen–Poiseuille equation, which was developed empirically as early as 1841.

- Mesoscopic balances can be derived by taking a shell or control volume with the differential length in the flow direction but the whole cross-sectional area in the perpendicular direction. Several examples were presented in the text.

- An important application of mesocopic balance is in the simulation of a countercurrent heat exchanger, which leads to a boundary-value problem. The temperature appearing in this model is the flow-weighted average or the cup mixing average, and this is a useful point to note. Boundary-value problems can be solved by a number of techniques, the simplest being the shooting method. For the problem in the text we could get an analytic solution.

- Upon study of this chapter, you will learn the methodology to analyze a given problem or process and develop a suitable model for that process. You will appreciate some common methods for solving such problems, irrespective of the mode of transport. Analysis and model formulation at various levels are the main focus of this chapter. The next four chapters are mainly devoted to flow analysis and to the derivation and solution of differential models for fluid dynamic problems.

Problems

1. Show that the steady-state exit concentration for a first-order reaction is $1/(1 + Da)$. What is the corresponding expression if the reactor is modeled as a plug-flow reactor? Produce a comparison plot of the exit concentration for these two cases as a function of Da. Show that if $Da \lesssim 0.3$ then the mixing behavior is not important and the two models give nearly the same conversion.

2. Bolus injection is a very useful and important tool in pharmacokinetic analysis. Repeat the analysis for a bolus injection of a tracer that undergoes a first-order reaction. Sketch typical exit-concentration vs. time plots for various values of the rate constant (expressed as Da). How is the time constant of the response affected by the rate constant of the reaction?

3. Consider a reactor modeled as N interconnected tanks in series. Derive an expression for the exit response for a bolus injection of a tracer.

   Show that the maximum value of the tracer concentration is given by the following expression:

   $$C_{\max} = \frac{N(N-1)^{N-1}}{(N-1)!} \exp[-(N-1)]$$

   and this occurs at a dimensionless time of $(N-1)/N$.

4. **IVP: eigenvalue representation.** Consider the system of IVP in the matrix form which is repeated here for convenience:

   $$\frac{dy}{dt} = \tilde{A}y + R$$

   Eigenvalue representation is a useful and important concept in linear algebra. This exercise walks you through the key step in solving the IVP by the eigenvalue method. First one has to evaluate the eigenvalues of the matrix $\tilde{A}$ and the eigenvectors for each of these eigenvalues. The MATLAB command eig (A) can be used for this purpose. The statement to use is

   ```
   [ V , L ] = eig (A)
   ```

   Show that the solution of the IVPs for $R = 0$ can then be represented in a compact vector–matrix notation as

   $$y = \tilde{V} \operatorname{diag}(\lambda t)\alpha$$

   where α is a vector of integration constants. This vector can be estimated by applying the initial conditions. If y_0 is the vector of initial conditions then

   $$\alpha = V^{-1} y_0$$

   Hence the solution can also be expressed as

   $$y = \tilde{V} \operatorname{diag}(\lambda t) V^{-1} y_0$$

   Simple MATLAB multiplication commands can be used to do the numerics:

   ```
   y = V *  diag( exp( diag( L * time) ) )  * inv(V) * y0
   ```

   Implement this in MATLAB, and test it on some sample problems in compartmental modeling.

Note that the exponential of a matrix is related to the modal matrix by

$$\exp(\tilde{A}t) = \tilde{V} \operatorname{diag}(\lambda t) V^{-1}$$

and hence the two representations are equivalent. Verify this in MATLAB.

5. Consider again the IVPs shown above.

If R is time-varying, i.e., $R(t)$, then show that the solution can be formally written in terms of the exponential matrix as

$$y = \exp(\tilde{A}t)y_0 + \int_0^t \exp[-\tilde{A}(\tau - t)]R(\tau)d\tau$$

where τ is the dummy variable for integration purposes.

This formula is useful for the case of a time-varying stimulus. You may want to write a MATLAB code to implement this and test it on a benchmark problem.

6. Consider the series reaction scheme in a constant batch reactor:

$$A \xrightarrow{k_1} B \xrightarrow{k_2} C \xrightarrow{k_3} D$$

Assuming all reactions are first order and irreversible, set up the governing equations in matrix form for A, B, and C. Assume the rate constants $k_1 = 2$, $k_2 = 1$, and $k_3 = 2$ and initial concentrations of $C_A = 1$, $C_B = 0$, and $C_C = 0$. Note that time and concentrations are in arbitrary units for illustrative purposes.

Solve the equations using matrix algebra and plot the concentration profiles.

If the same reaction were now carried out in a CSTR with a mean residence time of 2, what would the exit steady-state conversion be for the inlet concentrations of $C_A = 1$, $C_B = 0$, and $C_C = 0$.

7. The following data were found in response to a drug injected as a pulse:

Time	Concentration
0	1.0000
0.3000	0.4867
0.6000	0.2748
0.9000	0.1700
1.2000	0.1101
1.8000	0.0487
2.4000	0.0219

How good is the fit to a one-compartment model? What is the time constant?

How good is the fit to a two-compartment model? What are the time constants? To what degree of accuracy can you determine these, since the data are limited?

8. **CHEBFUN examples.** CHEBFUN helps to make the MATLAB codes easier, and allows you to work with solutions as though they were analytic functions. You will find it very useful once you start using these. I will walk you through several examples in this book. This example is a starting point.

What are CHEBFUNs? The answer is that they are a type of polynomial fitting to a data or function using Chebyshev polynomials. Once fitted, these functions have the feel of a mathematical function rather than a discrete set of data points.

How does one access these? For getting started you need to copy and paste the following in your MATLAB command window:

```
unzip('http://www.chebfun.org/download/chebfun_v4.2.2889.zip')
cd chebfun_v4.2.2889, addpath(fullfile(cd,'chebfun')), savepath
```

This will install the chebfun directory and also add the required path to the file. Once this has been installed you will have access to a number of new functions, which makes the MATLAB coding easier. The code will look much more compact with this functionality added to your MATLAB. Some sample codes provided in problems in this book use these compact functions. Let us do some function manipulations first and see how compact the codes become. The code lines to be entered in MATLAB are indicated below. You should type these and test them out with your own functions.

$x = chebfun$ ($'x'$, $[0, 1]$) defines x as a CHEBFUN in the interval 0 to 1, and, using x as a variable, any function of x can be created. For example

$$y = exp(x)$$

creates an exponential function.

The statement $plot(y)$ generates a plot of this function.

The statement $sum(y)$ gives directly the integral of the function in the given domain.

Infinite or semi-infinite domains can be easily handled. Thus, for example, the statement

```
X = chebfun ('X', [0. ,inf] ) and
t = exp(-X.^2)
```

creates a CHEBFUN $t = exp(-x^2)$ defined from 0 to ∞. Note the X. (dot) operator to show a function multiplication.

The sum (t) gives the result 0.8862, which is the integral of the function. This matches exactly the analytic value of $\sqrt{\pi}/2$.

A first-order differential equation where the separation of variables can be used is readily solved using the sum command.

For more details on the CHEBFUN and the mathematics involved in the construction of such a representation, the papers by Battles and Trefethen (2004) and Trefethen (2007) should be consulted.

9. **CHEBFUN wrapper for ODE45.** This code shows how CHEBFUN can be used with MATLAB ODE45 to give a "symbolic" look to your answer. The presence of the calling argument *domain* in the calling statement creates a Chebyshev polynomial for the results.

```
% consecutive reaction using ODE45 and CHEBFUN
%    A --> B --> C
   k1 = 1.0; k2 = 2.0;
fun = @(t,v) [-k1*v(1); k1*v(1)-k2*v(2)];
v = ode45(fun,domain(0,3),[1 0]); % solves
 plot(v) % generates the plot
c_max = max ( v(:,2) ) % finds what the maximum value is.
F = diff( v(:,2)) % takes the derivative
roots(F)  % finds where the maximum occurs
```

Note that additional operations on the results such as finding the maximum values etc. have become rather easier with this approach.

The maximum value of B is 0.25 and occurs at time 0.6931 for the parameters used in the example. These match the exact values from the analytical model.

10. The velocity profile in laminar flow in a pipe is a parabolic function of r and can be represented as

$$v_z = v_c[1 - (r/R)^2]$$

The parameter v_c appearing in this equation can be interpreted as the center-line velocity.

(a) Using the Hagen–Poiseuille equation, how is the center-line velocity related to other parameters of the problem.

(b) Calculate the following for the above velocity profile: $\langle v \rangle$, $\langle v^2 \rangle$, and $\langle v^3 \rangle$.

(c) What error would be introduced if the rate of momentum flow were based on $\langle v \rangle^2$ instead of $\langle v^2 \rangle$?

(d) What error would be introduced if the rate of kinetic energy flow were based on $\langle v \rangle^3$ instead of $\langle v^3 \rangle$?

(e) What is the value for the momentum correction factor for this velocity profile? (**Answer:** 4/3.)

(f) What is the value for the kinetic-energy correction factor for this velocity profile? (**Answer:** 2.)

11. The velocity profile in turbulent flow in a pipe is often approximated by a one-seventh power law:

$$v_z = v_c[1 - (r/R)]^{1/7}$$

(a) Calculate the following: $\langle v \rangle$, $\langle v^2 \rangle$, and $\langle v^3 \rangle$.

(b) What error would be introduced if the rate of momentum flow were based on $\langle v \rangle^2$ instead of $\langle v^2 \rangle$?

(c) What error would be introduced if the rate of kinetic energy flow were based on $\langle v \rangle^3$ instead of $\langle v^3 \rangle$?

12. A turbine discharges 40 m³/s of water and generates 42 MW of power. The rotational speed is 24 r.p.s. Fluid enters at a radius of 1.6 m with a tangential velocity of 9 m/s. The pump operates at 1200 r.p.m. The fluid exits the impeller in a radial direction near the center. Determine the velocity of the entering stream and the head of fluid required by the Bernoulli equation. Find the power output of the turbine.

13. A compressor delivers natural gas through a straight pipe of length L. Find the pressure at the exit. Include frictional losses in the pipe.

The energy balance should now include the frictional loss, and the appropriate form of the equation is

$$\delta(v^2/2) + \delta(p/\rho) = -\delta[\dot{W}_\mu/\dot{m}]$$

which is applied locally, i.e., in a mesoscopic sense rather than in the macroscopic sense. Assume adiabatic conditions. Then compare the results with those for if the compressor operated isothermally.

14. Water flows from an elevated reservoir through a conduit to a turbine at a lower level through a conduit of the same diameter. The elevation difference is 90 m. The inlet pressure is 2.2 atm and the exit pressure is 1.2 atm.

What is the minimum flow rate needed to generate 700 kW of turbine output?

15. **Momentum analysis for a rocket.** A rocket has a mass of m_R and carries fuel with mass m_f at a given instant of time. Thus the total mass of the system at the current time is $m_R + m_f$.

The fuel is consumed at a rate of $\dot{m}$. The initial velocity of the rocket is zero and the initial amount of fuel is m_{f0}.

The motion of the rocket is directly upwards, and the air resistance is represented as a force component F_y (in the negative vertical direction). The exit velocity from the nozzle is v_r relative to the rocket.

Derive an equation for the acceleration of the rocket.

16. Integrate the equation in the above problem analytically to find the velocity of the rocket as a function of time.

    Also write a MATLAB code to integrate the system numerically using the ODE45 solver. Test your solver on the following data.

    The total mass of the rocket at the time of launching is 600×10^6 kg, of which 70% is fuel. The fuel-consumption rate is 1640 kg/s. The exit velocity from the nozzle relative to the rocket is 3300 m/s.

    Neglect air resistance.

    Note that the integration should be done only up to the time of complete burning of the fuel. The results are meaningless after this time.

17. A water jet is deflected through 180° by a vane that is attached to a cart, which is free to move. Assume that the flow along the vane is frictionless and neglect gravity. The jet enters with a velocity v_0 of 30 m/s with the area of the jet being 0.0018 m^2.

    Find the force on the vane if the vane is not moving.

    Set up a differential equation for the motion of the vane if the vane is now allowed to move. **Hint:** the momentum balance is easier to do if one assumes a coordinate system moving with cart; in other words use relative velocity values to compute the force exerted on the cart by the water jet. The exercise is similar to Example 2.7 in the text except that the turning angle is now 180°.

18. Consider the flow in a U-bend in a pipe. Water is flowing at a rate of 80 cm^3/s in a pipe of diameter 10 cm. The inlet pressure is 1.2 atm. The friction loss is estimated as if the bend were equivalent to a straight pipe of length 75 times the diameter. The density is $\rho = 1000$ kg/m^3 and the viscosity is $\mu = 0.001$ Pa · s.

    Find the exit pressure.

    Find the force exerted by the flowing fluid on the pipe.

19. Consider the draining of a tank with three different flow arrangements at the exit as shown in Fig. 2.25. Develop an equation to calculate the level of liquid in the tank as a function of time for the three cases shown below.

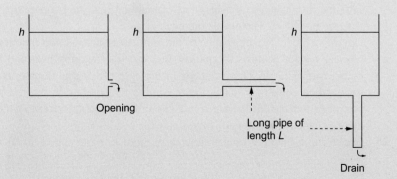

Figure 2.25 The diagram for Problem 19, showing the draining of a tank with a long exit pipe.

20. A compressor is to be designed to provide a compression ratio of 4. The inlet temperature is 27 °C. The C_p may be assumed to be a constant, 38.9 J/mol · K, and $\gamma = 1.3$.
    Find the power needed per unit mass, the temperature at the exit, the cooling needed, and the amount cooling water needed.
    The volumetric flow rate is 360 m³/h.

21. Show that for the windmill problem (Example 2.11) the maximum value of power for a constant wind incoming velocity is achieved when $v_2 = v_1/3$. Find the corresponding power and the efficiency of the windmill defined as the ratio of the power generated to the kinetic-energy flow rate in the incoming wind.
    The maximum efficiency is also called the Betz number. Show that the value is 59.3%.

22. Kundu and Cohen (2008) make the following statement which is counter-intuitive: *Friction can make the fluid go faster in an adiabatic flow in a channel of constant cross-section.*
    Develop a model for this system and comment on the above statement.

23. Consider a natural-gas pipe-line of inner diameter 60 cm and length 16 km. The pressure at the inlet is 7 atm and the temperature is 20 °C. The inlet velocity is 1.2 m/s. Find the exit pressure and the conditions at the exit.

    (a) Assume isothermal flow.
    (b) Assume adiabatic flow.

24. Derive the following equation for the conditions across a shock shown in Example 2.13.

$$\frac{v_1}{v_2} = \frac{(\gamma + 1)Ma_1^2}{(\gamma - 1)Ma_1^2 + 2}$$

$$Ma_2^2 = \frac{(\gamma - 1)Ma_1^2 + 2}{2\gamma Ma_1^2 + 1 - \gamma}$$

Ma_1 and Ma_2 are the Mach numbers before and after the shock.

25. An aluminum plate is at a temperature of 25 °C. It is then suddenly subjected to a uniform and continuing heating at the rate of 6 W/m². Develop a macroscopic model for the process. Find the time at which the surface temperature reaches a value of 100 °C. Find the total heat transferred up to this time.

26. A compressible gas of nitrogen is contained in a tank and is discharged thorough a small convergent nozzle. The pressure at the exit is maintained at p_2.
    Find the instantaneous discharge rate assuming that the temperature and pressure in the tank are p_1 and T_1. The nozzle diameter is d_o.
    Develop a model for the change of mass of gas in the tank as a function of time.

27. Nuclear reactor analysis. A nuclear fuel rod is generating heat and is in the shape of a cylinder of radius R. The surface is losing heat with a heat transfer coefficient of h to the surrounding coolant. What is the maximum temperature in the rod as a function of the system parameters? If there is a reduction in the coolant (water to generate steam) flow, is there a chance of meltdown? Investigate the conditions for meltdown.

28. Sketch the qualitative time–temperature profiles in the sphere for the following limiting cases.

    (a) Cooling is such that the Biot number is much less than one.
    (b) Cooling is such that the Biot number is large, say greater than 20.

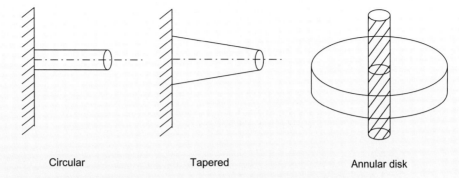

Circular   Tapered   Annular disk

Figure 2.26 Various arrangement of extended surfaces to be modeled in Problem 31.

29. A fluid is flowing in an annular region formed by two concentric cylinders. The inner cylinder has radius R_i and the outer cylinder has radius R_o. Develop an expression for the velocity profile and the volumetric flow rate for a fixed pressure drop G.

30. For the problem of heat transfer in a pin fin, solve the equations for the temperature profile for a fin at temperature T_0 at the base and with no heat loss at the edge of the fin.

31. Develop mesocopic models for heat transfer from extended surfaces of various shapes shown in Fig 2.26.

32. **A dynamic model for a tank with jacket cooling or heating.** The mathematical representation for the fluid in the tank is the same with T_S replaced by T_C in Example 2.15 in the text,

$$\rho c_p V \frac{d\bar{T}}{dt} = \dot{m} c_p (T_i - T_e) + UA(T_c - \bar{T})$$

where T is the average temperature in the tank.

An additional term due to heat generation can be added. This is $(-\Delta H)(-R_A)V$. However, this makes the model non-linear, and one also needs to set up a mass balance for the reacting species. The concepts and the method of analysis and model formulations are the same, except that more equations are involved and the numerical solution can be a challenging problem as well.

For the coolant we have

$$(\rho c_p V)_j \frac{d\bar{T}_j}{dt} = \dot{m}_j c_{p,j}(T_{ji} - T_{je}) - UA(T_c - \bar{T})$$

Note that the sign on the last term is now changed since the jacket is losing heat.

The output variables are the vessel and jacket temperatures. The fluid in the jacket is assumed to be well mixed, which might not be a good assumption.

Set up a case-study problem involving MATLAB simulation and test for some chosen parameter values. Consider both reacting and non-reacting systems.

3 Flow kinematics

Learning objectives

You will learn from this chapter the following concepts and various terminologies to characterize fluid motion:

- representation of motion in a stationary frame and in a frame moving with the fluid;
- the derivative following motion, the famous D/Dt operator;
- the expression for D/Dt when acting on a scalar and its form in common coordinate systems;
- the expression for D/Dt when acting on a vector and its form in common coordinate systems;
- representation of fluid acceleration;
- representation of change in volume of the fluid particle in compact vector notation; the physical meaning for divergence of a vector;
- derivation of the overall mass balance for a fluid in a compact vector form;
- the Reynolds transport theorem and its meaning;
- representation of local fluid rotation and the vorticity vector; the physical meaning for the curl of a vector;
- definition of the streamfunction; the streamfunction for axisymmetric problems;
- the meaning of the gradient of a vector; tensor quantities;
- the rate-of-deformation tensor and its physical meaning; and
- the Einstein summation convention and the index notation.

Mathematical prerequisites

- Divergence of a vector field.
- Curl of a vector field.
- The Reynolds transport theorem. This is an extension of Leibnitz rule to a 3D case. The Leibnitz rule is for differentiation (in, say, t) under the integral sign in one variable, say x, and the Reynolds theorem is an extension to a general 3D case, i.e., for a volume integral.

- The Green–Gauss theorem for converting a surface integration to a volume integration.
- The Stokes theorem for converting a line integration to an area integration (optional).
- The Einstein summation notation for vectors, tensors, and their derivatives.

Since the Green–Gauss theorem is widely used in the formulation of transport models, we state and indicate the meaning of this theorem in the introductory section of this chapter.

Fluid motion is caused essentially by the action of forces imposed on the flow system and the study of the effect of the forces on fluid motion is referred to as the flow **kinetics**. The motion itself can be characterized by the local velocity vector. Many properties of flow can be defined with reference to a given velocity field without any discussion on what caused the flow. This aspect of the study of motion is known as **kinematics** and is the focus of this chapter. In this chapter you will learn the various words used to describe fluid motion. You will also learn that the motion can be viewed and analyzed either using a stationary coordinate system or using moving coordinates and learn to relate the observations in the two frames. The definition and the meaning of the operator D/Dt, the time derivative following motion, will be discussed.

An important differential equation needed in the flow analysis, namely the overall mass conservation equation, will be derived. This equation is also called the continuity equation. Other concepts widely used in fluid dynamics, *viz.*, the streamfunction, vorticity, and rate of strain will be introduced. The rate-of-strain tensor will play an important role in relating the stresses acting on the control surface, similarly to stress–strain relations in solid mechanics.

The Green–Gauss theorem

This theorem states that a surface integral can be transformed into a volume integral. Let V be the volume bounded by a closed surface S. Then

$$\int_S \alpha \cdot n \, dS = \int_V \nabla \cdot \alpha \, dV$$

Here the quantity α can be a vector or a tensor. The theorem also applies if α is a scalar, in which case we use the following variation:

$$\int_S \alpha n \, dS = \int_V \nabla \alpha \, dV$$

Historical vignette. The Green–Gauss theorem was first discovered by Lagrange in 1762, then later independently rediscovered in 1813 by Gauss, in 1825 by Green, and in 1831 by Ostrogradsky, who also gave the first proof of the theorem. Subsequently, variations on the divergence theorem are correctly called Ostrogradsky's theorem, but also commonly Gauss's theorem, or Green's theorem (source: Wikipedia).

Students who may have some conceptual difficulty in understanding this theorem may wish to recall the following simple elegant formula from calculus:

$$f(b) - f(a) = \int_a^b \left(\frac{df}{dx}\right) dx$$

The divergence theorem is an extension of this from 1D to 3D! Do you see the analogy?

3.1 Eulerian description of velocity

The flow is described if one specifies the velocity vector, v, at various spatial locations (x, y, z) (or x_1, x_2, x_3) at each instant of time, t. Such a description is called a velocity field:

$$v = v(x, y, z, t) \tag{3.1}$$

The representation of velocity in the above form is known as the Eulerian form.

A compact indicial notation is $v_i = v_i(x, t)$, where the boldface x describes collectively the three coordinates x_1, x_2, and x_3. Cartesian coordinates are used mainly for discussion, and one can readily opt for other coordinate systems, depending on the geometry in which the flow problem is posed.

If the velocity field at any point does not depend on time, it is known as the steady-state field.

The following statement is worth noting and will become clearer later. *A control volume in Eulerian frame is fixed in space, and fluid can flow in or out from the control surfaces. The transport of any quantity due to flow has to be explicitly accounted for in the conservation laws as convection terms. In addition, for the unsteady-state situation, the local values of the quantities may change with time, leading to an accumulation term.*

3.2 Lagrangian description: the fluid particle

Here we focus attention on an infinitesimally small (but non-zero) region of a fluid, which is called a fluid particle. A fluid particle occupies a (vector) position ξ at time zero. The coordinates ξ are called the reference configuration or the marker coordinates of the particle. The particle itself is often referred to as the ξ particle. The motion can be described by a curve x traced out by this ξ particle. The mathematical equation for this path is

$$x = x(\xi, t) \tag{3.2}$$

which is called the Lagrangian representation. This is shown schematically in Fig. 3.1.

Note that the independent variables are the initial position vector given by ξ and time, while the dependent variable is the position x at any given time t. In contrast, in the Eulerian formulation the independent variables are any specified location x and time, and the dependent variables are the velocity components at that location at that particular instant of time.

Simulation of flow using the Lagrangian description involves tracking the trajectories of many material particles ξ. Note that ξ and time are the independent variables here. Time is a variable even for a steady-state flow in this formulation, since the particle ξ occupies different positions with the progress of time.

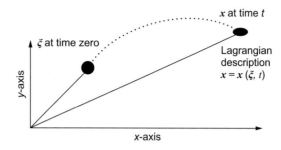

Figure 3.1 An illustrative picture for the Lagrangian description of motion. A particle initially at ξ moves to x in time t with the local velocity at each moment; the trajectory of the particle is the Lagrangian description of motion.

The following statement is worth noting and will become clearer later. *A control volume (and correspondingly the control surface in the Lagrangian frame) moves with the fluid and hence there is no fluid flow in or out from the control surfaces (as seen by an observer located within this moving control volume). Hence the convective transport is not explicitly accounted for and gets buried in the accumulation term.*

The conversion from the Eulerian description to the Lagrangian description and *vice versa* is a simple matter of integration or differentiation, as demonstrated by the following examples.

Example 3.1. From Eulerian to Lagrangian.
The position vector at any instant of time can be found by integration of the instantaneous velocity vector as

$$x = \int_0^t v(x, s)ds + \xi$$

which provides a method of converting from Eulerian description to Lagrangian. In the above equation, s is a dummy time variable. Also we use the initial condition that, at $t = 0$, the particle location is at ξ. The position terms x appearing in the integral term on the RHS can, in some cases, be eliminated in terms of ξ, thus providing an explicit description in terms of ξ and t alone for the position x. Otherwise the Lagrangian description is given by the above implicit form. The velocity vector appearing in the integral can be independent of time for steady-state flow, but time appears as a variable in the Lagrangian description even for this case.

Example 3.2. From Lagrangian to Eulerian.
Here we do the reverse. We differentiate the position vector to find the instantaneous velocity vector (keeping ξ fixed) as

$$v = \left(\frac{dx}{dt}\right)_\xi \tag{3.3}$$

which is a function of ξ. The derivative is taken with ξ fixed as a constant. The variables ξ appearing in the final result for v can be often expressed in terms of x to provide the Eulerian description in a more explicit form. Otherwise we get an implicit description.

3.3 Acceleration of a fluid particle

The velocity and the acceleration of the fluid particle (evaluated at the current location) can be found by differentiating the Lagrangian description with respect to time:

$$v = \frac{dx}{dt} \tag{3.4}$$

as indicated earlier in (Eq. 3.3), and

$$a = \frac{dv}{dt} = \frac{d^2x}{dt^2} \tag{3.5}$$

These equations look the same as those for a solid body's motion, but the interpretation of d/dt requires some discussion. The time derivative should be evaluated keeping the initial position (marker vector) constant since we are referring to a particle that started off at ξ. Hence Eq. (3.4) should be written as

$$v = \left[\frac{dx}{dt}\right]_\xi \tag{3.6}$$

where the marker position vector $\boldsymbol{\xi}$ is kept constant during differentiation since we are tracking the same particle and its initial position was fixed at $\boldsymbol{\xi}$.

Likewise Eq. (3.5) should be written as

$$a = \left[\frac{d^2x}{dt^2}\right]_\xi = \left[\frac{dv}{dt}\right]_\xi \tag{3.7}$$

The symbol d/dt in all the above equations implies that ξ is kept constant since we are referring to a given particle. However, to avoid confusion the symbol d/dt (with fixed ξ) is denoted by D/Dt in the fluid-dynamics literature and is called differentiation following motion or the substantive derivative. *The physical meaning of D/Dt is that it is the time rate of change of any property as one follows the motion.*

Note that this should not be confused with the usual partial derivative with respect to time, which is $\partial/\partial t$. The partial derivative involves differentiation keeping the position variables x, y, and z constant, and is therefore *not* a measure of the fluid acceleration of a particle at that given instant of time at a specified position.

3.4 The substantial derivative

It is of importance to know the relationship between the substantial derivative and the partial derivative with respect to time and the current velocity vector. This can be derived as follows. Consider a particle ξ that is currently at position x at time t. The particle would advance a distance Δx in time Δt. The acceleration of the particle at the current location can then be defined as follows:

$$a = \lim \frac{v_{(x+\Delta x)} - v_{(x)}}{\Delta t} \tag{3.8}$$

But, by the chain rule,

$$v_{(x+\Delta x)} = v_{(x)} + \frac{\partial v}{\partial t} \Delta t + \frac{\partial v}{\partial x} \Delta x + \frac{\partial v}{\partial y} \Delta y + \frac{\partial v}{\partial z} \Delta z \tag{3.9}$$

Using this in Eq. (3.8), we obtain

$$a = \frac{\partial v}{\partial t} + \frac{\partial v}{\partial x} \frac{\Delta x}{\Delta t} + \frac{\partial v}{\partial y} \frac{\Delta y}{\Delta t} + \frac{\partial v}{\partial z} \frac{\Delta z}{\Delta t} \tag{3.10}$$

The velocity components v_x, v_y, and v_z at the current location are defined by the following equations (velocity components = distance/time):

$$v_x = \frac{\Delta x}{\Delta t}; \quad v_y = \frac{\Delta y}{\Delta t}; \quad v_z = \frac{\Delta z}{\Delta t} \tag{3.11}$$

Using these results in Eq. (3.10), we obtain

$$\boxed{a = \frac{Dv}{Dt} = \frac{\partial v}{\partial t} + v_x \frac{\partial v}{\partial x} + v_y \frac{\partial v}{\partial y} + v_z \frac{\partial v}{\partial z}} \tag{3.12}$$

which is the expression for the acceleration of a fluid particle at the current position.

Note that the substantial derivative can be written in an operator form as

$$\boxed{\frac{D}{Dt} = \frac{\partial}{\partial t} + (v \cdot \nabla)} \tag{3.13}$$

The operator can act on any quantity, be it a scalar, a vector, or even a tensor. Thus, for example, DT/Dt represents the rate of change in temperature for an observer moving with the flow field.

The first term on the RHS of (3.13) represents the change in a property due to a change in time, while the second term represents the change in the property due to the fact that the particle itself has advanced to a new position during an elapse of time of Δt. Hence the second term is referred to as the convective term. Thus D/Dt includes both the effect of convection and the local accumulation. If the flow is steady, the first term is zero while the second term will still contribute to D/Dt.

The detailed form of the operator, $v \cdot \nabla$, is simple in Cartesian coordinates (see Problem 2), and the form of the operator acting on a scalar or a vector quantity is the same.

The expression for D/Dt in other coordinate systems is somewhat complicated, since the unit vectors do not often refer to a fixed direction, and some of them have derivatives with respect to the coordinate directions. The derivations of the expressions for this quantity in cylindrical and spherical coordinates are left as exercises, but the results for all these geometries are given in Tables 3.1 and 3.2. Also it may be noted that the expressions when D/Dt acts on a scalar are different from those when it acts on a vector except in Cartesian coordinates. (See Problems 3 and 4 for the form if the operator acts on a scalar.) These problems are important exercises for you to practice your vector calculus skills.

The following observations are important with regard to Tables 3.1 and 3.2 and should be noted.

Table 3.1. The convection operator $(v \cdot \nabla)v$ in a cylindrical coordinate system; componentwise equations

r component	$v_r \dfrac{\partial v_r}{\partial r} + \dfrac{v_\theta}{r}\dfrac{\partial v_r}{\partial \theta} + v_z \dfrac{\partial v_r}{\partial z} - \dfrac{v_\theta^2}{r}$
θ component	$v_r \dfrac{\partial v_\theta}{\partial r} + \dfrac{v_\theta}{r}\dfrac{\partial v_\theta}{\partial \theta} + v_z \dfrac{\partial v_\theta}{\partial z} + \dfrac{v_r v_\theta}{r}$
z component	$v_r \dfrac{\partial v_z}{\partial r} + \dfrac{v_\theta}{r}\dfrac{\partial v_z}{\partial \theta} + v_z \dfrac{\partial v_z}{\partial z}$

Table 3.2. The convection operator $(v \cdot \nabla)v$ in a spherical coordinate system; componentwise equations

r component	$v_r \dfrac{\partial v_r}{\partial r} + \dfrac{v_\theta}{r}\dfrac{\partial v_r}{\partial \theta} + \dfrac{v_\phi}{r\sin\theta}\dfrac{\partial v_r}{\partial \phi} - \dfrac{v_\theta^2 + v_\phi^2}{r}$
θ component	$v_r \dfrac{\partial v_\theta}{\partial r} + \dfrac{v_\theta}{r}\dfrac{\partial v_\theta}{\partial \theta} + \dfrac{v_\phi}{r\sin\theta}\dfrac{\partial v_\theta}{\partial \phi} + \dfrac{v_r v_\theta}{r} - \dfrac{v_\phi^2}{r}\cot\theta$
ϕ component	$v_r \dfrac{\partial v_\phi}{\partial r} + \dfrac{v_\theta}{r}\dfrac{\partial v_\phi}{\partial \theta} + \dfrac{v_\phi}{r\sin\theta}\dfrac{\partial v_\phi}{\partial \phi} + \dfrac{v_r v_\phi}{r} + \dfrac{v_\theta v_\phi}{r}\cot\theta$

1. The derivative with respect to θ always appears together with r.
2. Similarly, the derivative with respect to ϕ always appears together with $r\sin\theta$ in spherical coordinates.
3. The term $-v_\theta^2/r$ in the r-component in cylindrical coordinates is the centripetal acceleration.
4. The term $v_r v_\theta/r$ in the θ-component in cylindrical coordinates is the Coriolis acceleration.
5. Similar terms appear in spherical coordinates and should be noted.

3.5   Dilatation of a fluid particle

The dilatation of a fluid particle is defined as the rate of change of volume of the particle per unit volume:

$$\text{Dilatation} = \frac{1}{V}\frac{DV}{Dt} \tag{3.14}$$

where V represents the volume of the particle in the current configuration. The formula for calculation of the dilatation for a moving control volume of any general shape can be written vectorially as shown below.

Consider a differential area on the surface of a Lagrangian control volume with velocity v at that surface. Let n be the unit normal to the surface. Then $v \cdot n$ is the velocity component perpendicular to the surface (see Fig. 3.2).

The term $(v \cdot n)dA$ is the rate at which the surface is advancing with respect to time. This corresponds to the local time rate of change in volume of the Lagrangian control volume.

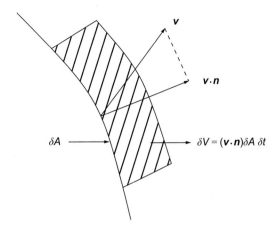

Figure 3.2 An illustration of a dilatation.

The total change in volume is obtained by integrating over the entire surface:

$$\text{Rate of change in volume} \ = \ \int_A (v \cdot n)dA$$

This can be converted to a volume integral using the Green–Gauss theorem:

$$\text{Rate of change in volume} \ = \ \int_V (\nabla \cdot v)dV$$

The dilatation is defined as the rate of change in volume per unit volume:

$$\text{Dilatation} \ = \ \frac{1}{V}\int_V (\nabla \cdot v)dV$$

For a point particle the volume integral can be approximated by an average value in view of the mean-value theorem. The point value is a good representation of the average value when $V \to 0$.

   Hence

$$\text{Dilatation} \ = \nabla \cdot v$$

In Cartesian coordinates this takes the usual form

$$\text{Dilatation} \ = \nabla \cdot v = \frac{\partial v_x}{\partial x} + \frac{\partial v_y}{\partial y} + \frac{\partial v_z}{\partial z} \tag{3.15}$$

The above equations provide physical meaning to the divergence of the velocity vector. *The divergence is a measure of the rate of change in volume of a fluid particle based on a unit volume of the particle.* The expressions in other common coordinate systems are also to be noted, and these are shown below. The derivations are left as exercise problems.

Divergence in other coordinate systems

The divergence in cylindrical coordinates is given by

$$\nabla \cdot v = \frac{1}{r}\frac{\partial}{\partial r}(rv_r) + \frac{1}{r}\frac{\partial v_\theta}{\partial \theta} + \frac{\partial v_z}{\partial z}$$

The divergence in spherical coordinates is

$$\nabla \cdot v = \frac{1}{r^2}\frac{\partial}{\partial r}(r^2 v_r) + \frac{1}{r\sin\theta}\frac{\partial}{\partial \theta}(v_\theta \sin\theta) + \frac{1}{r\sin\theta}\frac{\partial v_\phi}{\partial \phi}$$

3.6   Mass continuity

The mass of a "Lagrangian" fluid particle is conserved. The differential equation describing this is called the continuity equation, and we will derive this in a simple manner here.

Noting that mass = density × volume = ρV, the conservation requirement can be written as

$$\frac{D}{Dt}(\rho V) = 0 \tag{3.16}$$

Upon expanding the derivative and then dividing by the volume V, we have

$$\frac{D\rho}{Dt} + \rho\left[\frac{1}{V}\frac{DV}{Dt}\right] = 0 \tag{3.17}$$

Using Eq. (3.15) for the second (bracketed) term, the dilatation

$$\frac{D\rho}{Dt} + \rho\nabla \cdot v = 0 \tag{3.18}$$

which is one form of the continuity equation. An alternate form of this can be obtained by expanding the first term (the substantial derivative term):

$$\frac{\partial\rho}{\partial t} + v \cdot \nabla\rho + \rho\nabla \cdot v = 0 \tag{3.19}$$

The last two terms can be combined to yield

$$\boxed{\frac{\partial\rho}{\partial t} + \nabla \cdot (\rho v) = 0} \tag{3.20}$$

which is a common form of the continuity equation.

For a constant-density system (as in incompressible fluids) the equation simplifies to

$$\boxed{\nabla \cdot v = 0} \tag{3.21}$$

which is widely used in flow analysis. The equation ties up the three components of the velocity vector; the three components are tied together with the above condition and cannot take independent profiles. Also, as a general rule the compressibility effects can be ignored if the fluid velocity is less than 30% of the velocity of sound in the same fluid.

3.7 The Reynolds transport theorem

In order to appreciate the Reynolds transport theorem it is useful to review the Leibnitz rule for differentiation under the integral sign with one spatial variable. The rule is

$$\frac{d}{dt}\left[\int_{a(t)}^{b(t)} f(x,t)dx\right] = \left[\int_{a(t)}^{b(t)} \frac{d}{dt}f(x,t)dx\right] + \frac{db}{dt}f(b,t) - \frac{da}{dt}f(a,t)$$

Since db/dt is the velocity of the right side denoted as v_2 and da/dt is the velocity of the left side denoted as v_1, this can be written as

$$\frac{d}{dt}\left[\int_{a(t)}^{b(t)} f(x,t)dx\right] = \left[\int_{a(t)}^{b(t)} \frac{d}{dt}f(x,t)dx\right] + v_2 f(b,t) - v_1 f(a,t)$$

A graphical illustration of the theorem is shown in Fig. 3.3, which also captures the physical meaning of the theorem.

The extension of this to a 3D case with volume $V(t)$ and surface $S(t)$ leads to the Reynolds transport theorem: the change of any property α for a "particle" is given by

$$\text{Rate of change of } \alpha = \frac{D}{Dt}\left[\int_{V(t)} \alpha\, dV\right]$$

This can be written as

$$\text{Rate of change of } \alpha = \int_{V(t)} \frac{\partial \alpha}{\partial t} dV + \int_{S(t)} \alpha(v \cdot n)dS$$

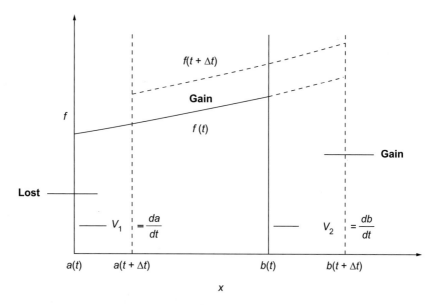

Figure 3.3 A graphical illustration of the Leibnitz theorem. The integral under a moving curve is related to the integral under the initial curve and the gain and loss due to the movement.

The first term represents the temporal rate of change of the property, while the second term deals with the change of the property due to the movement of the control surface itself. This is one version of the transport theorem.

The second term above can be converted to a volume integral by the use of the divergence theorem. Hence

$$\text{Rate of change of } \alpha = \int_{V(t)} \frac{\partial \alpha}{\partial t} \, dV + \int_{V(t)} [\nabla \cdot (\alpha v)] \, dV$$

The two terms can be combined to give

$$\text{Rate of change of } \alpha = \int_{V(t)} \left[\frac{\partial \alpha}{\partial t} + \nabla \cdot (\alpha v) \right] dV$$

which is a second version of the Reynolds transport theorem.

Mass conservation

For mass conservation, α is the density (mass per unit volume), and the term in the above equation is zero. Since V is arbitrary, the integrand itself is zero, leading to the equation of mass continuity given by Eq. (3.20).

Momentum change of a particle

If α is equal to ρv, the momentum per unit volume, then the term in the above equation is equal to the rate of change of momentum of the particle. Newton's law clicks in. The rate of change of momentum can in turn be equated to the body force acting on the particle, leading to an important equation (the equation of motion), which we shall derive in Chapter 5.

We now move on to study the meaning of the curl operator on the velocity vector.

3.8   Vorticity and rotation

The vorticity is defined as the curl of the velocity vector:

$$\omega = \nabla \times v \tag{3.22}$$

The Cartesian components of the vorticity vector (in 1–2–3 notation) can be represented by

$$\omega_1 = \frac{\partial v_3}{\partial x_2} - \frac{\partial v_2}{\partial x_3} \tag{3.23}$$

$$\omega_2 = \frac{\partial v_1}{\partial x_3} - \frac{\partial v_3}{\partial x_1} \tag{3.24}$$

$$\omega_3 = \frac{\partial v_2}{\partial x_1} - \frac{\partial v_1}{\partial x_2} \tag{3.25}$$

Plane flow

A plane flow is defined as the flow confined to two dimensions; say x_1 and x_2; i.e., there is a coordinate direction (x_3) along which the velocity component is zero. In that case the only surviving component of the vorticity vector is ω_3. This is simply referred to as vorticity for plane flow (ω) and not indicated as a vector. In general vorticity is, however, a vector for 3D flows.

The physical meaning of vorticity

The physical significance is that ω is a measure of the fluid rotation at any given point. To illustrate this, consider a rigid body rotating with an angular velocity of β. This can be represented as a vector, β, whose magnitude is the speed of rotation in rad/s and whose direction is the axis of rotation:

$$\beta = i\beta_1 + j\beta_2 + k\beta_3 \tag{3.26}$$

The linear velocity at a position r is then given by

$$v = \beta \times r \tag{3.27}$$

Expanding the cross product gives

$$v = i(\beta_2 z - \beta_3 y) + j(\beta_3 x - \beta_1 z) + k(\beta_1 y - \beta_2 x) \tag{3.28}$$

The curl of the above vector is given by

$$\nabla \times v = \begin{pmatrix} i & j & k \\ \partial/\partial x & \partial/\partial y & \partial/\partial z \\ \beta_2 z - \beta_3 y & \beta_3 x - \beta_1 z & \beta_1 y - \beta_2 x \end{pmatrix} \tag{3.29}$$

By performing the required algebra, it can be shown that

$$\nabla \times v = 2\beta \tag{3.30}$$

which shows that the vorticity vector (the curl of the velocity vector) is equal to twice the angular-velocity vector. If the vorticity is zero, then there is no rotational component of flow at that point. If the vorticity is zero everywhere, the flow is known as irrotational flow. The above discussion provides us with a physical meaning to the curl of the velocity vector: *the curl is a measure of the rotational motion of a fluid particle.*

ω is divergence-free

The vorticity vector has zero divergence. Thus

$$\nabla \cdot \omega = 0 \tag{3.31}$$

The proof is simple and follows by direct differentiation of the defining equation for the vorticity. Thus we note that both the velocity vector and the vorticity vector are divergence-free.

Vorticity plays an important role in turbulence modeling, and more details are left to a later chapter. It is useful to note the componentwise form of the curl operator in other coordinate systems, and these relations are presented below.

3.8.1 Curl in other coordinate systems

See Tables 3.3 and 3.4.

How will you derive these equations?

Some examples of vorticity calculations are now presented.

Table 3.3. Expressions for the curl of a vector in cylindrical coordinates

r-component	$\dfrac{1}{r}\dfrac{\partial v_z}{\partial \theta} - \dfrac{\partial v_\theta}{\partial z}$
θ-component	$\dfrac{\partial v_r}{\partial z} - \dfrac{\partial v_z}{\partial r}$
z-component	$\dfrac{1}{r}\left(\dfrac{\partial (r v_\theta)}{\partial r} - \dfrac{\partial v_r}{\partial \theta}\right)$

Table 3.4. Expressions for the curl of a vector in spherical coordinates

r-component	$\dfrac{1}{r\sin\theta}\left(\dfrac{\partial (v_\phi \sin\theta)}{\partial \theta} - \dfrac{\partial v_\theta}{\partial \phi}\right)$
θ-component	$\dfrac{1}{r\sin\theta}\dfrac{\partial v_r}{\partial \phi} - \dfrac{1}{r}\dfrac{\partial (r v_\phi)}{\partial r}$
ϕ-component	$\dfrac{1}{r}\left(\dfrac{\partial (r v_\theta)}{\partial r} - \dfrac{\partial v_r}{\partial \theta}\right)$

Example 3.3.

Find the vorticity in a simple shear flow. A simple shear flow is described by

$$v_x = V\frac{y}{d}$$

and can be generated by moving a plate at a **velocity** of V in a gap of dimension d (see Section 6.2.3). This is 2D or plane flow, and hence only the z-component of vorticity is non-zero:

$$\omega_z = \omega = -\frac{dv_x}{dy} = -V/d$$

The fluid trajectories are straight lines, but a fluid particle has a rotational motion. This shows that vorticity is a microscopic property, not a global property.

Example 3.4.

Find the vorticity for a free vortex. The velocity profile is

$$v_\theta = A/r$$

where A is a constant. This profile can, for example, be observed if there is a rotating stirrer at the center of a large pool of liquid. Upon applying the curl operator in polar coordinates we find that the vorticity is zero!

The fluid particles go round and round; the trajectory is a circle, but the fluid particle has no rotational motion. This shows that vorticity is a microscopic property, not a global property.

These two examples illustrate that vorticity is a microscopic property. The rotation itself might not be noticeable in a macroscopic sense, but is there in a microscopic sense if there is non-zero vorticity.

3.8.2   Circulation along a closed curve

This section may be omitted without loss of continuity.

A concept closely related to vorticity is that of the circulation. The concept of circulation can be grasped by considering a closed curve in two dimensions. Consider a simple closed curve, denoted by Γ, which encloses an area A as shown in Fig. 3.4. (A simple closed curve is one that does not intersect itself.)

The integral of the tangential component of the velocity along the closed curve is called the circulation over A and is closely related to the vorticity distribution over the area enclosing A. The definition of circulation, denoted by C, is

$$\text{Circulation} = C = \oint_{\Gamma} (v \cdot t)d\Gamma \tag{3.32}$$

where Γ is the perimeter of the closed curve. Note that t is tangent to the curve at any integration point, and hence $(v \cdot t)$ is the (tangential) component of the velocity along this curve. *The circulation around a closed curve is the area integral of the (normal component of) the vorticity enclosed by the closed curve.*

Stated mathematically, we have

$$C = \int \int_{A} \omega \, dA = \oint_{\Gamma} (v \cdot t)d\Gamma \tag{3.33}$$

Here ω is the z-component of the vorticity vector or the component in a direction perpendicular to the area. A simple example below demonstrates this property.

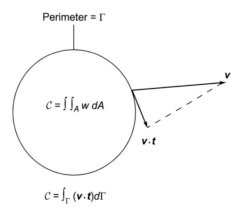

Perimeter $= \Gamma$

$C = \int \int_A w \, dA$

v

$v \cdot t$

$C = \int_{\Gamma} (v \cdot t)d\Gamma$

Figure 3.4 A simple closed curve with a perimeter of Γ enclosing an area A.

Example 3.5. Circulation for a rigid body

Consider a cylinder of radius R rotating about its axis with an angular velocity of β. Calculate the circulation and relate it to the vorticity.

Solution.

The closed curve is now a circle of radius R. The tangential velocity is $R\beta$ and is independent of the angular coordinate θ. Hence the integral in Eq. (3.32) defining the circulation is simply evaluated as

$$C = 2\pi R^2 \beta$$

Since the vorticity is twice the angular velocity, $\beta = \omega/2$. Hence the circulation can be expressed in terms of the vorticity as

$$C = \pi R^2 \omega = \omega A$$

which demonstrates the relation between circulation and vorticity, albeit for a simple problem.

3.9   Vector potential representation

A related key concept is the vector potential of velocity. It can be shown that the velocity field can be represented as the curl of a vector called the velocity potential vector, Ψ. This is defined as follows:

$$\nabla \times \Psi = v \tag{3.34}$$

This can be written in component form as

$$v_1 = \frac{\partial \Psi_3}{\partial x_2} - \frac{\partial \Psi_2}{\partial x_3} \tag{3.35}$$

$$v_2 = \frac{\partial \Psi_1}{\partial x_3} - \frac{\partial \Psi_3}{\partial x_1} \tag{3.36}$$

$$v_3 = \frac{\partial \Psi_2}{\partial x_1} - \frac{\partial \Psi_1}{\partial x_2} \tag{3.37}$$

The vector potential of velocity satisfies the continuity equation if the fluid is incompressible. This is because

$$\nabla \cdot \nabla \times \Psi = 0$$

for any vector. Hence, if we take the divergence on both sides of (3.34) then the divergence of velocity turns out to be zero.

The representation above is only a particular case of the famous Helmholtz decomposition theorem in vector calculus. According to this theorem any vector can be written as a sum of a gradient and a curl. Hence the velocity is represented as

$$v = \nabla \phi + \nabla \times \Psi \tag{3.38}$$

and we will use this for irrotational flow in a later chapter. The first term is known as the curl-free part of the vector, while the second term is known as the divergence-free part of the vector.

3.10 Streamfunctions

. . . .

3.10.1 Two-dimensional flows: Cartesian

Definition

If the flow is 2D, only the component Ψ_3 is meaningful. (Why?) Then Ψ_3 is represented as a streamfunction denoted by ψ. This function has the following properties:

$$v_1 = \frac{\partial \psi}{\partial x_2} \tag{3.39}$$

$$v_2 = -\frac{\partial \psi}{\partial x_1} \tag{3.40}$$

Note that the streamfunction automatically satisfies the continuity equation, i.e., $(\partial v_1/\partial x_1) + (\partial v_2/\partial x_2)$ is identically zero. This can be verified by direct substitution into the definitions for the velocity components.

The introduction of the streamfunction thus combines the variables v_1 or v_x, the x-component of velocity, and v_2 or v_y, the y-component, into a single variable and also satisfies the continuity equation. If the velocity components v_x and v_y are specified, the streamfunction can be calculated by integration:

$$\psi = \int v_x \, dy + F(x) \tag{3.41}$$

Similarly

$$\psi = -\int v_y \, dx + G(y) \tag{3.42}$$

Note that the streamfunction can be determined within an arbitrary constant for a given velocity field. For example, if ψ is a streamfunction, then a function $\psi + C$, where C is a constant, also yields the same velocity field.

Physical meaning

The physical significance of the streamfunction can be assigned by considering the flow across any arbitrary curve AB and calculating the rate of volume flow across it from left to right, as in Fig. 3.5.

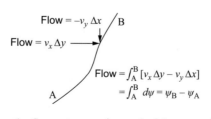

Figure 3.5 A sketch to show the flow rate as an integral of the streamfunction.

The flow across a differential element of length Δl is the sum of the flows in the x-direction and in the negative-y-direction, or $v_x \, \Delta y - v_y \, \Delta x$.

Hence the net flow (per unit width in the z-direction) is the line integral

$$\text{Net flow} = \int_A^B [v_x \, dy - v_y \, dx] \tag{3.43}$$

From the definition of the total derivative we have

$$d\psi = \frac{\partial \psi}{\partial x} \, dx + \frac{\partial \psi}{\partial y} \, dy \tag{3.44}$$

and, using the definitions for velocity in terms of ψ,

$$d\psi = -v_y \, dx + v_x \, dy \tag{3.45}$$

Using the third expression for net flow above, we get the following result:

$$\text{Net flow} = \int_A^B [v_x \, dy - v_y \, dx] = \int d\psi = [\psi(x,y)]_B - [\psi(x,y)]_A \tag{3.46}$$

Thus the difference between the values of the streamfunction at two points represents the flow across any curve connecting these points. If we consider another curve from A to B, the flow across it must be the same in order to satisfy the equation of continuity. Hence the line integral in Eq. (3.46) will be independent of path.

Lines of constant ψ are known as streamlines. These lines are always tangential to the local velocity vector. This follows from Eq. (3.45) on setting $d\psi$ equal to zero and re-arranging the equation in terms of the slope dy/dx, which turns out to be v_y/v_x, the slope of the velocity vector.

Example 3.6.

The velocity profile in a fluid is given as

$$v_x = A(1 - y^2)$$

Find the streamfunction. Find the flow across the system if the domain for y is from zero to one. Find the ratio of the average velocity to the maximum velocity.

Solution.

v_x is equal to $\partial \psi / \partial y$. We can derive an expression for the streamfunction by integration of the velocity with respect to y:

$$\psi = A(y - y^3/3)$$

where $\psi = 0$ is arbitrarily assumed for $y = 0$.

The flow across the system is obtained by taking the difference between the ψ values at 1 and 0. Hence the flow rate is $2A/3$.

The maximum velocity is at $y = 0$ and hence equal to A. The average velocity is therefore $2/3$ of the maximum velocity.

The relation to vorticity

Now consider the definition of vorticity ω in 2D flow:

$$\omega = \frac{\partial v_y}{\partial x} - \frac{\partial v_x}{\partial y} \tag{3.47}$$

where ω is the only non-vanishing component of the vorticity vector (the z-component) for a plane 2D flow in the x- and y-directions. If one substitutes for v_x and v_y in terms of the streamfunction one obtains

$$\boxed{\left(\frac{\partial^2 \psi}{\partial x^2} + \frac{\partial^2 \psi}{\partial y^2}\right) = \nabla^2 \psi = -\omega} \tag{3.48}$$

which is a kinematic relationship between the streamfunction and the vorticity for 2D flow.

Velocity as a scalar potential

If the vorticity vector ω is identically zero, then the flow is called irrotational, and the velocity can be represented as a potential of a scalar function defined by ϕ. The function ϕ is simply referred to as the velocity potential and is a scalar quantity,

$$v = \nabla \phi \tag{3.49}$$

3.10.2 Two-dimensional flows: polar

The 2D flows can be also represented using r and θ coordinates in lieu of x and y. The velocity components can be expressed in terms of the streamfunction as

$$v_r = \frac{1}{r}\frac{\partial \psi}{\partial \theta} \tag{3.50}$$

$$v_\theta = -\frac{\partial \psi}{\partial r} \tag{3.51}$$

The continuity equation is automatically satisfied.

3.10.3 Streamfunctions in axisymmetric flows

Although the streamfunction representation is defined for 2D flow (either Cartesian or polar), there are some cases where it can be used for 3D flows. These flows are axisymmetric flows in either cylindrical or spherical coordinates and the definitions are as follows.

Cylindrical coordinates

Flows where the velocity vector does not depend on the θ-direction in cylindrical coordinates are referred to as axisymmetric flows. For such flows the flow is symmetric around an axis passing through the line $r = 0$. The continuity equation then simplifies to

$$\frac{1}{r}\frac{\partial}{\partial r}(rv_r) + \frac{\partial}{\partial z}(v_z) = 0$$

This suggests the use of the following streamfunction for axisymmetric flow:

$$v_r = \frac{1}{r}\frac{\partial \psi}{\partial z}$$

$$v_z = -\frac{1}{r}\frac{\partial \psi}{\partial r}$$

Such a definition automatically satisfies the continuity equation, and the use of such a function is often useful for simplifying such problems.

Also note that this ψ is different from the ψ in polar coordinates for 2D flows for the case where the components of velocity are v_r and v_θ with $v_z = 0$. The flow in an axisymmetric system is 3D and can have three velocity components, the only requirement being that there is no θ-dependence. Hence such flows are sometimes called 2.5D flows.

Spherical coordinates

Flows where the velocity vector does not depend on the ϕ-direction in spherical coordinates are referred to as axisymmetric flows. For such flows the flow is symmetric around an N–S axis. Note that the designation N–S is arbitrary and represents a line joining a point at $\theta = 0$, the so-defined north pole, and a point at $\theta = \pi$, the south pole.

The continuity equation can then be represented as

$$\frac{1}{r^2}\frac{\partial}{\partial r}(r^2 v_r) + \frac{1}{r\sin\theta}\frac{\partial}{\partial\theta}(v_\theta \sin\theta) = 0$$

This suggests the use of the following streamfunction for axisymmetric flow in spherical coordinates:

$$v_r = \frac{1}{r^2\sin\theta}\frac{\partial \psi}{\partial\theta}$$

$$v_\theta = -\frac{1}{r\sin\theta}\frac{\partial \psi}{\partial r}$$

Such a definition automatically satisfies the continuity equation and is useful to simplify the flow representation for such problems. An application is in Stokes flow analysis (fluids with high viscosity) and will be studied in Chapter 15.

3.10.4   The relation to vorticity: the E^2 operator

There are kinematic relations between the vorticity and the streamfunction for axisymmetric flows similar to Eq. (3.48) for Cartesian cases. These are shown briefly here and the derivation is left as an exercise.

Cylindrical coordinates

For axisymmetric flows with v_θ equal to zero it can be shown that the only surviving component of the vorticity is ω_θ. Using the expression for this and substituting the velocity components in terms of the streamfunction, the following relation can be derived:

$$E^2\psi = -r\omega_\theta \tag{3.52}$$

where E^2 is an operator defined as

$$E^2 = \frac{\partial^2}{\partial r^2} - \frac{1}{r}\frac{\partial}{\partial r} + \frac{\partial^2}{\partial z^2} \tag{3.53}$$

Spherical coordinates

For axisymmetric flows with v_ϕ equal to zero it can be shown that the only surviving component of the vorticity is ω_ϕ. Using the expression for this and substituting the velocity components in terms of the streamfunction, the following relation can be derived:

$$E^2\psi = -r\sin\theta\,\omega_\phi \tag{3.54}$$

where E^2 is an operator defined as

$$E^2 = \frac{\partial^2}{\partial r^2} + \frac{\sin\theta}{r^2}\frac{\partial}{\partial\theta}\left(\frac{1}{\sin\theta}\frac{\partial}{\partial\theta}\right) \tag{3.55}$$

This operator is a key player in the analysis of irrotational flows.

3.11   The gradient of velocity

Consider two points A and B separated by a differential position vector of length δr. The x-component of the velocity at B $(r + \delta r)$ can be expressed in terms of the velocity at A (r) with a Taylor expansion term to a linear degree of approximation:

$$v_x(B) = v_x(A) + \frac{\partial v_x}{\partial x}\delta x + \frac{\partial v_x}{\partial y}\delta y + \frac{\partial v_x}{\partial z}\delta z$$

where all the partial derivatives are evaluated at the base point A.

The x-component of the relative velocity of B with respect to A is therefore equal to

$$\delta v_x = \frac{\partial v_x}{\partial x}\delta x + \frac{\partial v_x}{\partial y}\delta y + \frac{\partial v_x}{\partial z}\delta z$$

A similar expression holds for the velocity components δv_y and δv_z. All the three equations can be compacted into a vector–matrix form as follows. Let δv be a relative velocity vector with components δv_x, δv_y, and δv_z. Also let δr be a relative position vector with the components δx, δy, and δz. Then the three equations for the velocity components can be compacted as

$$\delta v = \nabla v \cdot \delta r \tag{3.56}$$

where ∇v is the gradient of the velocity. This is a tensor quantity with nine components as follows:

$$\tilde{\nabla}v = \begin{pmatrix} \partial v_x/\partial x & \partial v_x/\partial y & \partial v_x/\partial z \\ \partial v_y/\partial x & \partial v_y/\partial y & \partial v_y/\partial z \\ \partial v_z/\partial x & \partial v_z/\partial y & \partial v_z/\partial z \end{pmatrix} \tag{3.57}$$

What is the physical interpretation of the velocity gradient tensor?

Representation in other coordinates can be obtained by the vector operation rules shown in Problem 18.

3.12 Deformation and rate of strain

Equation (3.56) can be written as

$$\delta v = \frac{1}{2}(\nabla v + \nabla v^{\mathrm{T}}) \cdot \delta r + \frac{1}{2}(\nabla v - \nabla v^{\mathrm{T}}) \cdot \delta r \tag{3.58}$$

where ∇v^{T} is the transpose of the gradient of velocity obtained by interchanging the rows and columns of $\tilde{\nabla} v$. The diagonal terms are not affected and the cross diagonal terms get interchanged in making the transpose.

Note that, in the process, the first term on the RHS of Eq. (3.58) will be a symmetric tensor with six distinct components, while the second tensor is an anti-symmetric tensor with only three components (making it more like a vector). The two tensors are designated as the rate-of-strain tensor and the vorticity tensor. The relative velocity can hence be written as

$$\boxed{\delta v = \tilde{E} \cdot \delta r + \tilde{W} \cdot \delta r} \tag{3.59}$$

where $\tilde{E}$, the rate-of-strain tensor, is defined as

$$\tilde{E} = \frac{1}{2}(\nabla v + \nabla v^{\mathrm{T}})$$

and $\tilde{W}$, the vorticity tensor, is defined as

$$\tilde{W} = \frac{1}{2}(\nabla v - \nabla v^{\mathrm{T}})$$

The physical significance of these tensors is the following. The first term $\tilde{E} \cdot \delta r$ in Eq. (3.59) contributes to the pure strain or stretching of a line element δr, while the second term, $\tilde{W} \cdot \delta r$, in Eq. (3.59) contributes to a pure rotation of the line element. Figure 3.6

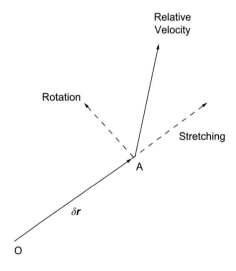

Figure 3.6 The relative velocity vector viewed as a sum of rotation and stretching vectors.

illustrates the stretching and rotation of a line element due to relative velocity at its end points.

The fact that the vorticity tensor contributes to pure rotation is shown by two methods in the following examples.

Example 3.7.
It can be shown by direct matrix algebra that

$$\delta r \cdot [\tilde{W} \cdot \delta r] = 0$$

Hence the part of the relative velocity $[\tilde{W} \cdot \delta r]$ is perpendicular to δr, indicating that we have a purely rotational motion.

Example 3.8.
It can be shown by direct matrix algebra that

$$\tilde{W} \cdot \delta r = (\omega \times \delta r)/2 = \beta \times \delta r$$

where ω is the vorticity vector and β is the angular velocity vector. The term $\beta \times \delta r$ represents the tangential velocity. Hence we verify that the contribution of $\tilde{W}$ is a pure rotation.

The concept of rotation and deformation is aptly illustrated by the following example for a fluid parcel placed in a simple shear flow.

Example 3.9.
Consider $v_x = \dot{\gamma} y$ and $v_y = 0$, where $\dot{\gamma}$ is a constant. This is a simple shear flow.

Figure 3.7 shows the relative velocity of a parcel in shear flow. The parcel is a square with the center at O. The line element OA has a purely the rotational component relative to O, while OB has both rotational and stretching components. The point C has no relative motion with respect to O. We note that the different points on the perimeter have different rotational and stretching tendencies, but on average the parcel undergoes both stretching and rotation.

We also compute the components of the various tensor quantities for clarity.

The components of the velocity gradient tensor are

$$\tilde{\nabla} v = \begin{pmatrix} 0 & \dot{\gamma} \\ 0 & 0 \end{pmatrix}$$

The rate-of-strain tensor is

$$\tilde{E} = \begin{pmatrix} 0 & \dot{\gamma}/2 \\ \dot{\gamma}/2 & 0 \end{pmatrix}$$

which is a symmetric tensor

The vorticity tensor is

$$\tilde{W} = \begin{pmatrix} 0 & \dot{\gamma}/2 \\ -\dot{\gamma}/2 & 0 \end{pmatrix}$$

which is an antisymmetric tensor. The magnitudes of the components are half the vorticity in the z-direction.

3.12.1   The physical meaning of the rate of strain

For understanding the physical meaning of the rate of strain the deformation of the Lagrangian control volume shown in Fig. 3.8 is useful. The control volume in the current configuration is shown as a square. Since the different points on the perimeter have different velocities relative to the center of the square, the control volume undergoes a deformation as shown in Fig. 3.8.

The deformation can be a change in linear dimensions (e.g., the line AB in Fig. 3.8), and such a deformation is called normal strain. It can also be changes to the angles (e.g., θ in Fig. 3.8), and such deformations are referred to as shear strain.

Normal strain

Consider a line element in the current configuration. Let us say that this line is oriented in the x-direction and has an original length of Δx. The rate of change in this line segment as

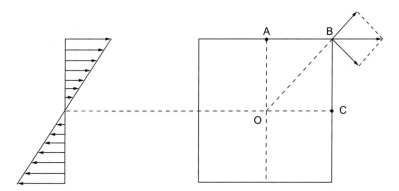

Figure 3.7 The relative velocity at various points on a parcel placed in simple shear flow: OA, pure rotation; OB, rotation plus stretching; and OC, no rotation and no stretching.

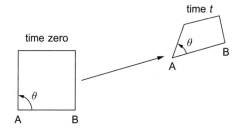

Figure 3.8 Elongational and shear deformation of a Lagrangian control volume and volume change of a fluid packet.

the particle moves can be shown to be equal to

$$\frac{1}{\Delta x}\frac{D}{Dt}(\Delta x) = E_{xx} \tag{3.60}$$

Similar equations hold for the other two directions. Thus E_{xx} (or E_{11} in index form) has the meaning of a rate of longitudinal strain or the normal strain rate. It is like the tension in a rod that is being stretched.

The diagonal terms of the rate of strain can therefore be interpreted as the change in length per unit time divided by the original length.

Example 3.10. Accelerating flow:

An accelerating flow is described by $v_x = Ax$, where A is a constant. Find the rate-of-strain tensor and interpret its physical meaning.

Solution.

Accelerating flows are called extensional flows. Such flows are encountered as a fluid enters an orifice or a contraction in a pipe. An example is shown schematically in Fig. 3.9.

The rate of strain in the x-direction is readily computed as A. Consider a line element OA in Fig. 3.9 oriented in the x-direction with a length Δx at time zero. Then its length will increase by $A\,\Delta x$ in a time interval Δt. This therefore represents the elongation of this line.

What about the y-direction? What is the velocity in the y-direction? What is the stretching of a line element oriented in the y-direction?

Volume change of a parcel

Upon extending the above analysis to three directions, we find that the volume change is given by

$$\text{Rate of volume change per unit volume} = \nabla \cdot v$$

which is the dilatation shown separately earlier.

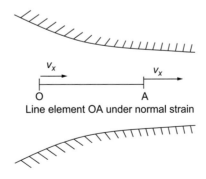

Line element OA under normal strain

Figure 3.9 An illustration of an extensional flow and the physical meaning of the rate of normal strain. The velocity at A is larger than that at O, causing OA to be under tension. This normal strain leads to a normal (viscous) stress.

Thus the diagonal terms of the strain-rate tensor will add up to zero for an incompressible fluid.

Shear rate

Similarly, consider two mutually perpendicular lines as indicated in Fig. 3.8. The angle between these lines is $\pi/2$ and will change as the particle deforms. The change in the angle can be shown to be

$$-\frac{D}{Dt}(\theta_{12}) = E_{12} \tag{3.61}$$

and hence the off-diagonal terms of the rate-of-strain tensor are the measures of shear deformation.

The concept of shear deformation is shown in Fig. 3.10

The stress and strain-rate relation

The strain-rate tensor is symmetric. For linear fluids (Newtonian fluids) the following constitutive equation is applicable if the incompressibilty condition holds:

$$\tilde{\tau} = 2\mu\tilde{E} \tag{3.62}$$

with μ being the fluid viscosity. Here $\tilde{\tau}$ is the stress tensor formally introduced and explained in Chapter 4.

This equation is referred to as the generalized version of Newton's law of viscosity.

For a general case of compressible fluids, a bulk viscosity correction is applied and the equation is modified to

$$\tilde{\tau} = 2\mu\tilde{E} - \frac{2}{3}\mu(\nabla \cdot \boldsymbol{v})\delta_{ij} \tag{3.63}$$

The correction applies only to normal stress terms, as is indicated by the presence of the Kronecker-delta term. For incompressible fluids the divergence of the velocity vector is zero and the bulk viscosity correction term disappears.

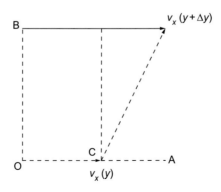

Figure 3.10 An illustration of a shear deformation: v_x changes in the y-direction, causing the angle BOA to change.

3.12.2   Rate of strain: cylindrical

In this section the expressions for the rate-of-strain tensor in cylindrical coordinates are provided. These are useful to relate the corresponding stress when working in this coordinate system (for a Newtonian fluid).

The normal strain rates are

$$E_{rr} = \frac{\partial v_r}{\partial r}$$

$$E_{\theta\theta} = \frac{1}{r}\frac{\partial v_\theta}{\partial \theta} + \frac{v_r}{r}$$

$$E_{zz} = \frac{\partial v_z}{\partial z}$$

The shear strain rates are

$$E_{r\theta} = E_{\theta r} = \frac{1}{2}\left(r\frac{\partial}{\partial r}(v_\theta/r) + \frac{1}{r}\frac{\partial v_r}{\partial \theta}\right)$$

$$E_{rz} = E_{zr} = \frac{1}{2}\left(\frac{\partial v_z}{\partial r} + \frac{\partial v_r}{\partial z}\right)$$

$$E_{\theta z} = E_{z\theta} = \frac{1}{2}\left(\frac{\partial v_\theta}{\partial z} + \frac{1}{r}\frac{\partial v_z}{\partial \theta}\right)$$

3.12.3   Rate of strain: spherical

In this section the expressions for the rate-of-strain tensor in spherical coordinates are provided. These are useful to relate the corresponding stress when working in this coordinate system (for a Newtonian fluid).

The normal strain rates are

$$E_{rr} = \frac{\partial v_r}{\partial r}$$

$$E_{\theta\theta} = \frac{1}{r}\frac{\partial v_\theta}{\partial \theta} + \frac{v_r}{r}$$

$$E_{\phi\phi} = \frac{1}{r\sin\theta}\frac{\partial v_\phi}{\partial \phi} + \frac{v_r}{r} + \frac{v_\theta \cot\theta}{r}$$

The shear strain rates are

$$E_{r\theta} = E_{\theta r} = \frac{1}{2}\left(r\frac{\partial}{\partial r}(v_\theta/r) + \frac{1}{r}\frac{\partial v_r}{\partial \theta}\right)$$

$$E_{r\phi} = E_{\phi r} = \frac{1}{2}\left(\frac{1}{r\sin\theta}\frac{\partial v_r}{\partial \phi} + r\frac{\partial}{\partial r}\left(\frac{v_\phi}{r}\right)\right)$$

$$E_{\theta\phi} = E_{\phi\theta} = \frac{1}{2}\left(\frac{\sin\theta}{r}\frac{\partial}{\partial \theta}\left(\frac{v_\phi}{\sin\theta}\right) + \frac{1}{r\sin\theta}\frac{\partial v_\theta}{\partial \phi}\right)$$

3.12.4 Invariants of a tensor

The following properties of a symmetric tensor are useful.

- The components of a tensor are dependent on the coordinate system used, but the tensor itself is a physical entity similar to a vector. One should not confuse a "tensor" with its components. In the words of Professor Aris (1962),

  a tensor is a tensor is a tensor.

- A symmetric tensor has real eigenvalues.
- The following "invariants" of a symmetric tensor (denoted by A here) can be defined:

$$I_1 = A_{11} + A_{22} + A_{33}$$

also known as the trace,

$$I_2 = A_{11}A_{22} + A_{22}A_{33} + A_{33}A_{11} - A_{12}^2 - A_{23}^2 - A_{13}^2$$

which is called the sum of the principal minors, and

$$I_3 = \det(A)$$

where det means the determinant of the matrix A.

- These are called invariants since they are the same irrespective of the coordinate system used.
- A symmetric tensor can be reduced to a diagonal form by a coordinate transformation.
- The coordinate system where the tensor takes a diagonal form corresponds to the eigenvectors of the tensor.
- For an incompressible fluid I_1 is equal to zero.

3.13 Index notation for vectors and tensors

In this notation a quantity without any subscript represents a scalar. The first rule is that a quantity with one index is a vector and a quantity with two indices is a second-order tensor. The indices are assumed to take values of 1, 2, and 3, which represent the components with respect to a Cartesian coordinate system. The subscript is also called a free index. Thus, for example, a_i is the index notation for a vector a with a_1, a_2, and a_3 as the components (referred to a Cartesian coordinate system). Similarly, τ_{ij} is the index notation for a tensor $\tilde{\tau}$ and as both the free indices i and j vary from 1 to 3 we generate a set of nine quantities, which are the components of $\tilde{\tau}$.

A second rule is that a repeated index implies summation. Thus $a_i b_i$ means $a_1 b_1 + a_2 b_2 + a_3 b_3$ and hence represents a dot product of vectors a and b. A repeated index is also called a dummy index.

The third rule is that differentiation is indicated by a comma notation. A comma followed by j means take derivatives with respect to each coordinate x_j. Therefore $T_{,j}$ is a compact way to write the gradient of a scalar T; the gradient is a vector quantity, as we already know, and hence the gradient operation increases the tensor order of the quantity.

Similarly, $v_{i,j}$ means differentiation with respect to x_j. Since both i and j are free indices, this represents a tensor, which is the gradient of velocity.

Further, $v_{i,i}$ means differentiation with respect to each x_i and also summation, since there is no free index. This operation thereby generates a scalar. It can be seen that $v_{i,i}$ is the divergence of the velocity.

The cross product is somewhat cumbersome, and needs the use of a quantity called the permutation tensor. This is defined with three subscripts, ϵ_{ijk}, which is a third-order tensor. It is defined such that it is zero if any of the indices are the same. It is defined as one if the order of the indices is 123, 231, or 312, and is defined as minus one otherwise. Thus the only non-zero components of this tensor are as follows:

$$\epsilon_{123} = \epsilon_{231} = \epsilon_{312} = 1$$

$$\epsilon_{132} = \epsilon_{213} = \epsilon_{321} = -1$$

The cross product of two vectors can then be represented as

$$a \times b = \epsilon_{ijk} a_j b_k$$

The summation with respect to both j and k is implied. Since there is one free index, i, this is a vector. You should verify that this is indeed the cross product by expanding (summing with respect to both j and k) and then using the definition of the permutation tensor.

Similarly, the curl of a vector is represented using the permutation symbol and the comma notation as

$$\nabla \times v = \epsilon_{ijk} v_{j,k}$$

The student should write out the various quantities. As an example, the rate-of-strain tensor would be

$$\tilde{E} = \frac{1}{2}(v_{i,j} + v_{j,i})$$

The operations of differentiation on tensor quantities become easy algebraic manipulations of the indices, as indicated in the following example.

Example 3.11. Derive an expression for the divergence of the rate of strain tensor.
We use the index notation. Let E_{ij} be the components of the rate of strain,

$$E_{ij} = (1/2)(v_{i,j} + v_{j,i})$$

Divergence is obtained by differentiation with respect to i.

The divergence is obtained as $E_{ij,i}$, where the repeated index means summation. Note that i becomes a repeated index and the final result is a vector. Hence

$$\nabla \cdot \tilde{E} = (1/2)(v_{i,j} + v_{j,i})_{,i}$$

The order of differentiation can be interchanged in the first term, leading to

$$\nabla \cdot \tilde{E} = (1/2)(v_{i,i,j} + v_{j,i,i})$$

Since $v_{i,i}$ is the divergence of the velocity, the first term above can be identified as the gradient of this quantity, i.e., $\nabla(\nabla \cdot v)$.

The second term can be identified as $\nabla \cdot (\nabla v)$, which is also written as $\nabla^2 v$. Hence

$$\nabla \cdot \tilde{E} = (1/2)[\nabla(\nabla \cdot v) + \nabla^2 v] \tag{3.64}$$

The form for the divergence of the rate of strain is needed in the equations of motion developed in later chapters. The expressions in different coordinate systems are useful and can be derived by using the vector operations indicated above.

Example 3.12. What form does the divergence of the stress tensor take for a Newtonian incompressible fluid?

Solution.
We use the generalized version of Newton's law of viscosity given by Eq. (3.62) and take the divergence on both sides:

$$\nabla \cdot \tilde{\tau} = 2\mu \, \nabla \cdot \tilde{E} \tag{3.65}$$

Now using the expression for the divergence of $\tilde{E}$ from the previous example and the incompressibilty condition leads to

$$\boxed{\nabla \cdot \tilde{\tau} = \mu \, \nabla^2 v} \tag{3.66}$$

which will be used in Chapter 5 to derive the equation for flow of a Newtonian fluid.

Summary

- Fluid motion can be represented in a fixed coordinate system (the Eulerian representation) or in a moving coordinate system (the Lagrangian representation). The representation and the equations are different in these two cases, but the final physical meaning and results are the same.

- The time derivative in a frame moving with the fluid velocity is called the substantial derivative. A general notation for this is D/Dt, and it represents the change in the property for a particle or control volume that flows with the fluid. Thus DT/Dt represents the change in temperature observed in a moving control volume.

- The substantial derivative of the velocity vector denoted as Dv/Dt represents the acceleration observed by a fluid particle. Its representation and detailed expressions in various coordinate systems are very useful and should be noted.

- Dilatation means the change in fluid volume (per unit volume) for a moving (Lagrangian) control volume. It is simply equal to the divergence of the velocity vector.

- The continuity equation is nothing but an overall mass balance for a control volume. The general equation is given by Eq. (3.20) in the text. For incompressible fluid it reduces to the condition that the divergence of the velocity vector is zero.

- A fluid particle can have a rotational component of motion viewed at a microscopic scale. A measure of this is the curl of the velocity vector, which is called a vorticity vector. A fluid motion with no circulation is called irrotational flow. A related concept is that of the circulation along a closed curve, which is the line integral of the tangential component of velocity.

- The streamfunction is defined in such a way that the continuity equation is satisfied. Lines of constant values of streamfunction are tangential to the velocity vector.

- The Laplacian of the streamfunction is equal to the negative of the vorticity for 2D flows. This is purely a kinematic property.

- The streamfunction is defined only for 2D flows. However, for flows in cylindrical and spherical coordinates that are axisymmetric it is possible to define a similar function. This function satisfies the continuity equation, and the definition is given in Section 3.10.3.

- The gradient of velocity is defined in a similar manner to the gradient of temperature. It therefore represents a change in velocity as one moves in a particular direction. However, it is a tensor quantity with nine components, not a vector.

- The gradient-of-velocity tensor can be split into an antisymmetric part (the vorticity tensor) and a symmetric part (the rate-of-strain tensor). The vorticity tensor is a measure of local rotation and does not cause any deformation of the control volume.

- The rate of strain is a measure of the deformation of the line elements (linear deformation) and angular dimensions (shear deformation) of the control volume. These cause stress and can be related to the stress tensor. If a linear relation between the two tensors holds, we have then an extension of Newton's law of viscosity introduced in Chapter 1 to the 3D case. Such fluids are called Newtonian fluids.

- The index-and-comma notation is a compact way of representing vector and tensor quantities. It is also called the Einstein notation.

Problems

1. Find the acceleration of a fluid particle for the following velocity profiles:
   (a) $v_x = A(1 - y^2); v_y = 0; v_z = 0$
   (b) $v_x = Ax; v_y = -Ay; v_z = 0$
   (c) $v_r = 0; v_\theta = Ar; v_z = 0$
   (d) $v_r = 0; v_\theta = A/r; v_z = 0$
   (e) $v_r = 0; v_\theta = 0; v_z = f(r)$

2. Derive an expression for DT/Dt in Cartesian coordinates.
   Also show that the same relation holds for a vector; i.e.,

   $$\frac{Dv}{Dt} = e_x \frac{Dv_x}{Dt} + e_y \frac{Dv_y}{Dt} + e_z \frac{Dv_z}{Dt}$$

   where Dv_z/Dt etc. have the same form as if they were acting on a scalar quantity.

3. In cylindrical coordinates, the temperature field (a scalar field) can be expressed as

   $$T = T(r, \theta, z, t)$$

   Based on this, derive an expression for DT/Dt in cylindrical coordinates.
   Answer:

   $$\frac{DT}{Dt} = \frac{\partial T}{\partial t} + v_r \frac{\partial T}{\partial r} + \frac{v_\theta}{r} \frac{\partial T}{\partial \theta} + v_z \frac{\partial T}{\partial z} \tag{3.67}$$

4. In spherical coordinates, the temperature field can be expressed as

$$T = T(r, \theta, \phi, t)$$

Derive from this an expression for DT/Dt in spherical coordinates.

   Answer:

$$\frac{DT}{Dt} = \frac{\partial T}{\partial t} + v_r \frac{\partial T}{\partial r} + \frac{v_\theta}{r} \frac{\partial T}{\partial \theta} + \frac{v_\phi}{r \sin \theta} \frac{\partial T}{\partial \phi} \tag{3.68}$$

5. Derive an expression for Dv/Dt in cylindrical coordinates.

   Hint: define the nabla operator in cylindrical coordinates and let it be dotted with a vector. Expand and use the chain rule, taking care to differentiate some of the unit vectors. The components in Table 3.1 would be obtained.

6. Derive an expression for Dv/Dt in spherical coordinates. Follow the same procedure as for cylindrical coordinates. Some unit vectors have derivatives. The components in Table 3.2 would be obtained.

7. Derive the expressions for the divergence of a vector in cylindrical and spherical coordinates. Thereby verify the results in the text.

8. Prove that $\nabla \cdot (\nabla \times A) = 0$; i.e., the divergence of the curl of a vector is zero. Use this result to prove that the vector potential of velocity satisfies the continuity equation.

9. Show that the necessary and sufficient condition for the flow to be irrotational is the existence of a scalar velocity potential.

10. Write the units for the (i) vorticity, (ii) circulation, (iii) velocity potential, and (iv) streamfunction.

11. By direct differentiation using Cartesian coordinates verify the following relations.

    (a) The divergence of the curl of any vector is zero.

    $$\nabla \cdot (\nabla \times A) = 0$$

    (b) The curl of the gradient of any scalar is zero

    $$\nabla \times (\nabla \phi) = 0$$

    Use these properties in the Helmholtz decomposition to confirm that any vector can be decomposed into two vectors, where one vector is divergence-free while the other is curl-free.

    Discuss the use of these in simplifying the description of the velocity vector for some limiting situations. In particular, indicate the proper representation for the following cases:

    (a) the divergence is zero, the flow is a general 3D flow
    (b) the curl is zero, the flow is 3D
    (c) the divergence is zero, the flow is 2D
    (d) both divergence and curl are zero, the flow is 3D
    (e) both divergence and curl are zero, the flow is 2D

12. Show by direct differentiation that the divergence of the vorticity vector is zero.

    $$\nabla \cdot \omega = 0$$

13. Derive expressions for the components of the curl operator in cylindrical polar coordinates. Using these relations, find the vorticity field for the flow described by the following

velocity field:

$$v_\theta = Ar + B/r$$

where A and B are constants. This field is a general representation of the "torsional flow" and occurs in many practical situations. What is the direction of the vorticity vector? What is the magnitude?

A free vortex can be considered to be a special case of the torsional flow of the above problem with the constant A equal to zero. Show that the vorticity field is zero for a free vortex. A rigid-body type of rotational motion is described by

$$v_\theta = Ar$$

Why is it called a rigid-body type of motion? Find the vorticity for this flow.

14. Show that for a 2D flow or plane confined to the (x, y) plane only the z-component of the vorticity is non-zero. This component, ω_z, is simply abbreviated as ω and treated like a scalar. Also verify by direct substitution that

$$-\omega = \nabla^2 \psi$$

for 2D flow, where ∇^2 is the Laplacian operator with only x and y terms included. Here ψ is the streamfunction for 2D flows.

Also derive the corresponding relations for axisymmetric flows.

15. Consider a 2D plane flow that is now represented in terms of the polar coordinates. The flow has then only v_r and v_θ components and no v_z component. How is the streamfunction defined here? Show that the continuity equation (in polar coordinates) is automatically satisfied by this function.

Also show that the vorticity of flow can be represented as

$$-\omega = \frac{1}{r}\frac{\partial}{\partial r}\left(r\frac{\partial \psi}{\partial r}\right) + \frac{1}{r^2}\frac{\partial^2 \psi}{\partial \theta^2}$$

or in other words

$$\omega = -\nabla^2 \psi$$

where ∇^2 denotes the Laplacian polar coordinates.

16. Write out the components of the vorticity tensor, $\tilde{W}$ in Cartesian coordinates. Show that the tensor is antisymmetric, i.e., $W_{ij} = -W_{ji}$, and has only three distinct components.

Show that the components of the vorticity tensor can be represented in terms of the components of the vorticity **vector** as

$$\begin{pmatrix} 0 & \omega_z & -\omega_y \\ -\omega_z & 0 & \omega_x \\ \omega_y & -\omega_x & 0 \end{pmatrix}$$

where ω_x etc. are the components of the vorticity vector $\boldsymbol{\omega} = \nabla \times \boldsymbol{v}$.

Now show by direct matrix expansion that

$$\tilde{W} \cdot \delta \boldsymbol{r} = (1/2)\boldsymbol{\omega} \times \delta \boldsymbol{r}$$

The RHS is the angular velocity of motion and hence this verifies that $\tilde{W} \cdot \delta \boldsymbol{r}$ is a purely rotational motion of the line element $\delta \boldsymbol{r}$.

17. Consider the simple shear flow described as $v_x = \dot{\gamma}y$ and $v_y = 0$, where $\dot{\gamma}$ is the rate of strain.

    Verify that the rate of strain has only shear components as shown in the text.

    Now consider a coordinate system that is rotated by an angle θ to the x-axis. Find the components of the vector v in this system. Then write the expression for the velocity-gradient, rate-of-strain, and vorticity tensors.

    Find the angle θ such that the rate of strain has only diagonal components. (Answer: $\theta = \pi/4$.) This represents a case where the strain is purely elongational. Show that the vorticity tensor remains unchanged by the rotation of the coordinates.

18. Derive expressions for the gradient of velocity in cylindrical and spherical coordinates. **Hint:** define the nabla operator in cylindrical/spherical coordinates and let it act on a vector. Use the chain rule, taking care to differentiate some of the unit vectors. Rearrange the results into a componentwise form.

    Using this and the definition of $\tilde{E}$, verify the expressions in Sections 3.12.2 and 3.12.3.

4 Forces and their representations

Learning objectives

You will learn the following important concepts:

- the pressure force and its representation;
- the equations of hydrostatics; calculation of the pressure profile in a column of fluid;
- the pressure variation on an immersed body and the famous Archimedes principle;
- representation of viscous forces; the stress tensor;
- calculation of the net viscous force on a control volume; the divergence of the stress tensor;
- forces at an interface; surface tension, contact angle, capillary rise; and
- definitions for integrated surface force on a macroscopic control volume; the lift and the drag.

Mathematical prerequisites

- The Green–Gauss divergence theorem for converting a surface integral to a volume integral introduced in Chapter 3.
- The dot product of a tensor and a vector. If $\tilde{\tau}$ is a tensor and n is a vector then the dot product of the two can be represented in a compact index notation as $\tau_{ij}n_j$, which is a vector. You should expand the index notation and be familiar with the components of this vector. Similarly the dot product $n \cdot \tilde{\tau}$ is equal to $n_i\tau_{ij}$, which is also a vector.
- The divergence of a tensor, which turns out to be a vector. In index notation this is $\tilde{\tau}_{ij,i}$, where the comma means differentiation with respect to the coordinates.

This chapter reviews the various types of forces acting on the control volume and control surface and their representation in a compact manner using the vector–tensor notation.

The forces are of two types: (i) body forces, which act on the volume, the most common example being gravity (also electromagnetic forces in some cases); and (ii) surface forces, which act on the control surface and may be viewed as resulting from the interaction of the surroundings with the control volume. Examples are the pressure and viscous forces. You will learn to represent the surface forces in a compact manner and will be introduced to the stress tensor. Volume forces, by contrast, are easier to represent.

In a static fluid the forces are the pressure and gravity, and a balance of these gives us an equation governing the pressure distribution in a static fluid. The buoyancy force on a body immersed in a fluid arises as a consequence of the variation in the pressure distribution on the surface of the immersed body. The resulting net force is enunciated in the famous Archimedes principle, which we will prove using vector calculus. A force and a torque are also felt on a surface in contact with a fluid. Once again we can use the pressure distribution to calculate these forces. Such calculations are important to civil engineers who design dams, retaining walls, etc., and are also useful in many other applications.

The surface tension is a force acting on an interface between, say, a gas and a liquid. A related concept is the contact angle. We will learn the nature of these effects and illustrate them with some simple examples. We can think of the surface tension as a hydro-static force acting on an interface.

In a moving fluid, an additional surface force acts, namely the viscous force. As noted in the previous chapter, a control volume is subject to deformation due to the variation of velocity on the control surface. The corresponding deformation can be either the normal strain or the shear strain. The deformation causes a surface stress and correspondingly a surface force. We will learn how to represent this force using a tensor quantity called the stress tensor.

For macroscopic balances the integrated value of the force over the entire control surface will appear in the model equations. Most commonly, we use the drag force (net force in the direction of flow) and the lift force (net force in a direction perpendicular to the flow) to describe the forces on a macroscopic control volume. Definitions of these forces will be given, and empirical representations in terms of the drag and lift coefficients will be introduced.

This chapter forms a prelude for the formulation of the differential equations for fluid dynamics discussed in next chapter.

4.1 Forces on fluids and their representation

When the conservation law is applied to momentum transport, the term generation will have the meaning of the (vector) sum of the forces acting on the control volume. We have already seen this in the context of macroscopic momentum balance in Section 2.3. It is obvious that a systematic methodology and notation to describe the forces acting on a control volume of the fluid are needed. This is the focus of this section. The forces acting on a fluid can be represented as

$$\text{Forces acting on the system} = \text{Gravity} + \text{Pressure} + \text{Viscous forces}$$

The pressure and viscous forces act over the control surfaces; gravity acts over the control volume. Since gravity is a volume force, the representation is straightforward:

$$\text{Gravity forces acting on the system} = \int_V \rho g \, dV$$

For a differential control volume ρ is assigned as the point value and hence

$$\boxed{\text{Gravity force per unit volume} = \rho g}$$

4.1.1 Pressure forces

We now examine the nature of pressure force in the following subsection.
    We pose the following question.

What is pressure and why is it a scalar quantity?

Before answering this we need to answer the following question.

What is a fluid?

A fluid by definition is a material that cannot resist a shear force (a force applied tangentially). It continues to flow under the action of a shear force. Hence in a static fluid only normal forces can exist; shear or tangential stresses cannot exist. The normal stress acting on a static fluid is defined as the fluid pressure. It acts equally in all directions, the famous Pascal principle, and hence it can be represented by a scalar field. It is also compressive in nature as indicated earlier in Chapters 1 and 2.
    The fact that pressure is a scalar and acts equally in all directions is demonstrated by balancing forces on a wedge-shaped control volume of the fluid shown in Fig. 4.1. We take the control volume in 2D for easy visualization. (The extension to 3D is straightforward.) Let the pressure on the face AC be p. Then the pressure force on AC is pAC per unit width and in the inward normal direction shown in Fig. 4.1. The component of this in the minus x-direction is $pAC \cos \alpha$.

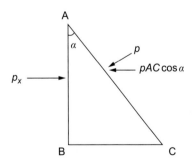

Figure 4.1 An illustration of a wedge- or prism-shaped control volume to examine the nature of the pressure force.

Let the pressure on the face AB be p_x. Then the pressure force on AB is $p_x AB$ per unit width and in the x-direction. For equilibrium the two pressure forces must balance:

$$pAC \cos \alpha = p_x AB$$

But $\cos \alpha = AB/AC$ and hence $p_x = p$.

Similarly, if one were to balance the forces in the y-direction one would write

$$pAC \sin \alpha + \rho g (AB)(BC)/2 = p_y BC$$

But $\sin \alpha = BC/AC$ and hence $p_y = p + \rho g AB/2$. The gravity term can be neglected as the line element AB tends to zero. Hence we can also show that $p_y = p$ for a differential volume in Fig. 4.1.

This proves the Pascal principle, namely that the pressure on any surface element (locally) is the same irrespective of the orientation of the control surface and that the pressure acts equally in all directions.

Although pressure is a scalar, the force due to pressure is a vector. If n is the unit normal pointing outward from the control surface then the pressure acts in the direction of $-n$. Hence the pressure force on a differential element of area dA is represented as a **vector**, $-pn\, dA$. The total pressure force is then a surface integral:

$$\text{Total pressure force over a surface } A = \int_A [-pn]\, dA$$

where the area integral is over the entire surface.

By Gauss's theorem, an area integral can be converted to a volume integral, by application of the divergence theorem:

$$\int pn\, dA = \int_V \nabla p\, dV \approx V \nabla p$$

Hence

$$\boxed{\text{Pressure force per unit volume} = -\nabla p} \qquad (4.1)$$

Explanation of the ∇p term

It is easier to understand and mark the surface forces in Cartesian coordinates. The control volume is in the shape of a box as shown in Fig. 4.2. The net pressure force in the x-direction can be shown to be $-\partial p/\partial x$ times the differential volume, $\Delta x\, \Delta y\, \Delta z$. This is obtained by

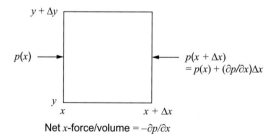

$y + \Delta y$

$p(x) \longrightarrow$

$p(x + \Delta x)$
$= p(x) + (\partial p/\partial x)\Delta x$

y

x $x + \Delta x$

Net x-force/volume $= -\partial p/\partial x$

Figure 4.2 An illustration of the pressure forces acting on a Cartesian control volume and the net pressure force on a control volume. Only the x-direction is shown in order to avoid clutter.

balancing the forces in the plane at x and the plane at $x + \Delta x$ as indicated in Fig. 4.2. Similar results hold for the y- and z-directions and hence vectorially; $-\nabla p$ is the net pressure force per unit volume as shown directly in the preceding discussion.

4.1.2 Viscous forces

Viscous forces also act on the surface but are somewhat different to describe. This is because the force can have tangential (shear) components and a normal component. Hence the direction of the force does not always coincide with the direction of the normal vector, unlike for the pressure force. This is the key difference between the representation of the viscous force and that of the pressure force. We will need a tensor field now. In order to proceed further the following lemma called the stress principle is useful.

Stress vector

The stress principle and the definition of the stress vector are as follows: *A vector function $f(n, x, y, z)$ exists at all points (x, y, z) in a body of fluid for all unit normals n such that the force exerted on a differential surface ΔA is $f\,dA$.*

Here the normal points outward from the control surface to the surroundings. The force defined above is the force exerted by the surroundings on the control surface. The above theorem is also called the Cauchy stress principle. The function f is called the stress vector or traction vector.

Figure 4.3 illustrates the definition of the traction vector.

Note that the vector f need not be in the direction of n. If the vector f is resolved in three perpendicular directions then f will have a normal component and two tangential components. The normal component represents the tension or compression acting on the surface and the tangential components represent the shear acting on the surface under consideration.

The stress vector at any fixed point is a function of the orientation of the normal and is therefore a function of n. We designate $f(n)$ as an abbreviation for the stress vector at a given point. We will need a formula to find f for a given n, which looks like a complicated problem. In this connection a useful property is the Cauchy lemma, which is nothing but Newton's third law of motion. Cauchy's lemma states that *the stress vectors acting upon opposite sides of the same surface at a given point are equal in magnitude and opposite in direction.*

Thus $f(n) = -f(-n)$. See Fig. 4.3, where the stress vectors on each side of a surface are shown. These have to balance according to Newton's third law.

Stress tensor

The stress vector and its components acting in the three principal coordinate directions have a special meaning.

For example, on a plane whose normal points along the x-axis the stress vector is $f(x)$. This can be expressed as

$$f(x) = e_x \tau_{xx} + e_y \tau_{xy} + e_z \tau_{xz}$$

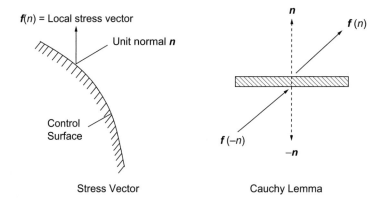

Figure 4.3 An illustration of the stress vector and the Cauchy lemma.

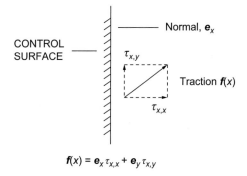

$$f(x) = e_x \tau_{x,x} + e_y \tau_{x,y}$$

Figure 4.4 The stress vector acting on the x-plane and its components in the x- and y-directions. The x-component is a normal stress, while the y-component is a shear stress.

Note that two suffices are necessary to represent the components. The first suffix indicates the direction of the normal, while the second suffix indicates the direction of the force component. See Fig. 4.4 for the definitions of these quantities

Similarly the stress vector acting on a plane that has a normal y is $f(y)$ and can be written in terms of the components as

$$f(y) = e_x \tau_{yx} + e_y \tau_{yy} + e_z \tau_{yz}$$

and for the third component direction

$$f(z) = e_x \tau_{zx} + e_y \tau_{zy} + e_z \tau_{zz}$$

These components τ_{xx}, τ_{xy}, etc. can be put into a matrix form called the stress tensor.

$$\tilde{\tau} = \begin{pmatrix} \tau_{11} & \tau_{12} & \tau_{13} \\ \tau_{21} & \tau_{22} & \tau_{23} \\ \tau_{31} & \tau_{32} & \tau_{33} \end{pmatrix} \tag{4.2}$$

Here we use the 1, 2, 3 notation for x, y, z for convenience later. Again we recall the physical meaning of the components: τ_{ij} represents the viscous stress acting in the j-direction on a plane whose outward normal points in the i-direction.

Properties of the stress tensor

The stress tensor has the following properties which should be noted.

Property 1. The stress components acting on the two opposing faces of a particular surface are equal and opposite. For example, $\tau_{-x,x}$ is equal to $-\tau_{x,x}$. Here $\tau_{-x,x}$ is the stress component in the x-direction on a plane whose normal is in the $-x$ direction, while $\tau_{x,x}$ is defined as above as the stress component in the x-direction on a plane whose normal is in the $+x$-direction. This property is a consequence of Cauchy's lemma.

Property 2. The stress tensor is symmetric. Thus $\tau_{x,y}$ is equal to $\tau_{y,x}$. This property is a consequence of the fact that for a tiny differential element the torque has to be equal to zero. An illustration of the torque due to the above two stress components is shown in Fig. 4.5.

The stress tensor has therefore only six independent components. The diagonal terms represent the normal stresses while the off-diagonal terms represent the shear stresses.

Representation of viscous forces

With all this background we are in a position to represent the viscous force acting on a surface. Here we ask a question. How does $f(n)$ change with n at a fixed point as n changes? The answer is given by the Cauchy stress theorem, which states that $f(n)$ at any point and specified direction n can be simply calculated if the value of the stress tensor at that point is known.

We first present the final relationship:

$$\boxed{f(n) = n \cdot \tilde{\tau}} \tag{4.3}$$

where n_j are the direction cosines of the normal. Thus $f(n)$ is a vector–tensor product of the normal vector and the stress tensor. Note that due to the symmetry of the stress tensor this product can also be written as $\tilde{\tau} \cdot n$.

It is also illustrative to write out the components of the viscous stress vector f_n,

$$f_1 = n_1 \tau_{11} + n_2 \tau_{21} + n_3 \tau_{31} \tag{4.4}$$

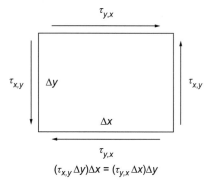

$$(\tau_{x,y}\,\Delta y)\Delta x = (\tau_{y,x}\,\Delta x)\Delta y$$

Figure 4.5 A sketch to illustrate the symmetry property of the stress tensor.

and similar relations for f_2 (or f_y) and f_3,

$$f_2 = n_1\tau_{12} + n_2\tau_{22} + n_3\tau_{32} \tag{4.5}$$

$$f_3 = n_1\tau_{13} + n_2\tau_{23} + n_3\tau_{33} \tag{4.6}$$

Proof of the stress relation given by Eq. (4.3)

The proof of the above relation is done using a stress prism as shown in Fig. 4.6, using Cauchy's lemma.

The x-component of $\boldsymbol{f}$ must balance the x-forces on the x- and y-axes:

$$\Delta s\, Zf_x = \Delta x\, Z\tau_{yx} + \Delta y\, Z\tau_{xx}$$

where Z is an arbitrary width into the plane of the paper. On dividing by Δs we have

$$f_x = \frac{\Delta x}{\Delta s}\tau_{yx} + \frac{\Delta y}{\Delta s}\tau_{xx}$$

The direction cosines of the normal vector are defined as

$$n_x = \frac{\Delta y}{\Delta s}$$

and

$$n_y = \frac{\Delta x}{\Delta s}$$

Hence we can write

$$f_x = n_y\tau_{yx} + n_x\tau_{xx} = n_x\tau_{xx} + n_y\tau_{yx} \tag{4.7}$$

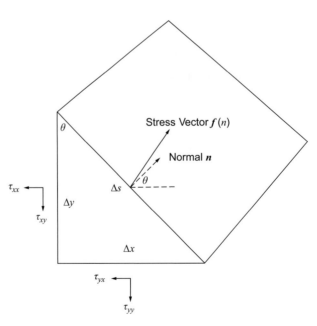

Figure 4.6 A figure showing the definitions of the stresses and the contact forces acting on any given surface; a 2D case is shown for simplicity.

or using the 1, 2, 3 notation for convenience of later compaction

$$f_1 = \sum_j n_j \tau_{j1} \tag{4.8}$$

Similarly for the y-direction:

$$f_2 = \sum_j n_j \tau_{j2} \tag{4.9}$$

Equations (4.8) and (4.9) can be compacted using the vector–matrix product notation as

$$\boldsymbol{f}(n) = \boldsymbol{n} \cdot \tilde{\tau}$$

which in index notation is $n_i \tau_{ij}$, with summation on i implied. Since j is a free index, this is a vector. The student should verify the veracity of this statement. This shows that the stress vector can be computed if the local value of the stress tensor is known and the direction of the normal is specified. Thus a tensor field is needed in order to represent the viscous forces.

The total viscous force on a control surface

The stress vector times the differential area is the local force on dA. Hence the total viscous force acting on a control volume is obtained by integration over the entire control surface A:

$$\text{Total viscous force} = \int_A (\boldsymbol{n} \cdot \tilde{\tau}) dA$$

The area integral can be reduced to a volume integral by use of the divergence theorem:

$$\text{Total viscous force} = \int_V \nabla \cdot \tilde{\tau} \, dV \approx V \nabla \cdot \tilde{\tau}$$

Hence

$$\boxed{\text{Viscous force per unit volume} = \nabla \cdot \tilde{\tau}} \tag{4.10}$$

Hence we have to use the divergence of a tensor, which is a vector, to represent viscous forces. The shorthand notation for this divergence is $\tau_{ij,i}$, where i means differentiation.

4.1.3   The divergence of a tensor

Since the divergence of a tensor plays such an important role, we will look at the forms of this in other coordinate systems as well.

Cartesian coordinates

The Cartesian form is rather easy. Note that the index notation applies in Cartesian coordinates and the divergence is τ_{ij}, i. Take $j = 1 = x$ for the x-direction and sum over $i = 1, 2, 3$ or x, y, z. The following result is obtained for the x-component of this vector representing the viscous force:

$$\frac{\partial}{\partial x}(\tau_{xx}) + \frac{\partial}{\partial y}(\tau_{yx}) + \frac{\partial}{\partial z}(\tau_{zx})$$

Similar expressions hold for the y- and z-components. The control volume (in 2D) and the force balance in the x-direction shown in Fig. 4.7 can be used to interpret the meaning of the divergence in a simple manner.

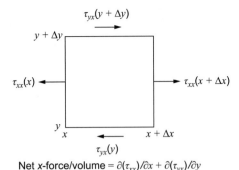

Net x-force/volume = $\partial(\tau_{xx})/\partial x + \partial(\tau_{yx})/\partial y$

Figure 4.7 An illustration of the calculation of the net viscous force. We show only the x-direction and a 2D case to avoid clutter.

$\nabla \cdot \tau$ in cylindrical coordinates

For other common coordinates the results are presented directly. The detailed derivation needs lengthy vector algebra manipulations and the concept of a metric of a coordinate system. The difficulty is due to the fact that the unit vectors do not always point in the same direction and have derivatives with respect to the coordinate variables. However, for flow analysis the equations presented here are sufficient and can be used directly. The detailed derivation is not needed.

For the r-component

$$\frac{1}{r}\frac{\partial}{\partial r}(r\tau_{rr}) + \frac{1}{r}\frac{\partial \tau_{\theta r}}{\partial \theta} - \frac{\tau_{\theta\theta}}{r} + \frac{\partial \tau_{zr}}{\partial z}$$

For the z-component

$$\frac{1}{r}\frac{\partial}{\partial r}(r\tau_{rz}) + \frac{1}{r}\frac{\partial \tau_{\theta z}}{\partial \theta} + \frac{\partial \tau_{zz}}{\partial z}$$

For the θ-component

$$\frac{1}{r^2}\frac{\partial}{\partial r}(r^2\tau_{r\theta}) + \frac{1}{r}\frac{\partial \tau_{\theta\theta}}{\partial \theta} + \frac{\partial \tau_{z\theta}}{\partial z}$$

$\nabla \cdot \tau$ in spherical coordinates

Expressions for the divergence of the tensor in spherical coordinates are presented below.

For the r-component

$$\frac{1}{r^2}\frac{\partial}{\partial r}(r^2\tau_{rr}) + \frac{1}{r\sin\theta}\frac{\partial}{\partial \theta}(\tau_{\theta r}\sin\theta) + \frac{1}{r\sin\theta}\frac{\partial}{\partial \phi}(\tau_{\phi r}) - \frac{\tau_{\theta\theta} + \tau_{\phi\phi}}{r}$$

For the θ-component

$$\frac{1}{r^3}\frac{\partial}{\partial r}(r^3\tau_{r\theta}) + \frac{1}{r\sin\theta}\frac{\partial}{\partial \theta}(\tau_{\theta\theta}\sin\theta) + \frac{1}{r\sin\theta}\frac{\partial}{\partial \phi}(\tau_{\phi\theta}) - \frac{\tau_{\phi\phi}\cot\theta}{r}$$

For the ϕ-component

$$\frac{1}{r^3}\frac{\partial}{\partial r}(r^3\tau_{r\phi}) + \frac{1}{r\sin\theta}\frac{\partial}{\partial \theta}(\tau_{\theta\phi}\sin\theta) + \frac{1}{r\sin\theta}\frac{\partial}{\partial \phi}(\tau_{\phi\phi}) + \frac{\tau_{\theta\phi}\cot\theta}{r}$$

Having characterized all the forces and learnt how to present them in a general vector–matrix form we take up some examples of problems in hydrostatics.

4.2   The equation of hydrostatics

In a stationary fluid, there are no viscous forces. Hence we need to consider only the pressure force and gravity. Hence

$$\text{Pressure force} + \text{Gravity forces} = \text{zero.}$$

From the representation of the forces, this leads to

$$\boxed{\nabla p = \rho g} \tag{4.11}$$

which is the equation of hydrostatics. This equation gives the local variation of the pressure at a given point. For an incompressible fluid, the density is treated as a constant. Then the following equation holds for the pressure variation in a column of, for example, a liquid:

$$p = p_0 + \rho g d \tag{4.12}$$

where d is the depth into the liquid (taken to be positive in the downward direction) and p_0 is the pressure at the surface of the liquid.

The above equation can be used for gases where the pressure changes are moderate so that an average density can be used. However, it is easy to combine this with the ideal-gas law to provide the following equation for the pressure variation in a column of gas:

$$\frac{p}{p_0} = \exp\left(-\frac{M_w z}{R_G T}\right) \tag{4.13}$$

The derivation is left out, but it involves using the density as a function of p in Eq. (4.11) followed by integration. Here z is the elevation or height measured from the surface or the datum plane with a base pressure p_0. In deriving this expression the variation of temperature with height is ignored. The effect of temperature variation is minor and is also left as an exercise for the reader.

An important application of the equation of hydrostatics leads to the famous Archimedes principle discussed in the following section.

4.2.1   Archimedes' principle

Consider a solid body immersed in a liquid as shown in Fig. 4.8. The pressure force acts on the surface of the body and is given by

$$\text{Pressure force} = \int\int_A (-p)n\, dA = \int\int\int_V -\nabla p\, dV$$

We now use the equation of hydrostatics as $\nabla p = \rho_l g$, where ρ_l is the liquid density. Hence

$$\text{Pressure force} = \int\int\int_V -\rho_l g\, dV = -\rho_l g V$$

We have

$$\text{Total force} = \text{Pressure force} + \text{Weight of the solid}$$

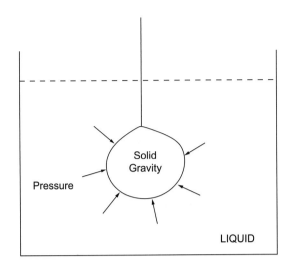

Figure 4.8 A diagram showing the forces on a solid immersed in a liquid, demonstrating Archimedes' principle. The net pressure force is upwards and reduces the weight of the immersed body.

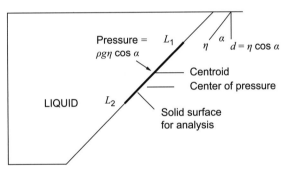

Figure 4.9 A diagram for force calculation on a plane surface. The surface between $\eta = L_1$ and $\eta = L_2$ is being analyzed.

Therefore

$$\text{Total force} = (\rho_s - \rho_l)gV$$

which is less than the weight of the body in the absence of the liquid, ρ_s being its density. The weight of the body (total force) measured in the liquid is decreased by $\rho_l gV$, which is equal to the weight of the liquid displaced by the body. This is the famous Archimedes principle, which is proved here using the equation of hydrostatics and vector calculus. Archimedes deduced it by intuition, but the formal proof had to wait for 1600 years until the invention of the vector calculus.

4.2.2 The force on a submerged surface: no curvature

In many applications of hydrostatics it is required to find the force on a submerged surface in contact with water. The calculation on a plane surface (surface with no curvature) is shown here as an illustrative example. See Fig. 4.9.

Let η be the coordinate measured from the surface of the liquid in the direction of the plate.

Let us consider the force on the plate between $\eta = L_1$ and $\eta = L_2$.

The pressure at any point is $\rho g d$, where d is the height measured from the top. Since $d = \eta \cos \alpha$ the pressure locally is equal to $\rho g \eta \cos \alpha \, d\eta$.

Hence the force on the plate between $\eta = L_1$ and $\eta = L_2$ is given by

$$\text{Pressure force} = \int_{L_1}^{L_2} \rho g \eta \cos \alpha \, d\eta$$

Upon integrating and after some minor algebra:

$$\text{Pressure force} = \int_{L_1}^{L_2} \rho g [(L_1 + L_2)/2] \cos \alpha (L_2 - L_1)$$

This can be interpreted as the pressure at the center $((L_1 + L_2)/2)$ multiplied by the area $(L_2 - L_1)$ of the section of the plate.

The moment of the force is equal to the force multiplied by the lever arm and is given as

$$M = \int_{L_1}^{L_2} \rho g \eta^2 \cos \alpha \, d\eta$$

Upon expressing this as

$$M = F \eta_{\text{c.p.}}$$

we find that the center of pressure $\eta_{\text{c.p.}}$ is located at

$$\eta_{\text{c.p.}} = \frac{2}{3} \frac{L_1^2 + L_1 L_2 + L_2^2}{L_1 + L_2}$$

This is a rather interesting result: the value of the pressure at the centroid is used to calculate the (net) pressure force, but the center of pressure (the point at which the net force can be presumed to act) is not at the centroid and is somewhat below the centroid.

4.2.3 Force on a curved surface

Consider a submerged surface immersed in contact with a fluid as shown in Fig. 4.10. The calculation of this force requires a vector integration. This is because the pressure forces at each location act in different directions. In other words, the vector n is not along a constant direction along the surface.

However, the calculations can be simplified by using what is called the free-body diagram in statics. This is illustrated in Figs. 4.10 and 4.11. The simpler "rules" for calculation of the horizontal and vertical components of the pressure force can be stated as follows.

Horizontal component
Horizontal component of force = Force on a vertical projection of the surface.

Vertical component
From a force balance in the vertical direction the following rule can be derived.

Vertical component of force = Weight of the fluid up to the free surface (real or hypothetical).

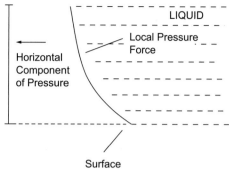

Horizontal component of pressure = pressure on a projection of the surface

Figure 4.10 A free-body diagram to calculate the horizontal component of the pressure forces on a curved surface.

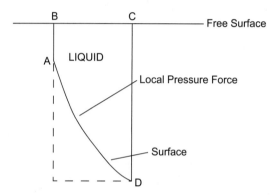

Vertical component of pressure = Weight of column of fluid, ABCD

Figure 4.11 A free-body diagram to calculate the vertical component of the pressure forces on a curved surface.

4.3 Hydrostatics at interfaces

At a fluid–fluid interface such as a liquid surface in contact with air, the surface tension forces can play a role in some cases. In such situations, it is necessary to consider the hydrostatics at a gas–liquid interface. These forces play a role where the interface has a curvature, e.g., gas bubbles, liquid films in contact with a solid, or when the surface tension varies strongly as a function of position due to temperature or concentration gradients at the surface.

First we discuss the nature of forces at an interface.

4.3.1 The nature of interfacial forces

The origin of the surface forces is the result of an intermolecular cohesive or attractive force similar to that shown in Fig. 1.23. For molecules in the interior of a liquid these attractive forces are balanced by the neighboring molecules, but those molecules near a gas interface lack these neighbours on the gas side and experience an unbalanced cohesive force directed

Table 4.1. Values of the surface tension of some common liquids in contact with air at 25 °C. The units are N/m or J/m^2. Note the high value for mercury.

Water	0.072
Benzene	0.0209
Methyl alcohol	0.023
Mercury	4.5

away from the interface. Thus the interface of a liquid is under tension and behaves like a stretched membrane or rubber band. The interfacial tension (more commonly referred to as surface tension) is defined as the force per unit perimeter of a line element and acts in a direction tangential to the line element. The units of surface tension are therefore N/m. Alternatively it can be viewed as the free energy per unit surface area in the same units but expressed as J/m^2. It can be viewed as the work required to change the surface area by one unit measure of area (m^2). Values for some common liquids in contact with air are given in Table 4.1.

We now look at some problems involving the effect of surface tension.

Example 4.1. A drop hanging on a pipette.
The problem is shown schematically in Fig. 4.12. Calculate the maximum size of the drop.

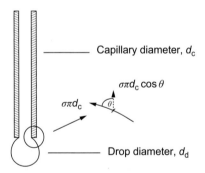

Figure 4.12 An illustration of a drop hanging on a pipette and the force balance to find the drop diameter.

The net force due to surface tension acts over a perimeter of πd_c and acts in the direction shown in Fig. 4.12. The magnitude of this force is equal to $\pi d_c \sigma$, where d_c is the capillary diameter. The force has a vertical component equal to $\sigma \pi d_c \cos \theta$. For the stable condition (with the drop just hanging in there) this has to be balanced by the downward force due to gravity, which is given by

$$\text{Gravity force} = \frac{\pi d_d^3}{6} \rho_l g$$

where d_d is the diameter of the drop.

The criterion for equilibrium can therefore be stated as

$$\pi d_c \sigma \cos \theta = \frac{\pi d_d^3}{6} \rho_l g$$

The maximum value of the LHS occurs when $\theta = 0$. Hence the maximum drop diameter can be written as

$$d_{d\,\text{max}}^3 = d_c \sigma / (\rho_l g) \qquad (4.14)$$

The equation can be rearranged into a dimensionless form by defining a dimensionless group (the Bond number) as

$$Bo = \frac{d_c^2 \rho_l g}{\sigma}$$

The physical significance of the Bond number is that it represents the ratio of the gravity force to the surface tension force acting on a drop. The maximum drop size in terms of this dimensionless group is therefore given by

$$\frac{d_d}{d_c} = (6)^{1/3} Bo^{-1/3} = 1.82 Bo^{-1/3}$$

This is in agreement with the experimental data except that the constant has a value closer to 1.6 rather than 1.82.

4.3.2 Contact angle and capillarity

Places where a fluid–fluid interface contacts a solid surface need special consideration as well. At such points, the liquid interface tends to meet the solid surface at a particular angle, referred to as the contact angle. The diagram shown in Fig. 4.13 illustrates this concept. The contact angle is in general given by the following equation

$$\sigma \cos \theta = \sigma_{SG} - \sigma_{SL}$$

where σ_{SG} is the surface tension or surface free energy per unit area for the solid–gas interface and σ_{SL} is the corresponding value for the solid–liquid side.

An application of the concept of contact angle together with a force balance is illustrated in the following example.

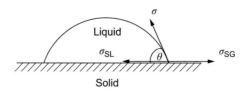

Figure 4.13 An illustration of the contact angle.

Example 4.2. The level rise in a capillary.

Consider a capillary of diameter d immersed in a fluid. For water, the liquid level in the capillary is higher than the level of the liquid in the beaker if the contact angle is less than $90°$. On the other hand, for mercury the level is lower, since mercury has a contact angle greater than $90°$. The liquid level can be calculated by a simple force balance.

A sketch of the basic situation is shown in Fig. 4.14. For a wetting liquid the surface tension force has a component of $\sigma \pi d_c \cos \theta$ in the vertical direction. Here d_c is the capillary diameter. This is balanced by the pressure force of $\rho g h (\pi d_c^2 / 4)$. The capillary rise is therefore given as

$$h = \frac{4\sigma \cos \theta}{\rho g d_c}$$

For non-wetting liquid there is a dip in the level as shown in Fig. 4.14.

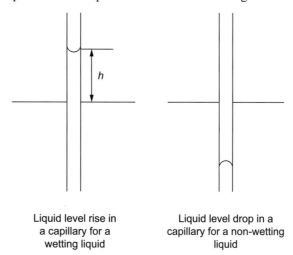

Liquid level rise in
a capillary for a
wetting liquid

Liquid level drop in a
capillary for a non-wetting
liquid

Figure 4.14 An illustration of the capillary-rise problem.

4.3.3   The Laplace–Young equation

Balance of static forces along an interface separated by two fluids leads to an important equation in hydrostatics of interfaces. The equation is called the Laplace–Young equation and may be stated as follows:

$$p_i - p_o = \sigma \left(\frac{1}{R_1} + \frac{1}{R_2} \right) \qquad (4.15)$$

where R_1 and R_2 are the radii of curvature of the interface, p_i is the inside surface pressure, and p_o is the pressure on the outer surface.

The derivation is as follows. Consider an interface that has a radius of curvature only in one direction as shown in Fig. 4.15.

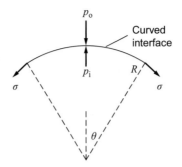

Figure 4.15 An Illustration of the derivation of the Laplace–Young equation.

Consider an arc of length $R\,\delta\theta$ as a representative control line. The surface tension then acts along the arc as shown in Fig. 4.15 and has a magnitude of σW, where W is an arbitrary width into the plane of the board. The component of this in the horizontal direction cancels out while the component in the vertical direction adds up. Each vertical component is equal to $\sigma W \cos(\pi/2 - \theta/2)$ or $\sigma W \sin(\theta/2)$.

Thus surface tension contribution pulling the surface inwards is 2 times $\sigma W \sin(\theta/2)$. To a first approximation $\sin(\theta/2)$ is equal to $\theta/2$. Hence the surface tension force acting downward is $\sigma W\theta$.

For a static situation this is balanced by the pressure difference $p_i - p_o$, which acts over an area of $R\theta W$. The balance leads to

$$p_i - p_o = \frac{\sigma}{R}$$

If the surface has curvature in the second direction as well, each contribution is additive and leads to the Laplace–Young equation shown in Eq. (4.15) indicated at the start of this section.

Note that the equation applies only to a static situation, and has to be modified for a moving fluid. The form for the moving fluid is indicated later and will appear as a normal-stress-balance boundary condition. This is due to the τ_{nn} term which arises due to the local normal strain at the given point. It may be also noted that in general surface tension forces will appear in the boundary conditions rather than in the governing equations for a differential model for fluid motion.

Example 4.3. Pressure inside a bubble.
Calculate the pressure inside a soap bubble floating in air.

Solution.
Consider the forces on a hemispherical part of a bubble as shown in Fig. 4.16. The surface tension acting downward is $\sigma(2\pi R)$, where R is the bubble radius. This has to be balanced by an upward pressure force of $(\Delta p)\pi R^2$. Note that the projected area πR^2 is used to calculate the pressure force. Upon equating the two forces we have

$$\delta p = \frac{2\sigma}{R}$$

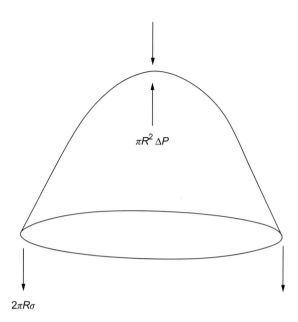

$\pi R^2 \Delta P$

$2\pi R\sigma$

Figure 4.16 An illustration of the force balance on the surface of a bubble. R is the bubble radius.

which is a measure of the excess pressure inside the bubble. The magnitude of this force is of the order of 100 Pa for typical liquids containing bubbles of radius 1 mm.

We now look at the forces generally used in macroscopic models together with a simple application.

4.4 Drag and lift forces

When macroscopic models are used the local forces are averaged. This leads to the drag and lift force on a surface with a fluid flowing past it. In this section we define these forces and show the relation between differential models and macroscopic models.

Consider a gas flowing past a solid surface as shown in Fig. 4.17. The surface force acting at any point can be computed from the local value of the stress tensor and the pressure:

$$\text{Local surface force } = [(\boldsymbol{n} \cdot \tilde{\tau}) - \boldsymbol{n}p)]dA$$

The first term on the RHS is the viscous force, while the second term is the force due to pressure.

The pressure can be included with the stress by using the identity matrix $\tilde{I}$, and the force can be written as

$$\text{Local surface force } = [\boldsymbol{n} \cdot (\tilde{\tau} - p\tilde{I})]dA$$

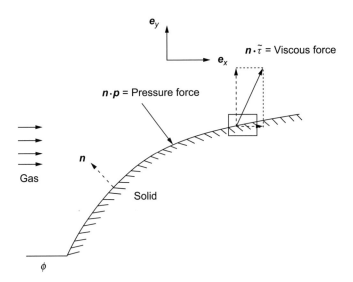

Figure 4.17 Illustration and definition of drag and lift forces. The components of the forces in the flow (x-) direction are added to get the drag force. Similarly, the components in the y-direction cause the lift.

The drag force

The component of this force in the direction of flow is obtained by taking the dot product with the unit vector flow direction. This force is called the drag force. Let e_x be the flow direction. Then

$$\text{Local drag force} = e_x \cdot [n \cdot (\tilde{\tau} - p\tilde{I})]dA$$

The total drag force is calculated by integrating over the area of the body:

$$\text{Total drag force} = \int_A e_x \cdot [n \cdot (\tilde{\tau} - p\tilde{I})]dA$$

There are two terms here: one is the viscous contribution, which is generally referred to as skin drag, and the second is the pressure contribution, called the form drag. It may be also noted that the viscous contribution is along the tangent to the surface and the normal viscous stress **right at** the solid surface can be shown to be equal to zero.

Drag is a macroscopic property, since it provides the total force. The above relation shows how the drag is related to microscopic or local forces. For complex shapes the drag force is correlated in an empirical manner using the drag coefficient, C_D. This is defined as

$$\text{Total drag force} = C_D A_p \rho v^2 / 2$$

Here v is the fluid approach velocity and A_p is the projected area of the body. The drag coefficient is similar to the friction factor for internal flows.

The lift force

Similarly the component of the surface force in a direction perpendicular to flow is called the lift force:

$$\text{Total lift force} = \int_A e_y \cdot [n \cdot (\tilde{\tau} - p\tilde{I})]dA$$

In practice it is convenient to use the lift coefficient values:

$$\text{Total lift force} = C_L A_p \rho v^2 / 2$$

An application of the drag force is for particle settling in a liquid, which is shown below.

Particle settling velocity

Consider a particle settling in a liquid. Systems of this type are encountered in filtration, sedimentation, etc. The system is characterized by the balance of gravity, pressure forces, and viscous forces. The pressure force is usually included as the buoyancy term. Hence the gravity + pressure (or buoyancy) force on the system is equal to $v_p g(\rho_p - \rho_l)$. Here v_p is the volume of the particle equal to $(4/3)\pi R^3$ for a spherical particle. The viscous force is usually represented by using a drag coefficient.

$$\text{Viscous forces} = C_D A_p \frac{1}{2} \rho_l v_t^2$$

where v_t is the relative velocity of the liquid and the solid and is equal to terminal velocity since the movement of the liquid phase is usually small. The area A_p is taken as the projected area in the direction of flow and is equal to πR^2 for spherical particles. Upon equating the forces we have

$$C_D \rho_l v_t^2 = (4/3) d_p g(\rho_p - \rho_l) \tag{4.16}$$

Equation (4.16) is rearranged to the following form using a Reynolds number based on the terminal velocity, Re_t, defined as $d_p v_t \rho / \mu$:

$$C_D Re_t^2 = \frac{4}{3} \frac{d^3 \rho_l g(\rho_p - \rho_l)}{\mu^2} \tag{4.17}$$

The Eq. (4.17) can be made dimensionless by defining a group (the RHS of this equation) called the Archimedes number, Ar,

$$Ar = \frac{d_p^3 \rho_l g(\rho_p - \rho_l)}{\mu^2}$$

and hence Eq. (4.17) can also be represented as

$$C_D Re_t^2 = \frac{4}{3} Ar \tag{4.18}$$

which is the equation for finding Re_t and therefore v_t. Note that C_D is a function of Re_t, and the corresponding relation has to be used in conjunction with the above equation.

The drag coefficient for a sphere

An illustrative plot of the drag coefficient for flow past a sphere is shown in Fig. 4.18.

The drag coefficient is a function of the Reynolds number and the relation is usually given in terms of charts or empirical correlations. A commonly used correlation valid for $Re_t < 1000$ is

$$C_D = \frac{24}{Re_t} \left[1 + 0.153 Re_t^{0.687} \right] \tag{4.19}$$

For Re_t greater than 1000 the drag coefficient is approximated as a constant equal to 0.44. This is known as the Newton regime since the settling is controlled mostly by inertia of the particle. Note that, for low Reynolds number,

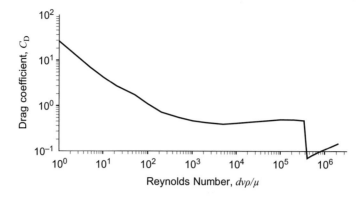

Figure 4.18 The drag coefficient for flow past a sphere as a function of the Reynolds number.

$$C_D = \frac{24}{Re_t} \qquad (4.20)$$

which is known as the Stokes equation. In Chapter 16 we will derive this equation from the theory of low-Reynolds-number flows.

Terminal velocity calculations

The calculation of the terminal velocity is essentially the calculation of Re_t from Eq. (4.18). Note that a trial and error calculation is needed since Re_t is not known. Only the product $C_D Re_t^2$ is known from the known quantities on the RHS of Eq. (4.18).

For $Re_t < 1$ the equation for C_D can be approximated as $24/Re_t$ as per Stokes' law. Hence Re_t can be directly calculated as

$$Re_t = \frac{Ar}{18} \qquad (4.21)$$

The corresponding value of the terminal velocity is

$$v_t = \frac{d_p^2 g(\rho_p - \rho_l)}{18\mu}$$

In the intermediate regime $1 < Re_t < 1000$, a trial and error calculation is needed since Re_t is not known. Only the product $C_D Re_t^2$ is known from the known quantities.

For Newton's regime a value of C_D of 0.44 is used, and trial and error is not needed. Note that the regime has to be checked after the calculation of Re_t, and a revised formula for drag may have to be applied if the assumed regime is not found to hold.

Summary

- Forces acting on a control volume can be grouped into two categories: body forces and surface forces.
- The most common example of a body force is the force of gravity. The electromagnetic or Lorentz force in conducting fluids subjected to a magnetic field is another example.

- Pressure and the forces caused by relative fluid motion (viscous forces) are examples of surface forces. In a stationary fluid there are no viscous forces and the force can be calculated if a scalar field, *viz.*, the fluid pressure, is known in the domain of interest.

- The pressure field in a static fluid is given by a differential equation that balances gravity and pressure forces. This is called the equation of hydrostatics.

- The equation of hydrostatics has many applications, including a formal proof of the famous Archimedes principle. An engineering application is the calculation of pressure forces acting on immersed surfaces, e.g., the surface of a dam. It may be noted in passing here that the pressure distribution in a flowing fluid is not as simple and can be a source of conceptual error as well as numerical error in computational fluid dynamics. The message is that the hydrostatic model for pressure is not applicable in a moving fluid.

- Viscous forces are due to relative motion of adjacent layers or surfaces of a fluid. They can be tangential to a surface (shear forces) or normal to a surface (tension or compression). The viscous forces are defined in terms of a stress vector or traction vector. This can be calculated if the stress tensor at a point is known. The viscous force (per unit area) is equal to the dot product of the normal vector and the stress tensor. The physical meaning of the stress tensor should be appreciated at this stage.

- At a gas–liquid interface, surface tension forces act. The phenomenon causes the fluid to rise in a capillary (for contact angles less than $\pi/2$). These forces are important in gas–liquid and liquid–liquid systems, and can be used to model bubble sizes and drop sizes.

- When macroscopic models are used, the local forces are averaged. This leads to the drag and lift forces on a surface with a fluid flowing past it. Empirical drag and lift coefficients are generally used here in lieu of the detailed models, and these are fitted as a function of the flow Reynolds number. A result that is simple but of wide applicability, which uses the correlation of the drag coefficient, is the terminal settling velocity of a solid in a liquid.

Problems

1. Is $n \cdot \tilde{\tau}$ equal to $\tilde{\tau} \cdot n$ in general? When will they be the same?
2. Indicate the direction and the plane over which the following stress quantities act.

   Rectangular: $\tau_{-x,x}$; $\tau_{-y,-y}$; $\tau_{z,-y}$.
   Polar: $\tau_{-r,\theta}$; $\tau_{\theta,\theta}$.
   Spherical: $\tau_{-\phi,\theta}$; $\tau_{-r,-\theta}$; $\tau_{r,\phi}$.

3. A stress tensor in two dimensions has the following components at a given point: $\tau_{x,x} = 3$, $\tau_{x,y} = 2$, and $\tau_{y,y} = 2$. Find the stress vector on a plane that is inclined at an angle of 60° with the x-axis. The plane is located at the same point where the stress tensor has the above reported values.
   Repeat for a plane oriented at 45°.
4. The operator ∇ can be considered to be a vector operator as defined in Chapter 1.
   Here τ is considered as a dyadic operator defined as

$$\tau = \sum_i \sum_j e_i e_j \tau_{ij}$$

With these definitions it is possible to write $\nabla \cdot \tau$ in terms of the spatial differentiation on τ. The equation can be simplified using the following property of the dot product of a unit vector and a dyad:

$$e_i \cdot (e_j e_k) = \delta_{ij} e_k$$

where δ_{ij} is the Kronecker delta.

Use these relations to derive the expression for the divergence of a tensor.

Similarly, by defining ∇ in cylindrical or spherical coordinates and carrying out the same operation on τ expressed as a dyad, you should derive expressions for $\nabla \cdot \tau$ cylindrical and spherical coordinates and verify the relations shown in the text in Section 4.1.3.

5. The average ocean depth is 2 km. Compute the pressure at this point. Assume a constant density.

   The change in density of water with pressure is small, and can be represented using the bulk modulus K defined by the following equation:

   $$K = \rho \left(\frac{\partial P}{\partial \rho} \right)_T$$

   Recalculate the pressure now using the density variation. $K \approx 2.2 \times 10^9$ Pa for water.

6. Derive Eq. (4.13) for pressure variation in the atmosphere with elevation. Find the pressure at Shangri-La, which is about 3000 m above the sea level. (In the Tibetan language Shangri-La means the Sun and Moon at heart.)

   The temperature variation with elevation is small, and is usually represented as a linear relation:

   $$T = T_0 (1 - \alpha z)$$

   Incorporate this into the calculation of density and then solve the hydrostatic equation to find a relation for the pressure variation with height including the above temperature correction.

   The change in temperature is about 6.4 °C for a change in height of 1000 m. Using this, find α defined above and recalculate the pressure at Shangri-La.

7. Consider a lighter solid of density ρ_s floating on the surface of a liquid of density ρ_l. Derive an expression for the volume fraction for the solid submerged inside the liquid by using the vector calculus. The result will verify the Archimedes principle for floating solids.

8. Consider a circular viewing port on an aquarium. This window has a radius of R, and the center of this port is at a depth $d + R$ from the water surface. Find the force and the center of pressure by direct integration of the differential pressure force.

9. A process requires the delivery of drops of volume 3.2×10^{-8} m^3. A liquid has a density of 900 kg/m^3 and a surface tension of 0.03 N/m. What size of capillary would you recommend to form these drops?

   In order to find the dripping rate, a simple criterion based on the Weber number can be used (Middleman, 1988b). The Weber number is defined as the dimensionless quantity $\rho v^2 d_c / \sigma$, where v is the flow velocity through the capillary. The value of the Weber number is suggested to be one at the onset of dripping. From this, calculate an estimate on the upper limit on the production rate from a single capillary in the units of kg of drops per hour. The drops are formed in air.

10. Calculate the settling velocity of a spherical particle of diameter 2.2 cm with a density of 2620 kg/m^3 in a liquid of density of 1590 and a viscosity of 9.58 millipoise. What is the regime under which the particle is settling?

11. Consider the motion of the particle in the initial stages, i.e., before it reaches the terminal velocity. Include the acceleration terms in the momentum balance of the particle and derive the following equation for the velocity as a function of time:

$$v(t) = v_t \left[1 - \exp\left(-\frac{9\mu t}{2\rho_p R^2} \right) \right]$$

Here v_t is the terminal velocity and the initial velocity is zero. Also the Stokes drag value is assumed to hold throughout the process. The time constant for settling is therefore the reciprocal of the constant in the exponential term above. This is usually small, and therefore it is usually assumed that the terminal velocity is reached instantaneously.

5 Equations of motion and the Navier–Stokes equation

Learning objectives

In this chapter you will learn

- the basic differential equations to describe the fluid motion in general (equations of motion),
- the equations for the velocity field for Newtonian fluids (the Navier–Stokes equations),
- common boundary conditions needed for flow simulation,
- important dimensionless variables for flow analysis and the principle of dynamic similarity, and
- how to understand the various fluid behaviors in general and how to model them with some constitutive relations.

Mathematical prerequisites

No additional mathematical tools are needed for the study of this chapter. You may wish to review the vector calculus, plus your knowledge of tensors and the various operations on tensor quantities.

This chapter starts with the development of the differential equation of motion, which is nothing but the statement of the momentum conservation principle. Thus we combine the fluid acceleration calculations derived in Chapter 3 and the representation of forces derived in Chapter 4. This leads to a general equation of motion that is based on the conservation principle alone. This is the basic model for the transport of momentum, which is also called the equation of (fluid) motion. As you may have guessed already, the divergence of the stress tensor will appear as a term, and the model is not in terms of the velocity alone. The model needs to be closed with appropriate constitutive relations between the stress and the strain rate. This requires knowledge of the rheological properties or the flow behavior of the fluid. Hence a discussion of common fluid behaviour is presented, and a classification of the rheological behaviors of fluids is presented next. Fluids obeying a linear relation are referred to as Newtonian fluids.

If a Newtonian model is used and the expression for the stress is substituted in terms of the velocity derivatives, the resulting equations constitute the classical **Navier–Stokes (N–S) equation**, which is a widely used equation in fluid dynamics. We will present these equations and examine the common boundary conditions which need to be used in conjunction with the N–S equations.

The dimensionless representation of the N–S equation is also presented in this chapter. This identifies the key dimensionless parameters which are important in fluid dynamics. Simple applications of dimensionless analysis to scaleup are illustrated.

The chapter concludes with a presentation of the common models used to characterize the stress–strain-rate behavior of non-Newtonian fluids. This provides a basic introduction to the study of flow problems involving such fluids, a few of which are examined in the next chapter.

This chapter forms the prelude for the study of prototypical flow problems illustrated in the next chapter.

5.1 Equation of motion: the stress form

5.1.1 The Lagrangian point particle

The equation of motion is readily obtained by applying Newton's second law to the Lagrangian "point" particle, which is defined as a differential packet of fluid moving with the fluid velocity. The velocity of the particle is represented as v at its centroid in its current configuration. The acceleration of the particle is therefore equal to Dv/Dt. Hence the rate of change of momentum is $\rho\, Dv/Dt$ per unit volume of the particle. Let $\sum F$ denote the total force per unit volume acting on the particle. Hence, by Newton's second law,

$$\rho \frac{Dv}{Dt} = \sum F$$

The expression for the sum of the forces was derived in Chapter 4, and is represented as

$$\sum F = \rho g - \nabla p + \nabla \cdot \tilde{\tau}$$

Note that additional forces such as electromagnetic forces can be added as an extra term, for example, for magnetohydrodynamic flows. In the absence of such forces, the second law of motion reads

$$\rho \frac{Dv}{Dt} = \rho g - \nabla p + \nabla \cdot \tilde{\tau} \tag{5.1}$$

Upon expanding the substantial derivative term we have

$$\boxed{\rho \left(\frac{\partial v}{\partial t} + (v \cdot \nabla)v \right) = \rho g - \nabla p + \nabla \cdot \tilde{\tau}} \tag{5.2}$$

which is the equation of motion in the stress form. Since no assumptions on the nature of the fluid have been invoked here, this expression is valid for all types of fluids, irrespective of whether or not they are Newtonian.

5.1.2   The Lagrangian control volume

The derivation shown above was for a point particle. A point particle is such that $V \to 0$ and hence the local values or the centroid values of the properties are representative of the entire particle. The derivation can also be done starting with an arbitrary control volume (not necessarily infinitesimally small). This requires the use of the Reynolds transport theorem discussed in Chapter 3. The final result is the same, but this derivation is mathematically more satisfying and will appeal to the purists.

The rate of change of momentum is an integral of the local values,

$$\text{Rate of change of momentum} = \frac{D}{Dt} \int_{V(t)} (\rho \boldsymbol{v}) dV$$

where $V(t)$ is based on the current configuration of the particle.

The forces are computed again using the integral of the local values:

$$\text{Sum of the forces} = \int_{V(t)} (\rho \boldsymbol{g} - \nabla p + \nabla \cdot \tilde{\tau}) dV$$

Combining the two terms in accordance with Newton's second law gives

$$\frac{D}{Dt} \int_{V(t)} (\rho \boldsymbol{v}) dV - \int_{V(t)} (\rho \boldsymbol{g} - \nabla p + \nabla \cdot \tilde{\tau}) dV = 0 \tag{5.3}$$

The first term on the LHS of Eq. (5.3) can be converted to a volume integral using the Reynolds transport theorem applied to vector quantities:

$$\frac{D}{Dt} \int_{V(t)} (\rho \boldsymbol{v}) dV = \int_{V(t)} \left(\frac{\partial (\rho \boldsymbol{v})}{\partial t} + \nabla \cdot (\rho \boldsymbol{v} \boldsymbol{v}) \right) dV$$

Hence we obtain

$$\int_{V(t)} \left(\frac{\partial (\rho \boldsymbol{v})}{\partial t} + \nabla \cdot (\rho \boldsymbol{v} \boldsymbol{v}) - \rho \boldsymbol{g} - \nabla p + \nabla \cdot \tilde{\tau} \right) dV = 0 \tag{5.4}$$

Note that the term $\rho \boldsymbol{v} \boldsymbol{v}$ is somewhat new to our vocabulary and is explained here briefly. The term $\rho \boldsymbol{v} \boldsymbol{v}$ is a dyadic product or a tensor, which is defined in index notation as $\rho v_i v_j$. This is a compact way of writing tensor quantities. Also see Problem 4 of Chapter 4. The divergence of this is defined in the usual way as the divergence of a tensor and is equal to $[\rho v_i v_j]_{,i}$.

Equation (5.4) is general and applies to any arbitrary control volume. A differential form can be obtained as follows: Since the volume V in Eq. (5.4) is arbitrary, the integrand has to be equal to zero to satisfy the above equation. This then leads to the pointwise representation of the equation of motion:

$$\boxed{\frac{\partial (\rho \boldsymbol{v})}{\partial t} + \nabla \cdot (\rho \boldsymbol{v} \boldsymbol{v}) = \rho \boldsymbol{g} - \nabla p + \nabla \cdot \tilde{\tau}} \tag{5.5}$$

which is one form of the equation of motion. The first term on the LHS is the transient term and the accumulation of momentum. The second term is the convective term. The terms on the RHS represent the gravity, pressure, and viscous forces, respectively.

The convective terms can be further simplified using the equation of continuity. The dyadic notation and index operations are quite handy for this simplification. The simplified version of Eq. (5.5) is the same as the equation presented in the previous section, *viz.*, Eq. (5.2). (see Problem 1).

5.1.3   The Eulerian control volume

The momentum balance can also be derived by considering the three components of the momentum vector in the three coordinate directions. Again, for simplicity, the control volume can be taken in the shape of a box as shown in Fig. 5.1. We will describe only the convection term here, since the balance of pressure forces was shown earlier in Fig. 4.2 and the balance of viscous forces was shown in Fig. 4.7.

The convective transport in the x-plane is $[\rho v_x \, \Delta y \, \Delta z]v$ evaluated at the take over point x. Note that the term in square brackets is the mass flow rate. This multiplied by the velocity vector gives the momentum flow into the plane, which is a vector quantity.

Similarly the convective transport in the $x + \Delta x$ plane is $[\rho v_x \, \Delta y \, \Delta z]v$ evaluated at the point $x + \Delta x$.

Net momentum efflux is the vector difference of these quantities.

This can be written as

$$\text{Net momentum, } x\text{-planes only } = \frac{\partial}{\partial x}(\rho v_x v)\Delta x \, \Delta y \, \Delta z$$

The convective term per unit volume is therefore

$$\text{Net momentum efflux, } x\text{-planes only, per unit volume} = \frac{\partial}{\partial x}(\rho v_x v)$$

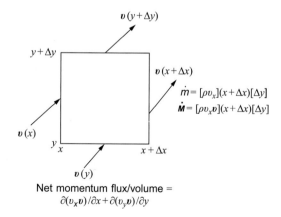

Figure 5.1 The Cartesian Eulerian control volume for the 2D case for the derivation of the momentum equation showing the convection terms.

Similarly the y-planes contribute a momentum efflux of

$$\text{Net momentum efflux, } y\text{-planes only, per unit volume} = \frac{\partial}{\partial y}(\rho v_y v)$$

A similar expression holds for the z-plane. On adding the contributions from all surface directions we get the net convective momentum transport. This can be shown to be the divergence of ρvv per unit volume. Using continuity, this turns out to be $\rho(v \cdot \nabla)v$ for an incompressible fluid.

Equations of motion: Cartesian coordinates

The component forms of the equation of motion in Cartesian coordinates are summarized below for ease of reference.

The x-momentum balance equation is

$$\rho\left(\frac{\partial v_x}{\partial t} + v_x\frac{\partial v_x}{\partial x} + v_y\frac{\partial v_x}{\partial y} + v_z\frac{\partial v_x}{\partial z}\right)$$

$$= -\frac{\partial p}{\partial x} + \left(\frac{\partial}{\partial x}(\tau_{xx}) + \frac{\partial}{\partial y}(\tau_{yx}) + \frac{\partial}{\partial z}(\tau_{zx})\right) + \rho g_x \tag{5.6}$$

The y-momentum balance is

$$\rho\left(\frac{\partial v_y}{\partial t} + v_x\frac{\partial v_y}{\partial x} + v_y\frac{\partial v_y}{\partial y} + v_z\frac{\partial v_y}{\partial z}\right)$$

$$= -\frac{\partial p}{\partial y} + \left(\frac{\partial}{\partial x}(\tau_{xy}) + \frac{\partial}{\partial y}(\tau_{yy}) + \frac{\partial}{\partial z}(\tau_{zy})\right) + \rho g_y \tag{5.7}$$

The z-momentum balance is

$$\rho\left(\frac{\partial v_z}{\partial t} + v_x\frac{\partial v_z}{\partial x} + v_y\frac{\partial v_z}{\partial y} + v_z\frac{\partial v_z}{\partial z}\right)$$

$$= -\frac{\partial p}{\partial z} + \left(\frac{\partial}{\partial x}(\tau_{xz}) + \frac{\partial}{\partial y}(\tau_{yz}) + \frac{\partial}{\partial z}(\tau_{zz})\right) + \rho g_z \tag{5.8}$$

In addition, the overall mass balance has to be satisfied. For an incompressible fluid this takes the following form:

$$\frac{\partial v_x}{\partial x} + \frac{\partial v_y}{\partial y} + \frac{\partial v_z}{\partial z} = 0 \tag{5.9}$$

Equations (5.6)–(5.9) form the starting point for the analysis of flow problems. Equations (5.6)–(5.8) are known as the equations of motion in terms of the stresses (also known as the stress divergence forms of the equations of motion). These equations are derived merely by application of the general principle of Newton's second law of motion and therefore are applicable to all fluids. The equations are, however, not complete, since additional equations for the stress distribution in the fluid are needed. These are provided by the constitutive equations for the fluid under consideration.

Cylindrical coordinates

The equation of motion in cylindrical coordinates can be written down in component form using the vector form as a basis. We have already seen the expressions and derived these separately. Hence the detailed form is not presented here. For easy reference the sections of the book where the expressions are shown are specified here.

The convective terms are given in Table 3.1. The gradient of pressure follows from the gradient operator applied to a scalar field, which is summarized in Section 1.4.4. The divergence of the tensor in cylindrical coordinates was presented in Section 4.1.3, which provides the viscous terms. Problem 2 in this chapter asks you to look at these and write out the equations of motion in cylindrical coordinates.

Spherical coordinates

The equation of motion in spherical coordinates can be written down in component form using the vector form as a basis. For easy reference the sections of the book where the expressions are shown are specified here.

The convective terms are given in Table 3.2. The gradient of pressure follows from the gradient operator applied to a scalar field, which is summarized in Section 1.4.4.

The divergence of the tensor in spherical coordinates was presented in Section 4.1.3, which provides the viscous terms. Problem 3 in this chapter asks to you look at these and write out the equations of motion term by term in spherical coordinates.

5.2   Types of fluid behavior

Fluids for which the strain rate and stress tensor show a linear relation are referred to as Newtonian fluids. Fluids showing more complex relations are referred to as non-Newtonian fluids. Some of these fluids exhibit an elastic (solid-like) behavior, as will be discussed later, and hence the non-Newtonian fluids can be further classified into viscoinelastic and viscoelastic fluids. The classification is shown in the tree diagram in Fig. 5.2.

5.2.1   Types and classification of fluid behavior

Fluid behaviour can further be classified as time-independent and time-dependent. The flow behavior of time-dependent fluids changes with time, whereas that of time-independent fluids does not depend on the history of motion.

Fluid behavior can be best understood by looking at flows in simplified geometries. One of the simplest geometries is that of simple shear flow, which is flow between two parallel

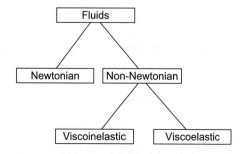

Figure 5.2 Classification of fluids in terms of their rheological behavior.

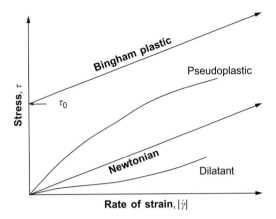

Figure 5.3 Stress vs. strain-rate diagram for common viscoinelastic behavior.

plates with no influence of pressure gradients. From our earlier discussion, we know that the velocity profile is linear and hence the velocity gradient dv_x/dy is a constant. This is the only component of the strain-rate tensor for this situation, and this will be denoted as $\dot{\gamma}$. The shear stress τ_{yx} denoted here simply as τ is also a constant across the system, and the ratio of the stress to the rate of strain is the viscosity of the fluid. A plot of $\dot{\gamma}$ vs. τ_{yx} is therefore indicative of the fluid behavior. An illustrative plot is given in Fig. 5.3, which shows the various types of fluid behavior.

The local slope of the curve can be defined as the apparent viscosity corresponding to a particular strain rate. For Newtonian fluids the slope is a constant as shown in Fig. 5.3. For some fluids, the slope (the apparent viscosity) can decrease with increasing shear. Such fluids are often called pseudoplastic fluids, although a more apt description as shear-thinning fluids is preferred. Examples of such fluids include polymer melts or solutions, and some solid suspensions. A common household example is hair-styling gel, which is difficult to pour (low shear) but flows easily when rubbed in the hair (high shear). It may be noted that polymer melts exhibit elastic behavior as well, and are not really viscoinelastic. A second type of behavior is shear thickening, where the viscosity increases with increasing shear rate. Such fluids are called dilatant. This type of behavior is not common. A household example is a mixture of cornstarch and water, which flows easily at low shear. At high shear the water gets squeezed out and the material does not easily flow.

For many materials a minimum stress (called the yield stress, τ_0) has to exist before the fluid can flow. An example is tomato ketchup, which may require many tappings before it can be poured out of the bottle. The flow behavior (as characterized by the apparent viscosity or the slope) can be linear once the yield stress is exceeded. Such fluids are referred to as Bingham plastics. Highly concentrated suspensions of solid particles exhibit such behaviour. In some cases the flow behaviour can be shear thinning once the yield stress is exceeded. These fluids are sometimes called Hershel–Bulkley fluids.

In some cases, the viscosity can change with time under a fixed shear rate and such fluids exhibit a time-dependent behavior. The structural rearrangement of molecules occurs in such fluids due to applied shear resulting in a change in stress with time (under constant-strain-rate conditions). The viscosity may increase with time, leading to a rheopectic fluid, while it may decrease with time for the so-called thixotropoic fluids. Some examples of rheopectic fluids are suspensions and emulsions in water, sols, etc., while examples of the thixotropic type include paint, ink, etc.

Now we discuss the nature of the elastic properties of the fluids. The normal stresses τ_{xx} and τ_{yy} are zero for simple shear flow for viscoinelastic fluids. But it is observed that the normal stresses are not zero for many non-Newtonian fluids. The fluid tends to experience a tension or compression similar to that in a stretched rubber band. Such fluids are called viscoelastic fluids. Many interesting and strikingly different behaviors are observed for such fluids. One effect is the rod-climbing effect. In a tank agitated with a rod there is a dip or vortex near the rod for Newtonian fluids, whereas in a viscoelastic fluid, the fluid moves upward around the rod, which is the rod-climbing effect. More discussion on these fluids and some common mathematical models for stress–strain relations for these fluids is provided later, in Section 6.11.

5.2.2 Stress relations for a Newtonian fluid

The linear stress–strain relation for incompressible fluids was indicated earlier:

$$\tilde{\tau} = 2\mu\tilde{E}$$

Expansion of the terms and the use of the symmetry property for both the stress and the rate of strain leads to the following constitutive equations:

$$\tau_{xx} = 2\mu\frac{\partial v_x}{\partial x} \tag{5.10}$$

$$\tau_{yx} = \mu\left(\frac{\partial v_x}{\partial y} + \frac{\partial v_y}{\partial x}\right) \tag{5.11}$$

$$\tau_{zx} = \mu\left(\frac{\partial v_z}{\partial x} + \frac{\partial v_x}{\partial z}\right) \tag{5.12}$$

These equations are for the stress components acting in the x-direction.

Similar equations can be written for the stress components in the y-direction (τ_{xy}, τ_{yy}, and τ_{zy}) and for the components in the z-direction. The symmetry of the stress tensor should also be noted here. Thus $\tau_{xy} = \tau_{yx}$, for instance.

The form of the stress tensor in other coordinate systems is obtained by using the definition of the rate-of-strain operator in these coordinates. Stress components are 2μ times the strain components. The relevant equations for strain components were shown in Section 3.12.1 for cylindrical coordinates and Section 3.12.2 for spherical coordinates, and Problem 4 should be done to complete these relations.

Constitutive relations for non-Newtonian fluids will be discussed later, in Section 5.7. Now we proceed with using the linear constitutive model in the equation of motion.

5.3 The Navier–Stokes equation

With all the background in place, it is a fairly trivial task to write down the Navier–Stokes equation. We note that

$$\nabla \cdot \tilde{\tau} = 2\mu \nabla \cdot \tilde{E}$$

where we assume a constant viscosity. From Example 3.12 we have shown that

$$\nabla \cdot \tilde{\tau} = \mu \, \nabla^2 v$$

On using this in the equation of motion, Eq. (5.2), given earlier, we obtain the Navier–Stokes equation, which is expressed vectorially as

$$\rho \left(\frac{\partial v}{\partial t} + (v \cdot \nabla)v \right) = \rho g - \nabla p + \mu \, \nabla^2 v \qquad (5.13)$$

This is supplemented with the continuity equation since the pressure field has to be computed as well.

For Cartesian coordinates the following equations hold and are shown for ease of reference.

For the x-component of motion,

$$\rho \left(\frac{\partial v_x}{\partial t} + v_x \frac{\partial v_x}{\partial x} + v_y \frac{\partial v_x}{\partial y} + v_z \frac{\partial v_x}{\partial z} \right) = \rho g_x - \frac{\partial p}{\partial x} + \mu \, \nabla^2 v_x \qquad (5.14)$$

For the y-component,

$$\rho \left(\frac{\partial v_y}{\partial t} + v_x \frac{\partial v_y}{\partial x} + v_y \frac{\partial v_y}{\partial y} + v_z \frac{\partial v_y}{\partial z} \right) = \rho g_y - \frac{\partial p}{\partial y} + \mu \, \nabla^2 v_y \qquad (5.15)$$

For the z-component,

$$\rho \left(\frac{\partial v_z}{\partial t} + v_x \frac{\partial v_z}{\partial x} + v_y \frac{\partial v_z}{\partial y} + v_z \frac{\partial v_z}{\partial z} \right) = \rho g_z - \frac{\partial p}{\partial z} + \mu \, \nabla^2 v_z \qquad (5.16)$$

together with continuity

The Laplacian of velocity takes a simple form in Cartesian coordinates, and the form is similar to that for a scalar.

For other coordinates we need the equations for the Laplacian of the velocity to complete the viscous terms. These equations are now summarized.

5.3.1   The Laplacian of velocity

The Laplacian of velocity: cylindrical

For the r-component,

$$\frac{\partial}{\partial r} \left(\frac{1}{r} \frac{\partial}{\partial r} (r v_r) \right) + \frac{1}{r^2} \frac{\partial^2 v_r}{\partial \theta^2} - \frac{2}{r^2} \frac{\partial v_\theta}{\partial \theta} + \frac{\partial^2 v_r}{\partial z^2}$$

For the θ-component,

$$\frac{\partial}{\partial r} \left(\frac{1}{r} \frac{\partial}{\partial r} (r v_\theta) \right) + \frac{1}{r^2} \frac{\partial^2 v_\theta}{\partial \theta^2} + \frac{2}{r^2} \frac{\partial v_r}{\partial \theta} + \frac{\partial^2 v_\theta}{\partial z^2}$$

For the z-component,

$$\frac{1}{r} \frac{\partial}{\partial r} \left(r \frac{\partial v_z}{\partial r} \right) + \frac{1}{r^2} \frac{\partial^2 v_z}{\partial \theta^2} + \frac{\partial^2 v_z}{\partial z^2}$$

The Laplacian of velocity: spherical

For the r-component,

$$\nabla^2 v_r - \frac{2}{r^2} v_r - \frac{2}{r^2} \frac{\partial v_\theta}{\partial \theta} - \frac{2}{r^2} v_\theta \cot\theta - \frac{2}{r^2 \sin\theta} \frac{\partial v_\phi}{\partial \phi}$$

For the θ-component,

$$\nabla^2 v_\theta + \frac{2}{r^2} \frac{\partial v_r}{\partial \theta} - \frac{v_\theta}{r^2 \sin^2\theta} - \frac{2\cos\theta}{r^2 \sin^2\theta} \frac{\partial v_\phi}{\partial \phi}$$

For the ϕ-component,

$$\nabla^2 v_\phi + \frac{2}{r^2 \sin\theta} \frac{\partial v_r}{\partial \phi} - \frac{v_\phi}{r^2 \sin^2\theta} + \frac{2\cos\theta}{r^2 \sin^2\theta} \frac{\partial v_\theta}{\partial \phi}$$

Here ∇^2 is the Laplacian operator for a scalar in spherical coordinates. See Eq. (1.58) for the expression for this term.

This completes our development of the Navier–Stokes equation. We now discuss how different types of boundary conditions can be applied to complete the model formulation.

5.3.2   Common boundary conditions for flow problems

We discuss first conditions for flow of a single fluid, followed by boundary conditions for two-phase flow at the common fluid–fluid boundary.

The domain of solution

In general, the boundary conditions are specified over the domain of the solution, namely the perimeter (for 2D simulation) or the bounding surface (3D) of a selected region of the fluid. The bounding region generally consists of the enclosing solid boundaries (also called the pipe walls), an inlet region for flow, and an exit region. Together they enclose the domain over which the solution is sought. In principle, two boundary conditions over each of these boundaries have to be specified for the case of 2D flow, one for each of the velocity components. Three per boundary are needed for 3D cases. An illustrative region over which the boundary conditions usually have to be prescribed for the case of 2D flow is shown schematically in Fig. 5.4.

The boundary conditions are usually specified as follows.

The no-slip condition

The velocity component normal to a solid boundary is set as zero, which is called the no-penetration condition; and, if viscous terms are included, the velocity tangential to a solid surface is also set to zero. Together these are known as the no-slip boundary conditions; i.e., the fluid velocity at the surface of the solid is set equal to the solid velocity, which is usually zero.

Porous wall

The tangential component of velocity is still zero for this case. If the wall is porous then the normal component of velocity is proportional to the seepage through the wall, which is assumed to be known. If mass transfer takes place to the surface, this condition is used even

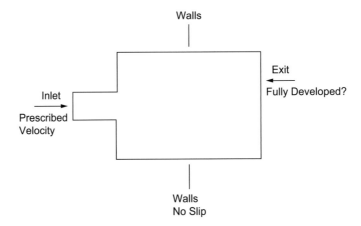

Figure 5.4 A typical region over which the boundary conditions are often specified for a 2D flow problem.

for a non-porous wall, since the mass transfer creates a velocity component proportional to the rate of mass transfer.

Inlet boundary
The boundary condition on the upstream boundary (i.e., where the fluid enters the system) is given by specifying the velocity vector components along this portion.

Exit boundary
The boundary condition on the downstream section is difficult to specify. Two common specifications are (i) the velocity profiles are allowed to reach a fully developed profile, i.e., the velocity in the direction perpendicular to the main flow is taken as zero; and (ii) no-traction boundary conditions, i.e., the normal and tangential stresses at the exit boundary are specified as zero. Other choices have been discussed in the book by Peyret and Taylor (1983). Note that the entrance and exit regions are often artificial boundaries, and are selected in order to avoid the need for simulation of the entire flow arrangement, including the tank from which the fluid enters and the mode by which it leaves the system. Hence the boundary conditions for the entrance and exit regions have to be judiciously selected on the basis of experience and intuition.

Pressure condition
The pressure is computed as an auxiliary variable for incompressible-flow problems, and hence specification of the pressure at one point is sufficient. No additional boundary condition for pressure is needed. The pressure field is allowed to evolve in such a manner that the continuity condition is satisfied.

The interface between two fluids
The tangential stress is taken to be the same at the interface between two fluids on either side of the interface. The total normal stress difference (viscous + pressure) is balanced by the surface tension forces. This condition is similar to the Laplace–Young equation, with the main difference being that the total stress jump is balanced instead of the pressure jump.

The phase velocity is assumed to be continuous. In general this is a free-interface problem and the interface shape has to be computed as a part of the domain.

5.4 The dimensionless form of the flow equation

Now we show the dimensionless form of the Navier–Stokes equation and identify the key dimensionless parameters needed for analysis of flow problems.

5.4.1 Key dimensionless groups

The dimensionless groups can be identified by rendering the governing differential equations into dimensionless form. This is done first by choosing reference values for all the independent variables:

Reference length, L_{ref};
Reference speed, v_{ref};
Reference pressure, P_{ref}; and
Reference time, t_{ref}.

Additionally, a reference temperature or temperature difference T_{ref} and a reference concentration C_{ref} are needed for heat and mass transfer problems.

The choice of some of the reference parameters, e.g., time, pressure, even temperature in some problems, might not be obvious so these are left undefined initially, and we let the differential equation guide us to the choice of these parameters. More discussion on this will be given later.

The next step is to define dimensionless variables for all the quantities by using the above reference scales.

Dimensionless distance: $x^* = x/L_{ref}$, and similar definitions for the y- and x- directions.
Dimensionless velocity: $v_x^* = v_x/v_{ref}$ etc., with v_y and v_z scaled in a similar fashion.
Dimensionless time: $t^* = t/t_{ref}$ etc. Note that t_{ref} is usually chosen later so that the differential equation containing the time term takes a simple form.
Dimensionless pressure: $p^* = p/p_{ref}$ etc. Note that p_{ref} is usually chosen later so that the differential equation containing the pressure term takes a simple form.

With these definitions, the differential equations can be made dimensionless by successive applications of the chain rule. Since all the dependent and independent variables are dimensionless, the quantities such as viscosity etc. will regroup and appear as dimensionless parameters. For example, the viscosity will appear in the Reynolds number and the effect of viscosity can be analyzed in terms of the effect of the Reynolds number.

Note that the boundary conditions must also be reduced to dimensionless from, and additional parameters may appear from the boundary conditions, for example, the Biot number in heat conduction problems and the Weber number for surface tension forces for study of gas–liquid interfaces.

Using these reference variables, the x-momentum balance takes the dimensionless form

$$St\frac{\partial v_x^*}{\partial t^*} + v_x^*\frac{\partial v_x^*}{\partial x^*} + v_y^*\frac{\partial v_x^*}{\partial y^*} + v_z^*\frac{\partial v_x^*}{\partial z^*} = -\frac{\partial p^*}{\partial x^*} + \frac{1}{Fr}\frac{g_x}{|g|}$$

$$+ \frac{1}{Re}\left(\frac{\partial^2 v_x^*}{\partial x^{*2}} + \frac{\partial^2 v_x^*}{\partial y^{*2}} + \frac{\partial^2 v_x^*}{\partial z^{*2}}\right) \quad (5.17)$$

Here Re turns out to be the Reynolds number defined in general manner as $L_{ref}v_{ref}\rho/\mu$.
Here p_{ref} is taken as ρv_{ref}^2 so that the leading coefficient in the pressure term is unity. The dimensionless pressure p^* is therefore defined as $p/(\rho v_{ref}^2)$. This group is known as the Euler number.

Fr is the Froude number defined as

$$Fr = \frac{v_{ref}^2}{gL_{ref}}$$

St is the Strouhal number defined in general as $L_{ref}/(v_{ref}t_{ref})$, where t_{ref} is the reference time or the observation time. Again this can be chosen depending on the problem. For example for start-up flow of fluid in a pipe, it is convenient to take t_{ref} as L_{ref}/v_{ref}, which makes St unity.

Similar equations can be written for the other components. For compactness it is useful to write the dimensionless form of the N–S equation in vector form,

$$\frac{1}{St}\frac{\partial v^*}{\partial t^*} + v^* \cdot \nabla_* v^* = -\nabla_* p^* + \frac{1}{Fr}g^* + \frac{1}{Re}\nabla_*^2 v^* \quad (5.18)$$

The vector operator terms such as ∇ are now applied with respect to dimensionless distance variables, and this is indicated by the subscript $*$ in these quantities.

Special forms of the dimensionless equations can be derived for various relative values of the Reynolds number, and these equations are now presented. The gravity term is ignored in order to focus on other terms.

5.4.2   The Stokes equation: slow flow or viscous flow

This simplified form of the above equation is applied for a slow or creeping flow where the inertial terms are negligible. The Reynolds number is small here.

Since $Re \to 0$ the RHS of Eq. (5.18) is much smaller than the LHS, and the equation reduces to

$$-\nabla_* \mathcal{P}^* + \frac{1}{Re}\nabla_*^2 v^* = 0 \quad (5.19)$$

The equation can be further simplified by defining

$$\mathcal{P}^{**} = Re\mathcal{P}^* = \frac{\mathcal{P}}{\mu v_{ref}/L_{ref}}$$

This scaling reduces the above equation to

$$-\nabla_* \mathcal{P}^{**} + \nabla_*^2 v^* = 0 \quad (5.20)$$

This is known as the Stokes equation of motion.

Note that the Reynolds number does not appear explicitly in the equation now. The equation has been reduced to the "bare bones". A rule in dimensionless analysis is that the equations should be reduced to the simplest possible form, and the above equation is an example of this rule. This was possible by redefining the reference pressure in this example. Recall that the reference pressure is now $\mu v_{ref}/L_{ref}$ rather than ρv_{ref}^2. The latter choice is better for inertia-dominated flows.

5.4.3   The Euler equation

The second limiting case is where the Reynolds number is large, resulting in the Euler equation of motion presented below. This is an example of inertia-dominated flow, also called inviscid flow, since the viscosity effects are neglected here.

The simplified form of the above equation for a flow where the viscous terms are negligible is obtained by setting the $1/Re$ term as zero:

$$\frac{\partial v^*}{\partial t^*} + v^* \cdot \nabla_* v^* = -\nabla_* \mathcal{P}^* \qquad (5.21)$$

This is known as the Euler equation of motion. Note that the order of the differential equation has been reduced by one since the viscous terms have been neglected. Hence the equation might not be able to satisfy all the specified boundary conditions, for example, the tangential component of the no-slip condition at the solid. (The Euler equation permits slip at the solid surface!) Nevertheless, it is a good approximation outside the region near the solid, e.g., outside the boundary layers near the solid.

Inviscid flows are treated in detail in Chapter 16 and will be taken up in more detail later.

Dimensionless groups are also useful in correlation of experimental data and for scaleup, and examples of such applications will now be discussed.

5.5   Use of similarity for scaleup

The key idea in scaleup is that the system behavior is the same, if the dimensionless groups are the same, irrespective of the size of the system. This property is known as dynamic similarity. This presumes that geometric similarity has been maintained, i.e., the ratios of all dimensions have been kept the same. For example, geometric similarity between a large agitated vessel and a small one implies that the tank-to-impeller diameter ratio must be the same, the impellers must be of the same type and scaled in similar proportion, and the ratio of the height of liquid to the diameter of the tank must be kept the same.

The use of dynamic similarity is best understood by taking some examples.

Example 1: flow past a cylinder
Flow past a cylinder is a well-studied problem in fluid mechanics. The parameters affecting the flow are

$$d, L, v_0, \rho, \text{ and } \mu.$$

The variables can be regrouped to the L/d ratio and the Reynolds number. Keeping L/d the same maintains the geometric similarity. Thus the effect of the Reynolds number on the flow can be studied and the results are general enough. No separate investigation of individual parameters is needed. For example, the pressure variation along the surface of the cylinder can be correlated by simply matching the Euler number since this is the other dimensionless group appearing in the dimensionless form of the N–S equation. Hence the data can be correlated as

$$Eu = f(Re, L/d)$$

Example 2: drag relations

Forces on submerged objects such as a cylinder due to the external flow of the above type are of importance in many applications. The force in the direction of the approach velocity is called the drag force, while the force in the direction perpendicular to the velocity is called the lift force. See also Section 4.4.

Note that force per unit area has the dimensions of pressure, which again has the same dimension as ρv_{ref}^2. The reference velocity is taken as the approach velocity v_∞ here. Hence we find that force per unit area, F/A, can be written as the following dimensionless group:

$$C_D = \frac{F/A}{\rho v_\infty^2/2}$$

The factor of 2 in the term in the denominator does not arise from dimensional considerations but is customarily used. The above group is referred to as the drag coefficient C_D. Since the pressure and viscous forces depend only on the Reynolds number and the geometric parameters such as L/d, the drag coefficient can be correlated as a function of the Reynolds number and the geometry. A sample plot for some cases was shown earlier for flow past a sphere (where only Re is needed) in Section 4.4. Data can therefore be presented as C_D vs. Re charts or by fitting empirical correlations for the same.

Hence the dimensional analysis gives us a clearer picture of why data are correlated in terms of certain particular dimensionless groups.

Example 3: unsteady-state flow in a pipe

Consider a pulsatile flow in a pipe caused by a pressure change that varies as a sinusoidal function of time. The change in dimensionless pressure is represented as

$$P^* = P_m^*[1 + \cos(\omega t)]$$

where P_m^* is a dimensionless mean value and ω is the frequency of oscillations. We consider slow flow, such that the convection terms are neglected. Hence the key groups are the Reynolds number and the Strouhal number. The governing equation in dimensionless terms is now represented for a slow flow (neglecting convection terms) as

$$\frac{1}{St}\frac{\partial v_z^*}{\partial t^*} = -\nabla_* P^* + \frac{1}{Re}\nabla_*^2 v_z^* \tag{5.22}$$

On multiplying by Re and using p^{**} as the dimensionless pressure we have

$$\frac{Re}{St}\frac{\partial v_z^*}{\partial t^*} = -\nabla_* p^{**} + \nabla_*^2 v_z^* \tag{5.23}$$

which eliminates one dimensionless group.

The time scale of pulsing is equal to the reciprocal of the frequency, $1/\omega$, and this can be chosen as the reference time. With this choice, the group Re/St can be shown to combine into a single group Wo^2:

$$\frac{Re}{St} = \frac{R^2\omega}{\nu} = Wo^2$$

Hence the equation reduces to

$$Wo^2 \frac{\partial v_z^*}{\partial t^*} = -\nabla_* p^{**} + \nabla_*^2 v_z^*$$

which is the equation for pulsatile flow for slow motion. The detailed solution and the effect of the parameter Wo will be examined later. Here we simply indicate that the correlation of the data should be based on dimensionless groups as

$$p^{**} = f(Wo, t^*)$$

with t^* defined as the dimensionless time ωt.

Example 4: pressure drop in a pipe (or other geometries such as a square pipe)

The key parameters are seen to be p^* or, equivalently, the Euler number, Eu, and the Reynolds number. Also the geometric parameter L/d_t will play a role. Hence the data should be correlated as

$$Eu = f(L/d_t, Re)$$

The dependence on L/d_t is usually linear in fully developed flow in pipes. Hence a combined parameter f defined as

$$f = \frac{Eu}{2} \frac{d_t}{L}$$

is used, and f is correlated as a function of the Reynolds number. This is nothing but the friction factor, and hence this analysis provides the justification for its widespread use in the correlation of pressure-drop data.

For laminar flow we have the classical result

$$f = \frac{16}{Re} \qquad \text{laminar flow} \qquad (5.24)$$

which also follows from theory and is the Hagen–Poiseuille equation. Note that $f \times Re$ is a constant, which is equivalent to saying that p^{**} is a constant. Note that p^{**} was the only key parameter in slow flow.

For turbulent flows, f and Re will appear as separate parameters and cannot be combined into one group. For turbulent flow the pipe roughness appears to play a role. A dimensionless group e/D can be used to characterize this. Here e is the average height of the protuberance at the wall. Hence it follows that

$$f = F\left(\frac{e}{d}, Re\right) \qquad \text{Turbulent flow} \qquad (5.25)$$

Some suggested correlations are as follows.

Friction factor correlations in turbulent flow

A correlation suitable for smooth pipes is the Blasius equation, which was shown in Chapter 1:

$$f = \frac{0.0791}{Re^{1/4}} \qquad (5.26)$$

for the range of Re from 10^4 to $< 10^5$ for smooth pipes.

Another equation is the modified Prandtl equation:

$$\frac{1}{\sqrt{f}} = 4.0 \log_{10}(Re\sqrt{f}) - 0.40 \tag{5.27}$$

This equation will be derived on a semi-theoretical basis for turbulent flow in Chapter 17.

The Colebrook and White equation is

$$\frac{1}{\sqrt{f}} = -4.0 \log_{10} \left(\frac{e}{d} + \frac{4.67}{Re\sqrt{f}} \right) + 2.28$$

This equation correlates the data well in turbulent flow for a wide range of Reynolds numbers. One minor disadvantage with the equation is that it is not explicit in f. The friction factor appears on both sides of the equation. Hence one needs to perform an iterative calculation with an assumed starting value for f. A starting value of 0.0075 for f appears to work well (Wilkes, 2006).

Rough pipe equation: for very rough pipes the dependence on the Reynolds number is unimportant and the following correlation is found to be applicable:

$$\frac{1}{\sqrt{f}} = -4.0 \log_{10} \left(\frac{e}{d} \right) + 2.28$$

The condition under which this will hold is when

$$\frac{e}{d} \gg \frac{4.67}{Re\sqrt{f}}$$

An example of use of similarity for scaleup calculations is now illustrated.

Example 5.1.

A new type of heat exchanger is to be used for a special nuclear application for a liquid with a viscosity of 2 cP. No data are available for the anticipated pressure drop in the system and experiments are therefore planned on a small scale, with a scale ratio of 1/10. Water ($\mu = 0.8$ cP) is to be used for safety reasons. Assume that the liquid densities are the same. What range of velocity is to be used if the commercial-scale velocities are 4–20 cm/s?

How would the measured pressure drop on the small scale translate to the large scale?

Solution.

The velocity on the small scale should be four times that for the large scale, i.e., 16–80 cm/s. (Match the Reynolds number and verify.)

The pressure drop will be 1/16 of the measured value for the large scale. (Match the Euler number and verify.)

Additional examples of the use of dimensionless gropes in scaleup are discussed in Chapter 14. Now we revert to the N–S equations. Various alternative forms of the N–S equations are also useful for flow analysis, and these are now discussed.

5.6   Alternative representations for the Navier–Stokes equations

5.6.1   Plane flow: the vorticity–streamfunction form

Plane flow is defined as flow in two dimensions (say, x and y). Here the z-dependence vanishes and $v_z = 0$. For this problem the N–S equations can be regrouped in terms of the following two variables: (i) the streamfunction ψ defined in (3.10) and (ii) the vorticity, ω. It may be noted here that in actuality ω is the z-component of the vorticity vector defined as $\nabla \times v$. In plane flow the x- and y-components of the vorticity vector are identically zero and only the z-component matters.

In terms of the above variables, ψ and ω, the N–S equations for steady-state flow can be reformulated as

$$\nabla^2 \psi = -\omega \tag{5.28}$$

$$\nu \nabla^2 \omega = \frac{\partial \psi}{\partial y}\frac{\partial \omega}{\partial x} - \frac{\partial \psi}{\partial x}\frac{\partial \omega}{\partial y} \tag{5.29}$$

It may be noted that the first condition is purely a kinematic condition, while the second condition is a kinetic condition (i.e., it incorporates the forces acting on the system).

The vorticity–streamfunction formulation has the advantage that the continuity equation is automatically satisfied. The exact satisfaction of the incompressibility condition is an advantage from a numerical-solution point of view. The disadvantage of this formulation is that the boundary conditions (see Section 5.3.2) are available in terms of the "primitive" variables v_x and v_y and not in terms of ψ and ω. Hence the boundary conditions for the latter must be deduced or derived.

5.6.2   Plane flow: the streamfunction representation

The N–S equations for plane flow can also be formulated in terms of the streamfunction alone. This is done by substituting for ω from Eq. (5.28) into Eq. (5.29). The resulting equation is a complicated fourth-order differential equation, which can be expressed compactly as

$$\nu \nabla^4 \psi = \frac{\partial(\psi, \nabla^2 \psi)}{\partial(x, y)} \tag{5.30}$$

where ∇^4 represents the biharmonic operator defined as

$$\nabla^4 = \frac{\partial^4}{\partial x^4} + 2\frac{\partial^4}{\partial x^2\,\partial y^2} + \frac{\partial^4}{\partial y^4} \tag{5.31}$$

The Jacobian notation is used in the RHS of Eq. (5.30). Thus

$$\frac{\partial(f, g)}{\partial(x, y)} = \begin{vmatrix} \partial f/\partial x & \partial f/\partial y \\ \partial g/\partial x & \partial g/\partial y \end{vmatrix} \tag{5.32}$$

If convective (inertia) terms are absent, i.e., if the RHS of Eq. (5.30) is negligible compared with the term on the LHS, then the streamfunction formulation reduces to the classic Stokes equation for a very slow flow of a viscous liquid in 2D:

$$\nabla^4 \psi = 0 \tag{5.33}$$

The condition for the applicability of the Stokes equation is that the flow Reynolds number should be less than unity. The Stokes flow is discussed in detail in a later chapter.

5.6.3   Inviscid and potential flow

In certain regions of the flow, the effect of fluid viscosity can be neglected. Here the N–S equations can be simplified by dropping the terms containing μ. The resulting flow is known as inviscid flow. The resulting equation is called the Euler equation of motion, which was shown earlier but is repeated here for ease of reference:

$$\rho \frac{Dv}{Dt} = \rho g - \nabla p \tag{5.34}$$

Potential flow

If the flow is irrotational as well, the velocity can be defined as a gradient of a scalar function: the velocity potential,

$$v = \nabla \phi \tag{5.35}$$

The velocity potential satisfies the Laplace equation. Thus the potential flow is

$$\nabla^2 \phi = 0 \tag{5.36}$$

Thus the flow field can be solved in terms of a single variable, namely the velocity potential, which is a scalar, and the velocity vector can be computed by taking the gradient of this scalar.

5.6.4   The velocity–vorticity formulation

The N–S equations can also be formulated in terms of the vorticity and velocity as the variables. The resulting equations are presented below in compact vector notation:

$$\nu \nabla^2 \omega + \nabla \times v \times \omega = \frac{\partial \omega}{\partial t} \tag{5.37}$$

$$\nabla^2 v = -\nabla \times \omega \tag{5.38}$$

where ω is the vorticity vector defined as usual by

$$\omega = \nabla \times v \tag{5.39}$$

The derivations of the equations are left as exercises.

5.6.5   Slow flow in terms of vorticity

If the viscous effects are large (a case of large ν in Eq. (5.37)) and if a steady state exists, then Eq. (5.37) reduces to

$$\nabla^2 \omega = 0$$

Further simplifications result for plane flows and flows with axisymmetry.

For plane flow we have $\nabla^4 \psi = 0$ using the kinematic relation for vorticity.

For axisymmetric flow we have $E^4 \psi = 0$.

5.6.6 The pressure Poisson equation

An important equation used in computational fluid dynamics and in the analysis of flow stability is the pressure Poisson equation. This is obtained by taking the divergence of the N–S equation. Since the divergence of a curl is zero, the viscous terms vanish. The result is a Poisson equation for the pressure field:

$$\nabla^2 p = -\rho \, \nabla \cdot [(v \cdot \nabla)v]$$

This is an elliptic equation (Poisson type), which permits the pressure field to be computed for a given velocity distribution. The use of this in CFD is widespread, since it is not easy to compute the pressure field.

5.7 Constitutive models for non-Newtonian fluids

In this section some common models to describe the behavior of non-Newtonian fluids are presented. The models are mainly applicable for shear flow, since they were developed mainly based on 1D fully developed flow.

The Ostwald–de Waale Model
In this model the stress is related in a power-law form to the strain rate:

$$\tau_{yx} = \Lambda \left| \frac{dv_x}{dy} \right|^{n-1} \frac{dv_x}{dy} \tag{5.40}$$

where Λ (units $Pa \cdot s^n$) and n are constants. This two-parameter model is also known as the power-law model. The parameter Λ is known as the consistency index, while n is known as the power-law index.

The local slope of τ_{yx} and the strain rate dv_x/dy is referred to as the apparent viscosity μ_{app}, which is given by

$$\mu_{app} = \Lambda \left| \frac{dv_x}{dy} \right|^{n-1} \tag{5.41}$$

If $n < 1$ the fluid is called pseudoplastic. The apparent viscosity decreases with shear rate in such fluids. Examples are aqueous solutions of macromolecular compounds such as carboxymethyl cellulose (CMC) and polyacrylamide (PAA).

If $n > 1$, the fluid is classified as dilatant. Here the shear stress increases with an increase in the strain rate.

There are some limitations to this simple model. It does not hold at high strain rates, where the apparent viscosity approaches a limiting asymptotic value.

The Carreau model
An illustrative plot of the strain rate vs. stress shows that there are two asymptotic values for the apparent viscosity. See Fig. 5.5. This is incorporated into the four-parameter Carreau model. The model is expressed as

$$\frac{\mu - \mu_0}{\mu_\infty - \mu_0} = \left[1 + (\Lambda \dot{\gamma})^2 \right]^{(n-1)/2} \tag{5.42}$$

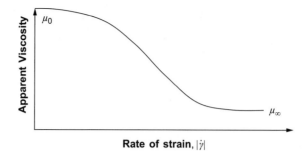

Figure 5.5 An illustrative plot of apparent viscosity for shear thinning fluids, showing the two asymptotic behaviors.

The Ellis model

The model is expressed in terms of an apparent viscosity, which is related to the stress by

$$\mu = \frac{\mu_0}{1 + \left(\tau/\tau_{1/2}\right)^{\alpha - 1}} \tag{5.43}$$

The parameters of the model are μ_0, α, and $\tau_{1/2}$. The last parameter may be interpreted as the shear stress at which the viscosity assumes half the asymptotic value of μ_0. Also α is an indicator of the shear thinning property of the fluid. The larger the value of α the larger is the shear thinning tendency of the fluid.

A typical plot of apparent viscosity as a function of strain rate is shown in Fig. 5.5.

The Bingham model

These fluids are characterized by a yield stress as discussed in Section 5.3. Thus a critical shear stress is needed for flow to occur. The model has been found to be suitable for many suspensions and pastes. The constitutive model proposed is of the following form:

$$\tau_{yx} = \mu \frac{dv_x}{dy} \pm \tau_0 \text{ if } |\tau_{yx}| > \tau_0 \tag{5.44}$$

and

$$\frac{dv_x}{dy} = 0 \text{ if } |\tau yx| < \tau_0 \tag{5.45}$$

The plus sign is used in Eq. (5.44) when τ_{yx} is negative and the negative sign when τ_{yx} is positive. When the applied stress is larger than the critical stress, the flow behavior is close to that of a Newtonian fluid.

An example is given below to illustrate the flow behavior of a Bingham fluid.

Example 5.2.

A fluid exhibiting Bingham flow character is contained in a tube of diameter d and height H. The tube is closed at the bottom in order to contain the fluid in the tube.

Now, if the bottom is removed, find the conditions under which the fluid will flow.

Solution.
The wall shear stress (taken as positive in the upward direction) can be estimated by balancing the force of gravity to the viscous force on the walls:

$$\tau_w(2\pi RL) = \rho g(\pi R^2 L)$$

or

$$\tau_w = \rho g R/2$$

This value has to be greater than the yield stress τ_0 of the material in order for the fluid to flow. Hence the fluid will flow only if

$$\rho g R/2 > \tau_0$$

Rearranging gives

$$R_{\text{critical}} > 2\tau_0/(\rho g)$$

in order for the fluid to flow. If the radius of the pipe is less than the above value the fluid will not flow (even when the bottom of the container is open!). This appears to be a strange result for common fluids.

The Casson model
Blood flow is often modeled by the Casson equation, which is as follows:

$$\sqrt{|\tau_{yx}|} = \sqrt{\tau_0} + s\sqrt{|E_{yx}|} \tag{5.46}$$

where s and τ_0 are the two model parameters. Here E is the rate-of-strain tensor. Note that the absolute values are to be used under the square-root sign. The parameters are sensitive to the hematocrit (volume fraction of red cells) and the fibrinogen content of the blood. It may be noted from the Casson model that blood flow is generally non-Newtonian. A yield stress and a power-law dependence on the strain rate are observed. Also no significant viscoelastic behavior is observed. This may be due to the fact that the Deborah number is small under the conditions of flow. The Deborah number is a measure of the ratio of the response time to viscous forces to elastic forces.

Summary

- The equation of motion is the basic equation describing fluid motion and is obtained by the application of Newton's second law to a differential control volume. The equation is general and applies to all types of fluids.

- The viscous stress component (or the divergence of the stress tensor) appears as a term in this equation and the equation is also known as the stress form of the equation of motion.

- In order to model and solve the equation in terms of the primary variables, a constitutive relation for stress as a function of the rate of strain (related to velocity gradients) must be used. This is the closure needed to link the continuum model to the molecular-level fluid behavior.

- For common fluids the relation is linear and the use of this in the equation of motion leads to the Navier–Stokes (N–S) equation. The equations are supplemented with the continuity equation. Finally, problem-specific boundary conditions are to be imposed.

- Equations are often made dimensionless and the dimensionless formulation has many advantages. The basic method of making the equations dimensionless should be understood. The key dimensionless groups emerge from the dimensionless representation of the governing differential equations. The various groups of importance in fluid dynamics are presented in the text. These groups and their physical meaning should be understood.

- Scaleup is often done by matching some key dimensionless groups. The process is called similarity analysis. Some illustrative examples are presented. The student will be able to use these in their own research or area of interest.

- Alternative formulations and simplified versions of the N–S equations are also useful, and the chapter presents these as well.

- Various models can be used to describe the stress–strain-rate behavior of non-Newtonian fluids. These include power-law models, the Carreau model, the Casson model, and the Bingham model, among many others.

Problems

1. Write the divergence of the dyad ρvv in index notation. Expand the derivatives using the chain rule.
   Write the continuity equation in index notation and use this in the expanded expression for the divergence of the above dyad. Simplify and show that the result is $(v \cdot \nabla)v$. Hence verify that Eq. (5.5) is the same as Eq. (5.2).

2. Write the equation of motion in cylindrical coordinates using the various vector operations. Also write the equation of continuity. These equations together are the needed equations for the solution of flow problems posed in cylindrical coordinates.

3. Repeat for spherical coordinates.

4. Write the stress vs. rate-of-strain relations in cylindrical and spherical coordinates for Newtonian fluids.

5. Unidirectional flows in 2D Cartesian coordinates (also known as channel flows) are defined as systems with only one velocity component, say v_x. How does the continuity simplify for such problems? Verify that v_x can be a function of y but not a function of x. How does the Navier–Stokes equation simplify for such problems? How does it compare with the equations derived from a basic shell balance in Section 2.5.3?

6. Unidirectional flows can also be posed as v_x as a function of y and z. A flow in a square duct (away from the entrance region) is an example. How does the Navier–Stokes equation simplify for such problems?

7. Simplify the Navier–Stokes equation for flows in a circular pipe with only the axial velocity v_z as the non-vanishing component.
   How does it compare with the equations developed in Section 2.6.3?

8. Determine the flow rate of water at 25 °C in a 3000-m-long pipe of diameter 20 cm under a pressure gradient of 20 kPa. Assume a relative roughness parameter of 2.3×10^{-4}. Use $v = 0.916 \times 10^{-6} \, \text{m}^2/\text{s}$ for the kinematic viscosity.

9. What form does the Colebrook–White equation take for a smooth pipe? Compare this with the Prandtl formula by plotting the values on the same graph.

10. Verify the steps leading to the vorticity–streamfunction formulation of the N–S equation for 2D flow given in Section 5.6.1.

   If the flow has no vorticity it is called irrotational flow. What form does this equation take for irrotational flow?

11. Derive the vorticity–velocity formulation of the N–S equation and discuss the advantages and disadvantages of using this formulation. **Hint:** take the cross product of the Navier–Stokes equation and simplify the resulting equation using standard vector identities.

12. Show that the vorticity transport equation can also be written as

$$\frac{D\omega}{Dt} = \nu\,\nabla^2\omega + [\omega \cdot \nabla]v$$

   How does it simplify for 2D flows? How does it simplify for slow flows?

13. Show all the steps leading to the pressure Poisson equation.

14. Derive the following form of the pressure Poisson equation shown in the book by Saffman (1993) on vortex dynamics:

$$\nabla^2 p = \rho(\tilde{W} : \tilde{W} - \tilde{E} : \tilde{E})$$

   where $\tilde{W}$ is the vorticity tensor and $\tilde{E}$ is the rate-of-strain tensor. The symbol : indicates a double dot product of a tensor. This is defined, for example, in index notation as $W_{ij}W_{ij}$. Double summation is implied with respect to both indices, and the resulting quantity is a scalar. You will need the double-dot notation in the energy-balance chapter (Chapter 7) as well.

6 Illustrative flow problems

Learning objectives

From this chapter you will learn

- solutions to flow problems in unidirectional flow and basic properties of such flows;
- the lubrication approximation and how the simple unidirectional flow equations can be extended to some more complex cases;
- basics of external flow past a solid, the concept of a boundary layer and boundary-layer separation;
- to set up and solve some flow problems with non-Newtonian fluid behavior;
- the Maxwell constitutive model for viscoelastic flow and a model for a simple channel flow involving flow of such a fluid; and
- how Lorentz forces can be included for flow of a conducting fluid in the presence of a magnetic field and the solution to the classical Hartmann flow problem.

Mathematical prerequisites

The ability to find general solutions to second-order differential equations is needed, as well as the procedure to find the integration constants and find solutions to specific problems.

The chapter aims essentially to illustrate the solution to some common and well-studied fluid-flow problems in laminar flow. The problems analyzed are mainly those where the Navier–Stokes (N–S) equations can be simplified and analytical solutions are possible. Steady-state flow problems are discussed. The types of problems examined here and a brief classification of flow problems are presented below.

Flows can be classified into external flows and internal flows. External flow is in a semi-infinite domain and the flow is always developing and of boundary-layer nature.

Thus there is a region near the solid surface called a boundary layer where the major changes in velocity can be anticipated. The thickness of this boundary layer increases as one moves along the flow direction. Further the flow is 2D (or even 3D) and both components of the velocity are needed to describe the flow correctly. The solution of the N–S equations are therefore more complex than those for internal flows, and for boundary-layer problems they can be classified as parabolic partial differential equations. Details of external flow are postponed until a later chapter. However, some key results that are also useful for study of heat and mass transfer in boundary layers are summarized.

The internal flow by contrast is simpler to describe, especially in regular geometries such as channels and pipes. The flow becomes fully developed after a certain entry length, and only one velocity component is of significance thereafter. Such flows are called unidirectional flows. This is the first class of problems which we study in this chapter. The N–S equations simplify and can be solved analytically. (Note that the flow is assumed to uncoupled, i.e., there are no heat transfer effects and the flow is laminar here). Solutions are presented both for channel flow (parallel plates or channels or films) and for axial flow in cylindrical geometry (pipes, annulus) in a general manner. Then solutions applicable to specific situations are examined and relations for important quantities such as the volumetric flow rate, pressure drop, and wall shear stress are presented.

Additional problems where only one component of velocity is important occur in tangential flows and radial flows. These are analyzed next.

Flow in other geometries such as tapered pipes can be analyzed as an approximation by extension of the unidirectional-flow equations. Such analysis is often called the lubrication approximation since it was and is being used in analysis of flow in lubrication bearings in shafts and similar mechanical equipment. Useful information on the pressure drop and other properties of flow can be obtained by lubrication analysis, and some examples are provided.

The next topic analyzed is some simple unidirectional flows involving non-Newtonian fluids. The stress form of the equation is used first here. Stress profiles are first computed. Then the constitutive equation for stress is substituted to get the velocity profile equation. This is solved for the velocity. Thus this is a two-step procedure, in contrast to that used for Newtonian fluids.

Viscoelastic fluid presents additional difficulties. Again we show a simple example where we develop the governing equations in detail for a simple channel flow. It is useful to study the methodology since it can be extended for more complex flows.

The final section deals with flows where electric/magnetic-field effects are felt. Such problems fall within the field of magnetohydrodynamics. A simple problem of channel flow is analyzed to indicate the method. This is the first or basic problem in the field of magnetohydrodynamics, and gives you an introduction to the field.

6.1   Introduction

6.1.1   Summary of equations

The equations of motion derived in the last chapter form the backbone of the field of fluid dynamics. The equations are repeated here in compact vector form for completeness and ease of reference:

$$\rho\left(\frac{\partial v}{\partial t} + (v \cdot \nabla)v\right) = \rho g - \nabla p + \nabla \cdot \tilde{\tau} \tag{6.1}$$

The equations are valid for all fluids and are known as equations of motion in stress form. For the non-Newtonian fluids discussed later in Section 6.10 the stress form is the starting point for flow simulation. But, if the fluid is assumed to be incompressible and Newtonian, we obtain the N–S equation as shown in the earlier chapter, which is reproduced here for ease of reference:

$$\rho\left(\frac{\partial v}{\partial t} + (v \cdot \nabla)v\right) = \rho g - \nabla p + \mu \nabla^2 v \tag{6.2}$$

These are solved in conjunction that the continuity condition

$$\nabla \cdot v = 0 \tag{6.3}$$

Use of modified pressure

One simplification is that the gravity term can be combined with the pressure term by using a "modified" pressure. Let

$$g = -\nabla\phi$$

where ϕ is a scalar function that will have the physical meaning of the gravitational potential. It can be shown that

$$\phi = gh$$

where g is the magnitude of the acceleration due to gravity (9.81 m/s^2) and h is the elevation measured from a datum plane. Therefore ϕ represents the potential energy per unit mass at a given elevation. The pressure and gravity are then combined into a modified pressure defined as follows. Let

$$\boxed{\mathcal{P} = p + \rho gh} \tag{6.4}$$

Then

$$-\nabla p + \rho g = -\nabla p - \rho \nabla\phi = -\nabla p - \rho g \nabla h = -\nabla\mathcal{P}$$

Hence the pressure terms and gravity terms can be combined into a single term and the N–S equations can be then written using the "modified "pressure as

$$\rho\left(\frac{\partial v}{\partial t} + (v \cdot \nabla)v\right) = -\nabla\mathcal{P} + \mu \nabla^2 v \tag{6.5}$$

This form is convenient for many problems since it applies to vertical, horizontal, or inclined flow, and the orientation of the system need not be explicitly considered.

6.1.2   Simplifications

The solution of the flow usually involves a number of assumptions. The common assumptions are stated here.

The common assumption is constant viscosity. This is valid for systems with no heat transfer or with limited effects of heat generation due to viscosity. If these assumptions do not hold then the problem becomes more complex and has to be treated as a set of coupled problems. The discussion on such problems is deferred to a later chapter.

The second assumption is that the system is isothermal or the change in density with temperature does not affect the flow. The change in density with temperature is usually small, but can be sufficient to generate convection current due to buoyancy effects. Such effects are called natural convection flows and analyzed later as coupled transport problems.

The difficulty in finding the solution to equations of motion or the N–S equation lies in the non-linear terms on the LHS. These can lead to bifurcations, multiple steady states, oscillatory flows, chaos, turbulence, etc. Hence the general and complete solution of the N–S equation is a formidable and perhaps impossible task. Hence we seek a simplified approach to tackle engineering problems

6.1.3   Solution methods

Solution methods can be classified into four types.

1. *Analytical solutions*

   Analytical solutions are common for unidirectional (1D) flows where the inertial (convection) terms vanish. These solutions provide exact solutions to the problem, and the analysis of such flows is the main theme of this chapter. The analytical solutions form an important part of the study since they are useful in many cases. For other cases, they provide a first-level solution for further analysis by perturbation methods or provide a benchmark solution to test the accuracy of the various numerical methods. The various analytical solutions to be presented in this chapter are also widely used in many engineering calculations or as input to macroscopic closures.

2. *Linear problems: analytic or semi-analytic solutions*

   For flows with low Reynolds number or highly viscous fluids the inertial terms can be neglected. Such flows are referred to as Stokes flow or creeping flow. Examples of Stokes flow are found in many applications such as micro-fluidic devices, etc. Analytical solutions can be obtained for flow in simple geometries, such as flow past a sphere. For other cases, numerical solutions may be needed, but, since the governing equations are linear, powerful numerical methods such as boundary element methods, singularity methods, and multipole expansion methods have been developed. An analysis of Stokes flow is presented in a later chapter.

3. *Reduction to ODEs; boundary-layer approximations*

   Flows in boundary layers are examples where a similarity solution is possible. In these cases certain simplifying assumptions can be invoked and then the problem is reduced from a PDE to an ODE. The resulting ODE is much simpler than the original N–S equations, and amenable to semi-analytical or numerical solutions.

4. *Numerical solutions*

Finally, for flows in complex geometry a numerical solution is needed. Here the full set of N–S equations is solved numerically by various techniques such as finite differences, finite-volume methods, finite-element methods, etc. Problems involving turbulent flow also need numerical solutions, except in some cases (1D) with some simple closure approximations (presented in Chapter 17 later).

The examples in this chapter are mainly from the first category. The other examples are taken up in subsequent chapters. Flows that are turbulent need a separate treatment. The N–S equations, *per se*, are valid but, due to the chaotic, time-dependent, and fluctuating nature of velocity, direct numerical solution is difficult and time averaging of the equations is needed. Averaging introduces additional modeling closure terms. In this chapter we consider only laminar flows and mainly unidirectional flows. Time averaging of N–S equations and simple closure models for turbulent flow are deferred to Chapter 17.

6.2   Channel flow

The simplest example is flow between two plates, or channel flow. The situation is illustrated schematically in Fig. 6.1.

6.2.1   Entry-region flow in channels or pipes

Consider a case where a fluid enters the channel at a uniform velocity. The no-slip condition applies at the wall, and hence there is a region near the wall where the velocity changes (from zero at the walls) to a uniform value (near the center). This region is called the boundary layer, and is shown schematically in Fig. 6.2.

The boundary-layer thickness increases with the axial length until it hits the channel or pipe center. The flow becomes fully developed at this point. The flow is unidirectional from this point onwards. The approximate length of the entry region is

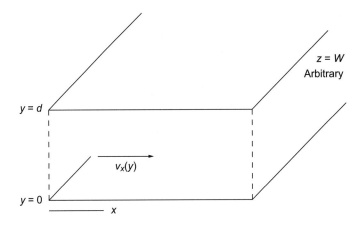

Figure 6.1 A description of channel flow between two plates, showing an example of unidirectional flow.

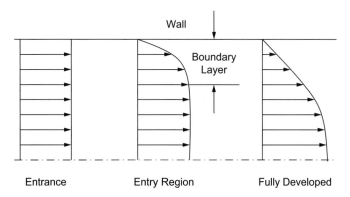

Figure 6.2 Flow in the entry region in a pipe or channel flow, showing the development of the fully developed profile.

$$L_e/d = 0.025 \frac{\langle v \rangle d}{\nu} = 0.025 Re$$

The flow is unidirectional after the entry length. An important simplification for a fully developed flow is that the acceleration (convection) terms are zero. This is verified in the following example.

Example 6.1.

Verify that the acceleration of a fluid particle is zero for a fully developed steady-state unidirectional flow in a channel.

Solution.

Let x be the main flow direction. Note that the acceleration term in Cartesian coordinates is

$$x\text{-component of the acceleration terms} = v_x \frac{\partial v_x}{\partial x} + v_y \frac{\partial v_x}{\partial y} + v_z \frac{\partial v_x}{\partial z}$$

The term v_x does not vary with x in unidirectional flows. Hence $\partial v_x/\partial x$ is zero, and the first term drops out. Also v_y and v_z are zero, and hence the second and third terms drop out. Hence the acceleration term is zero.

Other components of momentum are not important in 1D flows.

Hence the sum of the forces is equal to zero in fully developed flow. Taking only the force components in the x-direction,

$$\frac{\partial}{\partial y}(\tau_{yx}) = \frac{\partial p}{\partial x} - \rho g_x = \frac{\partial \mathcal{P}}{\partial x} = -G \tag{6.6}$$

Note that this can be verified also by writing a shell balance for the forces involved (see Chapter 2).

The quantity $G = -d\mathcal{P}/dx$ represents the pressure drop per unit length of pipe and can be considered to be a driving force for flow. G can be treated as a constant in unidirectional flow, which is verified in Example 6.2 below. This may be skipped to continue the main flow analysis.

Example 6.2.

Verify that the G can be assumed to be a constant for unidirectional flows.

Solution.

For this we need to examine the y-momentum balance. The forces acting in the y-direction are also zero, and this can be written as

$$\frac{\partial P}{\partial y} = 0 \tag{6.7}$$

which simply means that P can only be a function of x.

Now, inspection of Eq. (6.6) reveals that the LHS is a function of y only, while we require the RHS to be a function of x only in accordance with Eq. (6.7). The only way we can compromise is to assume that the pressure gradient is a constant along the flow direction.

This proves our assertion that the pressure gradient $-G$ can be treated as a constant for the purpose of integration of the governing equations for the case of unidirectional flows.

We now continue with the general solution to channel flow under fully developed flow conditions.

6.2.2 General solution

The force balance discussed above is restated here:

$$\frac{\partial}{\partial y}(\tau_{yx}) = -G \tag{6.8}$$

where G can be treated as a constant.

Up to this point the governing equations are valid both for Newtonian and for non-Newtonian fluids. Let us now proceed with the Newtonian fluid and compute the velocity distribution; non-Newtonian fluids are studied in a separate section.

Newton's law of viscosity for 1D flows,

$$\tau_{yx} = \mu \frac{dv_x}{dy} \tag{6.9}$$

used in conjunction with the momentum balance, yields the following differential equation for the velocity distribution:

$$\mu \frac{\partial^2 v_x}{\partial y^2} = -G \tag{6.10}$$

which is taken as the governing equation for the unidirectional flow.

A general solution for the velocity profile can be obtained by integration of the above equation twice:

$$v_x = -\frac{G}{2\mu}y^2 + C_1 y + C_2 \tag{6.11}$$

This general form of the solution is applicable to many problems in unidirectional flow in Cartesian coordinates. Particular solutions depend on the imposed boundary conditions, which are problem-dependent. This requires the evaluation of the integration constants C_1 and C_2 as per problem specifications.

Various common cases are presented here.

6.2.3 Pressure-driven flow

This flow is referred to as plane Poiseuille flow. The boundary conditions are now the no-slip boundary conditions: no slip at the boundaries; and $v_x = 0$ both at $y = 0$ and at $y = d$, where d is the gap width.

The student should use these conditions to evaluate C_1 and C_2, and back-substitute into the velocity distribution. The details are not presented here. The resulting solution for the velocity profile is

$$v_x = \frac{G}{2\mu} y(d - y)$$

It can be seen that the velocity distribution is a parabolic function. Other properties of flow of interest can be obtained as follows.

The average velocity is obtained by integration over the flow cross-section:

$$\langle v \rangle = \frac{1}{d} \int_0^d v_x \, dy = \frac{Gd^2}{12\mu}$$

The volumetric flow rate for a channel of arbitrary width W is therefore given as

$$Q = W \frac{Gd^3}{12\mu} \tag{6.12}$$

The stress field is given as

$$\tau_{yx} = \frac{G}{2} [d - 2y]$$

which is a linear function of y.

The vorticity field is obtained by taking the curl of the velocity and is a measure of the pointwise rotational tendency of the fluid. The component of vorticity in the z direction is given as

$$\omega = \nabla \times v = -\frac{G}{2\mu} [d - 2y] e_z$$

The components of vorticity in the x- and y-directions are zero.

6.2.4 Shear-driven flow

Top plate moving: no pressure gradient

The boundary conditions are now defined as $v_x = V$ at $y = d$, where V is the velocity with which the top plate is moving. The velocity is set as zero at the bottom plate. Also, since the pressure gradient is zero, the term G in Eq. (6.11) is set as zero. The solution can be shown to be

$$v_x = V \frac{y}{d}$$

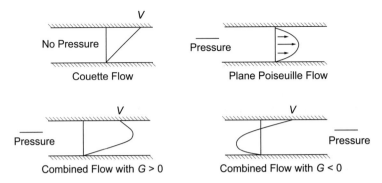

Figure 6.3 Combined plane Poiseuille flow and Couette flow, showing the various types of velocity distributions which can be generated.

The corresponding volumetric flow rate is

$$Q = W\frac{Vd}{2} \tag{6.13}$$

One obtains a linear velocity profile and flow is driven by shear. Hence these types of flow are called simple shear flows or Couette flows.

Top plate moving plus a pressure gradient is imposed

This is a combination of the above flows. The velocity profile turns out to be

$$v_x = \frac{G}{2\mu}y(d-y) + V\frac{y}{d}$$

This is called the principle of superposition of flows. The principle states that, if v_1 is the velocity profile to a problem 1 and v_2 is the profile for problem 2, then the flow for the combined problem is $v_1 + v_2$. Note that this applies only to linear problems. Since the unidirectional flow considered here is linear, we find that this principle holds and the flow is a combination of the plane Poiseuille flow and the Couette flow.

The various properties such as volumetric rate of flow can be computed. The details are left as exercise problems.

Flow reversal can occur at certain points in the flow if the pressure gradient and the direction of the plate movement are in opposite directions. The various velocity profiles which are possible are sketched in Fig. 6.3.

6.2.5   Gravity-driven flow

Gravity-driven film flow has many applications in engineering practice. The problem is shown schematically in Fig. 6.4. A liquid is flowing along a plate or wall as a thin film of thickness δ. A device of this type may be used to study mass transport, and the film flow occurs in condensation heat transfer as well.

Again we can use the general solution, privided that we define G appropriately. The parameter G is now the difference in modified pressure per unit height. This can be shown to be equal to ρg_x, where g_x is the component of gravity in the flow direction. For a vertical

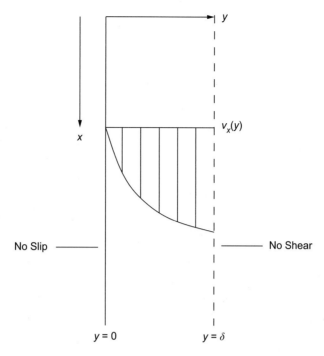

Figure 6.4 A schematic representation of film flow under gravity. x is marked downwards along the plate and y is taken normal to the plate. $y = 0$ is the location of the plate where the no-slip condition is applied.

wall this term is ρg, whereas for an inclined plate it is $\rho g \cos \alpha$, where α is the inclination of the wall to the vertical.

The boundary conditions also need a minor modification and are as follows.

There is no slip at $y = 0$, the wall, as for other problems. Hence v_x is set as zero at this point.

There is no shear at $y = \delta$, the gas–liquid interface. Hence τ_{yx} is set as zero at this point. This implies that the gradient of velocity is zero at this point. The physical interpretation is that there is no significant tangential force exerted by the gas on the liquid since the gas is assumed to be stagnant.

The result of using these boundary conditions in the general solution is

$$v_x = \frac{\rho g_x \delta^2}{2\mu} \left[\frac{2y}{\delta} - \left(\frac{y}{\delta}\right)^2 \right] \tag{6.14}$$

Later we will use the above velocity profile to model the mass transfer in a falling film.

The volumetric flow rate (per unit width of the plate) is given by

$$Q = \frac{\rho g_x \delta^3}{3\mu} \tag{6.15}$$

Later we will use the result to model the rate of condensation in a falling film. We find that the film thickness is proportional to $Q^{1/3}$.

The flow in a falling film tends to become unstable above a Reynolds number of 20. The model shown above is applicable only for stable laminar conditions. Above this Reynolds number the flow becomes wavy. The linear stability analysis discussed in Chapter 16 is a useful tool to study this transition and predict the evolution of the free surface. However, this is a challenging problem requiring a combination of analytical and computational tools. Pertinent references are provided in the later chapter on flow stability. At even larger Reynolds number the flow becomes turbulent.

The next class of problems amenable to analytic solutions is unidirectional axial flows posed in a cylindrical coordinate system.

6.3   Axial flow in cylindrical geometry

Pipe flow is a common example where v_z, the axial velocity, is the only non-zero velocity component. A second example is flow in an annular pipe, as shown schematically in Fig. 6.5. Another example is flow in an annulus caused by moving an inner solid rod, which is shown in Fig. 6.6. This is an example of a shear-driven flow in a cylindrical geometry. All these can be analyzed in a general manner, and we first provide the general solution for v_z using cylindrical coordinates.

The governing equation is obtained from the force balance, expressed now in cylindrical coordinates. We use here the z-component of the N–S equation, which takes the following form:

$$\frac{1}{r}\frac{d}{dr}\left(r\frac{dv_z}{dr}\right) = -\frac{G}{\mu} \tag{6.16}$$

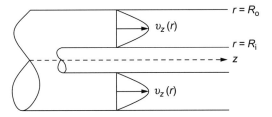

Figure 6.5 An example of axial flow in cylindrical geometry, namely flow in an annulus.

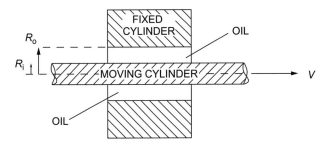

Figure 6.6 An example of shear flow in annular cylindrical geometry.

Note that the convective terms drop out and the equation simplifies considerably. The general solution is obtained by integrating this twice:

$$v_z = -\frac{Gr^2}{4\mu} + C_1 \ln r + C_2 \tag{6.17}$$

where the integration constants C_1 and C_2 can be found by applying the boundary conditions specified for the problem. The general model applies both to a circular pipe and annular flow in coaxial cylinders and to shear flow in an annulus.

6.3.1 Circular pipe

For flow in a circular pipe the constant $C_1 = 0$; otherwise the velocity becomes infinite at the center. The constant C_2 is found from the no-slip condition. The final expression for the velocity profile is

$$v_z = \frac{G}{4\mu}(R^2 - r^2) \tag{6.18}$$

and the cross-sectional average velocity is obtained as

$$\langle v \rangle = \frac{1}{\pi R^2} \int_0^R 2\pi r v_z(r) dr = \frac{GR^2}{8\mu} \tag{6.19}$$

The volumetric flow rate is

$$Q = \frac{\pi G R^4}{8\mu} \tag{6.20}$$

The equation is known as the Hagen–Poiseuille law. It was published in 1841 purely from empirical observations. We have already encountered this in Chapters 1 and 2.

The above equation is valid for $Re < 2100$. For higher Re, the streamlines become wavy, the flow starts to develop a random motion, and eventually, for $Re > 4000$, the flow becomes turbulent.

6.3.2 Annular pipe: pressure-driven

For an annular geometry driven by a positive pressure G we can show that

$$v_z = \frac{GR^2}{4\mu}\left[(1 - \xi^2) - (1 - \kappa^2)\frac{\ln \xi}{\ln \kappa}\right] \tag{6.21}$$

where $\kappa = R_i/R_o$, the ratio of the radius of the inner cylinder to that of the outer cylinder. The variable r has been expressed in dimensionless form as a variable ξ equal to r/R_o.

The volumetric flow rate can be calculated from the integral

$$Q_p = 2\pi \int_{\kappa R}^{R_o} v_z(r) r \, dr$$

where the subscript p is used to designate that this is for the pressure-driven flow case. After some algebra, we find

$$Q_p = \frac{\pi G R_o^4}{8\mu} \left[1 - \kappa^4 + \frac{(1 - \kappa^2)^2}{\ln \kappa} \right] \tag{6.22}$$

6.3.3 Annular pipe: shear-driven

A shear-driven flow can be generated by moving the inner cylinder with a velocity of V.

For a shear flow in annular geometry we use the general solution with the following boundary conditions. At $\xi = \kappa$, velocity $v_z = V$, the rod velocity. At $\xi = 1$, velocity $v_z = 0$, the zero velocity for the outer cylinder.

Using these in the general solution, we obtain the following velocity profile:

$$v_z = V \frac{\ln \xi}{\ln \kappa} \tag{6.23}$$

The volumetric flow rate is calculated as

$$Q_D = \frac{\pi R_o^2 V}{2} \left[\frac{2\kappa^2 \ln \kappa + \kappa^2 - 1}{\ln \kappa} \right]$$

Combined pressure- and shear-driven flow: the superposition

For a combined shear- and pressure-driven flow in an annular geometry, the flow can be computed from the superposition principle. Thus $Q = Q_p + Q_D$. This problem has applications in wire coating, where a viscous fluid is dragged by a wire and gets coated on the surface of the wire. The control of wire thickness by flow manipulations is a key problem, and the book by Middleman (1988b) discusses this in detail.

6.4 Torsional flow

Flows where v_θ is the only non-vanishing component are known as torsional flows. An example is shown in Fig. 6.7, where a viscous liquid is contained between two concentric cylinders. The flow is caused by rotating the inner or outer cylinder. Unidirectional flow holds, with v_θ as the only non-vanishing velocity, now a function of r only. The θ-momentum balance reduces to

$$\frac{d}{dr} \left(\frac{1}{r} \frac{d}{dr} (r v_\theta) \right) = 0 \tag{6.24}$$

The student should verify this result. Note that we have used the θ component of the Laplacian of velocity and simplified it for unidirectional conditions.

The equation can be integrated twice to yield the general solution for the velocity profiles in torsional flows,

$$\boxed{v_\theta = C_1 r + \frac{C_2}{r}} \tag{6.25}$$

where C_1 and C_2 are the integration constants as usual. These constants can be determined from the imposed boundary conditions.

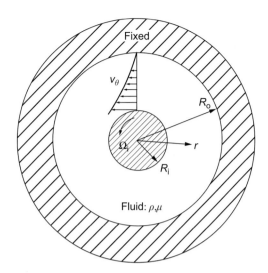

Figure 6.7 An example of torsional flow, with two concentric rotating cylinders. The velocity profile shown is for the inner cylinder rotating.

As an example, if one has a case where the outer cylinder (of radius R) is rotating at an angular speed of Ω_0 and the inner cylinder (of radius κR) is stationary, then the solution can be shown to be

$$v_\theta = R\Omega_0 \frac{\kappa R/r - r/(\kappa R)}{\kappa - 1/\kappa} \tag{6.26}$$

This turns out to be a good approximation for the velocity at low rotational speeds. At high speeds, secondary flow, (Taylor) vortices, and turbulence set in.

Once the tangential velocity profile is known, the radial pressure profiles can be calculated from the force balance in the radial direction:

$$\frac{\partial p}{\partial r} = \rho \frac{v_\theta^2}{r} \tag{6.27}$$

The physical interpretation of this equation is that the pressure forces in the radial direction are balanced by the centrifugal forces. However, it may be noted that the centrifugal acceleration has a tendency to produce secondary motions. If the viscous forces are not sufficient to balance these then the flow tends to become unstable.

The quantity of interest is often the torque needed to rotate the cylinders. The expression for torque can be derived as

$$T = 4\pi \mu \Omega_0 R^2 L \left(\frac{\kappa^2}{1 - \kappa^2} \right) \tag{6.28}$$

The measured torque can then be related to the viscosity of the liquid and hence an apparatus of this type can be used as a viscometer,

Forced vortex

Another example of torsional flow is the forced vortex. The flow is representative of the motion of a fluid contained in a tank, with the tank being rotated as shown in Fig. 6.8. The velocity profile can be shown to be a particular case of the general solution for torsional flow,

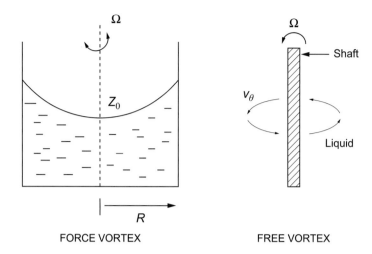

Figure 6.8 Examples of torsional flows.

$$v_\theta = r\Omega$$

where Ω is the rotation rate of the tank. The motion is as though the fluid were a solid body and hence is also referred to as a rigid-body rotation. The corresponding pressure profile is obtained by using this velocity profile in Eq. (6.27) and integration:

$$p - p_{atm} = -\rho g(z - z_0) + \frac{1}{2}\rho\Omega^2 r^2$$

where z_0 is the elevation of the free surface at the center of the tank. Since the free surface is at p_{atm}, the pressure profile causes the gas–liquid interface to assume a parabolic shape as shown in Fig. 6.8.

Free vortex
This is another example of torsional flow. The flow is representative of the motion of a large body of fluid being rotated by a central shaft of radius R as shown in Fig. 6.8. The velocity profile can be shown to be

$$v_\theta = R\Omega\frac{R}{r}$$

where Ω is the rotation rate of the shaft. The pressure profile can be obtained in a similar manner to that for a forced vortex.

6.5 Radial flow

The flow of a viscous fluid contained between two circular disks causes a radial velocity profile as shown in Fig. 6.9. We assume that the flow is directed radially outward from a central nozzle. The flow is fully developed after a short distance from the nozzle. The radial velocity is a function of z, since the velocity is zero at the top and bottom, i.e., $v_r(z = 0) = 0$ and $v_r(z = h) = 0$. However, the radial velocity is a function of r as well due to continuity.

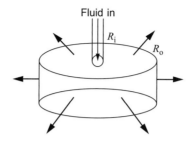

Figure 6.9 A description of radial flow between two parallel circular disks.

Hence we can postulate a solution of the form

$$v_r = \frac{F(z)}{r}$$

which satisfies the continuity condition. A differential equation for F can be obtained from the r-momentum balance:

$$-\frac{dP}{dr} + \frac{\mu}{r}\frac{d^2F}{dz^2} = -\rho\frac{F^2}{r^3} \tag{6.29}$$

You should verify this result before proceeding further.

An interesting characteristic is that one inertia term remains. No solution is possible if this term is retained. If this term is neglected, analytical solutions to both the pressure and the velocity profile can be obtained. In order to do this the Eq. (6.29) can be separated into two equations (after dropping the inertia term):

$$r\frac{dP}{dr} = \text{a constant denoted as } B$$

$$\mu\frac{d^2F}{dz^2} = \text{the same constant } B$$

We find from the first of the above equations that the pressure is a logarithmic function of r, while the F function is a parabolic function of z. Both the pressure profile and F function can be easily solved. Furthermore, the following expression holds for the volumetric flow rate in the limit of small Reynolds number:

$$Q = \frac{4\pi h^3 \Delta P}{3\mu \ln(R_o/R_i)} \tag{6.30}$$

The more complete problem has to be solved by a regular perturbation method whereby both components, v_r and v_z, are to be retained. (See Problem 24 in Chapter 14.) The solution shown above can be viewed as the leading term of the perturbation solution in the limit of Re tending to zero.

6.6   Flow in a spherical gap

This represents a problem posed in spherical coordinates. The problem is shown schematically in Fig. 6.10, where a fluid is flowing down a gap formed between two concentric spheres. The only velocity component is in the v_θ-direction now, and hence it resembles a

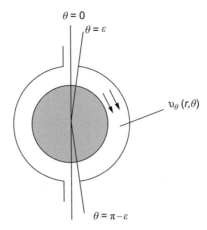

Figure 6.10 A schematic representation of flow in a spherical gap.

unidirectional flow. However, due to the changing cross-sectional area in the direction of flow, the continuity equation is needed and v_θ is not a constant in the flow direction (θ).

It can be shown using continuity that

$$v_\theta \sin \theta = F(r)$$

Using this in the θ-momentum balance, the following equation for F can be derived:

$$-\frac{1}{r}\frac{dP}{d\theta} + \frac{\mu}{r^2 \sin \theta}\frac{d}{dr}\left(r^2 \frac{dF}{dr}\right) = 0$$

Note that the additional inertia terms are neglected and the analysis is similar to that for radial flow.

The above equation can be separated into two equations, one for pressure variation and the second for the F function:

$$\sin \theta \frac{dP}{d\theta} = \text{a constant } B \tag{6.31}$$

which gives the pressure variation and

$$\frac{\mu}{r}\frac{d}{dr}\left(r^2 \frac{dF}{dr}\right) = \text{the same constant } B \tag{6.32}$$

which upon integration provides the F function and the corresponding velocity profile in the θ-direction.

6.7　Non-circular channels

Flow in a non-circular channel such as a square duct, a triangular channel, or an ellipsoidal tube is more conveniently formulated in Cartesian coordinates. Now let z be the main flow direction, and let x and y be the cross-flow directions. The z-momentum balance then reduces to

$$\frac{\partial^2 v_z}{\partial x^2} + \frac{\partial^2 v_z}{\partial y^2} = -\frac{G}{\mu} \tag{6.33}$$

which can be compacted and written as a "Poisson" equation:

$$\nabla^2 v_z = -\frac{G}{\mu} \tag{6.34}$$

where ∇^2 is the 2D Laplacian in Cartesian coordinates (i.e., with only x and y dependences). Recall that G is now the negative pressure gradient in the z-direction: $G = -d\mathcal{P}/dz$.

The boundary condition is that the velocity is zero along the channel walls. Solutions of certain simplified geometries can be obtained analytically, while for the general case a numerical solution is needed. A square duct is an example for which an analytical solution can be found.

Flow in a square channel

It may be useful to pursue the problem in a square channel briefly. The problem is also instructive in the sense that it can be solved by various methods. We present the series solution method here. A method based on boundary collocation is useful and left as a problem in Chapter 8. Purely numerical solution based on the finite-difference method can also be used. Here a technique of successive over-relaxation proves to be useful. This method is useful for a wide range of problems in transport phenomena. Thus three different solution techniques can be studied and tested on this simple problem and hence the problem provides a useful case study.

If L is the width of the channel, the distances can be scaled by L.

Assume that the velocity is scaled by a reference velocity v_{ref}. We do not know what to choose for this at this point. Let us leave it undecided. The dimensionless form of Eq. (6.34) is

$$\nabla_*^2 v^* = -\frac{GL^2}{\mu v_{\text{ref}}} \tag{6.35}$$

The * represents dimensionless quantities. The RHS can be simplified and set to minus one if we choose the reference velocity as

$$v_{\text{ref}} = \frac{GL^2}{\mu}$$

thereby providing an estimate of the order of the velocity in the system and also reducing the model to a bare-bones form:

$$\nabla_*^2 v^* = -1 \tag{6.36}$$

This is a partial differential equation in x and y with a non-homogeneous term. The boundary conditions are homogeneous but the governing equation is not. Therefore some pre-manipulation is needed before the PDE can be solved analytically using separation of variables. This is shown in the following paragraphs.

Series solution

A note on notation: for this paragraph we drop the * symbol for dimensionless quantities and assume that all variables here are dimensionless.

Equation (6.36) reads

$$\frac{\partial^2 v}{\partial x^2} + \frac{\partial^2 v}{\partial y^2} = -1$$

which is a prototype Poisson equation. The problem can be reduced to a homogeneous form by subtracting a particular solution. Let

$$v = v_p + v_h$$

where v_h is a particular solution. This solution is not unique and there are many choices. One such choice is

$$v_p = \frac{1}{2}(1 - x^2)$$

The homogeneous part of the problem, v_h, can then be solved by the method of separation of variables. The governing equation for v_h and the associated boundary condition are shown in Fig. 6.11. Note that the non-homogeneity is transferred to the boundary condition. Also it appears only in one boundary condition, and hence the overall problem can be solved by separation of variables. The complete solution can then be derived to be

$$v = \frac{1}{2}(1 - x^2) + \sum_{n=0}^{\infty} A_n \cosh(\lambda y)\cos(\lambda x)$$

where λ is an odd multiple of $\pi/2$ and is equal to $(n + 1/2)\pi$.

The series coefficient can be obtained by Fourier expansion of the boundary condition at $y = 1$:

$$A_n = -\frac{16h^2}{n^3\pi^3} \frac{\sin(n\pi/2)}{\cosh(n\pi/2)}$$

The average velocity can then be obtained by integration in the x- and y-directions, and the volumetric flow rate can be found to be

$$Q = \frac{4GL^4}{3\mu} \left(1 - 6\sum_n \frac{\tanh(\lambda_n)}{\lambda_n}\right)$$

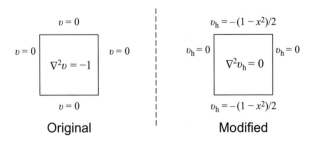

Figure 6.11 A schematic representation of flow in a square, showing the differential equations and the boundary conditions: (a) the original problem and (b) the modified problem. The variables are dimensionless here.

We now consider a variety of situations where the flow is actually 2D. But in many of these cases the 1D velocity profile can be used locally as an approximate solution. This is the famous and widely used lubrication approximation, and we illustrate this method with a few examples.

6.8 The lubrication approximation

6.8.1 Flow between two inclined plates

Consider the geometry shown in Fig. 6.12. The bottom plate is moving with a velocity of V while the top plate is stationary. The ends are at the same pressure. It is required to calculate the pressure profile and the volumetric flow rate in the system. The problem is also called the slider–block problem.

The volumetric flow rate for a channel (of arbitrary width W into the plane) is assumed to be given by

$$Q = W\frac{Gh^3}{12\mu} + W\frac{Vh}{2}$$

where we combine the pressure-driven, Eq. (6.12), and shear-driven, Eq. (6.13), flows. Upon expressing G as $-dp/dx$ and taking W as 1 since it is arbitrary, we have

$$Q = -\frac{dp}{dx}\frac{h^3}{12\mu} + \frac{Vh}{2} \tag{6.37}$$

We assume that this equation holds locally at any position x where the corresponding channel height is h (**the lubrication approximation**). Note that h varies with x. But the volumetric flow rate Q is constant for every x due to mass balance constraints. This implies that there has to be a pressure gradient in the system. The differential equation for the pressure gradient can be formulated by rearranging (6.37) as

$$\frac{dp}{dx} = \frac{6\mu V}{h^2} - \frac{12\mu Q}{h^3} \tag{6.38}$$

The functional dependence of h on x can be expressed as

$$h = h_0 - \alpha x \tag{6.39}$$

where α is the channel slope. Substitution into (6.38) gives

$$\frac{dp}{dx} = \frac{6\mu V}{(h_0 - \alpha x)^2} - \frac{12\mu Q}{(h_0 - \alpha x)^3} \tag{6.40}$$

Two boundary conditions are needed to complete the story. One condition is needed since Eq. (6.40) is a first-order differential equation. A second condition is needed since Q itself

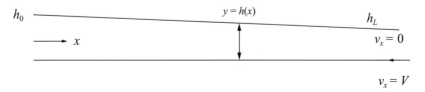

Figure 6.12 Flow between two plates with one plate inclined, namely slider–block flow.

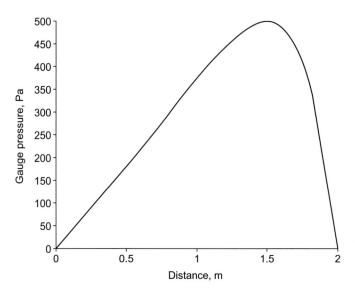

Figure 6.13 The pressure profile in the slider–block example. The parameters used are $h_0 = 3\,\text{mm}$, $h_L = 1\,\text{mm}$, $L = 2\,\text{m}$, $\mu = 0.01\,\text{Pa}\cdot\text{s}$, and $V = 0.1\,\text{m/s}$.

is an unknown and is to be solved as part of the problem. The two conditions are set as some pressure values at $x = 0$ and $x = L$. Since the system is at constant pressures in regions away from the lubrication zone, we can set these pressures as atmospheric values.

Upon integrating (6.40) from 0 to L we obtain the volumetric flow rate:

$$Q = V \frac{h_0 h_L}{h_0 + h_L}$$

Now, by integrating from $x = 0$ to any arbitrary position, we obtain the pressure distribution in the system as

$$p = p_{\text{atm}} + \frac{6\mu V \alpha}{h_0 + h_L} \frac{x(L - x)}{(h_0 - \alpha x)^2}$$

An illustrative sketch of the pressure profile is shown in Fig. 6.13. The solution has many interesting properties noted below. For $h_0 > h_L$, the pressure shows a maximum somewhere in the middle of the gap.

The y-component of the force (in excess of that due to atmospheric pressure) on the sloped surface is not equal to zero, unlike the case where the plates are both horizontal. Thus there is a lift force on the top plate. This lift force will be able to balance a load on the top plate. Egyptian engineers appear to have used this effect to move large weights of stone to build the pyramids. The calculation of the forces is left as an exercise.

6.8.2   Flow in a tapered pipe

Consider the flow in a tapered pipe shown in Fig. 6.14. It is required to develop a simple model to compute the volumetric flow rate in the system.

Locally we assume that the Hagen–Poiseuille equation holds, with the lubrication approximation. The conditions under which this assumption is applicable require a scaling

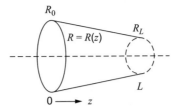

Figure 6.14 The tapered pipe analyzed in Section 6.8.2.

analysis, which will be discussed later. The following condition needs to be satisfied so that the flow can still be approximated as a unidirectional flow:

$$\frac{R_L}{L}\left[1 - \left(\frac{R_L}{R_0}\right)^2\right] \ll 1$$

We assume that the geometric parameters are such that the above condition is satisfied. The flow rate is given by

$$Q = \frac{\pi G R^4}{8\mu} \tag{6.41}$$

locally at the axial position z. The term G is $-dp/dz$ and can be written as

$$G = -\frac{dp}{dR}\frac{dR}{dz}$$

where R is the local radius at any z location.

Hence Eq. (6.41) can be written as

$$Q = \frac{\pi R^4}{8\mu}\frac{dp}{dR}\frac{dR}{dz}$$

Since Q is a constant at each z location this equation can be rearranged and integrated from R_0 to $R = R_L$:

$$Q\int_{R_0}^{R_L}\frac{dR}{R^4} = \frac{\pi}{8\mu}\frac{dR}{dz}\int_{p_0}^{p_L}dp$$

The term dR/dz is known from the geometry and, for a simple taper angle where the radius changes from R_0 at $z = 0$ to R_L at $z = L$, this term can be written as

$$\frac{dR}{dz} = \frac{R_L - R_0}{L}$$

Hence

$$Q\int_{R_0}^{R_L}\frac{dR}{R^4} = \frac{\pi}{8\mu}\frac{R_L - R_0}{L}\int_{p_0}^{p_L}dp \tag{6.42}$$

Upon integrating the LHS with respect to R and substituting the above expression for dR/dz on the RHS we obtain the following expression for the volumetric flow rate based on the lubrication approximation:

$$Q\frac{R_0^{-3} - R_L^{-3}}{3} = \frac{\pi}{8\mu}\frac{R_L - R_0}{L}(p_0 - p_L)$$

For a pipe with no taper the flow can be written as

$$Q_0 = \frac{\pi G_0 R_0^4}{8\mu}$$

where

$$G_0 = (p_0 - p_L)/L$$

and the inlet radius is used to define the geometry.

Hence the equation for the flow in a tapered pipe can be written as

$$Q = Q_0 F$$

where

$$F = \frac{3}{R_0^4} \frac{R_L - R_0}{R_0^{-3} - R_L^{-3}}$$

where F can be identified as a correction factor to account for the taper.

Additional examples of the application of the lubrication model are provided as Problems 9–13. Students will get a good feel for the method by working through these problems.

Now we take up briefly a second class of problems where the velocity has both components. Such problems are encountered in external flows, for example.

6.9   External flow

External flows have some important differences from internal flows. The main difference is that flow cannot be fully developed and is of boundary-layer type. The boundary-layer thickness increases with distance along the solid. Let us illustrate this from the study of flow past a flat plate, which is the basic problem studied in boundary-layer theory.

Flow past a flat plate

A schematic representation of flow past a flat plate is shown in Fig. 6.15. Here a fluid with a constant approach velocity v_∞ is flowing past a flat plate. All the features shown there should be carefully looked at and grasped. The velocity is zero at the plate itself. Near the plate a thin boundary layer of thickness δ develops, and the velocity increases from zero to the approach velocity in this boundary layer. The flow becomes turbulent at large distances from the contact point. The main features and the key equations are as follows.

Governing equations

A simplified version of the *x-momentum* can be shown to be applicable, and this will be shown in Section 14.2.4:

$$v_x \frac{\partial v_x}{\partial x} + v_y \frac{\partial v_x}{\partial y} = \nu \frac{\partial^2 v_x}{\partial y^2} \tag{6.43}$$

This is used together with the 2D version of the continuity equation:

$$\frac{\partial v_x}{\partial x} + \frac{\partial v_y}{\partial y} = 0 \tag{6.44}$$

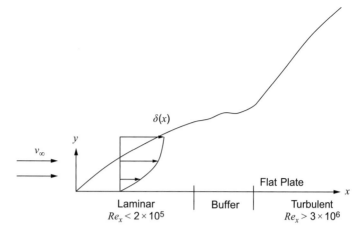

Figure 6.15 A schematic representation of velocity profiles in a boundary layer for flow over a flat plate. $Re_x = v_\infty \rho x/\mu$.

Thus we have two equations for the two unknown velocity components. The detailed justification leading to this simpler representation of the N–S equations is left to a later chapter, since the main goal here is introductory, in order to appreciate the key features of flow. Note that both v_x and v_y have to be solved together and the simplification due to unidirectional flow no longer applies. Detailed solutions are presented in Chapter 16.

The key results of engineering interest are presented here.

Summary of key results
The key results are as follows.

1. Boundary-layer thickness under laminar conditions:

$$\delta = 5\sqrt{\frac{\nu x}{v_\infty}}$$

or

$$\frac{\delta}{x} = \frac{5}{\sqrt{Re_x}}$$

where Re_x is a Reynolds number defined as xv_∞/ν.

2. Local shear stress exerted on the plate:

$$\tau_w = 0.332 v_\infty \mu \sqrt{\frac{v_\infty}{\nu x}}$$

3. Local drag coefficient under laminar flow conditions:

$$C_{\text{fx}} = \frac{0.664}{\sqrt{Re_x}}$$

4. Transition to turbulent flow; this occurs if $Re_x > 2 \times 10^5$. The flow becomes fully turbulent if $Re_x > 3 \times 10^6$.

5. Boundary-layer thickness under turbulent conditions:

$$\frac{\delta}{x} = \frac{0.376}{Re_x^{1/5}}$$

Note that the boundary-layer thickness increases dramatically as the flow transits to turbulence.

6. Drag coefficient under turbulent conditions:

$$C_{fx} = \frac{0.0576}{Re_x^{1/5}}$$

Flow past a cylinder

Now we examine the nature of the flow if the surface is curved. This is shown schematically in Fig. 6.16. The main aspect is that boundary-layer separation can occur.

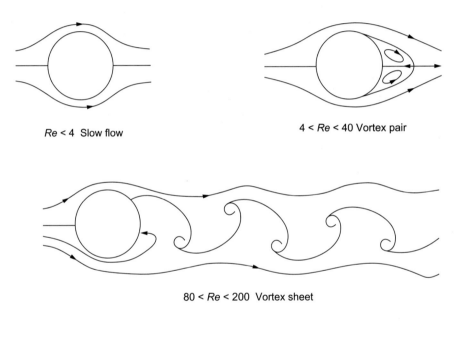

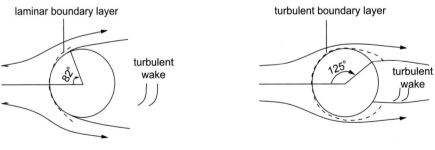

Figure 6.16 Flow over a cylinder, illustrating the various flow regimes as the Reynolds number is increased.

The observed flow pattern changes dramatically as the Reynolds number (defined as dv_∞/ν) is increased. At low Reynolds number the flow is smooth, but, at $Re \simeq 10$ a pair of vortices appears behind the cylinder and the streamlines separate from the cylinder surface. With further increase in Re the vortices appear in a regular manner. This regular pattern is known as the von Kármán vortex sheet. At even higher Reynolds number the flow becomes turbulent. Finally, at even higher Reynolds number, the wake region behind the cylinder narrows down considerably. Thus we see that the external flow past a curved object is difficult to model and shows a variety of flow patterns. The problem of simultaneous computation of heat or mass transfer from such an object is even more challenging. Often empirical representation of the drag coefficient as a function of the Reynolds number is used for practical calculations.

Now we switch our attention to non-Newtonian flows. Several aspects are discussed, with illustrations.

6.10   Non-Newtonian viscoinelastic fluids

6.10.1   A power-law model

Now we discuss some simple problems involving flow of Non-Newtonian fluids. We start with a power-law model. Pipe flow of a power-law fluid is discussed here. The stress relation for a power-law model was discussed earlier and is repeated here for convenience:

$$\tau_{rz} = \Lambda \left| \frac{dv_z}{dr} \right|^{n-1} \frac{dv_z}{dr} \tag{6.45}$$

where Λ (units $Pa \cdot s^n$) and n are constant, which can be considered as basic fluid properties.

Note that the equation is written as

$$\tau_{rz} = \Lambda \left(\frac{dv_z}{dr} \right)^n \text{ if } \frac{dv_z}{dr} \text{ is positive} \tag{6.46}$$

and

$$\tau_{rz} = -\Lambda \left(-\frac{dv_z}{dr} \right)^n \text{ if } \frac{dv_z}{dy} \text{ is negative} \tag{6.47}$$

for problem-solving purposes. This assigns the correct direction to the stress and at the same time avoids taking the root of a negative number. This is an important mathematical consideration for analysis of non-Newtonian flows.

The analysis of non-Newtonian fluids can be done by starting with the equation of motion in the stress form. For 1D flows in cylindrical coordinates, this reduces to

$$r\frac{dP}{dz} = \frac{d}{dr}(r\tau_{rz}) \tag{6.48}$$

The integrated form is

$$\tau_{rz} = -\frac{r}{2}G \tag{6.49}$$

where G is the pressure drop per unit length of pipe equal to $-dP/dz$.

Noting that dv_z/dr is negative in a pipe flow, the constitutive model can be written as (6.47). On combining (6.47) for the stress profile and (6.49), the constitutive model, we obtain

$$-\Lambda \left(-\frac{dv_z}{dr}\right)^n = -\frac{r}{2}G \tag{6.50}$$

Taking both sides to the power of $1/n$ gives

$$-\frac{dv_z}{dr} = r^{1/n}\left[\frac{G}{2\Lambda}\right]^{1/n} \tag{6.51}$$

which can be integrated for the velocity profile

$$v_z = -\frac{r^{1+1/n}}{1+1/n}\left[\frac{G}{2\Lambda}\right]^{1/n} + C_1 \tag{6.52}$$

The constant of integration, C_1, is obtained for applying the no-slip condition at the pipe wall. The final expression for the velocity profile is

$$v_z = \frac{n}{n+1}\left[\frac{G}{2\Lambda}\right]^{1/n}\left[R^{(n+1)/n} - r^{(n+1)/n}\right] \tag{6.53}$$

The maximum velocity is at the center and is obtained by setting $r = 0$:

$$v_{z,max} = \frac{n}{n+1}\left[\frac{G}{2\Lambda}\right]^{1/n}R^{(n+1)/n} \tag{6.54}$$

On taking the ratio, the velocity profile can also be expressed as

$$v_z = v_{z,max}\left[1 - \left(\frac{r}{R}\right)^{(n+1)/n}\right] \tag{6.55}$$

This reduces to the parabolic profile for $n = 1$. If $n < 1$, the profile is flatter, as shown in Fig. 6.17. The volumetric flow rate is given by

$$Q = \int_0^R 2\pi r v_z \, dr \tag{6.56}$$

Performing the integration gives

$$Q = \pi R^3 \frac{n}{3n+1}\left[\frac{GR}{2\Lambda}\right]^{1/n} \tag{6.57}$$

The following properties of the above equation for the volumetric flow rate should be noted.

1. The flow rate is now proportional to $R^{3+1/n}$. It simplifies to the Hagen–Poiseuille law for Newtonian fluids ($n = 1$).
2. It is proportional to the pressure drop G to the power of $1/n$. For Newtonian fluids the direct proportionality holds.

6.10.2   Flow of a Bingham fluid in a pipe

The flow of a Bingham fluid has to be analyzed somewhat differently since there is a yield stress τ_0. If the (magnitude of) stress is less than τ_0, then the velocity gradient is zero. The fluid flows as a plug with uniform velocity in the regions where this condition holds. This

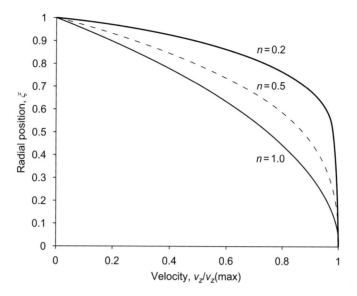

Figure 6.17 A plot of the velocity profile for power-law fluids. Note that the profile gets flatter as n decreases; ξ represents r/R.

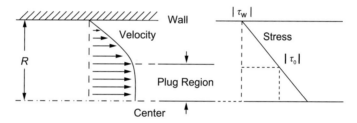

Figure 6.18 The stress profile and the velocity profile for the flow of a Bingham fluid in a pipe.

happens near the center of the pipe, since the stress values are smaller near the center. Thus the pipe has to be divided into two regions: (i) a center core, where the stress values are less than the yield stress; and (ii) a wall region, where a linear stress–strain relation holds. This is schematically shown in Fig. 6.18.

The detailed derivation of the velocity profile and the calculation of the volumetric flow rate by integration of the velocity profile are left as an exercise. The equation for Q is presented below, however, since this equation is the most often used one for design calculations:

$$Q = \frac{\pi R^3 \tau_w}{4\mu}\left[1 - \frac{4}{3}\left(\frac{\tau_0}{\tau_w}\right) + \frac{1}{3}\left(\frac{\tau_0}{\tau_w}\right)^4\right] \tag{6.58}$$

which is referred to as the Buckingham–Reiner equation. The equation holds only if $|\tau_w|$ is greater than τ_0. Otherwise the velocity is zero, and under such conditions the applied pressure difference is not sufficient to move the fluid. Note that the equation is more conveniently expressed in terms of the wall shear stress rather than the pressure drop. The student may wish to recall the relations between the two derived in the earlier chapter: $|\tau_w| = GR/2$.

6.10.3   The Rabinowitsch equation

The solution method described above is suitable for finding the velocity and the flow rate if the rheological properties of the system are known. However, in practice, the inverse problem is often important, i.e., given the measured flow rate vs. pressure-drop data, what type of model can be fitted to characterize the rheological properties of the fluid? In other words, the data should be recalculated as the wall shear-stress vs. rate-of-strain relation. The following analysis leading to the Rabinowitsch equation is useful for this purpose.

The wall shear stress is related to the pressure drop by the following equation irrespective of whether the fluid is Newtonian or not:

$$\tau_w = -\frac{GR}{2} \tag{6.59}$$

The volumetric flow rate is given by

$$Q = \int_0^R 2\pi r v_z \, dr \tag{6.60}$$

The RHS can be integrated by parts (treating r as the first function) to yield the following relation:

$$Q = \left[2\pi v_z r^2 / 2 \right]_0^R - \int_0^R \pi r^2 (dv_z/dr) dr \tag{6.61}$$

The first term on the RHS is zero due to the no-slip boundary condition at the walls. The second term is rearranged by using the local wall stress as the primary independent variable rather than r. First we note that the wall shear stress and radius are linearly related, which can be expressed as

$$\frac{r}{R} = \frac{\tau}{\tau_w} \tag{6.62}$$

Using this in Eq. (6.61) and also denoting dv_z/dr as $\dot{\gamma}$, which is the local strain rate, we get

$$Q\tau_w^3 = -\pi R^3 \int_0^{\tau_w} \dot{\gamma}\tau^2 \, d\tau \tag{6.63}$$

The expression can now be differentiated with respect to τ_w using the Leibnitz rule. The resulting expression can be rearranged to the following equation for $\dot{\gamma}$, the strain rate at the wall:

$$\boxed{\dot{\gamma} = -\frac{1}{\pi R^3} \left[3Q + \tau_w \frac{dQ}{d\tau_w} \right]} \tag{6.64}$$

This is the famous Rabinowitsch equation in rheology. This is a working equation to find the stress vs. strain-rate relations. Thus, if the flow-rate vs. wall-shear-stress data are available then the above equation can be used to compute the strain rate corresponding to each value of the wall shear stress. The procedure is illustrated by a worked example below.

Example 6.3.

The following data were measured for the flow rate as a function of pressure drop in a capillary viscometer of radius 1 cm. It is required to find the rheological behavior of the fluid.

Pressure drop (Pa/m)	Flow rate $Q \times 10^6 (m^3/s)$
1000	6.54
2000	26.18
3000	58.90
4000	104.72
5000	163.72
6000	235.62

Solution.

The first thing to do is to plot Q vs. ΔP. The plot is not a straight line, as can be verified by the student. The slope of the plot on a log–log scale is two. Assuming a power-law fluid, we find that the power-law index is two. This is one way of analyzing the data, but this assumes that the fluid behavior can be described by a power law. But in a more general framework we can construct the stress–strain-rate behavior for the fluid using Eq. (6.64). The procedure is as follows. For each value of pressure drop we calculate the wall shear stress. A plot of Q vs. τ_w is then constructed, and the local slope is computed at each data point. For each data point the strain rate can then be computed using Eq. (6.64). The results obtained are shown below and may be used to fit any chosen rheological model:

−G (Pa/m)	τ_w	Q $(10^6 m^3/s)$	$dQ/d\tau_w \times 10^4$	$\dot{\gamma}$ (1/s) (Eq. (6.64))
1000	5	6.54	0.1571	6.25
2000	10	26.18	0.3141	24.99
3000	15	58.90	0.4712	56.24
4000	20	104.72	0.6283	100.02
5000	25	163.72	0.7854	156.25
6000	30	235.62	0.9424	224.99

6.11 The effect of fluid elasticity

We briefly introduce the basic concepts needed to understand and model viscoelastic fluids. This section is introductory, but nevertheless will help you understand the basics and make it easy for you to understand more advanced textbooks or articles on this subject. For a detailed study on the topic, the book by Bird, Armstrong, and Hassager (1987) is useful.

Viscoelastic behavior, where a fluid exhibits both elastic and viscous effects, can be best understood by the spring-and-dashpot model sketched in Fig. 6.19.

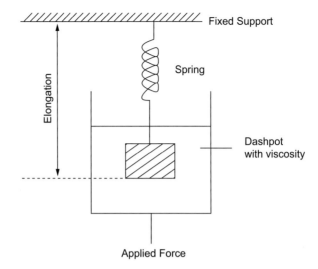

Figure 6.19 The basic spring-and-dashpot model used to represent a viscoelastic fluid.

The spring contributes to the elongation or pure strain, e, which is given by

$$e = \frac{\sigma}{2G} \tag{6.65}$$

where G is the modulus of elasticity and σ is the stress. This behavior is similar to Hooke's law for a solid under a load. Hence the rate of strain due to the spring is

$$\dot{\gamma} = \frac{de}{dt} = \frac{1}{2G}\frac{d\sigma}{dt} \tag{6.66}$$

The dashpot contributes viscous effects, and the rate of strain can be related to the stress by Newton's law of viscosity:

$$\dot{\gamma} = \frac{\sigma}{2\mu} \tag{6.67}$$

On combining the two effects, a constitutive model for the system can be represented as

$$\dot{\gamma} = \frac{\sigma}{2\mu} + \frac{1}{2G}\frac{d\sigma}{dt} \tag{6.68}$$

The extension of this to a 3D case forms the basis for the constitutive model for a Maxwell fluid. Here the stress tensor takes the place of σ. The substantial derivative is used instead of d/dt. The strain rate is replaced by the rate-of-strain tensor E_{ij}. Hence the spring–dashpot model leads to

$$2\mu E_{ik} = \tau_{ik} + \lambda \frac{D}{Dt}[\tau_{ik}] \tag{6.69}$$

where D/Dt is the material derivative in the convected coordinates. The constant λ is known as the relaxation constant.

Note that the material derivative for the stress tensor is computed by the following formula for the case where the flow field is not changing with time:

$$\frac{D}{Dt}[\tilde{\tau}] = (v \cdot \nabla)\tilde{\tau} - \nabla v^{\mathrm{T}} \cdot \tilde{\tau} - \tilde{\tau} \cdot \nabla v \tag{6.70}$$

The Maxwell model equations are now illustrated by two examples for a simple 2D plane flow.

Example 6.4.

What form does $(D/Dt)[\bar{\tau}]$ take for a plane 2D flow?

Solution.

The vector–tensor operations given by (6.70) are simple to execute since the unit vectors do not have any directionality. On performing the operations the following results are obtained. The student should verify these by doing the detailed algebra.

The expression for the normal stress in the x-direction is

$$v_x \frac{\partial \tau_{xx}}{\partial x} + v_y \frac{\partial \tau_{xx}}{\partial y} - 2\frac{\partial v_x}{\partial x}\tau_{xx} - 2\frac{\partial v_x}{\partial y}\tau_{xy} \tag{6.71}$$

The expression for the shear stress in the (x, y) plane is

$$v_x \frac{\partial \tau_{xy}}{\partial x} + v_y \frac{\partial \tau_{xy}}{\partial y} - \frac{\partial v_y}{\partial x}\tau_{xx} - \frac{\partial v_x}{\partial y}\tau_{yy} - \left(\frac{\partial v_x}{\partial x} + \frac{\partial v_y}{\partial y}\right)\tau_{xy} \tag{6.72}$$

The last (bracketed) term is taken as zero due to the incompressibility condition, and hence the expression for the convective transport of the shear stress is

$$v_x \frac{\partial \tau_{xy}}{\partial x} + v_y \frac{\partial \tau_{xy}}{\partial y} - \frac{\partial v_x}{\partial y}\tau_{xx} - \frac{\partial v_y}{\partial x}\tau_{yy} \tag{6.73}$$

Example 6.5.

Set up a model for the steady 2D flow of a Maxwell fluid.

A steady-state 2D flow of a Maxwell fluid can be characterized by the following set of equations. The equations are presented directly in dimensionless form. The starred notation usual for dimensionless groups is dropped for ease of reading.

Continuity:

$$\frac{\partial v_x}{\partial x} + \frac{\partial v_y}{\partial y} = 0$$

For the x-momentum:

$$Re\left(v_x \frac{\partial v_x}{\partial x} + v_y \frac{\partial v_x}{\partial y}\right) = -\frac{\partial p}{\partial x} + \frac{\partial}{\partial x}(\tau_{xx}) + \frac{\partial}{\partial y}(\tau_{yx})$$

For the y-momentum:

$$Re\left(v_x \frac{\partial v_y}{\partial x} + v_y \frac{\partial v_y}{\partial y}\right) = -\frac{\partial p}{\partial y} + \frac{\partial}{\partial x}(\tau_{xy}) + \frac{\partial}{\partial y}(\tau_{yy})$$

Three additional equations are needed for the stress field. Using the results for the previous example for the elastic part of the stress in the Maxwell model and adding the viscous contributions, the following equations are obtained.

The equation for the x-normal stress is

$$\tau_{xx} + We\left(v_x\frac{\partial\tau_{xx}}{\partial x} + v_y\frac{\partial\tau_{xx}}{\partial y} - 2\frac{\partial v_x}{\partial x}\tau_{xx} - 2\frac{\partial v_x}{\partial y}\tau_{xy}\right) = 2\frac{\partial v_x}{\partial x}$$

where We is the Weissenberg number defined below.

Similarly, we have the following equation for the y-normal stress:

$$\tau_{yy} + We\left(v_x\frac{\partial\tau_{yy}}{\partial x} + v_y\frac{\partial\tau_{yy}}{\partial y} - 2\frac{\partial v_y}{\partial y}\tau_{yy} - 2\frac{\partial v_y}{\partial x}\tau_{xy}\right) = 2\frac{\partial v_y}{\partial y} \tag{6.74}$$

and an equation for the shear stress in the (x, y) plane

$$\tau_{xy} + We\left(v_x\frac{\partial\tau_{xy}}{\partial x} + v_y\frac{\partial\tau_{xy}}{\partial y} - \frac{\partial v_x}{\partial y}\tau_{xx} - \frac{\partial v_y}{\partial x}\tau_{yy}\right) = \frac{\partial v_x}{\partial y} + \frac{\partial v_y}{\partial x}$$

In dimensionless form the fluid flow is then characterized by two dimensionless groups rather than one. The first is the usual Reynolds number, which represents the effect of viscosity, while the second is the Weissenberg number

$$We = \lambda v_{\mathrm{ref}}/L_{\mathrm{ref}}$$

This represents the ratio of the elastic forces to the inertial forces. A closely related dimensionless group is the Deborah number, which is defined in general as the ratio of the relaxation time to the observation time. The two are often used interchangeably in the literature, although there are some subtle differences in the definitions.

A numerical solution to the above problem, the channel flow of a Maxwell fluid, was obtained using COMSOL by Finlayson (2006); see this paper for further details. This paper also addressed some of the numerical instabilities that arise as the Weisenberg number is increased in the simulations. An earlier book by Crochet, Davies, and Walter (1984) is also a good source for numerical solutions and computational models of non-Newtonian flow.

More complex rheological models can be represented by combining the spring–dashpot arrangements in a series and parallel manner. Further details are not presented here, since the goal of this section is to provide an introduction to the basics of viscoelastic behavior, and Examples 6.4 and 6.5 give a flavor of the procedure.

Now we switch gears and look at examples where the effects of an electric field play a role in fluid motion. Again only simple examples are shown, so that the essential features can be understood. More complex problems are simply extensions of these concepts including higher dimensionality and some additional details.

6.12   A simple magnetohydrodynamic problem

Here we analyze a problem of channel flow where a magnetic field is applied. If the fluid is conducting (e.g., a molten metal) then an additional force (the Lorentz force) arises due to interaction of the electric conduction and the applied magnetic field. We wish to study the effect of this force on the velocity profile in the channel with an applied pressure gradient. The flow in the absence of the magnetic field is the classical plane Poiseuille flow, and a typical parabolic profile results for a Newtonian fluid. The same problem is analyzed here in the presence of an applied field. We examine a case where the field is applied in a direction

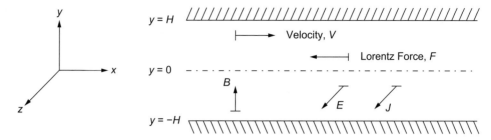

Figure 6.20 The geometry and problem analyzed for the effect of a magnetic field on flow.

perpendicular to the flow (the cross-flow direction). This is a basic problem in magneto-hydrodynamics and is known as the Hartmann flow problem. The problem analyzed is shown schematically in Fig. 6.20. Here x is the flow direction, the flow being caused by a pressure driving force G. A magnetic field of strength B is applied in the y-direction.

The x-momentum balance in the presence of additional body forces is

$$\mu \frac{d^2v}{dy^2} - \frac{\partial \mathcal{P}}{\partial x} + F_x = 0 \tag{6.75}$$

where v is used as a short form for v_x. The additional body force due to the magnetic field is denoted as F_x. An expression for this force is presented in the following section and used to compute the velocity profile.

Electric field

A moving charge in a magnetic field causes a current. This current in a stationary frame of reference is given by Ohm's law:

$$j = \sigma[E + v \times B] \tag{6.76}$$

where E is the induced electric field which is to be calculated under the requirement that the current continuity equation is satisfied. The charge conservation law is

$$\nabla \cdot j = 0 \tag{6.77}$$

The equation for the electric field is then obtained by taking the divergence on both sides of Eq. (6.76). Hence

$$\nabla \cdot E + \nabla \cdot (v \times B) = 0 \tag{6.78}$$

The electric field (a vector) can be more conveniently represented in terms of a scalar function ϕ, the electric potential:

$$E = -\nabla\phi \tag{6.79}$$

Using this definition in Eq. (6.78), the following equation is obtained for the electric potential:

$$\nabla^2\phi = \nabla \cdot (v \times B)$$

Thus the electric potential satisifies the Poisson equation. In the present case the only non-vanishing component of the electric field can be shown to be E_z, and hence we will directly solve the problem in terms of E_z rather than using the electric potential.

The Lorentz force

The additional body force is the Lorentz force, which is given as

$$F = j \times B \tag{6.80}$$

For the specific problem of duct flow analyzed here, we consider a constant applied magnetic field in the y-direction:

$$B = B_0 e_y$$

Also here v is the velocity vector, which in our case of unidirectional flow can be represented as $v = v(y)e_x$. Also E_x and E_y can be shown to be zero here, and the only non-vanishing component of electric field is E_z. All the information to perform the curl operations as per Eq. (6.76) is in our hands now. We find the following expression for j:

$$j = \sigma(E_z + v(y)B_0)e_z \tag{6.81}$$

or

$$j_z = \sigma(E_z + vB_0)$$

The current is therefore directed in the z-direction and varies as a function of y. The component to the Lorentz force is then calculated using (6.80). The only component is in the (minus) x-direction and can be shown to be equal to

$$F_x = -\sigma B_0 E_z - \sigma v B_0^2 \tag{6.82}$$

This contributes a force that opposes the flow. The first term in the above equation is sometimes referred to as the Laplace force, while the second term is referred to as the magnetic force.

The dimensionless equation for velocity

We now return to the flow equation, *viz.*, the x-momentum equation, which can be written, including the Lorentz force, as

$$\mu \frac{d^2 v}{dy^2} - \frac{\partial \mathcal{P}}{\partial x} - \sigma v B_0^2 - \sigma B_0 E_z = 0 \tag{6.83}$$

It is appropriate to cast this in dimensionless form:

$$\frac{d^2 v^*}{d\xi^2} - Ha^2 v^* + G^* - E^* = 0 \tag{6.84}$$

The various dimensionless quantities are defined as follows.

v^* is the dimensionless velocity, v/v_{ref}.

ξ is the dimensionless position, y/H, where H is half the gap width.

Ha is the Hartmann number, defined as

$$Ha^2 = \frac{\sigma B_0^2 H^2}{\mu}$$

G^* is the dimensionless pressure, defined as

$$G^* = \frac{GH^2}{v_{\text{ref}}\mu}$$

where G is the pressure drop per unit length which drives the base flow in the absence of the field effects. This is defined as earlier:

$$G = -\frac{\partial \mathcal{P}}{\partial x}$$

E^* is defined as

$$E^* = \sigma E_z B_0 \frac{H^2}{\mu v_{\text{ref}}}$$

The value of v_{ref} is chosen such that the term G^* is equal to one:

$$v_{\text{ref}} = \frac{GH^2}{\mu}$$

The term E^* is now equal to

$$E^* = \frac{\sigma E_z B_0}{G}$$

and represents the ratio of the Laplace force to the pressure force.

Hence the governing equation is

$$\frac{d^2 v^*}{d\xi^2} - Ha^2 v^* + 1 - E^* = 0 \qquad (6.85)$$

The analytical solution is

$$v^* = \frac{1 - E^*}{Ha^2}\left(1 - \frac{\cosh(Ha\xi)}{\cosh(Ha)}\right) \qquad (6.86)$$

where the no-stress condition is used at the center, $\xi = 0$, and the no-slip condition is used at the wall, $\xi = 1$.

Also the maximum velocity is given by

$$v^*_{\text{max}} = \frac{1 - E^*}{Ha^2}\left(1 - \frac{1}{\cosh(Ha)}\right)$$

On taking the ratio, the velocity profile can be expressed in a simple form as

$$\boxed{\frac{v}{v_{\text{max}}} = \frac{\cosh(Ha) - \cosh(Ha\xi)}{\cosh(Ha) - 1}} \qquad (6.87)$$

which is a classical result in this field. The result is also used as a benchmark to validate CFD-MHD numerical codes. The solution for large Ha shows a nearly plug-flow region near the center of the pipe and a thin boundary-layer type of region near the wall. This layer is called the Hartmann layer.

What is the result of the above equation as Ha tends to zero?

We have two parameters Ha and E^* that determine the maximum velocity and hence the velocity profile. The parameter Ha is known, since we know the applied magnetic field. The parameter E^* is the self-induced electric field, which is not known *a priori* and needs to be estimated by application of the total current considerations. Two limiting cases can be examined, as shown below.

For a conducting wall, the current can freely pass in the z-direction and hence there is no potential gradient induced in the system. We can set E^* as zero here. The dimensionless current is now given by

$$j^* = \frac{j_z B_0}{G} = Ha^2 v^*$$

For a perfectly insulating wall, no net current results and the current forms a circulating loop. Here an electric field is induced such that the net current across the cross-section is zero. The required mathematical condition is

$$\int_0^1 j^* \, d\xi = 0$$

In dimensionless form we have

$$E^* = -Ha^2 \int_0^1 u^* \, d\xi$$

Upon integrating we obtain the following analytical solution for the electric field for an insulating wall:

$$E^* = -\frac{Ha \coth(Ha) - 1}{Ha \coth(Ha)}$$

This can be used to find the velocity profile from Eq. (6.86). In general the velocity on the center is always less with a conducting wall than with insulating walls.

Summary

- Flow problems can be broadly classified into external flows and internal flows. An example of external flow would be the flow of air past an aircraft (with the coordinates attached to the aircraft). Such flows are of the boundary-layer type. There is a region near the solid surface where the velocity varies significantly, and there is an outer region where the viscous contribution to flow is relatively small. The thin region near the solid is called the boundary layer. The boundary-layer thickness increases with distance from the leading edge. The flow usually has 2D or 3D velocity components and is governed by the Navier–Stokes equation. The Navier–Stokes equation can be simplified in some cases, but a complete closed-form analytical solution is not possible. External flows are studied in a later chapter.

- Internal flows can be of the boundary-layer type in the entrance region, but the flow usually becomes fully developed as one moves away from the inlet. Such flows are unidirectional and have only one velocity component, which it is easy to model and solve. The Navier–Stokes equation reduces to a simple form for such cases, permitting an analytic solution for simple geometries such as channels and pipes.

- Unidirectional flows can be driven by pressure forces, shear forces, or gravity. A general solution can be obtained, and the application of suitable boundary conditions leads to solutions to specific problems. The velocity profile is a parabolic function in channel flows and pipe flows. The profile is a combination of a parabolic function and a logarithmic function for the general case of axial flow in a cylindrical geometry.

- Flows with a velocity component in the angular direction are referred to as torsional flows. An example of such a flow is flow in a fluid contained between a rotating shaft and a coaxial outer cylinder. Such flows find application, for example, in viscosity measurements, where the torque needed to rotate one cylinder can be measured. The viscosity of the fluid can be related to the torque by flow considerations, as shown in the text.

- Torsional flows find a number of other applications. For example, the flow in a tornado can be modeled as a combination of torsional flow in two regions: the inner rotating core and the outer free vortex core. The pressure distribution in the system can be predicted using the Navier–Stokes equations, and the results provide a simple but reasonable model for this complex system.

- Unidirectional flow in a non-circular channel can be described by the Poisson equation. Simple solution methods are available, although the equation itself is a partial differential equation in two variables.

- Although unidirectional flows are not applicable for tapered channels or pipes, the analysis can be extended by assuming that the unidirectional velocity profile is valid locally at a given flow location. Since mass is conserved across the cross-section, an equation for the pressure distribution can be set up and solved. This can be integrated and provides an approximate solution to this problem. This method is called the lubrication approximation. The pressure profiles in tapered channels are complex and do not follow the linear relation observed in regular (non-tapered) channels. The pressure shows a maximum at some intermediate flow location.

- Flow over a flat plate is the simplest example of external flow and is of the boundary-layer type. The boundary-layer thickness increases with the distance from the point of entry in a square-root manner. The velocity profiles (v_x and v_y) as well as local and average drag coefficients can also be computed from the boundary-layer theory. These results are needed for the calculation of heat and mass transfer from a flat plate in a later chapter. It may be noted also that after a certain distance from the entrance the flow becomes unstable, and eventually the flow becomes turbulent.

- External flow past a curved surface such as flow over a cylinder shows many complex patterns, depending on the Reynolds number. Boundary-layer separation, a wake region with vortices, vortex shedding, and turbulent wakes are commonly observed, and the flow simulation is usually done by numerical calculations.

- The Navier–Stokes equations are not applicable for non-Newtonian flows. A two-step procedure is needed to solve such flow problems. The first step is to get the stress profile using the equation of motion in the stress form. A constitutive model is then used to get an equation for the velocity profile, which can now be integrated. Examples of this procedure for unidirectional flow of common non-Newtonian fluids were discussed in the text.

- For finding the rheological behavior of a fluid we need the shear-rate vs. stress relations. The wall shear stress can be calculated from the pressure drop, but obtaining the strain rate is not so easy. The Rabinowitsch equation is useful in this context to calculate the strain rate at the wall from the measured volumetric flow-rate vs. pressure-drop data and finds applications in fitting constitutive models for non-Newtonian fluids.

- For viscoelastic fluids the calculation procedure is even more involved and requires the setting up of equations for the viscous normal stresses in addition to the shear stress. The stresses relax, and the constitutive equations are extended to include this effect. A common model is the Maxwell model here. A simple example of a channel flow gives you a good feel for the method. Even this simple flow needs six differential equations (two for velocity components, one for continuity and three equations for the stresses).

- Magnetohydrodynamics refers to the analysis of flows caused by an externally applied magnetic field. In conducting fluids, a force called the Lorentz force is created due to the interaction of the magnetic field and the fluid velocity. This is a volume type of force and is added as an extra term to the Navier–Stokes equation. For simple geometries such as channel flow and for a simple constant applied field an analytical solution to the velocity profile can be obtained, as demonstrated in the text. The additional consideration of the current distribution has to be included to complete the model. This depends on whether the channel walls are of conducting material or insulating material. The example in the text aptly illustrates the process and is the first or basic problem (the Hartmann problem) in this field.

ADDITIONAL READING

A good treatment of flow problems is the book by Wilkes (2006), who focuses in detail on chemical engineering applications.

A book devoted completely to non-Newtonian flow is that by Chabra and Richardson (2004), which provides a large number of examples of engineering applications.

A detailed study of the flow behavior of polymeric fluids is provided in the two-volume monograph by Bird, Armstrong, and Hassager (1987), who also treat the flow of viscoelastic fluids in detail.

Magnetohydrodynamic problems have been discussed in detail in the book by Davidson (2001).

Problems

1. Consider the case of channel flow where the top plate at $y = H$ is moving at a velocity of U_1 and the bottom plate at $y = 0$ is moving with a velocity of U_0. We also impose a pressure gradient in the system. Show that the velocity profile is given by

$$u = \frac{G}{2\mu}y(H - y) + U_1 + (U_2 - U_1)\frac{y}{H} \tag{6.88}$$

and that the volumetric flow rate per unit width of the channel, Q/W, is

$$Q/W = \int_0^H u\,dy = \frac{1}{2}(U_1 + U_2)H + \frac{GH^3}{12\mu} \tag{6.89}$$

2. **Flow in a triangular conduit.** Consider the flow in a conduit whose cross-section has the shape of an equilateral triangle. State the differential equations and the associated boundary conditions.

Show that the following expression satisfies the velocity profile in the x-direction (the flow direction):

$$v_x = \frac{G}{36\mu L}(2\sqrt{3}z + L)(\sqrt{3}z + 3y - L)(\sqrt{3}z - 3y - L) \tag{6.90}$$

where L is the length of each side of the equilateral triangle and the origin has been set at the centroid of the tube.

Find the flow rate for a given pressure drop. (**Answer:** $\sqrt{3}GL^4/(320\mu)$.)

3. **Flow in an ellipsoidal conduit.** Consider the flow in a conduit whose cross-section has the shape of an ellipse. State the differential equations and the associated boundary conditions. Show that the following expression satisfies the velocity profile in the x-direction (the flow direction):

$$v = \frac{G}{2\mu}\frac{a^2b^2}{a^2+b^2}\left(1 - \frac{y^2}{a^2} - \frac{z^2}{b^2}\right) \tag{6.91}$$

where a and b are the semiaxes of the ellipse corresponding to the y- and z-axes.

Find the flow rate, Q, for a given pressure drop. **Answer:**

$$Q = \frac{\pi G}{4\mu}\frac{a^3b^3}{a^2+b^2}$$

4. **Flow in a square conduit (adapted from BSL).** The following solution is presented for fully developed flow in a square duct that has a half width of B in the y- and z-directions.

$$v_x = \frac{GB^2}{4\mu}\left(1 - \frac{y^2}{B^2}\right)\left(1 - \frac{z^2}{B^2}\right)$$

The flow is in the x-direction. Is this solution correct? If not, state why.

The volumetric flow rate is presented in another paper as

$$Q = 0.563\frac{GB^4}{\mu}$$

What expression would you get if the above expression for the velocity profile were used.

5. **A model for a tornado.** A simple model for a tornado is a central core of radius R rotating at an angular velocity of Ω and an outer region. The flow is assumed to be tangential in both regions. Derive an expression for the tangential velocity and radial pressure distribution in the tornado. In particular, find an expression for the pressure at the eye of the tornado. This pressure will be found to be below atmospheric pressure and is responsible for all the damage caused by the tornado.

Hint: solve the problem as a two-region case, with an inner core with the solution domain of $0 \le r \le R$ and an outer core $R \le r \le \infty$. Also use the condition that the velocity and pressure are continuous functions of the radial position.

As a numerical example, consider a tornado with a core radius of 60 m with the maximum wind speed of 100 km/h. Find the pressure at the eye of the storm.

6. Verify the expression (6.28) for torque for flow between two cylinders with the outer cylinder rotating. What would the corresponding result be if the inner cylinder were rotating?

7. For flow in a square channel the shear stress is not a constant along the perimeter, unlike in a cylindrical channel. Obtain the stress distribution in a rectangular channel. At what point is the stress maximum?

Derive an expression for the volumetric flow rate for a square channel. Compare it with that for pipe flow for the same area of cross-section. Which geometry gives the higher flow rate? Why?

8. Show that the superposition of a radial velocity field and a torsional field results in a spiral flow.

Sketch typical streamlines. Spiral flows are good prototypes for tornadoes.

Explain why the pressure field cannot be superimposed. How would you compute the pressure field.

9. For the slider–block problem (Section 6.8.1) find the point where the pressure is a maximum and find the value of this pressure.

Calculate the tangential and normal stresses on the top plate.

Calculate the tangential stress on the bottom plate. Explain why the tangential stress on the top plate is not the same as that on the bottom plate.

10. Consider the flow of water in a pipe of length 2 m with an imposed pressure difference of 1000 Pa. Find the flow rate if (a) the pipe has a uniform cross-section of diameter 2 cm and (b) the pipe is tapered with an inlet diameter of 2 cm and an exit diameter of 3 cm. Assume the flow is laminar in both cases.

11. **Sqeezing flow.** Two parallel disks of radius R are separated by a distance H. The space between them is filled with an incompressible fluid. The top plate is moved towards the bottom at a constant velocity, causing the fluid to be squeezed out. Use the lubrication model and find an expression for the force needed to keep the squeezing motion.

12. **Film flow over a cone and over a solid sphere.** Consider the flow over a cone as shown in Fig. 6.21. Find the film thickness as a function of distance along the surface of the cone. Use the lubrication approximation. Repeat for the solid sphere in Fig. 6.21.

13. **The Reynolds equation for lubrication theory.** A more general scenario for the lubrication approximation is the problem where the fluid is contained in a gap of height h, which is assumed to be a general function of x, i.e., $h = h(x)$.

The lower surface is assumed flat and moves with a velocity v. The top surface can also move in the y-direction, and hence the gap height is changing at a rate of dh/dt. This is similar to the squeezing flow in the previous problem. The velocity profile locally is the superposition of the Poiseuille flow and the Couette flow:

$$v_x = \frac{1}{2\mu}(y^2 - hy)\frac{dp}{dx} + V\left(1 - \frac{y}{h}\right)$$

Develop an expression for the pressure distribution in the system using the lubrication approximation. For this the continuity equation is to be used in an integral sense rather than in a differential form:

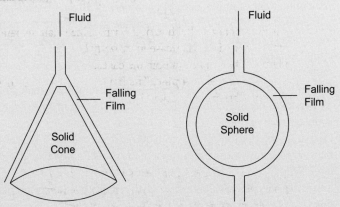

Figure 6.21 Film flow over complex shapes.

$$\int_0^{h(x)} \left[\frac{\partial v_x}{\partial x} + \frac{\partial v_y}{\partial y} \right] dy = 0$$

The final equation turns out to be

$$\frac{\partial h}{\partial t} = \frac{1}{\mu} \frac{\partial}{\partial x} \left(\frac{h^3}{12} \frac{\partial p}{\partial x} - \frac{v}{2} \frac{\partial h}{\partial x} \right)$$

Your task is to verify this. The above equation can then be used for a wide range of problems.

14. **A power-law fluid example.** A solution of 13.5% by weight of polyisoprene has the following power-law parameters: $\Lambda = 5000 \, \text{Pa} \cdot \text{s}^n$ and $n = 0.2$.
Consider the flow of such a solution in a pipe of internal diameter 1 cm and length 100 cm. Calculate and plot the volumetric flow rate as a function of the imposed pressure difference across the pipe.
If the pipe diameter is doubled, how does the volumetric flow change?

15. **The Reynolds number for a power-law fluid.** It is common to rearrange Eq. (6.57) to a friction-factor and Reynolds-number form similar to that for a Newtonian fluid. The friction factor is defined as usual by Eq. (1.22):

$$f = \frac{1}{4} \frac{p_0 - p_l}{\rho \langle v \rangle^2 / 2} \frac{d}{L}$$

If one wishes to express the results by the same formula as for Newtonain fluids,

$$f = \frac{16}{Re} \tag{6.92}$$

then how should the Reynolds number be defined for a power-law fluid?

16. **Power-law fluid: the effect of pipe diameter.** A power-law fluid has the following rheological constants: $n = 0.5$ and $\Lambda = 0.8 \, \text{Pa} \cdot \text{s}^n$.
It is pumped in a pipe of diameter 2 cm and length 5 m. The pipe is changed to one of diameter 4 cm. What is the change in flow rate if the same pressure drop is applied?
What is the pressure drop if the same flow is to be maintained?

17. **Power-law fluid: flow in parallel pipes.** A 1-m long pipe delivers a fluid with a power-law index of 0.5 at the rate of 0.02 m^3/s. The pressure drop of the pipe across the system is 3.5×10^4 Pa.
Now, if an additional parallel line of the same size is laid in parallel, what will the new flow rate be? How much is the increase in capacity?
Assume laminar flow holds in both the cases.

18. **Flow of a Casson fluid in a pipe.** The flow of blood is often described by the Casson fluid model described briefly in Section 5.7.
The model equation takes the following form for pipe flow:

$$\sqrt{-\tau} = \sqrt{\tau_0} + s\sqrt{-(dv_z/dr)} \tag{6.93}$$

since both the shear stress, τ, and the strain rate (dv_z/dr) are negative here.
Develop an equation for the velocity profile for flow of this type of fluid in a circular pipe and also an expression relating the volumetric flow rate to the pressure drop for this case.

19. **Case-study problem: flow analysis of an Ellis fluid.** The three-constant Ellis model (Eq. (5.43)) can describe a wide range of experimental data for many fluids. The paper by Matsuhisa and Bird (1965) provides a detailed analysis of this case. the various flow geometries were also analyzed in this paper, and hence this makes an interesting case study. Your goal is to reproduce the results and show some practical applications using rheological-property data for such a fluid.

20. Sketch the velocity profiles for the magnetohydrodynamic problem in Section 6.12 for the two cases of insulating and conducting walls, and compare the profiles.

7 The energy balance equation

Learning objectives

- To understand how the first law of thermodynamics can be used in the context of a moving (Lagrangian) or stationary (Eulerian) control volume to develop a general differential energy balance.
- To derive equations for the work done by the various forces acting on the control volume, in particular the work done by viscous forces, and to represent this work term in compact notation using tensor dot products.
- To derive an equation for the change in kinetic energy only for a control volume starting from the equation of motion.
- To derive the heat equation (the equation for the change in internal energy only) starting from the general energy equation and subtracting the kinetic-energy equation.
- To understand various simplifications in the heat equation, characteristic dimensionless variables, and common types of boundary conditions.
- To revisit macroscopic energy balances by volume averaging of the energy equation; to understand the formal definitions of the various terms in the macroscopic balance.
- To understand how the second law of thermodynamics can be used in the context of a moving or stationary control volume; to develop an entropy balance equation; to grasp the significance of this equation.

Mathematical prerequisites

No new mathematical prerequisites are needed for this chapter. However, the student may wish to refresh his or her knowledge of the following topics:

- tensor dot and double dot products
- index notation and the summation convention
- the Green–Gauss theorem

The objective of this chapter is to derive differential equations governing the flow of energy. The first goal is to derive the general energy balance which includes all the modes of energy. The internal and kinetic energy will appear as the important terms here.

A special equation known as the kinetic-energy equation can be obtained without the energy balance simply by starting with the equation of motion (just take the dot product). If this equation is subtracted from the general energy balance, an equation for transport of internal energy alone can be obtained. This equation can be "closed" by applying the thermodynamic relation which relates internal energy to temperature. Thus a differential equation for the temperature field can be obtained. This equation is the starting point in heat-transfer analysis.

The dimensionless form of the heat equation reveals the characteristic parameters which are important in heat transfer. We also state the various types of boundary conditions which are commonly used in heat transfer. The general dimensionless form together with these (problem-specific) boundary conditions will form the backbone for engineering analysis of heat transport problems.

A volume-averaging procedure applied to the general energy equation leads to a macroscopic model for energy transport. This model has many applications in practical engineering analysis and is derived in a systematic manner here. Some illustrative applications are demonstrated. The chapter has therefore a close parallel to the equation of motion developed in Chapter 5.

The chapter concludes with the derivation of an equation for the transport of entropy that can be used as a basis for a "rational" understanding of the second law of thermodynamics. Some examples, including the calculation of the famous Carnot principle, are shown.

7.1 Application of the first law of thermodynamics to a moving control volume

It is easier to use the substantial derivative to find the change in the energy of the system:

$$\rho \frac{D}{Dt}[\hat{u} + v^2/2] = \dot{Q} + \dot{W} + \dot{Q}_V \tag{7.1}$$

which is the restatement of the first law of thermodynamics applied to the control volume. Here $\hat{u}$ is the internal energy per unit mass and $v^2/2$ is the kinetic energy per unit mass, v being the speed or the magnitude of the velocity. The LHS therefore represents the change in the total energy content of the system, scaled per unit volume. The units of each term are $J/s \cdot m^3$ or W/m^3.

In the RHS of the above equation, $\dot{Q}$ is the heat added to the control volume from the surroundings per unit time and $\dot{W}$ is the working rate done **by** the surroundings on the system. The last term, $\dot{Q}_V$, is the energy generated within the volume due to various sources, e.g., microwave energy absorption, nuclear fission, chemical reaction, etc.

Heat flux and Fourier's law

The rate of addition of heat to the control volume can be related to the heat flux vector, which is defined as the heat leaving per unit area in the direction of maximum negative temperature gradient. The integral of this over the control surface gives the total heat lost from the control surface to the surroundings. This surface integral can be converted to a volume integral by application of the Gauss theorem. The heat added to the control volume from the surroundings is the negative of this quantity and hence the final result is

$$\dot{Q} = -\nabla \cdot \boldsymbol{q} \tag{7.2}$$

Using Fourier's law for $\boldsymbol{q}$, we have

$$\dot{Q} = \nabla \cdot (k \nabla T) \tag{7.3}$$

Now we look at the work terms.

7.2   The working rate of the forces

Note that the working rate is the dot product of the velocity and the force vector. The forces acting on the control volume are normally gravity, pressure, and viscous forces. The contributions of these forces to the work done are now calculated.

The working rate by gravity is equal to $\rho \boldsymbol{v} \cdot \boldsymbol{g}$.

This term is usually absorbed in the total energy term as the potential-energy change of the moving control volume.

Over a segment with an element of area dA, the pressure force is $-p\boldsymbol{n} \, dA$. The work done by this force on the control element is

$$\text{Pressure work rate, differential} = \boldsymbol{v} \cdot (-p\boldsymbol{n}) dA = -\boldsymbol{n} \cdot (p\boldsymbol{v}) dA$$

Integration over the control surface area and use of the Gauss divergence theorem yields that the working rate of the pressure forces per unit volume is equal to $-\nabla \cdot (p\boldsymbol{v})$.

Note that

$$\nabla \cdot (p\boldsymbol{v}) = \boldsymbol{v} \cdot \nabla p + p \nabla \cdot \boldsymbol{v} \tag{7.4}$$

which is a useful result to simplify the overall energy balance and will be needed later.

Also we can use the index notation and represent the work done by pressure forces as $(pv_i)_{,i}$. Using the chain rule, we can write this as $v_i p_{,i} + p v_{i,i}$, which is the same as the vector form given by Eq. (7.4).

The rate of working of viscous forces

Since the viscous forces are somewhat complex to represent, it is a good learning experience to derive the working rate by using the control volume in the shape of a box in Cartesian coordinates. We add the contributions of the surface forces acting on each of the six faces of the box. Later we can generalize using vector–tensor notation. This makes the understanding a little easier.

The working rate

The control volume in the form of a box and the forces acting in the x-direction are shown in Fig. 7.1. These forces are the contributions due to τ_{xx}, τ_{yx}, and τ_z.

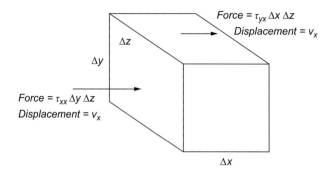

Figure 7.1 The control volume in the shape of a box to calculate the work done by viscous forces. The forces in the x-direction are shown on the x and $y + \Delta y$ faces. Note that τ_{zx} is not shown, in order to avoid clutter.

Consider the plane located at $x + \Delta x$ with the outward normal pointing in the positive-x-direction. Let us call it the $x + \Delta x$ plane. The area of the plane is $\Delta y \, \Delta z$. The stresses acting on this plane are as follows.

1. τ_{xx} in the x-direction, and the x-component of velocity is responsible for the work here. The working rate is therefore $[\tau_{xx} v_x]_{x+\Delta x} \, \Delta y \, \Delta z$.
2. τ_{xy} in the y-direction, and the y-component of velocity is responsible for the work here. The working rate is therefore $[\tau_{xy} v_y]_{x+\Delta x} \, \Delta y \, \Delta z$.
3. τ_{xz} in the z-direction, and the z-component of velocity is responsible for the work here. The working rate is therefore $[\tau_{xz} v_z]_{x+\Delta x} \, \Delta y \, \Delta z$.

Hence the work contribution to this face is

$$\text{Work rate, } x + \Delta x \text{ plane} = [\tau_{xx} v_x + \tau_{xy} v_y + \tau_{xz} v_z]_{x+\Delta x} \, \Delta y \, \Delta z$$

If one takes the corresponding face located at x (in the x-plane) this will have the normal in the $-x$ direction and hence the work done is

$$\text{Work rate, } x \text{ plane} = -\Delta y \, \Delta z [\tau_{xx} v_x + \tau_{xy} v_y + \tau_{xz} v_z]_x$$

Note the negative sign since the outward normal is in the $-x$-direction. The net contribution to the two planes (the $x + \Delta x$ plane and the x plane) is obtained by combining the two and using the Taylor expansion for the quantity at $x + \Delta x$:

$$\text{Net work rate from the two } x\text{-faces} = \Delta x \, \Delta y \, \Delta z \, \frac{\partial}{\partial x} \left(\tau_{xx} v_x + \tau_{xy} v_y + \tau_{xz} v_z \right)$$

Hence the work per unit volume is

$$\frac{\partial}{\partial x} \left(\tau_{xx} v_x + \tau_{xy} v_y + \tau_{xz} v_z \right)$$

The corresponding quantities for the $y + dy$ and y pairs and the $z + \Delta z$ and z pairs can readily be written down. Hence the rate of working of the viscous forces is

$$\dot{W} = \frac{\partial}{\partial x}\left(\tau_{xx}v_x + \tau_{xy}v_y + \tau_{xz}v_z\right)$$
$$+ \frac{\partial}{\partial y}\left(\tau_{yx}v_x + \tau_{yy}v_y + \tau_{yz}v_z\right)$$
$$+ \frac{\partial}{\partial z}\left(\tau_{zx}v_x + \tau_{zy}v_y + \tau_{zz}v_z\right) \tag{7.5}$$

A compact vector–tensor notation for the above quantity is $\nabla \cdot (\tilde{\tau} \cdot v)$.

Therefore we have the following result: the working rate of viscous forces per unit volume is equal to $\nabla \cdot (\tilde{\tau} \cdot v)$.

Index notation

In the index notation the working rate can be written as $(\tau_{ij}v_j)_{,i}$ where , in the subscript means differentiation. Note that $\tilde{\tau} \cdot v$ is the dot product of a tensor and a vector, and is therefore a vector. (The dot product reduces the dimensionality.) Taking the divergence on the resulting vector reduces the dimensionality further, reducing the final result to a scalar. Work is a scalar, and the final operation results in a scalar as required.

The work terms in (7.5) can be regrouped into two terms by using the chain rule as shown below in some detail. But it is easier to do this directly using the index notation:

$$(\tau_{ij}v_j)_{,i} = v_j(\tau_{ij})_{,i} + \tau_{ij}v_{j,i}$$

which is the final result shown below.

The viscous work term; splitting into two parts

By using the chain rule in Eq. (7.5), it can be seen that the terms with velocity in the front are of the form

$$v_x\left(\frac{\partial}{\partial x}\tau_{xx} + \frac{\partial}{\partial y}\tau_{yx} + \frac{\partial}{\partial z}\tau_{zx}\right) + v_y \text{ (etc.)} + v_z \cdots$$

A compact notation for this is $v \cdot (\nabla \cdot \tilde{\tau})$, which can be shown to be a scalar quantity. In the index notation this quantity reads as $v_j(\tau_{ij})_{,i}$.

The second group of terms in (7.5) is as follows:

$$\tau_{xx}\frac{\partial}{\partial x}v_x + \tau_{xy}\frac{\partial}{\partial y}v_y \text{ etc.}$$

which is written as $\tilde{\tau} : \nabla v$ or, in the index notation, as $\tau_{ij}v_{j,i}$.

The work term is the sum of the two terms and can be written in vector–tensor notation as

$$\text{Viscous work rate} = \nabla \cdot (\tilde{\tau} \cdot v) = v \cdot (\nabla \cdot \tilde{\tau}) + \tilde{\tau} : \nabla v \tag{7.6}$$

Working rate: summary

The contribution of the three forces to work are tabulated below.

Note that the work term is split into the sum of two terms: term 1 is the term with $v \cdot$ at the front and term 2 is the other term. Later we show that term 1 gets absorbed into the kinetic-energy change and term 2 gets absorbed into the internal-energy change.

Table 7.1. The rate of working of the forces acting on the control volume. Note that the work rate = Term 1 + Term 2 shown below.

Force	Work rate	Term 1	Term 2
Gravity	$\rho \boldsymbol{v} \cdot \boldsymbol{g}$	$\rho \boldsymbol{v} \cdot \boldsymbol{g}$	—
Pressure	$-\nabla \cdot (p\boldsymbol{v})$	$-\boldsymbol{v} \cdot \nabla p$	$-p \nabla \cdot \boldsymbol{v}$
Viscosity	$\nabla \cdot (\tilde{\tau} \cdot \boldsymbol{v})$	$\boldsymbol{v} \cdot (\nabla \cdot \tilde{\tau})$	$\tilde{\tau} : \nabla \boldsymbol{v}$

On collecting all terms the energy equation reads as

$$\rho \frac{D}{Dt}[\hat{u} + v^2/2] = -\nabla \cdot (\boldsymbol{q}) - \nabla \cdot (p\boldsymbol{v}) + \rho \boldsymbol{v} \cdot \boldsymbol{g} + \nabla \cdot [\tilde{\tau} \cdot \boldsymbol{v}] + \dot{Q}_V \qquad (7.7)$$

or, in expanded form,

$$\rho \frac{D}{Dt}[\hat{u} + v^2/2] = -\nabla \cdot \boldsymbol{q} - \boldsymbol{v} \cdot \nabla p - p \nabla \cdot \boldsymbol{v}$$
$$+ \rho \boldsymbol{v} \cdot \boldsymbol{g} + \boldsymbol{v} \cdot [\nabla \cdot \tilde{\tau}] + \tilde{\tau} : \nabla \boldsymbol{v} + \dot{Q}_V \qquad (7.8)$$

We now discuss how the various work terms get **budgeted** between the kinetic and internal energy.

7.3   Kinetic energy and internal energy equations

An equation for the rate of change of kinetic energy can be derived starting from the equation of motion, which is represented in substantial derivative notation as

$$\rho \left[\frac{D\boldsymbol{v}}{Dt} \right] = \rho \boldsymbol{g} - \nabla p + \nabla \cdot \tilde{\tau} \qquad (7.9)$$

On taking the dot product with the velocity vector, an equation for the rate of change of kinetic energy is obtained:

$$\boxed{\rho \left[\frac{D}{Dt}(v^2/2) \right] = \rho \boldsymbol{v} \cdot \boldsymbol{g} - \boldsymbol{v} \cdot \nabla p + \boldsymbol{v} \cdot [\nabla \cdot \tilde{\tau}]} \qquad (7.10)$$

The interpretation is that the kinetic energy is changed by work done by gravity and the interaction of the velocity with the pressure and viscous forces exerted on the control volume.

The total energy equation is given by (7.8). Now, by subtracting the kinetic-energy equation from the total energy equation, we obtain an expression for the rate of change of the internal energy:

$$\rho \frac{D}{Dt}[\hat{u}] = -\nabla \cdot \boldsymbol{q} - p[\nabla \cdot \boldsymbol{V}] + \tilde{\tau} : \nabla \boldsymbol{v} + \dot{Q}_V \qquad (7.11)$$

which is the equation for the change in internal energy alone. The term $\tilde{\tau} : \nabla \boldsymbol{v}$ is the viscous dissipation of heat and denoted as Φ_V for simplicity:

$$\Phi_V = \tilde{\tau} : \nabla \boldsymbol{v} \qquad (7.12)$$

Hence the internal-energy equation is written as

$$\rho \frac{D\hat{u}}{Dt} = -\nabla \cdot \boldsymbol{q} - p(\nabla \cdot \boldsymbol{v}) + \Phi_v + \dot{Q}_V \qquad (7.13)$$

The interpretation is that the internal energy is changed by heat added from the surroundings, by the contribution of the change in $\hat{u}$ due to volume compression, by generation of heat due to viscosity, and by heat generation due to internal sources.

An alternative form of the internal-energy balance is obtained by using the definition of the enthalpy. This is now presented.

7.4 The enthalpy form

It is now more convenient to use the enthalpy per unit mass, $\hat{h}$, defined as

$$\hat{h} = \hat{u} + p\hat{v} = \hat{u} + \frac{p}{\rho} \qquad (7.14)$$

where $\hat{v}$ is the specific volume, which is the reciprocal of the density.

Hence we can replace $\hat{u}$ in Eq. (7.13) by $h - p\hat{v}$, leading to the following equation for the LHS (after expanding the pressure term by application of the chain rule)

$$\rho \frac{D\hat{u}}{Dt} = \rho \frac{D\hat{h}}{Dt} - \rho\hat{v}\frac{Dp}{Dt} - p\rho\frac{D\hat{v}}{Dt} \qquad (7.15)$$

Note that $\rho\hat{v} = 1$. Also, it was shown in Chapter 3 that

$$\rho \frac{D\hat{v}}{Dt} = \frac{1}{\hat{v}} \frac{D\hat{v}}{Dt} = \nabla \cdot \boldsymbol{v} \qquad (7.16)$$

which is the dilatation (intrinsic rate of volume change) of a moving control volume.

Using these, Eq. (7.15) becomes

$$\rho \frac{D\hat{u}}{Dt} = \rho \frac{D\hat{h}}{Dt} - \frac{Dp}{Dt} - p\nabla \cdot \boldsymbol{v} \qquad (7.17)$$

Using this in (7.13), the energy balance in the enthalpy form is obtained:

$$\rho \frac{D\hat{h}}{Dt} = -\nabla \cdot \boldsymbol{q} + \frac{Dp}{Dt} + \Phi_v + \dot{Q}_V \qquad (7.18)$$

The energy balance in temperature form is obtained by relating the enthalpy to temperature, which is now addressed.

7.5 The temperature equation

The enthalpy is a function of temperature and pressure:

$$\hat{h} = \hat{h}(T, P)$$

Hence

$$d\hat{h} = \left(\frac{\partial \hat{h}}{\partial T}\right)_P dT + \left(\frac{\partial \hat{h}}{\partial P}\right)_T dP \tag{7.19}$$

From thermodynamics,

$$\left(\frac{\partial \hat{h}}{\partial T}\right)_P = c_p \tag{7.20}$$

and

$$\left(\frac{\partial \hat{h}}{\partial p}\right)_T = \hat{v} - T\left(\frac{\partial \hat{v}}{\partial T}\right)_P \tag{7.21}$$

Hence

$$\rho\frac{D\hat{h}}{Dt} = \rho c_p\frac{DT}{Dt} + \rho\left[\hat{v} - T\left(\frac{\partial \hat{v}}{\partial T}\right)_P\right]\frac{Dp}{Dt} \tag{7.22}$$

By using this in the LHS of (7.18) and doing some rearrangement, we obtain

$$\rho c_p\frac{DT}{Dt} = -\nabla \cdot \mathbf{q} + \left(\frac{\partial \ln \hat{v}}{\partial \ln T}\right)_P \frac{Dp}{Dt} + \Phi_v + \dot{Q}_V \tag{7.23}$$

Note that the term $\partial \ln \hat{v}/\partial \ln T$ at constant pressure is equal to one for ideal gases, and the energy equation can be simplified. Furthermore, the term Dp/Dt is dropped for incompressible fluids, which are under nearly constant (thermodynamic) pressure conditions, and the heat balance in terms of temperature is simplified to

$$\rho c_p\frac{DT}{Dt} = -\nabla \cdot \mathbf{q} + \Phi_v + \dot{Q}_V \tag{7.24}$$

Using Fourier's law, we obtain

$$\rho c_p\frac{DT}{Dt} = \nabla \cdot [k(T)\nabla T] + \Phi_v + \dot{Q}_V \tag{7.25}$$

which further reduces in the case of constant conductivity to

$$\rho c_p\frac{DT}{Dt} = k\nabla^2 T + \Phi_v + \dot{Q}_V \tag{7.26}$$

where ∇^2 is the Laplacian operator. This is the commonly used form of heat balance for transport problems. Note that we assume k to be a constant here, i.e., the temperature dependence of k is ignored. Otherwise the corresponding term should read $\nabla \cdot (k(T)\nabla T)$ as in Eq. (7.25).

Upon expanding D/DT we have the energy equation in temperature form in an Eulerian frame:

$$\boxed{\rho c_p\frac{\partial T}{\partial t} + \rho c_p(\mathbf{v} \cdot \nabla)T = k\nabla^2 T + \Phi_v + \dot{Q}_V} \tag{7.27}$$

The detailed form in Cartesian coordinates is obtained by expanding $\mathbf{v} \cdot \nabla$ as

$$\mathbf{v} \cdot \nabla = v_x\frac{\partial}{\partial x} + v_y\frac{\partial}{\partial y} + v_z\frac{\partial}{\partial z} \tag{7.28}$$

and then using this in the temperature equation. Hence the temperature equation can be written as

$$\rho c_p \frac{\partial T}{\partial t} + \rho c_p \left(v_x \frac{\partial T}{\partial x} + v_y \frac{\partial T}{\partial y} + v_z \frac{\partial T}{\partial z} \right) = k \nabla^2 T + \Phi_v + \dot{Q}_V \qquad (7.29)$$

in the Cartesian system. The forms in other coordinate systems are also useful for many problems.

They can be obtained by the corresponding expansion of $(v \cdot \nabla)T$ and $\nabla^2 T$ for the specified coordinate system. Students should write these out in detail for both cylindrical and spherical coordinates. See Problems 7 and 8.

7.6 Common boundary conditions

For heat transfer problems, the boundary conditions to be specified over the enclosing boundary or the perimeter of the region under consideration can be categorized into four types.

1. Dirichlet type. Here the value of the temperature on the boundary is specified. This is also known as the boundary condition of the first kind.
2. Neumann type. Here the normal gradient of the temperature is specified. If n is the direction of the outward normal at a given point in the perimeter, then values are assigned to $-k\,dT/dn$ for the Neumann case. This is the boundary condition of the second type. A special case of this is an insulated boundary; here dT/dn is set to zero.
3. Robin type. Here a rate of heat loss from the boundary is specified in accordance with Newton's law of cooling (by convective transport modeled by a heat transfer coefficient). Thus, the boundary condition takes the form

$$-k\frac{dT}{dn} = h(T - T_a) \qquad (7.30)$$

where h is the heat transfer coefficient for heat loss to the ambient whose temperature is T_a. The Robin boundary condition is more general and the particular case of the Dirichlet condition can be recovered from it as a limiting case of $h \to \infty$. The Robin condition is also known as the boundary condition of the third kind.
4. Radiation type. This type of boundary condition arises when a surface loses heat to an enclosure by radiation rather than by convection. See below for details.

The various boundary conditions and how they affect the temperature profile near a surface are shown schematically in Fig. 7.2.

Radiating surface
The boundary condition for the radiation case results in a non-linear condition. For the simple case of direct radiation the condition may be stated as

$$-k\frac{dT}{dn} = \epsilon\sigma(T^4 - T_a^4) \qquad (7.31)$$

where ϵ is the emissivity of the surface and σ is the Stefan–Boltzmann constant. T_a is the temperature of the ambient or surroundings, which emits some radiation back to the

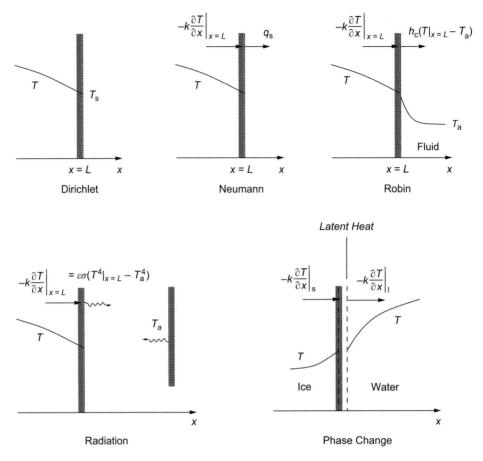

Figure 7.2 A schematic diagram showing the various types of boundary conditions.

surface. The boundary condition is of a non-linear type and the problem requires an iterative solution. In this context it is convenient to express this in a linearized form as

$$- k\frac{dT}{dn} = h_{\mathrm{R}}(T - T_{\mathrm{a}})$$

(7.32)

where h_{R} is an effective heat transfer coefficient for radiation defined by

$$h_{\mathrm{R}} = \epsilon\sigma(T + T_{\mathrm{a}})(T^2 + T_{\mathrm{a}}^2)$$

(7.33)

Since the surface temperature T is not known at the beginning of the calculation, a value of h_{R} is assumed to start off the calculation and this value is progressively updated using the calculated values of T. Note that if the surface loses heat both by radiation and by convection then the overall heat transfer coefficient for the surface is equal to $h + h_{\mathrm{R}}$.

Note that Eq. (7.31) is valid when both the surface and the surroundings see each other; for other cases a geometric parameter known as the view factor is needed. More details are provided in Chapter 18. Another small but important point is that the temperature has to be in absolute units (K) for radiation calculations.

Phase-change problems

The boundary conditions at a surface where a phase change occurs (e.g., a freezing or melting front) can be formulated by writing a heat balance at the surface. The temperature is continuous and equal to the freezing temperature. The heat flux is no longer continuous at this point. The heat flux into the solid minus that in the liquid has to be balanced by the enthalpy needed for freezing of the solid. More details of phase-change problems are given in a later chapter.

7.7   The dimensionless form of the heat equation

This is done first by choosing reference values for all the independent variables:

Reference length L_{ref},
Reference speed v_{ref},
Reference time t_{ref},
Reference temperature T_{ref}.

Note that in many cases a reference **temperature difference** is used for normalization. The choice of some of the reference parameters might not be obvious, e.g., time and even temperature in some problems and these are then left undefined initially and we let the differential equation guide us to the choice of these parameters. The rule is the "bare-bones" principle that the final equation should have as few parameters as possible.

The equation in dimensionless form is obtained from the dimensional equation (7.27), where we divide throughput by $\rho c_p v_{ref} T_{ref}/L_{ref}$. The resulting coefficients lead to various dimensionless groups. The resulting dimensionless equation is in general as follows:

$$\frac{1}{St}\frac{\partial \theta}{\partial t^*} + (v^* \cdot \nabla_*)\theta = \frac{1}{Pe}\nabla_*^2\theta + \Phi_v^* + \dot{Q}_V^* \tag{7.34}$$

Here * represents vector operation in dimensionless coordinates. The main dimensionless groups are the Péclet number, for convection

$$Pe = \frac{\rho c_p v_{ref} L_{ref}}{k} = \frac{v_{ref} L_{ref}}{\alpha}$$

for dimensionless viscous generation, an Eckert number

$$\Phi_v^* = \Phi_v k L_{ref}/(\rho c_p v_{ref} T_{ref})$$

and dimensionless heat generation,

$$\dot{Q}_V^* = \dot{Q}_V k L_{ref}/(\rho c_p v_{ref} T_{ref})$$

Thus the parametric dependence can be written as

$$\theta = \theta(x^*, t^*; Pe, \Phi_v^*, \dot{Q}_V^*)$$

The groups can be rescaled as needed, depending on the controlling mode of heat transfer. For convective heat transport problems the main parameter is the Péclet number, which is the role of convection relative to conduction. This presumes that the viscous generation of heat is small.

For problems involving purely conduction (for example, heat conduction in solids) the scaling is done differently, since there is no velocity in the system. It can be shown that the parametric representation for such systems will be of the form

$$\theta = \theta(x^*, t^*; \dot{Q}_V^*)$$

Additional dimensionless groups can come from the boundary condition. One common parameter is the Biot number in conduction heat-transfer problems. This group arises when a Robin type of boundary condition is applied at a solid–fluid interface.

Specific examples shown in Chapter 8 will make these things clear. The purpose of the above section was to show a general overview of putting the heat equation into dimensionless form.

7.8 From differential to macroscopic

It is instructive how the macroscopic balances for both internal energy and kinetic energy can be derived starting from differential balances and also to learn how the energy is partitioned between these two major forms of energy. We can start with either the kinetic-energy equation or the internal-energy equation.

Kinetic-energy changes

The kinetic-energy equation (repeated for convenience) reads

$$\boxed{\rho\left[\frac{D}{Dt}(v^2/2)\right] = \rho v \cdot g - v \cdot \nabla p + v \cdot [\nabla \cdot \tilde{\tau}]}$$

We take the steady-state version with no gravity to focus attention on the surface forces and steady state. The simplified version of (7.10) is

$$\nabla \cdot (\rho v v^2/2) = -v \cdot \nabla p + v \cdot \nabla \cdot \tilde{\tau} \qquad (7.35)$$

where we have expanded D/Dt and also moved ρv inside the ∇ operator, which is permissible in view of the continuity equation. From Eq. (7.4) we have

$$v \cdot \nabla p = \nabla \cdot (pv) - p\nabla \cdot v$$

Furthermore, using Eq. (7.6) rearranged as

$$v \cdot (\nabla \cdot \tilde{\tau}) = \nabla \cdot (\tilde{\tau} \cdot v) - \tilde{\tau} : \nabla v$$

the kinetic-energy equation above can be written as

$$\nabla \cdot (\rho v v^2/2) = -\nabla \cdot (pv) + p\nabla \cdot v + \nabla \cdot (\tilde{\tau} \cdot v) - \tilde{\tau} : \nabla v \qquad (7.36)$$

We now integrate this over a control volume V. The first term on the LHS, and the first and third terms on the RHS, i.e., all the terms where the $\nabla \cdot$ operator appears at the front, can be reduced to a surface integral. Hence we get the following macroscopic representation:

$$\int_S \rho(n \cdot v)(v^2/2)dS = -\int_S (n \cdot v)p\, dS + \int_S n \cdot (\tilde{\tau} \cdot v)dS$$

$$+ \int_V p(\nabla \cdot v)dV - \int_V (\tilde{\tau} : \nabla v)dV \qquad (7.37)$$

The following notations are used to simplify the equation:

$$\dot{W}_c = \int_V (p \nabla \cdot v) dV$$

$$\dot{W}_\mu = \int_V (\tilde{\tau} : \nabla v) dV$$

The macroscopic kinetic-energy equation is therefore

$$\int_S \rho(n \cdot v)(v^2/2) dS = - \int_S (n \cdot v) p \, dS + \int_S n \cdot \tilde{\tau} \cdot v + \dot{W}_c - \dot{W}_\mu \qquad (7.38)$$

The terms can be physically interpreted as: kinetic energy efflux from the control volume (LHS), work done by pressure forces, work done by viscous forces on the control surface S, volumetric generation due to work of compression, and loss in energy due to friction. Note that the work of compression is zero for an incompressible fluid since there is no change in volume. Also, if there are no viscous losses, the classical Bernoulli equation holds. The viscous loss term was added historically as an empirical term to account for the frictional losses, but here we have a formal way of accounting for this rather than in an ad-hoc manner. Hence this approach of going from differential to macroscopic via volume averaging gives a formal basis for the engineering Bernoulli equation.

Internal-energy changes

A similar analysis for the internal-energy equation gives

$$\int_S \rho(n \cdot v)\hat{u} \, dS = - \int_S (n \cdot q) dS - \dot{W}_c + \dot{W}_\mu + \dot{Q}_V(\text{total})$$

where the last term is the integral of the heat generation over the total control volume. The equation can be interpreted as saying that the change in internal energy is equal to the heat added from the control surface (the first term on the RHS), the change due to compression, and the gain to heat produced by viscous effects and the heat produced within the control volume. Note that $\dot{W}_c$ and $\dot{W}_\mu$ appear in both equations but with opposite signs. One man's gain is another man's loss.

7.9 Entropy balance and the second law of thermodynamics

7.9.1 Some definitions from thermodynamics

The primary property of the system is the internal-energy content of the system $\hat{u}$, which is a function of two state variables, the specific entropy $\hat{s}$ and the specific volume $\hat{v}$. Thus

$$\hat{u} = \hat{u}(\hat{s}, \hat{v}) \qquad (7.39)$$

This is a fundamental postulate of thermodynamics and known as the caloric equation of state or an equation of state for the internal energy.

The independent variables which are used here are the following: (i) the entropy and (ii) the specific volume or the density. These are referred to as the "canonical" variables or primary variables used to define an equation of state for the internal energy. Both variables are not directly measurable, and we need another two state variables. They happen to be the temperature and pressure, which we call "everyday" measurable variables. The change in internal energy due to changes in temperature and pressure can be written as

$$d\hat{u} = \left(\frac{\partial \hat{u}}{\partial \hat{s}}\right)_{\hat{v}} d\hat{s} + \left(\frac{\partial \hat{u}}{\partial \hat{v}}\right)_{\hat{s}} d\hat{v} \tag{7.40}$$

This can be written as

$$\boxed{d\hat{u} = T\,d\hat{s} - P\,d\hat{v}} \tag{7.41}$$

provided that we define the temperature and pressure as

$$T = \left(\frac{\partial \hat{u}}{\partial \hat{s}}\right)_{\hat{v}}$$

and

$$P = -\left(\frac{\partial \hat{u}}{\partial \hat{v}}\right)_{\hat{s}}$$

These definitions are strictly applicable to a static system at equilibrium. In transport phenomena we always deal with systems that are not at equilibrium. Hence, in order to use these definitions, we assume that the equations are valid locally at every point, although $\hat{u}$, $\hat{s}$, T, P, and $\hat{v}$ may be varying from point to point. Equation (7.41) is now used for a moving control volume, and hence the internal-energy changes can be related to changes in entropy and specific volume by

$$\rho\frac{D\hat{u}}{Dt} = \rho T\frac{D\hat{s}}{Dt} - P\rho\frac{D\hat{v}}{Dt} \tag{7.42}$$

This is the starting point for a microscopic entropy balance for a control volume. The second term on the RHS of the above equation can be related to the dilatation of the fluid particle, and can be shown to be equal to $-P\,\nabla \cdot \boldsymbol{v}$:

$$\rho\frac{D\hat{u}}{Dt} = \rho T\frac{D\hat{s}}{Dt} - P\,\nabla \cdot \boldsymbol{v} \tag{7.43}$$

The LHS is given by the internal-energy equation (7.13) derived in the earlier section, which is copied here for ease of following the derivations:

$$\rho\frac{D\hat{u}}{Dt} = -\nabla \cdot \boldsymbol{q} - p(\nabla \cdot \boldsymbol{v}) + (\tilde{\tau} : \nabla\boldsymbol{v}) + \dot{Q}_{\mathrm{V}} \tag{7.44}$$

On using this in (7.43), we obtain the following equation for the change in entropy of a moving fluid:

$$\rho\frac{D\hat{s}}{Dt} = -\frac{\nabla \cdot \boldsymbol{q}}{T} + \frac{\dot{Q}_{\mathrm{V}}}{T} + \frac{1}{T}(\tilde{\tau} : \nabla\boldsymbol{v}) \tag{7.45}$$

Note that we dropped the $(\nabla \cdot \boldsymbol{v})$ term, assuming an incompressible fluid for simplification to focus on the entropy.

The first term on the RHS of (7.45) is rearranged using the chain rule to

$$\frac{\nabla \cdot \boldsymbol{q}}{T} = \nabla \cdot \frac{\boldsymbol{q}}{T} + \frac{1}{T^2}\boldsymbol{q} \cdot \nabla T$$

The term $\boldsymbol{q}/T$ is designated as the entropy flux vector:

$$\boldsymbol{s} = \frac{\boldsymbol{q}}{T}$$

When this is substituted into the entropy balance we have the Jaumann relation for entropy:

$$\rho\frac{D\hat{s}}{Dt} = -\nabla\cdot\boldsymbol{s} - \frac{1}{T^2}\boldsymbol{q}\cdot\nabla T + \frac{1}{T}(\tilde{\tau}:\nabla\boldsymbol{v}) + \frac{\dot{Q}_V}{T} \tag{7.46}$$

which can be expressed as:

$$\boxed{\rho\frac{D\hat{s}}{Dt} = -\nabla\cdot\boldsymbol{s} + g_S + \frac{\dot{Q}_V}{T}} \tag{7.47}$$

where g_S is the second and third terms on the RHS of Eq. (7.46), which represents the contribution of transport fluxes of heat and momentum to entropy production:

$$g_S = -\frac{1}{T^2}\boldsymbol{q}\cdot\nabla T + \frac{1}{T}(\tilde{\tau}:\nabla\boldsymbol{v})$$

Equation (7.47) has the semblance of a "balance equation". The LHS is the change in entropy of the moving control volume. The second term can be viewed as the entropy crossing the bounding surface of the control volume. The last two terms can be viewed as the rates of production of entropy within the control volume due to the transport effects and internal heat sources. Let us examine the transport-effect term. Since $\boldsymbol{q}$ the heat flux vector has the opposite sign to ∇T vector this term is always positive. Similarly $\tilde{\tau}$ and $\nabla\boldsymbol{v}$ have the same sign and this term is always positive. We find that the rate of production of entropy due to transport is always positive, which can be viewed as a simple and general statement of the second law, namely internal generation of entropy is always positive or can be zero for a fully reversible process; it is never negative.

The implications are that the entropy production term is always positive. This is often called the second law of thermodynamics. The system entropy can increase or decrease depending on the direction of heat transfer and the magnitude of the heat source term $\dot{Q}_V$. But there is always a POSITIVE production of entropy associated with transport processes. This positive quantity can be viewed as a measure of the irreversibility of the process. A macroscopic version of the entropy balance can also be derived from the differential balance by volume averaging. The macroscopic version provides an elegant reinterpretation of some of the results from classical thermodynamics. Two examples are shown to illustrate the application of the entropy balance. The first example examines the irreversible losses in a process and the second example reiterates the Carnot efficiency of a cyclic process for work production. Both are classic examples from thermodynamics, but here we illustrate how transport-based models are used to arrive at the same results using the entropy balance.

Example 7.1. Entropy analysis of a compressor.
A compressor takes in gas at 300 K and 1 atm. The exit stream is at 500 K and 5 atm, and the system is completely insulated. Perform an entropy balance. Take C_p as 30 J/mol · K.

Is the compressor operating in a reversible mode? What is the amount of lost work?

Solution.
The macroscopic entropy balance in words is

Entropy in − entropy out + transport from boundaries + generation = 0

Here we assume that there is no transport from boundaries; say the compressor is insulated. Hence generation = entropy out minus entropy in. Therefore we need to calculate the entropy change at the exit condition. The change in entropy per mole is given by the following relation:

$$dS = C_p \frac{dT}{T} - \frac{RdP}{P}$$

or its integrated form

$$S_f - S_i = C_p \ln(T_2/T_1) - R \ln(P_2/P_1)$$

On substituting the numbers we find that the change in entropy is equal to +4.8 J/mol·K and this represents the internal entropy generation in the system. The compressor is operating in an irreversible mode. The measure of lost work is proportional to the entropy generation.

Example 7.2. Entropy analysis of an ideal engine (Carnot process).

A gas takes in Q_1 of heat at a temperature of T_1. It then does some work W on the surroundings and rejects a heat Q_2 to return to its original state. Find the efficiency of the process, i.e., the ratio W/Q_1. The system is shown schematically in Fig. 7.3.

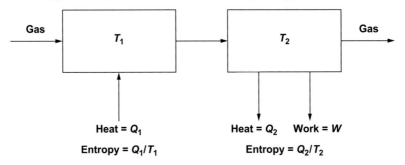

Figure 7.3 A schematic flowchart used to model an ideal engine.

Solution.
By energy balance

$$Q_1 = Q_2 + W \qquad \text{or} \qquad \frac{W}{Q_1} = 1 - \frac{Q_2}{Q_1}$$

By entropy balance

$$\text{In + generation} = \text{out}$$

or

$$\frac{Q_1}{T_1} + G_S = \frac{Q_2}{T_2}$$

where G_S is the total entropy production in the system. For an ideal case $G_S = 0$, which would be the case if the process were operated in infinitesimally small reversible steps. We cannot do better than this. Hence

$$Q_2 = Q_1 \frac{T_2}{T_1} \qquad \text{ideal case}$$

Using this in the energy equation, the work term can be calculated:

$$\frac{W}{Q_1} = 1 - \frac{Q_2}{Q_1} = 1 - \frac{T_2}{T_1}$$

This will be the maximum efficiency of an engine, which is called the Carnot efficiency. An expression for the Carnot efficiency can be derived from basic conservation laws!

Summary

- The first law of thermodynamics applied to a control volume (either Lagrangian as done in the text or Eulerian) gives the energy balance equation in the most general form. The heat flux term can be closed by applying the Fourier law of heat conduction. The working-rate term includes the contribution of gravity, the pressure force, and the viscous forces. These terms can be calculated by taking the dot product of the force and the velocity vector.

- The viscous work-rate term can be written in compact form using the compact tensor notation as $\nabla \cdot (\bar{\tau} \cdot \mathbf{v})$.

- An equation for the rate of change of kinetic energy for a moving control volume can be directly obtained starting from the Navier–Stokes (N–S) equations. All that you have to do is take the dot product of the N–S equation with the velocity vector! You can then rearrange it into a form that has the rate of change in kinetic energy. The contribution of the work terms to the kinetic energy should be noted as well.

- If the kinetic-energy equation is subtracted from the total energy equation, we get an equation for the change in internal energy alone. This equation is the starting point for heat-transfer analysis. One of the fascinating things to note about these two equations (kinetic and internal) is how the work terms gets allotted to each of these categories. It is also of interest to note that a part of the viscous work goes to change the kinetic energy while the other part gets dissipated as heat.

- Thermodynamic relations can be used in the internal-energy equation to get the energy equation in temperature form. Various forms can be derived but the most common one is that given by Eq. (7.27), which will be used in the later chapters.

- Common boundary conditions can be classified into three types: (i) Dirichlet or temperature-prescribed, (ii) Neumann or flux-prescribed, and (iii) Robin, an expression relating flux- and temperature-prescribed. The Robin condition is also known as the convection boundary condition.

- The dimensionless form of the temperature equation gives us the key dimensionless parameters of importance in heat transfer. The most widely used is the Péclet number in convective heat-transfer problems, which is equivalent to the Reynolds number for flow problems.

- Macroscopic energy balances can be obtained by volume averaging of the differential energy balance. This formal procedure gives us the same equations as we used in Chapter 2, but gives us a firmer definition and meaning for the various terms. It also provides a rationale to the engineering Bernoulli equation introduced there in an empirical manner.

- The entropy balance is an important equation in the subject of irreversible thermodynamics, and provides an illuminating new understanding of the second law of thermodynamics. The macroscopic version of the entropy balance is a useful starting point in the exergy analysis (tracking the lost work). Although the entropy balance is not central to our further study of heat transport, the conceptual understanding of this topic is essential.

Problems

1. Write out in detail all the terms for $\tilde{\tau} : \nabla v$ in rectangular Cartesian coordinates.

   Now assume a Newtonian fluid; use the generalized version of Newton's law of viscosity for the stress tensor $\tilde{\tau}$ and expand $\tilde{\tau} : \nabla v$. Thus derive the expression for the viscous dissipation rate in Cartesian coordinates.

   How does it simplify for a simple-shear-driven flow between two rectangular channels?

   How does it simplify for a pressure-driven flow between two parallel channels?

2. Verify that the viscous dissipation term is always positive, indicating that this is an **irreversible** conversion into internal energy.

3. Comment and elaborate on the following statement from the BSL book: *for viscoelastic fluids the term $\tilde{\tau} : \nabla v$ does not have to be positive since some energy may be stored as elastic energy.*

4. For fully developed flow in a pipe the contributions to viscous dissipation are from τ_{rz} and dv_z/dr. What is the form of the viscous generation term for a fully developed laminar flow of a Newtonian fluid in a pipe? How does it vary with radial position?

   Calculate this for (a) water and (b) crude oil flowing in a pipe of diameter 2 cm at a velocity of 5 cm/s.

5. Verify the following thermodynamic relation:

$$\left(\frac{\partial \hat{H}}{\partial P} \right)_T = \hat{V} - T \left(\frac{\partial \hat{V}}{\partial T} \right)_P \tag{7.48}$$

6. How does the temperature equation simplify if there is no flow? Write this out in detail in all of the three coordinate systems.

7. Write in detail the expression for $(v \cdot \nabla)T$ in cylindrical coordinates. Also write in detail the expression for the Laplacian, and thereby complete the temperature equation for cylindrical coordinates. What simplifications result for an axisymmetric case?

8. Write in detail the expression for $(v \cdot \nabla)T$ in spherical coordinates. Also write in detail the expression for the Laplacian, and thereby complete the temperature equation for spherical coordinates. What simplifications result for an axisymmetric case, i.e., when the temperature has no dependence on the ϕ direction?

9. A turkey is being roasted in a microwave oven. How would you calculate the internal heat generation term, $\dot{Q}_V$?

8 Illustrative heat transport problems

Learning objectives

The key learning objectives of this chapter are the following:

- temperature profiles for steady-state conduction in the slab, cylinder, and sphere;
- solution of problems with variable conductivity;
- appreciation of solution methods for multidimensional heat conduction;
- solution of problems involving heat generation with constant and linear heat sources;
- a solution method (p-method) for problems with non-linear heat sources;
- solution of problems with convection in the same direction as heat flow; the correction factor for heat transfer rate;
- a brief summary of equations for heat transfer for flow past a flat plate;
- area averaging and formulation of mesoscopic models; and
- volume averaging and formulation of a lumped model in heat transfer.

Mathematical prerequisites

No new mathematical tools are introduced in this chapter. Solution of second-order differential equations should be revisited.

Summary of equations

The starting point is the temperature equation derived in Chapter 7. This is reproduced here for ease of reference:

$$\rho c_p \frac{\partial T}{\partial t} + \rho c_p (\boldsymbol{v} \cdot \nabla)T = \nabla \cdot (k(T)\nabla T) + \Phi_{\mathrm{v}} + \dot{Q}_{\mathrm{V}} \qquad (8.1)$$

Various simplified versions of this equation are studied in this chapter.

First we examine steady-state conduction in solids with no generation of heat. Cases with constant conductivity are studied first, followed by problems with variable conductivity.

This chapter deals with the solution for the temperature profile and the heat flux calculations of some common problems in heat transfer. In many respects the chapter is similar to Chapter 6 on flow problems. This chapter focuses mainly on steady-state problems with and without internal heat generation. Transient heat flow problems with purely heat conduction as the mode of transport and general treatment of problems in convective transport are deferred to later chapters. A brief discussion of the type of problems we study here is given below.

First we examine steady-state problems with constant and variable thermal conductivity. We present analytical solutions for simple geometries such as the slab, long cylinder, and sphere. We will derive expressions for the heat flux and the rate of heat transfer and also show applications via some simple examples. The extension for multidimensional heat conduction will then be studied together with some simple computational tools.

The next type of problem examined is heat conduction with generation. The problem is classified into a number of types and analytical or semi-analytical solutions will be derived.

Convection in the direction of heat transfer is examined next. We will show that the convection can enhance or retard the heat transfer due to pure conduction and derive the factor to account for these effects. Heat transfer for flow past a flat plate represents a different type of convective problem. The direction of heat transfer is transverse to the flow direction here. Such problems are of a boundary-layer type and treated in more detail in Chapter 19. In this chapter we provide an overview and summarize the key results.

The area- and volume-averaging methods will then be applied to the heat equation to derive the mesoscopic and macroscopic balances, respectively. Illustrative applications of these problems will be shown.

Some problems of importance in steady-state heat transfer are thus examined in this chapter, enabling the student to acquire a good grasp of setting up and solving such problems.

8.1 Steady heat conduction and no generation

8.1.1 Constant conductivity

The governing equation (8.1) simplifies to the Laplace equation for the temperature field:

$$\nabla^2 T = 0$$

together with the associated boundary conditions on the surface of the solids. Boundary conditions may be of Dirichlet, Neumann, or Robin type as discussed in Section 7.6. The Laplace equation is a widely studied problem in physics, and appears in many fields such as heat transfer, potential flow, and electrostatics. The equations, being linear, are amenable to a number of powerful analytical or semi-analytical methods such as boundary integral methods, singularity methods, etc. These techniques are especially useful for solution of problems in general 3D cases. Here we present solutions to some simple heat conduction problems that are posed in terms of a single spatial variable. We consider three geometries, namely the slab, long cylinder, and sphere. These are well studied and classic problems in heat conduction and are also widely used in engineering analysis. The general solutions for the three geometries are presented below.

Slab geometry

The Laplacian takes a simple form here and is equal to d^2T/dx^2. Upon integration, the temperature profile can be shown to be a linear function of distance:

$$T = C_1 x + C_2 \tag{8.2}$$

where C_1 and C_2 are two integration constants. These constants can be determined if the boundary conditions are specified.

The corresponding heat flux q_x is given as $-kC_1$.

Cylindrical geometry

The Laplacian is given by

$$\nabla^2 T = \frac{1}{r} \frac{d}{dr} \left(r \frac{dT}{dr} \right)$$

where we assume that the conduction is mainly in the r-direction. This will hold, for example, for a long cylinder. Upon integration we find the general solution to the temperature. The profile is a logarithmic function of the radial position:

$$T = C_1 \ln r + C_2 \tag{8.3}$$

The corresponding heat flux q_r is given as $-kC_1/r$ and is an inverse function of radial position.

The heat crossing a cylindrical surface of length L at any location r, Q_r is given as $q_r(2\pi r)L$. This in turn is equal to $-kC_1(2\pi L)$ and is constant and is independent of radial position.

Spherical geometry

The Laplacian is given by

$$\nabla^2 T = \frac{1}{r^2} \frac{d}{dr} \left(r^2 \frac{dT}{dr} \right)$$

where r is the radial position in spherical coordinates.

Upon integration the following general solution for the temperature profile is obtained:

$$T = -\frac{C_1}{r} + C_2 \tag{8.4}$$

The corresponding heat flux q_r is given as kC_1/r^2 and is an inverse-square function of radial position.

The heat crossing a spherical surface at any location r, Q_r, is given as $q_r(4\pi r^2)$. This in turn is equal to $4\pi kC_1$ and is constant and independent of radial position.

Example 8.1. Heat transfer in a hollow cylinder.

The inner surface of radius R_i is maintained at T_i while the outer surface of radius R_o is maintained at a temperature of T_o. The temperature profile can be found by fitting these boundary conditions to the general solution:

$$T(r) = T_i - \frac{T_i - T_o}{\ln(R_o/R_i)} \ln\left(\frac{r}{R_i}\right)$$

The rate of heat flow across a hollow cylindrical shell of length L is given by

$$Q_r = \frac{2\pi kL}{\ln(R_o/R_i)}(T_i - T_o)$$

The rate of heat flow can be written in terms of a resistance in a manner analogous to Ohm's law. Note that the heat flow is the temperature difference divided by the resistance. Hence

$$\text{Resistance} = \frac{\ln(R_o/R_i)}{2\pi kL}$$

Example 8.2. Heat loss from a surface of a stagnant sphere.

Consider a sphere exposed to air. Assume that the air is stagnant and the heat transport from the surface to air is by conduction only. Find an expression for the heat loss from the sphere.

The boundary conditions are (i) at $r = R$, $T = T_s$, the surface temperature; and (ii) as $r \to \infty$, $T = T_\infty$, the air temperature. These conditions are used to fit the constants in the general solution. Hence the solution for the temperature profile is

$$\frac{T - T_\infty}{T_s - T_\infty} = \frac{R}{r} \tag{8.5}$$

The heat loss from the sphere is $Q_r = 4\pi kR(T_s - T_\infty)$.
If one defines a heat transfer coefficient as $Q_r = 4\pi R^2 h(T_s - T_\infty)$ then

$$h = k/R = 2k/d \tag{8.6}$$

The dimensionless version of this equation is

$$Nu = \frac{hd}{k} = 2 \tag{8.7}$$

where Nu is a dimensionless heat transfer coefficient, the Nusselt number. This sets the base value of the heat transfer coefficient and the Nusselt number for transport from a sphere to the surrounding fluid under stagnant conditions. In the presence of fluid motion an additional term is added. This term depends on the Reynolds number for flow.

The expression

$$Nu = \frac{hd}{k} = 2 + 0.2Re^{1/2}Pr^{1/3} \tag{8.8}$$

is a widely used empirical correlation for heat transfer to or from a sphere in a flowing fluid. For a slow flow under laminar conditions (Stokes flow) the following equation was derived by Acrivos and Taylor (1962) using matched asymptotic expansions:

$$Nu = 2\left(1 + \frac{1}{2}Pe + \frac{1}{2}Pe^2 \ln Pe + 0.414\,65Pe^2 + \frac{1}{4}Pe^3 \ln Pe\right)$$

The error in this expansion is of the order of (Pe^3). Note that the Péclet number Pe defined as $Pe = RePr$ is the key dimensionless group which appears in this model. This is a classic result in heat transfer under slow-flow conditions.

8.1.2   Variable thermal conductivity

Here we examine a case where k is some specified function of temperature. A slab-geometry case is analyzed; the other cases (cylinder and sphere) are left as problems.

The expression for the heat flux
For a slab the heat flux is constant and hence

$$q_x = -k(T)\frac{dT}{dx} \tag{8.9}$$

This equation can be separated and integrated across the slab from 0 to L since q_x can be treated as a constant:

$$\int_0^L q_x\,dx = q_x L = -\int_{T_0}^{T_L} k(T)dT \tag{8.10}$$

The RHS can be expressed as

$$\int_{T_0}^{T_L} k(T)dT = k_m(T_L - T_0)$$

by the use of the mean-value theorem of calculus, where k_m is a suitable mean value for the conductivity. Hence

$$q_x = k_m\frac{T_0 - T_L}{L} \tag{8.11}$$

where k_m is the mean value of thermal conductivity defined as

$$k_m = \frac{1}{T_L - T_0}\int_{T_0}^{T_L} k(T)dT$$

If now one knows the functional dependence of k on T the mean value can be evaluated and Eq. (8.11) can in turn be used to calculate the fluxes. Now, if k is a linear function of temperature the mean value is the arithmetic average of k at the two end-point temperatures. For other variations the mean value can be evaluated if the functional form of k vs. T is known.

Temperature profiles

To find the temperature distribution one needs to do a second integration, this time from 0 to any arbitrary x:

$$\int_0^x q_x \, dx = q_x x = - \int_T^{T_0} k(T) dT \tag{8.12}$$

Linear variation of k

A linear variation in k is often used. This can be expressed as

$$k(T) = k_0[1 + \beta(T - T_0)]$$

Here k_0 is the conductivity at a base or reference temperature T_0 and β is the coefficient for the variation of thermal conductivity with temperature. The β parameter can be either positive or negative, depending on the solid under consideration. The temperature profile for this case is given by the following non-linear implicit equation as a function of x:

$$x q_x = k_0(1 - \beta T_0)(T_0 - T) + \frac{k_0 \beta}{2}(T_0^2 - T^2)$$

where q_x is a constant that can be calculated from Eq. (8.11) as shown earlier.

An illustrative plot of the temperature profile for a linear variation of k is presented in Fig. 8.1. A linear profile is established for the base case of constant conductivity.

8.1.3   Two-dimensional heat conduction problems

In this section a brief discussion of 2D heat conduction problems and the solution to one prototypical problem are presented.

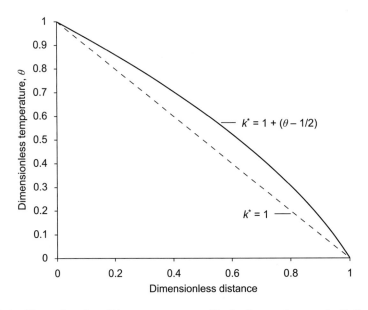

Figure 8.1 An illustrative plot of the temperature profile for the varying-conductivity problem for slab geometry.

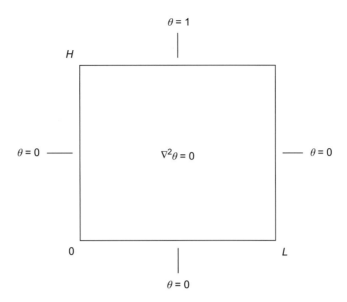

Figure 8.2 An illustration of 2D conduction in a square geometry.

The 2D problem is governed by the Laplace equation

$$\frac{\partial^2 T}{\partial x^2} + \frac{\partial^2 T}{\partial y^2} = 0$$

with appropriate boundary conditions on the perimeter.

As an example of a 2D application, consider the problem depicted in Fig. 8.2. Here the top edge $y = 1$ is maintained at T_s while the other edges are maintained at zero temperature. The problem can be solved by separation of variables. Here we simply present the solution and examine the nature of the solution. The solution is

$$T(x,y) = \sum_{n=1}^{\infty} C_n \sin(n\pi x)\sinh(n\pi y) \qquad (8.13)$$

where the series coefficients C_n can be shown to be equal to

$$C_n = \frac{2T_s}{\sinh(n\pi)} \left[\frac{1 - \cos(n\pi)}{n\pi} \right] \qquad (8.14)$$

The even values of the series coefficients can be shown to be zero and hence the summation is applied only for $n = 1, 3, 5, \ldots$. The calculated values of the first three series coefficients are $C_1 = 22.0498$, $C_3 = 0.01357$, and $C_5 = 1.53 \times 10^{-5}$. Note that the coefficient decreases rapidly as n increases. With these values in hand the temperature profiles can be generated. A contour plot for the case can easily be generated in MATLAB, and it can be shown that the corner point represents a singularity. The temperature is not uniquely defined at this point. These types of singularities are often present in 2D and 3D problems. The singularities can cause problems in numerical solutions since the grid near the singularity may not be sufficiently small to capture the steep changes in temperature. Care has to be exercised in interpretation of such results, especially for the values of the computed heat flux. The center temperature is found to be $T_s/4$, which is the mean value of the boundary temperatures. This property of elliptic equations is called the mean-value property.

The heat flux at the top surface can be calculated by applying Fourier's law at $y = 1$:

$$q_y(x, y = 1) = -2kT_s \sum_{n=1,3,\ldots}^{\infty} \sin(n\pi x)\coth(n\pi) \tag{8.15}$$

Again the heat flux is not defined at the corner singularity point.

The next class of problems studied is heat conduction with internal generation of heat.

8.2   Heat conduction with generation: the Poisson equation

The governing equation can be written as

$$k \nabla^2 T = -\dot{Q}_V$$

where the Laplacian depends on the geometry under consideration. This type of equation with a source term on the RHS is called the Poisson equation.

8.2.1   The constant-generation case

For the case of constant generation, the equations can be directly integrated. The solution to a cylindrical geometry is presented here. The other cases follow a similar solution procedure.

Solution for a long solid cylinder
Using the Laplacian for a long cylinder, the governing differential equation is

$$k\frac{1}{r}\frac{d}{dr}\left(r\frac{dT}{dr}\right) - \dot{Q}_V \tag{8.16}$$

It may be noted that this equation was derived by using a control volume or shell balance in Section 2.6.1. Here we show how the equation arises directly from simplification of the governing differential equation. Additional solution details are also presented in this section.

The boundary conditions are as follows.

At the center, $r = 0$, we use a symmetry condition $dT/dr = 0$. Alternatively we can state that the temperature must be finite at this point.

At the surface we can use the Robin condition in general. At $r = R$, the heat coming from the solid, $-k\, dT/dr$ must match the heat lost by convection, which is $h(T - T_a)$. Here T_a is the temperature of the ambient or the surroundings.

It is more convenient to solve the problem in dimensionless form. The governing equation in dimensionless form is

$$\frac{1}{\xi}\frac{d}{d\xi}\left(\xi\frac{d\theta}{d\xi}\right) = -1 \tag{8.17}$$

Here we use a dimensionless temperature θ, which is defined as

$$\theta = \frac{T - T_a}{\dot{Q}_V R^2 / k}$$

Thus $\dot{Q}_V R^2 / k$ serves as the reference temperature variable for the problem. You should verify that this quantity has the same dimension as temperature. Also note that the differential equation now takes a simple form with -1 on the RHS. The generation term has become equal to one with the above choice of the reference temperature!

The Robin boundary condition at the surface ($\xi = 1$) in dimensionless form is

$$\left(\frac{d\theta}{d\xi}\right)_1 = -Bi(\theta)_1$$

where Bi is defined as

$$Bi = \frac{hR}{k}$$

The solution is obtained as

$$\theta = \frac{1}{4}(1 - \xi^2) + \frac{1}{2Bi} \tag{8.18}$$

We note that the center temperature is equal to $1/4 + 1/(2Bi)$, while the surface temperature is equal to $1/(2Bi)$.

Position-dependent generation

For position-dependent generation, the RHS of Eq. (8.17) is some specified function of ξ, and the equation can be integrated twice to get the temperature profiles.

The case of temperature-dependent generation is more challenging and is examined next.

8.3   Conduction with temperature-dependent generation

. . . .

The temperature-dependent generation can be of two types: (i) linear and (ii) non-linear.

8.3.1   Linear variation with temperature

Assume that the source is a linear term represented as

$$\dot{Q}_V = \dot{Q}_{Vs}[1 + B(T - T_s)] \tag{8.19}$$

The parameter $\dot{Q}_{Vs}$ can be interpreted as the value of the heat generation at a base temperature T_s, while B is the coefficient of linear variation of heat generation with temperature.

The governing equation for slab geometry is

$$k\frac{d^2 T}{dx^2} = -\dot{Q}_{Vs}[1 + B(T - T_s)] \tag{8.20}$$

The case of a constant value of k is considered in the following analysis. The problem is then a linear second-order differential equation. We illustrate the solution for a Dirichlet condition at the surface. The slab thickness is taken as $2L$, which maintains the symmetry at $x = 0$, and a constant temperature of T_s is assumed at $x = \pm L$.

A dimensionless temperature is defined in general as

$$\theta = \frac{T - T_{base}}{T_{ref}}$$

The base temperature T_{base} can be taken as T_s. This makes the surface temperature equal to zero in dimensionless form. There is no particular reference temperature to choose since the internal temperature can vary widely. Hence we choose T_{ref} as $\dot{Q}_{Vs}L^2/k$ to reduce the differential equation to the simplest possible form.

The dimensionless temperature θ is therefore defined as

$$\theta = \frac{T - T_s}{\dot{Q}_{Vs}L^2/k}$$

The dimensionless distance ξ is chosen as x/L.

The governing equation in dimensionless form is

$$\frac{d^2\theta}{d\xi^2} = -[1 + \beta\theta] \tag{8.21}$$

The parameter β turns out to be equal to BT_{ref}. The differential equation (8.21) can be rearranged as

$$\frac{d^2\theta}{d\xi^2} + \beta\theta = -1 \tag{8.22}$$

and the solution can be written as the sum of the general solution to the homogeneous part (LHS) and a particular solution corresponding to the RHS:

$$\theta = C_1 \cos(\sqrt{\beta}\xi) + C_2 \sin(\sqrt{\beta}\xi) - 1/\beta$$

The solution satisfying the specified boundary condition is given by

$$\theta = \frac{1}{\beta}\left[\frac{\cos(\xi\sqrt{\beta})}{\cos(\sqrt{\beta})} - 1\right] \tag{8.23}$$

What happens in the limit $\beta \to 0$ to the above expression?

The heat flux at the surface $x = L$ can be calculated from Fourier's law applied at the surface:

$$q_x = -k\left(\frac{dT}{dx}\right)_{x=1} = (\dot{Q}_{Vs}L)\frac{\tan(\sqrt{\beta})}{\sqrt{\beta}}$$

The term $\dot{Q}_{Vs}L$ is the heat flux for a constant heat generation rate ($\beta = 0$), and hence the second term on the RHS can be viewed as the enhancement in the heat flux due to the linear generation. This formula should be noted since a similar formula will appear in mass transfer with chemical reaction and other related problems.

The solution has several interesting properties.

The temperature is maximum at the center, as expected, and the value of the maximum temperature is given by

$$\theta_{max} = \frac{1}{\beta}\left[\frac{1}{\cos(\sqrt{\beta})} - 1\right]$$

However, if the parameter $\sqrt{\beta}$ is equal to $\pi/2$ the value of the maximum temperature goes to infinity. The physical interpretation is that the rate of generation is now much larger than the rate at which the heat can be conducted out of the material. The result is that the system is not able to reach a steady state where the two rates can be balanced and the temperature tends to undergo "runaway". The prediction of runaway conditions is very important in a number of contexts, including the field of reactor design. This example provides a simple but illustrative demonstration of the phenomenon of runaway.

8.3.2   Non-linear variation with temperature

An exponential generation is often applicable, and this can be represented as

$$\dot{Q}_V = A \exp[-E/(R_G T)](-\Delta H) \tag{8.24}$$

which is similar to the Arrhenius equation. This is based on a zeroth-order reaction in the solid. The additional parameters are the heat of reaction (ΔH) and the activation energy E.

The model equation is then

$$k\frac{d^2 T}{dx^2} = -A \exp[-E/(R_G T)](-\Delta H)$$

The exponential term can be approximated as a linear term over a modest range of temperature, and hence approximate analytical solutions can be obtained.

A simple strategy to approximate an exponential function is shown in Fig. 8.3. One needs to estimate or guess the lower, θ_L, and the upper, θ_U, temperature values in order to construct this approximate linear model. (Note that this can be done in an iterative manner.) With the two profiles, one can also obtain an upper and lower bound on the generation rate and hence upper and lower bounds on the temperature as well. The analytical solutions for the linear case can then be directly used to bound the explosion limits. However, in

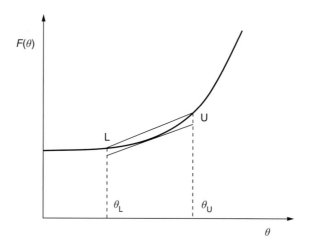

Figure 8.3 Two ways of approximating an exponential function by a linear function. (i) A linear profile with the expected end-point temperatures is constructed with a line joining the lower (L) and the upper (U) temperatures in the system. (ii) A linear profile with the slope evaluated at an intermediate temperature is constructed.

general and for more accurate results, a numerical solution is useful. The model problem is a boundary-value type of problem, and mathematical techniques are available to solve the problem. One method involves using the BVP4C solver in MATLAB, which is discussed in detail in a later chapter. A semi-analytical method of solution is the p-substitution method, which is also applicable to many related problems in mass transfer. Hence this method is presented as an example below.

Example 8.3.

Frank-Kamenetskii modeled the explosion of a solid as a zeroth-order chemical reaction with a high activation energy. The following differential equation is applicable for this process:

$$\frac{d^2\theta}{d\xi^2} = -\delta \exp(\theta)$$

with the boundary conditions of no flux at $\xi = 0$, the plane of symmetry, and the surface temperature of zero, i.e., $\theta(\xi = 1) = 0$. Note that all the variables are dimensionless here. The parameter δ is a measure of the heat generated by the chemical reaction.

Although the problem is non-linear, implicit analytical solutions can be obtained. For this purpose, we use the substitution $d\theta/d\xi = p$. Then

$$\frac{d^2\theta}{d\xi^2} = \frac{dp}{d\xi} = \frac{dp}{d\theta}\frac{d\theta}{d\xi} = p\frac{dp}{d\theta}$$

Note that the variable ξ has been eliminated and the differential equation can now be expressed as

$$p\frac{dp}{d\theta} = -\delta \exp(\theta)$$

which integrates to

$$\frac{p^2}{2} = -\delta \exp(\theta) + C$$

Here we use the boundary condition implicitly at $\xi = 0$. We take p, the gradient, as zero as needed, but leave the center temperature as an unknown parameter denoted as θ_c. Hence $C = \delta \exp(\theta_c)$. By using this and taking the square root, we get the following result for p:

$$p = \frac{d\theta}{d\xi} = -\sqrt{2\delta[\exp(\theta_c) - \exp(\theta)]}$$

where the negative sign rather than the positive sign has been kept for the square root since the expected temperature profile has a negative slope.

The above equation can be integrated a second time to obtain the following implicit integral equation for the center temperature. Note that we do a definite integration with the limits from 0 to 1 for ξ and θ_c to zero for θ:

$$\sqrt{2\delta} = \int_0^{\theta_c} \frac{du}{\sqrt{\exp(\theta_c) - \exp(u)}} \tag{8.25}$$

The integral can be evaluated analytically, leading to the following implicit expression for the center temperature:

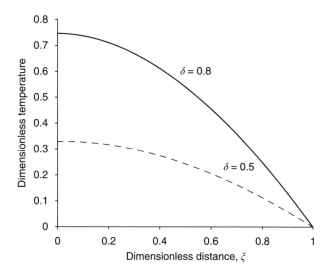

Figure 8.4 A plot of the temperature profile for the Frank-Kamenetskii problem as a function of the parameter δ.

$$\sqrt{2\delta} = \exp(-\theta_c/2)\cosh^{-1}[2\exp(\theta_c/2) - 1]$$

A non-linear root finder can be used to find θ_c, the center temperature. However, it is easier to evaluate a value for δ assuming a center-temperature value, and obtain the solution for a range of θ_c values as a parametric plot of δ vs. θ_c. Detailed temperature profiles can also be calculated using this value of θ_c and now doing a general integration of Eq. (8.25) from 0 to ξ rather than from 0 to 1. An illustrative plot of the calculated temperature profile is shown in Fig. 8.4. There exists a critical value of β above which no solution is possible, which is explored further in Chapter 16. This limit is the explosion limit and the behavior is similar to the linear heat-generation problem examined earlier.

8.3.3   Two-dimensional Poisson problems

Heat conduction with generation in a 2D domain is governed by the Poisson equation. Consider a square slab with a constant generation. The governing equation is

$$\frac{\partial^2 T}{\partial x^2} + \frac{\partial^2 T}{\partial y^2} = -\dot{Q}_V/k$$

Let us consider a case of constant surface temperature and a constant generation which makes the problem linear. Then the boundary conditions are set as zero temperature along the sides:

$$T(x = \pm L, y = \pm L) = 0$$

You should appreciate the analogy with the flow in a square duct. The solution can be represented as a sum of a particular solution and a homogeneous part. The latter can be solved by separation of variables.

The details are not shown here since the analogy with momentum transfer holds. The final result for the temperature distribution is

$$\frac{T(X, Y)}{\dot{Q}_V L^2/k} = \frac{1}{2}(1 - X^2) - 8 \sum_{n=0}^{\infty} \frac{(-1)^n}{\lambda_n^3} \frac{\cosh(\lambda_n Y)}{\cosh(\lambda_n)} \cos(\lambda_n X)$$

where $\lambda_n = (n+1/2)\pi$, i.e., odd multiples of $\pi/2$. The variables X and Y are dimensionless coordinates defined as x/L and y/L, respectively.

Other methods are the boundary-element method (Ramachandran, 1993), the method of boundary collocation, and, of course, numerical solution based on finite-difference approximation to the operators. The problems address some aspects of such computational methods.

8.4 Convection effects

8.4.1 Transpiration cooling

Consider a porous solid of thickness L and at temperatures of T_0 and T_L at the ends (see Section 8.5). The effect of blowing a gas across the system is to be analyzed. This problem is also known as the transpiration cooling problem.

Using Fourier's law, the heat transfer rate for no transpiration is

$$q_x^0 = k(T_0 - T_L)/L$$

The superscript is used to show that this is a base value in the absence of any flow across the system. If now a gas is blown across the system at a velocity of v_0 in the x-direction we need to find the heat transfer rate in the presence of convection. The problem is shown schematically in Fig. 8.5.

The governing equation should now include a convection term, which simplifies, for a constant velocity, to

$$\rho c_p v \cdot \nabla T = \rho c_p v_0 \frac{dT}{dx}$$

Hence the governing equation is

$$\rho c_p v_0 \frac{dT}{dx} = k \frac{d^2 T}{dx^2} \tag{8.26}$$

It is once again convenient to represent the problem in terms of dimensionless groups. Let $\theta = (T - T_L)/(T_0 - T_L)$ be the dimensionless temperature, and let $\xi = x/L$ be the dimensionless distance as commonly defined in this book. The temperature θ takes the values of 1 and 0 at the ends, with ξ equal to 0 and 1, respectively.

The dimensionless form of the model is then

$$\frac{d^2\theta}{d\xi^2} - Pe\frac{d\theta}{d\xi} = 0 \tag{8.27}$$

where Pe is a dimensionless group defined as

$$Pe = \frac{\rho c_p v_0 L}{k} = \frac{v_0 L}{\alpha}$$

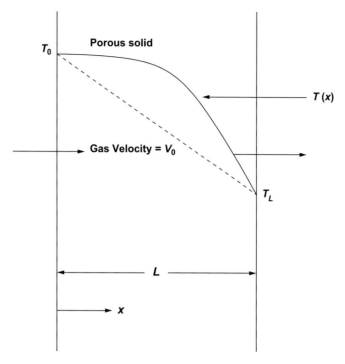

Figure 8.5 A schematic diagram for the problem of convection analyzed here.

This group is called the Péclet number and represents the ratio of the rate of heat transport by convection to that by conduction. The parameter α is the thermal diffusivity defined as $k/(\rho c_p)$. The solution is obtained from the roots of the characteristic polynomial equation $D^2 - PeD = 0$. The roots are 0 and Pe. Hence the general solution is

$$\theta = A + B\exp(Pe\xi)$$

where the constants of integration are found from the boundary condition. The final solution can be derived as

$$\theta = \frac{\exp(Pe\xi) - \exp(Pe)}{1 - \exp(Pe)}$$

The temperature profiles are plotted in Fig. 8.6.

Heat flux in the presence of convection

The effect of the flow on the rate of heat transfer is now examined.

The total heat flux in the presence of convection is defined as

$$q_x = -k\frac{dT}{dx} + \rho c_p v_0 T \tag{8.28}$$

The first term is due to conduction in accordance with Fourier's law and the second term is the additional heat transport due to convection. The reference temperature for the calculation of enthalpy can be taken as zero without the loss of generality.

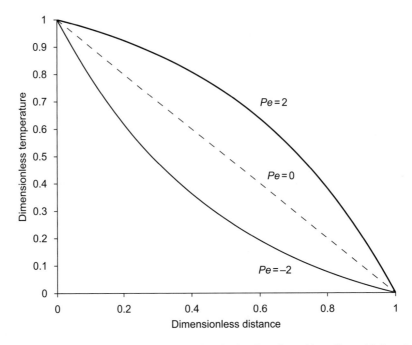

Figure 8.6 The temperature profile for convection in the direction of heat flow: (a) flow in the direction of heat flow, $Pe = 2$, and (b) flow opposite to the heat flow, $Pe = -2$; for the base case of no convection it is a straight line. Note the difference in profile for the two cases.

A dimensionless heat flux in the presence of convection can be defined as the ratio of the actual heat flux to the base value in the absence of convection:

$$q_x^* = \frac{q_x}{q_x^0}$$

Using the definition for q_x from Eq. (8.28), we have

$$q_x^* = -\frac{d\theta}{d\xi} + Pe\theta$$

Using the temperature profile, we find that the dimensionless heat flux is equal to

$$\boxed{q_x^* = \frac{Pe\exp(Pe)}{\exp(Pe) - 1}} \qquad (8.29)$$

We have that q_x^* is a constant, not a function of position ξ. Why?

In the heat-transfer literature this factor is referred to as the blowing factor or the augmentation factor, since it represents the augmentation of the heat transfer through blowing a gas across the system. Note that q_x^* tends to 1 as Pe tends to zero, which is the case with no blowing. The heat transport is now due to conduction alone. Such limiting cases should always be checked to verify that the results are in order.

If Pe is negative there is a reduction in heat transfer rate over that for pure conduction, whereas if it is positive then there is an increase. A negative value of Pe occurs when the gas is blown from right to left, i.e., in the direction opposite to the temperature gradient, leading to a reduction in the value of the overall heat transfer rate.

A simple worked example is shown below to illustrate the concepts.

Example 8.4.

Air at 1500 K is flowing past a porous plate which is to be maintained at 500 K. At a point under consideration, the base value of the heat transfer coefficient is 12.9 W/m^2 · K. A stream of air is blown through the bottom of the plate at a mass velocity of 0.0208 kg/m^{2}s. find the rate of heat transfer with and without blowing.

The rate of heat transfer in the absence of blowing is computed as $h_c(T_0 - T_L) = 12\,900$ W/m^2.

For the blowing case we find that the blowing parameter $Pe = -1.822$. It is negative since the blowing is against the direction of heat transfer. Correspondingly the augmentation factor is found as 0.3514 using Eq. (8.29). The heat transfer rate can be found by multiplying the base value by the augmentation factor.

Hence the heat transfer in the presence of blowing is calculated as 4534.0 W/m^2. This is the heat arriving at the plate surface from the hot gases. The contributions due to conduction and the blowing can also be separately calculated if needed, which can be useful in some design applications.

Augmentation or reduction of heat transport due to convection is important in many problems such as condensation of vapor, evaporative cooling, etc. These problems usually involve simultaneous heat and mass transfer, and are discussed more fully in a later chapter.

8.4.2   Convection in boundary layers

Here we show an example of convective transport to a surface where the flow is parallel to the surface. Such cases arise in flow in boundary layers. Only a simple example is shown here, while more detailed analysis is deferred to a later chapter.

The governing equation for temperature profile is as follows:

$$v_x \frac{\partial T}{\partial x} + v_y \frac{\partial T}{\partial y} = \frac{k}{\rho c_p} \frac{\partial^2 T}{\partial y^2}$$

The difference between this problem and the transpiration cooling problem in the earlier section is that in this case the flow is perpendicular to the primary direction of conduction heat transfer (a boundary-layer problem results). This leads to a PDE for the temperature profile as shown above. In the transpiration case the velocity is in the direction of heat transfer, leading to an ODE.

Note that the velocity distribution is needed for the solution of the boundary-layer heat equation, which has to be calculated from the momentum-transfer analysis. The detailed solution is left to a later chapter since the main goal here is introductory and to show correlations for heat transfer in boundary layers.

The key results are as follows.

1. As for momentum transfer, a thermal boundary layer develops along the plate. The thickness of the boundary layer can be shown to be

$$\delta_t = \delta/Pr^{1/3}$$

where Pr is the Prandtl number. This applies for a case where the plate is maintained at a constant temperature and for $Pr > 1$.

2. The local heat transfer coefficient is

$$h = \frac{0.332k}{x} Re^{1/2} Pr^{1/3}$$

for laminar flow conditions, where Re is the local Reynolds number.

The results are more commonly expressed in terms of a dimensionless number, namely the Nusselt number, which is defined as

$$Nu = \frac{hx}{k} = 0.332 Re^{1/2} Pr^{1/3}$$

This result is applicable for $Pr > 0.5$

3. For low-Prandtl-number fluids such as liquid metals, the following correlation is commonly used:

$$Nu = \frac{hx}{k} = 0.564 Re^{1/2} Pr^{1/2}$$

4. The local heat transfer coefficient under turbulent conditions and for $Pr > 1$ is

$$h = 0.036 Re^{0.8} Pr^{1/3}$$

For heat transfer with flow past complex geometries such as cylinders, numerical solutions are needed. These are, however, beset with their own difficulties, especially for turbulent flow. Hence the empirical correlations based on dimensionless groups are still popular in the design-oriented literature. However, computational simulation provides a method of rational analysis of empirical correlations, and the two approaches should be viewed as complementary.

Now we switch gears and show how the area and volume averaging can be applied to the heat equation. We illustrate that the meso and macro naturally evolve from the differential models via the averaging process. Two examples for mesoscopic model development are shown first, followed by one for a macroscopic model.

8.5 Mesoscopic models

Mesoscopic models are derived by area averaging over a cross-section. Two examples are provided here.

8.5.1 Heat transfer from a fin

The first case is conductive heat transfer from a solid exposed to a gas. The surface of the solid loses heat by convection to the surrounding fluid. The control volume analyzed is shown in Fig. 8.7. It is convenient in such cases to start with the heat equation in the heat flux form:

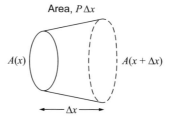

Area, $P\Delta x$

$A(x)$ $A(x + \Delta x)$

Δx

Figure 8.7 The control volume for derivation of a mesoscopic model by cross-sectional averaging.

$$\nabla \cdot q = 0 \tag{8.30}$$

Upon integrating over the control volume shown in Fig. 8.7, we have

$$\iiint_V (\nabla \cdot q)dV = 0 \tag{8.31}$$

Upon applying the Gauss theorem in reverse, this transforms to a surface integral:

$$\iint_S (n \cdot q)dS = 0 \tag{8.32}$$

The area S is the area surrounding the control volume. This can be decomposed into A_x, $A_{x+\delta x}$, and $P\,\Delta x$, and the integration can be done over these three areas. Here P is the local perimeter of the surface. Over the area A_x we can write this in terms of an average heat flux entering this cross-section. This average heat flux is defined as

$$q_x(\text{avg}) = \frac{1}{A} \iint_{yz} q_x \, dy \, dz \tag{8.33}$$

Similarly, an average heat flux at $x + dx$ can be defined. Finally, the heat loss from the surface $P\,\Delta x$, denoted here as S_e, can be related to the rate of convection heat transfer:

$$\iint_{S_e} n \cdot (k\,\nabla T)dS_e = -hS_e(T_s - T_a) = -hP\,\Delta x(T_s - T_a) \tag{8.34}$$

On collecting all terms together we have

$$[-q_x(\text{avg})A(x)]_x + [q_x(\text{avg})A(x)]_{x+\delta x} - hP\,\Delta x(T_s - T_a) = 0 \tag{8.35}$$

Taking the limit as $\Delta x \to 0$ we obtain

$$\frac{d}{dx}(q_x A(x)) + hP(T_s - T_a) = 0 \tag{8.36}$$

The average flux is related to the average temperature by Fourier's law leading to

$$\boxed{\frac{d}{dx}\left(A(x)k\frac{d\langle T\rangle}{dx}\right) = hP(T_s - T_a)} \tag{8.37}$$

This is the area-averaged representation of the Laplace equation. This was derived directly starting from a shell balance in Section 2.8.2. Here we show how it evolves naturally as a consequence of area averaging.

The mesoscopic model has to be solved with some lumping approximation. The commonly used one is that $T_s = \langle T \rangle$, which is a good approximation if the Biot number is small. This leads to

$$\frac{d}{dx}\left(A(x)k\frac{d\langle T \rangle}{dx}\right) = hP(\langle T \rangle - T_a) \tag{8.38}$$

which is the mesoscopic model widely used in the analysis of heat transfer from extended surfaces or fins.

The second example is for a single-stream heat exchanger, which was also studied by a shell balance in Section 1.4.

8.5.2   A single-stream heat exchanger

The governing equation can be shown, starting from the basic temperature equation, to be the following:

$$\rho c_p \frac{\partial (v_z T)}{\partial z} = k\left[\frac{1}{r}\frac{\partial}{\partial r}\left(r\frac{\partial T}{\partial r}\right)\right] \tag{8.39}$$

Note that we have included conduction only in the radial direction and have neglected conduction in the flow direction. This is usually a good assumption, since in the flow direction the convective heat-transfer term is more dominant. The temperature profile now depends on both r and z, and this is therefore a 2D problem. The goal of area averaging is model reduction in order to represent this as a 1D problem

The model reduction is achieved by multiplying both sides of the above equation by $2\pi r\,\delta r$ and integrating with respect to r from 0 to R. This procedure is known as cross-sectional averaging:

$$\rho c_p \int_0^R 2\pi r\frac{\partial (v_z T)}{\partial z}\,dr = 2\pi k \int_0^R\left[\frac{\partial}{\partial r}\left(r\frac{\partial T}{\partial r}\right)dr\right] \tag{8.40}$$

Since the derivative is with respect to z on the LHS, the integral can be pulled inside and the derivative sign can be outside. The RHS can be easily integrated, and the boundary conditions at $r = 0$ and $r = R$ can be applied. The result is

$$\rho c_p \frac{\partial}{\partial z}\left[\int_0^R 2\pi r v_z T\,dr\right] = 2\pi R\left[k\frac{\partial T}{\partial r}\right]_{r=R} \tag{8.41}$$

This LHS suggests that an average temperature should be defined as

$$T_b = \frac{\int_0^R 2\pi r v_z T\,dr}{\int_0^R 2\pi r v_z\,dr}$$

This is the cup mixing temperature used earlier, and its use is justified by this procedure. Note that the integral in the denominator can be expressed in terms of the volumetric flow rate since

$$\int_0^R 2\pi r v_z\,dr = \pi R^2 \langle v \rangle$$

Hence the integral term on the LHS of Eq. (8.41) is

$$\int_0^R 2\pi r v_z T \, dr = \pi R^2 \langle v \rangle T_b$$

The RHS can be related to q_R and is equal to $-2\pi R q_R$. Hence the 1D model for a single-stream heat exchanger can be written as

$$\rho c_p \pi R^2 \langle v \rangle \frac{dT_b}{dz} = -2\pi R q_R \qquad (8.42)$$

which can be interpreted as the 1D differential enthalpy balance for this system. Although it is possible to derive this directly by a heat balance as shown in Example 1.3, the interpretation of the type of average temperature to be used in the mesoscopic model is made clear by cross-sectional averaging. The cup mixing temperature defined above has to be used. The models for the constant-wall-flux and constant-wall-temperature cases were presented earlier, and we briefly reiterate the results here for ease of reading and completeness.

The constant-wall-flux case
This is the simple case and no additional closures are required. If q_R, the outward wall flux, is a constant, the mesoscopic model can be readily integrated. No additional parameters are needed. Equation (8.42) can be directly integrated with the inlet condition: at $z = 0$, $T_b = 0$.

It is nicer to present the result in dimensionless form as

$$\theta_b = 4\zeta$$

and ζ is a dimensionless axial distance defined as

$$\zeta = \frac{z}{R} \left(\frac{k/(\rho c_p)}{\langle v \rangle d} \right)$$

which is also called the contracted axial length. Here d is $2R$, the tube diameter. Also the dimensionless temperature is now defined as

$$\theta_b = \frac{T - T_i}{[-q_R]R/k}$$

We note that the cup mixing temperature varies linearly as a function of axial distance.

The constant-wall-temperature case
The constant-wall-temperature case is somewhat involved since the heat flux at the wall is not known ahead of time. We need to model this in terms of a heat transfer coefficient. Note that the information lost by averaging is now being supplemented by using this parameter. Let

$$q_R = h(T_b - T_w)$$

The mesoscopic model now reads

$$\rho c_p \pi R^2 v_{\text{avg}} \frac{dT_b}{dz} = -2\pi R \bar{h}(T_b - T_w) \qquad (8.43)$$

The equation can be integrated if h is known. It may be noted that h can vary along z, especially in laminar flow, as will be discussed later. However, the variation of h with z is not significant in turbulent flow. In design applications an average value of h denoted as $\bar{h}$ is used for integration. The integration leads to the results shown in Example 1.3. You may want to go back and review the final results and the dimensionless formula for temperature in terms of the NTU parameter.

8.6   Volume averaging or lumping

8.6.1   Cooling of a sphere in a liquid

Consider the problem of cooling of a sphere exposed to a cold liquid. This is a transient-heat-conduction problem. Again the equation in heat-flux form is useful here:

$$-\nabla \cdot \boldsymbol{q} = \rho c_p \frac{\partial T}{\partial t} \tag{8.44}$$

Now we integrate over the volume, the volume averaging procedure:

$$-\int_V \nabla \cdot \boldsymbol{q} \, dV = \int_V \rho c_p \frac{\partial T}{\partial t} \, dV \tag{8.45}$$

The volume integral on the LHS can be reduced to a surface integral by using the Gauss theorem in reverse:

$$\text{LHS} = -\iint_S (\boldsymbol{n} \cdot \boldsymbol{q}) dS \tag{8.46}$$

By using the boundary condition at the surface, this can be represented as

$$\text{LHS} = -\iint_S h(T_s - T_a) dS = -h(T_s - T_a)S \tag{8.47}$$

This assumes that the values of the surface temperature and the heat transfer coefficients are the same at every point in the surface. If this condition does not hold (h varies locally along the solid surface) then h and T_s are to be interpreted as some average values on the surface.

The RHS can be represented in terms of the average temperature of the solid using again the mean value theorem:

$$\text{RHS} = \int_V \rho c_p \frac{dT}{dt} \, dV = \rho c_p V \frac{d\bar{T}}{dt} \tag{8.48}$$

where $\bar{T}$ is an average temperature of the solid defined as

$$\bar{T} = \frac{1}{V} \int_V T \, dV \tag{8.49}$$

Hence the lumped model is

$$\rho c_p V \frac{d\bar{T}}{dt} = -h(T_s - T_a)S \tag{8.50}$$

The volume-averaged representation shown above is mathematically equivalent to the differential model. The difficulty is that the model involves two temperatures: the volume-averaged temperature of the solid and the (average of) the surface temperature of the solid.

If these are assumed to be the same, we obtain a simple lumped-parameter model for transient heat conduction:

$$\rho c_p V \frac{d\bar{T}}{dt} = -h(\bar{T} - T_a)S \tag{8.51}$$

This equation is often referred to as Newton's law of cooling, which was derived by a direct shell balance in Section 2.4.4.

The validity of the equation is established later as

$$\text{Biot number } Bi = hL/K < 1$$

The lumping analysis (first level) applies if the volume-average temperature of the material is the same as that of the fluid. The solution to the lumped model was also presented in Section 2.4.4 and you should review the details here.

8.6.2 An improved lumped model

The next example examines the validity of the lumped-parameter model and how it can be improved by adding a minor modification to the time constant.

The range of validity of the lumped model can be examined as follows. For this purpose one needs an estimate of the surface temperature and the average temperature. In order to do this, a quadratic approximation for the temperature is assumed for 1D problems such as the sphere, cylinder, and slab. The assumed temperature profile is

$$T(r) = a + br + cr^2$$

The temperature gradient dT/dr is taken as zero at the center, which requires the constant b to be zero since $dT/dr = b + 2cr$. Hence the temperature profile is simplified to

$$T(r) = a + cr^2$$

The convective heat-loss condition (Robin) from the surface is now applied:

$$-k(dT/dr)_{r=R} = h(T_{r=R} - T_a) = h(T_s - T_a)$$

A quadratic function that satisfies this requirement is

$$T - T_a = (T_s - T_a)\left[1 + \frac{Bi}{2} - \frac{Bi}{2}\left(\frac{r}{R}\right)^2\right] \tag{8.52}$$

The average temperature for a sphere is

$$\bar{T} = \frac{\int_0^R 4\pi r^2 T(r)dr}{(4/3)\pi R^3}$$

On using Eq. (8.52) and taking the average, we find

$$\bar{T} - T_a = (T_s - T_a)\left(1 + \frac{Bi}{5}\right)$$

Hence it is verified that the lumped model is valid if $Bi \ll 1$. Also for moderate Biot values the time constant can be adjusted by multiplying it by $1 + Bi/5$, and the lumped-parameter model can still be used as a first approximation.

A similar analysis done for a cylinder shows that the time constant has to be adjusted to be $1 + Bi/4$, and for a slab one must use $1 + Bi/3$. These derivations are left as an exercise in Problem 15.

Summary

- Steady-state heat transfer in solids is encountered in a number of situations. These problems are governed by the Laplace equation for the case of constant thermal conductivity. Since this is a linear partial differential equation, a number of powerful mathematical tools can be used to solve the problem. The method of boundary elements is one such method.

- A simple prototype of steady-state conduction is conduction with only one spatial dependence of temperature. Such problems can be posed in a 1D rectangular slab, a long cylinder, or a sphere. The solutions and the corresponding heat flux and rate of heat conduction across the material are basic to heat transfer and provide a number of important simple design formulas.

- The case of variable thermal conductivity adds a non-linearity to the problem. For simple geometries it is possible to integrate the equation twice to get analytic solutions for the temperature. However, the solution is implicit in temperature and not an explicit function of the spatial coordinate. This difference compared with the constant-thermal-conductivity case is worth noting.

- An interesting result for the varying conductivity is that the system can be integrated over the whole domain to get an expression for the heat flux without having to solve for the temperature profile. Often the flux is the quantity of interest and detailed temperature profiles may be needed only in specific cases. A further simplification occurs if the conductivity varies as a linear function of temperature. The heat flux can be directly calculated using a suitably defined average value of the thermal conductivity without the need for any model solutions.

- Illustrative results for 2D cases presented in the text are worth noting. Separation of variables can be used for the solution, provided that the geometry is regular (a rectangle). For irregular domains the method of boundary collocation is a useful technique.

- Steady-state problems with internal generation of heat belong to the class of Poisson equations. The equation can be either linear or non-linear depending on whether the heat source is linear or non-linear. Additional non-linearity arises if the conductivity is a function of temperature, sometimes referred to as material non-linearity.

- Analytical solutions can be obtained for 1D linear problems with internal heat generation. For non-linear (1D) problems a mathematical procedure called the *p*-substitution method reduces the problem to an integral. The solution of this integral gives the temperature profile and heat flux at the surface. A prototype of this problem is the famous Frank-Kamenetskii problem, where the heat generation is an exponential function of temperature. A (dimensionless) parameter appearing in this problem is a measure of the heat generated due to chemical reaction. There is a critical value of this parameter above which no solution exists to the problem. This can be physically regarded as the explosion limit.

- Problems with convection in the same direction as conduction are encountered in a variety of situations. Transpiration cooling by blowing a gas across a porous plate is an example. The problem has a simple mathematical structure and can be solved analytically. This is in contrast to the case of convection in a direction perpendicular to that of conduction which is treated in a later chapter. Such problems need a partial-differential-equation treatment. Some simple results for a flat plate were presented. These are useful for design calculations.

- For transpiration heat transfer, the rate of heat transfer can be either augmented or retarded by the blowing direction. A factor can be defined and used to account for this. This factor finds application in a number of areas, for instance, condensation heat transfer, as will be discussed in a later chapter.

- Mesoscopic models can be derived by area averaging of the heat equation. Often 2D models can be reduced to 1D models by such a procedure. Two example problems (i) heat transfer from an extended surface and (ii) a single-stream heat exchanger are demonstrated in the text.

- Volume averaging of the heat equation gives the lumped-parameter model. The validity of the lumped model was examined in the text by means of a simple problem involving the cooling of a sphere. A simple method to extend the range of validity of the lumped-parameter approach was also indicated.

Problems

1. Derive equations for the temperature in a slab if the thermal conductivity (a) is constant, (b) varies linearly as $k(T) = k_0 + a(T - T_0)$, and (c) varies as a quadratic function $k(T) = k_0 + a(T - T_0) + b(T - T_0)^2$.

   State how the heat flow should be calculated for each of these cases. How should the "mean" temperature on which to base the mean conductivity be defined for each of these cases?

2. Find the rate of heat flow in the radial direction through a spherical shell of inner radius r_i and outer radius r_o for the case where the thermal conductivity varies as a linear function of temperature,

$$k = k_0(1 + \beta T).$$

   The inside shell is at temperature T_i and the outside at temperature T_o. Compare this with a case where the thermal conductivity is treated as a constant evaluated at the average temperature.

3. Show that a variable transformation known as the Kirchhoff transformation,

$$F(T) = \int_0^T k(s)\,ds$$

   where s is a dummy variable, reduces the heat equation to $\nabla^2 F = 0$ for the variable conductivity case.

4. Consider the case of linear heat generation with a linear variable-thermal conductivity. Express the governing equation in terms of dimensionless form. What are the number of dimensionless parameters needed to characterize the model?

   The numerical solution based on BVP4C introduced in Chapter 10 would be useful here.

5. Consider heat conduction with generation in a slab and a sphere geometries. Derive the solution similar to Eq. (8.18) for slab and sphere cases.

   For all three cases (slab, cylinder, and sphere) find the heat flow from the surface to the fluid and show that the results satisfy an overall heat balance.

6. For the linear generation, use the Robin condition at the surface rather than the Dirichlet condition used in the text. Derive an expression for the temperature profile, the maximum temperature, and the stability condition.

7. Verify the solutions in the text for the temperature distribution in a square slab with constant generation of heat. Generate illustrative contour plots for the temperature profiles.

8. Second-order elliptic equations of the Laplace and Poisson type can be solved by central-difference-based finite-difference schemes. Thus, if we employ a square mesh, show that the temperature at a mesh point (i,j) is given by the averge of the adjacent mesh values:

$$T_{i,j} = \frac{1}{4}(T_{i+1,j} + T_{i-1,j} + T_{i,j+1} + T_{i,j-1})$$

This provides an iterative method of solving for the temperature field using the Gauss–Siedel iterative scheme.

Solve the temperature field for a 2D square geometry and compare your result with the analytic solution. Note that the scheme should not be used on the boundary nodes. The temperature is simply set as the boundary value at these points.

How would you modify this if there is also a heat generation in the slab?

9. The square problem can also be fitted using boundary collocation. For this we need functions that satisfy the Laplace equation in an exact manner. The solution can then be expanded in terms of these functions. Let us see whether we find suitable functions.

Prove that the real and imaginary parts of $F(z)$, where z is a complex variable, satisfy the Laplace equation.

We seek polynomial functions for simplicity. Thus we seek z^n, where $z = x + iy$, as the trial functions. Generate a set of functions for various values of n. For example, if $n = 2$ then show that $x^2 - y^2$ and xy are functions that satisfy the Laplace equation. In this way we can generate a whole set of basis functions.

For the square problem only even functions should be considered. Thus $x^2 - y^2$ is a good function, whereas xy is not. Similarly, so are some functions generated by z^4, but not those generated by z^3. Generate a set of, say, six such functions.

Use these functions to fit the boundary values of temperature in a least-square sense, which is the required solution by boundary collocation. Verify your answer with the analytic solution given in the text.

10. Show that, if purely Neumann conditions are imposed along the boundary, then the integral of the temperature gradient over the perimeter, Γ,

$$\int_\Gamma \frac{\partial T}{\partial n} \, d\Gamma = 0$$

has to be equal to zero in order to maintain a steady-state temperature profile inside the medium. This is called the consistency condition. Explain the physical meaning of this condition.

11. Consider the problem of a linear generation of heat examined in Section 8.3.1, but now we wish to analyze the case of a long cylinder.

State the governing equation and the dimensionless parameters needed to characterize the problem.

Verify that the temperature profile is given by

$$\theta = \theta_{max} \frac{J_0(\zeta\sqrt{\beta}) - J_0(\sqrt{\beta})}{1 - J_0(\sqrt{\beta})}$$

Show that it satisfies the differential equation and the boundary conditions. Also derive an expression for the maximum temperature, θ_{max}. Find the limit for no explosion. The answer is $\sqrt{\beta} < 2.4048$.

12. Repeat the above problem for a spherical geometry. Find the temperature profile: What is the limit for explosion for this case? The answer is $\sqrt{\beta} < \pi$.

13. Repeat the analysis for the above two problems if a Robin condition is applied at the surface.

14. A porous solid is in the shape of a hollow sphere and has temperatures of T_i and T_o at the radii of r_i and r_o, respectively. Find an expression for the heat loss from the sphere.
    In order to reduce the heat loss, a gas is blown through the porous hollow sphere at a velocity of v_0. Derive an expression for the reduction in heat loss from this system.

15. Apply the lumped model for cooling of a solid in the form of (i) a long slab and (ii) a long cylinder. Also apply the modified lumped model shown in the text, and show what correction is needed for the simple lumped model for an assumed quadratic variation in temperature of the solid.

9 Equations of mass transfer

Learning objectives

The main learning topics from this chapter are as follows:

- the concept of a concentration jump at an interface;
- the average velocity of a mixture and ways of averaging; correspondingly the definition of a mass (and molar) flux in a stationary frame and a moving frame;
- Fick's law for a binary system and its various forms;
- differential equations for mass transfer and its various forms; we focus here primarily on binary systems (multicomponent systems are deferred to a later chapter);
- averaging of differential models leading to mesoscopic and macroscopic models; the relation between these models; the need for a mass transfer coefficient;
- chemical potential as a driving force for mass transfer;
- the notion of pressure and thermal diffusion; and
- complexities associated with modeling diffusion.

Mathematical prerequisites

No new mathematical perquisites are needed for the study of this chapter. The student may wish to revise the Green–Gauss theorem for converting volume integrals to surface integrals for the derivation of macroscopic models. The notion of the substantial derivative must also be revisited.

Problems involving mass transport are important in many contexts and find applications in chemical, biological, and environmental processes. Some examples were discussed in Chapter 1, and more will follow as we progress further through the study of mass transport processes. We have already solved some illustrative problems involving mass transfer in Chapter 2. The approach demonstrated there employed a problem-to-problem basis. For a specified problem, we studied how the mass conservation principle followed by the constitutive equation can be used for setting up models for problems involving diffusion and in some cases diffusion with reaction. Simple Fick's law was used as the constitutive model for diffusion. The goal of this chapter is to provide a more formal setting and derive general differential equations for mass transfer. The chapter is organized as follows.

As a preliminary, the concentration and other necessary variables needed to characterize mass transfer are introduced. This is followed by a brief discussion on the conditions prevailing at a phase interface (a gas–liquid boundary, for example). At a phase interface, the concentration is always a discontinuous variable, unlike temperature (which is continuous at a phase boundary), and this is important in the analysis of mass transfer. Hence thermodynamic equilibrium considerations must always be applied at the interface.

In dealing with mass transfer problems, an important concept is the system average velocity and the frame of reference used to define the diffusion flux of various species. The system average velocity can be defined in a number of different ways, and the diffusion flux can be correspondingly defined in a number of different ways. Hence even Fick's simple law for binary diffusion takes various forms, depending on which frame of reference is used. These subtleties can be a source of confusion for the uninitiated. These concepts are then defined in Section 9.2. It will be pointed out that even in a seemingly stagnant system (such as alcohol evaporating from a wine glass) the system velocity is non-zero (although we do not see it, since it is too small), and hence convective-transport effects often need to be included. Thus convective flux gets superimposed on purely diffusional transport even in such "stagnant" systems. Careful attention to such details is, therefore, needed in many cases.

Differential equations describing mass transport are then derived in flux form. Using Fick's law for the flux, we then derive a set of differential equations of mass transfer in concentration form. These equations form the starting point in the analysis and modeling of a range of mass transport problems and have wide applications in many fields. The governing differential equations have almost a complete analogy with the heat transport equations derived in an earlier chapter, and hence similar solution methods are useful.

A volume average of the differential model provides the macroscopic model for mass transfer. These equations are formally derived starting from the differential models. Mesoscopic models can also be derived in a similar manner. An important application of the mesoscopic model is in simulation and design of gas–liquid contactors such as a packed-bed absorber.

Concentration gradients are primarily responsible for mass transfer and this phenomenon is often referred to as *ordinary* diffusion. However, this is not the only mechanism for mass transport, and the phenomena of diffusion can have many complexities. Chemical reaction can enhance the rate of mass transport, and is responsible for sustaining life itself due to the interaction of oxygen and hemoglobin. Mass transfer can also be caused by pressure gradients and thermal gradients. In addition, for charged species additional mass transfer occurs due to the migration of the charged species under the influence of an electric field. Diffusion in multicomponent systems cannot be described simply by Fick's law. In many biological systems mass transfer can occur even against a concentration gradient, and this phenomenon is referred to as active diffusion. It is useful to discuss the various driving forces for transport here, and hence the chapter concludes with a general overview of some complexities associated with the models for the diffusion processes and points out the constitutive equations commonly used for these cases. The detailed study of multicomponent and ionic transport is deferred to a later chapter.

9.1 Preliminaries

There are various ways of measuring concentration. Molar units or mass units can be used. These various definitions are summarized in this section. This is more of a review of well-known relations, but useful equations are summarized for ease of reference.

Mole units

The local value of the molar concentration denoted as C_A is defined as the number of moles of species A per unit volume of the system at a point in the fluid continuum. Thus, C_A has the units of mol/m^3, where mol refers to gm moles even in the SI system.

In the context of the continuum approximation, the local concentration may be defined as

$$C_A = \lim_{V \to 0} \frac{\mathcal{M}_A}{V} \tag{9.1}$$

where V is the volume element and $\mathcal{M}_A$ is the moles of species A present in that volume.

The total molar concentration of the mixture C (at any position and at any time) is defined as the summation of all the species concentrations:

$$C = \sum_{i=1}^{n_s} C_i \tag{9.2}$$

where n_s is the total number of species present in the mixture.

Very often it is necessary to refer to two or more phases, for example, a gas and a liquid. In such cases we use C_{AG} to represent the concentration of A in the gas phase and C_{AL} to represent the concentration in the liquid. Thus we use the first subscript to represent the species either by numbers $i = 1, 2$, etc., or simply by names such as A, B, etc., especially for binary mixtures. The second subscript will be used to represent the phase (or the location) where it is present.

The ratio of C_A to C is called the mole fraction. It is usually denoted by y_A for gas mixtures or x_A for liquid mixtures:

$$y_A \text{ or } x_A = C_A/C \tag{9.3}$$

The average molecular weight of the mixture $\bar{M}$ is calculated as

$$\bar{M} = \sum_{i=1}^{n_s} y_i M_i \tag{9.4}$$

where M_i is the molecular weight of species i.

Mass units

Another way of expressing the species concentration is to use mass units instead of moles. The mass concentration (also called the partial density) is designated by ρ_A and is defined as the mass of species A per unit volume of the mixture locally at a point in the fluid continuum. This quantity is related to C_A simply by the species' molecular weight:

$$\rho_A = M_A C_A \tag{9.5}$$

The summation of this for all the species is the mixture density, ρ:

$$\rho = \sum_{i=1}^{n_s} \rho_i \tag{9.6}$$

The ratio of ρ_i to ρ is the mass fraction of species i, which is often denoted by ω_i:

$$\omega_i = \rho_i / \rho \tag{9.7}$$

The average molecular weight of the mixture is calculated by the "reciprocal weighting rule" as

$$\frac{1}{\bar{M}} = \sum_{i=1}^{n_s} \frac{\omega_i}{M_i} \tag{9.8}$$

Partial pressure units

In gaseous systems, a measure of concentration is the partial pressure of a species. This follows from the ideal-gas law:

$$C_A = \frac{p_A}{R_G T} \tag{9.9}$$

or

$$p_A = C_A R_G T \tag{9.10}$$

where R_G is the gas constant (8.314 Pa $\cdot$ m^3/mol $\cdot$ K) and T is the temperature in K. The partial pressure is therefore a direct measure of concentration that is widely used for gaseous mixtures. Thus the partial-pressure difference can be used as a driving force for mass transfer instead of concentration differences for (ideal) gas-phase systems. The two sets of units can be interconverted easily if needed by using the above equations.

Number concentration

For diffusion in solids, the number concentration is often used. This is defined as the number of atoms per unit (local) volume of the system. Diffusion of dopants in semi-conductors is an example of a system where the atom unit is more useful than mole or mass units. Transport in ionized gases (plasma) is another example where this is useful. *To find the molar concentration simply divide by Avogadro's number.*

Concentration jumps at interfaces

Another important point to understand is that the concentration is not continuous at a phase boundary (e.g., a gas–liquid interface) unlike temperature. Consider the air–water system as an example.

Is the oxygen concentration on the air side of the interface the same as the oxygen concentration on the water side of the gas–liquid interface?

The answer is no. Oxygen has a poor solubility in water, and hence the oxygen concentration in the water phase is much lower than that in the air phase. The two concentrations are often related by a linear constant known as the Henry's-law constant. Various definitions are used, depending on the units used to measure the concentrations. The common form is as follows:

$$p_A = H_A x_A \tag{9.11}$$

under equilibrium conditions, where p_A is the partial pressure of the species in the gas phase and x_A is its mole fraction in the liquid. The constant H_A has the units of atm, Pa, or bars here. The values for some common gases are shown in Table 9.1.

The higher the value of H_A, the less soluble the gas is in the liquid phase. The effect of the total the pressure in the system is to increase the solubility in accordance with Henry's law. The effect of temperature is in general to decrease the solubility. Thus the Henry's-law constant is an increasing function of temperature. An exception is hydrogen, which shows a retrograde behavior. Here the solubility initially increases with an increase in temperature, then reaches a maximum, and decreases after that.

Other definitions

Other definitions for the Henry's-law constant are also common. Two common definitions are as follows:

$$p_A = H_{A,pc} C_A \tag{9.12}$$

where $H_{A,pc}$ is the Henry's-law constant (units of Pa $\cdot$ m^3/mol) with partial pressure and concentration units for the gas phase and liquid phase, respectively:

$$p_A H_{A,cp} = C_A \tag{9.13}$$

where $H_{A,cp}$ (units of mol/Pa $\cdot$ m^3) is the reciprocal of $H_{A,pc}$.

Table 9.1. Values of the Henry's-law constant (in atm) for some common gaseous species in water at 298.15 K

Gas	H (atm)
Hydrogen	7.099×10^4
Oxygen	4.259×10^4
Nitrogen	8.65×10^4
Ozone	4570
Carbon dioxide	1630
Sulfur dioxide	440
Ammonia	30

Also note that, since the concentration in the gas phase can be used instead of partial pressure for the gas phase, we have yet more definitions for the Henry's-law constant!

Example 9.1. Oxygen concentration in the water.

Find the dissolved oxygen concentration in water, assuming equilibrium conditions.

Solution.

Let A represent the oxygen species. The mole fraction in the liquid is then calculated using Henry's law: $x_A = p_A/H_A = 0.21$ atm$/4.259 \times 10^4$ atm, which is equal to 4.963×10^{-6}.

Hence the concentration in the liquid is $x_A C$, where C is the total concentration of the liquid mixture. For dilute systems C is simply the solvent concentration. For water, the density is 1000 kg/m^3 and the molecular weight is 18×10^{-3} kg-mass/g-mole.

Hence $C = 1000/(18 \times 10^{-3}) = 55\,560$ mol/m^3.

The oxygen concentration in the liquid is therefore 0.2739 mol/m^3. If air is bubbled for a long time through a batch liquid, the oxygen concentration in the entire liquid will reach the above value, which is the final equilibrium value.

To illustrate the concentration jump at the interface we also calculate the concentration of oxygen in air. This is calculated from the ideal-gas law:

$$C_{AG} = y_A P/(R_G T) = 0.21 \text{ atm} \times 1.01 \times 10^5 \text{ Pa/atm}/(8.314 \times 298)$$

$$= 8.467 \text{ mol/m}^3 \tag{9.14}$$

The ratio of $C_{AG,i}$ to $C_{AL,i}$ is 30.9, which is the magnitude of the concentration jump at the interface.

Note that mass transport usually occurs under non-equilibrium conditions. Hence Henry's law should be applied only at the interface, which is assumed to be at equilibrium. The third subscript i above indicates the interfacial values.

Vapor–liquid interfaces

Gases are defined as systems with $T > T_C$ (the critical temperature) while vapors are defined as systems where $T < T_C$. Here we consider the representation of the equilibrium at the vapor–liquid interface.

For liquid-phase species that are volatile, Raoult's law is often used to relate the interfacial concentrations. This states that, at the vapor–liquid interface, a pure liquid exerts a partial pressure equal to the vapor pressure of the liquid at that temperature. For a liquid mixture, the partial pressure at the interface is equal to the vapor pressure multiplied by mole fraction in the liquid. Hence the interfacial relation is

$$p_{A,i} = x_{A,i} p_{vap,A}$$

The vapor pressure varies as a function of temperature and is often correlated by the Antoine equation:

$$\log_{10} p_{vap} = A - \frac{B}{C + T} \tag{9.15}$$

We will need this equation later for simultaneous heat and mass transfer. Values of the constants are tabulated in many books and websites. The units are often in mm Hg for vapor pressure, and T is in °C rather than in standard SI units.

Liquid–liquid interface

A simple partition coefficient (denoted as m_A for species A) is often used to describe the interfacial equilibrium between two liquids:

$$y_{A,i} = m_A x_{A,i}$$

where $y_{A,i}$ is the interfacial mole fraction in one of the liquid phases, while $x_{A,i}$ is the interfacial mole fraction at the interface of the second liquid.

Gas–solid and liquid–solid systems are defined in a similar manner using an adsorption equilibrium constant that has the same status as the solubility or partition coefficient.

Non-linear equilibrium models

A note on the linear models described above is in order. More complex models are needed to describe equilibrium in (i) non-ideal liquid mixtures and (ii) liquid–liquid systems where there is a strong adsorbed layer due to a difference in the interfacial tensions of various species dissolved in the liquids. This layer often has completely different properties from those of the two liquids in contact, and is sometime referred to as a microphase. Models for systems exhibiting these complex behaviors are not considered in this book. However, it is worthwhile to note that deliberate creation of a microphase at a liquid–liquid interface can be used in many engineering applications. This field is known as microphase engineering. For more details, see Bhagwat and Sharma (1988) and related papers from their group.

On a similar note for fluid–solid systems, a linear adsorption equilibrium is often used, although these are in actuality non-linear and often follow the classical Langmuir equation. The linear relation is a good approximation for dilute systems.

9.3   The frame of reference and Fick's law

One subtle difference between heat transfer and mass transfer is that a motion need not be associated with a heat-conduction process. An example is heat conduction in a bulk solid. There is no solid motion (on a continuum scale), but there is heat transfer. In contrast, mass diffusion involves the transport or motion of a species from one location to another, hence involving movement of matter. Thus a motion is always associated with mass transfer, and the motion contributes to a convective flux in addition to the diffusion flux. The convective transport always needs to be looked into even when there is no external superimposed flow in the system. Often the convective effect is small (or even cancels out due to counter-movement of two species) and purely molecular transport of mass is the only mode of transport. Thus the mass transport effect can be analyzed as though the medium were stationary, and in these cases a one-to-one analogy between heat and mass transfer exists. But, to present the results in a general setting, there is a need to introduce a proper frame of reference to define more clearly the mass transport rate by diffusion. The existence of a velocity in a seemingly stationary medium is often not clear, as is illustrated in Fig. 9.1. A liquid evaporating in a glass looks stagnant, and the convective flow caused by evaporation is not evident.

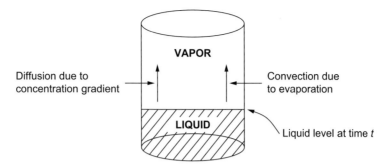

Figure 9.1 Liquid evaporating in a glass. Note that transport takes place both by diffusion and by convection, although the convective flow is not evident.

Another confusing aspect is that there are many ways of defining the average velocity of a mixture and there are many terminologies for the definition of diffusive fluxes. Two common ways are the mole-fraction-weighted average velocity and the mass-fraction-weighted average velocity. These are introduced now, and Fick's law for binary diffusion is revisited in this setting.

The mass-fraction-averaged velocity

Consider a plane that is oriented in the positive-x-direction at any given location. Let $n_{A,x}$ be the mass of A crossing this plane in unit time per unit area. Assume that the mass crossing is being measured by a stationary observer (an observer who is not a part of the system). Similarly, we can define $n_{A,y}$ and $n_{A,z}$ in the y- and z-directions. These components together form a local mass-flux vector:

$$n_A = e_x n_{A,x} + e_y n_{A,y} + e_z n_{A,z}$$

The velocity for species A, v_A, can be defined at this point, and this is related to the mass flux vector by the density of A:

$$\rho_A v_A = n_A$$

Similarly, the velocity of B can be defined (as can the velocities of other components in the system):

$$\rho_B v_B = n_B$$

The velocities v_A and v_B are again species velocities referred to a stationary observer.

The mixture velocity v is defined as a weighted average of these velocities. Now various definitions arise, depending on what weighting factor is used to define this average. Let us use here the mass fraction as the weighting factor. Thus the mixture velocity is defined as

$$v = \omega_A v_A + \omega_B v_B$$

for a binary mixture or in general

$$v = \sum w_i v_i = \left[\sum \rho_i v_i \right] / \rho \qquad (9.16)$$

where the summation is over all the species present in the system ($i = 1$ to n_s).

A flux with respect to a moving frame can now be defined. Let the moving frame have a velocity of v (the mass-averaged velocity), and hence let us define relative velocity and use this to define the fluxes. Let

$$j_A = \rho_A(v_A - v)$$

be defined as the **diffusion flux** (in mass units) of A. This quantity will then represent the mass flux vector seen by an observer who is also moving with the mixture velocity v.

The flux in a stationary frame is related to the flux in a moving frame as

$$n_A = \rho_A(v_A - v) + \rho_A v = j_A + \rho_A v \qquad (9.17)$$

or

$$\boxed{n_A = j_A + \rho_A v} \qquad (9.18)$$

The two fluxes n_A and j_A and the corresponding observers are shown in the cartoon in Fig. 9.2.

Since $\rho_A = \omega_A \rho$ and $\rho v = n_t$, the total flux (mass units), we can also write the flux as

$$n_A = j_A + \omega_A n_t \qquad (9.19)$$

Note that the total flux n_t is equal to $n_A + n_B$ in a binary mixture.

It turns out that

$$j_A = -\rho D_{AB} \nabla w_A$$

which is one version of Fick's law. The verification is left as an exercise. Here D_{AB} is the diffusivity of A in the the binary mixture of A and B.

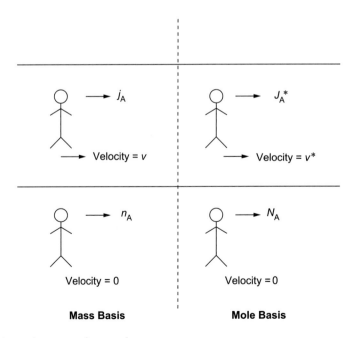

Figure 9.2 A moving vs. stationary observer.

Historical vignette. Fick was the son of a civil engineer and was born on September 3, 1829. Even as a child he was fascinated by advanced mathematics, but his father persuaded him to go to medical school, which he did. Later he felt that his passion for mathematics and medicine could be combined, and chose to become a biomedical engineer. He published extensively in this field, but is more famous for the law he postulated in 1855 concerning diffusion of matter. The law was proposed on the basis of an intuitive analogy with heat conduction.

The mole-fraction-averaged velocity

We start with the definition of the component of the molar flux of a species denoted by $N_{A,x}$ which is the number of moles crossing a differential area oriented in the x-direction per unit time per unit area.

Correspondingly, the molar flux vector of a species A over a differential control surface is equal to the number of moles crossing the surface per unit area per unit time:

$$N_A = C_A v_A \tag{9.20}$$

This flux is measured with respect to a stationary frame of reference. Obviously $n_A = M_A N_A$ since both the fluxes are with respect to a stationary frame.

Let v^* be the average (mole-fraction-weighted) velocity of the system defined as follows:

$$v^* = \sum_{i=1}^{n_s} y_i v_i \tag{9.21}$$

For an observer who moves with a system with a velocity of v^*, the flux observed will be

$$J_A^* = C_A(v_A - v^*) \tag{9.22}$$

Note that the observer has to move with the velocity v^* in order to observe this flux. This velocity is different from the bulk fluid velocity v discussed in momentum transfer. The bulk fluid velocity is a mass-fraction-averaged velocity defined by Eq. (9.16). In general the mass fraction is not equal to the mole fraction, and hence the two velocities are different.

Other definitions of diffusion flux are possible (see Problem 14). We use only j and J^* here. For further discussion the star superscript on J is dropped. We denote J_i^* as J_i in further discussions.

On rearranging Eq. (9.22) (with the * notation dropped as well) we have

$$C_A v_A = J_A + C_A v^* \tag{9.23}$$

and we find the following relation between the (molar) flux in a stationary frame and that in a moving frame:

$$N_A = J_A + C_A v^* \tag{9.24}$$

where the first term on the RHS is referred to as the diffusive flux while the second term is the flux arising out of the net fluid motion (the convective flux). The two fluxes N_A and J_A and the corresponding observers are shown in the cartoon Fig. 9.2.

The above equation can also be expressed in terms of the mole fraction as

$$\boxed{N_A = J_A + y_A N_t} \tag{9.25}$$

where N_t is the total molar flux (stationary observer). This follows from (9.24), since $v^* = N_t/C$.

It turns out that the diffusive flux in a binary system is proportional to the mole-fraction gradient, which is another version of Fick's law:

$$J_A = -CD_{AB} \nabla y_A \qquad (9.26)$$

Note that the mole-fraction gradient is used as the driving force rather than the usual concentration gradient. This is more general since the concentration can vary simply due to a local difference in temperature (as per the ideal-gas law). For instance, in a room with air (no mole-fraction difference) there is no diffusion, but the total concentration may be different at two points due to temperature differences between these points.

For systems with constant total concentration C, i.e., systems at a constant temperature and constant total pressure, we have

$$J_A = -D_{AB} \nabla C_A \qquad (9.27)$$

which is the traditional form of Fick's law introduced earlier, starting from Chapter 1.

Some properties of the diffusive flux are discussed below.

Property 1

The sum of the diffusive flux of all species is equal to zero. This is proved as follows. By summing the definition of diffusive flux of any species i given by Eq. (9.22) we have

$$\sum J_i = \sum (C_i v_i) - v^* \sum C_i \qquad (9.28)$$

Since the summation of C_i taken for all the species is equal to the total concentration C, the above equation can be written as

$$\sum J_i = \sum (C_i v_i) - v^* C \qquad (9.29)$$

We note that, from the definition of the mixture average velocity v^*,

$$v^* C = \sum_{i=1}^{n_s} y_i v_i C = \sum (C_i v_i) \qquad (9.30)$$

Hence the sum of diffusive flux taken over all the components is zero:

$$\boxed{\sum_{i=1}^{n_s} J_i = 0} \qquad (9.31)$$

For a binary we have therefore $J_A = -J_B$ and there is only one diffusive flux. In a similar manner, for a multicomponent system there can be only $n_s - 1$ independent diffusive fluxes.

Property 2

In a binary system a single parameter D_{AB} is sufficient to model the diffusive flux.

Let there be two parameters, D_A and D_B. Then the flux for each species is proportional to its mole-fraction gradient in the system. The fluxes are given by

$$J_A = -CD_A \frac{dy_A}{dz} \qquad (9.32)$$

and

$$J_B = -CD_B \frac{dy_B}{dz} \qquad (9.33)$$

For a binary system the diffusive fluxes of A and B are related by

$$J_A = -J_B$$

Also, for a binary system,

$$y_A + y_B = 1$$

or

$$\frac{dy_A}{dz} + \frac{dy_B}{dz} = 0$$

Hence it follows that

$$D_A = D_B = D_{AB}$$

Thus there is only one diffusion coefficient in a binary mixture, which is denoted as D_{AB}, the binary pair diffusion coefficient.

Property 3

In a multicomponent system $n_s(n_s - 1)/2$ parameters are needed to model the diffusive fluxes. This follows from Property 1 above. For a three-component system models for diffusion can be therefore set up with three basic binary parameters, D_{AB}, D_{BC}, and D_{AC},

Pseudo-binary diffusivity

Multicomponent diffusion does not follow Fick's law, and is discussed in detail in a later chapter. A simplified concept is that of pseudo-binary diffusivity, which is defined below:

$$J_A = -CD_{A\text{-m}} \nabla y_A$$

where $D_{A\text{-m}}$ is the diffusivity of A in the mixture defined as though the system were binary. Hence the name pseudo-binary diffusivity.

The multicomponent diffusion is then treated as though the binary version of Fick's law holds for each species taken individually. Note that this is only an approximation. This turns out to be a good approximation if species A is present in dilute concentrations and diffusing in a mixture of B, C, etc. A commonly used equation for such cases is the Wilke equation:

$$\frac{1 - y_A}{D_{A\text{-m}}} = \frac{y_B}{D_{AB}} + \frac{y_C}{D_{AC}} + \cdots \qquad (9.34)$$

9.4   Equations of mass transfer

The basic differential equation for species mass transfer will now be derived. We will show the derivations using the mass basis as well as the mole basis.

9.4.1   Mass basis

Eulerian control volume

For the Eulerian control volume shown in Fig. 9.3, the mass of A leaving the face $x + \Delta x$ is $n_{A,x}$ at $x + \Delta x$ times the area of the plane, which is $\Delta y \, \Delta z$:

$$\text{Mass leaving} = n_{A,x}(x + \Delta x)\Delta y \, \Delta z$$

By using the Taylor series we have

$$\text{Mass leaving} = n_{A,x}(x)\Delta y \, \Delta z + \frac{\partial}{\partial x}(n_{A,x})\Delta x \, \Delta y \, \Delta z$$

The first term on the RHS is seen to be the mass of A entering the plane at x.

Hence the net mass efflux from the two planes per unit volume is

$$\text{Net mass efflux per unit volume} = \frac{\partial}{\partial x}(n_{A,x})$$

Similar expressions hold for the y- and z-planes. Hence the net mass efflux is equal to $\nabla \cdot n_A$.

The accumulation of A in the control volume is $\partial \rho_A / \partial t$ per unit control volume.

Finally, the rate of production of A is r_A per unit control volume.

On putting all the terms in the mass balance

$$\text{In} - \text{out} + \text{generation} = \text{accumulation}$$

we have

$$-\nabla \cdot n_A + r_A = \frac{\partial \rho_A}{\partial t}$$

which is the basic differential species-mass-balance equation. Since $n_A = v\rho_A + j_A$ as per Eq. (9.18), the equation can also be written in terms of the diffusion flux as

$$\frac{\partial \rho_A}{\partial t} + \nabla \cdot (v\rho_A) = -\nabla \cdot j_A + r_A \tag{9.35}$$

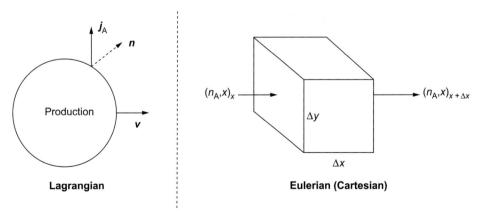

Figure 9.3 A schematic representation of the mass (or mole) balance for an arbitrary control volume.

The derivation of the equation in a Lagrangian control volume is now presented. The final result is the same equation as shown above, but showing the derivation by the two approaches has pedagogical value.

Lagrangian control volume

Consider a fluid particle and a frame of reference attached to this particle. Let the particle contain a multicomponent mixture of species A, B, C, etc. The mass balance of species A can then be stated in words as *the change in the substantial derivative of density of A is equal to the net flux due to diffusion from the bounding surface and the net mass-creation rate due to chemical reaction within the volume occupied by the particle.*

The net flux due to diffusion out of the particle is

$$\int_S (\boldsymbol{j}_A \cdot \boldsymbol{n}) ds = \int_V \nabla \cdot \boldsymbol{j}_A \, dV \approx V \nabla \cdot \boldsymbol{j}_A$$

This appears as a loss term in the mass balance since $\boldsymbol{j}_A$ is from the particle to the surroundings. Note that we have used the Green–Gauss theorem to convert the surface integral into a volume integral. Also note that the last term is written using the mean-value theorem in calculus and hence the quantity $\boldsymbol{j}_A$ is interpreted as some mean value within the particle volume. (A purist will use the Reynolds average theorem here.)

The rate of reaction is

$$\int_V r_A \, dV \approx V r_A$$

where r_A on the LHS is the **point** value for the rate of mass production per unit volume per unit time. Using these in the mass balance for A, the following equation can be written for the species mass balance:

$$\frac{D}{Dt}(\rho_A V) = -V \nabla \cdot \boldsymbol{j}_A + V r_A \tag{9.36}$$

Expanding the LHS gives

$$V \frac{D}{Dt}(\rho_A) + \rho_A \frac{D}{Dt}(V) = -V \nabla \cdot \boldsymbol{j}_A + V r_A \tag{9.37}$$

Using the dilatation for the volume-change term (the second term on the LHS) gives

$$\frac{D}{Dt}(V) = V \nabla \cdot \boldsymbol{v}$$

we obtain

$$\frac{D}{Dt}(\rho_A) + \rho_A \nabla \cdot \boldsymbol{v} = -\nabla \cdot \boldsymbol{j}_A + r_A \tag{9.38}$$

For an incompressible fluid $\nabla \cdot \boldsymbol{v}$ is zero, and hence the above equation can be simplified to

$$\frac{D}{Dt}(\rho_A) = -\nabla \cdot \boldsymbol{j}_A + r_A \tag{9.39}$$

Expanding the substantial derivative gives

$$\boxed{\frac{\partial \rho_A}{\partial t} + \boldsymbol{v} \cdot \nabla \rho_A = -\nabla \cdot \boldsymbol{j}_A + r_A} \tag{9.40}$$

This is the species-continuity equation in flux form. Note the similarity to the heat equation for the temperature given earlier. The equation is closed by providing an appropriate equation for flux as a function of the mass-fraction gradient (e.g., Fick's law for binary systems).

Note that Eq. (9.40) can also be expressed as

$$\frac{\partial \rho_A}{\partial t} + \nabla \cdot (v\rho_A) = -\nabla \cdot \boldsymbol{j}_A + r_A \tag{9.41}$$

which is a more convenient form for deriving the macroscopic balance equation. This is known as the conservative form of the species mass balance and holds for incompressible fluids.

The convective term and diffusive terms can be combined to give

$$\frac{\partial \rho_A}{\partial t} + \nabla \cdot \boldsymbol{n}_A = r_A \tag{9.42}$$

since $n_A = v\rho_A + j_A$ as per Eq. (9.18).

We now verify the overall continuity in the following example.

Example 9.2.

Sum the species mass balance equation over all the species and verify the overall continuity. We can use the following relationships.

1. The sum of the diffusion fluxes is zero: $\sum_i j_i = 0$.
2. The total mass produced by chemical reaction is zero:

$$\sum_i r_i = 0$$

If we now sum the species continuity equation over all the species and use the above-mentioned relations, then

$$\frac{D}{Dt}(\rho) + \rho \nabla \cdot v = 0 \tag{9.43}$$

For an incompressible fluid Eq. (9.38) reduces to $\nabla \cdot v$, which is the continuity equation.

9.4.2   Mole basis

The use of mole units is more suitable for reacting systems. Here we derive the species balance for A in terms of molar units. The procedure is the same as in Section 10.4.1 with the exception that a different average velocity (mole-fraction-weighted) is used here. Hence the "Lagrangian" particle is assumed to be moving with a velocity of v^* rather than v. The final equations are similar except for this difference, and hence some of the steps are not repeated here. For an incompressible fluid the species balance equation on a mole basis is therefore given by

$$\frac{D^*}{Dt}(C_A) + C_A \nabla \cdot v^* = -\nabla \cdot \boldsymbol{J}_A + R_A$$

which is analogous to (9.38). The substantial derivative is to be interpreted as

$$\frac{D^*}{Dt}(C_A) = \frac{\partial C_A}{\partial t} + v^* \cdot \nabla C_A$$

Hence

$$\frac{\partial C_A}{\partial t} + v^* \cdot \nabla C_A + C_A \nabla \cdot v^* = -\nabla \cdot J_A^* + R_A \tag{9.44}$$

The convective term and diffusive terms can be combined to give

$$\frac{\partial C_A}{\partial t} + \nabla \cdot N_A = R_A \tag{9.45}$$

since $N_A = v^* C_A + J_A$, which forms the starting point for derivation of the macroscopic models.

The overall continuity equation on the mole basis is shown in the following example.

Example 9.3.

Derive an overall continuity equation on the mole basis and point out the difference from that on the mass basis.

We still have that the sum of the diffusion fluxes is zero, $\sum_i J_i = 0$, but it should be noted that

$$\sum_i R_i \neq 0$$

By summing Eq. (9.44) over all the species we get the mole continuity equation as

$$\frac{\partial C}{\partial t} + v^* \cdot \nabla C + C \nabla \cdot v^* = \sum R_A \tag{9.46}$$

For constant total concentration, we have

$$\nabla \cdot v^* = \frac{\sum R_A}{C}$$

Note that $\nabla \cdot v^*$ need not be zero, unlike $\nabla \cdot v$. If there is no net change in the number of moles in the balanced chemical reaction, then $\sum_i R_i = 0$, but in general the above equation reduces to

$$\sum R_i = \left(\sum v_i\right) r$$

for a single reaction. Here v_i is the stoichiometric coefficient for the ith species and r is a rate function that is a measure of the rate of production (defined as the rate of formation of a product species with unit stoichiometry).

If $\sum v_i$ is equal to zero then there is no net change in the total number of moles and $\sum R_i$ is equal to zero.

9.4.3   Boundary conditions

The boundary conditions are similar to those for heat transfer and may be of Dirichlet, Neumann, or Robin type. The conditions for some common situations are indicated below.

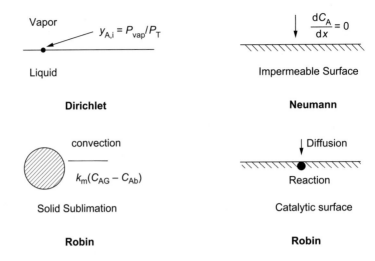

Figure 9.4 Common boundary conditions for mass transfer problems.

1. At the interface (such as a dissolving solid surface) the concentration may be prescribed. This leads to a Dirichlet type of condition.
2. At a plane of symmetry or an impermeable wall the flux is equal to zero, leading to a Neumann condition.
3. At a heterogeneous surface where a surface reaction occurs, a Robin condition can be used in general. The rate of diffusion is equated to the rate of surface reaction here:

$$N_A = -R_{A,s}$$

where $R_{A,s}$ is the rate of production of A by surface reaction defined as the number of moles produced per unit time per unit surface area.

4. A limiting case of heterogeneous reaction is fast reaction at the surface. The concentration is set as zero for this case. The Dirichlet condition is then used here. Note that the concentration cannot actually be equal to zero since then there would be no reaction. The zero value should be conceived as a limit when the rate constant tends to infinity.
5. At a gas–liquid interface the concentration-jump condition is used. Also the fluxes are continuous at this surface. Thus $N_{A,i}$ from the gas side is equal to $N_{A,i}$ into the liquid, while the concentration $C_{A,i}$ at the interface on the gas side is related to $C_{A,i}$ on the liquid side by an equilibrium constant or the Henry's-law constant.

An illustrative sketch of the various boundary conditions is shown in Fig. 9.4.

9.5 From differential to macroscopic

This section provides the derivation of the macroscopic model used in mass-transfer analysis by volume averaging of the differential equations of mass transport. Working models for the design of mass-transfer equipment are often based on such macroscopic models.

The starting point is integration of the differential equation given by (9.41), which is reproduced below for ease of reference:

$$\frac{\partial \rho_A}{\partial t} + \nabla \cdot (v\rho_A) = -\nabla \cdot j_A + r_A \tag{9.47}$$

The equation is integrated over a macroscopic control volume, which leads to the macroscopic species balance shown in Section 2.2. We show the representation term by term. The first term of Eq. (9.47) leads to the accumulation term, which is the time derivative of the total mass:

$$m_{A,tot} = \int_V \rho_A \, dV$$

The second term is now represented as a surface integral using the Green–Gauss theorem in reverse! Hence

$$\int_V \nabla \cdot (v\rho_A) dV = \int_S [n \cdot (v\rho_A)] dS \tag{9.48}$$

This gives the mass-flow terms from the mass crossing the boundary. (Diffusional crossing is ignored here.)

The first term on the RHS generates a term corresponding to the mass transferred to the walls:

$$\dot{m}_{AW,tot} = \int_{A_w} [n \cdot j_A] dA$$

gives the transport to the walls. (Convective transport is ignored here and can be added as a contribution that arises from (9.48) from the $n \cdot v$ at the wall.)

Finally, the last term gives the average rate of reaction in the macroscopic control volume:

$$\bar{r}_A = \frac{1}{V} \int_V r_A \, dV$$

On defining average concentrations and the average rate of reaction, we obtain the macroscopic species mass balance equation (2.4) presented earlier without any derivation in Chapter 2.

On a similar note, mesoscopic balances can be derived by taking a control volume of thickness Δx and area equal to the cross-section of the reactor. Such models are useful for tubular reactors and membrane transport systems.

Coupling of the macroscopic or mesoscopic description for the overall system (such as a reactor) with sub-models for local transport (such as of a catalyst particle) is used in a wide class of problems. The above procedure is a systematic way of deriving such coupled models.

9.6 Complexities in diffusion

Binary diffusion has been discussed so far, and binary diffusion in gases is well represented by Fick's law. However, many complexities are associated with the diffusion in other systems, and this section summarizes many of these observed effects.

1. Thermodynamic effects. The chemical-potential gradient can be considered to be the driving force for diffusion rather than the concentration gradients. Noting that the chemical potential can be represented as a function of mole fraction as

$$\mu_A = \mu_A^\circ + R_G T \ln a_A \tag{9.49}$$

the gradient in chemical potential can be written as

$$\nabla(\mu_A) = R_G T \frac{d\ln a_A}{dy_A} \nabla y_A \tag{9.50}$$

In these equations a_A is the activity of species A in the mixture. y_A is the mole fraction as usual.

Taking $y_A \nabla(\mu_A)/(R_G T)$ as the driving force, Fick's law takes the form

$$J_A = -CD_{AB} \frac{d\ln a_A}{d\ln y_A} \nabla y_A \tag{9.51}$$

For ideal solutions the term $d\ln a_A / d\ln y_A$ is equal to one, hence Fick's law holds in the form stated earlier. For non-ideal solutions Eq. (9.51) may be written in a form analogous to Fick's law as

$$J_A = -CD_{AB}^\circ \nabla y_A \tag{9.52}$$

where D_{AB}° is the representative observed diffusion coefficient, which is equal to

$$D_{AB}^\circ = D_{AB} \frac{d\ln a_A}{d\ln y_A} \tag{9.53}$$

However, D_{AB}° is now concentration-dependent rather than a constant, since the activity can be a function of concentration in non-ideal solutions.

A consequence of non-ideal thermodynamic considerations is that the diffusion coefficient is a function of the concentration of the species. These effects are common in many liquid–liquid systems. Another consequence is that the binary diffusivity in a mixture of a dilute A in a concentrated solution of B (denoted here as D_{A-B}) is not the same as D_{B-A}, the binary diffusivity in a mixture of dilute B in an excess of A. In many applications these complexities are ignored and absorbed in other fitting parameters. But one has to bear these considerations in mind for more accurate modeling of mass transport in liquid–liquid systems.

A suggested equation is the Darken relation, which allows description of the diffusivity in the mixture using the dilute-solution values, D_A and D_B:

$$D_{AB} = (x_B D_A + x_A D_B) \left[1 + \frac{\ln \gamma_A}{\ln x_A} \right]$$

with γ_A being the activity coefficient of species A in the liquid mixture.

2. Solute–solvent interaction effects. These are of importance in liquid–liquid systems where a solute associates with a solvent molecule and the associated pair diffuses together. The solute may associate with itself (a type of dimer), and the diffusion may be controlled by that of the dimer rather than the original solute. Also the solute may dissociate, and the two ions formed may diffuse as a pair together with the diffusion of any undissociated solute. Diffusion of acetic acid in water is an example:

$$CH_3COOH \rightleftharpoons CH_3COO^- + H^+$$

The diffusion rate of acetic acid is therefore determined by the diffusion of the undissociated species CH_3COOH and the diffusion of the ions CH_3COO^- and H^+.

A strong and complicated dependence on concentration and the pH of the solution can be expected in many cases (since the pH changes the extent of dissociation).

3. Multicomponent effects. For multicomponent systems the Stefan–Maxwell (S–M) model equation can be used for ideal-gas mixtures for an isobaric and isothermal system:

$$-\nabla y_i = \sum_{j=1}^{n_s} \frac{y_j N_i - y_i N_j}{CD_{ij}} \tag{9.54}$$

where n_s is the number of components and D_{ij} is the binary pair diffusivity in the gas phase. The equation relates the fluxes N_i to the concentration gradients as expected from a constitutive model. However, note that this is flux implicit. Fluxes are not provided directly as a function of concentration gradients and have to solved as a part of the solution procedure together with the species mass-balance equations.

A further complexity is that the fluxes are all coupled, which is evident from examination of the S–M equation, i.e., *the concentration gradient of any particular species is a function of fluxes of all the components present in the system.*

Some interesting and unexpected effects arise in multicomponent systems as a result of this coupling. These effects are not seen in ordinary diffusion (binary systems) and are briefly discussed below.

(a) Reverse diffusion. A species may diffuse in a direction opposite to its concentration gradient.

(b) Osmotic diffusion. A species diffuses even though the concentration gradient is zero.

(c) Barrier diffusion. A species does not diffuse even though there is a favorable concentration gradient.

It may be noted that these effects are not so common, but nevertheless they have been observed experimentally and validated in terms of theoretical considerations based on S–M equations.

The S–M model is not strictly applicable to dense gases or to liquid mixtures. However, this form is widely used even for such cases. One implication is that the diffusivity values are simply fitted values and do not have the meaning of binary pair diffusivity values for pair $i-j$. Also these fitted values are seen to be a strong function of concentration. Hence the analysis of diffusional effects in such systems is somewhat semi-empirical. Detailed analyses of problems using the S–M equations are provided in Chapter 21.

4. Reactive or facilitated transport. This is a complexity associated usually with transport in membranes. A chemical reaction binds the diffusing species and enhances its transport. Human life itself would not exist but for the oxygen–hemoglobin interactions. Similarly, many membranes (especially biological ones) have carriers, and the diffusing species binds to the carrier. Both the species and the carrier diffuse across the membrane and the transport is thereby "facilitated".

5. Restricted transport. Again this is of importance in transport in solids, in crystalline materials such as zeolites, membranes, etc. In these systems we encounter transport of small-diameter molecules. The size of these molecules is comparable to the size of the pores, and hence the diffusivity is restricted. The extent of the restriction depends on

the size of the solute in comparison with the size of the pores, and hence there is an engineering opportunity to design pores that are selective for particular solutes.

6. Thermal diffusion. A species diffuses due to the presence of a temperature effect. Additional mass flux resulting from thermal diffusion is modeled as

$$J_A = CD_{AB}k_T \nabla \ln T \qquad (9.55)$$

where k_T is the thermal diffusion factor.

In general, for binary systems, species with larger weight move to the colder region and *vice versa*.

Thermal diffusion is discussed further in Section 21.6, together with an illustrative example.

7. Pressure diffusion. A species diffuses as a result of a pressure gradient.

The contribution of pressure to diffusion is modeled as

$$J_A = -D_{AB}\frac{y_A}{R_G T}\left(1 - \frac{M_A}{\bar{M}}\right)\nabla p \qquad (9.56)$$

Pressure diffusion is discussed further in Section 21.5, together with an illustrative example.

8. Diffusion of charged species (e.g., ions). This is of importance in many systems such as electrochemical processes, fuel cells, etc.

See Chapter 22 for a complete discussion of this subject.

9. Active diffusion. Active transport refers to diffusion against a concentration gradient at the expense of some work done on the system. A common example is the Na^+–K^+ pump, which is common across all living systems. This "pump" involves transport of sodium ions against a concentration gradient using the hydrolysis of adenosine triphosphate (ATP) as the energy source. The free energy released in the hydrolysis is used to overcome the difference in chemical potential of Na^+ and a concentration difference of the order of 110 mol/m^3 exists for steady-state conditions across a cell membrane.

Summary

- At a gas–liquid interface (or for two phases in general) the concentration is discontinuous, with this effect being generally known as the jump condition at the interface. Thermodynamic equilibrium is assumed at the interface, which permits us to calculate or incorporate the concentration jump and relate the concentration at one side of the interface to that on the other side.

- In a multicomponent mixture, a velocity can be associated with each species. The velocity is used to find the fluxes in a stationary frame of reference. One can use mole or mass units freely here, and convert from one set to another.

- Averaging the species velocities gives a value for the velocity of the mixture as a whole. However, there is no unique way of averaging, thereby leading to many definitions for this average velocity.

- Two common ways of averaging are based on using either the mass-fraction or the mole fraction as the weighting factor. This leads to mass-averaged and mole-averaged velocities for the system. A flux can be defined on the basis of a coordinate system moving with either of these average velocities. This flux is called the diffusion flux. Thus we can have a diffusion flux based on mass-averaged velocity as a reference frame (j_A) or a diffusion flux based on a mole-averaged frame (J_A).

- In a binary mixture there is only diffusion flux. In a multicomponent system the sum of the diffusion fluxes taken for all the species is zero.

- Constitutive equations relate diffusion fluxes to the concentration gradient or more generally the mole (or mass) fraction gradient. Commonly Fick's law is suitable for binary systems or used as an approximation even for multicomponent systems. Now, depending on the frame of reference used to define diffusion flux, Fick's law can take different forms. The most common form relates J_{Ax} to dC_A/dx, which is used for isothermal and isobaric systems.

- Mass balance applied to a species leads to the differential equation for mass transfer. These equations can be written in either mass units or mole units. Mole units are more convenient for reactive systems.

- The boundary conditions used in conjunction with the mass-transfer equations are similar to those for heat transport. Thus the common conditions are of Dirichlet, Neumann, or Robin type. At the phase interface, fluxes are matched on either side, while a concentration-jump condition consistent with thermodynamic equilibrium is imposed for the concentrations.

- Integration of the differential models over an arbitrary control volume leads to the macroscopic mass-balance models discussed in Chapter 2.

- Binary diffusion caused by a concentration gradient is rather simple to describe, but in general many complexities are associated with diffusion. Diffusion can be caused by a pressure gradient, by thermal gradients, or by electric fields for charged species.

- Additional effects due to thermodynamic non-idealities are important in diffusion, mainly in diffusion in liquids. Here the nature of the species in solution (whether it is associated or dissociated etc.) can also affect the rate of diffusion.

- Fick's law is not suitable for multicomponent systems. A commonly used constitutive law is the Stefan–Maxwell equation. In some situations these equations cause strong coupling between the fluxes of various species. Complex effects such as reverse diffusion, osmotic diffusion, and barrier diffusion can arise due to these couplings.

- Reactive species present in the system can couple with the species being transported, causing a facilitated diffusion. In small pores the diffusion can be retarded if the size of the solute is relatively small compared with the pore size.

The application of the equations of mass transfer to many illustrative problems is taken up in the next chapter.

Problems

1. Show that mole fractions can be converted to mass fractions by the use of the following equation:

$$y_i = [\omega_i/M_i]\bar{M} \qquad (9.57)$$

Derive an expression for dy_i as a funciton of $d\omega_i$ values.

2. Show that mass fractions can be converted to mole fractions by the use of the following equation:

$$\omega_i = [y_i M_i]/\bar{M} \tag{9.58}$$

Derive an expression for $d\omega_i$ as a function of dy_i values.

3. At a point in a methane reforming furnace we have a gas of the composition 10% CH_4, 15% H_2, 15% CO, and 10% H_2O by moles.

Find the mass fractions and the average molecular weight of the mixture.

4. Express the Henry's-law constants reported in Table 8.1 as $H_{i,pc}$ and $H_{i,cp}$.

5. The Henry's-law constants for O_2 and CO_2 are reported as 760.2 l· atm/mol and 29.41 l· atm /mol.

What is the form of Henry's law used? Convert to values for the other forms shown in the text.

6. Solubility data for CO_2 are shown below as a function of temperature:

	Temperature (K)		
	280	300	320
H (bar)	960	1730	2650

Fit an equation of the type

$$\ln H = A + B/T$$

What is the physical significance of the parameter B?

7. The Antoine constants for water are $A = 8.071\,31$, $B = 1730.63$, and $C = 233.426$ in the units of mm Hg for pressure and °C for temperature.

Convert this to a form where pressure is in Pa and temperature is in K.

Also rearrange the Antoine equation to a form where temperature can be calculated explicitly. This represents the boiling point at that pressure.

What is the boiling point of water at Denver, CO (the so-called mile-high city)?

8. Given the Antoine constants for a species, can you calculate the heat of vaporization of that species?

9. Is n_A equal to $M_A N_A$? Why?

Is j_A equal to $M_A J_A$? Why?

10. Show the validity of the following relations between the species velocities v_A and v_B referred to a stationary frame of reference:

$$v_A - v_B = j_A \left(\frac{1}{\rho_A} + \frac{1}{\rho_B} \right)$$

and

$$v_A - v_B = J_A \left(\frac{1}{C_A} + \frac{1}{C_B} \right)$$

11. Show for a binary mixture the validity of the relation

$$j_A = \frac{M_A M_B}{M} J_A$$

12. Show the validity of the following form of Fick's law:

$$j_A = -\frac{C^2}{\rho} M_A M_B D_{AB} \nabla x_A$$

where x_A is the mole fraction.

13. Show the validity of the following equation based on Fick's law:

$$\nabla x_A = \frac{x_A N_B - x_B N_A}{C D_{AB}} \tag{9.59}$$

Show that the following equation holds as well:

$$\nabla x_A = \frac{x_A J_B - x_B J_A}{C D_{AB}} \tag{9.60}$$

Verify that the Stefan–Maxwell equation reduces to these equations when applied to a binary mixture.

14. **Velocity based on volume-fraction weighting.** If the volume fraction ϕ_a is used as the weighting factor, one can define an average velocity v^V as

$$v^V = \sum \phi_v v_i$$

Derive a form of Fick's law based on this definition of average velocity for a binary mixture. Note that the volume fraction of solute i is related to the partial molar volume of solute, $\hat{V}_i$, in the mixture by the following relation:

$$\phi_i = C_i \hat{V}_i$$

Also verify that the volume-fraction-averaged velocity is the same as the mole-fraction-averaged velocity when the total molar concentration (molar density) is constant, e.g., for a gas mixture.

Verify that volume-fraction-averaged velocity is the same as the mass-fraction-averaged velocity when the total mass density is constant, e.g., for a liquid mixture.

These observations led Cussler to conclude that the use of the volume-average velocity is more general, but most textbooks disguise the use of this averaging.

15. Knudsen diffusion is a phenomenon whereby transport takes place by gas–pore wall collisions rather than by gas–gas collisions. Knudsen diffusion is modeled using concepts similar to the kinetic theory of gases as $D_{KA} = (d_p/3)\bar{c}$, where $\bar{c}$ means the molecular velocity given earlier in Eq. (1.37). Use this to derive the following equation for the Knudsen diffusion coefficient:

$$D_{KA} = 4850 d_p \sqrt{\frac{T}{M_A}} \tag{9.61}$$

Calculate the value for silane diffusing in 10-μm-diameter pores at 900 K, which is representative of the deposition of solid silicon in thin tubes.

16. **Hindered diffusion** is a phenomenon of diffusion of a solute in narrow liquid-filled pores when the size of the pores is comparable to the size of the diffusing molecule. The hindered diffusion is often modeled by introducing two correction factors, F_1 and F_2, which are defined as follows:

$$F_1(\varphi) = (1 - \varphi)^2$$

which is simply a geometric factor known as the steric partition constant, and

$$F_2(\varphi) = 1 - 2.104\varphi + 2.09\varphi^3 - 0.95\varphi^5$$

which is known as the hydrodynamic hindrance factor. This is known as the Renken equation (Renken, 1954).

Here φ is the ratio of the solute's molecular diameter to the pore diameter. Find the factor by which the diffusivity is reduced for a large enzyme molecule of size 4 nm diffusing in a nanoporous liquid-filled membrane of diameter 30 nm.

10 Illustrative mass transfer problems

Learning objectives

- To understand the differences between the formulations for flowing and stagnant systems; to appreciate the role of convection in "stagnant" systems.
- To introduce the film theory for mass transfer from a surface and the two-film theory for gas–liquid mass transfer.
- To study the effects of mass transfer for systems with heterogeneous reactions at a catalyst surface.
- To study the effect of diffusion on the rate of reaction in a porous catalyst and to understand the concept of the effectiveness factor.
- To examine the effect of reaction on the rate of absorption of a gas into a liquid on the basis of the film model.
- To study the basic models for membrane transport in gaseous systems.
- To study mass-transport models for semi-permeable membranes for liquid systems.
- To study the effect of reaction due to a carrier for reacting membranes, and to study the numerical solution of boundary-value problems using the BVP4C solver.

Mathematical prerequisites

No new mathematical tools are needed for this chapter. Students should refresh their knowledge of solution methods for ordinary second-order differential equations.

This chapter demonstrates the use of equations of mass transfer for illustrative problems in mass transfer. This chapter is organized in a manner somewhat parallel to Chapter 8, so that the student can appreciate the commonalities in the analysis of heat and mass transfer problems. It would be appropriate at this point to classify the types of problems analyzed in this chapter so that the reader can get an overall perspective.

The chapter starts with steady-state problems with no superimposed external flow in the system. Such problems are often referred to as diffusion in stagnant media. The terminology stagnant can be misleading in many cases, which will become clear during the detailed discussion. Steady-state problems with no reaction are analyzed first, and problems posed in one spatial dimensions are considered. Such problems have applications, for instance, (i) in diffusivity measurements (the Arnold cell), (ii) in developing mass transfer models, and (iii) in membrane transport. Examples of such applications are presented.

We then proceed to analyze systems with chemical reactions. Two cases will be considered, namely (i) heterogeneous reactions and (ii) homogeneous reactions, and the difference in the model formulation will be clarified. For heterogeneous reactions (reactions occurring at a solid or catalyst surface), the rate term appears only as a boundary condition and does not appear in the differential equation. For a homogeneous reaction the rate appears in the differential equation. The equation has a similar form to a volumetric source/sink of heat in heat-conduction problems, and the various mathematical methods of solution for such problems studied in Chapter 8 can be put to effective use here. Important applications of such models are in diffusion with reaction in a porous catalyst and in gas–liquid reactions, and two detailed sections are devoted to these topics.

The cases of transport in membranes and semi-permeable systems are treated next. For semi-permeable membranes the concept of osmotic pressure is needed, and the applied pressure driving force has to be modified to account for the effect of osmotic pressures. Models for transport in such systems are developed and illustrated by examples. Finally, reactive membranes are analyzed by use of a diffusion–reaction model. Reactive membranes have a wide variety of applications in biological transport and some novel separation processes.

The problems analyzed in this chapter are quite important in practical applications, and provide a framework for design of a wide class of mass-transfer and reactor equipment. After a detailed study of this chapter, you will develop an ability to formulate and solve a wide range of mass transfer problems and also understand the basic methodology for the design of mass-transfer equipment.

10.1 Steady-state diffusion: no reaction

10.1.1 Summary of equations

A summary of equations is provided here for convenience.

Species mass balance
The basic balance equation which is widely used is the species balance derived in Chapter 9, which is repeated here for ease of reference:

$$\frac{\partial C_A}{\partial t} + \nabla \cdot N_A = R_A \tag{10.1}$$

For the present case of a steady state and no reaction this simplifies to

$$\nabla \cdot N_A = 0 \tag{10.2}$$

where the flux term N_A is the combined flux which can be expressed as a sum of the diffusion flux and a convection flux:

$$N_A = J_A + y_A N_T \tag{10.3}$$

The flux model

In this chapter we use Fick's law for the diffusion flux J_A, which is, strictly speaking, valid for ideal binary systems:

$$J_A = -C D_A \nabla y_A$$

Here D_A is the binary pair diffusivity, which is equal to D_{AB} for a truly binary mixture. For multicomponent mixtures D_A is interpreted as a pseudo-binary diffusivity D_{A-m} defined by the Wilke equation in Section 9.4, which is assumed to have a constant value in the domain of solution.

Using this in Eq. (10.3) the flux expression required in Eq. (10.2) is

$$N_A = -C D_A \nabla y_A + y_A N_T \tag{10.4}$$

Here N_T is the total molar flux, which for a binary system is

$$N_T = N_A + N_B$$

The total flux depends on the prevailing physical situation, and hence a constraint on N_T or some condition for this quantity has to be provided in order to complete the problem formulation. This is known as the determinacy condition, and examples will be provided later. Two common examples are as follows.

1. Diffusion of a species A in an inert or non-diffusing B. Here $N_A = N_T$ is used as the determinacy condition. Thus this is also often called unimolecular diffusion or UMD.
2. Diffusion of a species A with an equal counter-diffusion of species B. Here $N_T = 0$ is used as the determinacy condition. Thus this is also often called equimolecular diffusion or EMD.

Forced-convection systems

Note that Eq. (10.4) can also be expressed in terms of the system velocity as

$$N_A = -C D_A \nabla y_A + v^* C_A \tag{10.5}$$

which is useful for convection-dominated systems since the velocity is known from flow considerations for such systems. In such cases of forced-flow problems N_T is mainly due to the imposed external velocity in the system. This is usually specified from the flow conditions. The diffusion term is neglected in the flow direction in such cases. Problems where convection dominates are treated in a later chapter.

Stagnant systems

Here we consider systems with no superimposed flow, so-called stagnant systems. In such stagnant systems, N_T is generated by the diffusion process itself. Thus there may be a net velocity in a seemingly stagnant system. This is not often obvious, but can be of importance in some cases. In many problems this contribution has to be accounted for in addition to the diffusional contributions. We will study how to do this later in this chapter. In many cases, a simplifying assumption is made, and it is assumed that the contribution of the N_T term in (10.4) can be neglected. This is referred to as the low-mass-flux approximation.

The low-mass-flux model approximation

In this case

$$N_A \approx J_A = -CD_{AB} \nabla y_A \tag{10.6}$$

On combining Eqs. (10.6) and (10.2) we have

$$\nabla^2 y_A = 0 \tag{10.7}$$

For a system of constant concentration we can also write this as

$$\boxed{\nabla^2 C_A = 0} \tag{10.8}$$

Thus the concentration field satisfies the Laplace equation with the assumptions stated above. Note that we encountered the Laplace equation for steady-state heat transfer as well and looked at some illustrative solutions. Similar methods hold for mass transfer as well, and Example 10.1 discusses simple mass transfer in common geometries.

Example 10.1. Steady-state diffusion in three common geometries.

Three geometries, namely a slab, a long concentric cylinder, and a sphere, are considered here and shown schematically in Fig. 10.1. It is useful to represent the Laplace equation in a general way as

$$\frac{1}{\xi^s} \frac{d}{d\xi} \left(\xi^s \frac{dc_A}{d\xi} \right) = 0 \tag{10.9}$$

where s is called the shape factor and is equal to 0, 1, and 2 for a slab, a long cylinder, and a sphere, respectively. ξ is the distance coordinate in the direction of diffusion.

The concentration profile is identical to the temperature profile. The concentration profile in a slab is linear, in a hollow cylinder it is logarithmic, and for a spherical shell it is an inverse function of r.

The corresponding expressions for the numbers of moles transported across the system are also shown in Fig 10.1 and are useful in many applications.

The second example below shows the analysis for UMD.

Example 10.2. Liquid evaporating in an open container: the low-flux model.

The problem of evaporation in a tube was shown schematically in Fig. 9.1. Both diffusion and convection may need to be considered in determining the rate of mass transfer.

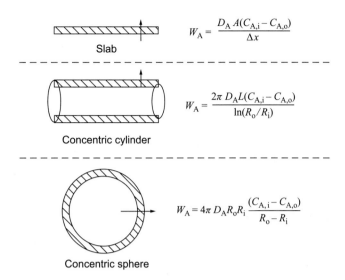

Figure 10.1 Basic steady-state diffusion problems with the corresponding expressions for the total number of moles transported across the system, which are important in many applications.

An apparatus of this type is called an Arnold cell. The level decreases as a function of time in the Arnold cell, and the change in level can be used to determine the diffusivity of the evaporating gas in the carrier gas (see Problem 2). It is in reality a transient diffusion problem, but for modeling purposes we focus on a particular instant in time at which the vapor space has a height H and assume a steady state. This assumption is called the pseudo-steady-state hypothesis and holds when the time scale for level change is much larger than the time scale for diffusion. The rate of mass transfer is now calculated under this assumption.

For the low-mass-flux case (N_T approximated as zero), the mole-fraction profile is linear as dictated by Eq. (10.7). Correspondingly the flux can be calculated as

$$N_A^\circ = \frac{D_A C}{H}(y_{A,s} - y_{A,b}) \tag{10.10}$$

We use the superscript o to show that the result is applicable to a low-mass-flux case. Corrections to this expression, including the convection contribution will now be presented.

Evaporation: the high-flux model

The evaporation causes a flow; thus N_T is not equal to zero here and the correction for this can be done with the high-flux model.

We assume that the component B has no flux in a stationary frame of reference. Hence $N_B = 0$ and $N_T = N_A + N_B = N_A$. This is the UMD model. The expression for the combined flux can now be written as

$$N_A = -CD_{AB}\nabla y_A + N_A y_A \tag{10.11}$$

On rearranging we have the flux expression for UMD:

$$N_A = -\frac{CD_{AB}}{1 - y_A} \nabla y_A \tag{10.12}$$

Note the presence of the $1 - y_A$ term in the denominator. This term is 1 in the low-flux model.

For 1D case, $\nabla y_A = dy_A/dz$, where z is the direction of diffusion. Hence the flux expression ($N_{A,z}$ abbreviated as N_A since flux is 1D) can be written, after separating the variables, as

$$N_A \, dz = -CD_{AB} \frac{dy_A}{1 - y_A} \tag{10.13}$$

Upon integrating across the system (keeping N_A constant), we get the rate of evaporation as

$$\boxed{N_A = -\frac{CD_{AB}}{H} \ln[(1 - y_{A,s})/(1 - y_{A,b})]} \tag{10.14}$$

where $y_{A,s}$ is the mole fraction at the evaporating surface $z = 0$ and $y_{A,b}$ is the mole fraction in the bulk gas, i.e., at $z = H$ and beyond.

Drift flux correction factor

The expression can be compared with the low-mass-flux model directly to ascertain the effect of "flow" in the system. The correction factor due to flow can be expressed as a drift flux factor and the following expression can be derived:

$$\text{Correction factor } \mathcal{F} = \frac{N_A}{N_A^\circ} = \frac{\ln[(1 - y_{A,s})/(1 - y_{A,b})]}{y_{A,b} - y_{A,s}} \tag{10.15}$$

The correction factor can also be written in terms of the mole fraction of B using the relation $y_B = 1 - y_A$ as

$$\mathcal{F} = 1/\log \text{ mean value of mole fraction of B}$$

where the log mean is defined as

$$\log \text{ mean } y_B = \frac{y_{B,b} - y_{B,s}}{\ln(y_{B,b}/y_{B,s})}$$

The calculation of the mole-fraction profile requires another integration and is illustrated in Example 10.3.

Example 10.3.

Find an expression for the mole-fraction profile in the evaporating system with a high-mass-flux model.

Integration of Eq. (10.13) is now done from 0 to any arbitrary position z rather than across the complete system:

$$N_A \int_0^z dz = -CD_{AB} \int_{y_{A,s}}^{y_A} \frac{dy_A}{1 - y_A} \tag{10.16}$$

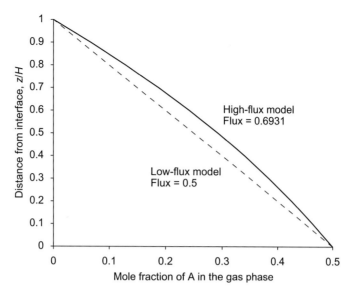

Figure 10.2 Evaporation in a tube, showing a comparison of concentration (mole-fraction) profiles with low- and high-flux models. Note that $y_{A,s} = 0.5$ and $y_{A,b} = 0$ are used. Flux is scaled by $D_A C/H$.

which yields

$$zN_A = CD_{AB} \, \ln[(1 - y_{A,s})/(1 - y_A)]$$

This can be rearranged to

$$y_A = 1 - (1 - y_{A,s})\exp\left(\frac{zN_{A,z}}{CD_A}\right)$$

We note that the profile is no longer linear, in contrast to the low-flux model (see Figure 10.2).

It will be illustrative to compare the effect of drift by means of a simple example.

Example 10.4.
A volatile liquid is in a tube with an exposed vapor height of 0.3 m. The vapor pressure is 0.15 atm under the given conditions of temperature 35 °C and pressure 1 atm. The diffusion coefficient is estimated as 8.8×10^{-6} m^2/s.

Find the rate of evaporation.

Solution.
The total concentration is $C = P/(R_G T)$, $= 39.56$ mol/m^3.

The flux from the low-flux model given by Eq. (10.12) is 1.04×10^{-4} mol/m$^2 \cdot$ s.

The flux from the high-flux model from Eq. (10.14) is 1.1318×10^{-4} mol/m$^2 \cdot$ s.

The correction factor is 1.1230; thus convection augments diffusive transport by about 12%.

The correction factor can also be found by calculation of the log mean mole fraction of the inert species, $(1 - 0.85)/\ln(1/0.85) = 0.9230$. The correction factor is equal to the reciprocal log mean mole fraction of the inert species. The two procedures give the same answer, as expected.

The equations, although derived for the evaporation of liquid, find wide application in the field of mass transfer. First we discuss the classical film theory for mass transfer where such equations find use.

10.2   The film concept in mass-transfer analysis

10.2.1   Fluid–solid interfaces

Near a solid surface there is a boundary layer where the velocity changes from zero to a bulk value over a small distance called the hydrodynamic boundary layer. Similarly, there is a region where the concentration changes, and this region is called the mass transport boundary layer. A typical concentration profile in the boundary layer is shown in Fig. 10.3. It is useful first to state some results from the boundary-layer theory to establish a background for the more simple-minded film model.

The flux at the surface can be calculated from the boundary layer model which is discussed later (Section 20.1.4). The concentration profile is often approximated as a cubic polynomial, and an expression for the boundary-layer thickness δ_m can be derived. The boundary-layer thickness is often correlated as a function of the momentum boundary-layer thickness and a Schmidt number defined as v/D_A.

$$\delta_m = \frac{\delta}{Sc^{1/3}} \tag{10.17}$$

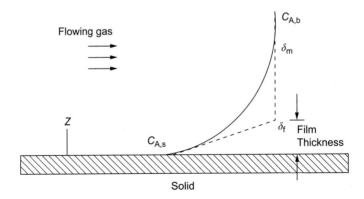

Figure 10.3 The concentration profile near a solid surface and its approximation by the film model; the solid line is a profile as envisaged by boundary-layer theory and the dashed line is an approximation according to the film model.

The mass transfer coefficient is predicted from this equation (for a cubic concentration profile) as

$$k_m = \frac{3}{2}\frac{D}{\delta_m}$$

This is a local mass transfer coefficient at any position x along the flat plate and varies along the plate since δ and δ_m are functions of x.

A more complete analysis is presented in a later chapter.

The film model

The film model provides a simple description of the mass transport. If a straight line is used to represent the concentration profile in Fig. 10.3 we find that there is a region of thickness δ_f over which the concentration profile changes, and, using Fick's law, the flux can be represented as

$$N_A^\circ = \frac{D_A}{\delta_f}(C_{A,s} - C_{A,b}) \tag{10.18}$$

Representation of the concentration distribution by a linear function and the concept of a film-thickness parameter δ_f form the basis for the film model for mass transfer. This was proposed as early as 1904 by Nernst. Although the film is a hypothetical concept, it has achieved widespread success in modeling mass transport processes, as we will observe during the course of this chapter.

Equivalently, a mass transfer coefficient can be defined to characterize the mass transfer near a solid with a flow past the solid,

$$N_A^\circ = k_m^\circ(C_{A,s} - C_{A,b}) \tag{10.19}$$

The mass transfer coefficient is therefore related to the film thickness by

$$k_m^\circ = \frac{D_A}{\delta_f} \tag{10.20}$$

The superscript o indicates that the coefficient is based on Fick's law and is therefore applicable to low-flux conditions.

If the low-flux mass transfer is not applicable, the flux can be defined as the low flux rate times a correction factor:

$$N_A = k_m^\circ(C_{A,s} - C_{A,b})\mathcal{F}_m \tag{10.21}$$

The correction factor depends on the prevailing situation, mainly on what other species are being transported. Note that the correction factor is given by Eq. (10.15) for UMD as before. It is equal to one for EMD, since N_T is zero in this case. An expression for the correction factor can be obtained in a rather general way as explained in the following discussion. Let us first see what the concentration profile looks like in the presence of (diffusion-induced) convection.

Example 10.5.

Derive an expression for the concentration profile for the film model which includes convection effects. Also relate the flux of A to the total flux in the system.

Solution.

The expression for the combined flux is used in the presence of convection. Thus we have

$$N_A = -CD\frac{dy_A}{dz} + y_A N_T$$

We can use CD/δ as a scaling parameter for the fluxes. Thus

$$N_A^* = -\frac{dy_A}{d\zeta} + y_A N_T^*$$

where ζ is the dimensionless distance in the film z/δ and the starred fluxes are dimensionless fluxes.

Integrate the above differential equation using the boundary condition y_{A0} at $\zeta = 0$ to obtain

$$y_A = \frac{N_A}{N_T} + \left(y_{A0} - \frac{N_A^*}{N_T^*}\right)\exp(N_T^*\zeta)$$

Now use the boundary condition at $\zeta = 1$, *viz.*, $y = y_{A\delta}$. Then rearrange. We obtain the following equation for N_A:

$$N_A^* = N_T^* \frac{y_{A\delta} - y_{A0}\exp(N_T^*)}{1 - \exp(N_T^*)} \qquad (10.22)$$

This expression can be used for a wide range of problems with a suitable condition specified for N_T, as shown below.

More on the determinacy condition for N_T

The conditions for N_T can be stated on a case-by-case basis. Let us review some common cases and state the relations.

1. Low-flux mass transfer.

   What is the limit of the LHS of the expression (10.22) for this case N_T^* tending to zero?

   Taking the limit of (10.22) as $N_T^* \to 0$, we have

   $$N_A^* = y_{A0} - y_{A\delta}$$

   which can be obtained by direct application of Fick's law.

2. UMD or evaporation of a pure liquid (also called Arnold diffusion). We use $N_T^* = N_A^*$ here, and rearrange Eq. (10.22):

   $$N_A^* = \ln\left(\frac{1 - y_{A\delta}}{1 - y_{A0}}\right)$$

3. EMD or equimolar counter-diffusion. Here $N_T^* = 0$, but the reasoning is different and the same result as for low flux is obtained.

4. Distillation of a binary mixture. Here a species A evaporates from a liquid interface to the vapor, and a species B condenses from the vapor. No heat is generally added in the distillation column. Hence the heat of vaporization must balance, which provides the determinacy condition.

$$N_A \,\Delta H_{vA} + N_B \,\Delta H_{vB} = 0$$

which relates N_A and N_B.

Note that if the heats of vaporization of A and B are equal then the equimolar counter-diffusion model holds.

5. Diffusion with a heterogeneous reaction at a surface. N_A and N_B are determined by the stoichiometry of the reaction. For example, for A → vB,

$$N_B = -vN_A$$

More discussion is given in Section 10.3.

We now extend the film model for gas–liquid interfaces. This necessitates consideration of two films.

10.2.2   Gas–liquid interfaces: the two-film model

Lewis and Whitman (1924) introduced the concept of two films at the interface, one on the gas side and one on the liquid side. The concentration profile according to this model is shown schematically in Fig. 10.4. Note that the profiles are linear, as shown in the figure, only for the low-flux model.

The mass transfer coefficients on each side can be defined as follows:

$$k_m = D_{AG}/\delta_G$$

where k_m is based on the concentration driving force. The flux on the gas side of the interface is therefore represented as

$$N_A = k_m(C_{AG} - C_{AG,i})$$

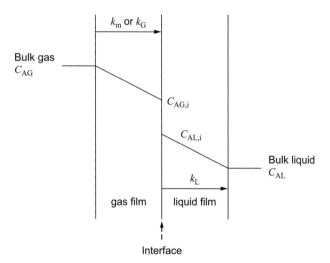

Figure 10.4 The two-film model. Two stagnant films are assumed, one on each side of the interface, and the resistance to mass transfer is assumed to be confined to these films.

Partial pressure units are more convenient and common for the gas phase, and we can also use a mass transfer coefficient k_G:

$$N_A = k_G(p_{AG} - p_{AG,i})$$

The two mass transfer coefficients are related by

$$k_G = \frac{k_m}{R_G T}$$

A similar definition can be used for the liquid film:

$$k_L = D_{AL}/\delta_L$$

where k_m is based on the concentration driving force:

$$N_A = k_L(C_{AL,i} - C_{AL})$$

The interfacial concentrations are related by a Henry's-law type of relation.

Let $p_{AG,i} = H_A C_{AL,i}$, where H_A is in pressure–concentration form.

Then an overall driving force can be derived and used. The overall driving force can be expressed in two ways: (i) as a partial pressure (gas-phase) driving force and (ii) as a concentration (liquid-phase) driving force.

Gas-phase driving force

Let $p^*_{AG} = H_A C_{AL}$ be a hypothetical partial pressure if the bulk liquid were to attain equilibrium. Then the driving force is $p_{AG} - p^*_{AG}$ and correspondingly an overall transfer coefficient can be defined as

$$N_A = K_G(p_{AG} - p^*_{AG})$$

It can be shown that

$$\frac{1}{K_G} = \frac{1}{k_G} + \frac{H_A}{k_L}$$

The two terms on the RHS may be interpreted as resistances, one on the gas side and the other on the liquid side.

Liquid-phase driving force

The overall liquid concentration driving force is defined as $C^*_{AL} - C_{AL}$, where $C^*_{AL} = p_{AG}/H_A$ is a hypothetical concentration corresponding to the bulk gas conditions. The rate of transfer is then defined using an overall coefficient K_L:

$$N_A = K_L(C^*_{AL} - C_{AL})$$

It can be shown that

$$\frac{1}{K_L} = \frac{1}{H_A k_G} + \frac{1}{k_L}$$

which relates the overall liquid-based mass transfer coefficient to the gas-side and liquid-side coefficients.

The controlling resistance

The controlling resistance may be on the gas side or the liquid side, depending on the value of H_A.

If H_A is small (highly soluble gas) then the controlling resistance is on the gas film. $K_G = k_G$ then.

If H_A is large (poorly soluble gas) then the controlling resistance is on the liquid film. $K_L = k_L$ then.

10.3 Mass transfer with surface reaction

10.3.1 Heterogeneous reactions: the film model

The low-flux model: first-order reaction

If now a chemical reaction occurs at the surface, then the mass transfer and reaction can be treated as two resistances in series. Let k_s be the rate constant for the reaction. The flux to the surface is equal to the rate of consumption by reaction Thus

$$N_A = k_m(C_{A,b} - C_{A,s}) = k_s C_{A,s} \qquad (10.23)$$

These expressions can be used to eliminate $C_{A,s}$, and the rate can be then expressed in terms of an overall rate constant k_o as

$$N_A = k_o C_{A,b} \qquad (10.24)$$

where k_o is given by

$$\frac{1}{k_o} = \frac{1}{k_m} + \frac{1}{k_s} \qquad (10.25)$$

This looks like two resistances being added. Thus the mass transfer and the reaction resistances operate in series, and the overall resistance is the sum of these two resistances.

These are the results of the simple film model for a heterogeneous first-order reaction.

Modification to account for flow in the direction of diffusion is now discussed, and the effects of counter-diffusion of products are accounted for in this revised model.

High flux: the effect of product counter-diffusion

Now consider a reaction scheme:

$$A \rightarrow (\nu_E + 1)B$$

where the stoichiometric coefficient for B is written in a funny way for ease of later algebra. The stoichiometry requires

$$N_B = -(\nu_E + 1)N_A$$

Hence

$$N_T = N_A + N_B = N_A - (\nu_E + 1)N_A = -\nu_E N_A$$

Hence the expression for the combined flux is

$$N_A = -\frac{CD_A}{1 + \nu_E y_A}\frac{dy_A}{dz}$$

It is also convenient to write this in dimensionless form as

$$N_A^* = \frac{N_A}{CD_A/\delta_f} = \frac{N_A}{k_m C}$$

and therefore the expression for the dimensionless combined flux is

$$N_A^* = -\frac{1}{1 + \nu_E y_A} \frac{dy_A}{d\zeta}$$

where ζ is the dimensionless distance in the film, z/δ_f.

Since N_A^* is a constant in the film, i.e., it is not a function of z, the expression may be integrated across the boundary layer to give

$$N_A^* = \frac{1}{\nu_E} \ln \left(\frac{1 + \nu_E y_{A0}}{1 + \nu_E y_{A\delta}} \right) \qquad (10.26)$$

What is the limit of the above expression for $\nu_E = 0$?

An additional condition is required for fixing $y_{A\delta}$. This comes from the rate of reaction at the interface. We can equate the flux at δ to the rate of surface reaction (per unit area). For a first-order surface reaction, the condition may be stated as

$$N_A = C k_s y_{A\delta}$$

It is also convenient to write this in dimensionless form as

$$N_A^* = (k_s/k_m)y_{A\delta} = Da_s y_{A\delta} \qquad (10.27)$$

where Da_s is a dimensionless number, the Damköhler number for surface reaction, which is defined as k_s/k_m for a first-order surface reaction.

The above two equations, Eqs. (10.26) and (10.27), provide the required model. Note that N_A (or N_A^*) has to be solved together with $y_{A\delta}$ and the simple addition of resistance concept is no longer applicable even for a first-order reaction.

For the limiting case of fast reaction, one can approximate $y_{A\delta}$ as zero, and the following limiting case of Eq. (10.26) holds:

$$N_A^* = \frac{1}{\nu_E} \ln(1 + \nu_E y_{A0}) \qquad (10.28)$$

For slow reaction $y_{A\delta} = y_{A0}$ and mass transfer resistance is no longer limiting the process.

10.4 Mass transfer with homogeneous reactions

We now discuss mass transfer with homogeneous reaction. The reaction taking place in a porous catalyst is often modeled in this manner since the diffusion and reaction take place in parallel here. First we discuss how diffusion in porous media is modeled.

10.4.1 Diffusion in porous media

Diffusion in a single cylindrical pore can occur by two mechanisms, as shown in Fig. 10.5. For small pores the transport occurs by pore-wall–gas collision and Knudsen diffusion (see Problem 15 in Chapter 9) is dominant. The criterion for small pores is that the pore diameter d_p should be much smaller than λ, the mean free path of the gas molecules. For large pores, gas–gas collision is the dominant mechanism and the diffusion can be modeled by the binary gas pair diffusivity D_{AB}. The overall diffusion is modeled by a combined model as

10.4 Mass transfer with homogeneous reactions

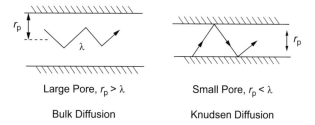

Large Pore, $r_p > \lambda$ Small Pore, $r_p < \lambda$

Bulk Diffusion Knudsen Diffusion

Figure 10.5 Mechanisms for diffusion in the pores of a catalyst.

$$\frac{1}{D_{pore}} = \frac{1}{D_K} + \frac{1}{D_{AB}} \tag{10.29}$$

where D_K is the Knudsen diffusion coefficient (see Problem 15 in Chapter 9) and D_{AB} is the gas phase (binary pair) diffusivity.

For a porous catalyst one defines an effective diffusivity that incorporates the pore structure:

$$D_e = D_{pore}\epsilon/\tau$$

where ϵ is the catalyst porosity and τ is a tortuosity factor. The factor ϵ accounts for the reduced area for diffusion, while the factor τ is the correction for non-straight pores.

10.4.2 Diffusion and reaction in a porous catalyst

Using the effective diffusivity, the model equation for reaction and diffusion in a porous catalyst can be set up.

The model equation is the diffusion–reaction equation of the form

$$D_e \nabla^2 C_A + R_A = 0$$

where C_A is the concentration of the diffusing species in the catalyst and D_e is an effective diffusivity of the species A in the pores of the catalyst. The assumptions are that we have a steady state and that the molecular diffusion is dominant. The convective flux terms are neglected here. Note that the problem is analogous to heat conduction with heat generation, so we can use similar solution methods. Again the method of solution depends on the kinetics of the reaction. For zeroth- and first-order kinetics analytical solutions can be obtained. Zeroth-order reactions have to be given special consideration, as will be shown later. Non-linear kinetics can be solved either by p-substitution for simple slab geometry or by numerical methods.

We first present results for a first-order reaction for the three common geometries where the Laplacian can be represented as a function of one spatial coordinate. These geometries are (i) slab, (ii) long solid cylinder, and (iii) solid sphere.

10.4.3 First-order reaction

For a first-order reaction,

$$R_A = -kC_A$$

The diffusion–reaction equation can be represented in dimensionless form for three common geometries as

$$\frac{1}{\xi^s}\frac{d}{d\xi}\left(\xi^s\frac{dc_A}{d\xi}\right) = \phi^2 c_A \tag{10.30}$$

The concentration c_A is $C_A/C_{A,b}$, where $C_{A,b}$ is the concentration in the bulk gas outside the surface, while ξ is a dimensionless spatial location equal to x/L or r/R. The parameter s is the geometry parameter as defined earlier.

The key dimensionless parameter in this problem is ϕ, which is called the Thiele modulus:

$$\phi^2 = kR^2/D$$

where R is the radius for the long cylinder and sphere and is taken as half the length of the slab (L) for the slab case. The Thiele modulus can be interpreted as the ratio of the characteristic time for diffusion R^2/D to the time for reaction $1/k$. It can also be interpreted as the ratio of the relative rate of reaction to the rate of diffusion within the pores.

The boundary condition at the plane of symmetry or the center is that, at $\xi = 0$, we impose $dc_A/d\xi = 0$.

The boundary condition at the surface ξ depends on whether external mass-transfer effects are accounted for or not. We postulate the existence of a thin film over a boundary layer near the surface in accordance with the film theory. This thin film offers all the resistance for the gas transport from the bulk to the surface of the catalyst. The balance of fluxes at the surface can then be represented as

$$k_m(C_{A,b} - C_{A,s}) = D_e\left(\frac{dC_A}{dx}\right)_{x=L}$$

where k_m is the external (gas film) mass transfer coefficient and the LHS above represents the number of moles of A transported from the gas to the surface. The RHS is the rate of diffusion into the catalyst surface. The dimensionless version of the condition is as follows and is of the third kind (Robin):

$$\left(\frac{dc_A}{d\xi}\right)_{\xi=1} = Bi[1 - (c_A)_{\xi=1}]$$

where Bi for external mass transfer is defined as

$$Bi = \frac{k_m R}{D_e}$$

The Biot number represents the ratio of the external transport rate to the internal transport rate and has the same significance as in heat transfer. For large values of Bi we would expect the problem to reduce to the Dirichlet type. Hence, if there are no transport resistances on the film near the catalyst then the Dirichlet condition $c_A = 1$ is applied at $\xi = 1$.

The parametric representation of the problem is therefore

$$c_A = c_A(\xi; \phi, Bi)$$

For large values of Bi there is no dependence on Bi and the solution depends only on ϕ. The solutions for the three geometries are now presented.

Slab solution: $s = 0$; Dirichlet

The solution for large Bi is as follows:

$$c_A = \frac{\cosh(\phi\,\xi)}{\cosh(\phi)}$$

The effectiveness factor is the quantity of interest in the design of a catalytic reactor and is defined in words as

$$\text{effectiveness factor} = \frac{\text{actual rate}}{\text{maximum rate}}$$

The maximum rate is the rate based on the external gas concentration. The actual rate can be calculated in two ways: (i) by taking the local rate and integrating over the whole catalyst or (ii) by calculating the rate of transport at the catalyst surface. In the first case we have to integrate the concentration profile, while in the second case we need the derivative of the concentration profile at the surface (in order to apply Fick's law here). Both cases lead to the same answer.

The effectiveness factor is calculated as the average rate of reaction in dimensionless units:

$$\eta = \int_0^1 c_A\, d\xi = \frac{\tanh(\phi)}{\phi} \qquad \text{Dirichlet problem}$$

The concentration profiles are shown in Fig. 10.6, together with illustrative values for the effectivness factor.

Slab: external transport effects

The solution for any general case where a Robin condition applies is slightly lengthy and is

$$c_A = \frac{Bi\cosh(\phi\,\xi)}{\phi\sinh(\phi) + Bi\cosh(\phi)} \tag{10.31}$$

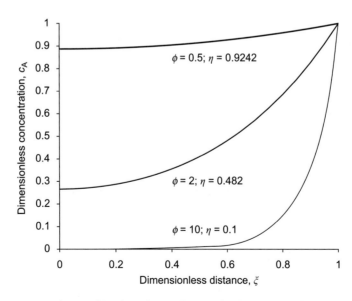

Figure 10.6 Concentration profiles for a first-order reaction in a catalyst in a slab shape.

The student should verify the algebra and also show that the Dirichlet limit is approached as $Bi \to \infty$. Such limiting cases should always be checked in general.

The effectiveness-factor expression for the above case can be derived as

$$\eta = \frac{\tanh(\phi)}{\phi} \left(\frac{Bi}{Bi + \phi \tanh(\phi)} \right)$$

The calculations can be automated in MAPLE, which reduces the algebraic burden. However, it may be good for students to derive these equations on paper without any computational crutches.

Long cylinder: Dirichlet problem

Here $s = 1$ and the differential equation is now the modified Bessel equation. This can be seen by expanding the Laplacian and rewriting Eq. (10.30) as

$$\xi^2 \frac{d^2 c_A}{d\xi^2} + \xi \frac{dc_A}{d\xi} - \xi^2 \phi^2 c_A = 0$$

The general solution is

$$c_A = A_1 I_0(\phi\xi) + A_2 K_0(\phi\xi)$$

The student should verify the general solution using MAPLE or using the information given in advanced engineering mathematics textbooks such as Kreyszig (2011).

The constants of integration are readily found for a solid cylinder. The function K_0 goes to infinity as $r \to 0$ and hence A_2 is set as zero so that the concentration will remain finite at the center. The second constant is then fitted from the boundary condition at $\xi = 1$, which may be of Dirichlet or Robin type. For the Dirichlet type (infinite Biot number) the solution is

$$c_A = \frac{I_0(\phi\xi)}{I_0(\phi)}$$

The effectiveness factor can be shown to be

$$\eta = 2 \int_0^1 \xi c_A \, d\xi$$

Why is this defined in such manner? Why the factor of two?

An alternative expression for the effectiveness factor is

$$\eta = \frac{2}{\phi^2} \left(\frac{dc_A}{d\xi} \right)_{\xi=1}$$

Students should verify that this will be the expression from flux considerations at the surface and that both expressions will lead to the same final answer, which is

$$\eta = \frac{2}{\phi} \frac{I_1(\phi)}{I_0(\phi)} \qquad (10.32)$$

The solution to the case of finite Biot number is left as an exercise.

Sphere

Here $s = 2$ and Eq. (10.30) needs some change of variables for integration purposes. A variable transformation $Y = \xi c_A$ reduces the equation to a form that can be integrated more easily (see Problem 10). The solution is

$$c_A = \frac{\sinh(\phi\xi)}{\xi\sinh(\phi)}$$

and the effectiveness factor is given by

$$\eta = \frac{3}{\phi^2}(\phi\coth(\phi) - 1)$$

10.4.4 Zeroth-order reaction

This is often important, especially in biological systems, since many metabolic reactions can often be approximated as zeroth-order processes. Although the model equation is straightforward, special considerations are needed, since the concentration value can actually become zero at some position in the diffusion path. This is unlike a first-order reaction where the concentration decays as an exponential function (which can, of course, become very small, but never becomes actually zero). For a zeroth-order reaction the concentration (a parabolic function as shown below) can actually become zero at some point within the porous catalyst, especially if diffusion is slow compared with the reaction rate. Hence we have to demarcate the position where the concentration actually becomes zero and analyze the problem in two regions separately. A simple example for a slab geometry is illustrated and other cases are left as exercises. Also we analyze an important problem in biomedical systems, *viz.*, oxygen transport in tissues, which is also known as the Krogh cylinder problem.

The slab analysis proceeds as follows with the basic diffusion–reaction equation:

$$D_e\frac{d^2C_A}{dx^2} - k_0 = 0$$

where the production rate of A, R_A, has been replaced by substitution of k_0, the zeroth-order rate constant for the metabolic process. The boundary conditions are set as follows

At $x = L$, $C_A = C_{A,s}$, a prescribed or known surface concentration.

At $x = 0$, the center of the slab, we use the no-flux condition $dC_A/dx = 0$.

The center concentration is assumed to be above zero so that the above condition can be applied. As we proceed we will find the condition under which the center concentration becomes exactly zero. The boundary conditions have to be modified beyond that case. The case where the concentration at the center becomes zero will be called the starvation condition.

The solution for the concentration profile is very simple:

$$c_A = 1 - \frac{\phi_0^2}{2} + \frac{\phi_0^2}{2}\xi^2 \tag{10.33}$$

where the parameter ϕ^2 is the square of the Thiele modulus and is defined as

$$\phi_0^2 = \frac{k_0 L^2}{D_e C_{A,s}} \tag{10.34}$$

Note that the Thiele modulus depends on the surface concentration, unlike in the case of a first-order reaction. This is true in general for any non-linear kinetics. The effectiveness factor is equal to one since the entire catalyst is exposed to the reactant and the rate of consumption of A per unit volume is k_0 and independent of the concentration, as expected for a zeroth-order reaction.

Onset of starvation and solution for larger Thiele modulus

The condition under which the concentration at the center becomes zero or remains positive is then obtained from Eq. (10.33) by setting $\xi = 0$ as

$$\phi_0 \leq \sqrt{2}$$

The solution given by Eq. (10.33) is valid if the above condition is satisfied. This can be taken as the condition for onset of starvation. If the Thiele modulus is larger than the above value then there will be a reactant-depleted zone in the slab. The analysis needs to be modified under these conditions.

Now assume that the concentration becomes zero at dimensionless location λ. The no-flux condition is now to be imposed at $\xi = \lambda$ since no mass of A crosses this point, rather than at $\xi = 0$. This leads to the following equation for the concentration profile

$$c_A = 1 - \frac{\phi_0^2}{2}(1 - 2\lambda) + \frac{\phi_0^2}{2}(\xi^2 - 2\xi\lambda)$$

The position λ is not known. Hence we impose an additional condition that the concentration c_A is also zero at $x = \lambda$. This provides the following condition for calculation of λ:

$$\lambda = 1 - \sqrt{\frac{2}{\phi_0^2}}$$

Note that only the region from $x = \lambda$ to $x = 1$ is effective for reaction. Hence the effectiveness factor is simply the length of the region and can be calculated as

$$\eta = \sqrt{\frac{2}{\phi_0^2}}$$

The rate of consumption of A per unit volume is now $k_0\eta$ and turns out to be

$$R_A = \sqrt{2D_e k_0 C_{A,s}}/L$$

with a square-root dependence on the surface concentration. Thus a zeroth-order reaction appears as a reaction of order one half due to the presence of diffusion effects. The concentration profiles for various ranges of the Thiele parameter are shown in Fig. 10.7.

10.4.5 Transport in tissues: the Krogh model

The Krogh (annular) cylinder model is often used to calculate the rate of oxygen transport in skeletal muscle and tissues. The conceptual basis of the model (Krogh, 1919) is shown in Fig. 10.8. Each capillary is assumed to be cylindrical with a radius R_C. A tissue region of radius R_0 is associated with each capillary. This radius R_0 is taken as half of the distance between the centers of two capillaries. It can be seen from the figure that the model fails to account for some area of tissue near R_0. But the assumption is needed so that a simple unit-cell concept can be used to simulate the problem. In practice the effect of this small region is expected to be minor. Also it may be noted that the model is not applicable to a

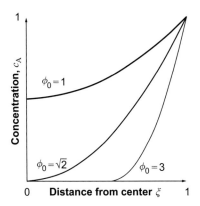

Figure 10.7 Concentration profiles for a zeroth-order reaction in a porous catalyst for three cases: (i) no starvation, (ii) onset of starvation, and (iii) strong diffusional gradients.

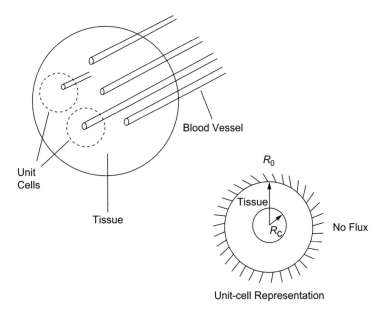

Figure 10.8 The Krogh cylinder model for oxygen transport in tissues.

complex arrangement of capillaries such as in brain tissue. The Krogh model assumes that there is a zeroth-order consumption of oxygen in the tissues, leading to

$$\frac{D_e}{r}\left[\frac{d}{dr}\left(r\frac{dC_A}{dr}\right)\right] = k_0 \tag{10.35}$$

with the boundary conditions of a saturation oxygen concentration of $C_{A,s}$ at $r = R_C$ and no flux of oxygen at $r = R_0$.

The dimensionless representation is achieved by defining the following reference scales: reference length R_0 and reference concentration $C_{A,s}$.

Then the parametric representation of the problem can be shown to be

$$c_A = c_A(\xi; \phi_0^2, \kappa)$$

where ϕ_0^2 is defined as $R_0^2 k_0/(D_e C_{A,s})$ and κ is the ratio R_C/R_0. The variable ξ is defined as r/R_0 and c_A is $C_A/C_{A,s}$. The solution can be obtained analytically for the concentration profile and the details are left as an exercise. The critical value of the parameter ϕ_0^2 at which the oxygen concentration drops to zero at $r = R_0$ is obtained as

$$\phi_0^2 \text{ critical } = \frac{4}{\kappa^2 - 1 - 2\ln\kappa}$$

For ϕ_0 values larger than this, there is a region of zero oxygen concentration referred to as an anoxic region. If such a region persists for a long time, the cells will die, leading to a condition known as necrosis. Although necrosis is rare under normal conditions, it can arise in tumors due to a higher rate of metabolism (higher k_0 values for tumor cells).

> **Historical vignette.** Krogh was a Danish physiologist interested in respiration and developed models for oxygen concentration and respiration rate. The classical model shown above was developed in 1920 with the aid of a mathematician friend, Erlang, who helped him to get the final solution. In 1920 he received the Nobel prize for medicine for this work. (Source: Truskey *et al.* (2004).)

A number of limitations to the Krogh model are now briefly indicated. The unit-cell model assumes a close-packed tissue and excludes about 21% of the tissues. It therefore underestimates oxygen consumption. Other unit-cell models are considered in later studies, such as a hexagonal arrangement of capillaries, but any such arrangement other than the cylinder requires a numerical solution. The model does not assume the discrete nature of the red cells. The role of myoglobin, which binds with oxygen in the tissue, is not explicitly accounted for. Details of the modeling of these issues are not considered here, and the interested student may wish to read the references cited in the textbook by Truskey *et al.* (2004) in this area. The key idea here was to show how some assumptions lead to a simplified and useful model for predicting the important features of the system. The idea of a unit-cell approach to model transport phenomena in heterogeneous systems has been used in many other contexts as well.

We now move to a non-linear mth-order reaction and illustrate a mathematical solution method for such problems. This is the familiar p-substitution method which we studied in heat transfer.

10.4.6 mth-order reaction

In this section we show the use of the p-substitution method for a non-linear diffusion–reaction problem. In order to illustrate the key points, we take an mth-order reaction and show the derivation directly in dimensionless form for a slab geometry. Also the Dirichlet condition is used at the surface.

Consider a model problem represented as

$$\frac{d^2 c_A}{d\xi^2} = \phi^2 c_A^m \tag{10.36}$$

where ϕ is now defined as

$$\phi^2 = L^2 k_m C_{A,s}^{m-1}/D_e$$

Note that ϕ is a function of the surface concentration $C_{A,s}$, unlike in a first-order reaction.

The order of the differential equation can be reduced by one by using the transformation $p = dc_A/d\xi$. You should verify that the following equation is obtained:

$$p\frac{dp}{dc_A} = \phi^2 c_A^m \tag{10.37}$$

The first integration and the use of the condition at $\xi = 0$ yields

$$p = \sqrt{2/(m+1)}\phi\sqrt{c_A^{m+1} - c_{A,c}^{m+1}} \tag{10.38}$$

where $c_{A,c}$ is the center concentration, which is unknown as yet.

On substituting for p and rearranging we can obtain an implicit integral to calculate $c_{A,c}$:

$$\int_{c_{A,c}}^{1} \frac{dc_A}{\sqrt{c_A^{m+1} - c_{A,c}^{m+1}}} = \sqrt{2/(m+1)}\phi$$

This equation may be integrated across the system to obtain an equation for $c_{A,c}$, which is the information needed for finding the flux into the catalyst.

The analysis is similar to that for the Frank-Kamenetskii problem in heat transfer.

For large values of the Thiele modulus, however, it is not necessary to evaluate $c_{A,c}$. Here the concentration drops significantly within the pellet and hence $c_{A,c}$ can be approximated as zero. The resulting solution is an asymptotic case for large ϕ. The flux at the surface is found directly from Eq. (10.38), by using this condition, as

$$p(\xi = 1) = \sqrt{2/(m+1)}\phi$$

which is the asymptotic solution for the mth-order for large Thiele parameter.

10.5 Models for gas–liquid reaction

Consider a gas A that dissolves into a liquid and reacts with a species B present in the liquid. The reaction scheme can be represented as

$$A(g \to l) + \nu B(l) \to \text{Products}$$

Examples of such a system include CO_2 absorption in various reacting solvents such as amine solutions and many other gas-treating processes.

The governing equations are as follows:

$$D_A\frac{d^2 C_A}{dx^2} = k_2 C_A C_B \tag{10.39}$$

$$D_B\frac{d^2 C_B}{dx^2} = \nu k_2 C_A C_B \tag{10.40}$$

Here x is the actual distance into the film, with $x = 0$ representing the gas–liquid interface.

Dimensionless representation

We now introduce the following dimensionless parameters:

$$c_A = C_A/C_A^*$$

where C_A^* is the equilibrium solubility of gas A in the liquid corresponding to the partial pressure of A in the gas phase.

This C_A^* is defined as

$$C_A^* = p_{AG}/H_A$$

where H_A is the Henry's-law constant for the species A in the units of atm m^3/mol. Note the units and the definition.

Similarly the dimensionless concentration for B is defined as

$$c_B = C_B/C_{BL}$$

where C_{BL} is the bulk liquid concentration of B.

Finally let ζ be the dimensional distance in the film ($\zeta = x/\delta$).

With these variables, the governing equations are the following:

$$\frac{d^2 c_A}{d\zeta^2} = Ha^2 c_A c_B \tag{10.41}$$

and

$$\frac{d^2 c_B}{d\zeta^2} = Ha^2 c_A c_B/q \tag{10.42}$$

where Ha and q are two dimensionless quantities:

$$Ha^2 = \delta^2 \frac{k_2 C_{BL}}{D_A} \tag{10.43}$$

and

$$q = \frac{D_B C_{BL}}{\nu D_A C_A^*} \tag{10.44}$$

The student should verify the algebra involved in going from the dimensional to the dimensionless formulation.

Note that the Hatta number (squared) represents the ratio of diffusion time to reaction time and hence a large drop in concentration in the film can be expected for such cases.

Noting that $k_L = D_A/\delta$, the Hatta number can also be expressed as

$$Ha^2 = \frac{D_A k_2 C_{BL}}{k_L^2} \tag{10.45}$$

Boundary conditions

The boundary conditions for species A in the first case, of no gas film resistance, are as follows.

At $\zeta = 0$, $C_A = C_A^*$ and hence $c_A = 1$.

At $\zeta = 1$, $c_A = C_{AL}$ is some specified value depending on the bulk processes. This value C_{AL} will depend on the extent of bulk reactions, convective and dispersive flow into the bulk, etc. But, even for moderately fast reactions, the bulk concentration of dissolved gas turns out be zero, and we take this value to be zero. Hence $c_A = 0$ at $\zeta = 1$ will be used as the second boundary condition. The validity of this assumption can be checked after a solution has been obtained and can be modified if needed.

In the second case, with gas film resistance included, a balance over the gas film provides the boundary condition:

$$k_G(p_{AG} - p_{A,i}) = -D_A \left(\frac{dC_A}{dx}\right)_{x=0} \tag{10.46}$$

Also note that $p_{A,i}$, the interfacial partial pressure of A, is related to the interfacial concentration of A in the liquid by Henry's law. Thus $C_A(x = 0) = p_{A,i}/H_A$. Using these, the boundary condition can be expressed in dimensionless form as

$$Bi_G[1 - c_A(\zeta = 0)] = - \left(\frac{dc_A}{d\zeta}\right)_{[\zeta=0]} \tag{10.47}$$

where $Bi_G = k_G H_A/k_L$, a Biot type of number for gas-side mass transfer. The boundary condition is now of the Robin type.

The boundary condition at $\zeta = 1$ is the same as before, taken as $c_A = 0$ if the reactions are reasonably fast; otherwise we need a mass balance for A in the bulk liquid as well.

The boundary conditions for species B are specified as follows.

At $\zeta = 0$ we use $dc_B/d\zeta = 0$, since B is non-volatile and the flux is therefore zero.

At $\zeta = 1$, we have $c_B = 1$. This completes the problem definition. Since the profiles of A and B are now coupled, this calls for a numerical solution. We will study the numerical solution using a MATLAB solver in a later section. But the essential details of the concentration profiles which can be expected are shown in Fig. 10.9.

The results are generally presented in terms of an enhancement factor E defined as

$$E = - \left(\frac{dc_A}{d\zeta}\right)_{\zeta=0} = -p_0$$

which is a measure of the flux at the interface. Here we use p for the concentration gradient $dc_A/d\zeta$.

Rather than solving numerically, we present here various limiting cases to give us a feel for the problem and the results which could be anticipated from the MATLAB solution.

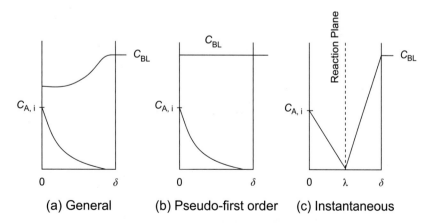

(a) General   (b) Pseudo-first order   (c) Instantaneous

Figure 10.9 Concentration profiles in the film for a gas–liquid reaction. Note the two limiting cases shown on the right. These are analyzed in detail in the text.

The invariance property of the system

The rate terms can be eliminated between Eq. (10.41) and Eq. (10.42) to give

$$\frac{d^2 c_A}{d\zeta^2} = q \frac{d^2 c_B}{d\zeta^2} \tag{10.48}$$

This can be integrated once to give

$$\frac{d c_A}{d\zeta} = q \frac{d c_B}{d\zeta} + p_0$$

where the integration constant has been assigned as the dimensionless flux of A at the interface. We note that the no-flux boundary condition for B at the interface is satisfied by this choice for the constant.

A second integration and use of the boundary condition at $\zeta = 1$ gives

$$c_A = q c_B + p_0 \zeta + A_1$$

From the boundary conditions at $\zeta = 1$ we can show that $A_1 = -p_0 - q$. Hence the concentration profile of B is related to that for A by

$$c_B = 1 + \frac{c_A}{q} + \frac{p_0}{q}(1 - \zeta)$$

The important quantity is the interfacial concentration of B, denoted as $c_{B,i}$, since it determines which of the regimes in Fig. 10.9 is likely to exist. This is obtained readily by setting $\zeta = 0$ in the above expression. This is an invariant of the system. All numerical results should show this result; otherwise the numerical procedure is wrong.

Since the enhancement factor E is equal to $-p_0$, the equation can be written in terms of E as

$$c_{B,i} = 1 + \frac{1}{q} - \frac{E}{q} \tag{10.49}$$

An illustrative plot of the enhancement factor as a function of the Hatta number for $q = 10$ is shown in Fig. 10.10. Note that the factor reaches an asymptotic value at large Hatta number that corresponds to the instantaneous regime.

It may be worthwhile to estimate the conditions under which $c_{B,i}$ is nearly unity. This will happen if $E \gg q$. But we do not know E. Let us proceed further assuming that the interfacial concentration of B is nearly unity. In this case Eq. (10.41) can be solved analytically. We call this a pseudo-first-order model.

10.5.1   Analysis for the pseudo-first-order case

In this case $c_{B,i} = 1$ in Eq. (10.41). The analytic solution for C_A is obtained as shown earlier for a first-order reaction. The corresponding enhancement factor is

$$E = \frac{Ha}{\tanh Ha} \tag{10.50}$$

The details leading to the above expression are left as an exercise.

Using this value of E in Eq. (10.49), we find that if $Ha \ll q$ then the concentration of B will be nearly one at the interface, leading to a pseudo-first-order approximation.

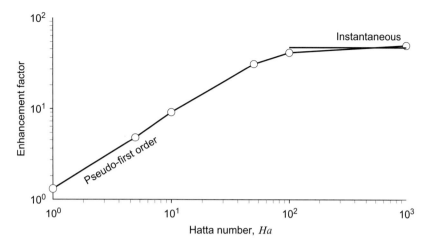

Figure 10.10 An illustrative plot of the enhancement factor as a function of the Hatta number; $q = 10$.

10.5.2 Analysis for instantaneous asymptote

The second limit can be analyzed easily as well. There are two ways of doing this. The first is that mathematically we set $c_{B,i} = 0$ in Eq. (10.49). We find

$$E = 1 + q \tag{10.51}$$

The second approach is based on a physical consideration postulating that there is a reaction plane λ that separates the A and B regions. In other words, we assume that A and B cannot co-exist in the film. The physical situation prevailing is shown in Fig. 10.9(c).

The rate of absorption and reaction is controlled by a balance of the rate of transport A from the interface to the reaction plane,

$$N_{A0} = \frac{D_A C_A^*}{\lambda}$$

and the transport of B from the bulk to the interface,

$$N_{A0} = \frac{N_B \lambda}{v} = \frac{D_B C_{BL}}{v(\delta - \lambda)}$$

By equating the two expressions for N_A above one can solve for both λ and N_A. From N_A the enhancement factor can be calculated. The result is the same as Eq. (10.51). The details leading to Eq. (10.51) are left as an exercise. One can likewise argue that if $Ha \gg q$ the instantaneous asymptote is reached.

10.5.3 The second-order case: an approximate solution

For $Ha \approx q$ we have a second-order reaction case where the two equations have to be solved together. This can be done numerically. However, a good analytical approximation was proposed by Hikita and Asai (1963). They claimed that the pseudo-first-order asymptote can still be used, provided that the Ha parameter is now based on the interfacial concentration of B rather than the bulk concentration. Thus we still use Eq. (10.50), but with a modified Hatta parameter defined as

$$Ha(\text{Modified}) = Ha\sqrt{c_{B,i}}$$

This leads to the following expression for E:

$$E = \frac{Ha\sqrt{c_{B,i}}}{\tanh[Ha\sqrt{c_{B,i}}]} \tag{10.52}$$

where $c_{B,i}$ is given by Eq. (10.49) earlier. Note that $c_{B,i}$ is a function of E and hence we have to solve for both E and $c_{B,i}$ simutaneously.

Example 10.6.

Carbon dioxide is absorbed into a solution of NaOH in a packed-column absorber. Locally at a given point in the absorber the concentration of NaOH is 1.0 M and the partial pressure of CO_2 is 1 atm. Find the rate of absorption. Neglect the effect of gas-side resistance.

Let us first make a list of the parameters needed. They can be classified into the physico-chemical parameters and the hydrodynamic parameters. The physico-chemical parameters needed are listed below together with the values specific to this problem. The parameter values are taken from the book by Danckwerts (1970): (i) the Henry's-law constant $H_A = 0.025$ atm at $20\,°C$; (ii) the diffusivities of gas A (CO_2) and liquid reactant B (NaOH) are $D_A = 1.8 \times 10^{-9}\,m^2 s$ and $D_B = 1.7 D_A$; (iii) the rate constant for the reaction is $k_2 = 10\,m^3/mol \cdot s$.

The hydrodynamic parameter is the liquid-side mass-transfer coefficient $k_L = 1 \times 10^{-4}\,m/s$.

Neglect the gas-side resistance.

Solution.

The dimensionless parameters needed are Ha and q. These are calculated from Eqs. (10.43) and (10.44), respectively as 42 and 21.25.

The next step is to solve Eq. (10.49) for the interfacial concentration of B. We find $c_{B,i} = 0.186$. Hence this represents a condition where there is a significant depletion of the reactant at the interface.

The enhancement factor is calculated as 18.3 using Eq. (10.52). Correspondingly the rate of absorption is $k_L C_A^* E$ and is equal to $0.0732\,mol/m^2 \cdot s$.

10.5.4 The instantaneous case: the effect of gas film resistance

The analysis of instantaneous reaction in Section 10.5.2 is now extended to include the gas film resistance. The concentration profile for various cases are shown in Fig. 10.11.

For the gas film we use

$$R_A'' = k_G(p_{AG} - p_{A,i})$$

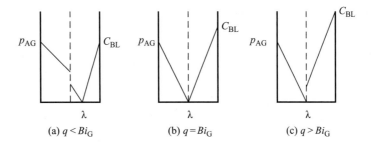

Figure 10.11 Concentration profiles of A and B for an instantaneous reaction with gas film resistance included; note the shift in the controlling regime.

where R_A'' is the interfacial flux (the same as N_{A0}), which is also equal to the rate at which the gas is being absorbed.

The interfacial concentration of A in the liquid side is related to the interfacial partial pressure by $C_{A,i} = p_{A,i}/H_A$. Now consider the transport of A from the interface to a distance λ and use the Fick's law. Then

$$R_A'' = \frac{D_A(p_{A,i}/H_A)}{\lambda}$$

Finally, consideration of transport of B from the bulk to λ gives

$$R_A'' = \frac{R_B''}{v} = \frac{1}{v}\frac{D_B C_{BL}}{\delta - \lambda}$$

We now have three equations for three unknowns, R_A'', $p_{A,i}$, and λ. The final result can be expressed in terms of the enhancement factor, which is now defined as

$$E = \frac{R_A''}{k_L(p_{AG}/H_A)}$$

It is easier to solve if the various expressions are in terms of $1/E$. From the transport equations of A (gas film plus from $x = 0$ to $x = \lambda$) we can show that

$$\frac{1}{E} = \frac{1}{Bi_G} + \frac{\lambda}{\delta}$$

where $Bi_G = k_G H_A/k_L$.

From the transport considerations for B we find

$$\frac{1}{E} = \frac{1}{q}\left[1 - \frac{\lambda}{\delta}\right]$$

where q is now defined as $D_B C_{BL}/(v D_A p_{AG}/H_A)$.

On equating the two expressions for $1/E$ and solving for λ, we obtain

$$\frac{\lambda}{\delta} = \frac{1}{1+q}\left[1 - \frac{q}{Bi_G}\right]$$

It should be noted here that the above equation is valid only if $q < Bi_G$, since λ cannot be negative (it can be no lower than zero, where the reaction plant shifts right at the interface). In such cases the value of E is simply equal to Bi_G; the process becomes controlled entirely by the gas film resistance.

Upon using this value of λ in either of the equations we find that the enhancement factor (for the case where $q < Bi_G$) can be calculated after some "minor" algebra as

$$E = \begin{cases} (1 + q)/(1 + 1/Bi_G) \text{ if } q < Bi_G \\ Bi_G \text{ if } q > Bi_G \end{cases}$$

It is useful to illustrate the calculations for a numerical problem, which is shown below.

Example 10.7. H_2S is absorbed in NaOH solution.

Hydrogen sulfide absorption in amine solutions may be assumed to be instantaneous. At a point in the absorber the total pressure is 20 atm and the gas contains 1% of H_2S. The ratio D_B/D_A is 0.64 and the amine concentration is 0.25 M. The mass transfer coefficients are $k_L = 0.03$ cm/s and $k_G = 6 \times 10^{-5}$ mol/cm^2 s atm.

The Henry's-law constant for H_2S is 1950 Pa m^3/mol.

Find the rate of absorption at this point.

Also find the amine concentration at which the process becomes entirely controlled by gas-side mass transfer.

Solution.

We need two dimensionless numbers q and Bi_G in order to use the model for instantaneous reaction with gas-side resistance. It is useful to convert Henry's law to atm to simplify the calculations.

We obtain

$$q = \frac{0.64 \times (0.25 \, \text{mol/l} \times 10^3 \, \text{l/m}^3)}{20 \, \text{atm} \times 0.01 \times (1950 \, \text{Pa m}^3/\text{mol} \times 10^5 \, \text{Pa/atm})} = 15.6$$

$$Bi_G = \frac{(6 \times 10^{-5} \, \text{mol/cm}^2 \, \text{s atm})(1950 \, \text{Pa m}^3/\text{mol} \times 10^5 \, \text{Pa/atm})}{0.03 \, \text{cm/s}} = 39$$

The enhancement factor is then calculated as $(1 + q)/(1 + 1/Bi_G)$ and is equal to 16.19. The rate of absorption is calculated as $k_L(p_{A,G}/H_A) \times E$ and is found to be 0.05 mol/m$^2 \cdot$ s.

If the concentration of amine is increased, the rate increases and eventually becomes gas-side controlled. This concentration is reached when $q = Bi_G$. The rate of absorption at this point is calculated as $k_G p_{A,G}$ and is equal to 0.12 mol/m$^2 \cdot$ s. If the amine concentration is increased further, the rate stays at this value since the reaction plane has already shifted to the interface. These are useful guidelines for choosing the optimum operating conditions depending on the mass transfer coefficient prevailing in the contactor.

10.6 Transport across membranes

10.6.1 Gas transport: permeability

An important manifestation of Fick's law occurs in transport across a membrane. This has applications in various fields of engineering, including biomedical engineering.

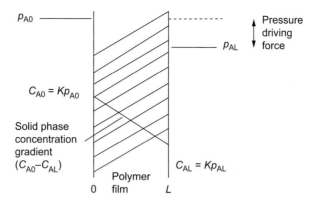

Figure 10.12 Concentration profiles of A across a membrane; note the concentration jump across the interface.

Consider a membrane of thickness L with the concentration of the diffusing solute being C_{m0} at $x = 0$ and C_{mL} at $x = L$. Then we can use the Fick's-law type of model for the transport across the membrane itself. See Fig. 10.12. We have

$$N_A = D_m \left(\frac{C_{m0} - C_{mL}}{L} \right) \tag{10.53}$$

where N_A is the flux of the diffusing species across the membrane. Note that the low-flux mass-transfer model is used here, and N_A is assumed to be the same as the Fick's-law flux. The concentration of the diffusing species in the membrane is usually not directly known. What is known is the concentration in the fluid phase on either side of the membrane. The concentration in the fluid is related to the concentration in the membrane by an equilibrium relationship. Thus, if C_{f0} is the concentration in the fluid at $x = 0$ then the concentration in the membrane C_{m0} is given as

$$C_{m0} = KC_{f0} \tag{10.54}$$

where K is an equilibrium constant also called the partition coefficient. A similar equation holds at $x = L$. Hence

$$C_{mL} = KC_{fL} \tag{10.55}$$

Using these "equilibrium" relations in (10.53), we obtain the flux in terms of the concentration driving force based on the fluid concentrations:

$$N_A = D_m K \left(\frac{C_{f0} - C_{fL}}{L} \right) \tag{10.56}$$

Usually the thickness of the membrane, L, is not precisely known. Also the partition coefficient K is usually not known. Hence the effects of these terms are combined into an overall term called the permeability P of the membrane:

$$P = D_m K / L \tag{10.57}$$

Hence the transport rate across the membrane can be described (in terms of the fluid phase concentration difference as the driving force) as

$$N_A = P(C_{f0} - C_{fL}) \tag{10.58}$$

Note that the permeability of a membrane, P, is a composite parameter that depends on the diffusion coefficient of the membrane itself, on the solubility or the partition coefficient of the solute in the membrane, and finally on the thickness of the membrane itself. Thus the values of P may vary from one type of membrane to another and cannot be correlated in a theoretical sense unless all of the three effects mentioned above can be quantified separately. However, in practice the permeability is often used as a matter of convenience. It is, however, not a basic parameter describing the transport phenomena in the membrane.

It should be noted here that in membrane-related publications the permeability is often reported in units of barrer. The conversion is as follows:

$$1 \text{ barrer} = 3.348 \times 10^{-16} \text{ mol} \cdot \text{m/m}^2 \cdot \text{s} \cdot \text{Pa}$$

The permeability is based on unit membrane thickness and the gas-phase partial-pressure difference across the membrane as the driving force. The unit barrer is named after R. M. Barrer, who published many of the early papers in this field, including a widely used monograph (Barrer, 1951). One reason for the persistence of this unit is that permeability of common gases for many membranes is in the range of 1–10 barrer.

Gases such as hydrogen can dissociate into H atoms in the membrane, and in such cases a non-linear transport model is used:

$$N_A = P\left(C_{f0}^{1/2} - C_{fL}^{1/2}\right) \tag{10.59}$$

This equation is called Sievert's law. It has been found to describe hydrogen transport in Pd-based membranes.

10.6.2   Complexities in membrane transport

The discussion in the earlier section was based on a simple Fick's-law concept with constant values of solubility and diffusivity. However, complexities in the transport law often arise due to local differences in both solubility and diffusivity, both of which can be functions of concentration. Stern (1994) has made the following classification.

1. Constant diffusion and constant partition constant K. This is the simplest case and the model in Section 10.6.1 is applicable. The diffusion of permanent gases in elastomers and many harder polymers is often described by such a model.
2. Concentration-dependent diffusivity D but constant K. This phenomenon is exhibited by gases with critical temperatures ranging from near the ambient to 200 °C, e.g., C$_4$ hydrocarbons in rubbery membranes.
3. Variable K and variable H. Gases with high critical temperatures, organic vapors, etc., are examples of systems showing this pattern.
4. Time-varying behavior: $D(c, t)$ and constant K, where t is the time. Here time- and history-dependent diffusion phenomena are observed. These are seen in polymers with longer relaxation times, e.g., organic vapor in ethyl cellulose.
5. Diffusivity as a function of position, e.g. membranes prepared from composite materials of two or more layers with different values of D for each layer.

Further discussion of these complex effects is not considered here. The main idea is for the student to be aware of these phenomena in practical applications. The models for all these cases can be set up using a methodology similar to that described in Chapter 8 for the variable-thermal-conductivity problem.

10.6.3   Liquid-separation membranes

Separation of liquid mixtures using membranes is similar to filtration (separation of a liquid–solid mixture). Hence similar terminology is often used, and the process is classified into three categories, depending on the pore size of the membrane: (i) microfiltration, (ii) ultrafiltration, and (iii) nanofiltration.

Microfilters have pore sizes in the range 0.05–10 μm and are used for recovery of particles in the size range 0.1–20 μm. These filters find applications in bio-separation of animal cells, yeasts, bacteria, etc., from fermentation broths. These essentially remove insoluble solids. Membranes for ultrafiltration processes have pore sizes in the range of 100 nm and are used to separate low-molecular-weight solutes such as enzymes from high-molecular-weight solutes such as virus particles. Nanofiltration uses membranes with pore sizes in the range of 10 nm to separate dissolved solutes. An example is purification of salt water to produce potable water. This process is also known as reverse osmosis. Separation of products like glucose, vitamins, etc., from fermentation broths can also be done by nanofiltration. Modeling of microfiltration and ultrafiltration can be done using concepts from flow in porous media, while for nanofiltration and reverse osmosis one needs the additional concept of osmotic pressure and how this influences the rate of transport. This is considered in the next section.

The transport mechanism depends on the relative magnitude of the solute size in comparison with the pore size. The various mechanisms are shown schematically in Fig. 10.13. For large pores the process is more of bulk flow and convection. The rate of flow is assumed to be proportional to the pressure gradient, which is referred to as Darcy's law. For simultaneous transport of two solutes there is hardly any difference in the rates, leading to there being no selectivity. For smaller pores the transport is controlled by diffusion. If now the solute size is comparable to the pore size the solute transport is restricted, leading to hindered diffusion. Selectivity is now achieved due to exclusion of smaller solutes by a diffusion barrier. Finally, if the pores are of nanoscale size, the species dissolves in the membrane and a high selectivity can be obtained owing to exclusion due to solubility. Here only one type of solute may be transported and the other may be excluded. The osmotic-pressure differences are created by the exclusion which is discussed in the following section.

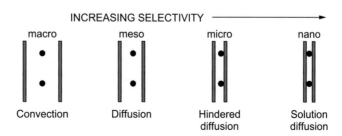

Figure 10.13 The effect of pore size on the transport mechanism of solutes in membranes.

10.7   Transport in semi-permeable membranes

What is a semi-permeable membrane? It is a membrane that permits transport of only one species (or a limited number of species), usually the solvent such as water. A basic concept in understanding transport in membranes is the osmotic pressure for semi-permeable membranes. Consider a semi-permeable membrane that allows the solvent (water) to diffuse but not the solute (salt). Assume that these (pure water and a salt solution) are separated by a membrane into two compartments. At equilibrium there is a pressure difference between the solution and solvent as shown in Fig. 10.14. This pressure difference is called osmotic pressure and can be calculated from either thermodynamic or kinetic considerations using the concept of dynamic equilibrium.

Thermodynamic derivation

The thermodynamic argument runs as follows. At equilibrium the chemical potential of the solvent is the same on both sides of the membrane. On one side we have pure solvent and let μ_w° represent its chemical potential (the subscript w is used since the solvent is very often water). On the other (solution) side the chemical potential is related to the concentration of the solvent and can be represented as

$$\mu_w = \mu_w^\circ + R_G T \ln a_w + \left(\frac{\partial \mu_w}{\partial P}\right)_{T,C} \Delta P \tag{10.60}$$

where the second term on the RHS is a correction due to concentration, while the third term is the effect of pressure on the chemical potential. The pressure difference is the pressure on the solution side minus that on the pure water side. From thermodynamics,

$$\left(\frac{\partial \mu_w}{\partial P}\right)_{T,C} = \tilde{V}_w$$

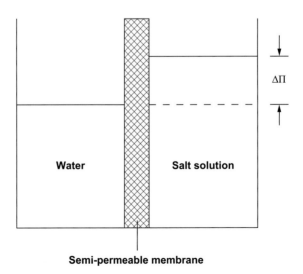

Figure 10.14 An illustration of the osmotic-pressure difference across a semi-permeable membrane.

where $\tilde{V}_w$ is the partial molar volume of water. At equilibrium the chemical potential of water on the solution side μ_w is the same as the chemical potential of pure water:

$$\mu_w = \mu_w^\circ \tag{10.61}$$

Hence from Eq. (10.60) we can find an expression for the difference in pressure across the membrane:

$$\Delta P = -\frac{R_G T}{\tilde{V}_w} \ln a_w = \Delta \Pi \tag{10.62}$$

This pressure difference is often represented by $\Delta \Pi$, the usual notation for osmotic pressure. Also for dilute solutions

$$\ln a_w = \ln x_w = \ln(1 - x_s) \approx -x_s$$

where x_s is the mole fraction of the solute. Hence Eq. (10.62) can be written as

$$\boxed{\Delta \Pi = \frac{R_G T}{\tilde{V}_w} x_s = C_s R_G T} \tag{10.63}$$

where $C_s = x_s/\tilde{V}_w$ is the concentration of the solute. This equation, known as the van't Hoff equation, has some resemblance to the ideal-gas law. It is also seen that the osmotic pressure is a colligative property, namely a property that depends only on the concentration of the solute and not on the nature of the solute. Correction factors are applied to the above equation for non-ideal solutions. Also, for solutes such as NaCl, a correction factor of two is used since NaCl exists as Na^+ and Cl^- ions in solutions.

With the expression for osmotic pressure in hand we can examine the effect of an applied pressure gradient across the membrane.

10.7.1   Reverse osmosis

Now, depending on the applied pressure difference compared with the osmotic pressure difference, three situations can arise, as shown in Fig. 10.15. If $\Delta P < \Delta \Pi$ there is osmotic flow with the solvent crossing (from the pure side) into the solute side. If the two are equal, we have a situation of equilibrium and there is no net flow across the membrane. The third case is when $\Delta P > \Delta \Pi$, which is the condition for reverse osmosis. Here the solvent flows across the membrane (from the solute side), leaving behind an enriched solute solution. This is the condition, for example, for desalination of sea water. The difference $\Delta P - \Delta \Pi$ may be considered to be the driving force for this case, and the flux of the solvent is represented as

$$J_w = K_w(\Delta P - \Delta \Pi) \tag{10.64}$$

which is the widely used equation for transport in osmosis and reverse osmosis. Here K_w is a permeability constant for the membrane in units of $mol/m^2 \, Pa \cdot s$.

This equation is applicable under conditions of complete solute rejection. Normally there is some solute diffusion as well, i.e., complete solute rejection is not achieved. In such cases the flux expression is modified to

$$J_w = K_w(\Delta P - \sigma \, \Delta \Pi) \tag{10.65}$$

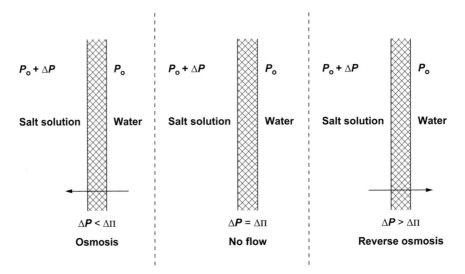

Figure 10.15 An illustration of the influence of the pressure difference across a semi-permeable membrane, for osmosis, no flow, and reverse osmosis.

where σ is a factor between 0 and 1. This is referred to as the Starling equation in the semi-permeable-membrane literature. The case of complete solute rejection corresponds to $\sigma = 1$, and the full impact of the osmotic pressure is left on the rate of transport. The other extreme is $\sigma = 0$, which is the case where the solute is also fully permeable. The parameter σ is known as the Staverman constant, since it was Staverman who introduced this first, in 1951, or simply the osmotic reflection coefficient (Staverman, 1951).

For the case where the Staverman constant is not equal to one, an additional equation for the rate of solute transport (across the membrane) is needed. This is modeled by the following equation for the combined flux:

$$N_s = -D_s \frac{dC_s}{dz} + C_s J_w (1 - \sigma)$$

The first term is the Fick's-law diffusion term for the solute, while the second term is the solute carried with the solvent flow.

We consider now a case with no solute transport ($\sigma = 1$) and show that the solute concentration at the membrane surface is larger than that in the bulk liquid.

10.7.2 Concentration-polarization effects

The convective flow of the solvent causes a corresponding solute flow on the liquid side of the membrane. However, the solute is not transported across the membrane (assuming complete rejection). This causes an increase in solute concentration near the membrane surface relative to that in the bulk liquid. The phenomenon is known as concentration polarization. This is illustrated in Fig. 10.16. The effect of concentration polarization is to increase the osmotic pressure at the surface of the membrane and thereby reduce the rate of solvent transport across the membrane. The effect can be modeled as follows by setting the net flux of solute equal to zero.

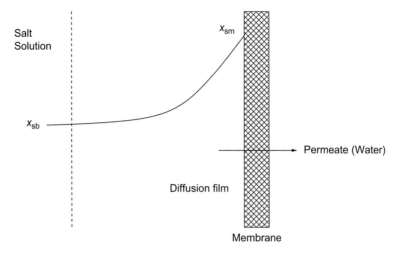

Figure 10.16 An illustration of the concentration-polarization effect. z is measured from right to left, with $z = 0$ at the membrane surface.

The solute transport (in the liquid phase) is the combination of diffusive and convective transport,

$$N_s = -D\frac{dC_s}{dz} - vC_s = 0$$

where v is the solvent velocity, which is equal to J_w/C. The minus sign is used in the convective term simply because the distance z is measured from the surface to the bulk. The direction of solvent flow is therefore in the minus-z-direction.

This equation can be written in terms of solute mole fraction as

$$-DC\frac{dx_s}{dz} - x_s J_w = 0$$

with the boundary conditions of x_{sm} at $z = 0$ and x_{sb} at $z = \delta$, where δ is the film thickness near the membrane surface. Solution of the above equation gives

$$x_{sm} = x_{sb}\exp[J_w\delta/(CD)]$$

The osmotic pressure for solvent transport should therefore be based on x_{sm} rather than x_{sb}. If $\Delta\Pi$ is the osmotic pressure based on the bulk, then the effective osmotic pressure is

$$\Delta\Pi(\text{Effective}) = \Delta\Pi\exp[J_w\delta/(CD)]$$

The transport equation for the solvent given by (10.64) is now modified by multiplying the osmotic pressure by the factor $\exp[J_w\delta/(CD)]$. The modified equation (for the case of complete solute rejection) now reads

$$J_w = K_w\{\Delta P - \Delta\Pi\exp[J_w\delta/(CD)]\} \tag{10.66}$$

which is a transcendental equation for J_w and can be solved by iteration.

Simplified form

For small values of $J_w\delta/(CD)$ the above equation can be simplified by expanding the exponential term and keeping only the linear term. The resulting equation can be explicitly solved for J_w. The final answer shown below has a rather interesting physical meaning:

$$J_w = \frac{(\Delta P - \Delta \Pi)}{1/K_w + \Delta \Pi \, \delta/(CD)}$$

The term in the denominator on the RHS can be viewed as the sum of two resistances. The reciprocal of this quantity can be viewed as the measured effective permeability:

$$K_w(\text{Effective}) = \left[\frac{1}{K_w} + \frac{\Delta \Pi \, \delta}{CD} \right]^{-1}$$

This effective permeability will be a function of the flow rate and other parameters (due to the film-thickness term) and hence will not be a true measure of the membrane permeability. This is again another effect where a mass-transfer effect masks the true parameter value as in diffusion in a porous catalyst, where the measured kinetic constant $k\eta$ is not necessarily the true kinetic constant k.

Returning to partial solute transport (which applies if σ is less than one), we present here another model for membrane transport.

10.7.3 The Kedem–Katchalsky model

The Kedem–Katchalsky model (Kedem and Katchalsky, 1958) is applicable when the solute rejection is not complete. It attempts to provide a simplified representation using the concept of irreversible thermodynamics and the Onsager reciprocal relations (Onsager, 1931) to decrease the number of model parameters. The model also finds important applications in transport in biological membranes.

The starting point is the entropy production due to transport of both solvent and solute, which can be represented as

$$g_S = J_w \, \Delta \mu_w + J_s \, \Delta \mu_s \tag{10.67}$$

The subscript w represents the solvent (usually water and hence the subscript w). $\Delta \mu_w$ is the chemical potential difference across the solution. The subscript s represents the solute, with the associated quantities defined in a similar manner. The changes in chemical potential both of the water and of the solute can be related to the activity and pressure difference across the membrane under isothermal conditions. Thus

$$\Delta \mu_w = \tilde{V}_w \, \Delta P + R_G T \, \Delta \ln a_w$$

and

$$\Delta \mu_s = \tilde{V}_s \, \Delta P + R_G T \, \Delta \ln a_s$$

For thermodynamically ideal solution, the activity can be related by concentration. Hence

$$\Delta \ln a_s = \Delta \ln C_s = \frac{1}{C_{s,(l.m)}} \, \Delta C_s$$

Here $C_{s,(l.m)}$ is the log mean concentration difference across the membrane. A similar expression holds for the solvent.

Using these relations, the entropy production in Eq. (10.67) can be written as

$$g_S = [J_w \tilde{V}_w + J_s \tilde{V}_s] \Delta P + R_G T [J_w \, \Delta C_{w,(l.m)} + J_s \, \Delta C_{s,(l.m)}]$$

Volumetric-flow-type quantities are now defined to simplify the expression for the entropy production:

$$Q_t = J_w \tilde{V}_w + J_s \tilde{V}_s$$

which is the total volumetric flux in the system, and

$$Q_s = J_{s,(l.m)} \tilde{V}_s - J_{w,(l.m)} \tilde{V}_w$$

which can be viewed as an exchange flow, i.e., volumetric flow of solute with respect to the volumetric flow of the solvent.

Hence

$$g_S = Q_t \, \Delta P + Q_s R_G T \, \Delta C_s$$

The quantity $R_G T \Delta C_s$ is equal to the osmotic pressure difference $\Delta \Pi$, and hence the equation can be written as

$$g_S = Q_t \, \Delta P + Q_s \, \Delta \Pi$$

Thus Q_t and Q_s can be considered as the fluxes, while ΔP and $\Delta \Pi$ are the associated driving forces. From irreversible thermodynamics, the flux is related to all the pertinent driving force and therefore the following constitutive models can be proposed:

$$Q_t = K_{11} \, \Delta P + K_{12} \, \Delta \Pi \qquad (10.68)$$

and

$$Q_s = K_{21} \, \Delta P + K_{22} \, \Delta \Pi \qquad (10.69)$$

Furthermore, Onsager reciprocity requires that the cross-coefficients are equal. Hence $K_{21} = K_{12}$, leading to a model with three parameters.

Equation (10.68) is rearranged by defining a parameter σ as $-K_{12}/K_{11}$ and denoting K_{11} as P_w. Thus

$$Q_t = P_w[\Delta P - \sigma \, \Delta \Pi] \qquad (10.70)$$

The parameter σ is the solute-rejection parameter introduced earlier: $\sigma = 0$ means complete accessibility for the solute, whereas $\sigma = 1$ means complete rejection.

The solute transport rate is likewise obtained, after some simplifications, as

$$Q_s = (1 - \sigma) C_{s,(l.m)} Q_t + \kappa \, \Delta \Pi \qquad (10.71)$$

where

$$\kappa = \frac{K_{22} K_{11} - K_{12}^2}{K_{11}}$$

In this version of the Kedem–Katchalsky model the three parameters K_{11} (or P_w), σ, and κ are used to model the membrane transport. The meanings of these parameters are as follows: P_w is a measure of the Darcy permeability; σ is a measure of osmotic permeability; and κ is a measure of diffusion of the solute in relation to convection.

A very brief introductory discussion on transport in biological membranes is now presented. A detailed discussion is outside the scope of this book. The book by Sten-Knudsen (2002) should be consulted for additional details.

10.7.4   Transport in biological membranes

The following characteristics of biological membranes such as a cell wall should be noted.

1. They are highly selective.
2. They are composed of lipid layers with narrow channels that can transmit only specific molecules. Polar molecules are often excluded since they have a low solubility (partition coefficient) in the lipid phase.
3. The channels of diameter of the order of 1Å are highly selective to transport of only one type of molecule.
4. Both active transport and passive transport can occur. Thus transport can take place against the concentration gradient. This is referred to as active transport, and an energy source is required in order for transport to occur.

10.8   Reactive membranes and facilitated transport

The models for systems where the species being transported reacts within the membrane are now analysed.

10.8.1   Reactive membrane: facilitated transport

Often the membrane has a carrier that promotes the chemical reaction

$$A + B \rightleftharpoons C$$

where C is a complex of A and B. An important example is hemoglobin in blood, which binds strongly to oxygen. By itself the solubility of oxygen in blood is small, and life as known for human beings cannot be sustained by this small solubility. Here is where the hemoglobin comes into action. Oxygen absorption in the lungs is enhanced by the carrier. The model equations for the carrier-mediated transport are formulated by adding a reaction term to the diffusion equation:

$$D_A \frac{d^2 C_A}{dx^2} = k_2(C_A C_B - C_C/K) \tag{10.72}$$

$$D_B \frac{d^2 C_B}{dx^2} = k_2(C_A C_B - C_C/K) \tag{10.73}$$

$$D_C \frac{d^2 C_C}{dx^2} = -k_2(C_A C_B - C_C/K) \tag{10.74}$$

The boundary conditions are set as follows.

For species A we can specify the concentration on both sides of the membrane, $C_A(x=0) = C_{A0}$ and $C_A(x = L) = C_{AL}$. Note that these are concentrations in the membrane phase, not in the external fluid near these sides. If the external concentrations are known, then the membrane-phase concentrations have to be obtained by multiplying the external concentrations by the partition coefficient.

Species B and C are bound to the carrier and hence they cannot diffuse out of the membrane. Hence the no-flux conditions are used for both these species at both $x = 0$ and $x = L$. Since the no-flux condition is specified at both sides, there is no base concentration to calculate the concentration profiles and the flux of A. Hence the problem is not unique. We have to augment the problem by an integral condition, which states that the total concentration of B in the free form and in the bound form is fixed and equal to some value C_{B0}:

$$C_{B0} = \frac{1}{L} \int_0^L [C_B(x) + C_C(x)]dx \tag{10.75}$$

Once C_{B0} has been specified by the above integral constraint, the problem has a unique solution. The above equation is simply a mass balance for the total carrier concentration within the membrane.

The above set of equations can be made dimensionless by using the following variables: $c_A = C_A/C_{A0}$, $c_B = C_B/C_{B0}$, and $c_C = C_C/C_{B0}$.

The dimensionless equations are as follows:

$$\frac{d^2 c_A}{d\xi^2} = M^2(c_A c_B - c_C/K^*) \tag{10.76}$$

$$\frac{d^2 c_B}{d\xi^2} = \frac{M^2}{q}(c_A c_B - c_C/K^*) \tag{10.77}$$

$$\frac{d^2 c_C}{d\xi^2} = -\frac{M^2}{q}(c_A c_B - c_C/K^*) \tag{10.78}$$

The dimensionless parameters in the model are defined as follows:

$$M^2 = k_2 C_{B0}(L^2/D_A)$$

a Hatta modulus parameter representing the ratio of the diffusion time to the reaction time;

$$q = \frac{D_B C_{B0}}{D_A C_{A0}}$$

a concentration and diffusivity ratio parameter; and

$$K^* = K C_{A0}$$

a dimensionless equilibrium constant.

Note that in the above representation we have used $D_C = D_B$, i.e., equal diffusivity for the free and bound carrier, for simplicity. Also note the similarity in the model to the gas–liquid reactions discussed in Section 10.5.

Illustrative solutions to the model are now presented in terms of the key dimensionless parameters.

We now look at two asymptotic features of the solution.

The fast reaction asymptote

By combining Eqs. (10.74) and (10.76) we find

$$\frac{d^2 c_A}{dx^2} + q \frac{d^2 c_C}{dx^2} = 0$$

Upon integrating once we find that the total flux of species A is

$$\frac{dc_A}{dx} + q\frac{dc_C}{dx} = \text{Constant} = \text{Dimensionless flux} \qquad (10.79)$$

The constant can be viewed as the total (dimensionless flux) of A diffusing in the free form and in the form of the solute–substrate complex. A second integration relates the flux to the concentration gradient of A in original and combined form:

$$\text{Dimensionless flux} = c_A(0) + qc_C(0) - c_A(1) - qc_C(1)$$

Further simplification can be made by using the equilibrium assumption $c_C = K^* c_A c_B$ and the total balance on the substrate $c_B = 1 - c_C$. These conditions lead to

$$c_C = \frac{K^* c_A}{1 + K^* c_A}$$

On substituting this into the expression for the combined flux we find

$$\text{Dimensionless flux of A} = 1 + \frac{qK^*}{1 + K^*}$$

The second term on the RHS can be viewed as the enhancement in transport due to the presence of the reacting carrier for the case of rapid reaction asymptote. An illustrative concentration profile for this case is shown in Fig. 10.17.

The slow reaction asymptote

When the diffusion is fast compared with the reaction a second asymptote (the slow reaction asymptote) is reached. The condition is that Ha is small. Again analytical solutions are possible. The concentrations of the substrate and of the complex are constant throughout the membrane and may be evaluated using an equilibrium condition based on the average concentration of A in the system. An illustrative concentration profile for this case is shown in Fig. 10.18.

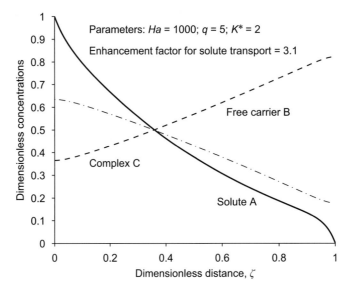

Figure 10.17 Concentration profiles in a reactive membrane for high Hatta number, $Ha = 1000$.

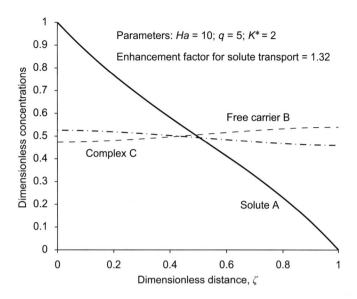

Figure 10.18 Concentration profiles in a reactive membrane for low Hatta number, $Ha = 10$.

10.8.2   Co- and counter-transport

Membranes with carriers exhibit many other complexities and some discussion is provided below. If two solutes compete for the same substrate, a phenomenon known as counter-transport is often observed. The reaction scheme for such cases can be schematically represented as

$$\text{Solute 1} + \text{Substrate S} \rightleftharpoons \text{Complex S+1}$$

$$\text{Solute 2} + \text{Substrate S} \rightleftharpoons \text{Complex S+2}$$

Both solute 1 and solute 2 can diffuse in and out of the membrane, while the substrate, complex 1, and complex 2 cannot leave the membrane. Under certain conditions, solute 2 can be transported against its own concentration gradient. Examples are common in biological systems. Thus oxygen and CO can both bind with hemoglobin, leading to counter-transport of oxygen and CO poisoning. On a similar note, the cure for methanol poisoning is ethanol, which displaces methanol from the liver enzyme.

Co-transport occurs when the following type of overall reaction scheme holds:

$$\text{Solute 1} + \text{Solute 2} + \text{Substrate S} \rightleftharpoons \text{Complex S+1+2}$$

In this case solute 2 can be transported by solute 1. Transport of solute 2 can occur even if there is no concentration gradient, or even against its own concentration gradient. Thus solute 1 carries solute 2 along with it, and the phenomenon is known as co-transport. An *in vitro* example would be a combination of an anion (solute 1) and a cation (solute 2) within a membrane with some amine as a carrier (substrate). Additional *in vivo* examples can be found in biological systems, as indicated in the book by Cussler (2009).

Detailed analysis can be done by setting up the differential models for each of the species, identifying the dimensionless groups, followed by computations. If the reactions can be assumed to be fast then the equilibrium assumption for each of the reactions can be invoked, leading to analytical solutions. A detailed discussion of such models is provided in the book by Cussler (2009).

We note that the problems of mass transport are rich and cover a wide range of practical situations. Non-linear rates and transport of multiple species are the norm, and hence many problems are best solved by numerical methods. We therefore now move to show some examples of using numerical programs using MATLAB.

10.9 A boundary-value solver in MATLAB

10.9.1 Code-usage procedure

Boundary-value problems are commonly encountered in steady-state modeling of transport problems. Hence it is useful and important to be familiar with the numerical solution methods for such equations. Here we present directly the use of a solver in MATLAB, which is called BVP4C. This section illustrates the use of this software as a "black box" and illustrates the application to some commonly encountered problems in transport phenomena. Although use of software as a black box without understanding the basic scheme which goes into the solution method is not in general a good idea and can even be dangerous, we still use it here in order to enable the student to get a feel for numerical solution. Details of numerical solution methods are covered in other courses, and there are many textbooks dealing with this subject.

The basic structure of 1D boundary-value problems can be represented as

$$\frac{d^2y}{dx^2} = f(x, y, p) \tag{10.80}$$

where $p = dy/dx$. Many transport problems can be reformulated to fit this category. In general y is a vector called the solution vector and f is a vector function for multiple differential equations.

The solution is needed in the region $a \leq x \leq b$, and one boundary condition at each end is provided. These can be stated in general terms for linear boundary conditions as

$$a_1p + a_2y = a_3 \text{ at } x = a$$

and

$$b_1p + b_2y = b_3 \text{ at } x = b$$

The solver BVP4C can be used for such problems, and it is easier to study this by taking an example problem as given below.

10.9.2 BVP4C example: the selectivity of a catalyst

Consider a model problem represented as

$$\frac{d^2c}{dx^2} = f(c) \tag{10.81}$$

which is a dimensionless form of the diffusion-with-reaction problem in a pore of a catalyst. Here c is the dimensionless concentration and x is the dimensionless distance. Here $f(c)$ is a measure of the reaction rate. For example, $f(c) = \phi^2 c$ for a first-order reaction, where ϕ is the Thiele modulus. The variable x is a dimensionless distance along the pore. The point $x = 0$ is taken as the pore mouth and $x = 1$ at the pore end. The boundary conditions are taken as follows.

At $x = 0$, the dimensionless concentration is $c = 1$ (Dirichlet).

At $x = 1$, the gradient of the concentration is $dc/dx = 0$ (Neumann).

We use the MATLAB program BVP4C to solve this problem. This requires that Eq. (10.81) be written as two first-order equations rather than as a single second-order differential equation. This can be done as follows. Consider a solution vector y with components y_1 and y_2 defined as follows:

$$y_1 = c \text{ and } y_2 = dc/dx \tag{10.82}$$

Equation (10.81) is then equivalent to the following two first-order equations:

$$\frac{dy_1}{dx} = y_2 = dc/dx \tag{10.83}$$

and

$$\frac{dy_2}{dx} = f(y_1) = d^2c/dx^2 \tag{10.84}$$

The boundary conditions are as follows:

$$\text{at } x = 0; y_1 = 1 \text{ (pore mouth)} \tag{10.85}$$

and

$$\text{at } x = 1; dy_1/dx = y_2 = 0 \text{ (pore end)} \tag{10.86}$$

The solution for the concentration profile and also for the local values of the gradient can then be obtained using the function BVP4C in MATLAB. The effectiveness factor is related to the gradient at the pore mouth and is calculated as

$$\eta = -y_2(0)/\phi^2 \tag{10.87}$$

The sample code for solving this problem is as follows.

```
% program calculates the concentration profiles and the effectiveness
% factor for mth-order reaction in  a slab geometry.
global phi m phi2 n
global eta

% user actions, specify parameters, a guess function for trial
    solution, a
% function odes to define the set of first-order differential
    equations
% and a function bcs to specify the boundary conditions.
phi = 2.0 ;    % thiele modulus for first reaction
m =2.0 ;       % order of reaction
nmesh = 21   % initial mesh
nplot = nmesh ;   % meshes for plotting the result.
```

```
% solution block.
x = linspace ( 0, 1, nmesh ) ;
solinit = bvpinit ( x , @guess ) % trial solution given by
    guess function
sol = bvp4c (@odes, @bcs, solinit) % bvp solved,
% post-processing. plot concentration profiles, find eta
 x = linspace ( 0, 1, nplot) ;
 y = deval(sol, x) ;
 y(1,:) % concentration profiles displayed
 plot ( x, y(1,:) ) ; % these are plotted.
 eta = -y(2,1) / phi^2
%_____
function yinit = guess (x)
% provides a trial solution to start off
global phi m phi2 n
y1= exp(-phi * x)
y2= 0. * y1
yinit = [y1
         y2 ] ;
%------------------------------------------------------------
function dydx = odes ( x, y )
% defines the rhs of set of first-order differential equations
global phi m phi2 n

dydx = [  y(2)
          phi^2 * y(1)^m
        ] ;
%------------------------------------------------------------
function res = bcs ( ya , yb)
% provides the boundary conditions at the end points a and b
res = [ ya(1) - 1
        yb(2) ] ;
------------------------------------------------------------------
```

It is fairly easy to extend the code to multiple reactions. As an example, consider a series reaction represented as

$$A \rightarrow B \rightarrow C$$

The governing equations are as follows, assuming both reactions to be first order:

$$\frac{d^2 c_A}{dx^2} = \phi_1^2 c_A \qquad (10.88)$$

and

$$\frac{d^2 c_B}{dx^2} = -\phi_1^2 c_A + \phi_2^2 c_B \qquad (10.89)$$

The boundary condition at $x = 0$ (pore mouth) depends on the bulk concentrations of A and B. The boundary condition at $x = 1$ (pore end) is the no-flux condition for both A and B.

Table 10.1. Results for a diffusion-with-reaction series simulated with BVP4C code in MATLAB

x	c_A	c_B
0.0	0.2658	0.0481
0.2	0.2874	0.0507
0.4	0.3557	0.0581
0.6	0.4811	0.0658
0.8	0.6851	0.0604
1.0	1.0000	0
$p(x = 1)$	1.9281	-0.5850

The solution vector y has a size of four and consists of

$$y = \begin{pmatrix} y_1 = c_A \\ y_2 = dc_A/dx \\ y_3 = c_B \\ y_4 = dc_B/dx \end{pmatrix} \tag{10.90}$$

The system is now formulated as four first-order ODEs for the four components of the solution vector and solved by BVP4c in exactly the same way.

A sample result for $\phi_1 = 2$ and $\phi_2 = 5$ is given in Table 10.1. You should also compare this with the analytical solution. A key parameter of design interest is the yield of B, which is a measure of the rate of the reaction A to B compared with that for B to C. This is defined as the ratio of the flux of B outward from the surface to the flux of A into the surface:

$$\text{Yield of B} = - \left(\frac{dc_B/dx}{dc_A/dx} \right)_{x=1}$$

In the present case the yield is equal to 0.3034. The yield can be increased by changing the ϕ parameters, which can be done by changing the catalyst size. The explanation is that if the size of the catalyst is small then B is able to diffuse out faster than it can react further, and hence the B to C reaction is reduced. The manipulation of the yield by choice of appropriate catalyst size and shape is an important inverse problem in reactor design.

Summary

- Analysis of mass transfer processes is done by solving the species mass balance equations with suitable boundary conditions. A constitutive model for diffusion is used in conjunction, usually Fick's law.

- Stagnant systems are defined as systems with no superimposed external flow. A flow can be generated in such cases due to diffusion itself. A common example is evaporation of a liquid in an inert gas. The evaporation causes a flow (barely noticeable), which enhances the rate of transport. The enhancement is often accounted for by a drift flux correction factor.

- The Arnold cell is a simple device to measure diffusion coefficients (in gas mixtures) by simple measurement of the change in level of a liquid due to evaporation. The model for such systems is formulated using the concept of a pseudo-steady state. The rate of evaporation is calculated as though the system were in (pseudo-)steady state at any given instant of time. The liquid level is then solved as a function of time using the (pseudo-steady-state) value for the evaporation.

- The assumption of a pseudo-steady state is used in many problems in mass transfer calculations. This assumption is reasonable whenever the time scale of diffusion is much smaller than the process time.

- Mass transfer near fluid–solid interfaces is often modeled by using a film model. This model assumes that there is a stagnant film near the interface within which all the resistance to mass transfer is contained. The rate of transport can then be calculated using the relation D/δ_f (or equivalently a mass transfer coefficient) multiplied by the overall concentration difference.

- The film model is useful to study mass transfer accompanied by a surface reaction. For a simple first-order reaction with no mole change, an overall rate constant can be defined and used. The reciprocal of this rate constant is the sum of the reciprocal of the mass transfer coefficient and the reciprocal of the reaction rate constant. Thus it is not the true or intrinsic rate constant but only a convenient parameter for representation of the data. The methodology is exactly equivalent to the law of addition of resistances.

- The above simple concept of resistances fails even for a first-order reaction where there is a change in the number of moles due to reaction. The counter-diffusion of the product can enhance or retard the mass transfer rates. For non-linear kinetics, even with no change in number of moles, the resistance concept does not hold. It may be noted that a non-linear or transcendental implicit equation for the rate is generally obtained.

- The film model when applied to gas–liquid systems needs the use of two films, one on the gas side of the interface and the other on the liquid side. Each film offers its own resistance to mass transfer. These resistances can be added (for low-flux mass-transfer cases) to get an expression for the overall resistance. Correspondingly, an overall mass transfer coefficient can be defined and used to calculate the mass transfer rate. The mass transfer rate may be governed by the gas-side resistance or the liquid-side resistance or both. Relative contributions can be evaluated, which is very useful information for the selection of the appropriate equipment and design.

- Mass transfer with reaction in a porous catalyst is an important problem in reaction engineering. The model (when applied to catalysts of simple shapes) leads to second-order ordinary differential equations of the boundary-value type.

- The concentration profile within the porous catalyst depends on a dimensionless parameter, the Thiele modulus. Small values of the Thiele modulus lead to an almost uniform profile in the catalyst, leading to complete utilization of all of the catalyst. For large Thiele modulus the concentration profile is confined to a thin region near the surface, and only this part of the catalyst is being used for reaction. An effectiveness factor can be defined to measure the rate of reaction divided by the rate based on the surface concentration.

- Mass transfer accompanied by a zeroth-order reaction has some special features. Here the rate is not affected by diffusion and remains constant up to a critical value of the Thiele parameter. Beyond this the rate is affected by diffusional gradients. The problem has important applications in oxygen transport in tissues and growth of microbial cells, which often follows a zeroth-order metabolism.

- Gas–liquid reactions are another area where the mass transfer analysis is needed and effectively used to understand and design such systems. The key parameter is the Hatta number, which is a measure of the rate of reaction relative to the speed of diffusion. A large Hatta number leads to reaction being confined to a thin region near the gas–liquid interface. Reactions can become "instantaneous" in which case the rate of reaction is solely determined by mass transfer considerations, i.e., how fast the reactant can reach the reaction plane. The chemical kinetics plays no role under these conditions.

- Membrane-based processes and separations are becoming important in a number of industrial applications, and an engineer should be able to analyze and evaluate these processes. Membranes are, of course, of importance in biological transport in cells and other organs as well.

- For separation of gaseous mixtures, polymer-based membranes are most commonly used. These are essentially non-porous, and hence the rate of transport is governed by diffusion. In view of the small size of these pores, Knudsen diffusion is often the mechanism operating in these systems.

- Membranes for gaseous separation can be modeled on the basis of the solution–diffusion model. The key equation is (10.56). It is seen that the permeability of the membrane is the product of the diffusivity and the solubility (divided by the membrane thickness). The rate of transport can be changed by changing the solubility rather than the diffusivity. Thus the solubility parameter plays a key role in the selectivity of the membrane.

- Semi-permeable membranes used in many liquid-phase separations permit the transport only of the solvent. Such membranes find application, for example, in water desalination. At a low pressure difference, the water flows from the solvent to the salt side (osmosis). At a critical pressure drop equal to the osmotic pressure the system is at equilibrium and no transport of water occurs. If the applied pressure is larger than the osmotic pressure then we have conditions for desalination (reverse osmosis).

- The rate of transport in a semi-permeable membrane is proportional to the applied pressure gradient **minus** the osmotic pressure of the solution. The transport equation is given by the Starling equation (10.65) with the parameter σ known as the Staverman constant, which has a value of one for a completely semi-permeable case. The case $\sigma = 0$ applies when the solute and solvent permeate equally. Thus the Starling equation can be used both for a semi-permeable and for a completely permeable membrane with the appropriate choice of the σ parameter.

- The solvent flow causes convection effects towards the membrane surface, and there is a transport of solute (salt) by convection towards the membrane surface. Since there is no net transport of salt across the semi-permeable membrane, the concentration of the salt at the membrane surface has to assume a higher value. Thus the convective transport of salt is balanced by the diffusive transport of salt in the opposite direction. This phenomenon is known as concentration polarization.

- The effect of concentration polarization is to increase the osmotic pressure locally at the surface, thereby reducing the driving force for salt transport. The equation for solvent transport is now given by Eq. (10.66). It is seen that the measured permeability under a strong concentration-polarization condition is not the true membrane permeability.

- The irreversible thermodynamics provides a platform for analysis of membrane transport, which leads to the Kedem–Katchalsky equation. This equation has close parallels to the Starling law, although the two models were developed independently.

- Membranes containing a reactive carrier are of importance in biological applications. Transport of the solute is now governed by the diffusion–reaction model. The model can be solved numerically with a modified integral boundary condition for the carrier. The asymptotic solution can be obtained analytically for instantaneous reaction. Variations of the system with different types of carriers lead to co-transport, counter-transport, and other complex mass transfer behaviors.

- You should review the summary and also examine to what extent you have understood the learning objectives stated at the beginning of the chapter. The analysis of mass transfer processes is important in very many contexts, and a thorough understanding of the examples presented in this chapter will go a long way towards your education.

ADDITIONAL READING

Diffusion in a porous catalyst is a part of every book on chemical reaction engineering. A concise and easy-to-read presentation is given in the book by Harriott (2003).

The classic references on gas–liquid reactions are the books by Astarita (1967) and Danckwerts (1970).

A complete source for separations is the book by Seader, Henley, and Roper (2011), *Separation Process Principles*. This gives a detailed coverage of membrane separations as well as a number of solved design examples.

Reactive and facilitated transport are treated in detail in the book by Cussler (2009).

Diffusion in polymeric systems (which is important also for the design of membrane separation processes) is treated in a recent book by Vrentas and Vrentas (2012) together with other aspects of polymer behavior, sorption in polymers, and properties of polymer–solvent systems. Technological aspects of membrane separation together with the governing principles of membranes are treated in a book edited by Noble and Stern (1995).

Problems

1. Verify that the concentration profile in a slab is linear, whereas that in a hollow cylinder is logarithmic, and for a spherical shell it is an inverse function of r for the three geometries in Fig. 10.1.

   Also verify the expressions for the numbers of moles transported across the system, $\mathcal{W}_A$.

2. Derive an expression for the fall in the liquid level during evaporation using a quasi-steady-state approach. Show that

$$\frac{dH}{dt} = \frac{M_A}{\rho_L} N_A$$

   where N_A is the instantaneous rate of evaporation, i.e., based on the current height of the vapor space H.

   Verify the following expression for the height change:

$$H^2 - H_0^2 = 2 \frac{M_A}{\rho_L} D_A C y_{A,s} \mathcal{F} t$$

where $\mathcal{F}$ is the drift correction factor. In this expression we assume that the bulk mole fraction of A is zero.

3. Benzene is contained in an open beaker of height 6 cm and filled to within 0.5 cm of the top. The temperature is 298 K and the total pressure is 1 atm. The vapor pressure of benzene is 0.131 atm under these conditions, and the diffusion coefficient is $9.05 \times 10^{-6}\,\mathrm{m^2/s}$.

   Find the rate of evaporation based on (a) the low-flux model, (b) exact solutions, and (c) the low-flux model corrected for the drift flux.

4. For the above problem find the time for the benzene level to fall by 2 cm. The specific gravity of benzene is 0.874.

   For this condition find the mole-fraction profile of benzene in the vapor phase and compare your answer with the linear approximation (which would be the prediction of the low-flux model).

5. A liquid is contained in a **tapered** conical flask with a taper angle of 30°. The radius in the flask for the liquid level at the bottom is 7 cm and the vapor height above this is 10 cm.

   Find an expression for the rate of evaporation and the mole-fraction profile in the vapor space and compare your answer with the case of a straight cylinder.

   Assume that the liquid is benzene under the conditions stated in Problem 3. The bulk gas is at zero concentration.

6. Two bulbs are connected by a straight tube of diameter 0.001 m and length 0.15 m. Initially one bulb contains nitrogen and the bulb at the other end contains hydrogen. The system is maintained at a temperature of 298 K and a total pressure of 1 atm. The volume of each bulb is $8 \times 10^{-6}\,\mathrm{m^3}$.

   Calculate and plot the mole-fraction profile of nitrogen in the first bulb as a function of time. Verify the validity of the quasi-steady-state approximation by calculating the diffusion time and the process time constants.

7. Derive an expression for the case of mass transfer with second-order surface reaction based on a low-flux model. The resistance concept does not hold, unlike for a first-order reaction.

8. An example of a problem where there is severe counter-diffusion of the products is the deposition of SiO_2 from tetraethoxysilane (TEOS) on a solid substrate. The reaction is represented as

$$SiO(C_2H_5)_4(g) \rightarrow SiO_2(s) + 4C_2H_4(g) + 2H_2O(g)$$

   Note that 6 moles have to counter-diffuse and hence this retards mass transfer unless the mole fraction of TEOS is small. Consider deposition from a gas at a temperature of 400 K and pressure 0.1 atm and a mole fraction of TEOS of 0.2. The diffusivity is $0.1\,\mathrm{cm^2/s}$. Calculate the rate of reaction and the film growth rate assuming that the surface reaction is very rapid.

9. At a certain point in a mass transfer equipment the bulk mole fractions are $y_A = 0.04$ in the gas phase and $x_A = 0.004$ in the liquid phase. The mass density of the liquid is nearly the same as water. The Henry's-law constant for A is reported as $7.7 \times 10^{-4}\,\mathrm{atm \cdot m^3/g\,mol}$. Is the species absorbing or desorbing?

   If the mass transfer coefficients are $k_G = 0.010\,\mathrm{g\,mol/m^2\,s \cdot atm}$ and $k_x = 1.0\,\mathrm{mol/\,s \cdot m^2}$ find the overall transfer rate. Here k_x is the mass transfer coefficient on the liquid side based on mole-fraction difference as the driving force. What percentage of the resistance is in either film?

10. Consider the diffusion–reaction problem represented in the three geometries.

Verify the analytical solutions shown in the text for the three geometries with the Dirichlet condition of $c_A = 1$ at $\xi = 1$ and a symmetry condition (Neumann) at $\xi = 0$.

Note that the solution for a sphere needs a small coordinate transformation ($c_A = f(\xi)/\xi$, which reduces the governing equation to a simpler one in f).

Find the average concentration in the system for the three cases which represents the effectiveness factor. Make a plot of the effectiveness factor vs. ϕ^* for all of the three cases, where ϕ^* is a shape-normalized Thiele modulus defined as

$$\phi^* = \frac{\phi}{s+1}$$

Thus ϕ^* is equal to $\phi/2$ for a cylinder and $\phi/3$ for a sphere. Show that the results for the three geometries are quite similar when η is plotted as a function of ϕ^*, which is referred to as the generalized Thiele modulus.

11. Consider the same problem with now a Robin condition at the surface:

$$\left(\frac{dc_A}{d\xi}\right)_1 = Bi[1 - (c_A)_1]$$

Derive an expression for the effectiveness factor as a function of Bi in addition to the ϕ parameter. Do the analysis for all three geometries.

12. Consider a second-order reaction in a catalyst in the form of a slab. Show that the differential equation can be expressed as

$$\frac{d^2 c_A}{d\xi^2} = \phi^2 c_A^2 \tag{10.91}$$

How is the ϕ parameter defined for this case? Use the p-substitution method and derive an implicit integral representation to the solution of this problem.

Calculate the concentration profiles and the effectiveness factor for ϕ equal to 1 and 3.

13. The sulfur compounds present in petroleum fractions such as diesel can be removed by contact with a porous catalyst containing active metals such as Mo in the presence of hydrogen. If the catalyst is 3 mm in radius and the concentration of sulfur in the liquid surrounding the catalyst is $10\,mol/m^3$ find the rate of reaction. Assume that the reaction is of second order in sulfur concentration with a volumetric rate constant of $5\,m^3/mol \cdot s$. Note that these numbers are arbitrary and for illustration purposes only. Use $D_e = 1 \times 10^{-9}\,m^2/s$.

14. Gas absorption in an agitated tank with a first-order reaction.

Oxygen is absorbed in a reducing solution, where it undergoes a first-order reaction with a rate constant of $40\,s^{-1}$. The conditions are such that the liquid-side mass transfer coefficient k_L is equal to $8 \times 10^{-3}\,cm/s$.

Will there be appreciable reaction in the film? Use $D_A = 1.5 \times 10^{-5}\,cm^2/s$ for dissolved oxygen?

What would the rate of reaction be if the oxygen partial pressure were 0.21 atm? Use the same Henry constant as that for oxygen for water.

15. Verify the expression for the critical Thiele modulus for the oxygen concentration to become zero at R_0 in the tissue as a function of κ, the ratio of the capillary diameter to the unit cell diameter of the tissue given in the text.

Apply the model to the following data (from Truskey *et al.* (2004)) to find the maximum intercapillary radius for there to be no anoxic region:

Metabolic rate $= 1 \times 10^{-7}\,\text{mol/cm}^3 \cdot \text{s}$
Capillary radius 1.5–4 µm
$D_{O_2} = 2. \times 10^{-5}\,\text{cm}^2/\text{s}$ in tissue
Oxygen concentration in blood at the given point $4.05 \times 10^{-8}\,\text{mol/cm}^3$

Answers: 26.1 µm to 31.5 µm for the radius range shown above.

16. The effective diffusion coefficient of H_2 in a mixture of H_2 and CO in a porous catalyst was found to be 0.036 cm^2/s at a temperature of 373 K and 2 atm total pressure. The catalyst has a monodispersed pore structure with pore void fraction of 0.3.
    What is the mean pore size of this material? Are the conditions such that this parameter can be calculated with a reasonable confidence?

17. Consider a porous catalyst with a series reaction represented as

$$A \rightarrow B \rightarrow C$$

Write governing equations for A and B. Express them in dimensionless form. How many dimensionless groups are needed?
Solve the equations for a case where the dimensionless concentrations at the catalyst surface are one and zero for A and B, respectively. Note that both concentrations are normalized with the concentration of A at the surface.
Calculate the dimensionless gradient for these two species at the surface. From these expressions calculate the yield parameter defined as

$$S = -(dc_B/d\xi)/(dc_A/d\xi) \text{ at } \xi = 1$$

What is the significance of this parameter? Plot it as a function of ϕ. For a high selectivity to B, would you operate with a low ϕ or a high ϕ.

18. A spherical capsule has an outer membrane thickness with inner and outer radii r_i and r_o, respectively. A solute is diffusing across this capsule. Consider the case where the diffusion coefficient is a function of concentration and can be represented in a general form as $D(C_A)$ and is not a constant.
    Derive an expression for the (total) rate of mass transport across this spherical shell. Use the low-flux model.

19. A pool of liquid is 10 cm deep, and a gas A dissolves and reacts in the liquid. The solubility of the gas is such that the interfacial concentration is equal to 2 mol/m^3 in the liquid and the diffusivity of the dissolved gas is $2 \times 10^{-9}\,\text{m}^2/\text{s}$.
    Sketch and plot the concentration profile if the gas undergoes a first-order reaction with a rate constant of $10^{-6}\,\text{s}^{-1}$ in the liquid.
    Repeat the analysis, if instead of being of first order, the reaction were a zeroth-order reaction with a rate constant k_0 equal to $10^{-6}\,\text{mol/m}^3 \cdot \text{s}$. Would you expect a depleted zone (gas-starved region) in the liquid? What should the minimum surface concentration be in order to avoid there being a starved zone?

20. CO_2 is absorbed into a liquid under conditions such that the liquid-side mass transfer coefficient is $2 \times 10^{-4}\,\text{m/s}$. The diffusion coefficient of CO_2 in the liquid is $2 \times 10^{-9}\,\text{m}^2/\text{s}$. The interfacial concentration of CO_2 can be found using Henry's law. The pressure is 1 atm and the temperature is 300 K. Assume that CO_2 reacts with a dissolved solute in

the liquid with a rate constant of $1\,s^{-1}$. Also assume that the bulk concentration of CO_2 is zero.

Find the Hatta number.

Find the flux of CO_2 at the interface.

Find the flux of CO_2 going into the bulk liquid.

What percentage of CO_2 reacts in the film itself?

21. A porous catalyst is used for CO oxidation, and the process is modeled as a first-order reaction with a rate constant of $2 \times 10^4\,s^{-1}$. The effective diffusion coefficient for pore diffusion was estimated as $4 \times 10^{-6}\,m^2/s$. The catalyst is exposed to a bulk gas at 600 K and 1 bar pressure with a CO mole fraction of 2%. The external mass transfer coefficient is assumed not to be rate limiting.

    Find the rate of reaction for a spherical pellet of diameter 3 mm. Express the rate as moles reacted per catalyst volume per second.

    Find the effectiveness factor.

    Find the effectiveness factor if the catalyst diameter were changed to 1 mm.

    Estimate the range of values of the external mass transfer coefficient for which this resistance can be neglected.

22. For the CO_2 absorption in NaOH problem in the text, examine the effects of (i) changing the partial pressure of CO_2 and (ii) the concentration of the liquid-phase reactant. State the range of conditions under which the reaction is expected to take place under pseudo-first-order conditions. State the conditions under which the reaction can be treated as instantaneous.

    Find the conditions (partial pressure of CO_2) for which the reaction becomes instantaneous. Find the rate under this condition.

    Find the conditions (partial pressure of CO_2) for which reaction occurs under pseudo-first-order conditions, i.e., the interfacial concentration of B is nearly the same as the bulk concentration. Find the rate under this condition.

23. Extend the analysis of gas absorption with reaction to the case of absorption of two gases with a common liquid-phase reactant. A detailed study of this topic has been published in an award-winning paper by Ramachandran and Sharma (1971). Simulate numerically some illustrative examples from this paper.

24. **Condensation rates for a binary vapor mixture (adapted from Taylor and Krishna (1993)).** Here we apply Eq. (10.22) to the condensation of a binary mixture of ethylene dichloride (A) and toluene (B) for the following conditions.

    Vapor-phase composition: $y_{A0} = 0.4$
    Liquid-phase composition at the start of condensation $x_A = 0.325$
    Interface equilibrium relation: $y = 2x/(1 + 1.14x)$
    Vapor-phase mass transfer coefficient under low-flux condition 0.054 m/s
    Pressure 101.325 kPa; temperature 400 K

    Assume that the fluxes are proportional to the liquid composition, i.e., the composition of the first drop is fixed by the relative rates of condensation. This fixes the determinacy condition as

    $$x_A = \frac{N_A}{N_t}$$

Use this condition in the relation for N_A given by Eq. (10.22). Rearrange the equation and show that the total mass flux of condensation is given by

$$N_t = \ln \left(\frac{x_{AL} - y_{A\delta}}{x_{AL} - y_{A0}} \right) \qquad (10.92)$$

Use the numerical values given and verify that $N_A = 0.47$ and $N_B = 0.98$ in the units of mol/m$^2 \cdot$ s.

Note that in practice the condensation rate is determined not only by the rate of mass transfer but also by the rate at which heat can be removed from the vapor. Hence this is a coupled problem in heat and mass transfer. The example above provides the solution to the mass-transport part of the overall problem.

25. Code for pore diffusion with CHEBFUN is shown below. Note how compact the code is compared with BVP4C. The backslash operator is used as an overloaded operator here.

```
k1 = 0.05; De = 1e-10; R = 1.0e-03/3.; C_As = 10; % system parameters
Thiele = R * (k1 * C_As/De)^0.5; % Thiele modulus (second order)
 N = chebop(0,1); % defines an operator
 u = chebfun ('u'); % defines solution as function
  rate =  Thiele^2 * u.^2;  % rate function
N.op  = @(x,u) diff(u,2) - Thiele^2 * u.^2 ; % defines the problem
N.lbc = @(u) diff(u,1); % left boundary condition
N.rbc = 1     % right boundary condition
u = N\0 ;  % solution using the backslash operator.
s = sum(u.^2)  % average rate (effectiveness factor)
plot (u)  % plots the solution;
```

Test the code which is applicable for diffusion with a second-order reaction in a porous catalyst. Then use the code to develop a model for concentration of two species reacting with each other in a porous catalyst.

26. A membrane separator is 3 mm in diameter, and the membrane permeability was estimated as 2×10^{-6} m/s. The solute being transported has a diffusivity of 2×10^{-9} m^2/s in the liquid.

Estimate the overall permeability. What fraction of the resistance can be attributed to mass transfer?

The above membrane separator showed 70% solute recovery under certain operating conditions. A new membrane with 10 times the permeability is being considered as a replacement. Estimate the solute recovery in this new case.

27. Develop a simple backmixed model to evaluate the performance of a gas-separation system. Here a feed gas enters a high-pressure chamber with a mole fraction of A of x_{Af} and leaves enriched with a mole fraction of x_{Ae}. The exit stream leaving the high-pressure chamber is often referred to as the retentate.

A membrane separates the high-pressure chamber from a permeate side maintained at a low pressure. A permeate stream leaves this section. Exit mole fractions and the permeate flow rates are to be calculated from the model. Generate equations to calculate these quantities.

28. **Pervaporation: a case-study problem.** Pervaporation refers to removal of the permeate as vapor and represents an intermediate case between purely gas transport and purely liquid transport in a membrane. The process is important in production of pure ethyl alcohol,

which can otherwise be done only by a more expensive azeotropic distillation. Additional information on this process may be found in the book by Seader *et al.* (2001).

The purpose of this problem is to establish a model for the pervaporation process. Develop a model assuming equilibrium conditions both on the upstream liquid side and on the downstream gas side. Also assume that the permeate diffusion is controlled by Fick's law. The effective partial-pressure driving force may be used for transport, but the upstream side is at higher pressure, and an activity correction needs to be used here. Develop a model based on these concepts.

11 Analysis and solution of transient transport processes

Learning objectives

You will learn from this chapter

- the method of separation of variables for standard problems,
- separation of variables for non-homogeneous problems,
- use of superposition to use simpler 1D solutions for multidimensional cases,
- solution of transient problems posed on a semi-infinite domain,
- the error-function solution for constant boundary temperature,
- transient mass transfer problems and illustrative solutions,
- the penetration model for mass transfer,
- the effect of chemical reactions on transient mass transfer,
- solution methods for transient problems with a periodic boundary condition, and
- numerical solution of PDEs with the PDEPE program in MATLAB.

Mathematical prerequisites

- Orthogonal functions; the Sturm–Liouville equation.
- Fourier expansion of a function and evaluation of the Fourier coefficients.
- The Laplace transform and inverse Laplace transforms.
- Complex representation of a sine or cosine function; manipulations of complex variables.

In Chapters 8 and 10 we examined a number of steady-state problems in heat and mass transfer. In this chapter we examine unsteady-state problems. Owing to the similar mathematical structure heat and mass transfer problems can be treated in a similar manner. The concentration or temperature is now a function of both time and the spatial coordinates, and therefore the governing equations are now partial differential equations. This adds additional mathematical complexity to the solution. The most general problem in 3D and time and with time-varying boundary conditions needs a numerical solution. However, there are many simpler but nevertheless important problems where analytic solutions are possible and useful. This chapter introduces and solves such problems and illustrates a number of important techniques to obtain the solutions.

For linear partial differential equations, analytic solutions based on the method of separation of variables are commonly used. It can be applied to problems where the differential equations and the boundary conditions are homogeneous (as will be explained in the text) and applies to problems posed in a finite spatial domain. This solution procedure is developed and demonstrated here more from a physical perspective rather than a rigorous mathematical viewpoint.

The method of separation of variables cannot be used directly for non-homogeneous problems and needs some modifications. We study such problems next, and then show a number of examples. A more general solution method is based on the eigenfunction operator and is known as the finite Fourier transform method. This can be applied to a wider class of (still linear) problems, but this method is not discussed in this text.

Many problems in transient diffusion are posed on a semi-infinite domain. For such problems a solution method based on similarity transformation is useful. This method is introduced and applications to a number of problems are shown.

The method involving the Laplace transform is another useful method for such problems. This is introduced briefly with an example of a diffusion-with-reaction problem.

In some cases a boundary condition that varies with time in a cyclic manner at the boundary appears in the problem specification. For such problems, a method based on complex variables is a useful solution method. We show how to use this for a few prototypical problems.

For non-linear problems numerical solutions are needed, and we discuss one program based on MATLAB code, PDEPE, as an illustrative and useful method for non-linear problems.

The chapter therefore provides an introduction to solution methods for transient heat and mass diffusion problems together with some useful worked examples.

Some examples of transient flow problems are also presented, since these can be solved by similar methods.

11.1 Transient conduction problems in one dimension

The governing equation for transient conduction for heat transfer follows from Chapter 7 for the case of constant physical properties:

$$\rho c_p \left(\frac{\partial T}{\partial t} \right) = k \nabla^2 T + \dot{Q}_V \tag{11.1}$$

Similarly we find that for mass transfer we have

$$\frac{\partial C_A}{\partial t} + \nabla \cdot N_A = R_A \tag{11.2}$$

If a low-flux mass-transfer model is used for N_A, which is then equal to J_A, and if Fick's law is used for diffusion we have

$$\frac{\partial C_A}{\partial t} = D_A \nabla^2 C_A + R_A \tag{11.3}$$

We find Eqs. (11.1) and (11.3) are very similar and can be solved by using similar mathematical tools. In fact for some problems the formulation looks exactly the same when expressed in dimensionless variables.

The Laplacian can be simplified for common 1D problems. These problems are 1D diffusion/conduction in a slab, diffusion in the radial direction only in a long cylinder, and diffusion under axisymmetric conditions in a sphere. See Fig. 11.1. These are important and useful problems that will be analyzed in detail in the following sections.

The equations for all of these three cases can be compacted and represented in a general manner as

$$\rho c_p \frac{\partial T}{\partial t} = \frac{k}{r^s} \frac{\partial}{\partial r} \left(r^s \frac{\partial T}{\partial r} \right) + \dot{Q}_V \tag{11.4}$$

where s is the shape factor defined as 0, 1, and 2 for a slab, a cylinder, and a sphere, respectively.

On defining $\alpha = k/(\rho c_p)$ as the thermal diffusion coefficient we can write this as

$$\frac{\partial T}{\partial t} = \alpha \frac{1}{r^s} \frac{\partial}{\partial r} \left(r^s \frac{\partial T}{\partial r} \right) + \dot{Q}_V/(\rho c_p) \tag{11.5}$$

which is a commonly studied partial differential equation (PDE).

Test for homogeneity
Classification of PDEs into homogeneous and non-homogeneous equations is useful for further discussion. The standard test used here is as follows.

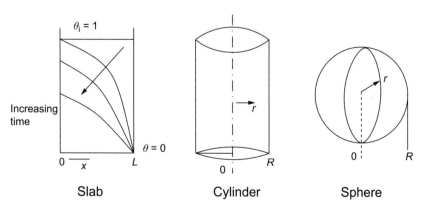

Figure 11.1 Three simple geometries where the transient problem can be treated in one spatial coordinate (and time).

A homogeneous differential equation is unaltered in form if T is substituted by another variable defined here as cT, where c is an arbitrary constant.

On applying this test to Eq. (11.5), we find that it is non-homogeneous since the coefficient $\dot{Q}_V$ becomes $\dot{Q}_V/c$. If this term were not present the equation would be homogeneous.

A similar test applies to boundary conditions. The test is as follows. If T is a solution then cT is also a solution with the same boundary conditions. Thus $T = $ constant at a boundary would make the boundary condition non-homogeneous, whereas $T = 0$ at that boundary would make it homogeneous.

Direct separation of variables is useful for homogeneous problems. For non-homogeneous problems a modification involving finding a partial solution is needed prior to using the separation of variables as discussed later.

Dimensionless representation

Let us now look at the dimensionless version of Eq. (11.4) which will be used for most of the solutions presented in this chapter.

The dimensionless distance is defined as $\xi = r/R$ or $\xi = x/L$ for slab geometry.

The dimensionless time τ can be defined as t/t_{ref}. It can be shown that $t_{ref} = R^2/\alpha$ is the most convenient choice. This follows from the differential equation. Hence

$$\tau = t/t_{ref} = t\alpha/R^2$$

The dimensionless temperature θ can be defined in a number of ways. The proper choice of which definition to use is problem-dependent. The key idea is to make the problem as simple as possible or make the equations and the boundary conditions homogeneous. The example provided will make this clear.

The governing equation in dimensionless form is then represented as

$$\frac{\partial \theta}{\partial \tau} = \frac{1}{\xi^s}\frac{\partial}{\partial \xi}\left(\xi^s \frac{\partial \theta}{\partial \xi}\right) + \dot{Q}_V^* \qquad (11.6)$$

where $\dot{Q}_V^*$ is the dimensionless heat generation rate. The common boundary conditions can be of the Dirichlet, Neumann, or Robin type. This is supplemented with an initial condition.

The parametric representation of the solution is therefore

$$\theta = \theta[\xi, \tau, \dot{Q}_V^*, Bi, \theta_i(\xi)]$$

where θ_i is the initial profile in dimensionless terms. The additional parameter is the Biot number, Bi, which comes into play if Robin-type boundary conditions are applied at one or both of the boundaries.

Having stated the problem in dimensionless form, we now examine first the problems where the separation of variables can be used as the solution method.

11.2   Separation of variables: the slab with Dirichlet conditions

Problem statement

No internal heat generation is considered in this section. The basic problem useful to understand the method of separation of variables is

$$\frac{\partial \theta}{\partial \tau} = \frac{\partial^2 \theta}{\partial \xi^2} \tag{11.7}$$

with the following boundary conditions.

At $\xi = 0$ we have symmetry and hence $\partial \theta / \partial \xi = 0$.
At $\xi = 1$ we fix a specified temperature and hence $\theta = 0$.
At $\tau = 0$ we assume a specified initial temperature profile $\theta = \theta_i(\xi)$ in general and $\theta = 1$
for a uniform starting profile.

The dimensionless temperature θ is defined here as $(T - T_s)/(T_i - T_s)$, where T_s is the surface temperature, which is maintained constant in the Dirichlet problem, and T_i is the initial temperature. This definition makes the boundary conditions homogeneous.

The form of the solution
Let us seek solutions of the form

$$\exp(-\lambda_n^2 \tau) F_n(\xi)$$

where λ_n is a constant. (The subscript n is needed since there are many values of λ, as will be shown later.)

The exponential dependence on time looks logical since there is a steady-state tempera-ture equation to zero. Also we note that locally at any point we start off with a temperature of one and end up with a final value of zero, and an exponential function of time would capture this behavior.

On substituting the assumed solution into Eq. (11.7) we find that the F function should satisfy the following ordinary differential equation (verify the algebra):

$$\frac{d^2 F_n}{d\xi^2} + \lambda_n^2 F_n = 0 \tag{11.8}$$

The general solution is

$$F_n = A_n \cos(\lambda_n \xi) + B_n \sin(\lambda_n \xi) \tag{11.9}$$

where A_n and B_n are the integration constants. Now we will use the boundary condition to evaluate these.

Use of the symmetry condition at $\xi = 0$ will lead to $B_n = 0$. Hence

$$F_n = A_n \cos(\lambda_n \xi)$$

Use of the condition at $\xi = 1$ will lead to

$$A_n \cos(\lambda_n) = 0 \tag{11.10}$$

A_n cannot be zero since that would lead to a trivial solution. Hence the above equation provides us with the means to find the values for λ_n. Since the cos function is zero at $\pi/2$, $3\pi/2$, $5\pi/2$ etc. we have many discrete values for λ_n:

$$\lambda_n = \left(n + \frac{1}{2} \right) \pi$$

where $n = 0, 1, 2, 3, \dots$. These are known as the eigenvalues. The F function is therefore of the form $F_n = A_n \cos(\lambda_n \xi)$. These are known as the eigenfunctions.

Hence a series solution that incorporates all the values of n seems logical, and therefore we can write the solution as an infinite series:

$$\theta(\xi, \tau) = \sum_{n=0}^{\infty} A_n \exp(-\lambda_n^2 \tau)\cos(\lambda_n \xi) \qquad (11.11)$$

The only thing missing is that the initial conditions are not yet satisfied. The series constants A_n can be found to satisfy the initial conditions as shown below.

Evaluation of the series coefficient

Using the initial condition we have

$$\theta_i(\xi) = \sum_{n=0}^{\infty} A_n \cos(\lambda_n \xi) \qquad (11.12)$$

Now, to find explicit expressions for A_n, we have to use the orthogonal property of the F function. Students often find this step difficult, and we illustrate this by the following example.

First we state the orthogonality property of the eigenfunctions:

$$\int_0^1 F_n(\xi)F_m(\xi)d\xi = 0 \text{ if } n \neq m$$

In our particular case of this problem,

$$\int_0^1 \cos([n + 1/2]\pi \xi)\cos([m + 1/2]\pi \xi)d\xi = 0 \text{ if } n \neq m$$

which can be verified from the table of integrals, MAPLE, or MATHEMATICA. But it is important to note that the orthogonality property is a general property for all eigenvalue problems of Sturm–Liouville type. If the differential operator is linear and symmetric then the eigenfunctions are orthogonal, which is a basic property of such operators. The interested student should refer to the book by Ramkrishna and Amundson (1985) on linear operator methods. We now proceed to show how to evaluate the series coefficients.

Example 11.1.

Show how the orthogonality property of the eigenfunction helps to unfold the series.

Solution.

Let us expand the series in Eq. (11.12) and write the first few terms:

$$\theta_i(\xi) = A_0 \cos(\lambda_0 \xi) + A_1 \cos(\lambda_1 \xi) + A_2 \cos(\lambda_2 \xi) + A_3 \cos(\lambda_3 \xi) + \cdots \qquad (11.13)$$

Consider finding A_1. If one multiplies both sides by $\cos(\lambda_1 \xi)$, one gets

$$\theta_i \cos(\lambda_1 \xi) = A_0 \cos(\lambda_0 \xi)\cos(\lambda_1 \xi) + A_1 \cos^2(\lambda_1 \xi)$$
$$+ A_2 \cos(\lambda_2 \xi)\cos(\lambda_1 \xi) + A_3 \cos(\lambda_3 \xi)\cos(\lambda_1 \xi) + \cdots \qquad (11.14)$$

Now integrate from 0 to 1 with respect to ξ. Only the A_1 term remains on the RHS, and upon rearranging we get A_1 as a ratio of two integrals:

$$A_1 = \frac{\int_0^1 \theta_i(\xi)\cos(\lambda_1\xi)d\xi}{\int_0^1 \cos^2(\lambda_1\xi)d\xi}$$

Using a table of integrals or by numerical integration, A_1 can be computed for any prescribed initial conditions. Note that the series unfolds upon integration due to the orthogonality property of the eigenfunctions. In general we have

$$A_n = \frac{\int_0^1 \theta_i(\xi)\cos(\lambda_n\xi)d\xi}{\int_0^1 \cos^2(\lambda_n\xi)d\xi}$$

For a particular case of a constant initial temperature $\theta_i = 1$, the series coefficients are

$$A_n = \frac{2(-1)^n}{(n+1/2)\pi}$$

which can be verified by performing the integration. This completes the solution to the Dirichlet problem in a slab.

The method holds for all similar problems, e.g., the other 1D geometries mentioned in Fig. 11.1. The only change will be in the eigenfunction, eigencondition, eigenvalues, and the series coefficient. If expressions for these are derived, the series solution is complete.

In fact, the solution to a number of problems examined in this and the following chapter can be represented in a general form similar to that given by Eq. (11.11). Here the term $\cos(\lambda_n\xi)$ will be replaced $F_n(\xi)$, the eigenfunction of the problem. Also the summation may be from $n = 1$, depending on how the eigenvalues are defined. The general form is noteworthy. One can therefore extend the method to study the Dirichlet problem in a cylindrical and spherical geometry, which is left as exercises. The solution for convective boundary conditions (Robin) at the surface will also have a similar form.

Let us return to the slab problem and look at some features of the solution.

11.2.1   Slab: temperature profiles

Plots of the solution can easily be generated for various values of time. An illustrative plot obtained by summing the series up to 50 terms in MATLAB is shown in Fig. 11.2.

The key points to note are the following.

Temperature profiles are rather sharp for small values of τ. The series will converge rather slowly here, and a large number of terms will be needed. An alternative is to model this as a semi-infinite slab as shown in Section 11.7. The following solution (derived in Section 11.7) will then be applicable:

$$\theta = \text{erf}\left(\frac{1-\xi}{2\sqrt{\tau}}\right)$$

where erf represents the error function. This solution is marked as * in Fig. 11.2. Note the excellent comparison with the above erf solution for $\tau < 0.05$.

A steady state is almost reached around a τ of one.

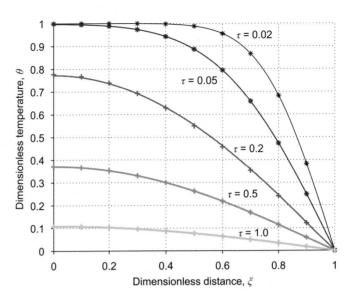

Figure 11.2 A time–temperature plot for the slab case with Dirichlet boundary conditions at the surface. The points marked * represent the semi-infinite solution for comparison; points marked + show the series solution computed with only one term.

The series converges rapidly for large values of τ. Just one term may be sufficient to get a reasonable estimate of the temperature profile.

Using the term approximation ($A_0 = 4/\pi$ here), the center temperature can be described as a function of time as

$$\theta_c = (4/\pi)\exp(-\pi^2\tau/4)$$

which is valid for $\tau > 0.2$ or so. The one-term solution is marked as + in Fig. 11.2 for comparison and we find that this is sufficient for $\tau > 0.2$.

11.2.2 Slab: heat flux

The expression for the heat flux at the surface is important in design calculations. This is obtained by applying Fourier's law at the surface. For a constant initial temperature of T_0, the expression can be shown to be

$$q_s = \frac{2k(T_0 - T_s)}{L} \sum_{n=0}^{\infty} \exp\left[-(n+1/2)^2\pi^2\tau\right] \tag{11.15}$$

The heat flux is very large at time near zero and decreases with time with a square-root proportionality.

11.2.3 Average temperature

The average temperature is defined in a general manner for all of the three geometries in Fig. 11.1:

$$\bar{\theta} = \frac{(s+1)}{1} \int_0^1 \xi^s \theta(\xi) d\xi$$

where s is the shape factor. For the slab case $s = 0$ and the average temperature is simply the integral under the curve in Fig. 11.2 for each time value. The integrated expression is

$$\bar{\theta}(\tau) = \sum_{n=0}^{\infty} A_n B_n \exp\left[-(n+1/2)^2 \pi^2 \tau\right] \tag{11.16}$$

where the term B_n arises from the spatial integration and is given by

$$B_n = \int_0^1 \cos(\lambda_n \xi) d\xi = \frac{\sin(\lambda_n)}{\lambda_n}$$

A related quantity is the fractional-energy-loss parameter, Φ, which is defined as

$$\Phi = 1 - \bar{\theta}$$

and represents the ratio of the heat that has been removed up to the given time to the total amount of heat that has to be removed.

11.3 Solutions for Robin conditions: slab geometry

If the surface is losing heat to the surroundings by convection, the surface temperature is unknown. What we know is the external or ambient temperature denoted as T_a here. For such problems the convective boundary condition should be used. This is obtained by taking a balance of the heat flux at the surface:

$$-k \left(\frac{\partial T}{\partial x}\right)_{x=L} = h(T_{x=L} - T_a)$$

The LHS is the heat arriving by conduction, with k defined as the thermal conductivity of the material, while the RHS is the heat leaving the surface by convection to the surroundings with h defined as the heat transfer coefficient. The dimensionless temperature θ is now defined as $(T - T_a)/(T_i - T_a)$. The boundary condition in dimensionless form at the surface is

$$\left(\frac{\partial \theta}{\partial \xi}\right)_{\xi=1} = -Bi\theta_{\xi=1}$$

where Bi is the Biot number defined as hL/k.

The differential equations and the center boundary condition remain the same. The initial condition is taken as unity for further analysis.

The series given by (11.11) is still used, but with the modified values for the eigenvalues, which now are specific to each Biot number. The series coefficients are also different now.

The following relations hold for this case, and the student should verify the details.

1. The eigenvalues are the solutions to

$$Bi \cos(\lambda) - \lambda \sin(\lambda) = 0$$

Note that the eigenvalues now depend on the Biot number and have to be calculated by solution of the above transcendental equation. They are not simple multiples of quantities like $\pi/2$, in contrast with the Dirichlet case.

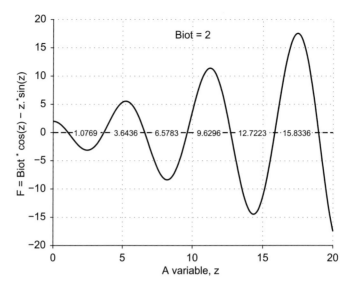

Figure 11.3 A graphical illustration of the calculation of the eigenvalues for the Biot boundary condition.

The eigenvalues can be found graphically by plotting $Bi\cos(\lambda) - \lambda\sin(\lambda)$ as a function of λ and noting the points of intersection of this curve with the x-axis. An illustrative plot is shown in Fig. 11.3.

The procedure to find eigenvalues can be automated in MATLAB or MAPLE. In particular the use of CHEBFUN reduces the coding and calculation efforts, and a sample code is shown as an exercise problem.

2. The eigenfunctions F_n are $\cos(\lambda_n\xi)$. These functions have the orthogonality property.
3. The series expansion coefficients for a uniform initial condition of one are

$$A_n = \frac{2\sin(\lambda_n)}{\lambda_n + \sin(\lambda_n)\cos(\lambda_n)}$$

The expression needed to find the average temperature is the same as before, as in Eq. (11.16), but the B_n values are now different and are Biot-number-dependent.

$$B_n = \frac{\sin(\lambda_n)}{\lambda_n}$$

Some properties of this solution are noted below.

1. For large Bi, the eigenvalues reduce to $\lambda_n = (n + 1/2)\pi$ and the Dirichlet solution is recovered for large Bi.
2. For small Biot numbers the temperature profiles are nearly uniform and a lumped-parameter approximation as discussed in Section 2.4.4 is sufficient. For a quick recap, the key assumption of the lumped model is that there are no significant internal temperature gradients. The average temperature of the solid predicted by this model can be stated as

$$\bar{\theta} = \exp(-Bi\tau)$$

You may wish to go back and review the section on the lumped-parameter model at this point.

Example 11.2. Biot problem: sample results for $Bi = 1$.
Sample results are provided for verification.

The solutions at the surface for various values of time are as follows.

Dimensionless time = 0.0, seven-term summation gives 9690; one term gives 0.7299. Note that even the seven-term summation does not converge well.

Time = 1; the solution is 0.3482; one term is sufficient. If a lumped model were used the solution would be 0.3679.

Time = 5; the solution is 0.018; one term. The lumped model gives 0.0067.

It can be seen that the lumped model is good even for a Biot number of 1.

Series expansion is also the solution method for the cylinder and sphere, and the final results are now presented.

11.4   Robin case: solutions for cylinder and sphere

Long cylinder

The solutions for a cylinder are in terms of the Bessel functions of the first kind and can be enumerated as follows.

1. The eigenvalues are the solutions to

$$Bi J_0(\lambda) - \lambda J_1(\lambda) = 0$$

2. The eigenfunctions F_n are $J_0(\lambda_n \xi)$.
3. The eigenfunctions are orthogonal to each other, with a weighting factor of ξ.
4. The series expansion coefficients for a uniform initial condition of one are

$$A_n = \frac{2 J_1(\lambda_n)}{\lambda_n [J_0^2(\lambda_n) + J_1^2(\lambda_n)]}$$

The coefficient needed in order to find the average temperature is

$$B_n = 2 \frac{J_1(\lambda_n)}{\lambda_n}$$

Sphere

The solution for a sphere belongs to a class of Bessel functions of the spherical kind. The results can be enumerated as follows.

1. The eigenvalues are the solutions to

$$\lambda \cos(\lambda) + (Bi - 1)\sin(\lambda) = 0.$$

2. The eigenfunctions F are $\sin(\lambda \xi)/(\xi \lambda)$
3. The eigenfunctions are orthogonal, with a weighting factor of ξ^2.
4. The series expansion coefficients for a uniform initial condition of one are

$$A_n = 2 \frac{\sin(\lambda_n) - \lambda_n \cos(\lambda_n)}{\lambda_n - \sin(\lambda_n)\cos(\lambda_n)}$$

Table 11.1. Coefficients for one-term approximation solution for the three common geometries for large Biot number, with the expressions $\theta_c \approx A_1 \exp(-\lambda_1^2 \tau)$ and $\bar{\theta} \approx A_1 B_1 \exp(-\lambda_1^2 \tau)$

	λ_1^2	A_1	B_1
Slab	2.467	1.273	0.6366
Cylinder	5.784	1.602	0.4317
Sphere	9.869	2.000	0.3040

The coefficient needed to find the average temperature is

$$B_n = 3 \frac{\sin(\lambda_n) - \lambda_n \cos(\lambda_n)}{\lambda_n^3}$$

Often one-term approximation can be used for long time. Usually we are interested in the temperature at the center, since the response is slowest at this point.

Similarly, the average temperature, another quantity of interest is expressed as

$$\bar{\theta} = \sum_n A_n B_n \exp(-\lambda_n^2 \tau)$$

The coefficients A_1 and B_1 have to be computed for each Biot number. The CHEB-FUN code shown in the exercise is handy for this purpose. These coefficients are given in Table 11.1 for the case of large Biot number, and these are useful for quick calculations for the long-term response and the average concentrations.

11.5 Two-dimensional problems: method of product solution

It is interesting to note that the 1D solutions can be extended to 2D (or 3D) problems by a (variation of the) superposition rule. In the context of transient problems this is known as the method of product solution. The method applies if the differential equation and the boundary conditions are homogeneous.

Consider a 2D problem shown schematically in Fig. 11.4.

The differential equation is the 2D transient heat-conduction equation

$$\frac{\partial \theta}{\partial t} = \alpha \left(\frac{\partial^2 \theta}{\partial x^2} + \frac{\partial \theta}{\partial y^2} \right) \tag{11.17}$$

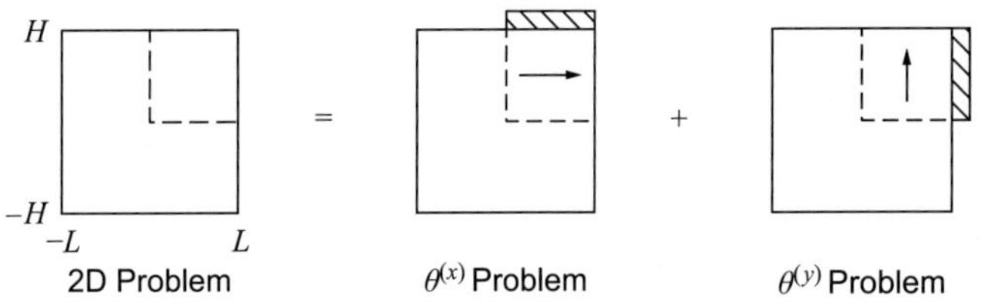

2D Problem $\theta^{(x)}$ Problem $\theta^{(y)}$ Problem

Figure 11.4 The product solution method: a problem in 2D transient conduction split into two simpler problems.

We claim that a product solution of the following type holds:

$$\theta(x, y, t) = \theta^{(x)}(x, t) \times \theta^{(y)}(y, t) \tag{11.18}$$

where $\theta^{(x)}(x, t)$ is the sub-problem with temperature varying in only the x-direction while $\theta^{(y)}(y)$ is the sub-problem posed in the y-direction only. (Superscripts with brackets are used here for clarity since often the subscripts denote differentiation and superscripts can also be mistaken for exponents.) The two sub-problems are shown schematically in Fig. 11.4.

The above claim can be verified by direct substitution in the governing equations. How about the boundary conditions? Since the boundary conditions on θ have been made homogeneous, we can also show that each sub-problem satisfies the boundary conditions for the sub-problems in the x- and y-directions.

Some caution has to be exercised in calculating the solutions to the sub-problems since the length scales in the x- and y-directions can be different. Let $2L$ be the length along the x-direction and $2H$ be the length along the y-direction. We know in general that the temperature profile depends on the dimensionless distance, time, and the Biot number. These have to be calculated separately for each sub-problem.

Thus, for the heat flow in the x-direction, we use the characteristic distance L and represent the sub-problem in the x- direction parametrically as

$$\theta^{(x)}(x) = \theta^{(x)}(x/L, \alpha t/L^2; hL/k) \tag{11.19}$$

i.e., dimensionless distance of x/L, dimensionless time of $\alpha t/L^2$, and the Biot number of hL/K should be used for the calculation of $\theta^{(x)}(x)$.

Similarly, for the y-problem the parametric representation is

$$\theta^{(y)}(y) = \theta^{(y)}(y/H, \alpha t/H^2; hH/k) \tag{11.20}$$

Once the temperatures have been calculated the average temperature can be shown to be

$$\bar{\theta}(t) = \bar{\theta}^{(x)}(t)\bar{\theta}^{(y)}(t) \tag{11.21}$$

and the fractional energy loss $1 - \bar{\theta}$ can be calculated as

$$\Phi = \Phi^{(x)} + \Phi^{(y)} - \Phi^{(x)}\Phi^{(y)} \tag{11.22}$$

where $\Phi^{(x)} = 1 - \bar{\theta}^{(x)}$, with $\Phi^{(y)}$ defined similarly.

Finite cylinder

For a finite cylinder the temperature profile can be found by the product solutions of the form

$$\theta(r, z, t) = \theta^{(r)}(r, t)\theta^{(z)}(z, t) \tag{11.23}$$

The r-sub-problem is to be computed by using the solution for transient conduction in a long cylinder. The radius of the cylinder is used as the characteristic length for this case. The z-sub-problem is to be computed using the 1D slab solution. Half the height of the cylinder is to be used as the reference length here.

11.6   Transient non-homogeneous problems

In this section, we consider how the heat equation can be solved for non-homogeneous cases. Clearly, using separation of variables directly is not possible, and we have to manipulate the equations somewhat before we can proceed further. We now state a general property and demonstrate it with two examples.

Fourier's lemma If a steady-state solution, $T_s(x)$, to the heat equation can be obtained with the prescribed boundary conditions, then the problem can be reduced to a homogeneous problem in terms of a new variable $Y(x, t)$ defined as

$$Y(x, t) = T(x, t) - T_s(x)$$

The variable Y can be interpreted physically as the deviation of the temperature from the steady state.

11.6.1   Subtracting the steady-state solution

Here we study the case of transient heat conduction with internal heat generation and demonstrate the method of subtracting the steady-state solution as per Fourier's lemma. The problem statement is as follows. A solid (a slab, cylinder, or sphere can be considered here) has an internal heat generation of $\dot{Q}_V$ per unit volume and is initially at a temperature of T_i. It is exposed to surroundings at a temperature of T_a, and is losing heat from the external surface with a convective heat transfer coefficient of h. It is required to track the temperature–time history of the solid as the system evolves to a final steady-state temperature.

The model equation is the same as Eq. (11.4) shown earlier in dimensional form:

$$\rho c_p \frac{\partial T}{\partial t} = \frac{k}{r^s} \frac{\partial}{\partial r} \left(r^s \frac{\partial T}{\partial r} \right) + \dot{Q}_V \tag{11.24}$$

We choose $T_{\text{ref}} = \dot{Q}_V R^2 / k$ so that the differential equations take a simple form. The solution can be represented as

$$\theta(\xi, \tau) = \theta_s(\xi) + Y(\xi, \tau) \tag{11.25}$$

where θ_s is the final steady-state part of the solution and Y is the transient part of the solution which can be obtained by separation of variables. Note that the problem cannot be directly solved by separation of variables, and that the above "superposition" approach is needed.

The steady-state solution for a long cylinder ($s = 1$) is obtained as

$$\theta_s = \frac{1}{4}(1 - \xi^2)$$

as shown earlier. The transient problem for Y is then set up. The governing equation is now the same as the transient conduction with no generation:

$$\frac{\partial Y}{\partial \tau} = \frac{1}{\xi^s} \frac{\partial}{\partial \xi} \left(\xi^s \frac{\partial Y}{\partial \xi} \right) \tag{11.26}$$

with the modified initial conditions

$$Y(0, \xi) = -\frac{1}{4}(1 - \xi^2)$$

The further details are left as an exercise. A point of caution is to note that one must change the initial conditions for the Y problem. The original conditions should not be inadvertently used.

Next we examine non-homogeneous problems for which no steady-state solutions exist. Very often an asymptotic solution valid for large values of time can be found and used to reduce the problem to a homogeneous form.

11.6.2 Use of asymptotic solution

The problem examined here is the solution to the transient heat conduction problem with the no-flux boundary condition at $\xi = 0$ and a flux of one at $\xi = 1$. Thus, at $\xi = 0, \theta_\xi = 0$, and at $\xi = 1, \theta_\xi = 1$, where the subscript ξ denotes differentiation. This represents a physical situation where one side of the slab is insulated while the other side is being heated by a constant heat flux. Since Neumann conditions are applied at both sides, this is also referred to as the Neumann–Neumann problem. The initial conditions are specified in a general manner as $\theta = \theta_i(\xi)$ for $\tau = 0$.

The differential equation is homogeneous, while the boundary conditions are non-homogeneous, and hence some adjustment is needed. However, the problem does not have a steady-state solution (why?) and hence the method proposed for the earlier problem does not apply. Fourier's lemma cannot be applied to separate out the steady-state solution here! A different approach is needed.

Here we propose that an asymptotic solution, θ_{asy}, exists for large time such that temperature varies linearly with time at any given position.

It can be shown that the following asymptotic form satisfies the PDEs and the boundary conditions:

$$\theta_{asy} = \tau + \frac{\xi^2}{2} - \frac{1}{6} \qquad (11.27)$$

This seems to have been pulled out of a hat, but Problem 11 shows the steps leading to it. Note that the initial conditions can be satisfied only in an integral average sense.

If the full solution is needed then a composite solution of the type

$$\theta(\xi, \tau) = \theta_{asy}(\xi, \tau) + Y(\xi, \tau)$$

where Y represents the initial transient part of the solution (the deviation from the asymptotic solution) can now be proposed. It can be shown that the Y part of the model is homogeneous and is amenable to solution by separation of variables. These details are not presented here but are left as an exercise. The solution for Y is

$$Y(\xi, \tau) = -\frac{2}{\pi^2} \sum_{n=1}^{\infty} \frac{(-1)^n}{n^2} \exp(-\lambda_n^2 \tau)\cos(\lambda_n \xi) \qquad (11.28)$$

where λ_n are the eigenvalues of the Y problem and are equal to $n\pi$ for this case, with n being an integer. Figure 11.5 shows how the temperature evolves as a function of time at various axial positions.

The most general case, in which these approaches might not be effective, can be handled by the eigenfunction expansion. The interested reader should refer to the excellent chapter in the book by in Deen (1998) or to mathematical texts on partial differential equations.

11.7 Semi-infinite-slab analysis

The series solution is valid for all values of time. However, at sufficiently short times the temperature changes are localized near the surface of the slab or solid. The heat does not

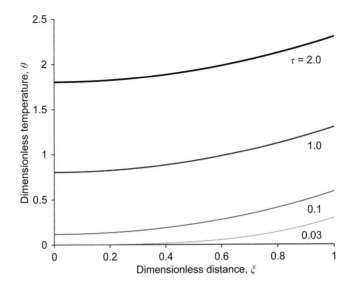

Figure 11.5 Temperature profiles as a function of time for the Neumann boundary condition at both sides.

penetrate far enough into the interior of the solid. For these small values of time ($\tau < 0.05$ is taken as the criterion), the series solution converges rather slowly, and many terms need to be included. In such situations, it is easier to treat the problem as one involving a semi-infinite medium and obtain separate closed-form analytical solutions. This is taken up in the following discussion. Solutions for three types of boundary conditions are studied.

11.7.1   Constant surface temperature

The dimensionless temperature for this problem is commonly defined as

$$Y = \frac{T - T_i}{T_s - T_i} = 1 - \theta$$

where T_i is the initial temperature of the slab and T_s is the surface temperature. Note that Y is $1 - \theta$, with θ as defined earlier. Note that this Y should not be confused with Y in Section 11.6.

The governing equation is

$$\frac{\partial Y}{\partial t} = \alpha \frac{\partial^2 Y}{\partial x^2} \tag{11.29}$$

Note that the distance and time have not been normalized. There is no obvious reference length scale or time scale for the problem. But we note that the following group of variables is dimensionless:

$$\eta = \frac{x}{2\sqrt{\alpha t}} \tag{11.30}$$

The factor of 2 in the denominator is there only for later numerical convenience.

You should verify that η is dimensionless.

Now the solution should be expected to be a function of η only. This is known as a similarity principle. Equation (14.13) can be shown to reduce to an ordinary differential equation of the following form:

$$\frac{d^2 Y}{d\eta^2} + 2\eta \frac{dY}{d\eta} = 0 \qquad (11.31)$$

The boundary conditions for the Y problem are as follows: At $\eta = 0$ (the surface), $Y = 1$. At $\eta \to \infty$, $Y = 0$, which applies at $t = 0$ or $x = \infty$. Note that the initial condition and the distant conditions merge into a single condition, which is a characteristic feature of problems with similarity transformations.

The first integration of Eq. (11.31) is straightforward and leads to

$$\frac{dY}{d\eta} = A \exp(-\eta^2)$$

with A being the integration constant. The second integration has no analytical closed-form solution and is written as

$$Y = A \int_0^{\eta} \exp(-u^2) du + B$$

where A and B are integration constants and u is a dummy variable.

Upon applying the boundary conditions, the final result can be expressed in a compact form using the "error" function. The final solution is

$$Y = 1 - \mathrm{erf}(\eta) = 1 - \mathrm{erf}\left(\frac{x}{2\sqrt{\alpha t}}\right) \qquad (11.32)$$

where erf represents the error function. This can be calculated directly in MATLAB or Excel. The statement employed to find the solution is $Y = erfc(eta)$, which is equal to $1 - \mathrm{erf}(\eta)$. This completes the solution for the temperature profile in a semi-infinite slab with a constant surface temperature.

The expression for the flux at the surface is obtained as

$$q_{x=0}(t) = \frac{k(T_s - T_i)}{\sqrt{\pi \alpha t}} \qquad (11.33)$$

An illustrative plot of the temperature profile is shown in Fig. 11.6. The profile is of a boundary-layer type, with a region near the surface coming under the influence of the surface change in temperature. This layer moves inwards as time progresses.

Results for other boundary conditions at the surface are now presented below without derivations.

11.7.2   Constant flux and other boundary conditions

Constant flux: Neumann

If the surface is exposed to a constant heat flux, the following results can be obtained:

$$T(x, t) - T_i = \frac{q_s}{k} \left[\sqrt{\frac{4\alpha t}{\pi}} \exp\left(-\frac{x^2}{4\alpha t}\right) - x \, \mathrm{erfc}\left(\frac{x}{2\sqrt{\alpha t}}\right) \right] \qquad (11.34)$$

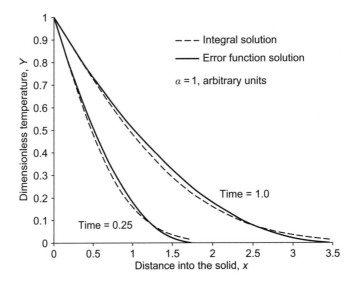

Figure 11.6 Temperature profiles in a semi-infinite slab; time is in arbitrary units.

Note that the surface temperature increases with time now, while the surface heat flux remains constant.

Convection: Robin
Here the surface is exposed to a bulk gas at a temperature of T_a and receives heat with a heat transfer coefficient of h. The solution for this case is

$$\frac{T - T_i}{T_a - T_i} = \text{erfc}\left(\frac{x}{2\sqrt{\alpha t}}\right) - \exp\left(\frac{hx}{k} + \frac{h^2\alpha t}{k^2}\right)\text{erfc}\left(\frac{x}{2\sqrt{\alpha t}} + \frac{h\sqrt{\alpha t}}{k}\right) \qquad (11.35)$$

Note that the surface temperature T_s is now a function of time and that the heat flux also varies with time. The heat flux can be calculated directly as $h(T_a - T_s)$.

Energy pulse
If a source of energy E per unit area is introduced at the surface for a very short duration of time this leads to a Dirac-delta condition at the surface. This can be done, for example, by switching on a laser pulse and then switching it off immediately. Assume that all the energy enters the body, with no loss from the surface. The temperature response is then given by

$$T(x, t) - T_i = \frac{E/(\rho c_p)}{\sqrt{\pi \alpha t}} \exp\left(-\frac{x^2}{4\alpha t}\right) \qquad (11.36)$$

The surface temperature is a maximum at time zero and decreases thereafter. The surface heat flux is zero at all times greater than zero. Examples involving use of the above equations are left as problems.

11.8   The integral method of solution

In this section we look again at the semi-infinite-domain problem and look at a method of solution based on integral balance of the differential equation. The integral methods involve integration of the governing equation with respect to x from 0 to an assumed heat-penetration depth δ. This sets up a macroscopic model for the process. The information on the detailed temperature profile is, however, lost in the process. Hence, in order to solve the problem an approximate temperature variation in the region has to be assumed (usually a simple quadratic relationship). Then an ordinary differential equation for the variation of δ with time can be derived. Integration of this equation gives an approximate solution to the problem and also the information on how the penetration depth varies with time. The details are as follows.

The governing equation is reproduced here for ease of reference:

$$\rho c_p \frac{\partial T}{\partial t} = k \frac{\partial^2 T}{\partial x^2} \tag{11.37}$$

We integrate both sides with respect to x from 0 to δ, the heat-penetration depth. We also note that δ is a function of t. Hence the integral representation of the problem is

$$\int_0^{\delta(t)} \rho c_p \frac{\partial T}{\partial t} \, dx = \int_0^{\delta(t)} k \frac{\partial^2 T}{\partial x^2} \, dx \tag{11.38}$$

This can be interpreted as a macroscopic heat balance. The LHS is the accumulation of enthalpy in the system, while the RHS is the input–output terms generated by conduction. The integral formulation can be simplified further. The LHS can be rearranged by use of the Lebnitz rule for differentiation under the integral sign:

$$\text{LHS} = \frac{d}{dt} \left(\int_0^{\delta(t)} \rho c_p T \, dx \right) - \rho c_p T_i \frac{d\delta}{dt} \tag{11.39}$$

where T_i is the temperature $x = \delta$, which is also equal to the initial temperature. The last term can be dropped if the initial temperature is zero, which can be done without loss of generality.

The RHS can be expressed as

$$\text{RHS} = \int_0^{\delta(t)} k \frac{\partial^2 T}{\partial x^2} \, dx = \left(k \frac{\partial T}{\partial x} \right)_{x=\delta} - \left(k \frac{\partial T}{\partial x} \right)_{x=0} = q_0 \tag{11.40}$$

where q_0 is the heat flux at $x = 0$. Note that $\partial T/\partial x$ at $x = \delta$ is taken as zero.

Hence the integral representation of the problem is

$$\boxed{q_0 = \frac{d}{dt} \left(\int_0^{\delta(t)} \rho c_p T \, dx \right) - \rho c_p T_i \frac{d\delta}{dt}} \tag{11.41}$$

The above expression is general and applies to all of the three (common) types of boundary conditions at the surface. Also note that the expression is an exact representation of the problem, not an approximate representation, and is a macroscopic model for transient conduction in a semi-infinite slab. However, no information on the detailed temperature profile can be extracted, since it is a lumped model. Specific solutions are obtained by invoking an approximate form for the temperature profile. The most common form is a quadratic:

$$T = A + Bx + Cx^2 \tag{11.42}$$

The solution using this temperature profile for the case of constant surface temperature is presented below. The solutions for other types of boundary conditions are left as problems. Applying the boundary condition for constant surface temperature case leads to

$$\frac{T - T_i}{T_s - T_i} = \left(1 - \frac{x}{\delta}\right)^2 \tag{11.43}$$

Substitution of this into the integral representation given by Eq. (11.41) and further algebra leads to the following differential equation for δ:

$$\delta \frac{d\delta}{dt} = 6\alpha \tag{11.44}$$

Upon integrating we obtain an expression for the variation of the penetration depth with time:

$$\delta = \sqrt{12\alpha t} \tag{11.45}$$

The corresponding temperature profile is now obtained from Eq. (11.46). The final expression can be written in terms of the similarity variable as

$$\frac{T - T_i}{T_s - T_i} = \left(1 - \frac{\eta}{\sqrt{3}}\right)^2 \tag{11.46}$$

The accuracy of the integral solution can be ascertained by comparison with the error-function solution given earlier. A comparison plot is shown in Fig. 11.6. Here the temperature profile obtained by the integral method with an assumed quadratic profile is shown as a dotted line and compared with the error-function solution shown as the solid lines. The error in heat flux is of the order of 3%.

The student should work out the details to understand the procedure and also work through the cases of constant flux (type 2 boundary condition) and specified convection rate to the surface (type 3 condition). It is easier to work out the algebraic details by taking the initial temperature T_i to be zero, which can be done without any loss in generality.

The integral approach to solving transport problems can also be used in boundary-layer theory (which will be discussed later), and for other problems as a quick approximate way to solve transport problems. Hence it may be useful to study the method in detail.

Important problems in transient mass diffusion with and without reaction are examined in the next section.

11.9   Transient mass diffusion

11.9.1   Constant diffusivity model

Finite-domain solutions

First we analyze transient mass transfer problems with no reaction and with constant diffusivity. The governing equation is given by Eq. (11.3) shown earlier, with R_A set as zero. Problems in the three common geometries shown in Fig. 11.1 are examined. Since the problem is identical in terms of dimensionless variables, the solutions for heat conduction can be used. The thermal diffusivity is replaced by D, the molecular diffusivity. In particular, the series solution given by (11.11) is useful, with the dimensionless concentration now defined as

$$c_A = \frac{C_A - C_{A,s}}{C_{A,i} - C_{A,s}}$$

and time defined as

$$\tau = \frac{t}{R^2/D_A}$$

R^2/D_A is the reference time, also called the diffusion time as noted earlier, and is approximately the time taken for internal gradients to reach a steady profile for a surface change in the concentration. The definition of the spatial variable ξ is obviously the same.

The equations for calculations of c_A for mass transfer and θ for heat transfer have identical forms and can be used interchangeably. The physical properties are replaced as follows.

The thermal diffusivity α in heat transport is replaced by the mass diffusivity D_A (D is used as an abbreviated notation in this section) for mass transport problems. In the calculation of the heat flux the quantity k appears as an explicit parameter. This is replaced simply by the diffusion coefficient in the mass transport problems and the term ρc_p in the heat-flux equation is set as one for the mass transport problem since the capacity for mass accumulation is one.

We illustrate this with an example problem of drug release.

Example 11.3. Drug release rate from a capsule.
A capsule has a radius of 0.3 cm and has an initial drug concentration of 68.9 mg/c.c. The diffusion coefficient of the drug in the porous matrix is 3×10^{-6} cm^2/s.

Calculate and plot the center concentration and the amount of drug released as a function of time.

Use one-term approximation as a simplification. (But note in general that this is not valid for short times.)

Solution.
First we estimate the time constant in order to know how long the process will last:

$$\text{Time constant} = \frac{R^2}{D_A} = \frac{(0.3^2)\,\text{cm}^2}{3 \times 10^{-6}\,\text{cm}^2/\text{s}} = 3 \times 10^4\,\text{s} = 8.33\,\text{h}$$

Hence we look for values of time in this range.

For $t = 1\,$h, which corresponds to $\tau = 1/8.33 = 0.12$, the center concentration is calculated from the one-term approximation to the series solution as 0.6119 using the value of A_1 from Table 11.1.

The average concentration is calculated as 0.1860 using the value of B_1 from Table 11.1. The fraction of drug released is $1 - \bar\theta$ and is 81.4%.

The quantities for other values of time can be calculated in a similar manner.

Semi-infinite region
The solution to transient diffusion in a semi-infinite region is identical to the heat transfer problem. We simply reiterate the key equations for ease of reading. Thus the concentration profile for a fixed concentration at the surface is

$$\frac{C_A - C_{A,i}}{C_{A,s} - C_{A,i}} = \text{erfc}\left(\frac{x}{\sqrt{4D_A t}}\right)$$

while the instantaneous flux into the surface is given as

$$N_{A,s} = [C_{A,s} - C_{A,i}]\sqrt{\frac{D_A}{\pi t}}$$

which is similar to Eq. (11.33). The surface flux decreases with a square-root-of-time dependence.

The average flux over a period of time t_E, the exposure time, is given by

$$\bar{N}_{A,s} = \frac{1}{t_E}\int_0^{t_E} N_{A,s}\, dt = (C_{A,s} - C_{A,i})2\sqrt{\frac{D_A}{\pi t_E}} \tag{11.47}$$

This expression is important in the context of mass transfer at interfaces exposed to a short contact time. The resulting model is known as the penetration model and is elaborated upon in Section 11.9.2.

The pulse response is also of interest in some applications, and the solution is

$$C_A - C_{A,i} = \frac{S}{\sqrt{\pi D_A t}}\exp\left(-\frac{x^2}{D_A t}\right) \tag{11.48}$$

which is analogous to Eq. (11.36) in heat transfer. Here S is the pulse strength, the quantity of solute placed on the surface at time zero.

Example 11.4.

Silicon is doped with phosphorus under (i) constant surface conditions and (ii) with a pulse introduced at time zero. Calculate the concentration profiles as a function of time. Assume that the initial concentration $C_{A,i}$ is zero and take the surface concentration as one, which is simply a scaling parameter.

Solution.

The diffusion coefficient of phosphorus in silicon is reported to be 6.5×10^{-13} cm^2/s at 1100 °C in the book by Middleman and Hochberg (1993).

The penetration depth is where the complementary error function takes a value of 0.01. This corresponds to an argument of η of 1.8. Hence at time $t = 3600$ s the phosphorus would have diffused up to $1.8\sqrt{2Dt}$, which is equal to 1.23 μm.

At a location of half this distance the parameter η is 0.6364. The corresponding value of erfc is 0.3173. Hence the phosphorus concentration at this location is 31.73% of the surface value.

The total quantity of phosphorus diffusing is equal to $\sqrt{4Dt/\pi}$, which is 5.45×10^{-5} per cm^2 times the surface concentration of phosphorus.

If the doping were done with a pulse of strength $S = 5.45 \times 10^{-5}$ then the concentration at 0.615 μm would be calculated from Eq. (11.48) as 0.1260. The concentration with a pulse source is lower than that with a step source and provides a means of controlling the junction thickness.

11.9.2   The penetration theory of mass transfer

In Section 10.2 we discussed a film model for mass transfer. This assumes that a steady-state concentration profile is established near the interface from where mass transfer is taking place. There are, however, many situations where the steady-state profile does not quite get established. For example, consider the case of mass transfer from a bubble to a liquid. Mass transfer can take place only when the bubble is in contact with the liquid, which receives mass only for this time, and hence the mass transfer is to be viewed as a transient process. The penetration model proposed by Higbie (1935) is an attempt to describe the mass transport using the transient diffusion model.

In this model the liquid element is assumed to be in contact with a bubble (or a gas phase) for a certain time t_E, the exposure time. Mass transfer takes place up to this time. At the end of this time, the eddies mix the liquid, a fresh liquid is exposed, and the process starts again. The mass transfer rate over this "exposure" time can then be calculated using the transient diffusion model shown above. The mass transfer coefficient is calculated by dividing the average flux Eq. (11.47) by the driving force and the following expression results for the mass transfer coefficient:

$$k_L = 2\sqrt{\frac{D_A}{\pi t_E}} \qquad (11.49)$$

11.9.3   The effect of chemical reaction

For a reacting system a source term is added to the transient diffusion equation, resulting in the following equation for a first-order reaction:

$$\frac{\partial C_A}{\partial t} = D\frac{d^2 C_A}{dx^2} - kC_A \qquad (11.50)$$

The solution depends on the domain under consideration. Let us start with a finite domain.

Finite domain
The dimensionless version is more useful in this context:

$$\frac{\partial c_A}{\partial \tau} = \frac{d^2 c_A}{d\xi^2} - \phi^2 c_A$$

where ϕ is the Thiele modulus.

The boundary conditions are the usual no-flux condition at $\xi = 0$ and a concentration of $c_A = 1$ at $\xi = 1$. The initial condition examined here is $c_A(\tau = 0) = 0$.

It should be noted here that the differential equation is homogeneous while the boundary condition is not. Hence the method of subtraction of the steady-state solution would apply.

The solution can be represented in dimensionless form:

$$c_A(\xi, \tau) = c_A^{(s)}(\xi) + Y(\xi, \tau)$$

where $c_A^s(\xi)$ is the solution and Y is the perturbation from steady state. The steady state was shown earlier:

$$c_A^{(s)}(\xi) = \frac{\cosh(\phi\xi)}{\cosh(\phi)}$$

The Y problem has the same form as the original equation:

$$\frac{\partial Y}{\partial \tau} = \frac{\partial^2 Y}{\partial \xi^2} - \phi^2 Y$$

except that the boundary condition at $\xi = 1$ is now $Y = 0$. The no-flux condition is still used at $\xi = 0$. These conditions render the problem homogeneous, and hence the series solution can be obtained for Y. The solution for Y proceeds by assuming a time-dependent function and a position-dependent function as before. The final result can be shown to be

$$Y(\xi, \tau) = \sum_n A_n \exp\left(-[\lambda_n^2 + \phi^2]\tau\right) F_n(\xi) \tag{11.51}$$

where λ_n is a constant. Note that the form is slightly different from the no-reaction case, and the term ϕ^2 is needed together with λ^2. The eigenvalues can be shown to be the same as before, $\lambda_n = (n + 1/2)\pi$.

The eigenfunctions are the same as before, $F_n = \cos(\lambda_n\xi)$.

The series coefficients are now different. The series coefficients can be evaluated as

$$A_n = \frac{\int_0^1 Y_i(\xi)\cos(\lambda_n\xi)d\xi}{\int_0^1 \cos^2(\lambda_n\xi)d\xi} \tag{11.52}$$

where Y_i is the initial condition for the Y problem. This initial condition is obtained by subtracting the steady-state solution from the prescribed initial condition for c_A. Thus, for a zero starting condition, the initial condition for Y is

$$Y_i(\xi) = -\frac{\cosh(\phi\xi)}{\cosh(\phi)}$$

The symbolic integration in MATLAB or MAPLE may be used here for Eq. (11.52), or closed-form analytical expressions can be derived for the series coefficients. Note that the series coefficients are functions of ϕ as well. The student may wish to fill in the details and complete the solution.

It may also be noted that the time constant is altered due to reaction. The time taken to reach steady state will be around $1/[\pi^2/4 + \phi^2]$. For low values of ϕ the time constant is not altered much by ϕ, although the final concentration profile is a function of ϕ. For large ϕ the response is governed more by the reaction rather than diffusion, and the time taken to reach steady state is a function of ϕ.

The solutions for the cylindrical case and for a sphere can be obtained in the same manner. The solution for a semi-infinite slab is discussed in the following section.

Semi-infinite domain

The solution was derived by Danckwerts and holds a prominent place in the chemical reaction literature. The model equation is the same as above, except that the domain is semi-infinite. The boundary conditions are again standard. The method of the Laplace transform is useful for such problems.

The Laplace-transform method

The starting equation is Eq. (11.50), and we define $\bar{C}(S)$ as the Laplace transform:

$$\bar{C}(S) = \int_0^\infty C_A(t) \exp(-st)dt$$

This reduces the PDE to an ODE:

$$D_A \frac{d^2 \bar{C}_A}{dx^2} - (s+k)\bar{C}_A$$

where we have used an initial condition of zero.

The boundary condition at the surface in the Laplace domain is $C_{A,s}/s$, which holds for $x = 0$. (Note that this is for a step change in the concentration at the surface, the problem being studied here.) The second boundary condition is at $x = \infty$, where $\bar{C}(s) = 0$. Hence the solution to Eq. (11.50) is

$$\bar{C}_A(x,s) = \left(\frac{C_{A,s}}{s}\right) \exp\left(-x \sqrt{\frac{s+k}{D_A}}\right)$$

The student should verify the solutions algebra leading to this equation at this stage.

All that remains is to find the inverse transform to find the solution in the time domain. That is easier said than done! The algebra is complex, and needs the concept of Cauchy integration and considerable work. However, tables of inverse transforms are available for a large class of problems, including this one, and can be directly used. Computer algebra based on MAPLE or MATHEMATICA can be used, and the book by White and Subramanian (2010) has a nice illustration of a step-by-step approach using MAPLE. The final solution in the time domain is

$$\frac{C_A(x,t)}{C_{A,s}} = \frac{1}{2} \exp\left(-x\sqrt{k/D_A}\right) \text{erfc}\left(\frac{x}{2\sqrt{D_A t}} - \sqrt{kt}\right)$$
$$+ \frac{1}{2} \exp\left(x\sqrt{k/D_A}\right) \text{erfc}\left(\frac{x}{2\sqrt{D_A t}} + \sqrt{kt}\right) \tag{11.53}$$

We can now sketch the profiles with and without reaction, which are illustrated in Fig. 11.7. The interesting behavior is that there is not much difference at small values of time between physical and chemical absorption. For large values of time the concentration front for the reactive system hardly moves and appears to stagnate. Here the profiles are close to a steady-state solution represented as

$$\frac{C_A(x,t)}{C_{A,s}} = \exp\left(-x\sqrt{k/D_A}\right)$$

In other words, the effect of the transient term dies out at large values of time, and the concentration profile is established by a balance of diffusion and reaction at each spatial location. These differences are important in analyzing experimental data using the penetration theory. Thus, by changing the contact time or time of exposure of the system, we can determine both the diffusion coefficient and the rate constant.

Flux and average rate of mass transfer

The flux of A at the surface in the presence of reaction is of more practical importance and can be calculated from the following formula derived from Eq. (11.53):

$$N_{A,s} = C_{A,s}\sqrt{D_A k}\left[\mathrm{erf}(\sqrt{kt}) + \frac{\exp(-kt)}{\sqrt{(\pi kt)}}\right]$$

A rather interesting result is the average flux over a time interval of t_E. This is calculated by integrating the above instantaneous flux with respect to time and then dividing by t_E. The result is

$$\bar{N}_{A,s} = C_{A,s}\sqrt{\frac{D_A}{kt_E^2}}\left[(kt_E + 1/2)\mathrm{erf}(\sqrt{kt_E}) + \exp(-kt_E)\sqrt{\frac{kt_E}{\pi}}\right]$$

which is the expression for the penetration-theory model for diffusion with a first-order reaction.

The enhancement factor is obtained by dividing by the no-reaction value given by Eq. (11.47).

Example 11.5.

Consider a gas in contact with a liquid for 1 s. Calculate and plot the concentration profiles for (i) no reaction, (ii) a first-order reaction with a rate constant of $k = 0.1\ \mathrm{s}^{-1}$, and (iii) a first-order reaction with a rate constant of $1\ \mathrm{s}^{-1}$.

Assume a diffusion coefficient of $2 \times 10^{-5}\ \mathrm{cm}^2/\mathrm{s}$.

Solution.

We should first establish the distances for which we want to calculate the solution. From the penetration model we know that, for a no-reaction case, the depth of penetration, δ, is approximately equal to $\sqrt{12 D_A t}$. Using the data given, we find $\delta = 0.0155\ \mathrm{cm}$. Hence values of x from 0 to 0.0155 are used in Eq. (11.53) and the dimensionless concentration $C_A/C_{A,s}$ is calculated. The plot for these two values of k is shown in Fig. 11.7. It is seen that for $k = 0.1$ the effect of reaction on the mass transfer rate is negligible. The time interval considered is simply not sufficient for any reaction to occur.

For $k = 1.0$ there is a significant effect of reaction, and the concentration front is now confined to a narrower region near the surface.

If k were 2 (results not shown to avoid clutter in the graph) the concentration front would have attained a nearly steady-state type of situation.

The penetration depth for a reacting system is of the order of $3\sqrt{D_A/k}$. Hence the concentration profile is expected to be confined to a region of thickness 0.0134 cm for $k = 1$ and 0.0424 cm for $k = 0.1$ for large values of time.

Extensions to other systems where two components diffuse and react with each other are not presented here, but follow an identical approach. Only the mathematics gets more involved. An illustrative sketch is, however, presented below to give a feel of the solution. Numerical solutions can be obtained using the PDEPE solver introduced later in this chapter. This will provide you with a useful tool to extend the analytical solutions to more practically relevant problems.

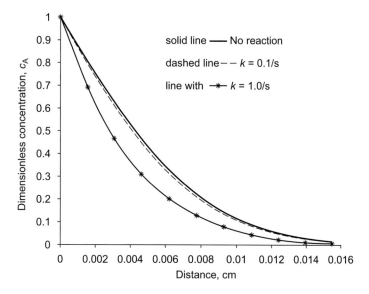

Figure 11.7 Illustrative concentration profiles for transient diffusion with reaction in a semi-infinite region; $D_A = 2 \times 10^{-5}$ cm^2/s and $t = 1$ s were used in calculations.

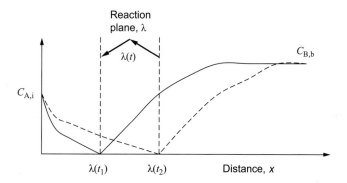

Figure 11.8 Concentration profiles for A and B for an instantaneous reaction based on the penetration model for mass transfer. The solid line is at time t_1 while the dashed line is at t_2.

A special case of the situation is shown in Fig. 11.8, for when the reaction is instantaneous. The solution is left as an exercise. The reaction front is assumed to be of thickness λ as in the film model. The main difference here is that the front moves inwards with time. The expression for the front movement can be taking derived by a mass balance at the interface, and the details are addressed in Problem 26.

11.9.4 Variable diffusivity

Problems where D is a function of concentration are of importance in many applications, including fabrication of micro-electronic devices. In this section we present some examples and introduce a mathematical method that is useful to solve these problems.

The problem can be stated in a slab geometry as

$$\frac{\partial C_A}{\partial t} = \frac{\partial}{\partial x}\left(D(C_A)\frac{\partial C_A}{\partial x}\right) \tag{11.54}$$

A power-law model for the diffusivity shown below is often used in semi-conductor device analysis:

$$D(C_A) = D_0[1 + BC_A^n]$$

where D_0 is the limiting value of diffusivity at low concentration, B is a constant for concentration dependence and n is an index. $n = 0$ corresponds to a constant diffusivity. Arsenic diffusion in Si is modeled with an index of one (Middleman and Hochberg, 1993). Solution in a semi-infinite region is often needed, and analytical solutions are rare.

The dimensionless form for a power law with first-order diffusion variation ($n = 1$) is

$$\frac{\partial c_A}{\partial t} = D_0\frac{\partial}{\partial x}\left((1 + \beta c_A)\frac{\partial c_A}{\partial x}\right) \tag{11.55}$$

where the diffusion coefficient is assumed to be a linear function of concentration. The coefficient of proportionality is β, and the constant-diffusivity model is represented with $\beta = 0$.

The boundary conditions are (i) $x = 0$, $c_A = 1$ and (ii) as $x \to \infty$, $c_A = 0$, which is also the initial condition.

The problem posed in a semi-infinite slab can be cast in terms of the similarity variable η defined as $x/\sqrt{4D_0 t}$.

Equation (11.55) reduces to an ordinary differential equation. Using the chain rule for differentiation on both sides we find

$$-2\eta\frac{dc_A}{d\eta} = \frac{d}{d\eta}\left((1 + \beta c_A)\frac{dc_A}{d\eta}\right)$$

which can be rearranged to the following second-order non-linear but ordinary differential equation:

$$\frac{d^2 c_A}{d\eta^2} + \frac{\beta}{1 + \beta c_A}\left(\frac{dc_A}{d\eta}\right)^2 + 2\frac{\eta}{1 + \beta c_A}\frac{dc_A}{d\eta} = 0$$

Unlike in the constant-diffusivity case, this equation cannot be solved analytically and a numerical solution is needed. The equation was solved by BVP4C and the plot is shown in Fig. 11.9. The key idea here was to demonstrate how to use the similarity transformation to simplify the computations even for a non-linear diffusivity, since the procedure reduces the PDE to an ODE. The ODE is easier to solve numerically than a PDE. Also note that the initial condition and the condition at ∞ merge into a single condition, which is needed in order for the similarity method to apply.

11.10 Periodic processes

Periodic variation in boundary conditions is often encountered in many processes. An example is the variation of the surface temperature of the soil due to the daily variation of temperature due to solar loading. This effect is often combined with another periodic component due to the annual variation of the air temperature. The distribution of temperature below the soil may be of interest in many applications. A second example is the wall of

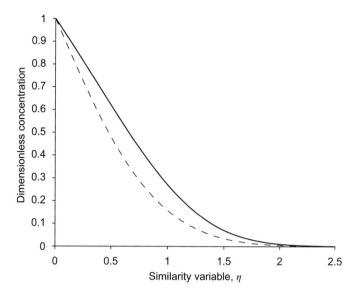

Figure 11.9 Concentration profile for non-linear diffusion as a function of the similarity variable. A linear variation is used and compared with the constant diffusivity case. The dotted line is the constant-D, erfc solution, whereas the solid line is the varying-diffusivity case, with $\beta = 1$.

a reciprocating combustion engine which is subjected to periodic temperature changes. This variation of the temperature causes thermal stresses, and such calculations are important in assessing the structural integrity of the engine.

This section analyzes the problems where the boundary conditions change as periodic functions of time. The solution procedure is also interesting and involves the use of complex variables. The problems posed in the geometries of a finite slab and a semi-infinite slab are both amenable to solution by the complex-variable method. The models for the two cases are quite similar. However, the solutions are different due to the differences in the spatial domain, and hence the two cases have to be treated separately.

11.10.1   Analysis for a semi-infinite slab

Consider a region extending from 0 to ∞ with the surface $x = 0$ exposed to a periodic variation of temperature represented by

$$T(x = 0) = \bar{T}_s + A\cos(\omega t) \tag{11.56}$$

Here $\bar{T}_s$ is the mean value of the temperature around which the surface temperature oscillates and ω is related to the oscillation frequency f. (Note that $f = \omega/(2\pi)$). The parameter A is the amplitude of the oscillations. Thus the surface temperature varies from $\bar{T}_s - A$ to $\bar{T}_s + A$ over a time interval of $2\pi/\omega$.

For a semi-infinite region there is no characteristic length scale, and it is therefore easier to pose the model in terms of actual variables rather than dimensionless variables. The governing equation is repeated for convenience:

$$\frac{\partial T}{\partial t} = \alpha \frac{\partial^2 T}{\partial x^2} \tag{11.57}$$

The initial temperature of the slab is denoted as T_i. We need to specify the boundary conditions as well. The first one at $x = 0$ is the periodic condition given by Eq. (11.56).

The effect of surface temperature variation is confined to a small region near the surface, and far away from the surface the temperature remains at the initial value. Hence the second boundary condition is as follows: at $x = \infty$, the temperature remains at T_i.

After an initial start-up period has elapsed, the temperature starts to oscillate everywhere in the slab in a sinusoidal manner. This state is often referred to as a periodic steady state. The terminology *steady state* can be misleading insofar as the temperature at each point changes with time. But the change occurs in a cyclic manner and hence the term periodic "steady" state can justifiably be used. The amplitude of the surface temperature oscillation is A, but the amplitude of the oscillation at the interior points will be smaller. In fact, for large distances from the surface the amplitude will be zero since the effect of the surface disturbance can be felt only for some distance into the surface. The solution at any point can therefore be represented in a periodic manner as

$$T(x) - T_s = F(x)\cos(\omega t) \tag{11.58}$$

where F is the amplitude of the oscillation at position x.

Equation (11.58) can be represented as a complex function:

$$T(x) - T_s = F(x)[\cos(\omega t) + i\sin(\omega t)] = F(x)\exp(i\omega t) \tag{11.59}$$

where $i = \sqrt{-1}$. The complex representation is used with the explicit understanding that only the real part of the equation is meaningful and representative of the solution.

By substituting the complex representation Eq. (11.59) into the governing differential equation (11.57) the following equation can be derived for $F(x)$:

$$\frac{\partial^2 F}{\partial x^2} = i(\omega/\alpha)F \tag{11.60}$$

The boundary conditions for F are as follows.

$F(x = 0) = A$. This follows from the boundary condition at $x = 0$.
$F(x = \infty) = 0$. This follows from the requirement that the surface perturbations are not evident at large distances.

The solution to F with these boundary conditions can be shown to be

$$F(x) = A\exp(-\sqrt{i\omega/\alpha}x) \tag{11.61}$$

Hence the solution for temperature in the complex space is

$$T(x) - T_s = A\exp(-\sqrt{i\omega/\alpha}x)[\cos(\omega t) + i\sin(\omega t)] \tag{11.62}$$

All that remains is to extract the real part of the above function. This is a simple but lengthy exercise in complex variables. The final result is

$$T(x) - T_s = A\exp(-\sqrt{\omega/(2\alpha)}x)\cos[\omega t - (\sqrt{\omega/(2\alpha)}x] \tag{11.63}$$

The amplitude of the oscillation at any position is therefore equal to $A\exp(-\sqrt{\omega/(2\alpha)}x)$ and decays with distance measured from the surface.

The depth of penetration of the disturbance can be found by setting the exponential term to, say, 3, which will be the point where the amplitude decays by 1%. Hence

$$\delta = 3\sqrt{\frac{\alpha}{\omega}}$$

Heat flux at the surface

The heat flux at the surface for the above case can be obtained by simply applying Fourier's law at the surface. Using the temperature profile given by Eq. (11.63) we obtain the following result for the heat flux:

$$q_s = -k(\partial T/\partial x)_{x=0} = k\sqrt{\omega/(2\alpha)}\, A \cos(\omega t + \pi/4)$$

It can be seen that the heat flux is also periodic, but $45°$ out of phase with the temperature variation.

The total heat flux over a complete cycle can be shown to be equal to zero, as expected.

11.10.2   Analysis for a finite slab

Consider a slab of thickness $2L$ subject to a variation of the surface $(x = \pm L)$ temperature in a periodic manner as described by the following equation:

$$T(\pm L) = \bar{T}_s + A \cos(\omega t) \tag{11.64}$$

where $\bar{T}_s$ is the mean value of the temperature and ω is related to the oscillation frequency. The slab is initially at a constant temperature of T_i and it is required to find the temperature profile in the slab, $T(x, t)$.

In view of the symmetry, only half the slab needs to be included in the analysis. Thus the distance parameter, x, varies from 0 to L. Also there is a characteristic length scale L, unlike in the case of the semi-infinite problem. Hence the governing equations are stated directly in dimensionless form:

$$\frac{\partial \theta}{\partial \tau} = \frac{\partial^2 \theta}{\partial \xi^2} \tag{11.65}$$

Recall the definitions of the dimensionless parameters as follows:

Dimensionless time $\tau = t(\alpha/L^2)$, where α is the thermal diffusivity.

Dimensionless distance $\xi = x/L$, where $\xi = 0$ represents the mid point in the slab in order to take advantage of the symmetry of the problem.

Dimensionless temperature is defined here as $\theta = (T - T_s)/A$. The initial temperature T_i is not included in the definition. It is expected that the effect of the initial conditions will die down after some time has elapsed, leading to a periodic steady-state type of profile.

The boundary conditions are $\partial \theta/\partial \xi = 0$ at $\xi = 0$ due to symmetry and $\theta(\xi = 1) = \cos(\omega t) = \cos(\omega L^2/\alpha)\tau$.

Denoting $\omega L^2/\alpha$ as ω^*, a dimensionless measure of the frequency, the above boundary condition is written as

$$\theta(\xi = 1) = \cos(\omega^* \tau)$$

The initial condition in dimensionless form is $\theta_i = (T_i - T_s)/A$ at time $\tau = 0$. This is needed only if the initial start-up is to be simulated and not needed if a time-periodic solution is to be computed.

After an initial start-up period has elapsed, the temperature starts to oscillate everywhere in the slab in a sinusoidal manner, and this state is often referred to as a periodic steady state. The effect of the initial condition will not be important for the periodic steady state. (This is why one defines the dimensionless temperature without including the initial temperature.) The amplitude of the surface temperature oscillation is one in dimensionless units but the amplitude of the oscillation at the interior points will be smaller. The solution at any point can be therefore represented in a periodic manner as

$$\theta(\xi) = F(\xi)\cos(\omega^*\tau) \tag{11.66}$$

where F is the amplitude of the oscillation at position ξ.

This is replaced by a complex variable solution as

$$\theta(\xi) = F\xi)\exp(i\omega^*t^*) \tag{11.67}$$

The governing equation for F can then be derived by using this in (11.65):

$$\frac{d^2F}{d\xi^2} = i\omega^*F \tag{11.68}$$

which is similar to Eq. (11.60). However the boundary conditions are different and are as follows.

$\partial F/\partial\xi = 0$ at $\xi = 0$. This follows from the symmetry condition at $\xi = 0$.
$F(\xi = 1) = 1$. This follows from the requirement that the dimensionless amplitude at the surface is equal to unity.

The solution to Eq. (11.68) subject to the above boundary conditions can be shown to be

$$F = \frac{\cosh(\xi\sqrt{i\omega^*})}{\cosh(\sqrt{i\omega^*})} \tag{11.69}$$

and hence the composite solution for the temperature in the complex space is

$$\theta(\zeta,\tau) = \frac{\cosh(\xi\sqrt{i\omega^*})}{\cosh(\sqrt{i\omega^*})}\cos(\omega^*\tau) \tag{11.70}$$

All that remains is to extract the real part of this *complex*-looking complex expression. This is sufficient to put the reader off further study of this approach and make her or him rush to the purely numerical methods discussed later in this chapter. However, help is on the way. We can use the MATLAB functions to extract the real part of the above expression and directly compute the solution for various values of time and position.

11.11   Transient flow problems

Transient flow problems arise due to various factors, e.g., (i) flow development from start-up, (ii) flow caused by a time-varying pressure gradient, of which blood flow in arteries due to periodic pumping of the heart is an example; (iii) flow caused by a moving wall (this is also known as peristaltic pumping); and (iv) flow in the transition region due to flow instability, wavy film flow, etc. It may be noted that the turbulent flows are essentially transient in nature, but a technique called time averaging discussed in a later chapter is commonly used to capture the averaged properties of the flow. Here we analyze some simple cases assuming laminar flow conditions. Also the flow problems posed here are for unidirectional flow. The model equations for these problems are governed by PDEs similar to those for heat and mass transfer problems.

11.11.1 Start-up of channel flow

For unidirectional channel flow the governing equation for steady-state flow (see Section 6.2.1) is now augmented by adding the accumulation terms in the balance:

$$\rho \frac{\partial v_x}{\partial t} = \mu \frac{\partial^2 v_x}{\partial y^2} + G \tag{11.71}$$

where G is the pressure drop per unit length (or the drop in the modified pressure $p + \rho g h$ if gravity is included).

Initial conditions are required in addition to the usual boundary conditions at $y = 0$ and 1. A zero velocity is usually taken as the initial condition for start-up flow analysis.

Clearly the equation is similar to transient heat transfer with internal generation. Note the presence of a non-homogeneous source term G. Hence the problem can be solved as

$$v_x(y, t) = v_s(y) + v(y, t)$$

where v_s is the steady-state solution and v is the deviation of the velocity from the steady state. The problem v will satisfy the homogeneity condition and hence can be solved by separation of variables. Further details are left as an exercise.

11.11.2 Transient flow in a semi-infinite mass of fluid

Consider a semi-infinite body of liquid occupying the region above the x-axis with $y = 0$ as a solid wall. The fluid is stagnant at time zero and the motion is caused by moving the solid wall at a constant velocity V. For a semi-infinite region there is no pressure variation and the model equation is therefore

$$\rho \frac{\partial v_x}{\partial t} = \mu \frac{\partial^2 v_x}{\partial y^2} \tag{11.72}$$

The problem is analogous to that for heat transfer in a semi-infinite region and the solution may be written by analogy as

$$\frac{v_x}{V} = 1 - \mathrm{erf}\left(\frac{y}{\sqrt{4vt}}\right) \tag{11.73}$$

The development of the velocity profile is shown in Fig. 11.10. A boundary layer in velocity is seen, and this increases as a function of time similarly to the heat front for the semi-infinite-slab problem.

11.11.3 Flow caused by an oscillating plate

Again we look at a semi-infinite region bounded by a solid wall at $y = 0$. In this case the flow is caused by moving the plate back and forth with a frequency ω. The governing equation is the same as (11.72) and only the boundary condition at $y = 0$ is different. A periodic variation gives

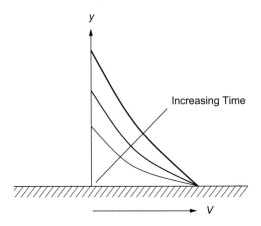

Figure 11.10 A schematic sketch of flow caused in a semi-infinite mass of fluid by a moving plate.

$$v_x(y = 0) = V \cos(\omega t)$$

The problem is similar to heat transfer with a periodic variation of the temperature. Hence the method of complex variables should be used here, and we alter the boundary condition to

$$v_x(y = 0) = \exp(i\omega t)$$

where we note that the amplitude of the oscillation can be taken as one without loss of generality. All the velocity values simply get scaled by V. The (scaled) velocity at any location is also assumed to be periodic, and is represented as

$$v_x = A(y)\exp(i\omega t)$$

where A is the local amplitude.

Using this in the differential equation, we can solve the problem in the complex domain. The details of the solution are omitted, and the final solution is directly given below by analogy with the similar heat-transfer problem:

$$v_x = \exp\left(-y\sqrt{\frac{i\omega}{\nu}}\right)\exp(i\omega t)$$

Separation of the solution into real and imaginary parts is done and the real part is the "real" solution. The final result is as follows:

$$\frac{v_x}{A} = \exp\left(-y\sqrt{\frac{\omega}{2\nu}}\right)\cos\left(\omega t - y\sqrt{\frac{\omega}{2\nu}}\right)$$

The solution is sketched in Fig. 11.11 for four values of the parameter ωt (0, $\pi/2$, π, and $3\pi/2$). This represents one cycle of the system.

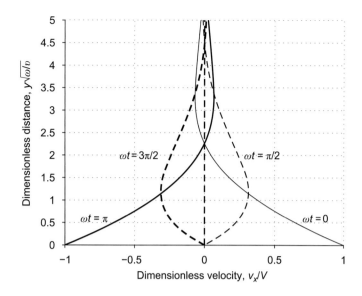

Figure 11.11 The response of a fluid to an oscillating plate.

11.11.4   Start-up of Poiseuille flow

The governing equation assuming a unidirectional velocity profile is

$$\rho \frac{\partial v_z}{\partial t} = G + \mu \frac{1}{r} \frac{\partial}{\partial r} \left(r \frac{\partial v_z}{\partial r} \right) \tag{11.74}$$

where G is an imposed pressure gradient.

The solution can be obtained by subtracting the steady-state parabolic profile and posing a homogeneous problem for the perturbation velocity. The following final analytical solution to the problem can be derived in terms of dimensionless variables:

$$v^* = \frac{1}{4}(1 - \xi^2) - \sum_{n=1}^{\infty} A_n \exp(-\lambda_n^2 \tau) J_0(\lambda_n \xi)$$

where the eigenvalues are the roots of $J_0(\lambda) = 0$ and the series coefficients are

$$A_n = \frac{2 J_0(\lambda_n)}{\lambda_n^3 J_1(\lambda_n)}$$

How are the dimensionless variables defined here?

You should verify this solution and then also use PDEPE to do numerical computations and compare the results. Note that the solution has two parts: (i) a steady-state parabolic function and (ii) a time-dependent function that decays to zero for large values of τ.

Pulsatile flow in a pipe

Problem statement

Here we examine a problem where the pressure driving force G varies as a sinusoidal function of time:

$$G = G_0 \cos(\omega t)$$

Here G_0 is the amplitude of the pressure gradient; in other words, the pressure changes from $+G_0$ to $-G_0$ and ω is the frequency. The governing equations are the same as Eq. (11.74) except that G is now a function of time.

Problems of this type occur in the context of blood flow in arteries, although it is an idealized picture. The problem was studied first by Womersley (1955) and is also known as the Womersley problem. The velocity profile is expected to vary as a periodic function of time after an early start-up period, and this periodic flow is of more interest in the context of blood flow and similar problems. We can immediately identify the similarity to periodic processes in heat transfer studied in Section 11.10.2, the difference being that there is a pressure forcing term, which leads to a non-homogeneous PDE. First let us cast the problem in dimensionless form and identify the parameters. The distance scale is readily identified as R and the dimensionless distance is $\xi = r/R$. The velocity scale must be such that the pressure term is simplified. This can be achieved by defining a reference velocity v_{ref} as equal to $G_0 R^2/\mu$. The dimensionless velocity v^* is therefore defined as $v_z \mu/(G_0 R^2)$.

Choice of reference time

The time scale can be R^2/v, the viscous diffusion time. This is the time scale on which the viscous effects reach some steady-state profile as seen in the previous section. The parameter τ is then tv/R^2 as in the previous subsection. In this case the governing equation is

$$\frac{\partial v^*}{\partial \tau} = \cos([\omega R^2/v]\tau) + \frac{1}{\xi}\frac{\partial}{\partial \xi}\left(\xi \frac{\partial v^*}{\partial \xi}\right) \qquad (11.75)$$

A dimensionless parameter $\omega R^2/v$ appears in the pressure gradient term. This parameter is the square of Wo, the Womersley number, which is defined as $Wo = \sqrt{\omega R^2/v}$. The parameter is the ratio of the viscous diffusion time to the oscillation time:

$$Wo^2 = \frac{R^2 \omega}{v} = \frac{\text{Viscous diffusion time}}{\text{Oscillation time}}$$

We will see that the magnitude of the Womersley number has a significant effect on the velocity profiles.

A second way to dimensionalize is to choose t^* as ωt:

$$t^* = \frac{\text{Observation time, } t}{\text{Oscillation time, } 1/\omega}$$

In this case the Womersley constant appears in front of the time derivative:

$$Wo^2 \frac{\partial v^*}{\partial t^*} = \cos(t^*) + \frac{1}{\xi}\frac{\partial}{\partial \xi}\left(\xi \frac{\partial v^*}{\partial \xi}\right) \qquad (11.76)$$

Which time scale should we choose? This depends on the value of Wo. If Wo is small, t^* is the scale to use. Then the LHS can be dropped as a first approximation. In this case, it means that the frequency of pulsations is sufficiently low that a parabolic velocity profile has time to develop during each cycle, and the flow will be very nearly in phase with the pressure gradient, and will be given to a good approximation by a parabolic Poiseuille profile, using the instantaneous pressure gradient at each t^*. The flow is of a quasi-steady-state nature. The pressure forces at each time are balanced by viscous forces and the inertia forces are negligible.

The second case is when Wo is large. The viscous diffusion time is now large, and hence a nearly steady-state parabolic profile cannot be achieved. The pressure has changed direction long before the time required in order to reach a quasi-periodic state has elapsed. The fluid does not have much time to respond to the pressure changes. The flow remains nearly where it was. Nearly plug flow is achieved, with some peaks near the wall. We now examine the nature of the solution for this case using a complex-variable method. We still use t^* (for both large and small Wo) as the dimensionless time since this is more mathematically convenient. The formulation is therefore given by Eq. (11.76). Now instead of using $\cos(t^*)$ we use $\exp(it^*)$, with the understanding that the real part of the solution is meaningful. The velocity is also assumed to have a periodic variation, which can be represented as

$$v^* = V_0(\xi)\exp(it^*)$$

By substitution we can derive the following differential equation for the local amplitude variable V_0:

$$\frac{1}{\xi}\frac{d}{d\xi}\left(\xi\frac{dV_0}{d\xi}\right) - iWo^2 V_0 = 1$$

The student should recognize that this is a Bessel equation with a non-homogeneous term. The solution can therefore be written as a sum of a particular solution and a homogeneous solution. The solution is

$$V_0(\xi) = \frac{1}{iWo^2} + \left(\frac{J_0(i^{3/2}Wo\xi)}{J_0(i^{3/2}Wo)}\right)$$

where the standard boundary conditions have been used. Recall that the velocity profile is the real part of the above expression. The separation of the solution into real and imaginary parts is readily accomplished in MATLAB without wasting any mathematical sweat equity. You should write a small piece of code to calculate this. You should run the code for various values of Wo and see what type of velocity profile is reached. For small Wo a nearly parabolic profile will be obtained, whereas for large Wo a nearly plug flow in the center region of the pipe, with a sharp boundary layer near the wall, will be obtained. Also you will observe that for small Wo the solution will be in phase with the pressure changes, whereas for large Wo the profiles will be out of phase. These limiting cases of Wo can also be examined by using the perturbation methods discussed in Chapter 14.

Now we discuss a method for numerical solution of these types of PDEs using an in-built MATLAB function.

11.12 A PDE solver in MATLAB

11.12.1 Code usage

The PDEPE routine solves the initial-boundary-value problems for parabolic-elliptic PDEs in 1D systems. Small systems of parabolic and elliptic PDEs in one space variable x and time t can be solved to modest accuracy, and this is a good practice tool with which to understand the behavior of such systems.

The general structure of the program is the solver for PDEs of the following type:

$$c\frac{\partial u}{\partial \tau} = \frac{1}{\xi^s}\frac{\partial}{\partial \xi}\left(\xi^s F\right) + S \tag{11.77}$$

Here s is equal to 0, 1, and 2 for the geometries of slab, infinitely long cylinder, and sphere, respectively. The solution is sought for u as a function of distance ξ and time τ. Note that multiple PDEs can be solved, and hence u is the "solution vector" with components u_1, u_2, etc.

The other terms in Eq. (11.77) are explained below. The coefficient in the time derivative, c, is a capacity term and is in general a function of ξ, τ, u, and p, where p is the gradient of u defined as $du/d\xi$. In general, c is a diagonal matrix when multiple equations are being solved. The diagonal elements of the matrix c are either identically zero or positive. An entry that is identically zero corresponds to an elliptic equation and otherwise to a parabolic equation. There must be at least one parabolic equation.

F is the heat-flux type of term and can be a general function of the quantities, τ, u, and p as well. Note that F can be defined using a local value of the diffusion coefficient. Hence variable-diffusivity cases can also be solved.

Finally, $S = S(\xi, \tau, u, p)$ is the source term, which can be any linear or non-linear function of the variables.

The calling statement is

```
SOL = PDEPE (shape, PDEFUN, ICFUN, BCFUN, XMESH, TSPAN)
```

The variables in the calling statement are defined as follows.

1. Shape is the parameter which defines the geometry and must be equal to either 0, 1, or 2, corresponding to slab, cylindrical, or spherical symmetry as earlier in Eq. (11.77).
2. PDEFUN is the function that evaluates the quantities defining the differential equation. The calling statement is

$$[C,F,S] = PDEFUN(X,T,U,DUDX)$$

The input arguments are scalars X (ξ) and T (τ) and vectors U and DUDX that approximate the solution and its partial derivative with respect to x, respectively. PDEFUN returns column vectors C (containing the diagonal of the matrix of the capacity term C(x,t,u,dudx)), F, and S (representing the flux and source term, respectively). For constant thermal conductivity, the "flux" term is usually written as F = dudx. Note that here we assume a dimensionless formulation and hence the thermal conductivity does not appear as an explicit term.
3. The ICFUN is the function which defines the initial conditions, and the structure of this function is as follows: U = ICFUN(X). When called with a scalar argument X, ICFUN evaluates and returns the initial values of the solution components at X in the column vector U.
4. The BCFUN defines the boundary conditions. These are defined as the left and right boundaries in the general form A + B F = 0, where A and B are in general functions of the values of x, t, and u, and B is a diagonal matrix.

11.12.2 Example general code for 1D transient conduction

A sample code that is useful for you to get started is shown below. You can modify this for a wide range of problems.

```
function pde1t
%  PDE1t is a solver for  a transient diffusion problem in
%  a slab, cylinder, or sphere using the Matlab solver "pdepe"
%  that solves a single PDE as given in the text.
alpha = 1;  % geometry
% meshes in the x-direction created with linspace function
xnodes = 21 ;
x = linspace (0,1,xnodes);
% time intervals at which solution is sought
t = [0   0.2  0.4  0.6   0.8 1.0  1.25 1.5 1.75 2.00];
% call the function to generate the solution.
sol = pdepe(alpha,@pdexpde,@pdexic,@pdexbc,x,t);
u1 = sol(:,:,1)  % solution displayed on the command window.
% surface values

nt = length(t);
ps = zeros(1,nt);
% seriesI = zeros(1,nt);
iok = 2:nt;
for j = iok
   % At time t(j), compute Du/Dx at x = 1.
   [dummy, ps(j)] = pdeval(alpha,x,u1(j,:),1);
   [ tcenter(j), dummy ] = pdeval(alpha,x,u1(j,:),0);
   %  seriesI(j) = serex3(t(j),L,d,eta,I_p);
end
ps  % surface gradient values
tcenter % center temperature values
% uncomment this section to generate a plot of the results
%   figure;
%   surf(x,t,u1);
%   title('u1(x,t)');
%   xlabel('Distance x');
%   ylabel('Time t');
% plot of surface gradient as  a function of time
% plot ( t, ps)
% plot of center temperature as  a function of time
% plot ( t, tcenter)

% PROBLEM DEFINITION -----------------------
function [c,f,s] = pdexpde(x,t,u,DuDx)
c = 1 ;   % capacity term
f =  DuDx;           % flux term is defined. (constant-k case)
```

```
s = 1.0 ;             % source term
% INITIAL CONDITIONS -----------------------
function u0 = pdexic(x);
u0 = 0.000 ;    % initial conditions
% BOUNDARY CONDITIONS----------------------
function [pl,ql,pr,qr] = pdexbc(xl,ul,xr,ur,t)
% Left boundary conditions (Neumann; symmetry)
pl = 0 ;
ql = 1 ;    % no-flux condition
% Right boundary condition (Robin type)
biot = 10. ;
pr = [ biot * ur(1)];
qr = 1 ;
% Right boundary condition for illustration (Dirichlet type)
% uncomment for Dirichlet type
% ts = 0.0  % surface temperature.
% pr = [  ur(1) -ts ];
% qr = 0 ;
```

Two applications are shown. The first concerns the transients in a long cylinder with internal generation of heat, which system was studied by analytical methods in Section 11.6.1. The problem to be solved is

$$\frac{\partial u}{\partial \tau} = \frac{1}{\xi} \frac{\partial}{\partial \xi} (\xi F) + Q_V^* \tag{11.78}$$

Here $F = \partial u / \partial \xi$. Also $s = 1$ for a cylindrical geometry.

The boundary conditions are symmetry at $\xi = 0$ and the Robin condition at $\xi = 1$:

$$\left(\frac{\partial u}{\partial \xi} \right)_{\xi=1} = -Bi u_{\xi=1}$$

The initial condition is taken as $u = 0$.

The steady-state solution should be

$$u = \frac{1}{4}(1 - \xi^2) + \frac{1}{2Bi}$$

and should be obtained by running the code for a large enough value of time.

The following result is obtained by running the code ($Q_V^* = 1$ and $Bi = 1$) (you should compare this with the analytical solution by using the modified separation of variables):

τ	ξ					
	0	0.2	0.4	0.6	0.8	1.0
0.2	0.4040	0.3988	0.3844	0.3602	0.3249	0.2767
0.5	0.5958	0.5872	0.5635	0.5248	0.4703	0.3992
1.0	0.7234	0.7125	0.6826	0.6343	0.5670	0.4806
2.0	0.7570	0.7455	0.7139	0.6631	0.5924	0.5020

The second example is transient diffusion in a sphere with a first-order reaction with $\phi = 2$. The solution is shown in the following table for c_A as a function of ξ and τ. The center concentration should be 0.5514 for large τ but the code gives some error. This is due to crude spatial discretization. We used only six points here and if you increase that to 101 points the exact solution is recovered. One should always perform such mesh-dependence studies and check the results. Testing it on a simpler problem having an analytical solution in hand is useful for such studies before attempting to deal with a more complex case.

τ	$\xi = 0$	0.2	0.4	0.6	0.8	$\xi = 1$
0	0	0	0	0	0	1.0000
0.5	0.2155	0.2445	0.3436	0.5035	0.7258	1.0000
1.0	0.5398	0.5546	0.6051	0.6887	0.8160	1.0000
2.0	0.5418	0.5564	0.6065	0.6897	0.8164	1.0000
5.0	0.5418	0.5564	0.6065	0.6897	0.8164	1.0000

Other similar problems can be coded with only minor modifications of the sample code.

Summary

- Transient problems lead to partial differential equations involving time and spatial variables. The simpler problems use only one spatial variable and are posed in the ideal geometries of the slab, long cylinder, and sphere. Further constant physical properties and a constant or linear rate of production are assumed, which leads to a linear PDE. The governing equations can be represented in a compact form in terms of dimensionless variables, as shown by Eq. (11.6) in the text.

- Linear PDEs with homogeneous boundary conditions can be solved by separation of variables. General properties of such systems are the eigencondition (the solution of which gives the eigenvalues of the problem), eigenfunctions, and representation of the solution as a series. The series coefficient can be obtained by Fourier expansion of the initial condition in terms of the eigenfunctions. The procedure can be applied to all three geometries. Obviously the eigenfunctions, etc., are geometry- and boundary-condition dependent, but the general procedure and the form of the solution are the same.

- Obtaining the eigenvalues for the Dirichlet problem is simple, but the eigenvalues for the Robin problem have to be obtained as the solution of a transcendental equation. This difference is worth noting. The eigenvalues for the Robin problem are functions of the Biot number. For small Biot number the temperature profile within the slab is nearly uniform and a lumped-parameter model (Section 2.4.4 from an earlier chapter) can be used as a simple model.

- Important properties of the series solution should be noted. The series converges rather slowly for small values of time, whereas for large values of dimensionless time just one term is sufficient. Since the series converges slowly, a semi-infinite model is more suitable for small values of time.

- Many 2D problems can be split into two 1D problems, and the results for the separated 1D problems can be combined and used for such cases. The result for 2D is a product of the separated 1D problems, and hence the method is also called the method of product solutions.

- Series solutions can be implemented with computer algebra tools such as MAPLE. This takes out the tedium of doing lengthy algebraic manipulations. These tools, however, can be more effectively appreciated and used only after a study of some simple problems.

- Many transient problems are posed in a semi-infinite domain. This approach is suitable when the depth of penetration of the temperature front is much smaller than the domain dimension. The mathematical problem can then be reduced to an ordinary differential equation in terms of the transformed variable. For the constant surface temperature the solution is an error function of the similarity variable. For a pulse input at the surface, the solution is a half-Gaussian function.

- Transient mass diffusion problems can be solved by analogy with heat transfer. One additional effect that is important may be the effect of chemical reaction. Examples were provided in the text to illustrate transient mass transfer with chemical reaction.

- Gas–liquid mass transport can be modeled by a transient diffusion equation if the contact time between the gas and the liquid is small. This leads to a model for interfacial mass transfer known as the penetration model. This has wide applications, especially for reacting systems, and is often used as an alternative to the film model.

- Problems involving varying boundary conditions are of importance in many applications. If a sinusoidally varying function is applied, then a mathematical tool based on complex variables is useful. Two examples in the text provide illustration of the method. The final solution requires complex algebra involving the separation of a function of a complex number into real and imaginary parts, which is readily accomplished with MATLAB.

- Pulsatile flow in a pipe is an interesting example of the approach using complex variables. There are two time scales (the viscous diffusion time and the oscillation time) here, and, depending on their relative magnitude, the velocity profile can take completely different shapes.

- MATLAB code PDEPE is a versatile tool for solving transient diffusion problems. Non-linear diffusion, non-linear source terms, and non-linear boundary conditions can readily be incorporated. The solution can be used for all of the three basic geometries by simply changing the shape parameter, s, in the code. The code can also handle multiple differential equations. The worked examples provide a learning tool for this case. The student will be in a position to analyze and solve a wide range of problems in transient transport problems after a study of the several examples in the text.

ADDITIONAL READING

The classic text on transient heat transfer is the book by Carslaw and Jaeger (1959). Transient diffusion problems are treated in a monograph by Crank (1975).

Problems

1. Consider expansion of a function in terms of a series $F_n(x)$ in the following form:

$$f(x) = \sum_n A_n F_n(x)$$

If the functions F_n are orthogonal then this property helps to unfold the series and permits us to find the series coefficients, one at a time.

State what is meant by orthogonality of two functions F_n and F_m.

Using this, show that the series coefficients can be calculated as

$$A_n = \frac{\int_0^1 f(x) F_n(x) dx}{\int_0^1 F_n^2(x) dx}$$

where the domain of solution is assumed to be from zero to one.

Apply this to the case of a slab with constant initial temperature ($f(x) = 1$). Show that the function $\cos[(n + 1/2)\pi x]$ is orthogonal in the interval from zero to one. Hence verify the following expression for the series coefficient given in the text for A_n:

$$A_n = \frac{2(-1)^n}{(n + 1/2)\pi}$$

for a constant initial condition equal to one.

What is the series coefficient if the initial temperature is a linear function of position?

Note that tools such as MATHEMATICA and MAPLE can be used to do the algebra and you may wish to use them.

2. The fact that the eigenfunctions are orthogonal can be verified easily using the symbolic calculations in MATLAB or MAPLE. But the underlying theory is based on the Sturm–Liouville problem. You may wish to explore this further by looking at some books.

State what a Sturm–Liouville equation is.

Prove that the eigenfunctions of this equation are orthogonal with the weighting factor.

State how to identify the weighting function directly from the Sturm–Liouville equation in its general form.

3. There is nothing magical about the eigenfunctions being orthogonal. This can be shown by integration by parts twice.

Consider two eigenfunctions F_n and F_m, both of which satisfy the following equations:

$$\frac{d^2 F_n}{dx^2} = -\lambda_n^2 F_n \tag{11.79}$$

and

$$\frac{d^2 F_m}{dx^2} = -\lambda_m^2 F_m \tag{11.80}$$

Obviously λ_n and λ_m are two distinct eigenvalues. Also assume that some homogeneous boundary conditions apply for both F_n and F_m at $x = 0$ and $x = 1$ (the end points assuming dimensionless variables).

Now consider the integral

$$\int_0^1 \left[F_m \frac{d^2 F_n}{dx^2} \right] dx = -\lambda_n^2 \int_0^1 F_m F_n \, dx \tag{11.81}$$

Show that this follows from Eq. (11.79).

Integrate this by parts twice. Then use the homogeneous boundary condition. Then use Eq. (11.80) to get rid of the second-derivative term for F_m. This involves a lot of algebra, but all of it is elementary. Finally show that the resulting expression is

$$(\lambda_m^2 - \lambda_n^2) \int_0^1 F_m F_n \, dx = 0 \tag{11.82}$$

Since λ_m is not equal to λ_n (the eigenvalues are distinct), we conclude that the functions F_m and F_n are orthogonal to each other. Generalization of this forms the basis of the Sturm–Liouville theory for eigenfunctions indicated in Problem 2 above.

4. Analyze the transient problem with the Dirichlet condition for a long cylinder and for a sphere. Derive expressions for the eigenfunctions, eigenconditions, and eigenvalues. Find the series coefficients for a constant initial temperature profile.

5. A concrete wall 20 cm thick is initially at a temperature of 20 °C, and is exposed to steam at pressure 1 atm on both sides. Find the time for the system to reach a nearly steady state. Find the rates of condensation of steam at various values of time.
The thermal diffusivity constant for concrete is needed. Use a value of $7.5 \times 10^{-7} \, \text{m}^2/\text{s}$.

6. A finite cylinder is 2 cm in diameter and 3 cm long and at a temperature of 200 °C, and is cooled in air at 30 °C. The convective heat transfer coefficient is estimated as 10 W/m$^2 \cdot$K. Calculate and plot the center temperature.
Use the following physical property values: $k = 50 \, \text{W/m} \cdot \text{K}$, $\rho = 2000 \, \text{kg/m}^3$, and $c_p = 1000 \, \text{J/kg} \cdot \text{K}$.

7. For the Biot problem in a slab by expanding the sin and cos term and keeping only terms up to λ^2 the following approximate relation can be obtained for the eigenvalues for small Biot number:

$$\lambda = \sqrt{Bi/(1 + Bi/2)}$$

Verify the result.
Find the value for a Biot number of 1, and estimate the error in the time constant if the above equation were used.
What are the corresponding expressions for a long cylinder and a sphere?

8. Consider the problem of transient heat transfer with a constant heat source in a slab. Show that the governing equation in dimensionless form is

$$\frac{\partial \theta}{\partial t^*} = \frac{\partial^2 \theta}{\partial \xi^2} + 1 \tag{11.83}$$

Identify and define the parameters in this problem stated in the above dimensionless form. State the boundary conditions for (i) constant surface temperature of the slab and (ii) convective heat loss from the surface.
Now consider the series solution to the case of constant surface temperature. The initial temperature is θ_i, which is assumed to be a constant for this problem. For simplicity take this as equal to the surface temperature.
Use the superposition to split the problem into two problems, (ii) a position-dependent steady-state part and (ii) a time-dependent part.
Solve for each of these parts for the case of constant surface temperature and find the composite solution.
Sketch and plot illustrative temperature distributions for different values of the Fourier number (t^*).

9. **Eigenvalues without pain: CHEBFUN code.** Eigenfunctions can be derived using the CHEBFUN with MATLAB since it has an overloaded eig function. The following code solves for the eigenfunctions of the Robin problem shown in the text in Section 11.3.

The code can also readily evaluate the series coefficient. Thus the whole procedure of the method of separation of variables can be automated with a code similar to this. You will be able to solve or verify many of the problems in this chapter with this code and also will be able to use it in your research.

```
xi = chebfun('xi',[0,1]);
A = chebop(0,1);
A.op = @(xi,u) diff(u,2) ;
% A.lbc = 'neumann';
% A.rbc = 'Robin';
A.lbc = @(u) diff(u,1)  % neumann
Biot = 1.0
A.rbc =  @(u) diff(u,1) + Biot* u
B = chebop(0,1);
B.op = @(xi,u) -u ;


%%
% Then we find the eigenvalues with eigs.
[F,L] = eigs(A,B)
omega = sqrt(diag(L)) % eigenvalues

% series coefficient can be evaluated
i = 1;
for i = 1:6
  center = F(0,i);
N1 = center * int ( F(:,i)  );
  N2 = int ( F(:,i).* F(:,i)  );
  B1(i) = int ( F(:,i)  )/center
  A1(i) = N1/N2 % Answer C1 =  1.119
end
```

On running the code you should get the following results for Biot = 1.
The eigenvalues are 0.8603, 3.4256, 6.4373, 9.5293, 12.6453, 15.7713, and 18.9024.
The series coefficients for Biot = 1 can be calculated as 1.1191, −0.1517, 0.0466, −0.0217, 0.0124, and −0.0080.
It is only a matter of summing the series for any chosen time to find the temperature profiles. You can write a small piece of code and generate time–temperature plots for any given Biot number.

10. A hot dog at 5 °C is to be cooked by dipping it in boiling water at 100 °C. Model the hot dog as a long cylinder with a diameter of 20 mm. Find the cooking time, which is defined as the time, for the center temperature to reach 80 °C. The heat transfer coefficient from the water to the surface is 90 W/m$^2 \cdot$ K.

The following data can be used: $k = 0.5$ W/m$\cdot$K, $\rho = 880$ kg/m^3, and $c_p = 3350$ J/kg$\cdot$K.

11. Consider the Neumann–Neumann problem in Section 11.6.2 once again. Assume that the asympotic solution has a time-dependent part, which is linear in time, and a position-dependent solution, which is an unknown function. Thus the solution should be of the following form:

$$\theta_{asy} = A\tau + F(\xi)$$

where A is a constant to be determined. The linear variation in time can be justified by physical arguments. Substitute into the equations, solve for F, and use the boundary conditions to show that the constant $A = 1$.
Hence the solution is

$$\theta_{asy} = \tau + \frac{\xi^2}{2} + B$$

The constant B is the integration constant, which has to be determined from the initial conditions. However, the effect of the initial conditions is lost in the asymptotic solution, and hence an overall heat balance from time zero to time τ is used as an alternative.
Find the average temperature of the solid at any time and the total energy content at this time. Equate this to the heat added from the start and show that $B = -1/6$ for a starting temperature of zero.

12. A slab has a thermal diffusivity of 5×10^{-6} m^2/s and is fairly thick. The initial temperature is 300 K and the surface temperature is raised to 600 K at time zero. Find the temperature 0.5 m below the surface after one hour has elapsed.
Also find the depth to which the heat front has penetrated.

13. Consider the problem of transient diffusion in a composite slab with two different thermal conductivities. Thus region 1 extending from 0 to κ has a thermal conductivity k_1, while the region 2 from κ to 1 has a conductivity of κ_2. The slab is initially at a dimensionless temperature of 1, and both ends are exposed to a zero temperature.
Set up the problem and solve by separation of variables.

14. A large tank filled with oxygen has an initial concentration of oxygen of 2 mol/m^3. The surface concentration on the liquid side of the interface is changed to 9 mol/m^3 and maintained at this value.
Calculate and plot the concentration profile for times 3600 s and 36 000 s.
Find the rates of oxygen transport at these times.
Find the total amount of oxygen transferred from the beginning to the end of the above time period.
The diffusion coefficient of oxygen is 2×10^{-9} m^2/s in water.
The penetration depth of oxygen is defined as $\sqrt{12D_A t}$, and is a measure of the depth up to which the oxygen concentration has changed. Calculate the values at the two times above.

15. Consider the drug-release problem in the text, but now the drug is in the form of a cube with sides 0.652 cm. Find the center concentration after 48 h and the fraction of drug released.

16. A porous cylinder, 2.5 cm in diameter and 80 cm long, is saturated with alcohol and maintained in a stirred tank. The alcohol concentration at the surface of the cylinder is maintained at 1%. The concentration at the center is measured by careful sampling and is found to drop from 30% to 8% in 10 h. Find the center concentration after 15 and 20 h.

17. A gas bubble of diameter 3 mm is rising in a pool of a liquid. What is the mass transfer coefficient if the diffusion coefficient is 2×10^{-5} cm^2/s?

Note that the contact time is needed for finding the mass transfer coefficient. One way of approximating this is

$$\text{Contact time} = \frac{\text{Bubble diameter}}{\text{Rise velocity}}$$

The rise velocity can be calculated using Stokes' law (discussed in Section 4.4). Small bubbles are spherical and Stokes' law is a reasonable approximation here.

18. Consider again the periodic variation of temperature in a semi-infinite domain analyzed in Section 11.10.1. Sketch or plot the temperature for three values of time. Show that a wave type of propagation of the temperature is observed. Find the distance between adjacent maxima in the temperature at any given instant of time (which is called the wavelength). Also find the speed of propagation of the thermal waves. This is defined as the distance between two adjacent peaks divided by the time needed for the crest to travel this distance.

19. In a flow distribution network that progresses from a large tube to many small tubes (e.g., a blood-vessel network), the frequency, density, and dynamic viscosity are (usually) the same throughout the network, but the tube radii change. Therefore the Womersley number is large in large vessels and small in small vessels. Find typical values for the aorta, major arteries, terminal artery, etc. from the literature. Sketch the anticipated velocity profile and indicate which forces are expected to be dominant in each of these systems.

20. Transient heat conduction in citrus fruits is of concern to farmers in Florida and other farmers as well. They must develop strategies to prevent freezing during cold weather. The thermal diffusivity data are needed for this, and one way to find this information is to insert a thermocouple in the center and dip the fruit in a stirred tank of iced water at 0 °C. From the following data for a 5.8-cm-diameter grapefruit estimate the thermal diffusivity and compare it with that of water at 10 °C.

t, min	0	5	10	15	20	25	30	35	40	45
T_C, °C	20	20	19.6	18.0	15.8	13.0	10.8	8.4	7.2	5.8

21. Use the thermal-diffusivity value estimated from the above problem to find out whether any part of a fruit of diameter 8 cm hanging in a tree will freeze if the ambient temperature suddenly drops to −6 °C, from an initial temperature of 15 °C. The heat transfer coefficient to the fruit surface is estimated as 10 W/m^2 · K. This is, of course, a function of the air velocity.

22. A membrane of thickness L separates two solutions. Both solutions and the membrane have initially a zero concentration of a permeable solute. At time $t = 0$ and thereafter one side is maintained at a concentration of C_{A0} of this solute. Solve the concentration profile in the membrane as a function of time and position.

23. For gas absorption with a semi-infinite region with reaction show that the following limiting values for the flux into the system can be derived starting from the detailed equation in the text:

$$N_{A,s} = C_{A,s}\sqrt{\frac{D_A}{\pi t}}(1 + kt) \text{ for } kt \ll 1$$

The solution error is 5% for $kt = 0.5$.
Find the total quantity of gas absorbed up to t_E for this case.
Find the average rate of absorption.

If the average rate is divided by $k_L C_{A,s}$ we can obtain an expression for the enhancement factor. Here k_L can be calculated using the penetration model.

24. Consider the above problem of gas absorption, but now for the limit for large values of kt, and show that

$$N_{A,s} = C_{A,s}\sqrt{D_A k} \text{ for } kt \gg 1$$

The solution error is 3% for $kt = 2.0$.
Find the total quantity of gas absorbed from time 0 to t_E for this case.
Find the average rate of absorption.
Compare your answer with the film model. State why there is no difference between the penetration model and the film model.

25. A gas stream with CO_2 at partial pressure 1 atm is exposed to liquid in which it undergoes a first-order reaction for 0.01 s. The total amount of gas absorbed during this time was measured as $1.5 \times 10^{-4}\,\text{mol/m}^2$. Estimate the rate constant for this reaction. Use $C_{A,s} = 30\,\text{mol/m}^3$ and $D_A = 1.5 \times 10^{-9}\,\text{m}^2/\text{s}$.

26. Extend the penetration model for two species reacting instantaneously. Assume that the solution on either side of the reaction front in Fig. 11.8 can be expressed in terms of error function which obviously satisfies the differential equation. Fit the boundary conditions for each side. Now use the flux balance at the reaction plane and derive an expression for λ as a function of time. How does this model compare with the film model?

27. **Transient channel flow.** Complete the solution to transient channel flow with a pressure gradient using the separation of variables. Also solve the problem numerically using the PDEPE or other solvers and compare the results.

28. **Transient Couette flow.** The problem is similar to Section 11.11.1 and is obtained by assuming the pressure gradient to be zero. Set up the problem and state the boundary conditions. Verify that the boundary conditions lead to a non-homogeneity. Obtain the transient solution by separation of variables after subtraction of the steady-state solution.

29. Pipe flow with periodic pressure variation. Verify the result for the velocity profile in the complex domain shown in Section 11.11.6. Write MATLAB code to find the real and imaginary parts. Use this code to plot the velocity profile for various values of Wo as a function of position and time. Also derive an expression for the volumetric flow rate. Plot this as a function of time and also superimpose on the plot the pressure variation with time. Show that there is a phase lag between pressure and flow for high values of Wo.

30. **A case-study problem: oscillatory flow of a Casson fluid.** The rheology of blood is non-Newtonian and is often represented by the Casson fluid model which was discussed in Section 5.7. Your project is to examine the blood flow with this model and examine the conditions under which the deviation from Newtonian fluid behavior may be expected. The mathematics is horrendous except for my students, and you may wish to refer to the paper by Rohlf and Tenti (2001), who studied the problem in detail.

12 Convective heat and mass transfer

Learning objectives

You will learn from this chapter the following main ideas:

- setting up problems in laminar-flow heat and mass transfer;
- dimensionless representation and the key dimensionless groups in convective transport;
- the separation-of-variables method for a constant wall temperature;
- a modified separation-of-variables method for constant wall flux;
- extraction of mesoscopic parameters, *viz.*, transfer coefficients, from detailed solutions;
- heat and mass transfer in the entry region; the similarity solution in terms of the incomplete gamma function;
- convective mass transfer with wall reaction;
- mass transfer rate in solid dissolution from a flowing film;
- mass transfer rate in gas absorption from a flowing film;
- a model for a laminar-flow chemical reactor; dimensionless representation;
- pure-convection and plug-flow models for a laminar-flow reactor;
- a mesoscopic model for a laminar-flow reactor; the role of the axial dispersion parameter; and
- setting up and solving convection problems in MATLAB.

Mathematical prerequisites

The chapter needs the use of two functions in mathematics:

- hypergeometric functions
- incomplete gamma functions

No additional prerequisites are needed.

The method of separation of variables and series solution of PDEs should be revised at this stage. Numerical solution of PDEs is another useful mathematical tool for the type of problems examined here.

In this chapter we examine problems in convective heat and mass transfer mainly for laminar internal flows. Owing to the similar mathematical structure of the problems both heat and mass transfer problems can be treated in a similar manner. The concentration or temperature is now a function of both coordinate directions, namely the axial distance and the radial (or cross-flow) direction, and therefore the governing equations are now partial differential equations (PDEs). Mathematically speaking, the PDEs are similar to that for transient heat transfer studied in the previous chapter, and hence the method of separation of variables can be used. But in the context of convective transport these PDEs, although often still linear, have variable (position-dependent) coefficients. This adds additional mathematical complexity to the solution, mainly in the form of the eigenfunctions and the calculations of the eigenvalues. These details will be examined in this chapter. The structure of this chapter is as follows.

Heat transfer in laminar flow under fully developed conditions will be the first problem examined. Both constant-wall-temperature and constant-wall-flux boundary conditions will be studied, and series solutions applicable for these cases will be derived. The nature of the series, convergence properties, the heat flux, and the predicted heat transfer coefficients will be examined, and illustrative examples are presented.

The series solution has poor convergence near the entrance, and hence an alternative method using the similarity variable is more useful here. The boundary-layer nature of the problem near the entrance region permits this approach. We will define the similarity variable and derive a closed-form analytical solution to the entry-region problem.

The next class of problems studied is those in convective mass transfer. Owing to the similar mathematical nature, the solutions can be obtained readily by analogy with heat transfer. Hence the mathematical treatment of this problem will not be presented in detail. Two common problems for mass transfer in pipe flow, namely (i) solid dissolution from pipe walls and (ii) transport to and heterogeneous reaction at the walls, will be presented. For the first problem the results will be found to be identical to that for heat transfer with a constant wall temperature. For the second case we will find that the boundary conditions are now of the third kind (Robin), necessitating some modifications in the eigenvalues' series coefficients.

Mass transfer into a liquid flowing as a film over a vertical (or inclined) surface is an important prototype problem in convective mass transfer since it simulates the flow in more complex industrial equipment such as packed beds. This is studied next. We examine two important cases here: (i) solid dissolving from the wall and (ii) gas being absorbed from the interface of the liquid film. We will explore the similarity solution approach to predict the mass transfer coefficients for the two cases. We will also compare and contrast the two problems.

The next problem examined in detail is the classical problem of the laminar-flow reactor. The model shown is simply an extension of the convective diffusion model with reaction terms added. We will show the key dimensionless groups and present solutions to some limiting cases. The mesoscopic model for a laminar-flow reactor will then be derived by cross-sectional averaging. It will be shown that an additional parameter, *viz.*, the dispersion coefficient, is needed in the context of mesoscopic models.

The final section of the chapter deals with the numerical solution of convective transport problems. Mainly the use of the PDEPE solver in MATLAB is demonstrated. Some analytical results will be validated numerically, and numerical results for some additional problems will be demonstrated.

This chapter provides an introductory basis for the analysis of convective transport problems. More discussion and analysis will be taken up in Chapter 18 for heat and Chapter 20 for mass, where we will consider transport in boundary layers and in turbulent flows.

12.1 Heat transfer in laminar flow

12.1.1 Preliminaries and the model equations

Consider a fully developed flow in a circular pipe. The fluid enters at $z = 0$ at a temperature of T_0 and the pipe walls are maintained at a constant temperature of T_w. We will write the differential equation for the temperature distribution as a function of r and z, and then express this in a dimensionless form and identify the important dimensionless parameters. Heat generation in the pipe due to viscous dissipation is neglected, and a Newtonian fluid is assumed. Also we neglect the changes in viscosity due to temperature variation. This makes it a one-way coupled problem. Otherwise momentum and heat equations have to be solved simultaneously.

The governing equation, neglecting viscous dissipation, is

$$\rho c_p v_z \frac{\partial T}{\partial z} = k \left[\frac{\partial^2 T}{\partial z^2} + \frac{1}{r} \frac{\partial}{\partial r} \left(r \frac{\partial T}{\partial r} \right) \right] \tag{12.1}$$

In this model, we assume that $z = 0$ is at a location in the pipe where the velocity profile is already fully established. This requires a dimensionless entry length to be given by the following equation:

$$L_e^* = \frac{L_e}{R} = 0.07 Re$$

where Re is the Reynolds number. If the heating starts at a position earlier than this length, the model has to be modified, since the velocity profile is not a parabolic profile in this region. Flow is of boundary-layer type here, and both v_z and v_r contribute to convective transport.

For the fully developed region, the velocity profile is given by

$$v_z = v_{max} \left[1 - \left(\frac{r}{R} \right)^2 \right] = 2\langle v \rangle \left[1 - \left(\frac{r}{R} \right)^2 \right]$$

where v_{max} is the center velocity and $\langle v \rangle$ is the average velocity.

It may also be noted that there is a thermal entry region near $z = 0$, and the temperature profiles become fully developed only after a certain thermal entry length. More on this is discussed later. The two entry lengths (hydrodynamic and thermal) are shown in Fig. 12.1

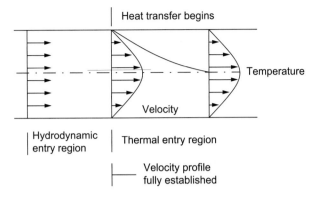

Figure 12.1 Hydrodynamic and thermal entry lengths in laminar flow in a heated or cooled pipe. Temperature changes only in the region near the wall in the thermal entry region.

for clarity. The thermal entry length is defined as a region where the temperature change is confined to region near the wall and the influence of the wall is not felt at the tube center.

Dimensionless form

In dimensionless form the following equation can be derived:

$$(1 - \xi^2)\frac{\partial \theta}{\partial z^*} = \frac{1}{Pe}\left[\frac{\partial^2 \theta}{\partial z^{*2}} + \frac{1}{\xi}\frac{\partial}{\partial \xi}\left(\xi\frac{\partial \theta}{\partial \xi}\right)\right] \tag{12.2}$$

Here z^* is the dimensionless axial distance defined as z/R and ξ is the dimensionless radial distance defined as r/R. The definition of the dimensionless temperature will depend on the boundary condition imposed at the wall ($\xi = 1$). Two common boundary conditions are (i) constant wall temperature and (ii) constant wall flux as illustrated in Fig. 12.2.

For the case of constant wall temperature θ will be defined as

$$\theta = (T - T_w)/(T_0 - T_w)$$

The dimensionless inlet temperature is 1, while the wall temperature is zero with this definition. This is convenient in order to make the problem homogeneous. The symmetry condition is applied at $\xi = 0$ as usual. The case of constant wall flux and the appropriate definition for θ will be taken up subsequently.

The Péclet number

The parameter Pe appearing in the dimensionless form Eq. (12.2) is defined as follows:

$$Pe = \rho c_p \langle v \rangle d_t / k = RePr$$

Re here is the Reynolds number based on the pipe diameter d_t, while Pr is the Prandtl number defined as $c_p \mu / k$ or equivalently as v/α. Also note that the Prandtl number is the ratio of the momentum diffusivity to the heat diffusivity.

The dimensionless group Pe, the Péclet number, has the following significance:

$$Pe = \frac{\text{Heat transport by convection}}{\text{Heat transport by conduction}}$$

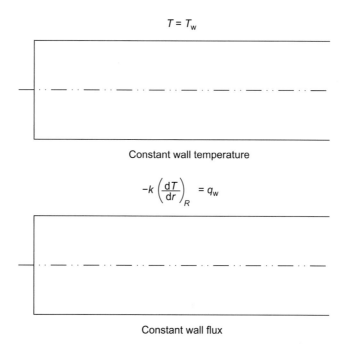

$$T = T_w$$

Constant wall temperature

$$-k\left(\frac{dT}{dr}\right)_R = q_w$$

Constant wall flux

Figure 12.2 Two common boundary conditions applied at the tube wall; the wall temperature is known in one case and the slope of the temperature at the wall is known in the other case.

which can also be realized by writing Pe as

$$Pe = \frac{\rho c_p \langle v \rangle d_t}{k} = \frac{\rho c_p \langle v \rangle \Delta T}{k \Delta T / d_t}$$

where ΔT is some representative temperature difference in the system.

The dimensionless temperature distribution depends only on Pe (neglecting the heat production due to viscous dissipation). Hence the parametric representation of the problem is

$$\theta = \theta(z^*, \xi; Pe)$$

Heat flux and Nusselt number

The quantity of engineering interest is usually the heat flux q_R or heat transfer rate to the wall. The heat flux at the wall is readily obtained by using the Fourier law at the wall,

$$q_R = -k\left(\frac{\partial T}{\partial r}\right)_{r=R} \tag{12.3}$$

The heat transfer coefficient h defined as

$$q_R = h(T_b - T_w)$$

is commonly used in lieu of the wall heat flux in mesoscopic models, design equations for heat exchangers, etc. In the above expression, T_b is the cup mixing or flow-averaged temperature defined as

$$T_b = \frac{\int_0^R 2\pi r v_z(r) T(r) dr}{\int_0^R 2\pi r v_z \, dr} \tag{12.4}$$

The heat flux to the wall or equivalently the heat transfer coefficient can be expressed in dimensionless form in terms of the Nusselt number defined as

$$Nu = \frac{2Rq_R}{k(T_b - T_w)} = \frac{hd_t}{k}$$

From the relation for q_R given by Eq. (12.3), the following expression for the Nusselt number is readily obtained:

$$Nu = -\frac{2}{\theta_b}\left(\frac{\partial\theta}{\partial\xi}\right)_{\xi=1} \tag{12.5}$$

where θ_b is the dimensionless cup mixing average temperature defined as

$$\theta_b(\zeta) = 4\int_0^1 \xi(1 - \xi^2)\theta\, d\xi \tag{12.6}$$

for a parabolic velocity profile.

Thus, once the detailed radial variation of the temperature profile has been calculated at any chosen axial position, all the quantities needed to find the local Nusselt number by use of Eq. (12.5) are known. The Nusselt number can be related to the Péclet number, and this relation can be used for design or simulation purposes in conjunction with the mesoscopic models. This is the main idea of modeling convection effects in heat transfer.

Simplification for large *Pe*

If $Pe > 10$ the axial conduction term will be smaller than the radial conduction, and the model can be reduced considerably. This can be shown by a detailed scaling analysis, which is left as an exercise.

The differential equation now reduces to

$$(1 - \xi^2)\frac{\partial\theta}{\partial\zeta} = \frac{1}{\xi}\frac{\partial}{\partial\xi}\left(\xi\frac{\partial\theta}{\partial\xi}\right) \tag{12.7}$$

where ζ is now a dimensionless axial length defined as z^*/Pe. This is often referred to as a contracted (dimensionless) axial distance. Note that the parameter Pe has been absorbed into the contracted axial distance ζ. Thus the parameter dependence can be written simply as

$$\theta = \theta(\zeta, \xi) \tag{12.8}$$

Note that there is no free parameter, which is a characteristic of a well-scaled problem.

12.1.2   The constant-wall-temperature case: the Graetz problem

The associated boundary conditions for a constant-wall-temperature case for Eq. (12.7) are as follows:

$$\text{Entrance } \zeta = 0, \qquad \theta = 1$$
$$\text{Wall } \xi = 1, \qquad \theta = 0$$
$$\text{Center } \xi = 0, \qquad \partial\theta/\partial\xi = 0$$

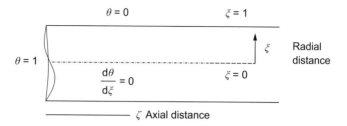

Figure 12.3 A schematic representation of the boundary conditions for the classical Graetz problem.

This is the classical Graetz problem which has been well studied in the heat-transfer field. The variables and the boundary conditions are shown schematically in Fig. 12.3. Note that only one condition (the entrance condition) is needed for ζ, which would not be the case if the axial conduction term were retained in the model.

> **Historical vignette.** The historical development of laminar forced convection in circular ducts was initiated by Graetz in 1883 and by Nusselt in 1910. These independent investigations led to a classical problem usually referred as the Graetz–Nusselt problem in the literature. The solution to the problem has been investigated in considerable detail as well, and there is a large body of literature dealing with many methods for the solution of this problem.

Solution methods for the Graetz problem

The solution can be obtained by separation of variables in view of the fact that the equation is homogeneous in the ξ boundaries. In some sense the problem is similar to transient heat transfer, the only difference being that the $1 - \xi^2$ term on the LHS of Eq. (12.7) makes the problem somewhat more complicated than the simpler transient heat conduction problem.

The solution can be represented as a series of the form

$$\theta(\zeta, \xi) = \sum_{n=1}^{\infty} C_n \exp(-\lambda_n^2 \zeta) F_n(\xi) \tag{12.9}$$

where λ_n are the eigenvalues to the associated eigenproblem and F_n are the eigenfunctions. The difficulty lies in the fact that the eigenvalues and eigenfunctions cannot be readily computed, unlike the situation for transient heat-conduction problems. The eigenfunctions are no longer simple trigonometric functions or even the more complicated Bessel functions. They are hypergeometric functions.

The differential equation for F can be derived by substituting the proposed solution Eq. (12.9) into Eq. (12.7), and you should verify that the following eigenvalue problem is obtained:

$$\frac{d}{d\xi}\left(\xi \frac{dF_n}{d\xi}\right) = -\lambda_n^2 \xi (1 - \xi^2) F_n \tag{12.10}$$

Table 12.1. Constants needed for the solution to the Graetz problem with constant wall temperature

	$n = 1$	$n = 2$	$n = 3$	$n = 4$	$n = 5$
Eigenvalues λ_n	2.704	6.6790	10.6733	14.6710	18.6698
Coefficient C_n	0.9774	0.3858	−0.2351	0.1674	−0.1292
$F_n(\xi = 0)$	1.5106	−2.0895	−2.5045	−2.8426	−3.1338
$F_n'(\xi = 1)$	−1.5322	−2.8192	3.9379	−4.9631	5.9256

and the solutions to this equation are the eigenfunctions to the Graetz problem. It can be shown by a series expansion that the eigenfunctions are

$$F_n(\xi) = \exp(-\lambda_n \xi^2/2) M\left(\frac{1}{2} - \frac{\lambda}{4}, 1, \lambda_n \xi^2\right)$$

where M is the confluent hypergeometric function or the Kummer function. These functions are power series in ξ similar to an exponential function. (See the book by Abramowitz and Stegun (1964) for more details on these types of functions.)

The functions represented above have the symmetry property at $\xi = 0$ since they are even functions. Hence the boundary condition at $\xi = 0$ is satisfied. Now, using the boundary condition at $\xi = 1$, we obtain the eigencondition;

$$\exp(-\lambda/2) M\left(\frac{1}{2} - \frac{\lambda}{4}, 1, \lambda\right) = 0 \tag{12.11}$$

The solution of this algebraic equation provides us with the eigenvalues. These can be readily computed in MATLAB since it has a built-in hypergeometric function calculator. The first few eigenfunctions are given in Table 12.1.

The series coefficients can be calculated from the orthogonality property of the F function:

$$C_n = \frac{\int_0^1 \xi(1 - \xi^2) F_n(\xi) d\xi}{\int_0^1 \xi(1 - \xi^2) [F_n(\xi)]^2 \, d\xi}$$

Note that the eigenfunctions F_i and F_j (with $i \neq j$) are not orthogonal with themselves but need a weighting function $\xi(1 - \xi^2)$ to make them orthogonal. Hence this term is also needed in the integral for the calculation of the series coefficient. A few values of the series coefficients are also given in Table 12.1 together with the corresponding eigenvalues. The formula above is applicable for an inlet condition of one.

The center temperature profile and the cup mixing average temperature profiles are shown in Fig. 12.4 using five terms to sum the series. The solution generated numerically by PDEPE is also shown for comparison. Note that the five-term series solution is not accurate for $\zeta < 0.05$. More terms are needed here for the series to converge. An alternative method is the asymptotic solution discussed in Section 12.2.

The leading term in the solution for the center temperature is therefore

$$\theta(\zeta, 0) \approx 0.9774 \exp(-2.704^2 \zeta) F(0)$$

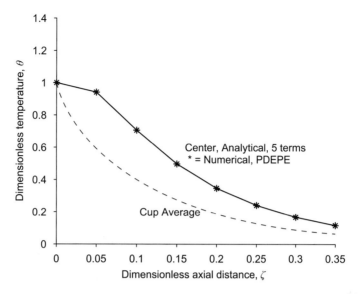

Figure 12.4 The temperature profile for the Graetz problem; the plot compares the analytical and numerical solutions as well.

which is the one-term approximation. The values of $F(0)$ are also tabulated in Table 12.1 for quick calculations of the center temperature. We find that the heat transfer is almost complete at $\zeta = 0.5$. For this length the value of the center temperature is 0.0381.

Heat flux and Nusselt number

The heat flux to the wall and the corresponding local Nusselt number can be calculated by taking the derivative of the F function at the pipe wall and also by evaluating θ_b (see Eq. (12.6)). The result for the local Nusselt number, after some clever algebra that allows one to avoid integration of the Kummer functions, is

$$Nu(\zeta) = \frac{\sum_{n=1}^{\infty} C_n F'_n(1)\exp(-\lambda_n^2 \zeta)F'_n(1)}{\sum_{n=1}^{\infty} 2C_n[F'_n(1)/\lambda_n^2]\exp(-\lambda_n^2 \zeta)}$$

The first five numerical values of F'_n are also shown in Table 12.1 as a quick aid to calculation of the Nusselt number. For one-term approximation the value of Nu reduces to $\lambda_1^2/2$. Hence the Nusselt number takes an asymptotic value of 3.657. The reader may wish to note that the entire calculation can easily be done in MATLAB.

Note that the local Nusselt number will be a function of the axial position ζ. The quantity of interest, especially for the 1D or mesoscopic model discussed in Section 1.7.4 is the average Nusselt number over a length L^* of the pipe.

This is defined as

$$\bar{Nu}_L^* = \frac{1}{L^*} \int_0^{L^*} Nu \, d\zeta$$

The average Nusselt number over a length L of pipe can be calculated by integration as

$$Nu_L = 3.66 + \frac{0.0668 Pe/L^*}{1 + 0.04[Pe/L^*]^{2/3}}$$

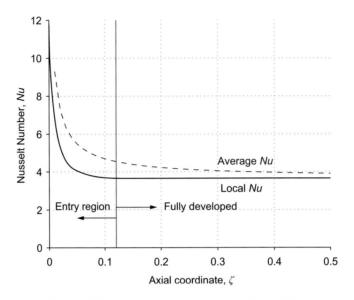

Figure 12.5 A plot of the local Nusselt number as a function of the axial distance.

where $L^* = L/d_t$. As L^* becomes large the average Nusselt number approaches an asymptotic value of 3.66.

A plot of the local and average Nusselt number as a function of the axial coordinate ζ is shown in Fig. 12.5.

12.1.3 The constant-flux case

The wall boundary condition for a constant heat flux of q_w at the walls is

$$k \left(\frac{dT}{dr} \right)_{r=R} = (-q_w)$$

Here $(-q_w)$ is the inward-directed prescribed wall heat flux. This quantity is positive for wall heating and represents the common case where the pipe is heated by an electrical coil or by direct radiation impinging on the pipe wall.

How should the dimensionless temperature be defined for this case? It should be defined so that the boundary condition at the wall takes a simple form. The dimensionless temperature should be now defined as

$$\theta = \frac{T - T_0}{(-q_w) R/k}$$

The wall boundary condition takes a simple form:

$$\frac{\partial \theta}{\partial \xi} = 1 \text{ at } \xi = 1$$

The symmetry condition is used at the center $\xi = 0$ as usual, and the inlet temperature is now zero.

A direct series solution is not possible due to the non-homogeneous boundary condition, but for large distances from the entrance an asymptotic solution can be obtained.

The essential assumption for getting this solution is that the temperature profile is now a linearly increasing function of the axial position. Hence a solution of the following form is postulated:

$$\theta(\zeta,\xi) = A\zeta + F(\xi)$$

where $F(\xi)$ is a function of ξ to be evaluated by matching the model equations and the boundary conditions.

The inlet condition at $\zeta = 0$ can no longer be used here since the above (asymptotic) solution is not valid near the inlet. This is not really a problem since the cup mixing dimensionless temperature θ_b is known at any axial position from an overall (macroscopic) heat balance. From Section 8.5.2 we know that $\theta_b = 4\zeta$ and hence this condition is used in lieu of the inlet condition.

The asymptotic solution for the constant flux case is then readily obtained which is valid for large distance from the entrance (usually for $\zeta > 0.1$).

$$\theta(\zeta,\xi) = 4\zeta + \xi^2 - \frac{1}{4}\xi^4 - \frac{7}{24} \tag{12.12}$$

Note that the solution has a similar structure to transient heat conduction with a constant heat flux. The value of Nusselt number based on pipe diameter can be derived to be equal to 48/11 which is a classical result. Details are left as an exercise in Problem 4.

12.2   Entry-region analysis

12.2.1   The constant-wall-temperature case

Application of the series solution described in Section 12.1.2 requires a large number of terms for small values of ζ, i.e., near the entrance region. An alternative method is to seek a similarity solution similar to the error-function solution discussed for transient heat transfer. An entrance-region solution will now be presented. The starting differential Eq. (12.7) is rewritten by expanding the Laplacian as

$$(1 - \xi^2)\frac{\partial\theta}{\partial\zeta} = \frac{\partial^2\theta}{\partial\xi^2} + \frac{1}{\xi}\frac{\partial\theta}{\partial\xi} \tag{12.13}$$

The distance from the wall defined as $s^* = 1 - \xi$ is now more convenient. The velocity-profile term $1 - \xi^2$ is now approximated as $2s^*$, i.e., we keep only the linear term. Also we keep only the first term for the Laplacian. Thus the second term on the RHS is dropped. This amounts to ignoring the curvature effects and representing the geometry by simpler rectangular coordinates.

The differential equation now is

$$2s^*\frac{\partial\theta}{\partial\zeta} = \frac{\partial^2\theta}{\partial s^{*2}} \tag{12.14}$$

We find that $2s^{*3}/\zeta$ (or its cube root) should be used as a similarity variable since there is no well-defined length scale in the entry region. Note that the radius of the pipe is not a relevant length scale here since the temperature changes are now confined to a thin region near the wall. It is convenient now to use a similarity variable defined as

$$\eta = s^* \left(\frac{2}{9\zeta} \right)^{1/3} \quad \text{or } \eta = s^* \left(\frac{2Pe}{9z^*} \right)^{1/3}$$

The factor 2/9 is included for convenience in later algebra.

The differential equation (12.14) now reduces to an ordinary differential equation in terms of the similarity variable η:

$$\frac{d^2\theta}{d\eta^2} + 3\eta^2 \frac{d\theta}{d\eta} = 0 \tag{12.15}$$

The boundary conditions to be used now are as follows. (i) At $\eta = 0$, the wall, $\theta = 0$. (i) At $\eta \to \infty$, the center or the inlet, $\theta = 1$. We also note that the three conditions merge into two, which is a characteristic of similarity solution methods.

The solution is found by integrating the above equation twice. The first integration gives an expression for the temperature gradient $p = d\theta/d\eta$. A second integration gives the temperature profile. The details are left as an exercise problem. The final answer is in terms of an incomplete gamma function:

$$\boxed{\text{Entry region: } \theta = \frac{1}{\Gamma(4/3)} \int_0^\eta \exp(-t^3)dt} \tag{12.16}$$

where η is a similarity variable defined earlier and t is a dummy variable for integration. Equation (12.16) is also known as the Lévêque solution.

The results can be expressed compactly as an incomplete gamma function as $\theta = $ gammainc($\eta^3, 1/3$). A plot of an incomplete gamma function is shown in Fig. 12.6 as an aid to quick calculations.

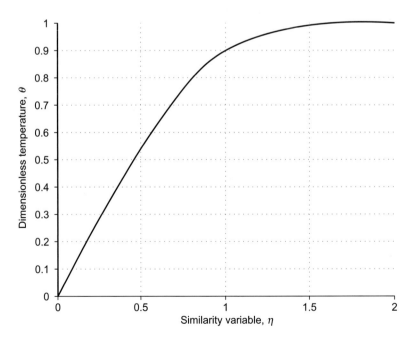

Figure 12.6 A plot of the temperature profile for the entry region as a function of the similarity variable. This is a plot of gammainc($\eta^3, 1/3$).

The corresponding local heat flux is calculated as

$$q_R \frac{R}{k} = \left[\frac{(2Pe/z^*)^{1/3}}{9^{1/3}\Gamma(4/3)} \right] (T_w - T_0) \tag{12.17}$$

On rearranging and basing the Nusselt number on diameter we have

$$Nu = 1.077(2Pe/z^*)^{1/3} \tag{12.18}$$

Note that the driving force for heat transfer is taken as $T_w - T_0$ rather than $T_b - T_0$ in the above definition of the Nusselt number. The bulk of the fluid is still at the inlet temperature, and hence this simple choice may also be the appropriate choice.

The Graetz number defined as $Gz = \dot{m}c_p/(kz)$, where $\dot{m}$ is the mass flow rate, is used in the literature on transfer. This laminar heat turns out to be equal to $Gz = (\pi/4)(2Pe/z^*)$. Using this in Eq. (12.18), the Nusselt number in the entry region can also be expressed in terms of the Graetz number as

$$Nu = 1.877Gz^{1/3} \tag{12.19}$$

What is the dependence of the heat transfer coefficient on the mass flow rate in this region?

12.2.2   The constant-flux case

The solution for constant flux in the entry region can be shown to be

$$\theta = \sqrt[3]{9\zeta/2} \left[\frac{\exp(-\eta^3)}{\Gamma(2/3)} - \eta\,\mathrm{gamainc}(\eta^3, 2/3) \right] \tag{12.20}$$

Now we study illustrative mass transfer problems. Problems that can posed in Cartesian coordinates are shown first, and then we examine a problem in cylindrical coordinates.

12.3   Mass transfer in film flow

An important class of problems involve mass transfer to or from a liquid flowing as a thin film over a vertical plate. The velocity profile was derived in Section 6.2.5 and is reproduced here for ease of reference:

$$v_z = \frac{\rho g}{\mu} \left[\delta y - \frac{y^2}{2} \right]$$

as shown earlier, neglecting entry effects. Also the film is assumed to flow in a laminar flow. (This requires a Reynolds number based on film thickness and an average velocity of less than 20.)

The film thickness δ can be computed from an overall integral mass balance (integral from 0 to y) and is related to the liquid flow rate Q/W (per unit width of the wall):

$$\delta = \sqrt[3]{\frac{3\mu Q/W}{\rho g}}$$

Now we analyze the mass transport of a species dissolving from the wall or absorbed from the interface into the film.

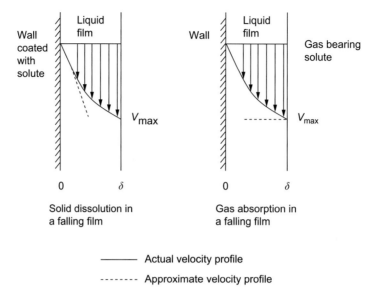

Actual velocity profile

-------- Approximate velocity profile

Figure 12.7 A schematic representation of mass transfer in film flow. For solid dissolution the region near the wall is important. For gas absorption the region near the interface is important.

12.3.1   Solid dissolution at a wall in film flow

The governing equation is

$$v_z(y)\frac{\partial C_A}{\partial z} = D\frac{\partial^2 C_A}{\partial y^2} \tag{12.21}$$

Here we neglect diffusion in the flow direction. Thus the LHS represents convection in the direction of flow (z) and the RHS represents diffusion in the cross-flow direction (y). Sc defined as v/D is the key parameter affecting the mass-transfer penetration distance. For a liquid Sc is large, and we expect a boundary-layer type of concentration distribution and the solute concentration changes are confined to just a small part of the hydrodynamic film thickness.

Anticipating a concentration boundary layer near the wall, we use a linearized form as an approximation for the velocity. The approximation is shown in Fig. 12.7. Here

$$v_z = \beta y$$

where β is the slope of the velocity profile near the wall, which is given by

$$\beta = \frac{\rho g}{\mu}\delta$$

This will be the velocity profile "seen" by the diffusing species, and hence the convection–diffusion model (Eq. (12.21)) reduces to

$$\beta y\frac{\partial C_A}{\partial z} = D\frac{\partial^2 C_A}{\partial y^2} \tag{12.22}$$

The domain of the solution is set as semi-infinite since the concentration profile is assumed to be confined to a thin region near the wall. Hence the similarity solution method can be used. Following the heat transfer problem studied in Section 12.2.1, we find that y^3/z should be a scaling parameter. The cube root of this, with some constants added, is defined as the similarity variable η,

$$\eta = y \left(\frac{\beta}{9Dz} \right)^{1/3}$$

The solution to Eq. (12.22) can be derived as

$$\frac{C_A}{C_{A,s}} = 1 - \frac{\int_0^\eta \exp(-\eta^3)d\eta}{\Gamma(4/3)} \tag{12.23}$$

The detailed derivation of the equation is not presented here but left as an exercise. The solution can be readily computed using the built-in function in MATLAB as

$$\frac{C_A}{C_{A,s}} = 1 - \text{gammainc}(\eta^3, 1/3) \tag{12.24}$$

which gives the dimensionless concentration as a function of η. The local rate of mass transfer is obtained by using Fick's law at the wall:

$$N_A = \frac{DC_{A,s}}{\Gamma(4/3)} \left(\frac{\beta}{9Dz} \right)^{1/3}$$

The average rate of mass transfer over a length L is given by

$$\bar{N}_A = \frac{2DC_{A,s}}{\Gamma(7/3)} \left(\frac{\beta}{9DL} \right)^{1/3}$$

If one divides this by the driving force $C_{A,s}$, one can calculate the average value of the mass transfer coefficient over a length L:

$$k_{sl} = \frac{2}{\Gamma(7/3)} D^{2/3} \beta^{1/3} L^{-1/3}$$

The following parametric effects can be deduced by substituting for β from the above equation. (i) The mass transfer coefficient varies as diffusivity to the power of 2/3. (ii) It increases with film thickness to the power of 1/3, since β is proportional to δ. (iii) It decreases with the length with an exponent of 1/3. (iv) It is proportional to the 1/6 exponent of the liquid flow rate per unit width.

12.3.2   Gas absorption from interfaces in film flow

Here we consider the mass transfer of a species present in the gas into the liquid film. Again a concentration boundary layer is proposed near the interface. The velocity profile seen is now flat and equal to v_{max} rather than a linear profile, as in the previous subsection. The diffusing species now sees a flat velocity profile, and this is also illustrated in Fig. 12.7. The convection–diffusion equation now simplifies to

$$v_{max} \frac{\partial C_A}{\partial z} = D \frac{\partial^2 C_A}{\partial y^2} \tag{12.25}$$

The maximum velocity in turn is obtained from Eq. (6.14) by setting $y = \delta$, and is given by

$$v_{\max} = \frac{\rho g \delta^2}{2\mu}$$

Note that the LHS of Eq. (12.25) has no functional dependence on y and therefore the problem is similar to the transient diffusion problem rather than being a convection–diffusion problem. In fact, $z/v_{\max}$ is a time-like variable and represents the exposure time of an element of fluid moving down the interface. The concentration profile is now represented in terms of an error function:

$$\frac{C_A}{C_{A,s}} = 1 - \mathrm{erf}\left(\frac{y}{\sqrt{4Dz/v_{\max}}}\right) \tag{12.26}$$

Here y is now defined as the distance measured from the interface.

The local rate of mass transfer is obtained by using Fick's law at the wall. The result is the same as in the penetration-theory model discussed earlier:

$$N_A = C_{A,s}\sqrt{\frac{D v_{\max}}{\pi z}}$$

with t replaced by the exposure time, which is equal to $z/v_{\max}$. The average mass transfer coefficient over a length L is given by

$$\boxed{k_L = 2\sqrt{\frac{D v_{\max}}{\pi L}}} \tag{12.27}$$

which is the same as in the penetration-theory model with the contact time set as $L/v_{\max}$.

The following parametric effects can be deduced by substituting for $v_{\max}$ from the above equation. (i) The mass transfer coefficient varies as the diffusivity to the power of 1/2. (ii) It increases with film thickness to the power of 1/2. (iii) It decreases with the length with an exponent of 1/2. (iv) It is proportional to the 1/6 exponent of the liquid flow rate per unit width.

An important application of convective mass transfer with chemical reaction is the laminar-flow reactor, which will now be studied in some detail.

12.4 Laminar-flow reactors

12.4.1 A 2D model and key dimensionless groups

Consider fully developed flow in a circular pipe of a Newtonian fluid. The fluid has a soluble compound A at a concentration of C_{A0} at the inlet. This compound undergoes a chemical reaction in the system with a rate constant of k. The reaction is assumed to be of first order. Also, the walls are coated with a catalyst and the compound reacts at the wall with a surface reaction constant of k_s.

Dimensionless model equations

The differential equation for the concentration distribution of species A as a function of r and z as given by the convection–diffusion–reaction (CDR) model is

$$2\langle v\rangle\left(1-(r/R)^2\right)\frac{\partial C_A}{\partial z}=D\left[\frac{\partial^2 C_A}{\partial z^2}+\frac{1}{r}\frac{\partial}{\partial r}\left(r\frac{\partial C_A}{\partial r}\right)\right]-kC_A \qquad (12.28)$$

where a parabolic velocity profile has been used, which is valid for a Newtonian fluid.

If the reaction is not of first order the rate term kC_A is replaced by some "rate" function of C_A, and the differential equation becomes non-linear. We consider only the linear case, and the non-linear case is left as a computational exercise.

The following dimensionless variables are introduced.

The dimensionless radial position is $\xi = r/R$.

The dimensionless concentration is $c_A = C_A/C_{A0}$.

The dimensionless axial length is $\eta = z/L$. Here L is some specified length of the reactor, and the variable η goes from 0 (at the entrance) to 1 (at the exit of the reactor).

The axial diffusion term in Eq. (12.28) (the first term within the square bracket on the RHS) can usually be neglected. As in the case of heat transfer, this is justified if the Péclet parameter (defined later) here is greater than, say, 10. In dimensionless form the following equation will then be obtained:

$$2(1-\xi^2)\frac{\partial c_A}{\partial \eta}=\frac{L}{R}\frac{1}{Pe_R}\left[\frac{1}{\xi}\frac{\partial}{\partial \xi}\left(\xi\frac{\partial c_A}{\partial \xi}\right)\right]-Dac_A \qquad (12.29)$$

The dimensionless numbers appearing here are (i) the Damköhler number Da defined as

$$Da = kL/\langle v\rangle$$

and (ii) a Péclet number Pe_R defined as

$$Pe_R = \langle v\rangle R/D$$

Note that the common definition is to use the tube diameter to define the Péclet number. In this section we use the tube radius as the representative length to define the Péclet number, which is something of a departure from tradition. This turns out to be useful in the context of the mesoscopic model in the next subsection. The subscript R on Pe is used to distinguish the two.

Time constants and their ratios

It is also useful to define the various time constants and write these dimensionless groups as the ratios of time constants.

The dimensionless group Pe_R has the following significance:

$$Pe_R = \frac{\text{Mass transport by convection}}{\text{Mass transport by conduction}}$$

Likewise Da can be written as

$$Da = \frac{L/\langle v\rangle}{1/k} = \frac{\text{Residence time}}{\text{Reaction time}}$$

Finally, the group appearing at the front of the RHS of Eq. (12.29) denoted as B can be shown to be

$$B = \frac{L}{R}\frac{1}{Pe_R} = \frac{L/\langle v\rangle}{R^2/D} = \frac{\text{Residence time}}{\text{Radial diffusion time}}$$

and will play an important role in determining the importance of the radial diffusion. If this parameter is large, the radial diffusion time is small and the concentration is likely to equalize in the radial direction. Plug-flow behavior can be expected.

Boundary conditions

At the inlet $\eta = 0$ we set $c_A = 1$. At the center $\xi = 0$ we have a symmetry, and therefore $\partial c_A/\partial \xi = 0$.

The boundary condition at the wall in the presence of a heterogeneous wall reaction is of the Robin type and can be expressed as

$$\left(\frac{\partial c}{\partial \xi}\right)_{\xi=1} = -Da_w(c)_{\xi=1} \tag{12.30}$$

Thus an additional dimensional parameter, the wall Damköhler number, $Da_w = k_s R/D$, arises for this problem. This group can be shown to signify the ratio of the radial diffusion time to the surface reaction time. The no-flux condition can be used in the case of no wall reaction, leading to a Neumann condition. Furthermore, if the surface reaction is extremely rapid the concentration at the wall can be set as zero, leading to a Dirichlet problem. Hence it is interesting to note that all of the three common conditions can be applied to this problem, depending on the wall condition.

Parametric representation

Hence the result of the dimensionless formulation leads to the following parametric representation of the problem:

$$c_A = c_A(\eta, \xi, B, Da_R, Da_w)$$

The results are usually represented in terms of a cup mixing average concentration, which is defined as

$$c_{A,b}(\eta) = 4 \int_0^1 \xi(1 - \xi^2)c_A(\xi)d\xi \tag{12.31}$$

Usually the value of the cup mixing concentration at the exit ($\eta = 1$) is the quantity of design interest. Note that the conversion of the reactant A equals $1 - c_{A,b}$ at $\eta = 1$.

The effect of radial diffusion

The parameter appearing at the front of the radial diffusion term can be written as

$$B = \frac{L}{R}\frac{1}{Pe_R} = \frac{L}{\langle v \rangle}\frac{D}{R^2} = D\bar{t}/R^2$$

which is the ratio of the mean residence time to the diffusion time in the radial direction. Equation (12.29) can be written as

$$2(1 - \xi^2)\frac{\partial c_A}{\partial \eta} = B\left[\frac{1}{\xi}\frac{\partial}{\partial \xi}\left(\xi\frac{\partial c_A}{\partial \xi}\right)\right] - Da\,c_A \tag{12.32}$$

Some limiting cases, depending on the value of the parameter B, will now be analyzed.

An illustrative result is shown in Fig. 12.8 as a plot of $c_{A,b}$ (exit) as a function of B for $Da = 3$. The results were simulated numerically using PDEPE as shown in Section 12.6, but the consideration here is useful in understanding the effect of radial diffusion. We now discuss the limiting cases of small B and large B.

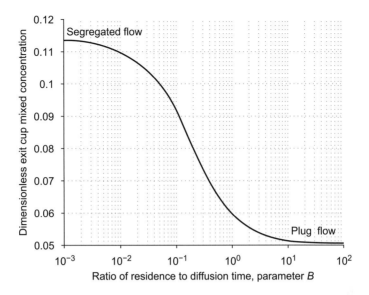

Figure 12.8 The effect of radial diffusion on the performance of a laminar-flow reactor. The results are for $Da = 3.0$.

12.4.2   The pure convection model

If the parameter B is small then the radial diffusion term can also be dropped, leading to a model called the pure convection model, which is also known as the segregated flow model.

The segregated model has a simple representation:

$$2(1 - \xi^2)\frac{\partial c_A}{\partial \eta} = -Dac_A \tag{12.33}$$

for a steady-state case.

The model can be readily solved and integrated over the cross-section to model the reactor performance. The concentration at any radial position is obtained as

$$c_A(\xi) = \exp\left(-\frac{Da\eta}{2(1 - \xi^2)}\right)$$

The reactor exit is at $\eta = 1$. Knowing the radial distribution of the concentration, it remains to find the cup mixing concentration by radial integration as per Eq. (12.31). The final result can be expressed as an exponential integral:

$$c_{A,b}(\text{exit}) = \frac{Da^2}{4}\text{expint}(Da/2) + (1 - Da/2)\exp(-Da/2) \tag{12.34}$$

The plug-flow model

The second limiting case, that for large B, is the plug-flow model. The result after some model manipulations leads to the classical result for a plug-flow reactor:

$$c_{A,b}(\text{exit}) = \exp(-Da) \tag{12.35}$$

Figure 12.9 shows a comparison of these limiting cases for various values of Da. The differences start to appear at larger values of Da, i.e., for large reactor conversions.

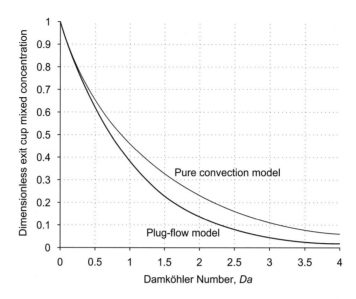

Figure 12.9 A comparison of the pure convection model and the plug-flow model.

The axial dispersion model

If the parameter B is in the intermediate range, neither the segregated model nor the plug-flow model will be reasonable. There is some smearing of concentration in the radial direction. The full 2D model should be used for such cases. However, the model can be reduced to a 1D or mesoscopic model by radial averaging, which is taken up in the following section. The radial averaging needs additional closure terms, as shown below. A common way of bringing about closure is the Taylor dispersion model, and the resulting 1D model for the reactor is known as the axial dispersion model.

Example 12.1.

Find the exit concentration and the conversion for $k = 0.5\,\mathrm{s}^{-1}$ and a residence time of 6 s.

For this case $Da = 3$. The concentration at various radial positions for the segregated model are computed from the equation given in Section 12.4.1, and upon averaging these or using Eq. (12.34) directly we find $c_b(1) = 0.1135$.

If the reactor were operating in plug-flow we mode would have a concentration of $\exp(-Da) = \exp(-3) = 0.0498$. The conversion is reduced compared with that in plug flow, due to the velocity profile and the spread in the residence time.

12.5   Laminar-flow reactor: a mesoscopic model

12.5.1   Averaging and the concept of dispersion

We again start with the differential model given by Eq. (12.29), and average it over the cross-section, which is standard practice to get the mesoscopic model. For this purpose, a cross-sectional average $\langle c \rangle$ (with the subscript A dropped) can be defined as

$$\langle c \rangle = 2 \int_0^1 \xi c_A \, d\xi$$

Equation (12.29) can be averaged by multiplying it by 2ξ and integrating with respect to ξ from 0 to 1. The boundary condition of no flux is applied at $\xi = 0$ and $\xi = 1$, which is applicable to a case of no wall reaction. The result is

$$\frac{\partial}{\partial \eta} \left(\int_0^1 4(1 - \xi^2)\xi c_A \, d\xi \right) = -Da \, \langle c \rangle \tag{12.36}$$

The integral appearing on the LHS is the cup mixing concentration $c_{A,b}$, and hence the averaged model can be written as

$$\frac{dc_{A,b}}{d\eta} = -Da \, \langle c \rangle \tag{12.37}$$

This looks like a simple plug-flow model, but looks can be deceptive. The LHS involves the cup mixing concentration, while the RHS involves the area-averaged concentration. An approximation is in order. Let

$$c_{A,b} = \langle c \rangle - H \frac{d\langle c \rangle}{dx}$$

where H is a dispersion factor. Equation (12.37) reduces to

$$H \frac{d^2 \langle c \rangle}{d\eta^2} - \frac{d\langle c \rangle}{d\eta} - Da \, \langle c \rangle = 0$$

which is referred to as the dispersion model.

Taylor derived the following result:

$$H = \frac{1}{48B} \tag{12.38}$$

See Section 20.5 for the derivation.

The reciprocal of H is defined as the dispersion Péclet number,

$$Pe^* = \frac{1}{H} = \frac{\langle v \rangle L}{D_E}$$

where D_E is called the axial dispersion coefficient.

Using Taylor's value for H, a reasonable value for D_E for laminar flow is

$$D_E = \frac{1}{48} \frac{\langle v \rangle R^2}{D}$$

which is the classical result derived by Taylor (Eq. (20.55)).

In terms of the dispersion Péclet number the model is written as

$$\frac{1}{Pe^*} \frac{d^2 \langle c \rangle}{d\eta^2} - \frac{d\langle c \rangle}{d\eta} - Da \, \langle c \rangle = 0 \tag{12.39}$$

Danckwerts boundary conditions are commonly used in the context of the dispersion model. At the inlet we allow some drop in concentration due to dispersion: at $\eta = 0$,

$$\langle c \rangle - \frac{1}{Pe^*} \frac{d\langle c \rangle}{d\eta} = 1 \tag{12.40}$$

and at the exit the stream is allowed to blend smoothly with the outlet fluid: at $\eta = 1$,

$$\frac{d\langle c \rangle}{dz^*} = 0 \qquad (12.41)$$

The first condition applies for a closed–open system with no dispersion upstream of the reactor. The second condition applies at an open–open section at the exit and assumes that the flux due to dispersion is zero at the exit.

The solution can be obtained analytically for a first-order reaction and numerically for other cases.

The result for a first-order reaction using the boundary condition stated above is as follows:

$$\langle c \rangle(1) = \frac{4\alpha \exp(Pe^*/2)}{(1+\alpha)^2 \exp(\alpha Pe^*/2) - (1-\alpha)^2 \exp(-\alpha Pe^*/2)} \qquad (12.42)$$

where

$$\alpha = \sqrt{1 + 4Da/Pe^*}$$

12.5.2 Non-linear reactions

For non-linear reactions one can perform the averaging as done earlier. However, one obtains the average reaction rate as a term. The model can be represented as

$$\frac{dc_{A,b}}{d\eta} = -Da\langle r^*(c) \rangle \qquad (12.43)$$

where r^* is the rate model in terms of the dimensionless concentration. The value of $\langle r^*(c) \rangle$, the cross-sectional average of the rate of reaction, is not the same as $r^*\langle c \rangle$ the rate based on the average concentration, unless the reaction is of first order. For example, the average of c^2 is not the same as the square of the average of c. This introduces yet another uncertainty into the dispersion model. One has to assume that

$$\langle r(c) \rangle \approx r(\langle c \rangle)$$

in order to close the model as a first approximation, or one must use some modified closure equations.

12.6 Numerical study examples with PDEPE

The PDEPE program shown in the previous chapter is a useful tool to simulate these classes of problems as well. Several illustrative examples are shown here.

12.6.1 The Graetz problem

This problem is readily handled by PDEPE by simply defining the capacity term as a function of the radial position. In the notation used in the sample code in Section 11.12.2, x has the meaning of the dimensionless radial position, ξ, while t has the meaning of the dimensionless axial position η. The capacity term is now defined as

$$C = 1 - x^2$$

Table 12.2. Case 1

ζ	θ_c	q_{wall}	θ_{cup}	Nu
4.9999997×10^{-2}	6.0493886×10^{-2}	1.158698	0.4212878	4.004401
9.9999987×10^{-2}	0.2988414	0.7331823	0.6047488	3.709956
0.1500000	0.5081392	0.5012615	0.7264640	3.665050
0.2000000	0.6581963	0.3469469	0.8103118	3.658075
0.2500000	0.7628176	0.2406021	0.8684155	3.656999
0.3000000	0.8354517	0.1669046	0.9087162	3.656829
0.3499999	0.8858457	0.1157870	0.9366729	3.656788
0.3999999	0.9208059	8.0326751	0.9560673	3.656806

The value from the one-term solution for $\zeta = 0.4$ is 0.9207 for the center temperature, which matches well with the PDEPE-generated solution shown in Table 12.2.

Table 12.3. Case 2

ζ	t_c	t_{wall}	Nu
0.05	0.0174	0.6021	4.972193
0.10	0.1392	0.8429	4.513927
0.15	0.3169	1.0539	4.404197
0.20	0.5107	1.2569	4.374793
0.25	0.7089	1.4578	4.366726
0.30	0.9085	1.6580	4.364493
0.35	1.108	1.8580	4.363877
0.40	1.308	2.0580	4.363703

The results can be verified by matching with the asymptotic solution given by Eq. (12.12) and are in almost exact agreement.

This should not be confused with the concentration variable. A sample code graetz.m can be written with only a minor modification of the code pde1t.m listed in the previous chapter.

Some results are given below so that the student can benchmark the programs and verify the results.

Case 1: constant wall temperature

Note that the results for wall temperature in Table 12.2 are based on a dimensionless wall temperature of one and the inlet temperature of zero, and hence the results are one minus the analytical value given in the earlier section.

Case 2: constant wall flux

The results generated numerically are shown in Table 12.3 and you should run the code and get the results. You should also verify the solution with the analytical solution.

The cup mixing temperature is readily calculated analytically here as

$$\theta_{cup} = \theta_{in} + 4\zeta$$

Table 12.4. Case 3A

Distance η	c_{center}	c_{wall}
0	1.0000	1.0000
0.1000	0.7906	0.5557
0.2000	0.5725	0.3707
0.3000	0.4006	0.2536
0.4000	0.2778	0.1748
0.5000	0.1921	0.1207
0.6000	0.1328	0.0834
0.7000	0.0918	0.0576
0.8000	0.0635	0.0398
0.9000	0.0439	0.0275
1.0000	0.0303	0.0190

You should vary the parameter B and show how the solutions for the segregated model and the plug-flow model are obtained for small B and large B, respectively.

Table 12.5. Case 3B

Distance η	c_{center}	c_{wall}
0	1.0000	1.0000
0.1000	0.7765	0.2570
0.2000	0.5110	0.1447
0.3000	0.3171	0.0867
0.4000	0.1943	0.0527
0.5000	0.1187	0.0322
0.6000	0.0725	0.0196
0.7000	0.0443	0.0120
0.8000	0.0271	0.0073
0.9000	0.0165	0.0045
1.0000	0.0101	0.0027

This result should be verified after running the program by numerical integration of the radial temperature profiles.

Case 3: laminar-flow reactor

Some sample results for the laminar-flow reactor are presented so that the student can verify the results

For Case 3A, with no wall reaction, $B = 0.5$ and $Da = 2$; see Table 12.4.

For Case 3B, with wall reaction included, $B = 0.5$, $Da = 2$, and $Da_w = 0.5$; see Table 12.5.

The case of wall reaction is solved simply by changing the boundary condition at $\xi = 1$. A sample result is shown for testing the code. The effect of wall reaction can be examined by changing the parameter Da_w.

Summary

- Heat transfer from the walls to a fluid in a pipe under laminar-flow conditions is the most widely studied and prototype problem in convective heat transfer. Various assumptions are usually employed to simplify the problem, such as constant viscosity and fully developed flow. Even then the problem is mathematically complex and results in a partial differential equation in r and z with a variable coefficient term. The variable coefficient arises due to the velocity variation in the r direction. It is useful to understand the mathematical structure of the model equation.

- The boundary conditions at the wall can be of constant-temperature (first kind or Dirichlet) or of constant-wall-flux type (Neumann) or wall-reaction type (Robin). Series solution can be obtained for both cases; it is straightforward for the constant-wall-temperature case, but the eigenfunctions are not simple well-known common functions but belong to a class of functions called hypergeometric functions. These functions can, however, be readily evaluated in MATLAB or MAPLE.

- The constant-wall-flux case is a non-homogeneous PDE and hence the series solution cannot be used directly. One has to obtain an asymptotic solution and represent the solution as a sum of the asymptotic solution and a perturbation solution. The latter can be expressed as a series.

- The key result from an engineering application point of view is the local wall heat/mass transfer coefficient. This can be predicted from the detailed temperature/concentration profiles with the differential model. The results are expressed in terms of a dimensionless group, the Nusselt number. It is seen that the Nusselt number is a function of the axial distance and decreases with distance, reaching an asymptotic value for large distance from the entrance.

- The convergence of the series solution is very rapid for large distances from the entrance. Only one term of the series is sufficient to predict the temperature profile and the heat flux. Using the heat flux and a radially averaged cup mixing temperature, the Nusselt number reaches a constant value of 3.67 for constant wall temperature and a value of 4.1 for constant wall flux.

- The convergence of the series is rather poor near the entrance. In this region there is a thermal boundary layer near the wall, and hence it is more convenient to obtain the similarity solution. Closed-form similarity solutions can be obtained both for the constant-wall-temperature case and for the constant-wall-flux case.

- Convective mass transport problems can be analyzed by using the analogy with heat transfer. The simplest case is dissolution of solid from the wall, which is identical to heat transfer with a constant wall temperature since the dissolution leads to a constant-concentration condition at the wall. A more involved case is that of heterogeneous reaction at the walls, which leads to a boundary condition of the third kind. The series solution applies, but the eigencondition now involves the (dimensionless) wall reaction rate constant.

- Mass transfer into a falling film is an important problem in convective mass transfer. Dissolution of solid at the walls and gas absorption at the interface are the two important prototype problems. The length of wall (absorber) is usually small, and hence the asymptotic entry-region type of model is generally used for these cases. The local and average values of the mass transfer coefficient can be predicted from such a model. It is important to note the dependence on the diffusion coefficient. In the case of solid dissolution the dependence is to the power of 2/3, whereas in the gas absorption case it is 1/2.

- An important application of the convective transport equations is in the simulation of the laminar-flow reactor. A 2D model shows that the Damköhler number and Péclet number (B parameter) are the key dimensionless groups which influence the reactor performance.

- The laminar flow model can be simplified in some cases where the radial diffusion term is small. This leads to a model called the pure convection model. If the velocity variation is neglected the reactor can be modeled as a plug-flow model. These two models provide two limiting cases for the complete 2D laminar reactor model.

- Mesoscopic models can be obtained by cross-sectional averaging of the differential model. The mesoscopic model for a laminar-flow reactor leads to a formulation where both the area average and the cup mixing averaging appear in the model, and an additional closure relating these is, therefore, needed. A widely used concept for this closure is the concept of dispersion. This leads to the dispersion models for reactors. Dispersion models can be viewed as a correction to the plug-flow model and have been used in many reactor applications as well.

- The PDEPE software studied in the last chapter is a useful tool for convective mass transport problems as well. The usefulness of this for a number of test problems is shown in the text.

Problems

1. **Stretching transformation.** Consider the differential equation given by Eq. (12.2). Use a coordinate transformation $\zeta = z^* Pe^a$, where a is some index to be chosen suitably. Show that in the transformed equation there are no free parameters in the convection term and the radial conduction term if the index a is chosen as one. Also verify that the axial conduction term has a leading coefficient of $1/Pe$. Hence justify the neglect of the axial conduction term if Pe is, say, greater than 10.

2. **Eigenfunctions of the Graetz problem.** Show that the differential equation for F given in the text can also be written as

$$\xi \frac{d^2 F}{d\xi^2} + \frac{dF}{d\xi} + \lambda^2 \xi (1 - \xi^2) F = 0 \tag{12.44}$$

Show that this belongs to a class of Sturm–Liouville problem. Verify that the F functions are orthogonal with a weighting factor of $\xi(1 - \xi^2)$. This permits the calculation of the series coefficients by the integrals shown in the text.

The solution can be expressed in terms of the hypergeometric functions shown in the text, but some transformations are needed in order to reduce it to a standard Kummer equation. Mathematically motivated students may wish to pursue all the details leading to the hypergeometric functions.

3. **Eigenvalues and solutions for laminar flow heat transfer with CHEBFUN.** This exercise demonstrates how to find the eigenfunctions using CHEBFUN. The procedure is the same as that for the transient conduction prolbem. The first step is to declare ξ as a CHEBFUN and write Eq. (12.44) as a CHEBOP operator with an LHS and RHS such that the RHS is the term with λ^2. Please fill in the code details.

Now we can use the *eig* funciton in MATLAB directly, but in our case this is an overloaded operator and computes the eigenvalues of the differential equation rather than for a set of of algebraic equations. But the syntax is the same. Thus [F, L] = eig (LHS, RHS) gives the first six eigenvalues and the corresponding eigenfunctions. These functions can be manipulated in a symbolic manner.

Apply this to the Graetz problem to find the eigenfunctions and eigenvalues.

The series coefficient and the complete solution can be coded, and the method provides a simple and efficient method to solve convective transport problems in internal flow.

4. **Constant heat-flux analysis for pipe flow.** Show all the steps leading to (12.12) and verify that the Nusselt number has a value of 48/11 for this case.

5. **Constant heat-flux analysis for flow in a channel.** Consider laminar flow between two parallel plates. The plates are electrically heated to give a uniform inward wall flux $-q_w$.

Set a mesoscopic model to find the cup mixing average temperature in the system as a function of the flow direction. This will be needed for the temperature profile calculations in the asymptotic region.

Now propose that the temperature is the sum of a linear function of flow direction and some unknown function of the cross-flow direction. Use this in the governing differential equation and solve for the temperature profile in the asymptotic region.

Show that the Nusselt number is 140/17.

6. Repeat the analysis if only one plate is exposed to the constant flux while the other plate is kept insulated. What is the value of the Nusselt number for this case?

7. Show all the steps leading to Eq. (12.15).

Integrate once, using the substitution $p = d\theta/d\eta$.

Now integrate a second time to get the temperature distribution given by Eq. (12.16) in terms of the incomplete gamma function.

Verify the expression for the Nusselt number given by (12.18).

8. Consider a liquid flowing down a vertical wall at a rate of $1 \times 10^{-5} \, \text{m}^2/\text{s}$ per meter unit width. Find the concentration at a height 25 cm below the entrance for a dissolving wall such as a wall coated with benzoic acid as a function of perpendicular distance from the wall. Also find the local mass transfer coefficient.

Find the average mass transfer coefficient for a wall of height 50 cm. Assume $D = 2 \times 10^{-9} \, \text{m}^2/\text{s}$, and for other properties take those of water. Use $C_{A,s} = 20 \, \text{mol/m}^3$ and the physical properties of water.

9. Repeat the analysis if instead a gas such as oxygen is being absorbed into the liquid film.

10. A pipe of diameter 2 cm with an oil flow rate of 0.02 kg/s heated by a wall temperature of 400 K. The inlet temperature is 300 K.

Find and plot the radial temperature profile at a distance of 5 cm from the entrance.

Find and plot the radial temperature profile at a distance of 100 cm from the entrance.

If an outlet temperature of 380 K is needed, what is the length of the heat exchanger needed? Use an average Nusselt-number value of 3.66.

The following values for the physical properties are applicable: $\nu = 4 \times 10^{-5}$, $c_p = 2120$, $k = 0.14$, and $\rho = 1000$. All are in SI units.

11. Consider the oil flowing in a pipe in Problem 10. The inlet temperature is 300 K and the pipe is now heated electrically at a rate of 76 W/m². Plot the wall temperature, the center temperature, and the bulk temperature as a function of position using the asymptotic solution for large length.

Verify that the cup mixing temperature in dimensionless units is equal to 4ζ.

12. **The effect of viscous dissipation: the Brinkman problem.** Heat transfer in laminar flow with internal generation of heat is called the Brinkman problem. The additional dimensionless group needed here is the Brinkman number. Numerical solutions can be readily obtained using MATLAB functions or other numerical methods. Set up and solve the problem, for example using PDEPE. Also set this up using CHEBFUN as an eigenvalue problem, and solve by the method of series solution.

13. Derive Eq. (12.34) leading to the exponential integral solution of the segregated flow model.

14. Find the exit concentration and the conversion for a laminar-flow reactor under the following conditions using the segregated model: radius 1 cm, length 500 cm, mass flow rate 0.1 kg/s, density 1000 kg/m^3, and first-order rate constant $k = 0.1\,\text{s}^{-1}$.

   Also model the reactor (a) as a plug-flow reactor and (b) as a completely backmixed reactor.

15. **A laminar reactor with a power-law non-Newtonian fluid.** Extend the analysis of the laminar-flow reactor in Section 12.3 for a power-law fluid. Perform some computations using the PDEPE solver, and show how the power-law index affects the conversion in the reactor.

16. **Mass transfer in oscillating flow: a case-study problem.** Mass transfer can be enhanced by flow oscillation. Your goal in this study is to review the literature and, in particular, examine the following model problem, *viz.*, mass transfer in pipe flow with a dissolving wall. Assume that the flow oscillation is caused by a sinusoidal variation of the inlet pressure. Perform a scaling analysis based on the key time constants to find conditions under which the enhancement is likely and also investigate the problem using numerical tools. How does the time-average Sherwood number vary with the dimensionless amplitude of oscillation?

13 Coupled transport problems

Learning objectives

You will learn a lot from this chapter, mainly from a problem-solving basis. The concepts and the problems examined here are listed below.

- One-way and two-way coupling problems.
- The Boussines approximation for buoyancy-driven flows.
- Velocity profiles in natural convection in a fluid contained between two plates: the role of the Grashof number and the Rayleigh number.
- The velocity profile in a pipe due to a radial temperature gradient; its effect on the rate of heat transfer.
- The temperature profile due to viscous heat generation; its effect on the velocity profile.
- Wet-bulb and dry-bulb temperatures; the role of the Lewis number.
- Augmentation of the heat transfer rate due to mass transfer.
- A model for condensation of a mixture of vapor and an inert gas.
- A model for solution of temperature and concentration profiles in a porous catalyst; the key dimensionless groups needed to characterize this problem. Estimation of the magnitude of the temperature rise in a porous catalyst.

Mathematical prerequisites

No additional prerequisites are needed for this chapter.

The study of all of the three transport mechanisms individually was presented in the earlier chapters. The goal of this chapter is to analyze problems where two or more modes of transport have to be analyzed together. Such problems are generally referred to as coupled problems. In industrial practice coupled problems are more common, and hence it is important to get a good understanding of the modeling, solutions, and computations of such problems. Therefore I have devoted a whole chapter to this discussion, and will illustrate the methodology with a number of examples.

It is useful to distinguish between one-way coupling and two-way coupling. In many problems, the information from one transport problem is needed for the solution of the other transport problem, but the solution for the first transport problem can be done separately without consideration of the second problem. Such problems are referred to as one-way coupling. In a sense, we have already seen one example of this, *viz.*, convective heat and mass transfer. Information on the velocity profile is obtained from momentum transfer considerations only and then used in heat and mass transfer. The velocity profile is in turn assumed not to be affected by heat or mass transfer. The assumption will not be valid if there is a strong variation of the fluid viscosity with temperature, and in such cases the velocity profile is distorted by heat transfer. Hence this becomes a two-way coupled problem. One of the goals of this chapter is to look at some two-way problems, in addition to some additional problems involving one-way coupling.

An important example of a coupled problem is in natural convection heat transfer. This is an interesting problem in the sense that it can be treated as a one-way coupling as an approximation or as a two-way coupled problem. If a temperature profile is assumed (e.g., in a closed enclosure) the effect of the temperature-induced buoyancy on flow can be calculated as a one-way coupled problem. If the geometry is of external flow type, a boundary layer develops near the plate for both heat and mass transfer, and this becomes a two-way coupled problems. Here we study the one-way coupled problem in a closed enclosure, and the external-flow two-way coupled problem is taken up subsequently in Chapter 21.

Another useful problem to study is convective heat transfer with viscous generation of heat. This is called the Brinkman problem, and we have already seen a numerical solution for a pipe geometry (Exercise 12, Chapter 12). If the velocity profile is assumed (calculated from a purely flow problem), the problem can be solved as a one-way coupled problem; the heat-generation term is calculated from the velocity and simply used as a source term in the heat balance. The temperature rise can, however, lead to viscosity variation, leading to a two-way coupling. Two-way coupled problems require simultaneous solution of heat and momentum equations and are more difficult.

Mass transport problems with simultaneous heat transfer can lead to quite a number of interesting coupling effects. The most common applications are in the areas of humidification, condensation of a vapor, and evaporation of a liquid. Diffusion in a porous catalyst with an exothermic reaction is another example in which there are many complex effects such as multiple solutions, etc. These coupled effects will be studied in some detail, and useful examples will be provided to illustrate the methodology.

13.1 Modes of coupling

13.1.1 One-way coupling

Problems where the momentum transport can be solved and used in heat and mass transport are the most common examples of one-way coupling. Velocity profiles are always needed to characterize the convective transport of heat and mass, and hence the momentum transport is solved for velocity profiles and used in heat and mass transfer problems. Velocity profiles are also needed to estimate the viscous dissipation term in the energy equation. Such problems can be solved by one-way coupling.

13.1.2 Two-way coupling

Two-way coupling refers to problems where the momentum transport is affected by heat or mass transport and hence cannot be solved in isolation. Examples include natural convection problems, high-mass-flux problems, strong changes in viscosity due to temperature, etc. In addition, mass transfer may be coupled to heat transfer owing to the generation of heat due to reaction. Industrially important problems usually are two- or three-way coupled, and an example is the polymerization reactor alluded to in Chapter 1.

13.2 Natural convection problems

. . . .

Problems involving natural convection are generally more complex since both momentum and heat transfer have to be solved together. The feedback needed for solving such problems is illustrated schematically in Fig. 13.1. This is in general a two-way coupled problem. The temperature field is needed in order to calculate the buoyancy force, which is then used to generate the flow field. The flow-field solution generates the convection terms from which to calculate the temperature field.

Now we take a simple problem where the solution to the heat transfer problem is assumed and used to solve the momentum transfer problem, making it a one-way coupled problem. This example also gives us an appreciation of dimensionless grouping of variables for natural convection problems. A two-way coupled natural convention problem will be studied later in Section 18.4.

13.2.1 Natural convection between two vertical plates

The problem analyzed is shown schematically in Fig. 13.2. This is the simplest problem in natural convection. A temperature field is assumed, thereby allowing the problems to be decoupled.

The key issues we address are the following.

- What are the model equations after making some simplifying assumptions?
- What are the key dimensionless groups?

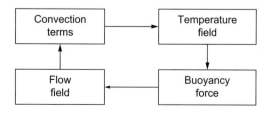

Figure 13.1 An illustration of coupling for natural convection problems.

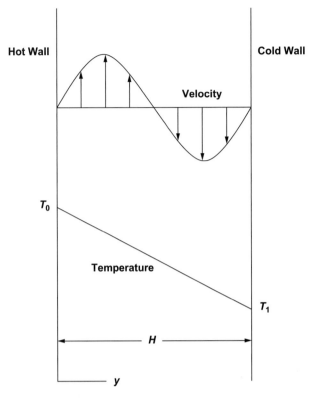

Figure 13.2 Natural convection between two vertical plates. A sketch of the temperature and velocity profiles.

- What is the order of velocity by scaling arguments?
- What is the shape of the velocity profile?

The simplifying assumptions used at the first level of modeling are as follows. (i) Inertial terms are ignored in the Navier–Stokes (N–S) equations. (ii) The pressure profile in the vertical direction is hydrostatic, being based on a mean density value. (iii) There is a linear temperature profile across the two plates. This uncouples the heat transport and reduces the problem to a one-way coupling.

Model equations

The governing equation for z-momentum with the inertia term excluded is

$$-\frac{dp}{dz} + \rho g_z + \mu \frac{d^2 v}{dy^2} = 0 \tag{13.1}$$

Here z is the vertical direction and y represents the coordinate across the gap width. The notation v is used for v_z for simplicity.

Since $g_z = -g$, we have

$$\mu \frac{d^2 v}{dy^2} = \frac{dp}{dz} + \rho g \tag{13.2}$$

To account for density variation with temperature we expand the density in a Taylor series as a function of temperature:

$$\rho = \rho_m + \left(\frac{\partial \rho}{\partial T}\right)_m (T - T_m) \tag{13.3}$$

where the temperature T_m is an appropriately defined mean temperature in the system and ρ_m is the fluid density evaluated at this mean temperature.

The coefficient of volume expansion β is defined as follows:

$$\beta = -\frac{1}{\rho}\left(\frac{\partial \rho}{\partial T}\right)_P \tag{13.4}$$

and represents the change in density with temperature.

In general β is a function of temperature and decreases with an increase in temperature. However, water shows an anomalous behavior and the β value becomes negative below 4 °C. At this point the density of water reaches a maximum. Typical values for some fluids are as follows:

water: 0.000214 (1/K) at 20 °C,
ethyl alcohol: 0.00109 (1/K), and
oil: 0.00070 (1/K).

For ideal gases

$$\beta = \frac{1}{T}$$

which follows from the ideal-gas law.

The coefficient of volume expansion is a function of temperature, but in the present case it will be evaluated at the mean temperature as an approximation.

Hence the density expansion Eq. (13.3) can be expressed as

$$\rho = \rho_m - \rho_m \beta_m (T - T_m) \tag{13.5}$$

where β_m is the value of β evaluated at the mean temperature.

Hence Eq. (13.2) can be expressed as

$$\mu \frac{d^2 v}{dy^2} = \frac{dp}{dz} + \rho_m g - \rho_m \beta_m g (T - T_m) \tag{13.6}$$

The pressure variation can be assumed to be hydrostatic, being based on the mean density. Hence we can set

$$\frac{dp}{dz} + \rho_m g = 0 \tag{13.7}$$

Note that a precise value was not assigned to the mean temperature earlier, and the above equation gives us the flexibility to define it and simplify the problem. It turns out here that the mean temperature is simply the arithmetic average of the plate temperatures. The governing equation for the velocity then simplifies to

$$\mu \frac{d^2 v}{dy^2} = -\rho_m \beta g (T - T_m) \tag{13.8}$$

The heat equation is uncoupled by assuming a linear temperature profile:

$$T = T_0 + (T_1 - T_0)\frac{y}{H} \tag{13.9}$$

Dimensionless form

Upon introducing a dimensionless temperature,

$$\theta = \frac{T - T_1}{T_0 - T_1} \tag{13.10}$$

and a dimensionless distance,

$$\xi = y/H \tag{13.11}$$

we have

$$\theta = 1 - \xi \tag{13.12}$$

and the mean dimensionless temperature is 1/2. Hence

$$(T - T_m) = (T_0 - T_1)(\theta - 1/2) \tag{13.13}$$

Hence the y-momentum is

$$\mu \frac{d^2 v}{dy^2} = -\rho_m \beta_m g (T_0 - T_1)(\theta - 1/2) \tag{13.14}$$

or, since $\theta = 1 - \xi$,

$$\frac{d^2 v}{d\xi^2} = \frac{\rho_m \beta_m g H^2 (T_0 - T_1)}{\mu}(\xi - 1/2) \tag{13.15}$$

The leading term appearing on the RHS has the units of velocity and is defined as v_C, the characteristic velocity for natural convection flow,

$$v_c = \rho_m \beta_m g H^2 (T_0 - T_1)/\mu \tag{13.16}$$

This is the representative velocity generated by the temperature gradients and provides an order-of-magnitude estimate of the velocity generated by natural convection.

The dimensionless velocity v^* can therefore be defined as v/v_c, and the differential equation takes a bare-bones form:

$$\frac{d^2 v^*}{d\xi^{*2}} = \xi - 1/2 \tag{13.17}$$

which is the dimensionless representation of the problem.

Dimensionless groups

It is also customary to define a dimensionless groups similar to a Reynolds number based on the characteristic velocity v_c defined above. This dimensionless group is known as the Grashof number and represents the ratio of buoyancy effects to viscous effects in flows dominated by natural convection:

$$Gr = \frac{v_c H \rho}{\mu} = \frac{\rho_m^2 g \beta_m H^3 (T_0 - T_1)}{\mu^2} \tag{13.18}$$

Velocity profile

Equation (13.17) may be integrated twice to get the velocity profile with no-slip conditions at the two ends. A cubic profile is obtained:

$$v^* = \xi^3/6 - \xi^2/4 + \xi/12 \tag{13.19}$$

The average velocity is computed as the integral from 0 to 0.5 of the above equation. This is a measure of the net upflow or circulation Q (per unit width into the plane). This is also equal to that net downflow near the colder region obtained by integrating the velocity from 0.5 to 1. The result is

$$Q = \int_0^{H/2} v \, dy = H v_{\mathrm{c}}/384 \tag{13.20}$$

The maximum or minimum velocity is given as

$$v_{\max} = 0.032 v_{\mathrm{c}}$$

The points at which the maximum and minimum velocities occur are obtained by finding the ξ at which $dv^*/d\xi$ is equal to zero. This is at $1/2 \pm 1/\sqrt{12}$. These locations are 0.2113 (maximum) and 0.7887 (minimum) in dimensionless distance units.

13.2.2   Natural convection over a vertical plate

This problem is usually solved as a two-way coupled problem since both the temperature and the velocity profiles are of the boundary-layer type. The natural convection of a vertical wall (assuming laminar conditions) is amenable to analytical solution. Two limiting cases are well studied in the literature: (i) constant-wall-temperature and (ii) constant-wall-flux conditions. The first case was studied by Ostrach (1953), who developed an expression for the local Nusselt number and the second problem was studied by Sparrow and Gregg (1956).

The detailed solution is shown later in Section 18.4, and here we present the key results which are useful for finding the heat transfer rate from a vertical wall. The heat transfer data are correlated in terms of the dimensionless parameter Nu_x defined as hx/k, with h being the local value of the heat transfer coefficient at location x:

$$Nu_x = \left(\frac{Gr_x}{4}\right)^{1/4} f(Pr)$$

where Gr_x is the local Grashof number and $f(Pr)$ is a function that corrects for the effect of the Prandtl number.

This correlation is applicable under laminar-flow conditions for constant wall temperature. The flow shifts from a laminar to a transition region if $GrPr$ is above 10^9. The product of Gr and Pr is known as Rayleigh number. The linear stability analysis discussed in Sections 16.4 and 16.5 is useful to identify the onset of transition. Flow becomes fully turbulent if the Rayleigh number is greater that 10^{12}. A wide range of data was collected and correlated by Churchill and Chu (1975), and their equation for the Nusselt number is widely used for calculation of natural convection heat-transfer rates in turbulent conditions. A book completely dedicated to buoyancy-driven flow is that by Gebhart et al. (1988), which is an excellent starting source for further study of this field.

Aqueous systems

Buoyancy effects can be complex in aqueous systems and solutions. Owing to the anomalous density variation of water, the natural convection can be complex at low temperatures. Buoyancy reversal, bidirectional velocity distribution, and total convective inversion can be expected as a consequence, as discussed in the book by Gebhart et al. (1988).

13.2.3   Natural convection: concentration effects

Concentration variation can also lead to natural convection flows.

Similarly to the analysis for heat transfer, we can define a coefficient

$$\beta_c = -\frac{1}{\rho_m}\left(\frac{\partial \rho}{\partial \omega_A}\right)_T \tag{13.21}$$

which is the mass transfer analogue of β, the coefficient of thermal expansion.

The additional term which needs to be added to the momentum equation is $\rho_m g \beta_c \times (\omega - \bar{\omega}_m)$.

The analysis for simple geometries can be done in a similar fashion to heat transfer. The key dimensionless group which arises as a result of such models is the Grashof number for diffusion:

$$Gr_m = \frac{g\beta_c \, \Delta\omega \, L^3}{\nu^2} \tag{13.22}$$

The concentration-induced buoyancy is an order of magnitude smaller than the thermally induced buoyancy. We can also use a composite value of the Grashof number combining the two effects. Such an approach is common in empirical correlations for Nu and Sh. In some cases, the free-convection heat transfer provides a source for forced-convection mass transfer. This represents an example of coupling of all three transport mechanisms.

Now we examine problems in heat transfer with viscous generation of heat.

13.3   Heat transfer due to viscous dissipation

The viscous generation of heat can be viewed as a source term of the form $\tilde{\tau} : \nabla v$ as shown in Chapter 7, Eq. (7.12), for a Newtonian fluid. If the velocity profile is calculated assuming a constant viscosity then this term is known and simply added as a source term to the temperature equation. This is then a one-way coupled problem. We provide two examples here as representative problems. The effect of variable viscosity is treated in Chapter 14 in the context of perturbation methods.

13.3.1   Viscous dissipation in plane Couette flow

Consider the problem where a highly viscous fluid is contained between two parallel plates of gap width H with temperature of zero at the top and bottom. The top plate is moving with a velocity of V, and this generates heat in the system due to viscous dissipation. The problem, shown schematically in Fig. 13.3, is to develop an equation for the temperature distribution in the system.

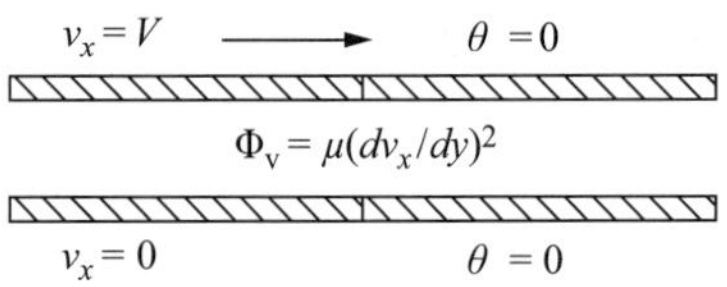

Figure 13.3 Shear flow with viscous dissipation.

If we assume a constant viscosity then the velocity profile is given as a linear function of y:

$$v_x = V\frac{y}{H}$$

The governing equation for the temperature profile simplifies to

$$k\frac{d^2T}{dy^2} = -\Phi_v$$

where Φ_v is the heat generation due to viscous generation and k is the thermal conductivity. For the simplified velocity profile for Couette flow, this term is

$$\Phi_v = \mu\left(\frac{dv_x}{dy}\right)^2 = \mu(V/H)^2$$

Hence

$$k\frac{d^2T}{dy^2} = -\mu(V/H)^2$$

If the top and bottom plates are maintained at temperature T_0 the integrated form of the equation is

$$\frac{T-T_0}{T_0} = \frac{1}{2}Br\left[\frac{y}{H} - \left(\frac{y}{H}\right)^2\right]$$

where the dimensionless group is the Brinkman number defined as

$$Br = \frac{\mu V^2}{kT_0} \tag{13.23}$$

The maximum temperature is seen to be $Br/4$, and provides us with a direct estimate of the heat generation due to viscous effects. Thus viscous effects in the temperature equation may be neglected if $Br \ll 1$. The Brinkman number may be interpreted as the relative ratio of heat generation to heat conduction.

If Br is large then there is a large increase in temperature due to viscous dissipation. This will change the viscosity and the linear velocity profile no longer holds. The problem has to be treated as a coupled problem, and an elegant method of solution is the regular perturbation analysis described in Chapter 14.

13.3.2   Laminar heat transfer with dissipation: the Brinkman problem

Here we revisit the problem of heat transfer in laminar flow but now include the effect of viscous dissipation. A constant-wall-temperature case is analyzed here. Also we assume constant viscosity, leading to a one-way coupling. This is known as the classical Brinkman problem The viscous dissipation adds an extra heat-source term to the governing equation:

$$\Phi_v = \mu\left(\frac{dv_z}{dr}\right)^2$$

For a parabolic profile this term is equal to

$$\Phi_v = 16\mu\langle v\rangle^2 r^2/R^4$$

This additional term is now added to the heat equation and the resulting equation is presented below in terms of dimensionless groups:

$$(1 - \xi^2)\frac{\partial \theta}{\partial \zeta} = \frac{1}{\xi}\frac{\partial}{\partial \xi}\left(\xi \frac{\partial \theta}{\partial \xi}\right) + 16 Br\xi^2 \tag{13.24}$$

where ζ is the dimensionless axial length defined as z^*/Pe as before. This is often referred to as a contracted (dimensionless) axial distance, as noted earlier. Note that the parameter Pe has been absorbed into the contracted axial distance ζ.

The additional dimensionless group is the Brinkman number defined as

$$Br = \frac{\mu \langle v \rangle^2}{k(T_0 - T_w)} \tag{13.25}$$

which can be interpreted as

$$Br = \frac{\text{Heat generated by viscous dissipation}}{\text{Heat transport by conduction}}$$

Thus the parameter dependence can be written simply as

$$\theta = \theta(\zeta, \xi; Br) \tag{13.26}$$

The associated boundary conditions for a constant-wall-temperature case are as follows:

$$\text{entrance } \zeta = 0, \qquad \theta = 1$$
$$\text{wall } \xi = 1, \qquad \theta = 0$$
$$\text{center } \xi = 0, \qquad \partial\theta/\partial\xi = 0$$

These are the same as for the classical Graetz problem. The governing PDE belongs to a non-homogeneous type, and hence direct solution using separation of variables is not

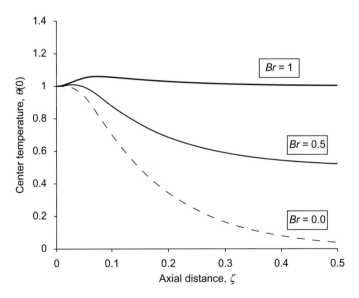

Figure 13.4 Temperature profiles in heat transfer in laminar flow with viscous generation of heat.

possible. A partial solution has to be found, and the remaining part of the PDE can be reduced to a homogeneous type and solved analytically. The details are left as an exercise. Numerical solution using the PDEPE solver is also useful and sample results generated using this method are shown in Fig. 13.4 for the center temperature. Note that the center temperature does not decay to zero, the wall temperature, unlike in the Graetz problem. The radial conduction balances the viscous generation, setting up a pseudo-steady-state type of profile.

13.4 Laminar heat transfer: the effect of viscosity variations

In this section we revisit the problem of heat transfer in (forced-convection) laminar flow, but now we explore the effect of variation in viscosity with temperature. The effects of viscosity variation on the velocity profiles are different for heating and cooling. Here we present a simple example assuming a prescribed temperature distribution, rather than providing a complete solution. In order to appreciate the finer details, the qualitative aspects are discussed first.

Case A: heating
Here the walls are hotter and hence μ_w is less than μ_b (for liquids). Note that for gases the opposite trend will hold. Liquid can flow fairly easily near the wall now, and the velocity profile is flattened. The center velocity is smaller than that for a constant-viscosity case for the same flow rate. The convective transport is increased compared with the conductive transport in the wall region. This corresponds to an effective increase in the local Pe, and a corresponding increase in the heat transfer coefficient. An empirical correction factor to the Nusselt number $(\mu_b/\mu_w)^{0.14}$ known as the Sieder–Tate factor is often used.

Case B: cooling
Here the walls are cooler and hence μ_w is greater than μ_b for liquids. Fluid motion near the wall is somewhat retarded, leading to a bulge in the velocity profile in the center. The center velocity is larger than that for the isothermal case. The convective transport near the wall is decreased compared with the conductive transport. This causes an effective decrease in the heat transfer coefficient, which is again captured by the Sieder–Tate factor.

The velocity profiles for the two cases are sketched qualitatively in Fig. 13.5. For more quantitative analysis, numerical simulation is needed. We now present a simplified model to calculate the velocity profiles.

A model for variable viscosity
A simple model for the prediction of the velocity profile in a tube under varying viscosity conditions is presented. The starting equation is the stress profile, which remains the same for variable viscosity since it is only a force balance. The stress is related to the pressure gradient parameter G, assuming a unidirectional flow locally at any axial position. Hence

$$\tau_{rz} = -G\frac{r}{2}$$

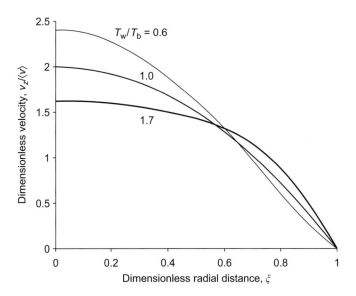

Figure 13.5 A plot of velocity profiles for variable viscosity values.

where G is the pressure gradient. Assuming a Newtonian fluid,

$$\mu(r)\frac{dv_z}{dr} = -G\frac{r}{2}$$

The integrated form is obtained by treating μ as a function of r:

$$v_z = G\int_r^R \frac{s\,ds}{2\mu(s)} = GI(r) \tag{13.27}$$

where s is a dummy variable for integration and $I(r)$ is defined as

$$I(r) = \int_r^R \frac{s\,ds}{2\mu(s)}$$

Here G is the local pressure gradient whose value can be obtained from the mass flux rate. The pressure gradient varies with z in this case, but a local value is used for flow analysis and G is treated as a constant.

If just the dimensionless velocity is needed, then we can write the following expression for the average velocity (mass flow rate) and eliminate the pressure-gradient term:

$$\dot{m} = \langle v \rangle \rho \pi R^2 = \int_0^R 2\pi r\rho v_z\,dr = -G\rho \int_0^R 2\pi rI(r)dr \tag{13.28}$$

On taking the ratio of Eqs. (13.27) and (13.28) we have the following expression for the dimensionless velocity:

$$\frac{v}{\langle v \rangle} = \frac{R^2 I(r)}{\int_0^R 2rI(r)dr}$$

and the pressure gradient need not be explicitly calculated. Local values of the pressure gradient can also be calculated from Eq. (13.28) if needed.

The calculated v_z can then be used in the thermal model, and the problem can be solved by a sequential iterative manner. Models of this type are typically applied in simulation of polymerization reactors, as indicated in the book by Neumann (1987). The alteration of the velocity profile can affect the performance and even the operability of such reactors. Generally speaking, a flattened profile is closer to plug flow and will improve the performance, whereas an elongated profile will hurt the performance. An elongated profile is typical of strong cooling or significant polymerization near the wall.

In this discussion we have assumed that v_r is small enough that it does not affect the calculation of v_z, which is based on the unidirectional-flow assumption. In a sense this is equivalent to the lubrication approximation discussed in Section 6.8. It does not mean that v_r is zero. Furthermore, the radial convection term $v_r \, \partial T / \partial r$ in the temperature equation can have a significant and cumulative effect on the temperature profile and should be included in the heat balance at the next level of modeling. The local balance of v_r is obtained from continuity:

$$ v_r = -\frac{1}{r\rho} \int_0^r \rho s \frac{\partial v_z}{\partial z} \, ds $$

and can be used in the temperature and concentration equations. These equations can then be solved by suitable discretization in the r-direction and axial marching in the z-direction. More details can be found in the paper by McLaughlin *et al.* (1986). Those authors also discuss the role of radial velocity in affecting the temperature profiles.

13.5 Simultaneous heat and mass transfer: evaporation

13.5.1 Dry- and wet-bulb temperatures

Wet-bulb and dry-bulb thermometers provide a means to measure the moisture content of air. In the simplest form a wet-bulb thermometer is an ordinary thermometer covered with a wetted wick. The evaporation of water causes the wet bulb to cool, and the temperature recorded by this thermometer is different from that measured by one with a dry bulb. In this section we model the system in terms of simultaneous heat- and mass-transfer considerations and show how the temperature difference between the dry and wet bulbs is related to the moisture content of the air. The system modeled is shown schematically in Fig. 13.6.

The mass transfer rate (liquid to gas) is calculated using a mass transfer coefficient, k_m°, assuming a low-flux model. The mole fraction of vapor corresponds to the saturation

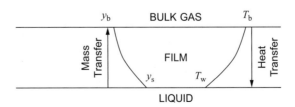

Figure 13.6 A schematic representation of heat and mass transfer from an evaporating surface. Note that the profiles would be modeled as linear functions for the low-flux mass-transfer case.

condition at the surface denoted by y_s and the mole fraction in the gas is y_b. Hence the mass transfer or evaporation rate is

$$N_A = k_m^o C(y_s - y_b)$$

Here y_s is the equilibrium mole fraction at the interface, which depends on T_s. Note $y_s = P_{vap}(T_s)/P$, where P_{vap} is the vapor presssure and P is the total pressure.

Heat transfer from gas at a temperature of T_b to the liquid–gas interface at T_s is modeled using a heat transfer coefficient h,

$$q = h(T_b - T_s)$$

where q is the heat flux towards the interface. Assume an adiabatic process such that the latent heat of evaporation ΔH_v comes from the heat transfer in the gas film. Then

$$q = N_A \, \Delta H_v$$

On equating the two expressions, we get an expression for the surface temperature:

$$k_m^o C(y_s - y_b)\Delta H_v = h(T_b - T_s) \tag{13.29}$$

The heat transfer coefficient and mass transfer coefficients are related by the film model. We have $k_m = D/\delta_f$, while $h = k/\delta_t$, where δ_f and δ_t are the film thicknesses for mass and heat transfer, respectively. From a scaling analysis of boundary-layer equations for heat and mass transfer theory (which will be discussed later) we can use the following relations:

$$\delta_m \propto Sc^{-1/3} \text{ and } \delta_t \propto Pr^{-1/3}$$

Hence

$$\frac{h}{k_m} = \frac{k}{D}\left(\frac{Sc}{Pr}\right)^{-1/3}$$

Also $k/D = \rho c_p (Sc/Pr)$. Hence

$$\boxed{\frac{h}{k_m} = \rho c_p \left(\frac{Sc}{Pr}\right)^{2/3}} \tag{13.30}$$

The ratio Sc/Pr is referred to as the Lewis number Le:

$$Le = \frac{Sc}{Pr} = \frac{\alpha}{D} = \frac{k}{\rho c_p D} = \frac{\text{Thermal diffusivity}}{\text{Mass diffusivity}}$$

Therefore

$$\frac{h}{k_m} = \rho c_p Le^{2/3}$$

On using this in Eq. (13.29) and rearranging, we have

$$(y_s - y_b) = (T_b - T_s)\frac{\rho c_p}{C \, \Delta H_v} Le^{2/3}$$

since ρ/C is equal to the molecular weight of the gas and since $M_W c_p$ is equal to C_p, the molar specific heat, this equation can be written as

$$(y_s - y_b) = (T_b - T_s)\frac{C_p}{\Delta H_v} Le^{2/3} \tag{13.31}$$

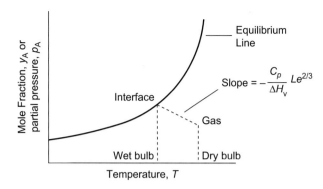

Figure 13.7 A representation of a humidification process in terms of an equilibrium curve and operating curve.

This permits us to calculate the humidity in the vapor phase from the temperature of the evaporating liquid. If the wet-bulb and dry-bulb temperatures are known, the calculation of the moisture content is straightforward. If the unknown is the wet-bulb temperature, an iterative calculation is needed, since y_s is based on the wet-bulb temperature T_s, which is not yet known.

The system can be plotted in terms of an operating line and an equilibrium line, as is typically done in mass transfer operations. An illustrative plot is shown in Fig. 13.7. We note that the Lewis number is an important parameter in determining the slope of the operating line.

Problems of this type are of importance in a number of situations, including fog formation in environmental applications.

A simple numerical example is shown below to get an appreciation for the numbers.

Example 13.1.

Find the gas composition and relative humidity if the (dry) gas temperature is 31.8 °C and the wet-bulb temperature is 26.8 °C. The total pressure is 1.01 bar.

Average properties at 300 K may be used.

The following values of physical properties evaluated at 300 K are used: $\rho = 1.1769 \text{ kg/m}^3$, $c_p = 1006 \text{ J/kg} \cdot \text{K}$; $\nu = 1.57 \times 10^{-5} \text{ m}^2/\text{s}$, and $Pr = 0.713$.

The vapor pressure of water at 26.8 °C is computed using the Antoine equation and is equal to 26.35 mm Hg. Hence y_s has a value of 26.34/760 or 0.0347. Using Eq. (13.31), y_b is found to be 0.0313.

The results are normally reported in terms of the relative humidity, which is defined as follows:

$$\text{Relative humidity} = \frac{\text{Mole fraction in air}}{\text{Mole fraction if the air were saturated}}$$

In the above example, if the air were saturated at the temperature of 31.8 °C, it would exert a vapor pressure of 35.17 mm Hg. Hence the mole fraction would be 35.17/760 or 0.0463, while the actual value is 0.0313. The relative humidity is therefore 0.0313/0.0463 or 67.6%.

13.5.2   Evaporative or sweat cooling

A related application of the concept of the wet-bulb temperature is evaporative cooling, which is also called sweat cooling. This is an alternative to air conditioners based on a compression–refrigeration cycle and is suitable for dry climates where the relative humidity of air is rather low. Thus it finds extensive applications in the western USA and places like New Delhi in India, for example. Evaporative cooling is not suitable for Mumbai, where the relative humidity is in the range of 85%.

The air is cooled to a temperature in between the dry-bulb and wet-bulb temperatures which depends on the efficiency of the cooler. The efficiency depends on the air flow rate and other design parameters.

13.6   Simultaneous heat and mass transfer: condensation

13.6.1   Condensation of a vapor in the presence of a non-condensible gas

Consider a liquid condensing on a cold wall. In the presence of a non-condensing gas, concentration and temperature profiles are established near the gas film adjoining the liquid. This is shown schematically in Fig. 13.8. The rate of condensation depends on the mass transfer rate across the gas film and on the heat that can be transported across the gas film and through the condensed liquid film. Simultaneous transport considerations are needed.

The mass flux from the bulk gas to the interface is given by

$$N_A = \frac{CD_A}{\delta_f} \ln\left(\frac{1 - y_{A,i}}{1 - y_{A,g}}\right) \tag{13.32}$$

from our earlier analysis for mass transport in a stagnant film. Here we use a high-flux model since the concentration ranges are usually high for condensing vapors. It is more convenient to use the mass transfer coefficient k_m° defined as D_A/δ_f and use

$$N_A = Ck_m^\circ \ln\left(\frac{1 - y_{A,i}}{1 - y_{A,g}}\right) \tag{13.33}$$

The mole fraction in the gas phase, $y_{A,g}$, is usually known, while the mole fraction in the gas at the interface, $y_{A,i}$, depends on the vapor pressure at the interface in accordance with

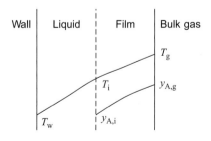

Figure 13.8 Concentration and temperature profiles in the vapor film adjacent to a condensed liquid.

Raoult's law. This in turn depends on the interface temperature, which is unknown, and hence a coupling with heat transport is needed.

The rate of heat transport if there were no significant mass flux is given by Fourier's law using a linear temperature profile:

$$q^\circ = k\frac{T_g - T_i}{\delta_t} = h_g^\circ(T_g - T_i)$$

where h_g° is the gas-side heat transfer coefficient ignoring the effect of mass transfer.

Since there is a convection due to mass transfer the heat transfer rate is augmented by introducing the ϕ parameter as shown in Section 8.4.1:

$$q = h_g^\circ(T_g - T_i)\mathcal{F}_h \qquad (13.34)$$

where $\mathcal{F}_h$ is the augmentation factor for convection. This is given by the heat transfer model for the film including convection (caused by mass transfer) and is calculated as

$$\mathcal{F}_h = \frac{\phi\exp(\phi)}{\exp(\phi) - 1} \qquad (13.35)$$

where ϕ is the blowing factor (a type of Péclet number), which is defined as

$$\phi = \frac{N_A C_{pg,A}}{h_g^\circ}$$

Note the following limiting cases for $\mathcal{F}_h$.

1. At low ϕ the augmentation factor is equal to one. The mass transport in the gas film does not enhance the rate of heat transfer to any significant extent. The conduction in the gas film controls the rate of heat transport, i.e.,

$$q = h_g^\circ(T_g - T_i)$$

2. At high ϕ the augmentation factor is equal to ϕ. The mass transport in the gas film dominates the rate of heat transfer. The rate of heat transport is simply equal to the change in enthalpy on going from T_g to T_i, i.e.,

$$q = N_A C_{pg,A}(T_g - T_i)$$

The general expression given by Eq. (13.35) takes care of the intermediate case between these two limiting scenarios.

The total heat released at the interface can now be calculated. This is equal to the rate of heat transport in the gas film plus the energy released at the interface due to the latent heat of condensation. The latter depends on the rate of mass transfer, by which heat is removed. The total heat released is therefore equal to

$$q^i = q + N_A \Delta H_c \qquad (13.36)$$

where ΔH_c is the latent heat of condensation.

The heat transferred at the interface is then equal to the heat transfer in the liquid to the walls of the condenser:

$$q^i = h_l(T_i - T_w) \qquad (13.37)$$

Hence the interfacial temperature is determined by equating the two equations for q^i:

$$h_g^o(T_g - T_i)\mathcal{F} + N_A \Delta H_c = h_l(T_i - T_w)$$

Usually the wall temperature is known or specified but the interfacial temperature is not known, and the latter has to be solved in an iterative manner. This is illustrated in the example shown later in the text. Before showing this example, it is useful to give some additional comments.

Additional Notes

The analysis does not include the sensible heat change of the condensed liquid from T_i to T_w. An additional term has to be added to account for this. The term can be neglected as a first approximation. More precisely, the calculation of the sensible heat loss requires the calculation of the average temperature of the liquid film. A simple way to account for this is to augment the value of the latent heat of vaporization by adding a correction term as shown below:

$$\Delta H_c^a = \Delta H_c + (3/8)C_{pl}(T_i - T_w)$$

The augmented value, ΔH_c^a, is then used in place of ΔH_c to account for the sensible heat change in the liquid. The factor 3/8 can be viewed as the contribution from the average temperature of the liquid.

Again the wall temperature might not be known and only the temperature of the cooling liquid may be known. In this case Eq. (13.37) is replaced by

$$q^i = U_i(T_i - T_c) \tag{13.38}$$

where U_i is the overall heat transfer coefficient from the interface to the cooling liquid and T_c is the (local) temperature of the coolant at the point under consideration.

Example 13.2.

Water is condensing on a surface at 310 K. The gas mixture has 65% water vapor and is at a temperature of 370 K. The total pressure is 1 atm.

Calculate the rate of condensation.

The data needed are of two types: (i) physical properties and (ii) transport properties or values for the transport coefficients. We use the following values for illustration of the calculation method.

(i) Physical properties.

The specific heat of water vapor is $C_{pg,1} = 37.06$ J/kg·K
The specific heat of liquid water is 75.13 J/kg·K
The heat of condensation is $\Delta H_c = 43\,000$ J/mol
The vapor pressure as a function of temperature is given by the Antoine equation (Eq. (9.15)), and the constants for water are $A = 18.3036$, $B = 3816.44$, and $C = 46.13$ with temperature in °C.

(ii) Transport properties or coefficients. In general these have to obtained by using suitable empirical correlations or detailed CFD modeling. Our purpose is to illustrate the calculation procedure, and hence we use directly the following values for the transport parameters. These values are in the range of expected values of the parameters. Sensitivity analysis to these parameters is provided as an exercise.

The gas-side heat transfer coefficient is $12 \, \text{W/m}^2 \cdot \text{K}$.

The liquid-side heat transfer coefficient is $400 \, \text{W/m}^2 \cdot \text{K}$.

The gas-side mass transfer coefficient is $0.1 \, \text{m/s}$.

Note that the liquid-side mass transfer coefficient is not needed since the liquid is a pure component.

Solution.

The interface temperature is dictated by the relative rates of heat and mass transfer. This information is unknown. Once the interface temperature has been found the problem is essentially solved. Hence the solution procedure amounts to simply solving for the interface temperature in an iterative manner. The steps are outlined here.

1. We start with an assumed interface temperature, T_i. Let $T_i = 314 \, \text{K}$.
2. The vapor pressure at the interface is calculated using the Antoine equation. Note that the units are mm Hg, which needs to be converted to pascals. $P_{vap} = 58.325$ mm Hg. This sets the mole fraction of water at the gas side of the interface:

$$y_i = p_{vap}/P = 58.325/760 = 0.0768$$

3. The mass transfer rate across the gas film can now be calculated using Eq. (13.33). The total concentration needed here is calculated using the ideal-gas law at an average film temperature of $T_f = (T_i + T_g)/2$. Hence

$$C = \frac{P}{R_G T_f} = 35.61 \, \text{mol/m}^3$$

The total concentration based on the bulk gas temperature would be $32.93 \, \text{mol/m}^3$, Hence no significant errors would be expected in using any of these values for the temperature. Now, using Eq. (13.33), we have

$$N_A = 0.2979 \, \text{mol/s} \cdot \text{m}^2$$

4. The heat transfer to the interface can now be calculated using (13.34) and (13.35). The following values are found, and the student should verify them: Péclet factor $\phi = 0.92$, $\mathcal{F}_h = 1.5296$, and finally $q = h(T_g - T_i)\mathcal{F}_h = 1050 \, \text{W/m}^2$.
5. The rate of total heat release at the interface can now be calculated. This is the sum of the rate of heat transfer to the interface plus the heat released by the latent heat of condensation. The latter is proportional to the mass transfer rate. We have

$$q(\text{total}) = 1698 \, \text{W/m}^2$$

The heat released at the interface is equal to the heat transferred across the liquid film to the cold walls:

$$q(\text{total}) = h_L(T_i - T_w)$$

The interface temperature can therefore be calculated as

$$T_i(\text{calcuated}) = T_w + \frac{q(\text{total})}{h_L}$$

The calculated value is 314.2, which differs from the assumed value by only 0.2 °C. A new guess for the interface temperature can be taken as the average of the assumed and calculated values:

$$T_i(\text{new}) = (T_i(\text{calculated}) + T_i(\text{assumed}))/2$$

The calculations can be repeated until convergence is reached.

In the problem we neglected the subcooling of the liquid. This is usually accounted for by using a modified vaporization value. For the converged solution above, the added value increases by 2% the total heat flux at the interface. The calculations can be repeated for further improvement in the values. The added accuracy might not be commensurate with the errors introduced by the uncertainties in the transfer coefficient values (see the sensitivity analysis exercise).

13.6.2   Fog formation

What leads to fog formation? Vapor cools below the dew-point temperature, leading to supersaturation. The vapor can then condense as mist or fog rather than condensing into a liquid surface. The temperature and concentration profiles during the process of cooling of a vapor–inert-gas mixture are shown in Fig. 13.9. Three possibilities can arise: (i) the p_A vs. T values in the vapor film can touch the equilibrium line tangentially, which occurs for $Le = 1$; (ii) the line can be beneath the equilibrium curve for $Le < 1$ and no fog formation can be expected; and (iii) the curve can intersect and cross the equilibrium line before the interface point is reached. In case (iii) there is a region where the vapor is supersaturated, and this can lead to fog formation in the vapor film near the interface. This happens for $Le > 1$.

13.6.3   Condensation of a binary gas mixture

The problem of condensation from a binary mixture is considered next.

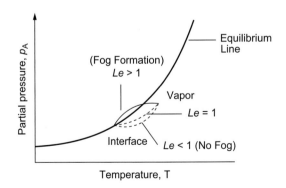

Figure 13.9 Concentration and temperature profiles during cooling of a vapor, indicating the conditions for fog formation.

The mass transport part is now solved as two-component diffusion.

The liquid-phase composition at the vapor–liquid interface is fixed by the relative rates of condensation. Thus

$$x_{A,i} = \frac{N_A}{N_A + N_B} = \frac{N_A}{N_T} \tag{13.39}$$

The mole fraction of B is equal to $1 - x_{A,i}$ for a binary system.

This assumption is often referred to as the unmixed film model assumption (Colburn and Drew, 1937; Taylor and Krishna, 1993) and implies a large resistance to mass transfer on the liquid side. Since the interfacial concentration is directly related to the condensation rates, the value of the liquid-side mass transfer coefficient is not needed. An alternative assumption is completely mixed film. Here $x_{A,i}$ is taken as the bulk liquid value $x_{A,l}$ and the bulk liquid concentrations are specified or calculated from the overall mass balance in the bulk liquid. This would imply no resistance on the liquid side for mass transfer. Again the values of liquid-side mass transfer coefficients are not needed. We proceed with the unmixed film assumption here and derive an equation for the calculation of the condensation rates from equilibrium considerations at the interface.

We assume that the interfacial temperature is known. Actually, it is not known but needs to be solved iteratively, since usually the coolant or wall temperature is known. But once the interface temperature has been specified the game is over and the problem is essentially solved. Therefore we demonstrate the calculation procedure assuming that the interfacial temperature is known, since in principle it can be iteratively calculated as shown later in this section.

What are the maximum and minimum values which the interfacial temperature can take for a given bulk gas composition?

First the K values are calculated from the vapor pressure data corresponding to the interfacial temperature. The vapor composition at the interface is related to the interfacial liquid concentration by a suitable equilibrium relation. Raoult's model is used here:

$$y_{A,i} = K_A x_{A,i} \tag{13.40}$$

This is valid for ideal solutions, whereas for non-ideal systems an activity coefficient correction is used:

$$y_{A,i} = \gamma_A K_A x_{A,i}$$

where γ_A is the activity coefficient, which is a function of the liquid composition. The activity coefficients are usually calculated by the Margules equation with the needed parameters obtained from Gmehling and Onken (1984) or other data sources. For more details standard books on thermodynamics (Smith and van Ness, Sandler etc.) should be consulted. We use the ideal solution for our discussion.

On applying Eq. (13.40) to species B, we have

$$y_{B,i} = K_B x_{B,i} \tag{13.41}$$

The sum of the mole fractions is equal to one. Hence

$$K_A x_{A,i} + K_B x_{B,i} = 1$$

or, since $x_{B,i}$ is equal to $1 - x_{A,i}$, the above expression reduces to

$$x_{A,i} = \frac{1 - K_B}{K_A - K_B} \tag{13.42}$$

All the compositions are known now.

From the mass transfer model discussed earlier (Eq. (10.92), Problem 24, Chapter 10) we have

$$\exp(N_T) = \left(\frac{x_{A,i} - y_{A,i}}{x_{A,i} - y_{A0}}\right)$$

where $y_{A,i}$ is substituted as the equilibrium value $K_A x_{A,i}$.

The individual fluxes can be calculated as $x_{A,i} N_T$ and $x_{B,i} N_T$.

We illustrate the calculation of the condensation fluxes by a simple example where we assume that the interface temperature is known.

Example 13.3.

A mixture of methanol and water at 90 °C containing 40% by mole of methanol is in contact with a cold wall. Find the rate of condensation and the composition of the condensate formed at this point.

Assume an interface temperature of 85 °C.

The data needed are of two types: (i) physical properties and (ii) transport properties or values for the transport coefficients. We use the following values for illustration of the calculation method.

(i) Physical properties.

The specific heats of components in the gas phase (methanol = 1 and water = 2) are $C_{p,g1} = 45$ J/mol · K and $C_{p,g2} = 34$ J/mol · K.

The heats of condensation are $\Delta H_{gl,1} = 36\,000$ J/mol and $\Delta H_{gl,2} = 43\,000$ J/mol.

The Antoine constants with pressure in pascals and temperature in °K are $A_1 = 23.402$, $B_1 = 3593.4$, $C_1 = -34.92$, $A_2 = 23.196$, $B_2 = 3816.4$, and $C_2 = -46.13$.

(ii) Transport properties or coefficients.

The gas-side heat transfer coefficient 60 W/m² · K

The gas-side mass transfer coefficient 0.08 m/s.

The liquid-side mass transfer coefficient is not needed for this model.

The liquid-side composition at the interface is fixed by assuming that the mole fraction in the liquid for each component is proportional to its condensation rate, i.e.,

$$x_1 = N_1/N_T$$

This is referred to as the completely unmixed assumption for the condensate. The assumption implies an infinite resistance on the liquid side.

Solution.

This problem involves making guesses both for the interfacial temperature and for the molar flux of either of the two species. If these guesses are made, one cycle of iteration can be done. The iterations are then repeated until convergence for both parameters is achieved. Thus it is an iteration problem in two variables. The steps are outlined below.

1. Using the Antoine equation, the vapor pressures are computed as
   $$p_{\text{vap1}} = 2.153 \text{ bars}$$
   $$p_{\text{vap2}} = 0.575 \text{ bars}$$
2. The corresponding K values are
   $$K_1 = 2.1528$$
   $$K_2 = 0.5745$$
3. The liquid composition at the interface can now be calculated:
   $$x_1 = 0.27786$$
   $$x_2 = 1 - x_1 = 0.7221$$
4. The composition in the vapor at the interface is calculated using Raoult's law and an ideal solution:
   $$y_1 = 0.5904$$
   $$y_2 = 0.4096$$
5. All the quantities needed to find the total flux are known. Thus, using Eq. (10.92), we have

$$N_{\text{T}}^* = 0.9394$$

The individual fluxes are

$$N_{\text{A}}^* = 0.2610$$

$$N_{\text{B}}^* = 0.6784$$

The actual fluxes are calculated by multiplying by $k_{\text{m}}C$, the scaling factor. Here C is the total concentration calculated as 34.75 mol/m^3.
The species fluxes are

methanol 0.7057 mol/m^{2}s
water 1.8344 mol/m^{2}s

Note that the methanol flux is lower, as expected. Note that methanol accumulates at the vapor side of the interface. The mole fraction is close to 0.6, compared with that in the vapor of 0.4.

Water accumulates on the liquid side of the interface. The mole fraction is 0.72, compared with 0.6 in the bulk gas phase.

Iterative calculations of the interface temperature

All the calculations above are based on an assumed interfacial temperature. This is fixed by the heat-balance equation. Hence the following liquid-phase heat-balance equation can be used to solve for the interfacial temperature in an iterative manner. This part of the calculation is similar to that for condensation of a single component in the vapor. Only the mass transport part is different for a binary condensing system. To reiterate, the equation for calculation of the interfacial temperature for a binary system is

$$h_{\text{g}}^{\circ}(T_{\text{s}} - T_{\text{i}})\mathcal{F} + N_{\text{A1}}\,\Delta H_{\text{c1}} + N_{\text{A2}}\,\Delta H_{\text{c2}} - h_{\text{l}}(T_{\text{i}} - T_{\text{w}}) = 0 \qquad (13.43)$$

where the correction factor for heat transfer augmentation ($\mathcal{F}$) is based on ϕ, which is now defined as

$$\phi = \frac{N_A C_{pgA} + N_B C_{pgB}}{h_g^\circ} \tag{13.44}$$

If just the coolant temperature is known rather than the condensing wall temperature, then the last term in the above equation is replaced by $U(T_i - T_w)$, where U is the overall (condensate side + condensing wall) heat transfer coefficient.

13.7 Temperature effects in a porous catalyst

The simultaneous transport of heat and mass in a porous catalyst can be represented by the following set of equations:

$$D_e \frac{d^2 C}{dx^2} = f(C, T) \tag{13.45}$$

and

$$k_e \frac{d^2 T}{dx^2} = (\Delta H) f(C, T) \tag{13.46}$$

where k_e is the effective conductivity of the catalyst and ΔH is the heat of reaction. See Fig. 13.10. The local rate (of disappearance) $f(C, T)$ is now a function of both the pore concentration and the temperature at any point in the pellet.

The boundary conditions are specified in a general manner as follows. The no-flux condition is imposed at the pore end ($x = 0$) for both heat and mass:

$$dC/dx = 0; \quad dT/dx = 0$$

At the pore mouth ($x = L$) the flux into the pellet must match the transport through the gas film. This leads to the following Robin-type boundary conditions.

For mass transport

$$D_e \left(\frac{dT}{dx} \right)_{x=L} = k_m (C_g - C_{x=L})$$

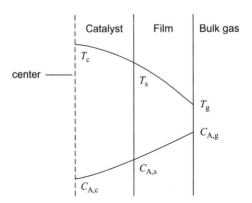

Figure 13.10 A schematic diagram of concentration and temperature profiles near and in a porous catalyst.

For heat transport

$$k_e \left(\frac{dT}{dx} \right)_{x=L} = h(T_g - T_{x=L})$$

The equations can be put in terms of dimensionless variables by scaling the concentration and temperature by reference values. The bulk concentration and temperature can be used as the reference values if the particle-scale model is being analyzed. For reactor-scale models, the inlet feed conditions can be used as reference values.

The dimensionless mass and heat transport equations are then as follows:

$$\frac{d^2c}{d\xi^2} = \frac{f(C_{ref}, T_{ref})L^2}{D_e C_{ref}} [f^*(c, \theta)] \tag{13.47}$$

where ξ is the dimensionless distance, x/L, and f^* is a dimensionless rate defined as

$$f^* = f(C, T)/f(C_{ref}, T_{ref})$$

The first term on the RHS of (13.47) can be identified as a Thiele parameter defined in the following manner:

$$\phi^2 = \frac{f(C_{ref}, T_{ref})L^2}{D_e C_{ref}}$$

Then the mass balance equation is

$$\frac{d^2c}{d\xi^2} = \phi^2 f^*(c, \theta) \tag{13.48}$$

The heat balance equation can be represented as

$$\frac{d^2\theta}{d\xi^2} = -\beta\phi^2 f^*(c, \theta) \tag{13.49}$$

where β is the dimensionless parameter (thermicity group) defined as

$$\beta = \frac{(-\Delta H)D_e C_{ref}}{k_e T_{ref}} \tag{13.50}$$

Note that β takes a positive value for exothermic reactions and a negative value for endothermic reactions.

The dimensionless rate has to be appropriately defined, including the effect of temperature on the rate constant. For example, using the Arrhenius equation, the dimensionless rate for a first-order reaction can be expressed as

$$f^* = c \exp[\gamma(1 - 1/\theta)] \tag{13.51}$$

where $\gamma = E/(R_G T_{ref})$. E is the activation energy for the reaction. The parameter γ is a dimensionless group sometimes referred to as the Arrhenius number. The governing equations and the boundary conditions shown in the earlier section are summarized for convenience here. The species mass balance is

$$\frac{d^2c}{d\xi^2} = \phi^2 c \exp[\gamma(1 - 1/\theta)] \tag{13.52}$$

The heat balance is

$$\frac{d^2\theta}{d\xi^2} = -\beta\phi^2 c \exp[\gamma(1 - 1/\theta)] \tag{13.53}$$

Boundary conditions

The boundary conditions can be normalized as

$$dc/d\xi = 0; \quad d\theta/d\xi = 0$$

at $\xi = 0$, and the conditions at $\xi = 1$ can be related to the Biot numbers for heat and mass:

$$\left(\frac{dc}{d\xi}\right)_s = Bi_m(c_g - c_s)$$

and

$$\left(\frac{d\theta}{d\xi}\right)_s = Bi_h(\theta_g - \theta_s)$$

where the subscript s is used for $\xi = 1$, i.e., the surface values. Here we define

$$Bi_m = k_m L/D_e$$

and

$$Bi_h = hL/k_e$$

This completes the problem formulation. The problem requires a numerical solution in general. For the particle problem (local particle-scale model) we set c_g and θ_g to one.

We find that, even for this simple first-order reaction, the overall effectiveness factor is a function of five parameters: ϕ, β, γ, Bi_m, and Bi_h. The results are often shown in terms of an effectiveness factor as a function of the Thiele modulus defined as

$$\eta = \int_0^1 c_A \exp[\gamma(1 - 1/\theta)]d\xi$$

An alternative definition is based on the flux at the interface:

$$\eta = \frac{1}{\phi^2}\left(\frac{dc_A}{d\xi}\right)_{\xi=1}$$

The two expressions are theoretically equivalent, but the flux form may be less accurate from a numerical point of view since the fluxes are calculated to a lesser precision than the concentration. You may wish to explore both ways of computing the effectiveness factor.

Let us now examine some invariants and see whether we can calculate the maximum temperature rise (in the exothermic case) even without solving the full set of equations.

Internal temperature gradients

The magnitude of the internal temperature gradient can be assessed by combining the above two equations without even actually solving the equations. The mass and heat transport equations (13.52) and (13.53) can be combined as

$$\frac{d^2\theta}{d\xi^2} + \beta\frac{d^2c}{d\xi^2} \tag{13.54}$$

Hence the integration of the above equation twice leads to

$$\theta - \theta_s = \beta(c_s - c) \tag{13.55}$$

which is one form of the invariant of the system.

The maximum value of the temperature occurs for an exothermic reaction when the center concentration drops to zero. This occurs in the strong pore diffusion resistance for mass transfer. Hence an estimate of the internal gradients is

$$\text{Max } \theta_c - \theta_s = \beta c_s = O(\beta) \tag{13.56}$$

where θ_c is the center temperature and the symbol O means "of the order of".

External gradients

Also the magnitudes of the external gradients can be estimated by the following procedure. Here we introduce the overall effectiveness factor η_0 defined as the actual rate divided by the rate based on reference conditions.

The rate of transport to the surface is equal to $S_p k_m (C_g - C_s)$.

The rate of reaction is $V_p f(C_{ref}, T_{ref}) \eta_0$.

By equating the two and using the dimensionless groups we find an estimate of the external concentration gradients:

$$c_g - c_s = \frac{\phi^2 \eta_0}{Bi_m} \tag{13.57}$$

Note that $\phi^2 \eta_0$ is a combined group known as the Weisz modulus that can be directly calculated if the measured data are available. The value of η_0 need not be explicitly evaluated in order to use the above equation for diagnostic purposes. We have

$$\phi^2 \eta_0 = (-R_{measured}) L^2 / (D_e C_{ref}) \tag{13.58}$$

A similar analysis equating the heat transport in the film to the total heat released from the reaction leads to an estimate of the film temperature difference:

$$\theta_s - \theta_0 = \frac{\beta \phi^2 \eta_0}{Bi_h} \tag{13.59}$$

The above relation enables a rapid determination of the relative importance of external and internal temperature gradients. It can, for instance, be used to determine whether the pellet is operating under near-isothermal conditions or whether a temperature correction to the data is necessary.

The temperature rise in the pellet can be assessed as follows. On combining (13.55) and (13.57), we find

$$\theta_c - \theta_s = c_g \beta \left(1 - \frac{\phi^2 \eta_0}{Bi_m} \right) \tag{13.60}$$

which is a closer estimate of the pellet temperature rise than the earlier equation based on the surface concentration.

Detailed numerical results for various combinations of parameters can be generated, for example, using BVP4C with MATLAB. The solution can show multiple steady states for certain ranges of parameters, and some additional study of this problem is taken up in Section 16.2.2.

Summary

- For natural convection flow between two vertical plates, a first level of modeling involves assuming a temperature profile. This is used in the Navier–Stokes equation to solve for the velocity profile. A truncated Taylor series (up to linear terms) for the density variation as a function of temperature is used, the so-called Boussinesq approximation.

- The Grashof number is the key dimensionless parameter for natural convection problems and represents the ratio of buoyancy forces to viscous forces. The velocity profile for flow between a hot plate and a cold plate shows a recirculating pattern with the fluid flowing downwards near the cold wall.

- Natural convection in external flows (on a heated wall for example) is of boundary-layer type. The results can be expressed in terms of the Rayleigh number, which is a product of the Grashof number and the Prandtl number. The boundary-layer thickness increases with the height. The flow can become oscillatory or turbulent beyond a certain height. More details on this will be presented in Section 18.4.

- Natural convection can also be caused by concentration gradients. The effect, though small, can influence the rate of mass transfer.

- For heat transfer in laminar flow, the heat transfer coefficient for cooling can be different from that for heating. The viscosity variation and the resultant modification of the velocity profile are responsible for this. As a simple rule the values are corrected by the $\mu_{\mathrm{w}}/\mu_{\mathrm{b}}$ ratio to the power of 0.14.

- Heat generation due to viscous dissipation can be of importance in the processing of highly viscous liquids. The key dimensionless group is the Brinkman number. The problem can be solved as a one-way coupled problem as a simple approximation to get the magnitude of the temperature profile due to heat generation. For a more accurate model it can be solved as a two-way problem. The perturbation method of analysis discussed in the next chapter is a useful method to study such two-way coupled problems.

- Condensation of a vapor with an inert non-condensing species is an example of the problem of simultaneous heat and mass transfer. The heat transfer rate is augmented by the mass transfer due to condensation and can be accounted for by an augmentation factor that depents on the mass transfer rate. The interface temperature is needed for calculation of these rates, and this is calculated iteratively from the rate of overall heat transport from the vapor to the coolant. The analysis in the text can also be extended to the case of two or more species condensing from the vapor.

- Diffusion with an exothermic reaction is an example of simultaneous mass and heat transfer in a porous catalyst. For the steady state an invariant relation between the temperature and concentration can be established without solving the complete set of differential equations. This relation, together with transport considerations in the external film, permits the estimation of the maximum temperature which can be generated within the catalyst. Such relations are useful for interpretation of experimented data and also for finding the regions of safe operation.

- A liquid exposed to a gas containing a vapor will reach a lower temperature than the gas due to evaporative cooling. This temperature is called the wet-bulb temperature, while the temperature of the gas itself is called the dry-bulb temperature. The values can be calculated on the basis of a film model with simultaneous heat and mass transfer. Such calculations are important in the design of humidifiers, cooling towers, air-conditioning equipment, etc.

- The student will be able to analyze problems where simultaneous transport involving two or more modes of transport is important. Several practical problems are solved in this chapter to familiarize the reader with this approach. Mainly problems for which analytical solutions are possible are considered. For some complex problems, numerical results are presented to get a feel for the numbers and the key results.

ADDITIONAL READING

The book by Gebhart *et al.* (1988) is a treatise on natural convection with an emphasis on engineering applications.

The book by Taylor and Krishna (1993) has more details on the condensation of multicomponent mixtures.

The two-volume monograph by Aris (1975) is a scholarly treatise on diffusion with reaction in porous catalysts.

Problems

1. **Example: natural convection in water.** Water is contained between two vertical plates with a gap width of 2 cm. The temperatures of the plates are 25 and 75 °C. Find the velocity profile and plot the velocity profile as a function of distance. Also find the maximum velocity in the system.

   Use the following physical properties, the values of which are evaluated at 50 °C, the average temperature in the system: density $\rho = 988 \, \text{kg/m}^3$, viscosity $\mu = 565 \times 10^{-6} \, \text{Pa} \cdot \text{s}$, and coefficient of volume expansion $4.54 \times 10^{-4} \, \text{K}^{-1}$.

2. **Natural convection: air.** Show that $\beta = 1/T$ for an ideal gas.

   Repeat the example above for air instead of water.

   Which fluid generates more circulation? Suggest a reason for this.

3. **Heat generation due to viscosity: the effect of boundary conditions.** Consider again the problem of a highly viscous fluid contained between two parallel plates of gap width d. The top plate is moving with a velocity of V and this generates heat in the system due to viscous dissipation.

   Develop an equation for the temperature distribution in the system if the upper plate is at T_0 while the lower plate is insulated. Assume constant viscosity.

   What happens to the problem if both plates are insulated?

4. **The Brinkman problem: analytical solutions.** Owing to the non-homogeneous term in Eq. (13.24), direct separation of variables is not possible. A partial solution has to be found, and the problem has to be solved by the modified method of separation of variables. Develop this analytical model. Plot typical values of the temperature profiles and compare your results with the numerical solution generated by PDEPE.

   Repeat the analysis for flow in a channel with viscous heat generation.

5. **The Brinkman problem with constant wall flux.** Reexamine the problem of laminar flow with heat generation. Now do the analysis for a constant heat flux at the walls.

   How should the dimensionless temperature θ be defined here? How is the Brinkman problem defined for this case?

   Does the problem have an asymptotic solution for large ζ similar to Eq. (12.12) in Section 12.1.3? If so, derive this result and find the value of the Nusselt number.

   Show some numerical results and compare them with the asymptotic solution.

6. **Viscosity variation.** Extend the analysis in Section 13.4 to a power-law fluid. Compare the changes in the nature of the profile with varying power-law index.

7. The thermodynamic wet-bulb temperature is defined as the temperature at which water evaporates and brings the air to equilibrium conditions. Show that the thermodynamic wet-bulb temperature is the same as the wet-bulb temperature shown in the text, which is derived from transport considerations, if the Lewis number is equal to one. What is the difference between the values computed by using the two definitions if the Lewis number is not unity?

8. **A dryer for solid with benzene.** At a point in a dryer benzene is evaporating from a solid. The air temperature is 80 °C and the pressure is 1 atm. The relative humidity of benzene in air is 65%. Find the wet-bulb temperature.

9. A wall subject to intense radiative and convective heating is to be protected by sweat cooling. For this purpose water is injected onto the surface through a porous stainless steel plate at a rate sufficient to keep it wetted and to keep the surface temperature at 360 K. Bone-dry air at 840 K and 1 atm flows past the surface and causes a convective transport to the surface. In addition there is a radiation of 461 W/m^2 impinging on the surface. The heat and mass transfer coefficients are estimated from the boundary layer. We use the following values: $k_m = 0.02$ m/s and $h = 250$ W/m$^2 \cdot$ K.

   (i) Calculate the water flow rate needed to keep the surface at the desired temperature.
   (ii) If the water comes from a cold reservoir, find the water temperature in this reservoir which would be required in order to maintain a steady state.

10. **Condensation of a pure vapor.** Water is condensing on a surface at 310 K. The gas mixture has 65% water vapor and is at a temperature of 370 K. The total pressure is 1 atm.

    Calculate the rate of condensation. Look for and use numerical values of physical properties from the web or other books.

    Use the following values for the transport coefficients.

    The gas-side heat transfer coefficient is 12 W/m$^2 \cdot$ K.

    The gas-side mass transfer coefficient is 0.009 m/s.

    Neglect the heat-transport resistance in the condensing film.

11. Examine the sensitivity of the results of Problem 1 to changes in the transport parameters by $\pm 20\%$ on either side and tabulate the results for the predicted values of the condensation rate, heat transfer rate, and interface temperature. Which of the three transport parameters is the most influential?

12. **Condensation of the binary gas mixture with an inert species.**

    The condensation of a binary mixture A and B in the presence of an inert species C can be analyzed in the same manner. The liquid side has only A and B, and hence the composition on the liquid side is calculated by the same set of equations.

    The composition on the vapor side should now include the presence of the inert species.

Extend the analysis of Section 13.6.3 for condensation of a ternary mixture of A, B, and C.

13. **Condensation with reaction.** Now consider the above problem for a case where a reaction between the condensing vapor A and the relatively non-condensing gas B can take place in the liquid according to

$$A + B \rightarrow C$$

Model the system and study the effect of reaction on the condensation rate.

14. **The temperature effect in a porous catalyst.** The effectiveness factors for a porous catalyst in the presence of significant temperature gradients can be larger than one. Explain why.

Your goal is to generate such a plot of η as a function of ϕ for the following set of parameters: $\beta = 1/3$, $\gamma = 27$, $Bi_m = 300$, and $Bi_h = 100$.

Vary ϕ as a parameter gradually and generate a plot of the effectiveness factor. Careful continuation of parameters based on a good starting solution is needed in order to track the profile. An alternative method is the arc-length continuation discussed in Section 16.2.2.

14 Scaling and perturbation analysis

Learning objectives

The study of this chapter will enable you to

- use linear algebra methods to group variables into dimensionless numbers;
- identify the scales and do a scaling analysis on a differential equation; learn what to keep and what to throw away (i.e., which terms are important and which terms can be neglected);
- use the results from scaling to suggest suitable correlating equations for a given set of experiments;
- recognize whether a problem can be solved by the perturbation method; identify whether the problem is of singular- or regular-perturbation type; and
- derive analytical solutions to non-linear problems using perturbation theory. Tools such as MAPLE and MATHEMATICA may come to your rescue in doing the lengthy and tedious algebra.

Mathematical prerequisites

- Linear algebra and the notion of the echelon form of a matrix.
- The concept of expansion in series in terms of a given (usually small) parameter or in terms of gauge functions.
- The concept of a singular perturbation problem.
- Matching and patching between the inner and outer solutions.

Some of these mathematical topics will be explained in the chapter. For additional reading on asymptotic and perturbation methods the book by Bender and Orszag (1978) is a good starting reference.

We have presented by now a detailed introduction to the analysis and modeling of transport phenomena. Basic differential equations were developed and examples were shown for all three modes of transport and some coupled problems in transport. We also studied how models at higher levels can be developed by either cross-sectional averaging or volume averaging. The goal of this chapter is to look into the details of some mathematical underpinning of the transport equations as well as some novel methods for solving transport equations. The chapter examines a number of important techniques for the analysis of transport models. Three tools are examined here.

1. Dimensionless analysis. This section expands on the dimensionless analysis discussed earlier. We learn how to form dimensionless groups from a set of variables using matrix algebra. Dimensionless analysis is an important tool in correlation of experimental data for transport processes and for assessing the impact of scaleup on the performance of larger equipment.
2. Scaling analysis. Given a differential equation for a process or system, the basic features of the system can be understood by examining the order of magnitude for the variables. A systematic way of doing this is the scaling analysis. We also learn which terms are important in a given differential equation and which terms can be neglected. Often important and useful results can be obtained simply from the scaling analysis without even solving the set of equations.
3. Perturbation analysis. Given a solution to a base set of parameters, we can expand the region of validity of the solution to a wider range of parameters by a technique similar to a series expansion. This is the basis of the perturbation method for solution of transport models.

Perturbation methods are based on a series type of expansion of the problem in terms of a small parameter. Depending on where this parameter occurs in the governing equation, this method can be classified into two general categories:

(1) regular perturbation, where a small parameter appears with the lowest-order term in the governing differential equation; and
(2) singular perturbation, where the small parameter multiplies the term with the highest-order derivative.

You will learn the difference in the solution method for these two classes of problem. A number of simple examples will be presented and references to key books on this subject will be pointed out for those of you who wish to study this fascinating topic in further detail.

14.1 Dimensionless analysis revisited

This method of dimensionless analysis was developed by E. Buckingham in 1914 and is useful when there is no simple governing differential equation that applies particularly to the problem or when the geometry is complicated. Often the effects of only a few chosen parameters are to be studied for data analysis and scaleup. An example of a complex model is in multiphase flow with turbulence, where the governing equations are complicated and not fully established even today. An empirical model useful for correlation of data and to examine the scale changes is often sufficient, and dimensionless analysis comes in handy for many such problems.

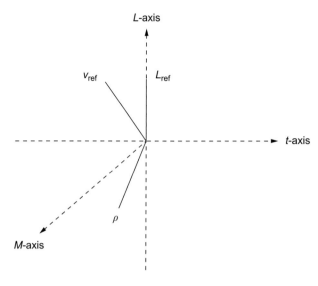

Figure 14.1 Dimensionless analysis viewed as a mapping in vector space. The illustration is for momentum transfer problems; the dotted lines are the M, L, and t coordinates. The solid lines are the new coordinates with L_{ref}, v_{ref}, and ρ as the coordinate directions.

14.1.1 The method of matrix transformation

The Buckingham method for forming the dimensionless groups can be reinterpreted as a matrix transformation problem and can be solved by using MATLAB or other software for dealing with matrices. We follow such a modern approach here rather than the classical "exponent" approach found in traditional books. The principle can be understood by looking at the background of basis vectors. The rule is rather simple and can be stated succinctly as follows. *A three (N)-dimensional space has three (N) basis vectors that can be chosen at will. Any other vector belonging to this space can be expanded as a linear combination of these three (N) basis vectors. The requirement is that the basis vectors have to be linearly independent.*

Dimensionless analysis can be viewed as changing the basis vectors and is therefore similar to coordinate transformation. A schematic representation of the change in basis is shown in Fig. 14.1.

We shall now see this rule and mapping in action to form dimensionless groups.

14.1.2 Momentum problems

The notion of core variables
Consider flow past a cylinder with a large L/d. The variables of interest are density, velocity, diameter, viscosity, and the force acting on the cylinder. It is necessary to find dimensionless groups that can represent experimental results obtained in the system. The following procedure based on matrix algebra can be used.

Density can be considered as a vector in three dimensions consisting of basis vectors of mass e_m, length e_L, and time e_t. (Of course, density has no time component in its unit.) We can write density as

$$\rho = (1)e_m - (3)e_L + 0(e_t) \tag{14.1}$$

Similarly, the length or diameter and the velocity can be represented as

$$d = (1)e_L \tag{14.2}$$

and

$$v = (0)e_m + (1)e_L - (1)e_t \tag{14.3}$$

All other quantities can be expressed in a similar manner. For example, the viscosity unit is as follows:

$$\mu = (1)e_m + (-1)e_L + (-1)e_t \tag{14.4}$$

The units for the magnitude of a force acting on a particle can be represented as a vector:

$$F = (1)e_m + (1)e_L + (-2)e_t \tag{14.5}$$

Note that, since there are only three basic vectors, the basic vectors (mass, length, and time) can themselves be written in terms of three reference quantities chosen from the system. Let us choose the length scale d, a reference velocity v, and density ρ as the new basis vectors and denote them by e_d, e_v, and e_ρ. In matrix transformation theory, this is referred to as a change in the basis. The three chosen variables for defining the new basis are also known as the core variables.

Transformation to core basis

It can be shown that the original basis vectors (length, mass, and time) can be expressed in terms of the new basis vectors of core variables as

$$e_L = (1)e_d \tag{14.6}$$

$$e_m = 3e_d + 1e_\rho \tag{14.7}$$

$$e_t = 1e_d + (-1)e_v \tag{14.8}$$

All other quantities can be expressed in terms of these bases. For example, it can be shown by direct substitution of the new basis into the viscosity vector given by Eq. (14.4) that

$$\mu = (1)e_d + (1)e_v + (1)e_\rho \tag{14.9}$$

This indicates that viscosity has the same units as $dv\rho$. Hence a dimensionless group can be formed as $dv\rho/\mu$, which is the Reynolds number.

Likewise the force can be shown to have the same units as $d^2v^2\rho$ in the new basis of the core variables. Hence $F/(d^2v^2\rho)$ is another dimensionless group. This is one form of the drag coefficient. Hence the dimensionless groups in this set of variables are two, viz., the Reynolds number and the drag coefficient.

The entirety of the experimental data can be compacted as a plot or fit of the drag coefficient as a function of the Reynolds number.

MATLAB calculations

The whole calculation can be done using matrix algebra. We form first a table of variables and their units in the three basic units of M, L, and t. The matrix for the present case will look like this:

	d	v	ρ	μ	F
m	0	0	1	1	1
L	1	1	-3	-1	1
t	0	-1	0	-1	-2

This can be expressed as a 3×5 matrix. Let us call it a. In MATLAB it will be written as

```
a = [ 0    0    1    1    1
      1    1   -3   -1    1
      0   -1    0   -1   -2 ] ;
```

The MATLAB operation *rref*(a) reduces this matrix to the "echelon" form.

What is the echelon form of a matrix?

```
         [ 1    0    0    1    2
rref(a)=   0    1    0    1    2
           0    0    1    1    1 ]
```

which is the echelon form of the matrix "a". The result may be interpreted as shown in the following table:

	d	v	ρ	μ	F
d	1	0	0	1	2
v	0	1	0	1	2
ρ	0	0	1	1	1

This shows that μ is a linear combination of vectors d, v, and ρ with coefficients of 1, 1, and 1, respectively. Hence $\mu/(dv\rho)$ is a dimensionless group, the reciprocal of the Reynolds number.

Likewise we find that F has coefficients 2, 2, and 1 with the basis vectors, and therefore $F/(d^2v^2\rho)$ is another dimensionless group. This is one representation of the drag coefficient, although the representation $(F/A_p)/(\rho v^2/2)$ is more commonly used.

Matrix theory also tells us the following rule, which is also referred to as Buckingham's theorem:

Number of dimensionless groups = Total number of variables minus the number of basic dimensions needed to describe the system.

In the present case we have five variables. Three basic dimensions (M, L, and t) are needed in order to describe the system. Hence the number of dimensionless groups is five minus three = two, *viz.*, the Reynolds number and the drag coefficient.

An additional example is provided for further illustration.

Example 14.1.

The mass m of drops formed by discharging a liquid by gravity from a vertical tube is a function of the tube diameter, liquid density, surface tension, and acceleration due to gravity. *How many independent dimensionless groups can be formed?* Determine these independent groups which would allow surface tension effects to be analyzed. Neglect the effect of viscosity.

Solution.
There are five variables and three basis vectors (core variables), and hence two dimensionless groups can be formed.

Since the mass of drops is to be correlated as a function of the surface tension, do not use this as a basis vector. Hence the basis vectors are chosen as the tube diameter, liquid density, and acceleration due to gravity. By matrix algebra, we can show that $m/(\rho d^3)$ and $d^2 \rho g / \sigma$ are the two dimensionless groups. Please verify this by forming the "a" vector and using the rref function in MATLAB.

The experimental data can therefore be correlated as

$$\frac{m}{\rho d^3} = f\left(\frac{d^2 \rho g}{\sigma}\right) \tag{14.10}$$

The group on the RHS is known as the Bond number, Bo:

$$Bo = \frac{d^2 \rho g}{\sigma} = \frac{\text{Gravity forces}}{\text{Surface tension forces}}$$

Experimental data for liquids of varying surface tension and different tube sizes can be analyzed together in this general framework, and the need to perform extensive experiments can be avoided by use of dimensionless groups.

Also, since the mass m of the droplet is equal to

$$m = (\pi/6) d_d^3 \rho$$

we have the following correlation for the droplet size relative to the nozzle size:

$$\frac{d_d}{d} = f(Bo) \tag{14.11}$$

The following correlation appears to work well with a large set of experimental data:

$$\frac{d_d}{d} = 1.82 Bo^{-1/3} \tag{14.12}$$

as indicated in the book by Middleman (1988a). See also Example 4.1 earlier.

A note of caution is that the viscosity has not been included in the analysis. Thus the method will not lead to a good fit of data if viscous effects are important for the situation under consideration. Thus the method of dimensionless analysis requires, in general, making some assumptions on the important variables of the problem. More variables can be included (e.g., viscosity for the above case), leading to additional dimensionless groups, but the problem becomes complex with too many groups. Hence the analysis is often simplified by including only the important groups for the given problem. Some prior experience and intuition is useful in this context.

14.1.3 Energy transfer problems

The basic groups (m, L, and t) are now augmented by including the temperature (T) and heat transfer rate (W) as additional dimensions. Note that the rate of heat transfer (in watts) can be expressed in basic dimensions as mL^2t^{-3} but is retained as such as an independent dimension in heat transfer problems, since it permits quantities such as the heat transfer coefficient to have simple dimensions. Thus the number of basic dimensions is five in heat transport problems, and five core variables are chosen. A simple application is shown below.

Example 14.2.

The heat flux, q, from a hot sphere exposed to a fluid at a velocity of v is to be correlated as a function of ambient temperature and other parameters. Find the groups.

Solution.

The expected parameters are the diameter of the sphere, the temperature difference between the surface and the ambient fluid, ΔT, and the fluid's physical properties ρ, c_p, μ, and k. There are eight parameters. Five basic dimensions (m, L, t, T, and W) are used. Hence a 5×8 matrix can be formed as shown below.

Use of the Buckingham theorem then tells us that three ($8 - 5$) independent dimensionless groups can be formed. The matrix algebra can be used to form these groups. First we set up the matrix in terms of the dimensions as shown below:

	d	v	ρ	ΔT	c_p	μ	k	q
M	0	0	1	0	-1	1	0	0
L	1	1	-3	0	0	-1	-1	-2
t	0	-1	0	0	1	-1	0	0
T	0	0	0	1	-1	0	-1	0
W	0	0	0	0	1	0	1	1

The operation *rref* on the above matrix results in

	d	v	ρ	ΔT	c_p	μ	k	q
d	1	0	0	0	0	1	1	0
v	0	1	0	0	0	1	1	1
ρ	0	0	1	0	0	1	1	1
ΔT	0	0	0	1	0	0	0	1
c_p	0	0	0	0	1	0	1	1

Three dimensionless groups can be identified and extracted from the above matrix:

(i) $dv\rho/\mu$, the Reynolds number;
(ii) $dv\rho c_p/k$, the Péclet number; and
(iii) $q/(v\rho c_p \Delta T)$, which is the Stanton number, St.

Thus the heat transfer data would be correlated as

$$St = f(Re, Pe)$$

The Péclet number, Pe, is often written as the product $RePr$, and hence the data could be correlated as

$$St = f(Re, Pr)$$

where Pr is the Prandtl number defined as $c_p \mu / k$. The Stanton number, St, can also written as $Nu/(RePr)$, where Nu is the Nusselt number defined as hd/k. Thus the data could be equally well correlated as

$$Nu = f(Re, Pr)$$

Note that all these forms are equivalent, but there are only THREE independent dimensionless groups. A new group formed by combining other groups is not a new independent group, and should be used in lieu of one of the other groups.

14.1.4   Mass transfer problems

The dimensionless analysis of mass transfer problems follows the same pattern. We illustrate this by taking a specific problem.

Consider the flow through a bed packed with catalytic particles. Such problems are of importance, for example, in automobile exhaust converters, methanol production from synthesis gas, etc. It is required to correlate the mass transfer coefficient as a function of system parameters. The mass transfer coefficient is defined as

$$k_m = N_A / \Delta C_A$$

and has the units m/s as indicated earlier. Here N_A is the molar flux and ΔC_A is a suitably chosen driving force for mass transfer. The mass transfer coefficient is expected to be a function of the system parameters as shown:

$$k_m = f(d_p, v, \rho, \mu, D_{A\text{-}m})$$

It is also a function of the bed porosity, ϵ_B, which is defined as the empty volume of the bed divided by the total volume of the bed. Note that the bed porosity or bed holdup is one minus the catalyst holdup. This is already dimensionless and hence not included in further analysis.

We find six variables in total and three independent dimensions. Hence, according to the Buckingham theorem (or using the concept of the rank of a matrix), we expect three $(6-3)$ dimensionless groups.

Choosing the particle diameter, density, and velocity as the set of core bases of vectors results in the following three groups from MATLAB.

1. The dimensionless mass transfer coefficient k_m/v, which is the Stanton number St for mass transfer.
2. The Reynolds number $Re = d_p \rho v / \mu$.
3. A group $\Pi_3 = D_{A\text{-}m}/(dv)$.

A group $1/(Re\Pi_3)$ is used rather than Π_3. This group is the Schmidt number,

$$Sc = 1/(Re\Pi_3) = \mu/(\rho D_{A\text{-}m})$$

and represents the ratio of momentum diffusivity to mass diffusivity. This group is a physical property and does not depend on the operating parameters.

Hence one way of correlating the mass transfer data in packed beds is

$$St = f(Re, Sc, \epsilon_B)$$

The group $StReSc$ is often used instead of St. This group is the Sherwood number defined as $k_m d / D_{A\text{-}m}$.

Hence mass transfer data in packed beds can also be correlated in terms of the Sherwood number:

$$Sh = f(Re, Sc, \epsilon_B)$$

A large set of correlations for mass transfer data in a variety of process equipment has been correlated in the above manner, and the dimensionless analysis provides the foundation for the form of the correlations.

Now we show two examples of the use of the method for analysis of process equipment.

14.1.5   Example: scaleup of agitated vessels

Agitated vessels are used in a number of chemical processes. Power consumption is an important parameter, and the effect of the agitator size on this quantity may be correlated by the dimensional analysis.

Three core variables are needed, and these are normally chosen as follows.

1. The characteristic length is taken as the impeller diameter

$$L_{ref} = d_I$$

2. The characteristic velocity is taken as the tip speed of the impeller. The tip speed in radians per second is equal to $2\pi R\Omega$ or $\pi d_I \Omega$, where Ω is the number of revolutions per second. The factor π is not used since it is merely a scaling factor. Hence

$$v_{ref} = d_I \Omega$$

which provides the second core variable.

3. The liquid density can be chosen as the third core variable. We could have chosen the viscosity instead. This choice is therefore arbitrary.

The remaining variables can be cast into dimensionless groups in terms of the above basis vectors. These groups are as follows.

1. The power number. The quantity to be correlated is the power consumed (P), which has units of ML^2/T^3. On transforming the basis the power has the units of $\rho \Omega^3 d_I^5$. Hence the dimensionless group is

$$\Pi_1 = \text{Power number} = \frac{P}{\rho \Omega^3 d_I^5} = Po$$

which is called the power number and denoted as Po.

2. The Reynolds number. A second group is based on the viscosity, which leads to the Reynolds number for agitated vessels:

$$Re = \frac{d_I^2 \Omega \rho}{\mu}$$

3. The Froude number. A third group can be formed by using gravity as the variable, which leads to a form of the Froude number. However, for well-baffled vessels there is no vortex formation at the surface and hence this variable is not so significant and this group is not normally considered to be important.

Hence the data for power consumption can be correlated as

$$Po = f(Re)$$

For large Reynolds number ($Re > 10^5$), the power number becomes independent of the Reynolds number. The explanation is that the viscous effects well captured by the Reynolds number are not so important under these conditions. An asymptotic value of 6.3 is often observed for the power number. Hence the power consumption can be calculated as

$$P = 6.3\rho\Omega^3 d_I^5$$

This gives a useful scaleup rule. We find that the power consumption is proportional to d_I^5, while the dependence on the agitation speed is of the third power.

The rule is not valid for low Reynolds number; here the power number is found to be a linear function of the inverse of the Reynolds number:

$$Po = \frac{A}{Re}$$

where A is a constant that depends on the geometry. This suggests forming a new dimensionless group $PoRe$ that remains constant on scaleup:

$$Po^* = PoRe = A; \text{ a constant}$$

Thus the dependences of the operating variables are different for the two regimes, and the scaleup may be erroneous if the proper flow regime is not known or identified. This is a general statement applicable to many complex systems. It may be noted here that the laminar regime usually exists if $Re < 1000$ in agitated tanks.

Note that, when scaling up, geometric similarity is needed and maintained. The matching of the dimensionless groups applies only when the geometric similarity is maintained.

14.1.6   Example: pump performance correlation

The performance of a pump is the pressure difference developed by the pump and the volumetric flow of the fluid Q which the pump delivers. The power consumption P is of importance as well. The pressure head can therefore be expected to be a function of the following variables:

$$\Delta P = f(\rho, d_I, \Omega, \mu, Q, P)$$

The dimensionless parameters needed to characterize the performance of the pump assuming geometric similarity are illustrated in this section. We have seven quantities and three basic dimensions. Hence four groups are important. The core variables are chosen as ρ, d_I, and Ω. The remaining variables are $\Delta P, \mu, Q$, and P.

Prior experience tells us that μ can be put in a Reynolds type of group. The tip speed $d_I\Omega$ is used as the reference velocity v_c. Hence the first group is

$$\Pi_1 = \frac{d_I^2\Omega\rho}{\mu}$$

which is the Reynolds number.

The second group for the pressure drop can be scaled in terms of ρv_c^2, which has the same dimension as pressure. Hence the second group is

$$\Pi_2 = \frac{\Delta P}{\rho d_I^2\Omega^2}$$

This is expressed conveniently as the head of fluid, h, developed by the pump:

$$\Delta P = \rho g h$$

Hence the above group is rearranged as

$$\Pi_2 = \frac{gh}{d_I^2\Omega^2}$$

This is known as the head coefficient in the pump jargon.

The third group involves the volumetric flow rate, Q. Noting that Q/d_I^2 has the same dimension as the velocity a third group can be formed as

$$\Pi_3 = \frac{Q}{d_I^2 v_c} = \frac{Q}{d_I^3\Omega}$$

This third group is called the flow coefficient.

The fourth group will involve the power consumed and is the power number

$$\Pi_4 = \frac{P}{\rho\Omega^3 d_I^5}$$

which was the same group as derived for an agitated tank.

These four groups are maintained to be the same for small and large pumps that are geometrically similar. An example is provided below to illustrate the matching of the groups.

Example 14.3.

A pump has an impeller diameter of 15 cm and provides a flow rate of 0.05 m^3/s when operating at 1200 r.p.m. A geometrically similar pump is needed to deliver 1 m^3/s with operation at 600 r.p.m.

Find the pump diameter needed. Find the head ratio and the ratio of power consumed for the two pumps.

Solution.

The flow number is first matched since all the quantities required in order to calculate this in the smaller pump are known:

$$\Pi_3 = \frac{Q}{d_I^3\Omega} = 0.05 \times 0.15/(1200/60) = 0.74$$

The dynamic similarity (matching the dimensionless numbers) requires that Π_3 should have the same value in the larger pump. Hence d_I for the larger pump can be calculated as 0.51 m = 51 cm. Now other dimensionless numbers can be matched.

The head is proportional to $\Omega^2 d_I^2$ as indicated by the dimensionless group Π_2. Hence the ratio of the head in the larger pump to that in the smaller pump is calculated by taking the ratio of large-pump values to small-pump values, and the result is 2.9.

Similarly, the power needed is proportional to $\Omega^3 d_I^5$, as indicated by the dimensionless group Π_4. Hence the ratio of the power in the larger pump to that in the smaller pump is calculated as 58.5.

14.2 Scaling analysis

Scaling analysis refers to a technique whereby the orders of magnitude of various terms are compared. In a properly scaled problem the order of each term should be the same. Terms that are much smaller can be neglected. Results of engineering interest can often be obtained by scaling analysis without the need to solve the full equations. Again the method is best learned by application, and we now show some simple examples to illustrate the scaling analysis.

14.2.1 Transient diffusion in a semi-infinite region

Consider the transient heat conduction in a semi-infinite medium described by the equation

$$\frac{\partial T}{\partial t} = \alpha \frac{\partial^2 T}{\partial x^2} \tag{14.13}$$

The initial temperature is zero and the surface is exposed to a higher temperature of T_s and maintained at that level. We can define a depth of penetration δ over which the temperature changes are non-zero, and this front moves with time. An estimate for this quantity is to be obtained using the scaling analysis.

The length scale x of interest is taken as δ. The temperature scale is T_s. Hence the RHS of Eq. (14.13) is scaled as $\alpha T_s / \delta^2$. The LHS scales as T_s / t, where t is the time for which the surface is exposed to a higher temperature. Since the transient and conduction terms in Eq. (14.13) are equally important, the scale of the LHS can be equated to the scale of the RHS. On rearranging we obtain an estimate of the penetration depth.

$$\delta \propto \sqrt{\alpha t}$$

This can be written, introducing a constant of proportionality, as

$$\delta = A\sqrt{\alpha t}$$

where A is a numerical constant that should of the order of 1. The scaling analysis provides the correct time dependence, i.e., square root in time for the penetration depth.

Scaling analysis, however, does not provide a value for this constant. Also note that there is no definitive value for this constant, since it depends on where the cut-off value of temperature is taken. Usually 0.01% of the final value is taken as the cut-off. The "exact" solution is then

$$\delta = 6\sqrt{\alpha t}$$

using the error-function representation in Section 11.3.2.

14.2.2 Example: gas absorption with reaction

Consider the penetration model for gas absorption. The governing differential equation is

$$\frac{\partial C}{\partial t} = D\frac{\partial^2 C}{\partial x^2} - kC \qquad (14.14)$$

In the absence of reaction we have

$$\delta = A\sqrt{Dt}$$

If the reaction is included, the following analysis can be performed.

The following relation is then established by scaling analysis:

$$\frac{1}{t_E} = A\frac{D}{\delta^2} - Bk$$

where δ is the penetration depth, and A and B are constants that again should be of the order of one.

On rearranging in terms of δ, we find

$$\delta = \sqrt{\frac{ADt_E}{1 + Bkt_E}}$$

Since the instantaneous rate of absorption is inversely proportional to δ, we find

$$N_A = (D\,\Delta C)\sqrt{\frac{1 + Bkt_E}{ADt_E}}$$

The expression reveals that a plot of N_A^2 vs. $1/t_E$ should be a straight line. Data from the literature confirm this trend (see Problem 11) Hence the above equation (obtained purely by scaling arguments) is useful as a simplified way of fitting and interpreting the data to estimate the diffusion coefficient and the rate constant.

14.2.3 Kolmogorov scales for turbulence: an example of scaling

This section can be omitted for now and studied in conjunction with the turbulent flow analysis of Chapter 19, but this is a good example of scaling analysis.

Turbulence is characterized by eddies of different sizes, and the energy transfer between the eddies and the viscous dissipation is an important property of the turbulent flow. Kolmogorov proposed a model for energy cascade whereby energy is simply transferred from large eddies to smaller eddies. But finally eddies of a particular small size and their characteristic velocity lose kinetic energy by viscous dissipation. The size of these eddies and the corresponding velocity at which this occurs are called Kolmogorov scales. More details of this are presented in Section 17.10. Here we show how scaling analysis can be used to estimate these scales.

A representative term in the viscous dissipation per unit volume is given by

$$\Phi_v = \mu \left(\frac{dv_x}{dx} \right)^2$$

An estimate of this term is therefore

$$\Phi_v \approx \mu v^2 / L_K^2$$

where v is the velocity scale at which viscous dissipation occurs and L_K is the length scale of these eddies.

If $\epsilon = \Phi_v / \rho$ is defined as the energy dissipation per unit mass then the scaling analysis provides an estimate for this quantity as

$$\epsilon = v v^2 / L_K^2 \qquad (14.15)$$

At the Kolmogorov scale the inertial forces and viscous forces have a similar order of values. Hence the Reynolds number (the ratio of inertial to viscous forces) based on these scales should be nearly unity. Hence

$$\frac{L_K v}{v} \approx 1$$

or

$$v(\text{Kolomogorv scale}) \approx v / L_K \qquad (14.16)$$

Using the estimate of the velocity from (14.15) and rearranging and solving for L_K, we have

$$L_K = \left(\frac{v^3}{\epsilon} \right)^{1/4}$$

which provides the estimate for the smallest size of eddies. Now, using this in Eq. (14.16), we obtain an estimate for the velocity scale for these eddies:

$$v = (v\epsilon)^{1/4}$$

We note that both the length scale and the velocity scale at which energy dissipation occurs can be predicted purely from scaling considerations. These Kolmogorov scales are very important in turbulent transport processes since diffusive transport in turbulent systems begins only at this or smaller scales. More on this is provided in a later chapter; our main goal here was to show the application of the scaling analysis for such a classical problem.

14.2.4   Scaling analysis of flow in a boundary layer

External flow past a flat plate is the most well-studied problem in fluid dynamics next only to the Hagen–Poiseuille equation for internal flow in a pipe. The classical work of Prandtl showed that there is a thin layer near the plate where the viscous effects are important and the flow becomes uniform outside this layer. Here we show how the scaling analysis is a powerful tool to estimate the thickness of the boundary layer. Scaling analysis also shows which terms are important in the Navier–Stokes equation and how these equations can be simplified for this problem.

Continuity equation scaling

We start with the continuity equation

$$\frac{\partial v_x}{\partial x} + \frac{\partial v_y}{\partial y} = 0$$

and do an order-of-magnitude analysis.

The distance x can be scaled with L, an arbitrary length along the plate, and the velocity by V, the approach velocity.

Hence the first term in the continuity equation scales as

$$\frac{\partial v_x}{\partial x} \approx \mathcal{O}\left(\frac{V}{L}\right)$$

where the symbol $\approx \mathcal{O}$ should be read as "approximately of the order of".

The second term in the continuity equation should be of the same order of magnitude as the above term. Thus

$$\frac{\partial v_y}{\partial y} \approx \mathcal{O}\left(\frac{V}{L}\right)$$

A scale for the distance y is the boundary-layer-thickness width δ. By comparing the orders of the two terms, we can estimate a scale for v_y:

$$v_y \approx \mathcal{O}\left(\frac{\delta}{L}V\right) \tag{14.17}$$

This provides the estimate for v_y. Since δ/L is expected to be much smaller than unity, we conclude that the velocity component in the y-direction is much smaller than that in the x-direction.

These estimates can be used in the x- and y-momentum balances to see the orders of magnitude of the various terms in the Navier–Stokes equation.

x-Momentum scaling

The x-momentum balance (divided by ρ on both sides) is

$$v_x\frac{\partial v_x}{\partial x} + v_y\frac{\partial v_x}{\partial y} = -\frac{1}{\rho}\frac{\partial p}{\partial x} + v\frac{\partial^2 v_x}{\partial x^2} + v\frac{\partial^2 v_x}{\partial y^2}$$

Let us do the order-of-magnitude analysis term by term:

$$v_x\frac{\partial v_x}{\partial x} \approx \mathcal{O}\left[V\left(\frac{V}{L}\right)\right] = V^2/L$$

Also

$$v_y\frac{\partial v_x}{\partial y} \approx \mathcal{O}[(\delta/L)V] \cdot [V/\delta] \approx \mathcal{O}\left(\frac{V^2}{L}\right) \approx V^2/L \tag{14.18}$$

Thus we find that the two convective terms are of the same order of magnitude. Although v_y is much smaller than v_x, v_y contributes as much to convection as does the v_x term.

Let us now look at the viscous terms.

The first term, momentum diffusion in the x-direction, scales as

$$v\frac{\partial^2 v_x}{\partial x^2} \approx \mathcal{O}\left(\frac{vV}{L^2}\right) \tag{14.19}$$

The y-"diffusion" of momentum term scales as

$$v\frac{\partial^2 v_x}{\partial y^2} \approx \mathcal{O}\left(\frac{vV}{\delta^2}\right) \tag{14.20}$$

Hence the ratio of the two terms (x-term divided by y-term) scales as δ^2/L^2. This indicates that the x-diffusion of the momentum can be neglected. Hence

$$\frac{\partial^2 v_x}{\partial x^2} \ll \frac{\partial^2 v_x}{\partial y^2}$$

Further, the scale of the viscous term in the y-direction (Eq. (14.20)) must be comparable to the convective terms, for example that in Eq. (14.18). Hence the orders of magnitude of these two terms can be equated:

$$\frac{vV}{\delta^2} \approx \frac{V^2}{L}$$

or

$$\delta^2 \approx \frac{vL}{V} \tag{14.21}$$

Since L is arbitrary, it can be replaced by x at any position along the plate. Using this change in notation, an estimate of the boundary layer thickness is therefore

$$\boxed{\delta = A\sqrt{\frac{vx}{V}}}$$

where A is a fitting constant. Later we show by a detailed analysis that A is nearly equal to 6, but for the time being we note that the scaling analysis has completed its job.

y-Momentum scaling

Now the ordering analysis is done for the y-momentum balance, which is

$$v_x\frac{\partial v_y}{\partial x} + v_y\frac{\partial v_y}{\partial y} = -\frac{1}{\rho}\frac{\partial p}{\partial y} + v\frac{\partial^2 v_y}{\partial x^2} + v\frac{\partial^2 v_y}{\partial y^2}$$

The first term scales as $V[(\delta/L)V]/L$ or as $(\delta/L)(V^2/L)$ and hence is of the order of δ/L.

The second term scales as $[(\delta/L)V][(\delta/L)V]/\delta$ or as $(\delta/L)(V^2/L)$. Thus the two terms have the same scale and appear to be small compared with the terms in the x-momentum.

Also it is easy to show that

$$\frac{\partial^2 v_y}{\partial x^2} \ll \frac{\partial^2 v_y}{\partial y^2}$$

and hence this term can be neglected.

Finally, show that

$$v\frac{\partial^2 v_y}{\partial y^2} \approx \mathcal{O}v[(\delta/L)V]\delta^2$$

Use the estimate of δ^2 from Eq. (14.21) to show that this term scales as $(\delta/L)V^2/L$:

$$\mu\frac{\partial^2 v_y}{\partial y^2} \approx \mathcal{O}(\delta/L)(V^2/L)$$

Thus all the terms in the y-momentum except the pressure term are of the order of δ/L or even smaller.

Hence we conclude that

$$\frac{\partial P}{\partial y} \approx \mathcal{O}(\delta/L) \approx 0$$

The pressure variation across the boundary layer is small. This is an important result, which can now be used in the x-momentum balance. Hence the pressure gradient in the x-momentum balance can be approximated by the external pressure gradient at the edge of the boundary layer:

$$\frac{\partial p}{\partial x} = \frac{dp}{dx} = \frac{dp_{\text{ext}}}{dx}$$

For a flat plate there is no variation of the velocity in the external flow and hence, by application of the Bernoulli equation, the above term is zero. Note that this term is not zero for plates inclined to the direction of flow or for flow over curved surfaces.

Hence the simplified x-momentum for the flat plate can be represented as

$$v_x\frac{\partial v_x}{\partial x} + v_y\frac{\partial v_x}{\partial y} = v\frac{\partial^2 v_x}{\partial y^2}$$

which needs to be solved together with the continuity equation. The y-momentum balance is no longer needed, which is an important conclusion gained from the scaling analysis. This is the starting point of boundary-layer models, which will be taken up in further detail in Section 15.5.

Wall shear stress

A correlating form for the wall shear stress can also be obtained from the scaling analysis. The wall shear stress is given by

$$\tau_w = \mu\left(\frac{\partial v_x}{\partial y}\right)_{y=0} \approx \frac{\mu V}{\delta}$$

On substituting into this the estimate for δ we find

$$\tau_w = A\mu V\sqrt{\frac{V}{vx}}$$

The constant A can be fitted to the data or, in this case, estimated from a detailed analysis. Its value is found to be equal to 0.332 by detailed analytical solution. The key point to note here is that the scaling analysis provides a reasonable method for correlation of quantities of engineering interest.

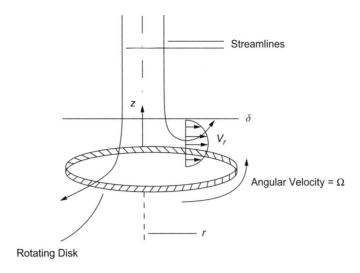

Figure 14.2 A schematic representation of flow over a rotating disk

14.2.5   Flow over a rotating disk

This is a classic and well-studied fluid flow problem that has important applications in mass transfer in electrochemical systems, deposition of coatings on surfaces, and similar classes of problems. Figure 14.2 illustrates the features of the problem. A disk of large radius is rotating with the angular speed Ω in still fluid. The fluid at the disk has to satisfy the no-slip condition. The centrifugal effect due to rotation causes the fluid to leave the disk radially near the disk. The flow above the disk must replace this flow through a downward spiralling flow. A cylindrical coordinate system is needed for description with all of the three velocities non-vanishing. Detailed analysis requires the use of all of the three momentum balances and is taken up in a later chapter. In this section we show how the scaling analysis of this problem leads to some key results directly without having to carry out a detailed solution of the problem. The region of influence of the disk is confined to a thin boundary layer near the disk, and we wish to predict the parametric dependence of the boundary-layer thickness by scaling arguments.

The tangential velocity v_θ will have a value of $r\Omega$ at the disk itself. Hence we expect this component of velocity to scale in proportion to this:

$$v_\theta = O(r\Omega)$$

To estimate a scale for v_r let us compare terms in the r-momentum balance (Table 3.1) of the Navier–Stokes equation. We find that the radial convective term scales as

$$v_r \frac{\partial v_r}{\partial r} = O(v_r v_r / r)$$

The scale of the centrifugal force term in the r-momentum is

$$\frac{v_\theta^2}{r} = O[r\Omega)^2 / r] = O(r\Omega^2)$$

Since the radial flow is driven by the centrifugal forces these two should be of comparable order of magnitude. Hence we compare the above two terms and we find that v_r also scales as $r\Omega$:

$$v_r = O(r\Omega)$$

Let the distance in the z-direction scale as some boundary-layer thickness δ, which will be the region of influence of the rotating disk. The terms in the continuity equations can now be scaled and compared with each other. We find

$$\frac{v_z}{\delta} \approx \frac{v_r}{r} \approx (r\Omega)/r \approx \Omega$$

Hence v_z must be of the order of

$$v_z \approx \delta\Omega$$

We can now use the result in the z-momentum to get an estimate of δ as shown below.
The viscous term in the z-direction is then

$$\nu\frac{\partial^2 v_z}{\partial z^2} = O(\nu v_z/\delta^2) = O(\nu\delta\Omega/\delta^2) = O(\nu\Omega/\delta)$$

Comparing this with the v_r term in the z-momentum convection gives

$$v_r\frac{\partial v_z}{\partial r} = O((r\Omega)(\delta\Omega)/r) = \mathcal{O}(\delta\Omega^2)$$

By matching the above two terms we get an estimate for δ, the boundary-layer thickness near the spinning disk, namely

$$\delta^2 = \mathcal{O}\left(\frac{\nu}{\Omega}\right)$$

or the following relation can be proposed:

$$\boxed{\delta = A\sqrt{\frac{\nu}{\Omega}}} \tag{14.22}$$

Thus the estimate of the boundary-layer thickness near a rotating disk can be obtained by scaling analysis without having to solve a single differential equation. A more detailed solution is presented later in Section 15.7, which will give us a theoretical estimate for the constant A.

It may be noted that the model applies, strictly speaking, to an infinite disk, but can be applied to a finite disk of radius R, provided that the condition $R \gg \delta$ or equivalently $R \gg \sqrt{\nu/\Omega}$ is satisfied. The "edge" effects are expected to be small under these conditions. The numerical solution and the experimental values of the torque exerted by the fluid agree very well under these conditions. The flow is expected to become unstable at $Re > 10^5$ as revealed the stability analysis. The experimental conditions for mass transfer studies can then be appropriately chosen by taking these considerations into account.

Mass transfer coefficient
The mass transfer coefficient from the disk to the fluid can be estimated approximately as D/δ for $Sc = 1$, since the hydrodynamic and mass-layer boundary-layer thicknesses are

expected to be of similar order. But in general a correction for the effect of Sc is applied with a 1/3 exponent as shown later. Hence

$$k_L \propto \frac{D}{\delta} Sc^{1/3}$$

Hence in a purely scaling analysis a correlating expression for mass transfer coefficient is

$$k_L \propto \Omega^{1/2} D^{2/3} \nu^{-1/6}$$

which fits the mass transfer data in this equipment well. We find a constant mass transfer coefficient at all radial positions, which is a special feature of a rotating disk. The disk is equi-accessible to mass transfer and hence the correlation of reaction data (for a reaction at the disk surface) becomes easier, and the deposition rate is also expected to be uniform. These are some of the advantages of the rotating disk contactor.

Now we move to a different method of analysis, $viz.$, the perturbation methods.

14.3 Perturbation methods

Perturbation methods are useful when a small parameter appears as a coefficient in the differential equation. These methods can be classified into two categories.

1. Regular perturbation. This method is used when the small parameter appears in the lowest-order term in a differential equation. An asymptotic series in terms of the small parameter can be obtained, and the series has a non-vanishing radius of convergence. Often a power-series form in terms of the small parameter is used, but this can be changed to suit the problem requirements.
2. Singular perturbation. This method is used when the small coefficient multiplies the term with the highest derivative. Here the solution is often of the boundary-layer type. The solution need not be of the power-series type here.

14.3.1 Regular perturbation

Example 1: diffusion with first-order reaction
Consider for illustration the diffusion–reaction problem given by

$$\frac{d^2 c}{dx^2} = Mc \tag{14.23}$$

Here x is the dimensionless parameter and c is the dimensionless concentration of the diffusing species. The parameter M is the square of the Thiele modulus and equal to ϕ^2 in Section 10.4.2. The notation is changed here for convenience in some algebraic manipulations.

We use the same boundary conditions as before: $dc/dx = 0$ at $x = 0$ and $c = 1$ at $x = 1$. We look for a perturbation solution for small values of M. Since the parameter M appears in the lowest-order term, the problem is of a regular perturbation. A solution for $M = 0$ is needed to start off the perturbation. The solution for $M = 0$, is of course, $c = 1$.

Now we propose that, for small values of M, the solution for c can be represented as

$$c = 1 + M f_1(x) + M^2 f_2(x) + O(M^3) \tag{14.24}$$

where the first term on the RHS represents the solution for $M = 0$ while the second term on the RHS is a product of an undetermined function $f_1(x)$ and the parameter M, and so on for other terms. The term $O(M^3)$, means that the neglected term has magnitude of M^3, and this is standard practice in perturbation analysis to indicate the magnitude of the error.

On substituting the perturbation approximation for c into the differential equation we have

$$Mf_1'' + M^2 f_2'' + \cdots = M + M^2 f_1 + M^3 f_2 + \cdots \tag{14.25}$$

where the prime indicates differentiation as usual. Upon equating the first power of M on both sides of the above equation, we find

$$f_1'' = 1 \tag{14.26}$$

The boundary conditions on f are as follows. At $x = 1$, $f = 0$, since the first term of the assumed solution already satisfies the condition $c = 1$ at $x = 1$. The second boundary condition is $\partial f / \partial x = 0$ at $x = 0$.

The solution to Eq. (14.26) with the above boundary condition is

$$f_1 = \frac{1}{2}(x^2 - 1) \tag{14.27}$$

Now, by equating the second power of M on either side of Eq. (14.25), a differential equation for f_2 is obtained:

$$f_2'' = f_1 = \frac{1}{2}(x^2 - 1) \tag{14.28}$$

The boundary conditions for f_2 are the same as those for f_1. The solution gives

$$f_2 = \frac{1}{2}\left(\frac{x^4}{12} - \frac{x^2}{2} + \frac{5}{12}\right) \tag{14.29}$$

The procedure can be continued to get higher-order terms. The next term can be shown to be

$$f_3 = \frac{x^6}{720} - \frac{x^4}{48} + \frac{5x^2}{48} - \frac{61}{720}$$

A comparison of the solution with the exact solution for $M = 0.5$ is shown in Table 14.1. The above procedure to obtain a regular perturbation solution is standard and can be applied to other problems of a similar nature.

Example 2: parallel flow with viscous dissipation

Here we consider a steady laminar flow between two parallel plates. The top plate is moved with a velocity V while the bottom plate is stationary. The viscous dissipation causes a temperature rise in the fluid, which in turn changes the viscosity of the fluid and affects the flow. Hence it represents a coupled problem in transport phenomena.

The problem was schematically shown in Fig. 13.3 in Section 13.3.1, where the base flow solution (the constant-viscosity case) was also calculated. We wish to determine the effect of varying viscosity on the velocity and temperature profiles. This can be done by a regular perturbation analysis as shown below.

Table 14.1. A comparison of the regular perturbation solution with the analytical solution for diffusion with the first-order-reaction problem with $M = 0.5$

| x | Analytical | Perturbation | |
		Three terms	Four terms
0.0	0.7933	0.8021	0.7915
0.2	0.8012	0.8096	0.7995
0.4	0.8252	0.8324	0.8238
0.6	0.8658	0.8709	0.8647
0.8	0.9236	0.9263	0.9231
1.0	1.0000	1.0000	1.000

The governing equations are given directly in terms of the dimensionless variable, here

$$\frac{d}{d\xi}\left(\exp(-\alpha\theta)\frac{dv}{d\xi}\right) = 0 \tag{14.30}$$

and

$$\frac{d^2\theta}{d\xi^2} + Br\exp(-\alpha\theta)\left(\frac{dv}{d\xi}\right)^2 = 0 \tag{14.31}$$

Here the term α is the coefficient of viscosity, which is represented as an exponential function of temperature. The parameter Br is the Brinkman at base conditions.

The boundary conditions for the dimensionless velocity v are as follows: at $\xi = 0$, $v = 0$ and at $\xi = 1$, $v = 1$.

The boundary conditions for θ are problem-specific, and could be of the three kinds discussed earlier. Here we show the solution procedure by taking the following simple Dirichlet conditions: both at $\xi = 0$ and at $\xi = 1$ the dimensionless temperature $\theta = 0$.

For the case where the parameter Br is zero, the problem has simple analytical solutions. Let v_0 and θ_0 represent these solutions. These "zero" perturbation solutions are readily obtained as $v_0 = \xi$ and $\theta_0 = 0$.

This is the base flow, a linear velocity profile, and the temperature profile is constant since there is no generation.

Now we seek the solution as an expansion in terms of Br parameter as

$$v = v_0 + Brv_1 + Br^2v_2 + O(Br^3) \tag{14.32}$$

and

$$\theta = \theta_0 + Br\theta_1 + Br^2\theta_2 + O(Br^3) \tag{14.33}$$

In order to obtain higher-order solutions, it is necessary to expand the non-linear terms in terms of the parameter Br. The exponential term in Eqs. (14.30) and (14.31) is expanded as

$$\exp(-\alpha\theta) = 1 - \alpha\theta + (\alpha\theta)^2/2 + O(\alpha^3)$$

On using Eq. (14.33) for θ and rearranging, we obtain the following expansion up to Br^2:

$$\exp(-\alpha\theta) = 1 - Br\alpha\theta_1 + Br^2([\alpha\theta_1/2]^2 - \alpha\theta_2) + O(Br^3)$$

which is valid for the specific case of Dirichlet conditions at both ends, with θ_0 of zero as the base solution.

Similarly the $(dv)(d\xi)^2$ term is expanded using Eq. (14.32) as

$$\left(\frac{du}{d\xi}\right)^2 = v_0^2 + 2Br v_0 v_1 + 2v_0 v_2 Br^2 + O(Br^3)$$

These expansions are now substituted into the equations, and terms of Br to the power of one can thus be matched. These lead to equations for u_1 and θ_1 as follows:

$$\frac{d}{d\xi}\left(\frac{du_1}{d\xi} - \alpha u_1 \frac{du_0}{d\xi}\right) = 0$$

and

$$\frac{d\theta_1}{d\xi} + \left(\frac{du_0}{d\xi}\right)^2 = 0$$

The homogeneous boundary conditions are to be used here. The solutions are obtained as

$$\theta_1 = \frac{1}{2}\xi(1-\xi)$$

and

$$v_1 = -\frac{\alpha}{2}(\xi - 4\xi^2 + 3\xi^2)$$

The solutions to the next order of approximation are obtained by matching the terms with Br to the power of two. The detailed derivations are left as an exercise. The second-order solutions are obtained as follows:

$$\theta_2 = -\frac{\alpha}{24}(\xi - 2\xi^2 + 2\xi^3 - \xi^4) \tag{14.34}$$

and

$$v_2 = -\frac{\alpha}{120}(\xi - 5\xi^2 + 10\xi^3 - 10\xi^4 + 4\xi^5) \tag{14.35}$$

which you should verify using computer algebra such as MAPLE. Higher-order solutions can be obtained by continuing the procedure.

14.3.2   The singular perturbation method

We demonstrate the basic ideas with simple examples and move to somewhat more involved cases.

Consider again the diffusion–reaction problem given by Eq. (14.23), but now with a large value for M. The equation can now be rearranged as

$$\epsilon \frac{\partial^2 c}{\partial x^2} = c \tag{14.36}$$

where $\epsilon = 1/M$. If one uses the perturbation approach blindly, one is tempted to put $\epsilon = 0$ (since large M implies small ϵ). This implies that $c = 0$ everywhere in the domain. This is not the correct approach, because by doing so the order of the differential equation is reduced, i.e., the second derivative term vanishes and we cannot satisfy the boundary condition at $x = 1$. Thus neglecting a small coefficient that multiplies the largest-order derivative leads to an incomplete solution to the problem. The difficulty can be overcome by saying that there is a thin region near $x = 1$ where the concentration changes sharply from unity to zero, which holds outside this region. This thin region is known as the boundary layer. The problem is now of a singular perturbation type.

The solution $c = 0$ holds for most of the region except near $x = 1$, where the boundary condition was not satisfied. Let us explore this thin region by introducing a stretched coordinate system that magnifies this region.

Let a new variable y be defined, namely

$$y = \frac{1 - x}{\epsilon^a} \tag{14.37}$$

where a is a constant to be determined.

This transformation (known as the stretching transformation because it magnifies or stretches the boundary layer) reduces the differential equation to

$$\epsilon^{1-2a} \frac{\partial^2 c}{\partial y^2} = c \tag{14.38}$$

We can now choose a as $1/2$, which removes the term containing ϵ, leading to

$$\frac{\partial^2 c}{\partial y^2} = c \tag{14.39}$$

The boundary condition at $x = 1$ implies that $c = 1$ at $y = 0$. Furthermore, the domain for y can be taken to be infinite since the boundary layer is now stretched and looks large. We require that the solution obtained from this equation for $y = \infty$ should match the inner solution for $x = 1$, i.e., $y \to \infty$, $c \to 0$. Thus the solution in the boundary layer is seen to be

$$c = \exp(-y) \tag{14.40}$$

or

$$c = \exp(-\sqrt{M}(1 - x)) \tag{14.41}$$

which can also be shown to be a good approximation to the exact solution near the diffusing surface.

14.3.3 Example: catalyst with spatially varying activity

We now consider a modified problem where the catalyst activity varies as a function of position. The model equation now is

$$\epsilon \frac{\partial^2 c}{\partial x^2} = cx \tag{14.42}$$

The inner solution is $c = 0$ as in the previous example. The boundary-layer part is now expressed in terms of the stretched variable $y = (1 - x)/\sqrt{\epsilon}$ as

$$\frac{\partial^2 c}{\partial y^2} = c(1 - \sqrt{\epsilon}y) \tag{14.43}$$

The solution is obtained now as an expansion in $\sqrt{\epsilon}$. Let

$$c = c_0 + \sqrt{\epsilon}c_1 + O(\epsilon)$$

to a first approximation. The differential equation for c_0 is obtained as

$$\frac{\partial^2 c_0}{\partial y^2} = c_0$$

The solution is the same as for a catalyst with constant activity:

$$c_0 = \exp(-y)$$

The c_1 differential equation is obtained by equating the terms with $\sqrt{\epsilon}$. You should verify that the following equation is obtained:

$$\frac{\partial^2 c_1}{\partial y^2} = c_1 - yc_0$$

On substituting the previously obtained solution for c_0 into this we have

$$\frac{\partial^2 c_1}{\partial y^2} = c_1 - y \exp(-y)$$

The boundary conditions for the c_1 problem are now $c_1 = 0$ at both $y = 0$ and $y = \infty$. The solution for c_1 is the sum of a homogeneous part and a particular solution to the non-homogeneous term (the second term on the RHS). This is readily obtained in MAPLE as

$$c_1 = \frac{1}{4}y(1 + y)\exp(-y)$$

The combined solution matches reasonably well with the numerical solution.

14.3.4   Example: gas absorption with reversible reaction

Consider a case of gas absorption with a first-order reversible reaction represented by

$$A \rightleftharpoons B$$

with the rate given by

$$\text{Rate} = k(A - B/K)$$

where k is the first-order rate constant and K is the equilibrium constant for the reaction. We will model this assuming a film model for mass transport. The problem is to set up the model equations and solve for the concentration profiles for the case of a large value of k, which leads to a singular perturbation problem. The problem has an analytical solution, which helps us to evaluate the accuracy of the perturbation results. Also we assume B to be a non-volatile species and assume equal values for the diffusion coefficients for A and B.

The problem can be modeled by a set of diffusion–reaction equations which is directly given in terms of dimensionless form below:

$$\frac{d^2 c_A}{d\xi^2} = Ha^2(c_A - c_B/K) = -\frac{d^2 c_B}{d\xi^2} \tag{14.44}$$

Distance is scaled by the film thickness while the concentrations are scaled by the interfacial concentration of species A. The parameter Ha is the Hatta number defined as $k\delta^2/D_A$. Also equal diffusivity is assumed for simplicity, since we wish to illustrate the application of the perturbation procedure.

The following simplified boundary conditions are used for illustration purposes. The concentration $c_A = 1$ at $\xi = 0$ while $dc_B/d\xi$ is taken as zero here, since we assume B to be a non-volatile species.

The concentration of A is taken as zero in the bulk liquid, $\xi = 1$. This, for example, would apply for a case where the bulk liquid is recirculated with a fresh solution, or at the top of an absorption column where the inlet liquid has no dissolved A.

Before proceeding with the perturbation analysis it is useful to examine the invariance property of the system and reduce the set of equations to a single equation for the concentration profile of A.

Invariance

Integration of the first and the last pair of equations in (14.44) and the use of the boundary conditions listed above provide a relationship for c_B in terms of c_A. This can be expressed in terms of dimensionless concentration as

$$c_B = -c_A + p_0(\xi - 1) \tag{14.45}$$

and p_0 is the flux $dc_A/d\zeta$ at the interface:

$$p_0 = \left(\frac{dc_A}{d\xi} \right)_{\xi=0}$$

The student should derive and verify the above invariance relation, Eq. (14.45). Invariance is simply a mathematical statement of the overall mass balance for A and B across the film.

Equation (14.45), when substituted into the first pair of equations in Eq. (14.44), yields the differential equation with the unknown coefficient, viz., p_0. The equation in terms of dimensionless variables is then

$$\frac{d^2 c_A}{d\xi^2} = \beta^2 \left[c_A - p_0 \frac{\xi - 1}{1 + K} \right] \tag{14.46}$$

where

$$\beta^2 = Ha^2 \frac{1 + K}{K} \tag{14.47}$$

The undetermined parameter p_0 appearing in the above differential equation is the slope of the concentration profile at $\xi = 0$. In fact this is often the primary quantity of interest, since it determines the rate at which gas A is absorbed in the liquid.

Problems with undetermined parameters are difficult to solve numerically. A paper by Ramachandran (1992) addresses some of these issues and provides a detailed study of the problem of gas absorption. The MATLAB routine BVP4C now has a built-in feature whereby ODEs with unknown parameters can be solved. An illustrative numerical solution is shown in Table 14.2, which can be used to compare the various solution methods, including the perturbation solution, which will now be elaborated upon.

Is it a singular or a regular perturbation problem?

Perturbation solution

We wish to examine the solution for large β and hence use $\epsilon = 1/\beta$ as the coefficient in the higher-order term of the differential equation. The problem to be solved then is

Table 14.2. Detailed concentration and gradient profiles for gas absorption followed by a first-order reversible reaction as discussed in Section 14.3; with $Ha^2 = 10^6$ and $K = 100$

x	c	p
0.0	1.0000	−91.782
0.001	0.9422	−33.266
0.005	0.9058	−2.327
0.01	0.8997	−0.952
0.03	0.8815	−0.915
0.05	0.8623	−0.909
0.075	0.8406	−0.909
0.1	0.8178	−0.909
0.2	0.7269	−0.909
0.3	0.6361	−0.909
0.4	0.5452	−0.909
0.5	0.4543	−0.909
0.7	0.2726	−0.909
1.0	0.0000	−0.909

$$\epsilon^2 \frac{d^2 c_A}{d\xi^2} = \left[c_A - p_0 \frac{\xi - 1}{1 + K} \right] \tag{14.48}$$

with the conditions $c_A(\xi = 0) = 1$ and $c_A(\xi = 1) = 0$.

The base solution in the outer region is obtained by setting ϵ to be zero. This generates a linear concentration profile

$$C_A^i = p_0 \frac{\xi - 1}{1 + K}$$

which also satisfies the boundary condition at $\xi = 1$. The boundary condition at $\xi = 0$ is not satisfied, and hence we conclude that there is a boundary layer near $\xi = 0$ and the problem is of singular perturbation type. In the boundary layer a stretched coordinate is called for, namely

$$y = \xi / \epsilon$$

which reduces the equation to

$$\frac{d^2 c_A}{dy^2} = c_A - p_0(y\epsilon - 1)/(1 + K) \tag{14.49}$$

In a further simplification, neglecting the term $y\epsilon$ can be justified, which is the first approximate perturbation solution:

$$\frac{d^2 c_A}{dy^2} = c_A + p_0/(1 + K) \tag{14.50}$$

The solution satisfying the boundary condition at $y = 0$ is

$$c_A = [1 + p_0/(1 + K)]\exp(-y) - p_0/(1 + K)$$

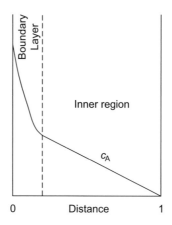

Figure 14.3 A diagram showing a constant-gradient inner solution.

The matching condition that is automatically satisfied is

$$c_A(BL)(y \to \infty) = c_A(\text{outer})(x = 0) = -p_0/(1 + K)$$

There is a slope discontinuity between the two solutions, which can be resolved by taking a higher-order solution within the boundary layer.

The unknown parameter p_0 is obtained by taking the derivative of the boundary-layer solution:

$$p_0 = -\frac{(1 + K)Ha}{1 + K + Ha}$$

For the data in Table 14.2 we have $p_0 = -91.7897$, and hence we conclude that the above perturbation solution is in good agreement both with the analytical (-91.8596) and with the numerical (-91.782) solution. Higher-order terms can be included if necessary, but the first approximation is usually sufficient for fast reactions of the type considered here.

Note that the boundary layer is of a constant-gradient type rather than a constant-concentration type. Thus the concentration varies in the inner layer, but the variation is linear, resulting in a constant-gradient inner solution as depicted in Fig. 14.3.

14.3.5   Stokes flow past a sphere: the Whitehead paradox

Here we discuss the problem of flow past a sphere under low-Reynolds-number conditions. We indicate that a seemingly regular perturbation problem is actually a singular perturbation problem. The Navier–Stokes equation in dimensionless form is used and stated below without the * notation:

$$\nabla^2 v = \nabla p + Re(v \cdot \nabla)v$$

where Re is the Reynolds number. All the quantities in the above equation are in dimensionless units with the distance scaled by R and the velocity scaled by the far stream velocity V. Hence Re is defined as RV/ν using the particle radius as the scale.

How is the dimensionless pressure defined in the above equation?

At low velocity or for highly viscous fluids we expect that Re is small and the problem may be approached as a regular perturbation problem. The base flow is set for $Re = 0$ and leads to

$$\nabla^2 v_0 = \nabla p_0$$

where v_0 is the solution to the problem with the inertia terms neglected and p_0 is the corresponding pressure field. This is known as the Stokes flow approximation. The detailed solution derived by Stokes in 1851 will be examined in a later chapter. Here we examine the qualitative features and the approximation for finite non-zero values of Re. For the latter case we may expect that the solution can be expanded as a regular perturbation problem as

$$v = v_0 + Re v_1 + Re^2 v_2 + \cdots$$

Whitehead attempted to solve the problem in this manner and failed in 1898. No higher-order solutions (v_1 etc.) that could match the far-field boundary condition could be obtained, which is called the Whitehead paradox. The failure to find the solution may have persuaded Whitehead to leave engineering analysis and pursue more exciting fields in philosophy and logic. The problem is of a singular perturbation nature as shown in the following scaling analysis and needs the theory of matched asymptotic approximation. This theory was not established at that time and hence this resource was not available to Whitehead to complete the solution.

The key problem is the choice of the radius of the sphere as the scaling parameter. Far away from the sphere the local radial position is much larger than the radius and hence the radius is not the correct choice for the scaling. Let us leave the scale as some arbitrary r and compare the magnitudes of the viscous and inertial terms.

The viscous terms scale as $\mu V/r^2$.

The convective terms scale as $\rho V^2/r$.

The ratio of the two terms is therefore

$$\frac{\text{Inertia}}{\text{Viscous}} = \frac{\rho r V}{\mu} = Re \frac{r}{R}$$

The inertia terms are small near the sphere (small r) but significant at large r. Hence a uniform (regular) perturbation solution is not possible. Neglecting the inertia terms in the far-field region is inappropriate! The matched asymptotic solution is needed, with a near-field solution matched with a far-field solution. This approach was applied first by Oseen and subsequently by Pearson and others. The details are not presented here, and the student may wish to study the book by Leal (2007), who provides an excellent detailed analysis of the solution that is recommended for the mathematically challenged. The two regions and the form of the solution for these two regions are shown qualitatively in Fig. 14.4.

Two-dimensional Stokes flow

The situation is even worse for 2D flow in that a base solution for $Re = 0$ that matches the far-field boundary condition cannot be found. This led Stokes to conclude that no such flows can exist, which conclusion is sometimes referred to as the Stokes paradox. The true picture is that even for $Re = 0$ the problem has to be solved as a singular perturbation problem. For 3D flow it turns out that the inner solution has a term that matches with the first term in the outer solution and further terms in the outer solution are not needed for a first approximation. Thus the Stokes flow could be solved for 3D but not for 2D flows.

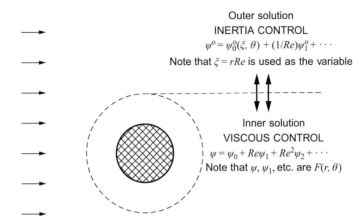

Figure 14.4 Matching and patching; figure showing two solutions for the Stokes problem. The crosshatched region is a sphere of radius R; ψ is the axisymmetric streamfunction defined in Section 3.10.3.

The heat transfer paradox

One may note that the analogous problem in heat transfer can be formulated in the same manner. The governing equation in dimensionless form is

$$\nabla^2 \theta = Pe(\boldsymbol{v} \cdot \nabla)\theta$$

The Péclet number replaces the Reynolds number here. Again an uniformly convergent solution with Pe as the expansion parameter of the form

$$\theta = \theta_0 + Pe\theta_1 + \cdots$$

is not possible, which is sometimes referred to as the Whitehead paradox for heat transfer. The solution for θ_0 (the pure conduction case) is simple and leads to a classical result that $Nu = 2$. A matched asymptotic solution procedure is once again called for in order to seek the higher-order solutions, for θ_1, θ_2, etc. Again, for more details the book by Leal (2007) is worthy of careful study.

14.4   Domain perturbation methods

This refers to a class of methods where the boundary is perturbed and the effect of small changes in the boundary is examined rather than the study of the changes of a parameter in the differential equation. The solution to a base case (e.g., for a simple shape of the boundary) is assumed to be known. The method is best learned by working through examples. Hence a simple example is provided; qualitative aspects are indicated and no detailed solutions are provided. The goal is mainly to introduce the topic since some of you will find some interesting research problem where this tool can be applied. For an excellent discussion on domain perturbation and many applications of the method, the book by Leal (2007) is again a good source.

Example 14.4.

Consider flow in a rectangular channel. Let the walls be located at $y = \pm h$, with $y = 0$ being the center of the walls. A pressure gradient G is applied in the x-direction and we have the classical plane Poiseuille flow problem with a parabolic velocity profile. This represents the base problem. Let us now consider the flow in a channel with a corrugated boundary with some sinusoidal corrugation in the z-direction. The geometry is shown in Fig. 14.5.

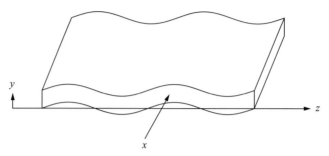

Figure 14.5 A figure showing the flow geometry considered in the domain perturbation problem. The gap width y is assumed to vary as a sinusoidal function of z. The flow direction is x.

The no-slip boundary condition is to be applied at

$$v = 0 \text{ at } y = \pm h[1 + \epsilon \sin(2\pi z/L)] \tag{14.51}$$

rather than at $y = h$.

The channel is corrugated in the z-direction and the flow is in the x-direction as indicated in Fig. 14.5. The symmetry condition holds at $y = 0$. In view of the symmetry, only the domain $0 \leq y \leq h$ need be considered.

Solution.

The assumption of unidirectional flow is used, so we consider only v_x as the variable. We denote v_x as v here. However, due to the corrugated boundary, v now depends on y and z. However, we wish to use the 1D equation

$$\mu \frac{\partial^2 v}{\partial y^2} = -G \tag{14.52}$$

and modify the boundary conditions so that they are applicable at $y = \pm h$ rather than at the corrugated boundary. Using approximate boundary conditions of this type is the crux of the boundary perturbation method.

We propose a solution in the form

$$v = v_0 + \epsilon v_1 + \epsilon^2 v_2 + \cdots \tag{14.53}$$

where v_0, v_1, etc. are functions of y and z. It is assumed that each of these functions v_0, v_1, etc. satisfies the governing equation (14.52). We need matching boundary conditions for each of these functions. The key to the domain perturbation method is to replace the exact boundary condition Eq. (14.51) with an approximate boundary condition at $y = h$

that is asymptotically equivalent to the exact boundary condition in the limit $\epsilon \to 0$. For this purpose we assume that v can be expanded near $y = h$ as in power series of ϵ:

$$v(y) = v(h) + (y - h) \left(\frac{\partial v}{\partial y} \right)_{y=h} + \frac{(y - h)^2}{2} \left(\frac{\partial^2 v}{\partial y^2} \right)_{y=h}$$

which, in view of the condition at $y = h[1 + \epsilon \sin(2\pi z/L)]$, becomes

$$v = v(h) + \epsilon \sin(2\pi z/L) \partial v/\partial y + \epsilon^2 \sin^2(2\pi z/L)/2 + \cdots$$

Now, on substituting the expansion for v from Eq. (14.53) into this, we have the following expression for v near the boundary:

$$v = [v_0 + \epsilon v_1 + \cdots]_{y=h} + \epsilon \sin(2\pi z/L) \left(\frac{\partial}{\partial y} [v_0 + \epsilon v_1 + \cdots] \right)_{y=h} + O(\epsilon^2)$$

and this needs to be set to zero. Because ϵ is an arbitrarily small parameter, each term in powers of ϵ can be set to zero. Hence we generate a sequence of boundary conditions for v_0, v_1, v_2, etc. Each problem is solved in succession in a similar manner to the regular perturbation method.

We find $v_0 = 0$ at $y = h$, and the use of this leads to the solution of the classical plane Poiseuille flow. Furthermore, we find

$$v_1(h) = - \sin(2\pi z/L) \left(\frac{\partial v_0}{\partial y} \right)_h$$

as the needed boundary condition for the perturbed problem for the first approximation. Thus the v_1 problem can now be solved. Using this, the boundary condition for the v_2 problem can be generated, and this problem can be solved then, and so on. Further details are left as an exercise.

Summary

Three methods of analysis of the transport phenomena problems and the corresponding governing equations are studied in this chapter, which have a wide range of applicability.

- Dimensionless analysis using linear algebra provides a means of finding the dimensionless numbers for a given set of parameters, and is useful when the governing differential equation or the system to be modeled is too complicated or not formulated properly due to lack of adequate closure terms. This is a tool of importance in the correlation of experimental data. It also provides some ideas of scaleup of complex systems utilizing the concept of dynamic similarity. The student will be able to identify the dimensionless groups for a given problem and use these in the interpretation and correlation of experimental data.

- Scaling analysis is a tool for order-of-magnitude estimation of differential equations. It permits us to identify unimportant terms and thereby permits simplification of the equations. Often some correlations can be suggested purely by scaling analysis without the need for a complete solution. The student should be able to apply these to new problems and also appreciate the simplicity of the procedure in resolving many complex problems.

- Perturbation analysis is a powerful mathematical technique that permits analytical or semi-analytical solutions to some difficult linear as well as non-linear problems. The boundary-layer theory and many other related topics in transport phenomena are heavily based on this technique. The student should be able to distinguish between regular and singular perturbation analysis, and be able to apply this to some common problems in transport phenomena.

ADDITIONAL READING

Dimensional analysis is covered in the book by Zlokarnik (1991), who provides also an introduction to scaleup of various process equipment using the dimensionless groups.

Scaling analysis is discussed in detail in the excellent treatise by Krantz (2007). The review chapter by Ruckenstein (1987) is a classic where he demonstrates the application to a variety of complex problems. His article is recommended because of the simplicity and originality with which he treats complex cases. These sources provide many examples of the application of this powerful method.

The book by Bender and Orzsag (1978) is the starting source for asymptotic expansion and perturbation analysis.

The classical books on perturbation methods are those by Cole (1968), van Dyke (1975), and Nayfeh (1981). The book by Johnson (2005) is an excellent source for singular perturbation methods with applications to a variety of problems in engineering and science.

The perturbation methods in general and analysis of Stokes flow in particular have been treated lucidly by Leal (2007).

Problems

1. **Simple pendulum: time of oscillation.** The time of oscillation t of a simple pendulum is expected to be a function of the mass of the pendulum, its length L, and the gravitational constant. How many dimensionless groups can be formed? Show that the key dimensionless group is $t\sqrt{g/L}$. Hence conclude that t should be proportional to $\sqrt{L/g}$ but is independent of the mass of the pendulum from purely dimensional considerations without invoking any physics.

   Note that the constant of proportionality is 2π, but this value cannot be determined from dimensionless analysis. A physics model is needed for this.

2. **Energy release from an explosion.** The eminent British fluid dynamicist G. I. Taylor deduced the energy release E from the first atomic explosion simply by dimensional analysis. This was strictly classified information, but Taylor showed that those well versed in dimensional analysis have access to such confidential information!

   A shock wave of radius R is produced by the explosion and R increases as a function of time, t. Taylor showed that

$$R = CE^{1/5}t^{2/5}\rho^{-1/5}$$

   Verify his result, which was based on dimensionless analysis. The energy release was classified information but from the pictures of the explosion R vs. t could be plotted on a log–log scale and the plot should have a slope of 2/5. The estimate of the energy release can then be obtained from the intercept of this graph. Taylor assumed the constant C above to be nearly one (this value was based on a similar model problem) and estimated the energy release. His estimate was in close agreement with the classified value.

3. In correlating the diameter of drops formed at the orifice, the viscosity is to be included as the core group rather than density as done in the text. Form groups with d_o, ρ, and μ as the core groups and surface tension and drop diameter as the variables.

4. **Natural convection past a vertical plate.** A vertical plate of height L at a temperature of T_s is immersed in a fluid at a temperature of T_a. Local density variation due to temperature causes a flow, which causes an enhancement in the heat transfer rate.

   Note that the heat flux depends now on the following variables:

   $$q = f(\Delta T, \beta, g, \rho, \mu, k, L)$$

   where β is the volumetric coefficient of thermal expansion.

   (a) What is the expression for β based on the ideal-gas law? What are its units?
   (b) Identify the dimensionless groups needed to correlate the data on heat transfer in such systems using the method of matrix algebra.
   (c) Suggest expressions to correlate data on natural convection heat transfer.

5. **Roasting of a turkey: scaling analysis.** It is required to find the cooking time of a large turkey. Experimental data are available for a small turkey of mass m_s, and the time is t_s. Suggest the cooking time, t_L, for a larger turkey of mass m_L by dimensionless and scaling analysis.

6. Apply scaling analysis for gas absorption with first-order reaction based on the film model. Derive a relation for the depth of penetration of the gas for a fast reaction. Using this relation, verify the following model prediction:

   $$R_A = C_{A,s}\sqrt{D_A k_1}$$

7. **The film-penetration model.** Gas-absorption systems are usually modeled by assuming a film thickness and steady-state diffusion. In an attempt to modify this picture one can assume a finite film but allow for a transient diffusion in the film. This model is known as the film-penetration model. Perform a scaling analysis for this system and develop a useful relationship to correlate the data on absorption.

8. How does the power in an agitated vessel depend on the impeller speed and the impeller diameter in (a) the laminar regime and (b) the turbulent regime?

9. **Scaleup of agitated systems for equal mass transfer.** The mass transfer coefficient in agitated gas–liquid systems is often proportional to the power per unit volume of the reactor rather than the total power P dissipated in the system. On this basic derive a scaleup criterion for equal-mass transfer for a small and a large reactor.

10. For gas–liquid dispersions in agitated vessels, the gas flow rate Q_G is also important. Show that an additional group, namely the flow number defined as $Q_G/(\Omega d_i^3)$ is needed in order to correlate the data. Hence the power number is then correlated as a function of the Reynolds number and the flow number. Search the literature and find the common correlations suggested for this.

11. The following data were obtained by Sharma and Danckwerts (1963) for CO_2 absorption in a laminar jet of solution in which the gas underwent a first-order reaction:

Exposure times	78.5	113	167	360
Total rate (mol/cm^2)	16.5	15.0	13.0	11.2

Interpret the data on the basis of the scaling model described in Section 14.2.2.

12. For turbulent flow of water in a pipe of diameter 5 cm with $Re = 10^5$ estimate the magnitude of the length and velocity scales at which viscous dissipation becomes important.

13. An agitated tank has an impeller diameter of 10 cm and operates at a speed of revolution of 10 r.p.s. The tank has a diameter of 20 cm and is filled up to a height of 20 cm with liquid. Find the following: (i) the power consumed assuming turbulent conditions, (ii) the length scale at which viscous dissipation takes place, and (iii) the corresponding velocity scale.

14. **Droplet distribution in oil dispersion: a case-study problem.** Treating an oil spill by application of a chemical dispersant is a useful means for dispersion of oil. Knowledge of the droplet size formed as a function of the energy dissipation rate is useful to scale the laboratory data to oceanic applications. A dimensional and scaling analysis was done by our group and published in a paper by Mukherjee *et al.* (2012). This paper presents a compelling case study of the methods discussed in this chapter.

Identify the important variables and perform a dimensionless analysis. Show that the Reynolds number, Weber number, and viscosity ratio of water to oil are the key parameters which affect the dispersed drop size.

The form for the correlation cannot be established from dimensionless analysis. Hence a scaling analysis where the order of magnitude of surface tension and the restoring viscous forces is compared with the inertial forces leading to drop break-up is needed. Perform such an analysis based on the discussion in the above paper and show that, depending on the relative values of the dimensionless groups, either the viscous or the surface-tension forces could be important as the restoring forces. Derive criteria to distinguish these effects and show practical implications.

15. **Regular perturbation for diffusion with second-order reaction.** Consider for illustration the diffusion–reaction problem given by

$$\frac{d^2 c}{dx^2} = Mc^2 \tag{14.54}$$

with the same boundary conditions as before.

Use the expansion for c in terms of M as in Eq. (14.24) for small values of M, which is a regular perturbation problem. Show that by substituting into the differential equation we obtain

$$Mf_1'' + M^2 f_2'' + \cdots = M + 2M^2 f_1 + O(M^3) \tag{14.55}$$

By equating the powers of M, write out the governing equations for f_1 and f_2. Integrate the equations to obtain the following results:

$$f_1 = x^2 - x$$

and

$$f_2 = (x^4/4 - x^3 + 2x)/3$$

Derive an expression for the effectiveness factor, and show that it can be approximated as $1 - 2M/3$ to an approximation of $O(M^2)$. Derive the next approximation.

Also solve the problem numerically using BVP4C and compare the results.

16. **Heat generation in a slab with variable thermal conductivity.** Solve the problem of heat generation in a slab with a variable thermal conductivity. Show that the problem can be represented as

$$\frac{d\theta}{d\xi}\left[(1 + \beta\theta)\frac{d\theta}{d\xi}\right] = -1$$

with the boundary condition of no flux at the center and convective heat loss at the surface, where β is the coefficient in the thermal conductivity relation. Thus β equal to zero is the base case of constant conductivity.

Obtain a regular perturbation solution with β as the expansion parameter.

17. **The convection–diffusion problem.** Solve the following problem:

$$\frac{d^2c}{dx^2} - Pe\frac{dc}{dx} = 0$$

Use the boundary conditions $c(0) = 1$ and $c(1) = 0$.

Solve both for large Péclet number, Pe, where this is a singular perturbation problem, and for small Pe, where this is a regular perturbation problem. Compare your answers with the analytical and numerical solutions. Where is the boundary layer for large Pe? Is it at $x = 0$ or at $x = 1$?

18. **Catalyst with varying activity.** Perform the order-of-ϵ^2 approximation for the problem in Section 14.3.3 of diffusion with reaction in a catalyst of variable activity. Compare the flux with that obtained from the BVP4C solver in MATLAB. Perform computations for ϕ of 10, 20, and 30. The results are 9.7313. 19.7416, and 29.745, respectively.

Now consider the problem of varying catalyst activity with the maximum activity at the center. The diffusion–reaction model is now given by

$$\epsilon\frac{\partial^2 c}{\partial x^2} = c(1 - x)$$

Perform a perturbation analysis for the problem. Use MAPLE or other programs to solve the c_0, c_1, etc. problems. You will note that the c_0 problem is an Airy function. Derive an approximate formula for the effectiveness factor.

Verify the following results obtained from BVP4C:

Thiele, ϕ	10	20	30	40
Flux, p	3.3837	5.3714	7.0384	8.5259

Note that ϵ is equal to $1/\phi^2$. The flux, p, is at the surface of the catalyst.

19. **A singular reaction–diffusion problem with two boundary layers.** Solve the following singular perturbation problem:

$$\epsilon\frac{d^2 c_b}{d\xi^2} = \xi c_b - 1$$

which arises in a model for a consecutive reaction in a liquid film (Deen, 2011). Here c_b is the concentration variable, ξ is the distance variable, and ϵ is a small parameter.

The boundary conditions are that there is no flux for c_b at both ends, viz., $\xi = 0$ and $\xi = 1$. Show that there are boundary layers at both sides (0 and 1), and that a different scaling is needed for each side. Obtain these solutions together with that for the inner region to the first needed level of approximation.

Plot your results for $\epsilon = 0.01$.

Answers from a numerical solution generated by BVP4C are provided for comparison: $c_b(0) = 5.9782$ and $c_b(1) = 1.1439$. You may also wish to run the BVP4C code to verify these results for extra practice.

20. **Gas absorption with a reversible reaction.** Consider again the problem discussed in Section 14.3.4. Your task is to obtain the solution by analytical and numerical methods and compare it with the perturbation solution given in Section 14.3.4.

Verify the following analytical solution for the concentration distribution and the gradient at the gas–liquid interface:

$$c = c_1 + \left(1 - c_1 + \frac{p_1}{K+1}\right) \frac{\sinh[\beta(1-x)]}{\sinh \beta} + \frac{p_1}{K+1}(x-1) \qquad (14.56)$$

and

$$p_1 = -\frac{(1-c_1)(1+K)}{1 + (K/\beta)\tanh \beta} \qquad (14.57)$$

These equations can be used to compare the numerical results as well as perturbation results. Perform such a comparison for values of M and K in the text and verify the results in Table 14.2.

21. **Gas absorption with rapid reaction.** The following problem arises in gas absorption with fast reaction. The notations have been slightly changed from Eq. (10.5) earlier:

$$\epsilon^2 \frac{d^2 c_A}{dx^2} = c_A[1 + q^* c_A - q^* p_{x=0}(x-1)] \qquad (14.58)$$

where q^* is defined as the reciprocal of q in Eq. (10.44). The parameter ϵ is the reciprocal of Ha. The boundary conditions for the diffusing gas c_A are $c(x=0) = 1$ and $c(x=1) = 0$. The problem has two interesting characteristics. (i) The solution is of a boundary-layer type for $Ha \to \infty$. (ii) An undetermined parameter $p_{x=0}$, the slope of the concentration gradient, appears. This parameter is not known at the start and can only be estimated *a posteriori*.

Here we wish to examine the solution for large Ha. For fast reaction Ha is large and hence ϵ is a small parameter, leading to a singular perturbation problem. Show that the solution for ϵ equal to zero leads to an instantaneous reaction profile and develop the next approximation, which is approximate to $O(\epsilon)^2$. Find the enhancement factor. Then use the results and find the next-higher-order solution.

22. **Pulsatile flow in a pipe.** Here we revisit the problem of pulsatile flow in a pipe discussed in Section 11.11.5. The problem was solved analytically using complex variables. The perturbation method is also suitable for this problem and provides additional physical insight into the solution.

The model equation in dimensionless form is restated here for ease of reference:

$$Wo^2 \frac{\partial v^*}{\partial t^*} = \cos(t^*) + \frac{1}{\xi} \frac{\partial}{\partial \xi} \left(\xi \frac{\partial v^*}{\partial \xi}\right) \qquad (14.59)$$

The term $\cos(t^*)$ is the pressure oscillation, and the response of the pressure oscillation to changes in velocity is to be determined.

The three terms in the above equation may be interpreted as inertia, pressure, and viscous forces. Recall that Wo^2 is defined as $\omega R^2/v$, where ω is the frequency. The parameter Wo^2 is then ratio of the viscous diffusion time to the period of oscillation.

The first boundary condition is that there is no shear at $\xi = 0$, and hence $\partial v^*/\partial \xi = 0$ here. The second boundary condition is that of no slip at $\xi = 1$ and here $v^* = 0$. We are looking for solutions for the periodic steady state, and hence there is no initial condition in t^* applied for the time variable.

We wish to seek a perturbation solution to this problem for two limiting cases of small and large Wo. The case of large Wo is examined in the next problem.

For small Wo the velocity is expressed as

$$v = v_0 + Wo^2 v_1 + \cdots$$

Show that the base solution when Wo is zero is

$$v_0 = \frac{1}{4} \cos(t^*)(1 - \xi^2)$$

which is the laminar flow under quasi-steady-state conditions. The inertia terms are nearly zero here, and the pressure forces are balanced by viscous forces. The flow is in phase with the pressure oscillation.

Then use this to show that v_1 is

$$v_1 = Wo^2 \left[\frac{3}{16} - \frac{3\xi^2}{4} + \frac{\xi^4}{16} \right] \sin(t^*)$$

This is the correction to the Poiseuille solution, and the contribution is out of phase with the pressure oscillation. Higher-order terms can be obtained if needed. The algebra is lengthy, but MAPLE can be used to ease the burden.

23. **Pulsatile flow at large Wo.** For large values of Wo the problem is of singular perturbation type. Verify that the inner solution is of a plug-flow type. The pressure is balanced by inertial forces here. Also show that the inner solution does not satisfy the boundary condition at the wall. A singular perturbation solution is therefore needed. Perform the necessary analysis and compare your answer with the PDEPE solution and the analytical solution. Show that the base solution is

$$v^* = \frac{1}{Wo^2} \sin(t^*)$$

We note that the profile is plug flow and the flow is completely out of phase with the pressure oscillation. The boundary condition at $\xi = 1$ is not satisfied, and hence this is a singular perturbation problem. Find the boundary-layer-type solution which is useful near the walls.

24. **Radial flow between parallel disks.** Consider again the radial flow between two parallel disks examined in Section 6.5. The case of low Reynolds number was examined there, and a solution in which the non-linear terms were ignored was obtained. Here we wish to examine the role of inertia by performing a regular perturbation analysis. Expand v_r and v_z as perturbation expansions of Re, and show that the base solution is the one derived in Section 6.5. Derive the next approximation which includes the inertia term, which was neglected in that section.

25. Complete the solutions for v_1 in Example 14.4 for domain perturbation in the text. What are the boundary conditions for the v_2 problem?

26. **Potential flow past a perturbed circle: an example of domain perturbation.** The problem of potential flow past a circular object is a well-studied problem in fluid mechanics governed by the Laplace equation. We wish to study the flow around a circle which is

slightly perturbed. The boundary is now described as

$$r = R(1 + \epsilon \cos \theta)$$

The solution for the streamfunction for the base problem where the boundary is at $r = R$ is that for a circle studied in Section 15.4.2. Now set up the domain perturbation analysis for this problem with the above perturbed boundary and solve for the modified domain.

15 More flow analysis

Learning objectives

The learning objectives of this chapter are the following:

- to understand the basic properties of Stokes flow; to understand how the forces and torques on a particle suspended in a body can be related to the velocity and rotational speed;
- to understand the mathematical tools useful for solution of Stokes flow; to study some additional examples of low-Reynolds-number flows, such as bubble flow and Brownian motion;
- to understand the properties of irrotational flow and potential flow; to derive the Bernoulli equation for irrotational and inviscid flow and indicate the differences;
- to study problems where the velocity field can be computed as the gradient of a scalar field; to review some mathematical tools for such computations;
- to review the simplification of the N–S equations for the boundary layers by a scaling analysis discussed in Section 14.2.1;
- to show how the external pressure profile gets imposed on the boundary-layer momentum balance and acts as a source term for momentum within the boundary layer;
- to find the velocity profiles over a flat plate by the integral method and to find an expression for the drag coefficient;
- to provide a more general analysis of boundary layers using the Falkner–Skan equations; and
- to examine some additional boundary-layer-type flows, such as Hiemenz flow, flow over a rotating disk, and other common geometries.

Mathematical prerequisites

- Basic eigenfunctions of PDEs.
- Fundamental solutions of PDEs.
- Legendre functions.
- Harmonic functions and general solutions to the Laplace equation.

Chapter 6 provided some illustration of the analysis of flow problems. The problems studied were usually 1D flow with the velocity changing in the cross-flow direction. The focus of this chapter is on studying somewhat more complex flow problems in further detail. The flow problems studied here are mathematically more involved and involve non-zero velocity components in two or more directions. The problem looks formidable at first sight, but can be broken down into a number of common types, each of which can be solved by a different set of simplifications. Hence it is useful to classify the types of problems which will be studied in this chapter and look at the main features of each of these types of flow. The mathematical method for solving these problems will also be briefly summarized, together with some computational procedures.

1. **Low-Reynolds-number flows or Stokes flow.** This applies for flow of a highly viscous liquid at relatively low velocity. This has a wide range of applications in particle settling, colloidal suspension flow, and other fields. The viscous effects dominate, and one can reduce the Navier–Stokes (N–S) equation simply to the Stokes equation as discussed in Section 5.4.2 by setting the Reynolds number to zero. Here we continue the study of this class of flow in some detail. First we show some properties of low-Reynolds-number flow and present some illustrative results. Then the section following these preliminaries will review the mathematical features and the solution methods for Stokes flow briefly.

2. **Flow at large Reynolds number or potential flow.** This is the opposite of the Stokes flow and the viscous terms are not considered in this model. The N–S equation reduces to the Euler equation of motion for this case. The case of external flows past solids with high Reynolds number is one such example where it is observed that, outside a thin region near the solid, the equations of flow can be simplified. The solution for the velocity potential (a scalar variable) is sufficient to characterize the flow and the details are presented. This is then followed by a brief section on the mathematical underpinning of the potential flow.

3. **Flow in boundary layers.** The potential flow does not apply near the solid surface and we have to include the effect of viscosity in this region. Such flows come under the theory of boundary-layer flows. We start off with the simplest and most well-studied problem in boundary-layer theory, *viz.*, the flat plate. Analytical solutions briefly presented earlier are revisited and more details on the flow analysis are provided. A number of additional flow examples will be studied.

Boundary-layer flows cannot be studied in isolation, and one requires information from potential flow to couple it with the region outside the viscous layer. The potential flow provides the information on the pressure gradient which drives the flow within the boundary layer itself. The numerics of this coupling will be examined. Additional complexities associated with the boundary-layer separation are briefly discussed. Furthermore, the flow may become turbulent at some distance from the starting point, which adds another level of complexity to the problem.

15.1 Low-Reynolds-number (Stokes) flows

The simplified version of the N–S equation where the inertial terms are neglected is known as Stokes flow or creeping flow. This simplified representation applies when the flow Reynolds number is small. The situation is encountered in a number of important applications such as the study of (i) micro-circulation, (ii) suspensions and emulsions and their rheological behavior, (iii) colloidal dispersions, (iv) flow in micro-fluidic devices, (v) flow in porous media, etc.

First, we present some basic properties of Stokes flow. Then general solutions are presented, and we show how the solution to complex flows can be constructed by the superposition of these general solutions. The application to the basic problem of Stokes flow past a sphere is presented. The derivations and other mathematical details pertinent to the general solutions are not presented here, since this is an introductory treatment of Stokes flow. Some of these are left as exercises.

15.1.1 Properties of Stokes flow

The differential equation describing Stokes flow can be represented as

$$\boxed{\mu \, \nabla^2 v - \nabla p = 0} \tag{15.1}$$

and

$$\nabla \cdot v = 0 \tag{15.2}$$

The term p can be treated as the modified pressure $\mathcal{P}$ to make the problem general; the notation p is used here for simplicity.

In the above simplification we note that the convective terms are dropped because the inertia effects are small due to the small value of the Reynolds number, while the time derivative is dropped since the viscous diffusion time in this case is small compared with the observation time (the Strouhal number is small). This amounts to making a pseudo-steady-state assumption for the time variable. For example, consider a particle oscillating around its mean position with a frequency ω. Then the corresponding coefficient term in the time derivative is the Wormersley number, which is expected to be small for highly viscous fluids; hence the time derivative can be dropped without loss of accuracy for such cases.

The following properties of the Stokes flow are worth noting. The proofs of some of these properties are left as exercises.

1. The pressure field satisfies the Laplace equation:

$$\nabla^2 p = 0 \tag{15.3}$$

2. The velocity field satisfies the biharmonic equation:

$$\nabla^4 v = 0 \tag{15.4}$$

3. The vorticity field satisfies the Laplace equation:

$$\nabla^2 \omega = 0 \tag{15.5}$$

4. The streamfunction, ψ, for 2D Stokes flow satisfies the biharmonic equation

$$\nabla^4 \psi = \nabla^2(\nabla^2 \psi) = 0 \tag{15.6}$$

for planar flow. Here ∇^2 is the Laplacian operator.

The above relation follows from using the kinematic relation $\omega = -\nabla^2 \psi$ in Eq. (15.5) above. This kinematic condition was derived in Section 3.10, and you should review this at this stage. You may wish to review also the definition of the streamfunction from Section 3.10.1 at this stage.

Why is ω treated as a scalar here?

5. The corresponding equation for axisymmetric flows is somewhat more complicated and is stated as

$$E^4 \psi = E^2(E^2 \psi) = 0 \tag{15.7}$$

and E^2 is the operator which was defined as

$$E^2 = \frac{\partial^2}{\partial r^2} + \frac{\sin \theta}{r^2} \frac{\partial}{\partial \theta} \left(\frac{1}{\sin \theta} \frac{\partial}{\partial \theta} \right) \tag{15.8}$$

for the spherical coordinate system. Note that the E^2 operator is not the same as the ∇^2 operator. The operator E^4 in spherical coordinates is also known as the Stokes operator. You may wish to recall the definition of the streamfunction for an axisymmetric sphere from Section 3.10.3 at this stage.

6. Stokes flow is a linear problem and hence the superposition principle holds. This property can be effectively used where the solutions to simple problems can be combined to construct the solution to a more complex flow situation. For example, a sphere which is translating, rotating, and subject to a shear flow (a case of far-field linear velocity) can be modeled as the combination of these basic flows.

7. The total force acting on an irregularly shaped control surface S_1 bounded by a control volume V_1 is equal to the force on a larger regularly shaped surface S_2 bounding the original control volume. This property is useful because the calculation of the force on the irregular shape S_1 can be reduced to a surface integration on a simple shape S_2 such as a sphere.

8. The torque acting on a control surface S_1 bounded by a control volume V_1 is equal to the torque on a larger regularly shaped control surface enclosing the original control volume.

Another interesting property is the reversibility of flow discussed briefly below.

Reversibility of Stokes flow

It may be noted that, if the velocity is replaced by $-v$ and the pressure gradient is also reversed, i.e., p can be replaced by $-p$, then the flow remains essentially the same. This property is referred to as the reversibility property of the Stokes flow and is not shared by the full N–S equation where the convection terms are retained. Thus, upon a reversal of the approach velocity, the particle will stop immediately (since the time term is absent there is no deceleration) and start to move back, tracing a route along exactly the same streamline. Such effects have indeed been observed in micro-circulation.

Note that the reversibility is to be considered as a kinematic reversibility, not a thermodynamic reversibility. The energy lost in viscous dissipation cannot be recovered by reversing the flow direction.

The reversibility leads to some interesting consequences, The drag on a body and the drag on a mirror image of the body are the same. A neutrally buoyant particle in laminar flow will have no lateral motion and will track the streamlines without drifting. Thus there is no lateral force on this particle. The absence of the lateral force means that there is no lift force on a body (such as a baseball) rotating in a simple shear flow. In real life the lift is caused by inertial effects and the non-uniform boundary layers on the top and bottom of, for example, a spinning ball. This is a very interesting topic that is discussed in detail in the book by Kundu and Cohen (2008) but is not pertinent to Stokes flow.

15.2 The mathematics of Stokes flow

15.2.1 General solutions: spherical coordinates

In this section, we take problems posed in spherical coordinates with ϕ symmetry. General solutions are obtained starting from the equation

$$E^4 \psi = 0$$

We first look for solutions for the E^2 operator defined by Eq. (15.8) using separation of variables and obtain general solutions. These are combinations of power functions in r and Legendre functions P_n in terms of $\cos \theta$. Then we can use these to get solutions for E^4. On combining all such functions the final result can be expressed as

$$\psi(r, \eta) = \sum_{n=1}^{\infty} [A_n r^{n+3} + B_n r^{n+1} + C_n r^{2-n} + D_n r^{-n}] Q_n(\eta) \qquad (15.9)$$

where $\eta = \cos \theta$ and the functions Q_n are defined as integrals of the Legendre polynomials:

$$Q_n = \int_{-1}^{\eta} P_n(s) ds$$

where s is a dummy variable.

Two values for Q_n are presented below for use in future discussions:

$$Q_1 = \sin^2 \theta$$

and

$$Q_2 = \frac{1}{2} \sin^2 \theta \, \cos \theta$$

Equation (15.9) is a general solution that satisfies the governing differential equations in an exact manner. The solution to many specific problems can be found by finding the coefficients which satisfy the boundary conditions. The differential equation is satisfied already and we need only fit the boundary conditions. This is one variation of what can generally be called the boundary collocation method.

The drag force equation

An interesting and useful result is that the drag force in the direction of the axis is simply related to the coefficient C_1 in Eq. (15.9):

$$F_z = 4\pi \mu C_1 \tag{15.10}$$

where z denotes the direction of the symmetry axis. Thus only the constant C_1 appears in the drag relation and other coefficients are irrelevant. The origin for the coordinate system is taken as the center of mass of the body. The expression for drag is simply stated here, and other sources such as Leal (2007) may be consulted for the proof.

Let us now apply the general solution to a particular case of flow past a sphere, which is the classical problem studied by Stokes.

15.2.2   Flow past a sphere: use of the general solution

The governing equation is $E^4 \psi = 0$ and the solution given by Eq. (15.9) already satisfies the differential equation. The problem is reduced to simply stating the boundary conditions and tuning the solution to fit these boundary values.

Boundary conditions

We need four boundary conditions in terms of ψ. Recall the definition of the streamfunction for spherical coordinates:

$$v_r = \frac{1}{r^2 \sin\theta} \frac{\partial \psi}{\partial \theta} \tag{15.11}$$

and

$$v_\theta = -\frac{1}{r \sin\theta} \frac{\partial \psi}{\partial r} \tag{15.12}$$

Since the surface of the sphere is a streamline, we can set $\psi(r = R) = 0$. The choice of zero for the streamfunction at this point is arbitrary.

Since there is no tangential velocity at the surface, i.e., v_θ is zero, we set $\partial\psi/\partial r$ as zero at $r = R$.

The boundary conditions far away from the sphere are as follows.

As $r \to \infty$, $v_z = v_\infty$, $v_y = 0$, and $v_x = 0$. Also the approach velocity v_∞ is uniform in the z-direction. These translate in terms of v_r and v_θ into $v_r = v_\infty \cos\theta$ as $r \to \infty$. On using this expression for v_r in Eq. (15.11) and integrating we find

$$\psi(\infty, \theta) = \frac{1}{2} v_\infty r^2 \sin^2\theta$$

which is the boundary condition for large distances from the sphere. Note that, if v_θ were used to find ψ instead of v_r, the result would be the same, and hence only one condition results for ψ in the far-field region. The boundary conditions are shown schematically in Fig. 15.1.

Examination of the general solution and the boundary condition specified above shows that for a sphere only the $n = 1$ solution remains since the term $\sin^2\theta$ in Q_1 matches the boundary condition stated above. Q_2, Q_3, etc. are not needed.

Matching the boundary condition at infinity requires B_1 to be $v_\infty/2$. Also the term A_1 should be made zero since it has the r^4 dependence and there is no such matching term in the far-field boundary condition. Matching the other two conditions at $r = R$ gives

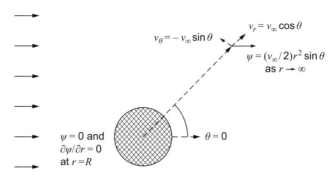

Figure 15.1 A diagram showing flow past a sphere and the boundary conditions for ψ.

$$C_1 = -3v_\infty R/4$$

and

$$D_1 = v_\infty R^3/4$$

The fitting of the boundary conditions has been done!
The final result is

$$\psi(r,\theta) = v_\infty r^2 \sin^2\theta \left[\frac{1}{2} - \frac{3}{4}\frac{R}{r} + \frac{1}{4}\frac{R^3}{r^3} \right] \tag{15.13}$$

The velocity components can then be obtained by using the definition of the streamfunction:

$$v_r = v_\infty \cos\theta \left(1 - \frac{3R}{2r} + \frac{1}{2}\frac{R^3}{r^3} \right) \tag{15.14}$$

and

$$v_\theta = -v_\infty \sin\theta \left(1 - \frac{3R}{4r} - \frac{1}{4}\frac{R^3}{r^3} \right) \tag{15.15}$$

The pressure distribution is given by integrating the relation $\nabla p = \mu \nabla^2 v$. The result is

$$p = p_\infty - \frac{3\mu v_\infty R}{2r^2} \cos\theta$$

The pressure distribution on the surface of the sphere is obtained from the above expression by substituting $r = R$:

$$p(R) = p_\infty - \frac{3\mu v_\infty}{2R} \cos\theta \tag{15.16}$$

The pressure is a maximum at the forward stagnation point ($\theta = \pi$) and is a minimum at the rear stagnation point ($\theta = 0$).

The drag force on the fluid can be calculated directly from (15.10) using the constant C_1 as

$$F_z = 4\pi \mu C_1 = 6\pi \mu v_\infty R \tag{15.17}$$

However, it is more educational to calculate this by integration of the differential forces on the surface of the sphere, as shown below.

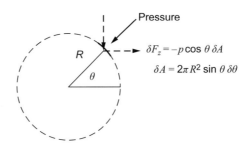

Figure 15.2 Resolution of the pressure force in the flow direction. δA is the differential element of surface at location θ.

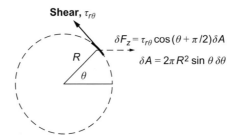

Figure 15.3 Resolution of the shear force in the flow direction. δA is the differential element of surface at location θ.

The pressure contribution to drag

Let us examine the contribution to drag due to the pressure distribution on a differential surface of the sphere.

The force due to pressure on the solid is given by $-pe_r\, dA$, where $dA = 2\pi R^2 \sin\theta\, d\theta$, an element of area in spherical coordinates integrated over ϕ. The net force due to pressure has a component in the flow direction. This becomes clear on resolving the pressure in directions along the flow and perpendicular to the flow (see Fig. 15.2).

The components perpendicular to the flow cancel out due to symmetry and the net component is in the direction of the flow. This component can be calculated as

$$\Delta F_z(\text{pressure}) = -p[e_r \cdot e_z](2\pi R^2 \sin\theta\, d\theta)$$

over a differential element of area. The variation of p as a function of θ is given by Eq. (15.16), and this is used in the above equation. We then integrate over θ from 0 to π and we find

$$F_z(\text{pressure}) = 2\pi \mu v_\infty R$$

The contribution of the force on the solid due to the pressure distribution is known as the form drag.

The shear contribution to drag

In a similar manner the force due to viscous stress can be calculated. See Fig. 15.3. The shear stress along the surface of the sphere in the θ-direction is equal to $\tau_{r\theta}$. This can be evaluated as $2\mu E_{r\theta}$. Using the expression for $E_{r\theta}$ given in Section 3.12.2, the stress is given by

$$\tau_{r\theta}(\text{at } r = R) = \mu \left[r \frac{\partial (v_\theta / r)}{\partial r} + \frac{1}{r} \frac{\partial v_r}{\partial \theta} \right]_{r=R} = \frac{3\mu v_\infty R}{\cos \theta}$$

The component of the force in the flow direction is $\tau_{r\theta}[e_\theta \cdot e_z]dA$ and the net force in the z-direction can be calculated again by taking the integral on θ. The final result after a few steps is:

$$F_z(\text{shear}) = 4\pi \mu v_\infty R$$

This is known as the friction drag. Note that the viscous normal stress τ_{rr} is zero at the sphere surface and hence it does not contribute to the drag. It may be noted that the normal stress is not zero at locations other than $r = R$. Also it may be noted that the normal stress is zero for all incompressible Newtonian fluids at a fixed solid surface.

The total force exerted by the fluid on the sphere is the sum of the pressure and shear forces and is therefore given by

$$F_z(\text{total}) = 6\pi \mu v_\infty R$$

which is in conformity with the earlier result using C_1. It is customary to write this in terms of a drag coefficient defined as

$$F_z = C_D A_p (\rho v_\infty^2 / 2) = C_D \pi R^2 (\rho v_\infty^2 / 2)$$

where the projected area A_p is taken as πR^2. Upon introducing the drag coefficient into the expression for F_z and the Reynolds number for the other terms we find the classical result for Stokes drag:

$$C_D = \frac{24}{Re}$$

This relation was introduced in an empirical manner in Section 4.4, where we also demonstrated the use of this equation to find the terminal velocity of a sphere falling freely under gravity through a fluid. This section provides the theoretical foundation to the model.

An important application of the concept of Stokes law is to the motion of bubbles and drops, which is discussed in the following subsection.

15.2.3   Bubbles and drops

The Stokes flow can also be applied to bubbles and drops rising in a pool of liquid at small velocities. In this case there may be an internal motion within the bubble that modifies the analysis. The internal circulation is suppressed if the surface has some adsorbed surface-active agents. In the absence of internal motion the bubble may be treated in the same manner as a rigid solid and the usual relation for the drag coefficient applies:

$$C_D = \frac{16}{Re_b} \text{ for } Re_b < 2$$

where Re_b is the Reynolds number based on the bubble diameter. Only the viscous contribution to the drag is included, since the pressure force inside the bubble balances that on the outside. This leads to a factor of 16 rather than 24 for solid spheres.

The simple explanation for the absence of internal circulation when surface-active agents are present is that the surface-active agents accumulate on the surface. When the drop moves through the continuous medium the surface-active agents accumulate on the rear

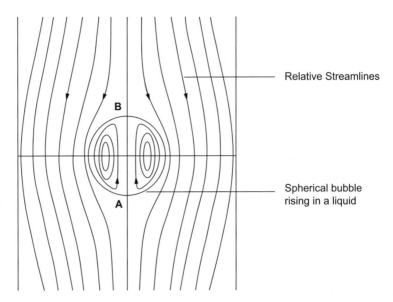

Figure 15.4 A schematic representation of the streamlines within a bubble for the case in internal circulation. Surface-active agents accumulate at the rear, point A, causing a flow to point B preventing the internal ciculation.

side (point A in Fig. 15.4), leaving the front (point B) uncontaminated. This sets up a surface-tension gradient, which induces a Marangoni flow at the surface. This surface-tension-induced flow retards the internal circulation, making the drop or bubble act as a rigid sphere.

Bubbles free of surface-active agents undergo a toroidal circulation as shown qualitatively in Fig. 15.4.

This phenomenon is referred to as Hadamard–Rybczyński internal circulation after those who developed the model; see Rybczyński (1911) and Hadamard (1911). The Stokes flow both in the external region and in the region interior to the bubble has to be analyzed for such a case. The problem is solved as a two-region problem with matching boundary conditions at the bubble surface. The velocity and the tangential stress are matched at the interface.

The drag on a sphere based on the Hadamard model is as follows:

$$F_z = 6\pi \mu v_\infty R \left(\frac{2\mu + 3\mu_p}{3(\mu + \mu_p)} \right)$$

where μ is the liquid viscosity and μ_p is the viscosity of the particulate phase (gas bubbles or the second liquid present as drops).

A more detailed expression is sometimes used attributed to Boussinesq (1903), who extended the analysis of Hadamard to include the interfacial stresses resulting from motion. These stresses were attributed to the combined effects of surface tension and a "dynamic" increment that varies over the surface. The combined effects were captured by assigning a surface viscosity parameter μ_s. The final result for the drag force was

$$F_z = 6\pi \mu v_\infty R \left(\frac{\mu_s + \alpha(2\mu + 3\mu_p)}{\mu_s + 3\alpha(\mu + \mu_p)} \right)$$

where α is a parameter that is indicative of the relative influence of surface viscosity and the tangential stresses at the interface. If $\alpha = 0$ the Stokes drag is recovered, and this corresponds to a case where the interface is rigid, whereas if α is large the Hadamard mode applies. Thus the two limiting cases of zero and complete circulation are thereby combined here in a single expression. Scriven (1960) extended the model to include material surfaces with arbitrary curvatures. Sternling and Scriven (1959) also investigated the role of Marangoni flows, and their paper can be considered as the start of surface rheology. A mathematical analysis by Bothe and Prüss (2010) provides additional information on this, and the books by Slattery, Sagis, and Oh (2007) and Edwards, Brenner, and Wasan (1991) are also good resources for understanding the interfacial phenomena in more detail.

Note that the normal-stress balance and the role of surface tension are not considered in the analysis of Hadamard. The assumption is that the surface tension is large and any differences in normal stress are balanced by the surface-tension forces such that the bubble or drop maintains its spherical shape. If the drop is perturbed, it will, however, deform, and whether it returns to its spherical shape or not will depend on the magnitude of the surface tension. A linear stability analysis based on boundary perturbation on the base flow is a useful method, and details are given in the book by Leal (2007). The qualitative conclusion is that a drop or bubble will restore its spherical shape if the surface tension is large; otherwise it will elongate to an ellipsoidal shape. The key dimensionless parameter is the capillary number Ca defined as $\mu v_\infty / \sigma$, and if $Ca \ll 1$ the drop will remain spherical. More information on this is also given in the book by Clift, Grace, and Weber (1978).

15.2.4   Oseen's improvement

Oseen (1910) improved the Stokes solution by including the inertial terms in an approximate manner. Again consider a situation where a body is moving with a steady velocity of v_p in a fluid and the fluid velocity far from the fluid v tends to zero. The local velocity at any point is expressed as

$$v = v_d + v_p$$

where v_d is a deviation velocity in the fluid medium due to the motion of the solid. The deviation velocity tends to zero at infinity, and v here is therefore equal to the negative of the particle velocity at the surface. The inertia terms are

$$v \cdot \nabla v = v_d \cdot \nabla v_d + v_p \cdot \nabla v_d$$

Oseen neglected the first term on the RHS, thereby keeping the linear nature of the problem. The Stokes equation is then modified to

$$\rho v_p \cdot \nabla v_d = \mu \nabla^2 v_d - \nabla p$$

The particle velocity was assumed to be one-directional and was represented as $v_p = v_p e_z$ The following was obtained by Oseen for the perturbation streamfunction:

$$\frac{\psi}{v_p R^2} = \left[\frac{r^2}{2R^2} + \frac{1}{4}\frac{R}{r} \right] \sin^2\theta - \frac{3(1 + \cos\theta)}{Re} \left[1 - \exp\left\{ -\frac{Re}{4}\frac{r}{R}(1 - \cos\theta) \right\} \right]$$

Note that in the above expression the point occupied by the center of the sphere at any instant of time is taken as the origin. The limiting case for $Re(r/R) \ll 1$ can be shown to be the Stokes solution.

The drag coefficient obtained from this expression is

$$C_D = \frac{24}{Re}\left(1 + \frac{3}{16}Re\right)$$

Proudman and Pearson (1957) extended Oseen's work by matching different series expansions for the flow regimes near and far away from the sphere. The relation obtained for the drag coefficient in their work was as follows:

$$C_D = \frac{24}{Re}\left(1 + \frac{3}{16}Re + \frac{9}{160}Re^2 \ln(Re/2)\right)$$

15.2.5   Viscosity of suspensions

An important application of Stokes flow is to compute the viscosities of dilute suspensions. The suspension is replaced by a hypothetical one-phase system, and the velocity and stress components are replaced by volume-averaged values. A Newtonian type of relation between the (volume-averaged) stress and the rate of strain is then proposed. The resulting coefficient is the effective viscosity of the suspension.

The problem was first considered by Einstein in 1905 as part of his Ph.D. thesis. He assumed the suspension to be dilute. Each particle behaves independently, but the collective behavior is equivalent to a disturbance in the far-field approach velocity. If a linear approximation is assumed for the far-field velocity this leads to a problem of axisymmetric elongational flow under low Reynolds-number conditions. See Problem 8 for a brief description of this problem. The velocity profile, strain rate, and stress profiles can then be solved analytically, and volume-averaged values can be obtained. For details the book by Leal (2007) may be consulted. A Newtonian type of constitutive model was proposed for the volume-averaged variable, leading to the following result for the viscosity of a dilute suspension:

$$\frac{\mu_{\text{eff}}}{\mu} = 1 + \alpha\frac{\mu + (5/2)\mu_p}{\mu + \mu_p}$$

Here α is the concentration of the particles by volume. The equation is valid only for dilute suspensions, and the condition $\alpha < 0.02$ is normally used as the range of validity.

For concentrated suspensions, the interparticle interactions become appreciable. Various modifications have been suggested to correct for these. A commonly used form is the Graham equation:

$$\frac{\mu_{\text{eff}}}{\mu} = 1 + \frac{5}{2} + \frac{9}{4}[\psi(1 + \psi/2)(1 + \psi^2)]^{-1}$$

where

$$\psi = 2\frac{1 - (\alpha/\alpha_{\max})^{1/3}}{(\alpha/\alpha_{\max})^{1/3}}$$

which is based on cell theory. Here $\alpha_{\max}$ is the volume fraction corresponding to close-packed spheres.

15.2.6 Nanoparticles: molecular effects

If the particle size is small ($\lesssim 10\,\mu m$ or so) then the effect of random molecular fluctuations will be felt on the particle. The phenomenon is known as Brownian motion. It was first observed in 1828 by a botanist, R. Brown, who studied the motion of small pollen particles in water. The motion of fine dust particles moving randomly is commonly observed when sunlight shines on a room full of dust. The conditions under which Brownian-motion effects become important can be stated in terms of a Knudsen number or as $d_p \approx \lambda$. This criterion is similar to that for Knudsen diffusion discussed earlier, and in fact the ratio of the above two length scales is used as a correction. Thus the first correction to Stokes drag is generally taken as $1 + 2Kn$, where Kn is the Knudsen number defined as λ/d_p. More details may be found in the book by Green and Lane (1957).

Einstein modeled the effect using the Stokes drag equation, and derived the following equation for diffusion of the solid particles:

$$D = \frac{\kappa T}{f} = \frac{\kappa T}{6\pi \mu R}$$

where f is the friction coefficient of the particle. This in turn can be obtained from the Stokes law as F_z/v_∞ and is therefore equal to $6\pi\mu R$. Also it can be seen that the fact that the relation $D\mu/T$ is a constant that is applicable for diffusion of a solute in a liquid phase follows from this model.

A more general model of particle motion under such conditions is given by the Langevin equation. This is essentially a stochastic differential equation.

$$m\frac{dv}{dt} = -6\pi \mu Rv + F(t)$$

The first term on the RHS is the Stokes drag, while the second term is a stochastic term resulting from the irregular force due to random fluctuations in the liquid due to molecular motion. The equation cannot be solved in the usual (deterministic) manner due to the stochastic term $F(t)$. One generally seeks information on the probability functions for such problems, rather than deterministic values. Solution methods for stochastic differential equations are reviewed, for instance, in the book by Oksendal (2010). In the present case (and in general for stochastic problems), a probability function W can be extracted, where W is the probability that at time t the particle is in a specific position range. For large times this function approaches the Gaussian form and hence the particle motion can be described by a variant of Fick's law based on the variance of the distribution function. The model itself is rather interesting in the sense that the position is calculated on a continuum basis using the averaged velocity, but the velocity is modeled on a molecular level and includes an interaction term given by $F(t)$. Thus models at two scales are combined.

Thermophoresis

Thermophoresis is the phenomenon of drift of small-sized particles in the presence of a temperature gradient. The gradients are small, such that the change in the viscosity of the suspending gas is small. Hence the drift is then essentially a manifestation of molecular effects (and not that due to the effect of variable viscosity). Gas molecules in the hotter region have larger velocities and hence transfer momentum to gas molecules in the cooler region. The result is a net force in the direction of $-\nabla T$. The force due to this is generally modeled as

$$F_{\text{thermo}} = -\frac{32}{15}\frac{R^2}{v_{\text{th}}}k_{\text{t}}\nabla T$$

where k_{t} is the translational contribution to the thermal conductivity and v_{th} is the thermophoretic velocity. A detailed analysis of nanoparticle motion has been published by De Bleecker *et al.* (2005) and can be used as an interesting case study of the modeling of such systems.

15.3   Inviscid and irrotational flow

Here we examine flow at high Reynolds number where the viscous effects may be neglected. The flow is usually of boundary-layer type and can be analysed in two regions as explained in the introduction to this chapter. Here we show the solution for the outer region, and in Section 15.5 we couple it with flow in the inner region, the boundary layer. First we examine the properties of inviscid and irrotational flow, and contrast the two. (The contrast is not clearly indicated in many textbooks, and the two cases are often taken as being the same.) This is followed by solution methods for irrotational flows and then some illustrative examples.

15.3.1   Properties of irrotational flow

It is useful to review Section 3.10 at this stage. Let us start with some definitions and some important relations.

1. If the vorticity vector $\boldsymbol{\omega}$ is identically zero, then the flow is called irrotational and the velocity vector must be curl-free.
2. The curl-free part of the Helmholtz representation (Eq. (3.38)) of the velocity is sufficient to represent the flow. Thus the velocity can be represented as a gradient of a scalar function defined by ϕ,

$$\boldsymbol{v} = \nabla\phi \tag{15.18}$$

and such flows are called potential flows.
3. Potential flows are irrotational, i.e., the vorticity vector for such flows is equal to zero. This follows automatically from the definition above on applying the curl on both sides of Eq. (15.18). Since the curl of a gradient is always zero, the vorticity is identically zero for a potential flow, and hence a potential flow is also an irrotational flow.
4. The scalar potential satisfies the Laplace equation:

$$\boxed{\nabla^2\phi = 0} \tag{15.19}$$

This can be easily demonstrated by taking the divergence on both sides of Eq. (15.18) and using the continuity $\nabla \cdot \boldsymbol{v} = 0$. Thus the Laplace equation is representative of irrotational flow (also known as potential flow), since the velocity can be described now as a gradient of a scalar potential function. The computation of the flow reduces to the computation of a scalar field subject to suitable boundary conditions. We also note that the Laplace equation was also encountered in steady-state heat transfer problems. Many powerful methods are available for the solution of the Laplace equation, and some of these techniques were indicated in the context of heat conduction in Chapter 8.

5. The linearity of the Laplace equation permits superposition of the solutions. Thus two solutions ϕ_1 and ϕ_2 can be added to construct a solution to another potential flow problem.

6. For 2D flows the streamfunction also satisfies the Laplace equation:

$$\boxed{\nabla^2 \psi = 0 \quad \text{2D plane flow only}} \tag{15.20}$$

Very often the streamfunction form is more useful, since it is easier to visualize flows by constructing streamlines. However, this is not valid for 3D cases since there is no simple streamfunction representation for such flows.

7. The viscous terms vanish if the flow has zero vorticity. This is an important property. The proof requires some vector identities that are shown in Example 15.1.

8. The Bernoulli equation is satisfied everywhere in the flow field. This is shown in detail in Section 15.3.2.

Example 15.1.

Show that the viscous terms vanish if the flow has zero vorticity.

Solution.

We start off with the N–S equation in vector form represented as

$$\rho \left[\frac{\partial v}{\partial t} + (v \cdot \nabla)v \right] = -\nabla \mathcal{P} + \mu \nabla^2 v \tag{15.21}$$

where $\mathcal{P}$ is the modified pressure equal to $P + \rho g h$.

The viscous (diffusive) terms in (15.21) are rearranged using the following vector identity:

$$\nabla^2 v = \nabla(\nabla \cdot v) - \nabla \times (\nabla \times v) \tag{15.22}$$

The first term on the RHS is zero for an incompressible flow. Hence

$$\nabla^2 v = -\nabla \times (\nabla \times v) = -\nabla \times \omega \tag{15.23}$$

This term is zero if the vorticity is zero. Hence both terms in Eq. (15.22) are zero for a flow field whose vorticity, ω, is zero if the fluid is also incompressible. This means that the viscous terms in the N–S equation automatically vanish for an irrotational flow.

Inviscid flow is defined as flow where the viscous terms can be neglected and μ can be set to zero, which is the inviscid-flow assumption. We find that there is a subtle difference between inviscid flow and irrotational flow. The vorticity can be non-zero in inviscid flow, but is always zero in irrotational flow. Also the Bernoulli equation is satisfied only along a streamline in inviscid flows, not everywhere in the fluid. This is shown in the next section.

15.3.2   The Bernoulli equation revisited

The Bernoulli equation states that the sum of the kinetic energy, the pressure energy, and the potential energy in a flowing fluid is constant in a frictionless fluid along a streamline.

We demonstrate this property now, and show further that for irrotational flow this sum is a constant at any two points in the fluid, not necessarily along a streamline:

$$\frac{1}{2}\rho v^2 + p + \rho gh = \text{constant} \tag{15.24}$$

The last two term can be combined into the definition of a modified pressure $\mathcal{P}$. A common form is in terms of the head of the fluid, which is obtained by dividing throughout by ρ. Hence

$$\frac{1}{2}v^2 + \frac{\mathcal{P}}{\rho} = \text{constant} \tag{15.25}$$

is valid for a system with constant density.

Example 15.2.
Show that, for a potential flow, the Bernoulli equation holds throughout the flow field.

Solution.
For inviscid flow the convective terms are rearranged using the following vector identity:

$$(v \cdot \nabla)v = \nabla\left(\frac{1}{2}v^2\right) - v \times \omega \tag{15.26}$$

Note that v^2 is a scalar equal to $v \cdot v$; i.e., the square of the speed.
    Hence the N–S equation (with no viscous term) can be written as

$$\rho\left[\frac{\partial v}{\partial t} + \nabla\left(\frac{1}{2}v^2\right) - v \times \omega\right] = -\nabla\mathcal{P} \tag{15.27}$$

The equation is applicable for inviscid flow (flows where μ is set as zero, as an approximation). Note that the $v \times \omega$ term remains, and this need not be zero for an inviscid flow.
    Further simplification results for irrotational flow, since the vorticity term can now be set as zero. Hence we find

$$\rho\left[\frac{\partial v}{\partial t} + \nabla\left(\frac{1}{2}v^2\right) + \nabla\left(\frac{\mathcal{P}}{\rho}\right)\right] = 0 \tag{15.28}$$

Since $v = \nabla\phi$ for potential flow, the above equation can be written as

$$\nabla\left[\frac{1}{2}v^2 + \frac{\mathcal{P}}{\rho} + \frac{\partial\phi}{\partial t}\right] = 0 \tag{15.29}$$

or

$$\left(\frac{1}{2}v^2 + \frac{\mathcal{P}}{\rho} + \frac{\partial\phi}{\partial t}\right) = F(t) \tag{15.30}$$

The form of $F(t)$ can be shown to have no physical significance, and hence can be taken as a constant. Since the velocity is the gradient of the potential, this will not alter the results.
    Hence

$$\frac{1}{2}v^2 + \frac{\mathcal{P}}{\rho} + \frac{\partial\phi}{\partial t} = \text{constant} \tag{15.31}$$

For steady-state flow

$$\frac{1}{2}v^2 + \frac{\mathcal{P}}{\rho} = \text{constant} \tag{15.32}$$

Thus, for an irrotational flow, the Bernoulli equation is satisfied everywhere in the flow field.

Note that, for an inviscid flow, the $v \times \omega$ term on the RHS of (15.27) remains. However, one can show that the Bernoulli equation holds along a streamline (but not along every curve in the field). This is due to $v \times \omega$ becoming zero along a streamline, since ω is perpendicular to v.

15.4 Numerics of irrotational flow

Computations of irrotational flow are done by solving the Laplace equation for the velocity potential subject to suitable boundary conditions. We now state what boundary conditions are generally used for these computations.

15.4.1 Boundary conditions

The ϕ formulation

The problem formulation and the boundary conditions to be applied are best illustrated by an example of external flow past a body or array of bodies. The common problem is flow past a body with a fluid approaching at a specified or constant velocity far from the body. Since this is an infinite-domain problem, it is convenient to split ϕ into two functions, one corresponding to the flow of the fluid alone and the second corresponding to a perturbation potential due to the presence of the body. We take a case where the far-field velocity is a constant. Let v_∞ represent the far-field velocity in the x-direction. Then the potential corresponding to this is

$$\phi_\infty = v_\infty x \quad \text{or} \quad v_\infty r \cos \theta$$

where the last term is the representation in polar coordinates, which is more useful for many problems. The potential is given by

$$\phi = \phi_\infty + \phi_s$$

where ϕ_s is the perturbation in the potential due to the presence of ths solid. Obviously

$$\nabla^2 \phi_s = 0 \tag{15.33}$$

which is now the governing equation to be solved. The first boundary condition is set as the solid surface. If n is the direction of the normal from the solid, the no-penetration condition requires v_n to be zero. Since $d\phi/dn$ is zero along a solid surface we can set a Neumann condition for ϕ_s on the solid surface:

$$\frac{d\phi_s}{dn} = -\frac{d\phi_\infty}{dn} \quad \text{on the surface}$$

This, together with the boundary condition that ϕ_s is zero far from the solid, i.e., as $r \to \infty$, completes the problem formulation. Note that the tangential component of velocity along the surface of the body, v_t, cannot be set to zero. There are no more boundary conditions to set. Hence the potential flow permits slip along a solid surface. Since the no-slip condition is not satisfied, the tangential stress is also zero, which leads to the following paradox.

D'Alembert's paradox states that the drag over a solid for a potential flow is zero. This is contrary to physical experience and is resolved by postulating that there is indeed a thin layer near the solid surface where the vorticity is non-zero. This thin layer is called the boundary layer, and the concept of the boundary layer permits the flow to be analysed in two regions: (i) a viscous region near the solid and (ii) an irrotational region as discussed in Section 15.5.

The ψ formulation

The governing equation for 2D cases in terms of the streamfunction requires the Laplacian to be zero. Since the solid surface is a streamline, ψ can be set to zero here. The second condition is set at $r \to \infty$. Since $v_x = d\psi/dy$ we can set this as v_∞ here. In polar coordinates (which is convenient for many problems) we would set $d\psi/d\theta$ equal to rv_r here. Since v_r in turn is equal to $v_x \cos\theta$, which is equal to $v_\infty \cos\theta$, we have

$$\frac{d\psi}{d\theta} = rv_\infty \cos\theta$$

or

$$\psi = rv_\infty \sin\theta \text{ at } r = \infty \tag{15.34}$$

as the second boundary condition.

15.4.2   Solutions using harmonic functions

Harmonic functions were introduced earlier in the context of heat conduction (Chapter 8, Problem 8). Here we demonstrate the usefulness of these functions for potential flows.

Functions that satisfy the Laplace equation are called harmonic functions. Four general harmonic functions in polar coordinates that are widely used are

$$r^n \cos(n\theta); \qquad r^n \sin(n\theta)$$
$$(1/r^n)\cos(n\theta); \qquad (1/r^n)\sin(n\theta) \tag{15.35}$$

where n takes positive integer values. These arise naturally if the method of separation of variables is used to find general solutions to the Laplace equation in polar coordinates.

Combinations of any of these functions also satisfy the Laplace equation by the principle of superposition. Thus a general solution can be proposed as a linear combination of these functions. The boundary conditions at specified points can be used to find the constants.

Flow past a circle

Let us apply this for flow past a circle. Here we will find that only $n = 1$ is needed (to accommodate the boundary condition at infinity) and only the sine function is needed. This is because the streamfunction for the far field is the $r\sin\theta$ term and there are no other r^n terms. Hence the solution is of the form

$$\psi = A_1 r \sin\theta + A_2 \frac{\sin\theta}{r}$$

The boundary condition at $r = \infty$ given by (15.34) requires $A_1 = v_\infty$, while the boundary condition at $r = R$ requires A_2 to be $-v_\infty R^2 \sin\theta$.

Hence the solution is

$$\psi = v_\infty r \sin\theta \left[1 - \frac{R^2}{r^2} \right]$$

The velocity can be computed from the definition of ψ, and, using this, the pressure profiles can be obtained by using the Bernoulli equation.

Similar fitting can be done for ϕ as well, and the result can be shown to be

$$\phi = -v_\infty r \cos\theta \left[1 + \frac{R^2}{r^2} \right]$$

Let us take a second example.

Example 15.3.

Consider the flow represented by the potential function ϕ given by

$$\phi = r^n \cos(n\theta)$$

Here r and θ represent the location of a point in polar coordinates. Also n is a specified constant. Find a geometry where such a flow field can be used.

Solution.

Since the function is harmonic, we find that the Laplace equation $\nabla^2 \phi = 0$ is satisfied and is therefore a solution to a potential problem. To find what that problem is, let us calculate the velocities along some fixed θ values.

The velocity components are obtained from

$$v = \nabla\phi$$

The operator ∇ is again taken in polar coordinates. Thus

$$v_r = \frac{\partial\phi}{\partial r} = nr^{n-1}\cos(n\theta)$$

and

$$v_\theta = \frac{1}{r}\frac{\partial\phi}{\partial\theta} = -nr^{n-1}\sin(n\theta)$$

The θ-component becomes zero if $\sin(n\theta) = 0$, i.e., at $\theta = 0$ and $\theta = \pi/n$. These can be then considered as walls since the normal component of velocity is v_θ. Hence the above function represents the potential flow in a corner. The corner angle is given as π/n.

To visualize flow it is better to find the streamfunctions. It is not difficult to show the following result:

$$\psi = r^n \sin(n\theta)$$

We find $\theta = 0$ and $\theta = \pi/n$ are lines of constant streamfunction, and these can be considered as contours of the solid body.

The pressure field for this example is calculated using the Bernoulli equation. Consider the $\theta = 0$ or the $\theta = \pi/n$ line. The velocity along this line is then nr^{n-1}. The pressure variation along this line is

$$-\frac{1}{S}(\rho)\frac{dP}{dr} = v_r\frac{dv_r}{dr} = n^2(n-1)r^{2n-3}$$

The pressure variation is important in coupling the potential flow to flow in the boundary layer, which is applicable to a thin zone near the solid boundary. Boundary-layer flows and transport will be taken up in Section 15.5.

A second method of solution is superposition of basic flows due to some singularities in the domain. This is the focus of the next section.

15.4.3 Solution using singularities

In this section the flow due to some singularity in the domain is described. These singularities may be sources (a single-layer potential), doublets (source–sink combinations), or vortices. The physical interpretation of these singularities is also useful and will be presented. Singularities have electric and magnetic analogies, for example, a point source of charge, a bar magnet of infinitesimal length but with a finite dipole moment, etc. These analogies are also useful for acquiring a better understanding of the model development. Once these "basic" flows are known, the flow around any object can be computed as a superposition of the flow generated by these basic singularities. This method of computation is known as the singularity method or the method of fundamental solutions. A variation of this method is the boundary integral method.

Flow due to a point source

Consider a long, thin porous pipe with a liquid leaking from its circumference. The resulting flow will be axisymmetric and hence radially outward. See Fig. 15.5. In the limit of the pipe radius tending to zero, we can imagine a point source of mass located at the origin of a 2D plane. The velocity can be obtained by a mass balance. The magnitude of the velocity times the area available for flow must be a constant in order for the mass balance to hold. Thus, at any radial position r, we have

$$2\pi r v_r = S \tag{15.36}$$

where v_r is the component of the velocity in the r-direction and S represents the flow rate per unit length of the pipe (dimension L^2/T). Here S is known as the source strength (flow rate per unit length of the leaking pipe). The velocity is then given by

$$v = v_r e_r = \frac{S}{2\pi r} e_r \tag{15.37}$$

The value of v_r is not defined at the point $r = 0$ and hence this point represents a singularity. For further discussion, let the source strength be unity ($S = 1$); the velocity field is then given by the above equation with $S = 1$. The potential, K, corresponding to this velocity field can be calculated by using the equation

$$\frac{\partial K}{\partial r} = v_r \tag{15.38}$$

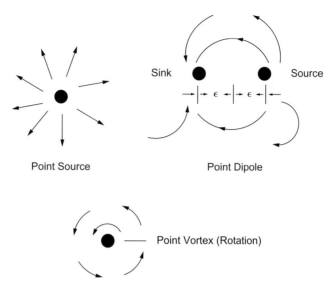

Figure 15.5 Representations of flow resulting from a point source of mass, a point dipole, and a point vortex.

On performing the integration of Eq. (15.38), we obtain the potential K_S resulting from a unit source:

$$K_S(r) = \frac{1}{2\pi} \ln r \qquad (15.39)$$

where the constant of integration is set as zero. An arbitrary constant can be added to the RHS of the above equation since the potential is not uniquely defined.

Since we need to consider several sources, we need to generalize the source location. Hence the flow due to a point source can be represented as

$$K(\boldsymbol{x}, \boldsymbol{\xi}) = K(x_1, x_2, \xi_1, \xi_2) = \frac{1}{2\pi} \ln r_{\xi x} \qquad (15.40)$$

where x is the observation point, ξ is the source point, and $r_{\xi x}$ is the distance between these points.

The corresponding result for the 3D case is obtained as

$$K(\boldsymbol{x}, \boldsymbol{\xi}) = -\frac{1}{4\pi \, r_{\xi x}} \qquad (15.41)$$

where $r_{\xi x}$ is the distance in 3D coordinates given by

$$r_{\xi x} = \left[(x_1 - \xi_1)^2 + (x_2 - \xi_2)^2 + (x_3 - \xi_3)^2\right]^{0.5} \qquad (15.42)$$

The result is easy to show and left as an exercise in Problem 14.

Flow due to a point vortex

Consider a point source of a vortex at the origin. A physical representation would be a shaft of very small diameter rotating at a huge speed of, say, Ω. This induces a rotation of the surrounding fluid. The velocity is now entirely tangential and is given by

$$v_\theta = \kappa \Omega \qquad (15.43)$$

The scalar potential corresponding to this tangential velocity field can be derived as

$$\boxed{K_v = \frac{\kappa}{2\pi}\theta} \qquad (15.44)$$

Here κ is the circulation defined as $2\pi r^2 \Omega$. The angle θ should be viewed as the angle between some (arbitrary) x-axis and the line joining the location of the vortex and the observation point.

Source/sink doublet or dipole

A combination of a source and sink placed side by side is called a doublet. A doublet and the geometric parameters used to derive the correponding potential are shown schematically in Fig. 15.6. The source and sink are separated by a distance 2ϵ, and the unit vector pointing in the direction joining the sink and the source is designated as n (the unit vector along the x-axis for the case shown in Fig. 15.6).

The strengths of the source and sink are numerically the same, and designated as S.

An expression for the velocity potential resulting from the combined action of the source and sink is now derived. For simplicity the source and sink are both located on the x-axis near the origin separated by a distance 2ϵ. Extension to arbitrary orientation of the line joining the source and the sink can be easily done. For this geometric arrangement the source location is at the point $(\epsilon, 0)$, while the sink location is at $(-\epsilon, 0)$.

Consider an observation point located at (x, y). Let this point be located at a distance r_1 from the source. Then the potential at this point due to the source is $S \ln r_1/(2\pi)$. Similarly, if the distance between the sink and the observation point is r_2, then the potential at x due to the sink is $-S \ln r_2/(2\pi)$. (Note the minus sign due to the action of the sink.) Now, by application of the superposition principle, the combined potential, K_D, at the observation point due to the source–sink doublet is

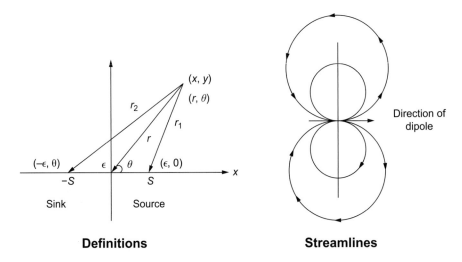

Figure 15.6 A schematic representation of a doublet.

$$K_D = \frac{S}{2\pi}\ln r_1 - \frac{1}{2\pi}\ln r_2 = \frac{S}{2\pi}\ln(r_1/r_2) \qquad (15.45)$$

where the subscript D is used for doublet.

Now the following relations hold from geometric considerations. The distance r_1 is given by

$$r_1 = \sqrt{(x-\epsilon)^2 + y^2} \approx \sqrt{x^2 + y^2 - 2x\epsilon} \qquad (15.46)$$

In the last expression the square term in ϵ is neglected.

This can be written as

$$r_1 = \sqrt{r^2 - 2x\epsilon} = r\sqrt{1 - 2(x/r^2)\epsilon}$$

where $r^2 = x^2 + y^2$, the distance of the observation point from the origin.

Similarly,

$$r_2 = \sqrt{r^2 + 2x\epsilon} = r\sqrt{1 + 2(x/r^2)\epsilon} \qquad (15.47)$$

On substituting the above expression into (15.45), we obtain

$$K_D = \frac{S}{4\pi}\ln\left(\frac{1 - 2(x/r^2)\epsilon}{1 + 2(x/r^2)\epsilon}\right) \qquad (15.48)$$

where the square root has been taken out of the logarithm (now giving a factor of 4 instead of 2).

The limit of this expression as ϵ tends to zero is needed. For this purpose the approximation

$$\ln(1 - a\epsilon) = -a\epsilon$$

can be used. Hence

$$K_D = -\frac{1}{\pi}S\epsilon(x/r^2) = -\frac{1}{\pi}S\epsilon\cos\theta/r \qquad (15.49)$$

since $x = r\cos\theta$.

The source strength of a dipole S is defined by a composite quantity λ defined as $S\epsilon$:

$$\lambda = S\epsilon$$

Hence the expression for the potential due to a dipole is

$$K_D = -\lambda\frac{\cos\theta}{r} \qquad (15.50)$$

Generalization to the case where the dipole is oriented in any arbitrary direction is straightforward. On noting that $\cos\theta = e_r \cdot e_x$, we can use $\cos\theta = e_r \cdot n$ for an arbitrary direction of the dipole. Hence the potential due to a dipole is

$$\boxed{K_D = -\lambda\frac{e_r \cdot n}{r}} \qquad (15.51)$$

Flow in complex geometries can be constructed by superposition of the above basic flows. Note that the potential, being scalar, can be added. A flow with a uniform velocity, say, in the x-direction, is often superimposed, and the potential due to this is $r\cos\theta$ for a unit velocity.

Example 15.4. Flow past a cylinder with rotation.

The potential for flow past a cylinder was derived in Section 15.4.2. It can be seen that this is a superposition of a dipole placed at the origin and a uniform flow in the x-direction. We now extend this for a rotating cylinder. We show that a lift force (a force in the vertical direction) is associated with this system. We simply add the potential corresponding to a point vortex placed at the origin. The combined potential satisfying all these three combinations can be expressed as

$$\phi = V_\infty \left(r + \frac{R^2}{r} \right) \cos\theta + \frac{\kappa\theta}{2\pi} \tag{15.52}$$

where κ is the circulation around the cylinder, which is defined as

$$\kappa = -\oint v_t \, d\Gamma = \int_0^{2\pi} R\Omega \cdot R \, d\theta = 2\pi R^2 \Omega$$

since the tangential velocity is equal to $v_\theta = R\Omega$.

The tangential velocity at the surface is

$$v_\theta = -2v_\infty \sin\theta + \kappa/(2\pi R)$$

The pressure profiles are then computed from the Bernoulli equation. There is a net lift force in the system, and this can be computed by integrating the upward component of the pressure force along the perimeter:

$$\text{Lift force per unit perimeter} = -\rho v_\infty \kappa$$

The lift force is proportional to the approach velocity and the circulation around the object. It turns out that this is a general result for bodies of arbitrary shape in 2D flows, where the lift turns out to be related to the circulation. The lift force is downward when the rotation is counterclockwise and upward when the rotation is clockwise. (The fluid is assumed to flow in the usual x-direction, i.e., from left to right.)

15.5 Flow in boundary layers

In many situations in transport processes, one can postulate the existence of a boundary layer where rapid change in a variable can be expected. The effect is due to a small parameter multiplying the highest-derivative term in the differential equation. For regions outside the boundary layer simpler solutions can be expected, where the small parameter can be totally neglected. Thus the complete solution can be constructed by a combination of the outer and inner solutions. These concepts were discussed and elaborated in Chapter 14. Here we demonstrate the application to fluid-dynamic problems, especially for external flows.

Boundary layers usually arise in three types of flow situations: (i) external flow past a solid object; (ii) flow in the developing region of a pipe or a closed conduit (such flows are known as internal boundary layers); and (iii) flow in a jet of gas issuing from a nozzle. Our discussion will be primarily for external flows past solid objects.

15.5.1   Relation to the vorticity transport equation

The physical concepts involved in the boundary layer analysis can be grasped by reviewing the vorticity transport equation. This is reproduced below from Section 5.6.4. For a 2D flow we can write the vorticity equation as

$$(v \cdot \nabla)\omega = \nu \nabla^2 \omega$$

This has the same form as the heat equation. The LHS represents the convection term by the velocity v, while the RHS represents the diffusion of the vorticity, the diffusion coefficient being the kinematic viscosity. Vorticity is generated at the solid surface due to viscous effects, and this diffuses into the main flow stream and is carried away by the velocity. The analogy with a plate at a higher temperature is obvious. The heat conducts from the plate and is carried away by convection. The influence of the heat (or analogously the vorticity) is confined to a thin region near the plate, and this represents the boundary layer. Outside this region the vorticity remains constant. The incoming fluid has a uniform velocity and hence zero vorticity, leading to there being an outside region with no vorticity. Thus the concept of the boundary layer permits the flow to be analyzed separately for the two regions.

Equations for the boundary layer: results from scaling analysis

The scaling analysis of flow in boundary layers was presented in Chapter 13. One of the key results was that there is no pressure variation in the y-direction within the boundary layer. Hence the pressure gradient in the x-momentum balance can be approximated by the external pressure gradient at the edge of the boundary layer:

$$\frac{\partial P}{\partial x} = \frac{dP}{dx} = \frac{dP_{\text{ext}}}{dx}$$

The external flow can be modeled using the potential-flow theory, and the variation of pressure in the external flow is related to the velocity by the Bernoulli equation:

$$\frac{1}{\rho}P_{\text{ext}} + \frac{1}{2}v_{\text{ext}}^2 = \text{constant}$$

where v_{ext} is the velocity in the external flow at the solid surface (i.e., just outside the boundary layer). Note that this can be calculated from the potential-flow model which applies for external flow, i.e., outside the boundary layer.

On using this in the x-momentum we get the equation for the x-component of velocity within the boundary layer:

$$v_x\frac{\partial v_x}{\partial x} + v_y\frac{\partial v_x}{\partial y} = v_{\text{ext}}\frac{dv_{\text{ext}}}{dx} + \nu\frac{\partial^2 v_x}{\partial y^2} \tag{15.53}$$

which needs to be solved together with the continuity equation. The y-momentum balance is not needed any more. Note how the external flow pressure term drives flow in the boundary layer via the P_{ext} or v_{ext} term in the above equations. This can be viewed as a source term for flow within the boundary layer.

We now look at flow past a flat plate, and consider how the x-momentum equation can be solved together with the continuity equation.

15.5.2　Flat plate: integral balance

Governing equations

The simplest problem to analyze is that of flow over a flat plate. The pressure-gradient term is now zero, or we can say that v_{ext} is not a function of x. The governing equations are therefore the continuity and simplified x-momentum balance with the pressure term set as zero. These equations have to be solved together:

$$\frac{\partial v_x}{\partial x} + \frac{\partial v_y}{\partial y} = 0 \tag{15.54}$$

and

$$v_x \frac{\partial v_x}{\partial x} + v_y \frac{\partial v_x}{\partial y} = v \frac{\partial^2 v_x}{\partial y^2} \tag{15.55}$$

The x-momentum can also be written, using the continuity equation, as

$$\frac{\partial (v_x^2)}{\partial x} + \frac{\partial (v_x v_y)}{\partial y} = v \frac{\partial^2 v_x}{\partial y^2} \tag{15.56}$$

which is an easier form from which to obtain the integral momentum balance. This is known as the conservation form of the momentum balance. We now formulate an integral form of this equation.

Integral form

The integral balance form is derived by integrating (15.56) with respect to y from 0 to δ, i.e., across the boundary layer. The result is

$$\int_0^\delta \left[\frac{\partial (v_x^2)}{\partial x} \right] dy + [v_x v_y]_0^\delta = v \left[\frac{dv_x}{dy} \right]_0^\delta \tag{15.57}$$

This is equivalent to a mesoscopic model for the boundary layer.

The following "boundary" conditions are applied.

1. $v \, dv_x/dy$ is set equal to τ_w/ρ at $y = 0$, where τ_w is the shear stress on the wall or the plate.
2. $v \, dv_x/dy$ is equal to zero at the edge of the boundary layer $y = \delta$; i.e., we assume that the boundary layer blends into the main flow in a smooth manner.
3. $(v_x v_y)$ is zero at $y = 0$ (no-slip condition).
4. $(v_x v_y)$ is $(v_e v_{y\delta})$ at $y = \delta$, where v_e is the external velocity (which is the same as v_{ext} at the edge of the boundary). This is the same as v_∞, the approach velocity, in the context of a flat plate.

   Note that v_e will not be the same as v_∞ for inclined plates and for flow past a curved shaped object. The pressure variation in the external flow has to be considered for such cases, which are taken up later.

The value of $v_{y\delta}$ can in turn be obtained by integrating the continuity equation (15.54) across the boundary layer:

$$v_{y\delta} = -\int_0^\delta \left(\frac{\partial v_x}{\partial x}\right) dy \tag{15.58}$$

Hence the integral balance reduces to

$$\int_0^\delta \left[\frac{\partial(v_x^2)}{\partial x}\right] dy - v_e \int_0^\delta \left(\frac{\partial v_x}{\partial x}\right) dy = -\frac{\tau_w}{\rho} \tag{15.59}$$

We need to apply the Leibnitz rule to the terms on the LHS. But the results turn out to be fairly simple, since the $d\delta/dx$ term cancels out:

$$\boxed{\frac{d}{dx}\left(\int_0^\delta (v_x^2 - v_e v_x) dy\right) = -\frac{\tau_w}{\rho}} \tag{15.60}$$

The above equation is known as the von Kármán integral, or simply as the momentum integral.

The above is an EXACT representation of the overall x-momentum balance in the boundary layer. It is equivalent to a mesoscopic momentum balance. It is instructive to note that a direct derivation of this equation can be obtained using the mesoscopic control volume of size Δx which spans the full boundary layer in the y-direction.

The pointwise velocity distribution cannot, however, be obtained from the integral balance. This is understandable since the averaging smoothes things out and destroys local information. In fact, a velocity distribution must be ASSUMED in order to proceed further. A cubic profile is normally assumed, and an equation for the boundary layer thickness can be then obtained, as discussed now.

15.5.3   The integral method: the von Kármán method

Von Kármán assumed the following velocity distribution, the so-called trial functions:

$$\frac{v_x}{v_e} = \frac{3}{2}\eta - \frac{\eta^3}{2} \tag{15.61}$$

where η is defined as y/δ. The equation is applicable for η of 0 to 1, i.e., within the boundary layer. For $\eta > 1$ the velocity v_x is simply equal to v_e.

The above profile satisfies certain boundary conditions for the velocity profile. See Problem 25 for the conditions leading to the above expression for the velocity profile.

If one substitutes this velocity profile into the integral balance, the following equation is obtained:

$$(39/280)\rho v_e^2 \frac{d\delta}{dx} = \frac{3\mu}{2}\frac{v_e}{\delta} \tag{15.62}$$

This can be integrated to obtain δ as a function of x:

$$\frac{\delta^2}{2} = 10.78\frac{\nu x}{v_e} \tag{15.63}$$

where the integration constant has been set to zero since $\delta = 0$ at $x = 0$ is zero. Using a local Reynolds number defined as

$$Re_x = \frac{x v_e \rho}{\mu}$$

the result can be expressed in a simple dimensionless formula as

$$\frac{\delta}{x} = \frac{4.64}{\sqrt{Re_x}}$$

Knowing δ, we can find the APPROXIMATE velocity profile in the boundary layer which is given by Eq. (15.61).

The drag force on the plate

The wall shear stress can also be obtained by differentiating the velocity profile, and the resulting expression is

$$\tau_w = \frac{3}{2}\frac{\mu v_e}{\delta} = 0.332\sqrt{\frac{\mu \rho v_e^3}{x}} \qquad (15.64)$$

The above equation shows that the drag coefficient is a function of x. An inverse-square-root dependence on x holds.

Noting that the local drag coefficient is defined as $\tau_w/(\rho v_e^2/2)$, the following expression for the drag coefficient is also obtained:

$$C_{Dx} = \frac{0.646}{\sqrt{Re_x}}$$

15.5.4   The average value of drag

A local value is not always needed, and an average value for a given length of plate L is more useful. Hence a mean drag coefficient is defined as

$$C_{DL} = \frac{1}{L}\int_0^L C_{Dx}\, dx$$

The mean drag coefficient is given upon integration as

$$C_{DL} = \frac{1.292}{\sqrt{Re_L}}$$

Hence the force on a flat plate of length L and an arbitrary width W in the z-direction is given by

$$F = LWC_{DL}\rho v_e^2/2 \qquad (15.65)$$

Note that the flow has to be laminar until the edge of the plate for the above equation to hold. The flow is laminar up to a local Reynolds number of 3×10^5 to 3×10^6 depending on the uniformity of the approaching condition. The flow is turbulent beyond these Reynolds-number values. Turbulent boundary layers will be analyzed briefly later (Section 17.7).

15.5.5   Non-flat systems: the effect of a pressure gradient

For a flat plate v_e is constant and not a function of x, and is equal to the approach velocity. Also the pressure gradient dp/dx is zero here. Here we show the form of the von Kármán analysis for flow systems where the external velocity v_e changes with x. Note that the direction along the body is taken as x, while that perpendicular to the body is taken as y, for non-flat surfaces.

The additional term needed is $v_e \, dv_e/dx$, which can also be written as $\frac{1}{2} d(v_e^2)/dx$.

The following integral results from this term:

$$\int_0^\delta \left[\frac{\partial (v_x^2)}{\partial x} \right] dy - v_e \int_0^\delta \left(\frac{\partial v_x}{\partial x} \right) dy = -\frac{\tau_w}{\rho} + \frac{1}{2} \frac{d(v_e^2)}{dx} \delta \qquad (15.66)$$

On using the Leibnitz rule and rearranging some terms, we have

$$\boxed{\frac{d}{dx} \left(\int_0^\delta (v_x^2 - v_e v_x) dy \right) = -\frac{\tau_w}{\rho} + \frac{dv_e}{dx} \left(\int_0^\delta (v_e - v_x) dy \right)} \qquad (15.67)$$

The last term is the correction for the change in external flow with x. This term is assumed to be known from the potential-flow analysis for the specified geometry. The solution of the integral equation proceeds in the same manner once v_e has been calculated as a function of x from potential-flow theory.

The pressure variation in the external flow can in turn be obtained from the potential-flow theory or experiments conducted outside the boundary layer. The sign of the pressure gradient has a significant effect on the boundary-layer flows. For a plate that is inclined upwards, the velocity increases with the distance along the plate. The pressure gradient is therefore negative here. The second derivative of velocity is proportional to the pressure gradient, as can be seen from the x-momentum balance. The second derivative is positive and does not change sign for the entire boundary layer. The boundary layer remains attached to the plate here. The situation is shown as case (a) in Fig. 15.7. The negative pressure gradient is also called a favorable gradient. For a flat plate the pressure gradient is zero (case (b) in Fig 15.7).

For a plate that is inclined downwards, the fluid decelerates and the pressure gradient is positive. The second derivative is positive at the wall now. Also the second derivative approaches zero at large y from the negative side. Hence the second derivative becomes zero at some point in the boundary layer. This is associated with an inflexion point in the velocity profiles. Thus the boundary layer can separate from the plate if the pressure gradient is positive.

For flow past a cylinder, the pressure gradient is negative for 0 to $\pi/2$ and positive thereafter for $\pi/2$ to π. Hence the boundary layer can separate from the solid at some point where $\theta > \pi/2$.

A more detailed analysis of the boundary-layer flow in inclined plates is now discussed.

15.6   Use of similarity variables

In this section we show how the use of a similarity variable is useful in simplification of the boundary-layer equation. The new form of the equation is also useful in computations for such flows in objects of arbitrary shapes.

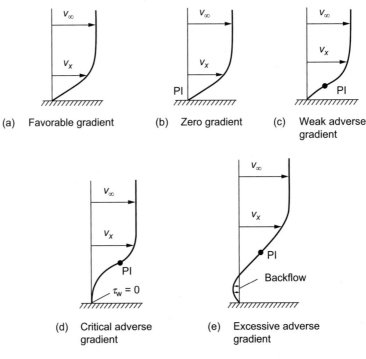

Figure 15.7 Velocity profiles for various values of the pressure gradient; PI indicates the point of inflexion.

First we eliminate the continuity equation by use of the streamfunction ψ defined earlier. The x-momentum balance Eq. (15.53) using the streamfunction becomes

$$\frac{\partial \psi}{\partial y}\frac{\partial^2 \psi}{\partial x \partial y} - \frac{\partial \psi}{\partial x}\frac{\partial^2 \psi}{\partial y^2} = v_e\frac{dv_e}{dx} + \frac{\mu}{\rho}\frac{\partial^3 \psi}{\partial y^3} \tag{15.68}$$

We now introduce a new stretched coordinate in the y-direction. This coordinate η is defined as

$$\eta = \sqrt{\frac{\rho}{\mu}\left(\frac{v_e}{x}\right)}^{1/2}\frac{y}{2} \tag{15.69}$$

We also introduce a dimensionless stream function f defined by

$$f = \frac{\psi\sqrt{\rho/\mu}}{(v_e x)^{1/2}} \tag{15.70}$$

Furthermore, a dimensionless pressure gradient parameter m is defined as

$$m = \frac{x}{v_e}\frac{dv_e}{dx} \tag{15.71}$$

With these substitutions Eq. (15.68) reduces to the following form:

$$f''' + (m+1)ff'' + m[1 - (f')^2] = x\left[f'\frac{\partial f'}{\partial x} - f''\frac{\partial f}{\partial x}\right] \tag{15.72}$$

where the prime $'$ refers to differentiation with respect to η.

The boundary conditions in terms of the primitive variables are

$$v_x = 0, \qquad v_y = 0 \tag{15.73}$$

at $y = 0$, and

$$v_x = v_e(x) \tag{15.74}$$

at $y = \delta$, where δ is the edge of the boundary layer.

The boundary conditions in terms of the variables f and η are

$$f' = 0, \qquad f = 0 \tag{15.75}$$

at $\eta = 0$, and

$$f' = 2 \tag{15.76}$$

at $\eta = \infty$.

Note that the boundary condition at $y = \delta$ is equivalent to $\eta = \infty$ since the Reynolds number is large and the boundary layer looks infinitely large in terms of the stretched coordinate, η. In computations, however, the boundary condition at $\eta = \infty$ is usually applied at a finite location equal to η_∞, which is usually in the range of 8–10. The quantity η_∞ is a measure of the local boundary-layer thickness.

The equations have been transformed into the form $f = f(\eta, x)$. The presence of x in (15.72) shows that a complete similarity has not been achieved. It appears that nothing has been gained by going from x–y coordinates to x–η coordinates. However, the advantage is that the solution is easier using η rather than y, and distance marching along x can be used for any arbitrarily shaped geometry. The range of η is nearly constant, unlike that of y. A simple computational scheme based on this concept is as follows.

15.6.1   A simple computational scheme

The procedure for the solution of the boundary-layer equation is then as follows.

1. Obtain the solution for the velocity potential for the external flow which is governed by

$$\nabla^2 \phi = 0$$

   The boundary condition to be used here is that $\partial\phi/\partial n$ is zero; i.e., the normal component of the velocity at the solid surface is zero. The second condition is that the far-field velocity should be obtained for large distances from the plate.

2. From the solution for ϕ, determine the v_e profile from the equation

$$v_e(x) = \frac{\partial\phi}{\partial x}$$

   at the solid surface $y = 0$.

3. Use the v_e profile, to obtain m as a function of x by using Eq. (15.71).

4. With the m profile in hand, solve Eq. (15.72) as a third-order differential equation after approximating the terms, $\partial f'/\partial x$ and $\partial f/\partial x$, by first-order finite differences.

5. Solve for f as a function of η.

6. From the solution for f, f', and f'', determine the required flow parameters such as velocity profiles in the boundary layer, wall shear stress, etc.

7. Proceed to the next x.

A case where Eq. (15.72) simplifies and leads to a complete similarity representation is the wedge flow discussed in the following section.

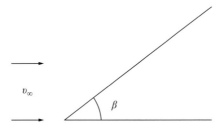

Figure 15.8 A schematic diagram of flow in a wedge.

15.6.2 Wedge flow: the Falkner–Skan equation

Wedge flow is shown schematically in Fig. 15.8. Here the fluid is blowing past a plate inclined at an angle β to the approach direction.

If m is a constant, then the flow does not depend explicitly on x. Such flows are called similar flows, since the velocity profiles look the same when plotted in terms of η. Similar flows are defined as flows where f is a function of η only (and not a function of x). This requires m to be a constant or the external flow V_e to vary in a power-law manner with respect to x,

$$V_e = Cx^m \tag{15.77}$$

where C is a constant. It can be shown that the above equation applies to this problem with $m = \beta/(\pi - \beta)$, β being **half** the wedge angle. For similar flows the Falkner–Skan equation is a third-order ordinary differential equation for f as a function of η only:

$$\boxed{f''' + (m+1)ff'' + m[1 - (f')^2] = 0} \tag{15.78}$$

where m is a constant with the boundary conditions given by Eqs. (15.75) and (15.76). The boundary condition at η_∞ is usually applied at a finite value of η, the usual range being 8–10 as indicated by Cebeci and Bradshaw (1977).

15.6.3 Blasius flow

The case with $m = 0$ refers to flow over a flat plate,

$$f''' + ff'' = 0 \tag{15.79}$$

the solution of which generates the velocity profile for flow over a flat plate (Blasius, 1910). The solution for Blasius flow for f, f', and f'' as a function of η is shown in Table 15.1. The velocity components can be calculated from the f function using the following equations:

$$v_x = \frac{v_\infty}{2}f'$$

and

$$v_y = \frac{1}{2}\left(\frac{vv_\infty}{x}\right)^{1/2}[\eta f' - f]$$

The solutions for other values of m are shown in Table 15.2 for the wall shear stress. Note that the case of negative values of m represents a positive (or adverse) pressure gradient and

Table 15.1. Solution to Blasius flow using the program THIRD.FOR; note that $m = 0$ for this case of a flat plate

η	f	f'	f''
0.00	0.0000	0.0000	1.3284
0.20	0.0266	0.2655	1.3260
0.40	0.1061	0.5294	1.3096
0.60	0.2380	0.7876	1.2664
0.80	0.4283	1.0336	1.1867
1.00	0.6500	1.2596	1.0670
2.00	2.3058	1.9110	0.2570
3.00	4.2796	1.9980	00096
4.00	6.272	2.0000	0.0000
5.00	8.2792	2.0000	0.0000

Table 15.2. Values of wall shear stress for various values of m for flow in a wedge

m	1	0	-0.05	-0.0904
$f''(\eta = 0)$	1.2325	0.3317	0.2128	6.35×10^{-3}

Uniform velocity far from solid

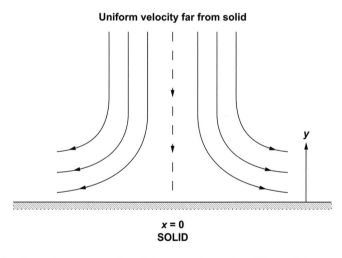

$x = 0$
SOLID

Figure 15.9 A schematic representation of the stagnation-point (Hiemenz) flow.

hence a possibility of flow separation. At the point of flow separation, the wall shear stress becomes zero.

15.6.4   Stagnation-point (Hiemenz) flow

The problem is sketched schematically in Fig. 15.9. A fluid is approaching a plate, and the flow diverts in two directions around the origin. The problem is to investigate the nature of the flow and of the boundary layer that develops around the plate (Hiemenz, 1911).

The vorticity distribution near the plate can also be obtained. The analysis proceeds in two steps: (i) the potential flow is calculated and (ii) this flow is used as the boundary condition for the viscous flow near the plate.

The flow in the irrotational region is readily obtained from the analysis presented in Example 15.3 (with $n = 1$) as $v_x^o = kx$ and $v_y^o = -ky$, where k is a constant that depends on the speed of the flow. The superscript o is used to indicate that this is the outer flow solution.

The form of the above velocity distribution suggests that the flow in the boundary layer may be represented by

$$v_y = -kyf(y)$$

where f can be considered a function that accounts for the effect of the boundary layer. It is obvious that this function should be equal to 1 as $y \to \infty$ and should be zero at the plate itself.

Using continuity we find that

$$v_x = kxf'(y)$$

For later algebra it is convenient to introduce a dimensionless length $\eta = y\sqrt{k/v}$. Using these results in the N–S equation, we find

$$f''' + 2ff' = (f')^2 - 1 = 0$$

which is an ordinary third-order differential equation for f as a function of η. It can be seen that this is a special case of the Falkner–Skan equation with $m = 1$. The solution can be obtained numerically with the following boundary conditions: (i) at $\eta = 0, f = 0$ and $f' = 0$; and (ii) at $\eta \to \infty, f = 1$. The solution can be obtained numerically.

The thickness of the boundary layer is found from the numerical solution to have the following value:

$$\delta = 2.4\sqrt{v/k}$$

We find an interesting property, namely that the thickness of the boundary layer is independent of x This property is also shared by flow over a rotating disk discussed in the following section.

15.7 Flow over a rotating disk

We continue the analysis for flow over a rotating disk. The scaling analysis was presented in Section 14.2.5. Here we present the similarity solution in detail. From the results of the scaling analysis, the following form of the solution can be anticipated:

$$v_r = r\Omega F(\eta)$$

$$v_\theta = r\Omega G(\eta)$$

$$v_z = \sqrt{v\Omega}H(\eta)$$

where F, G, and H are functions of η, which is defined as $z/\sqrt{v/\Omega}$. Note that η here is a measure of the dimensionless boundary-layer thickness and should not be confused with the similarity variable η used in the last section.

From the continuity we can relate F to H and eliminate one of these variables. Thus

$$F = -\frac{1}{2}\frac{dH}{dz}$$

The precise differential equations for F, G, and H can be obtained by back-substituting into the N–S equation. The solution of the three ordinary differential equations then gives us the complete information without our having to solve a bunch of partial differential equations. The governing equations can be shown to be the following:

$$\frac{1}{2}H''' + \frac{1}{4}H'^2 - \frac{1}{2}HH'' - G^2 = 0$$

and

$$G'' + GH' - G'H = 0$$

The boundary conditions are

$$H = H' = 0; \qquad G = 1$$

at $\eta = 0$, and

$$G = 0$$

at $\eta = \infty$.

The problem can be readily integrated using the BVP4C routine nowdays, but an accurate numerical solution was obtained by Cochran (1934) long before the advent of computers.

The model applies strictly to an infinite disk, but can be applied to a finite disk of radius R, provided that the condition that $R > \delta$ or equivalently $R > \sqrt{v/\Omega}$ is satisfied. The "edge" effects are expected to be small under these conditions. The numerical solution and the experimental values of the torque exerted by the fluid agree very well under these conditions. The thickness of the boundary layer may be defined as the point where the F function drops to a small value, say, 0.01. From the numerical results we can show that

$$\delta = 5.4\sqrt{\frac{v}{\Omega}} \tag{15.80}$$

Observation and stability analysis show that the flow is stable up to Re of 285, where Re is defined as $R(\Omega/v)^{1/2}$, where R is the disk radius. Above this value the flow develops spiral vortex pattterns. The flow is expected to become unstable at $Re > 10^5$.

Summary: Stokes flow

- The pressure field in Stokes flow satisfies the Laplace equation. The streamfunction satisfies the biharmonic equation for 2D flows given by Eq. (15.6). For axisymmetric flows the streamfunction satisfies the Stokes biharmonic equation, which is slightly different from the biharmonic equation (see Eq. (15.8)).

- Solution methods can therefore be based on the fundamental solutions to the Stokes operator (for axisymmetric flows). The coefficients in the expansion can be fitted to satisfy the boundary condition on the solid surface. For a sphere the series collapses to a one-term expansion and analytical solutions can be found. Using the analytical solution the drag on the particle can be calculated. This is an important result in particle–liquid motion.

- For deformable particles such as gas bubbles and liquid drops, the internal circulation within the particle affects the velocity field and the drag. Internal circulation is reduced in the presence of surface-active agents.

- Oseen's improvement is an attempt to extend the Stokes analysis to higher Reynolds numbers. The problem in general has to be solved by using the theory of matched asymptotic expansions.

- For nanoparticles, the effect of molecular motion of the gas phase is felt, leading to a phenomenon called Brownian motion. In general a stochastic model is needed to analyze these systems. A related effect is thermophoresis, namely particle motion induced by temperature gradients.

Summary: potential flow

- Flows with zero vorticity are called irrotational flows. The viscous terms in the N–S equations vanish for such flows, and therefore the problem can be treated as an inviscid flow. The Bernoulli equaion holds throughout the flow field in such cases. If the flow were treated as inviscid but not irrotational then the Bernoulli equation would hold only along a streamline.

- A scalar-valued function, the velocity potential, exists in such cases. The Laplacian of the scalar potential is zero. The velocity is simply given as the gradient of the potential.

- Since the harmonic function satisfies the Laplace equation, any harmonic function is a solution to some potential flow. The solution satisfying the required boundary condition can be fitted as a superposition of harmonic functions. This bears a resemblance to the boundary collocation discussed in Problem 9 of Chapter 8.

- The potential flow can also be constructed from the velocity induced by basic flows due to (i) a point source, (ii) a point dipole, and (iii) a point vortex. The superposition of the flows due to these singularities can be used to generate solutions to a wide class of potential-flow problems.

Summary: boundary-layer theory

- Scaling analysis provides us with the tools to simplify the N–S equations within the boundary layer. The x-momentum balance is the only equation needed. The pressure gradient in the boundary layer is the same as the pressure gradient in the external flow. The latter can be found simply by applying the Bernoulli equation for the external flow. The final expression is Eq. (15.53) in the text. The x-direction can be interpreted as the direction along the solid for non-flat geometries.

- The integration of the boundary-layer momentum balance together with the equation of continuity provides an integral representation to the problem, the famous von Kármán integral (Eq. (15.56) in the text). If an assumed profile for the velocity is used, the equation can be solved, providing us with the information on the boundary-layer thickness and the velocity profile. The solution procedure is discussed in Section 15.5.3 and should be reviewed, since the method is quite versatile and has applications in heat and mass transfer as well.

• For flat plates the problem can be solved in an "exact" manner by the Blasius method. Here, using a similarity-variable method, the equation is reduced to a single ordinary differential equation. The solution then gives the velocity profiles. The generalization of the method can be used for many other cases, and the resulting equation is the Falkner–Skan equation (Eq. (15.78) in the text). The equation can also be solved numerically by marching in distance and provides a simple method for the solution of boundary-layer equations in complex geometries.

ADDITIONAL READING

The classic text on the subject of low-Reynolds-number flows is the book by Happel and Brenner (1983).

Microhydrodynamics and slow motion of suspension of particles are treated in detail in the book by Kim and Karrila (1991).

The dynamics of flow around a single particle, bubbles, and droplets has also been discussed in greater detail in the book by Brennen (2005). This book also provides discussions on many fundamental aspects of multiphase flow in a textbook style and is useful for students to get started on more advanced study of multiphase flow.

The classic book on the boundary layer is the book by Schlichting and Gersten (2000), which is a reworking of the original classic.

The dynamics of the vorticity field and its relation to velocity have been succinctly explained in the book by Majda and Bertozzi (2002).

The book by Warsi (1999) is a detailed study of theoretical and computational methods useful for flow analysis.

Problems

1. **Properties of the Stokes equation.** Verify that both the pressure and the vorticity field satisfy the Laplace equation for Stokes flow.

   Verify that the velocity field satisfies the biharmonic equation

   $$\nabla^4 v = 0 \qquad (15.81)$$

2. **Forces and torques in Stokes flow.** Show that the force acting on a control surface of any arbitrary control volume is equal to the force on a larger regularly shaped control volume enclosing the given body. In order to do this you need a control volume that is multiply connected by surfaces S_1 and S_2. Then apply the divergence theorem in reverse to show that the forces on S_1 and S_2 are the same.

   You should also write the Stokes equation in the following stress-divergence form:

   $$\nabla \cdot \tilde{\sigma} = 0$$

   where $\tilde{\sigma}$ is the total stress tensor defined as $-p\tilde{I} + \tilde{\tau}$.

   Show that the torque acting on any arbitrary control volume is again given as the torque on a larger regularly shaped control volume enclosing the given body.

3. **The E^4 operator in spherical coordinates.** Confirm that the definition of the streamfunction given in the text satisfies the continuity equation.

Use this in the N–S equation and derive the form for the E^4 operator in spherical coordinates with ϕ symmetry.

4. **The E^4 operator in axisymmetric cylindrical coordinates.** Consider the flow in cylindrical coordinates with no dependence on θ. How is the streamfunction defined? What is the non-vanishing component of vorticity. Show that the E^2 operator takes the form

$$E^2 = \frac{\partial^2}{\partial r^2} - \frac{1}{r}\frac{\partial}{\partial r} + \frac{\partial^2}{\partial z^2} \qquad (15.82)$$

and $E^4\psi = 0$ for Stokes flow.

5. **General solution for ψ.** Derive the general solution for ψ given in the text (Eq. (15.9)).

6. **Dimensional consistency.** What are units for C_1 in Eq. (15.10)? Confirm that the units agree on both sides.

7. **A rotating sphere.** A sphere of radius R is rotating in an infinite fluid at an angular velocity of Ω. Derive the following expression for the velocity field:

$$v_\phi = r\Omega\sin\theta\,\frac{R^3}{r^3}$$

Find an expression for the torque exerted by the sphere on the fluid. **Answer:** the torque is $8\pi\mu\Omega R^3$.

Hint: assume a solution of the form $v_\phi = f(r)\sin\theta$, where $f(r)$ is a function just of r to be determined from the ϕ-momentum balance.

8. **Stokes flow with a far-field elongational flow.** Find the solution to Stokes flow past a sphere where the far-field velocity satisfies the elongational flow defined as $v_x = \dot{\gamma}x$, $v_y = \dot{\gamma}y$, and $v_z = -2\dot{\gamma}z$.
   First verify that the continuity equation is satisfied. Then transform the velocity to spherical coordinates to show that the flow is symmetric if z is chosen as the direction of symmetry. Now set up the boundary conditions for ψ. In particular, show that only the Q_2 term remains for the above far-field velocity. Complete the solution for ψ by boundary-condition fitting. What is the force on the sphere?

9. **The general solution to Stokes flow in 2D Cartesian coordinates.** For the 2D case the governing equation is $\nabla^4\psi = 0$. The operator ∇ may be applied either in Cartesian $(x,\ y)$ or in polar (r,θ) coordinates. In either case it would be appropriate to seek a general form of the solution to this biharmonic operator.
   The problem of finding a solution to the biharmonic operator can be broken down into two sub-problems:

$$\nabla^2\omega = 0$$

and

$$\nabla^2\psi = -\omega$$

The first problem is similar to the case of potential flow in the 2D case and admits a class of solutions of the following type:

$$r^n\cos(n\theta); \qquad r^n\sin(n\theta)$$

The second problem is then the solution to a Poisson equation with the non-homogeneous terms corresponding to each of the above functions. Solve these equations to derive a general solution to Stokes flow in 2D Cartesian coordinates.

10. **Stokes flow past a cylinder.** The case of flow past a cylinder of infinite length normal to the axis was also studied by Stokes. In view of the 2D nature of the problem, it is more convenient to work in r, θ coordinates. The governing equation is $\nabla^4 \psi = 0$.

    What are the boundary conditions that can be imposed on ψ? Assume that the far-field approach velocity is a constant.

    In view of the far-field condition, the following solution looks reasonable:

    $$\psi = v_\infty \left(\frac{A}{r} + Br + Cr \ln r \right) \sin \theta$$

    Try fitting the constants, and show that a uniformly valid solution cannot be obtained. (This is called the Stokes paradox.)

11. Verify the vector identity

    $$\nabla^2 v = \nabla(\nabla \cdot v) - \nabla \times (\nabla \times v) \tag{15.83}$$

12. Show that, for irrotational axisymmetric flows in cylindrical coordinates, the streamfunction satisfies the following equation:

    $$\frac{\partial^2 \psi}{\partial r^2} - \frac{1}{r} \frac{\partial^2 \psi}{\partial \theta^2} + \frac{\partial^2 \psi}{\partial z^2} = 0$$

13. Indicate the type of flow given by $\phi = \sqrt{r} \cos(\theta/2)$. Calculate and plot typical streamlines.

14. Consider a source of flow in 3D of strength S (dimensions L^3/T), Then the flow is axisymmetric and independent of the θ-direction in the spherical coordinate system. Show by a mass balance that

    $$S = 4\pi r^2 v_r$$

    and hence derive Eq. (15.41) for the potential due to a unit source in 3D.

15. Consider again the problem of flow past a cylinder with rotation considered in Example 15.4. Calculate and plot the location of the stagnation point on the cylinder surface for $0 < \kappa^* < 4$. Here κ^* is the dimensionless circulation defined as $\kappa/(v_\infty R)$. Plot typical streamlines.

    Also calculate the locations for larger values of κ^*. These are now along the $\theta = -\pi/2$ line but not on the surface of the cylinder.

16. Verify the principle of superposition, which states that if ϕ_1 and ϕ_2 are solutions to potential flow then $\phi = c_1\phi_1 + c_2\phi_2$ is also a solution.

    Also show that $\phi_1\phi_2$ is NOT a solution.

17. Verify by direct substitution that the functions $r^n \cos(n\theta)$ etc. shown in the text satisfy the Laplace equation in polar coordinates.

    Calculate the streamfunction for the above-mentioned potential function and construct a contour-plot of streamlines for $n = 1/2, 1$, and $3/2$. MATLAB or MATHEMATICA may be used to make these contour-plots.

    Also construct a contour-plot of equipotential lines for any one of the above cases and show that these lines are perpendicular to the streamlines.

    Answer for streamfunction: $\psi = r^n \sin(n\theta)$. A constant can be added to the streamfunction without altering the results for the flow field.

    Hence verify that the complex function $F(z) = z^n$ is a solution to a potential problem where the real part becomes the velocity potential while the imaginary part represents the streamfunction.

18. State the form of the Laplace equation in axisymmetric spherical coordinates.
    Verify that the following functions satisfy this equation:

    $$r\cos\theta; \qquad \cos\theta/r^2$$

    A linear combination is also a solution by superposition. Thus the following solution for ϕ obtained by taking the combination represents the potential flow around a sphere of radius R:

    $$\phi = v_\infty \left[r\cos\theta + \frac{R^3}{2} \frac{\cos\theta}{r^2} \right]$$

    Verify that the impermeability condition is satisfied at $r = R$, the radius of the sphere, by showing that $v_r = \partial\phi/\partial r$ is zero at these locations.

19. A cylinder of diameter 1.2 m and length 7.5 m rotates at 90 r.p.m. with its axis perpendicular to an air stream with an approach velocity of 3.6 m/s.
    Plot the tangential component of the velocity along the circumference of the cylinder.
    Plot the pressure distribution along the circumference.
    Find the lift force on the cylinder.

20. **Flow in porous media.** Model equations similar to those for potential flow arise in flow in porous media, which has a wide variety of applications, e.g., in groundwater treatment, water-purity remediation, filtration, flow of a drug into a tissue such as a tumor, etc. The mathematical structure of the problem is presented here.
    The model used is the Darcy equation, which states that the velocity is proportional to the pressure gradient,

    $$v = -\frac{\mathcal{L}^2}{150\mu} \frac{dP}{dz} \tag{15.84}$$

    where $\mathcal{L}$ is a characteristic length parameter of the system.
    Equation (15.84) has the form of the flux (velocity is a volumetric flux) as a function of a driving force (the pressure gradient) and is known as Darcy's law. Since all the parameters appearing on the RHS cannot be estimated exactly for porous media, they are combined into a constant κ defined as the Darcy permeability and the equation is written in a simple flux-driving force formula:

    $$v_z = -\frac{\kappa}{\mu} \frac{dP}{dz} \tag{15.85}$$

    This equation is known as Darcy's law. The term κ is called the permeability of the porous medium. The equation may be generalized for three dimensions as

    $$v = -\frac{\kappa}{\mu} \nabla P \tag{15.86}$$

    The vector v satisfies the continuity equation, i.e.,

    $$\nabla \cdot v = 0$$

    Hence, by applying the divergence operator on both sides of Eq. (15.86), verify that the pressure field is given by

    $$\nabla^2 P = 0 \tag{15.87}$$

Hence we find that the pressure field in a porous medium satisfies the Laplace equation. This equation has the same form as that for steady-state conduction, and hence similar solution methods can be used to compute the pressure field. Once the pressure field has been computed, the velocity field can be recovered by applying Eq. (15.86).

21. One way of cleaning up environmentally polluted underground water is to pump it out of the ground, treat it with some equipment (by catalytic or photochemical oxidation, for example), and then pump it back into the ground. This is referred to as the pump-and-treat process in the EPA jargon.

   Here we address a part of the design calculations, namely the problem of the pressure-drop calculation. In particular, the problem we wish to solve is that of determining an estimate of the pressure drop between the point of pumping and the point of discharge The system is modeled as a line source (along a vertical line) at a point A (the pump location) and a line sink at a point B (the location of the discharge point). Also the pressure variation in the vertical direction is ignored. This makes the problem 2D, and the theory for potential flow developed in this chapter can be used. Note that once the 2D case has been solved corrections can be applied for the pressure variation in the vertical direction. Your goal is to use some Laplace solver and simulate pressure profiles in this medium. The corresponding velocity profiles can then be computed by applying Darcy's law.

22. **Boundary-layer flows: the effect of cross-stream velocity.** Consider the flow past a flat plate. Here we examine the effect of a superimposed velocity v_{y0} at the plate surface. This can be done by the integral method with only minor changes in the equations. Verify that the velocity in the y-direction is now given by

$$v_{y\delta} = v_{y0} - \int_0^\delta \left(\frac{\partial v_x}{\partial x} \right) dy \qquad (15.88)$$

   This adds an extra term $v_{y0}v_\infty$ to the integral form of the x-momentum balance. The resulting equation can be derived as

$$\frac{d}{dx} \left(\int_0^\delta (v_x^2 - v_\infty v_x) dy \right) + v_{y0}v_\infty = -\frac{\tau_w}{\rho} \qquad (15.89)$$

   Apply a cubic profile and integrate the above equation to get an (implicit) relation for the boundary-layer thickness. Derive this equation and discuss how the boundary-layer thickness changes with the blowing or suction velocity v_0.

23. **Falkner–Skan flows.** Find the value of m for which the wall shear stress is independent of the principal flow direction.

   Find the value of m for which the boundary-layer thickness is a constant. (The answer is $m = 1$.)

24. **Flow with an interfacial traction.** Consider the problem of a semi-infinite fluid subject to a constant shear at the interface. This can be caused, for instance, by a surface-tension gradient. Show that the following differential equation is applicable for the boundary layer:

$$3f''' + 2ff'' - (f')^2 = 0$$

   State the boundary conditions. Use the program BVP4C to simulate the flow.

25. Von Kármán assumed a cubic profile for the integral momentum analysis over a flat plate. Since a cubic has four constants, four conditions were used.

   (i) $V_x = 0$ at $y = 0$.
   (ii) $V_x = V_e$ at $y = \delta$.
   (iii) $dV_x/dy = 0$ at $y = \delta$.
   (iv) $d^2 V_x/dy^2 = 0$ at $y = 0$.

   Show that the use of these conditions in a cubic profile leads to the representation given by Eq. (15.61).
   Justify condition (iv) above using the x-momentum balance applied at $y = 0$ together with the no-slip boundry condition.

26. Consider a boundary layer with no external pressure gradient (a flat plate). Then, from the Prandtl boundary-layer equation, deduce that

$$\mu \frac{d^2 v_x}{dy^2} = 0 \text{ at } y = 0$$

   Now derive an expression for the second derivative of velocity at the surface for the case where there is a finite pressure gradient. In particular, show that

$$\mu \left(\frac{d^2 v_x}{dy^2} \right)_{y=0} = -\rho v_e \frac{dv_e}{dx}$$

   where v_e is the x-component of the velocity at the edge of the boundary layer.
   Using the above result, show that the following velocity profile can be used in the presence of a finite pressure gradient:

$$\frac{v_x}{v_e} = \frac{3}{2}\eta - \frac{\eta^3}{2} + \frac{\delta^2}{4v} \frac{dv_e}{dx}\left[\eta - 2\eta^2 + \eta^3\right]$$

   where η is defined as y/δ. Also $v = \mu/\rho$ as usual. This can then be used in conjunction with the von Kármán integral momentum balance to find the boundary-layer thickness and the velocity profiles for non-flat surfaces.

27. A cubic approximation is commonly used in conjunction with the von Kármán momentum integral. An alternative form is the sine function:

$$v_x = \alpha \sin(by)$$

   What should the constants α and b be in order to satisfy some conditions on v_x for a flat-plate geometry? Repeat the von Kármán analysis using the above expression, and derive a differential equation for $d\delta/dx$. Integrate this equation to obtain the following expression for δ:

$$\delta = \sqrt{\frac{vx}{v_\infty} \frac{2\pi^2}{4 - \pi}}$$

   The mathematics is horrendous except for my students in Lopata.
   Also find an expression for the local drag coefficient.

28. There is a fluid evaporating from the surface. Here $v_x = 0$, but $v_y \neq 0$ at the plate surface. Derive the von Kármán momentum relation for this case.

29. Follow up the derivations leading to the Blasius equation in Section 12.5 leading to

$$f''' + ff'' = 0$$

A useful routine to solve this is BVP4C in MATLAB. Solve the Blasius equation using this routine, and compare your answer with the results in Table 15.1. In particular, use the boundary condition of $f' = 2$ at $\eta = 5$ rather than at infinity as an approximation. The other conditions at $\eta = 0$ are $f = 0$ and $f' = 0$.

You may wish to extend this for Falkner–Skan flow and show the effect of the m parameter on the wall shear stress.

16 Bifurcation and stability analysis

Learning objectives

Upon completion of this chapter you will be able to

- understand the basic characteristics of dynamical systems, linear stability analysis, and limit cycles;
- develop methods to analyze stability by using the method of normal modes;
- derive and solve, for simple cases, the classical Orr–Sommerfeld equation, and apply this to do flow stability analysis; and
- track the solution as a function of a specified parameter by the arc-length-continuation method, and apply this to common non-linear diffusion–reaction problems.

Mathematical prerequisites

- Taylor-series expansion and the notion of the Jacobian.
- Differential equations for dynamic systems and their matrix representation.
- Complex-variable representation of trigonometric functions.
- The concept of an eigenvalue of a differential operator and eigenfuncitons.

Non–linear differential equations exhibit a wide range of complexities such as multiple steady states, bifurcations, and chaos. The problem is particularly important since the Navier-Stokes (N–S) equations have the non-linear term $(v \cdot \nabla)v$ which lies at the root of chaos and turbulent flow behavior. The first goal of this chapter is to introduce mathematical tools to analyze the stability of dynamical systems. A common method is linear stability analysis. Here the base flow or a base steady state is perturbed by a small amount and modeled as a dynamical system. Then we can examine whether the flow returns to the original steady state (i.e., whether the perturbations decay with time) or whether the flow remains transient. When this method is applied to the N–S equations, we can derive a widely used differential equation, *viz.*, the Orr–Sommerfeld equation, which is useful for analysis of the stability of the flow. This chapter provides an introductory treatment of these topics.

The mathematical techniques are then illustrated by a number of examples of well-studied problems in flow stability. Some of these discussions are simply qualitative but still span some important common situations and also provide a glimpse of the historical developments in this field. As a further aid to learning, a sample code for the solution of the basic Orr–Sommerfeld equation is also provided. These sections and the sample code will go a long way towards providing a good understanding and stimulating the urge for additional learning on this topic.

The non-linear behavior is also common in heat and mass transfer and in many related systems with non-linear source terms. We also present some analysis of such systems and discuss a mathematical technique, *viz.*, the arc-length continuation, to track the dependence of the solution on a particular parameter. This is a useful method and has applications in heat and mass transfer. Although the chapter is mainly focused on flow stability, this is one additional topic that you will explore in this chapter. We revisit the Frank-Kamenetskii problem as an illustrative example and see how the nature of the solution changes as a critical parameter, δ (the thermicity parameter), is varied. We also examine again the problem of simultaneous heat and mass transfer in a porous catalyst.

16.1   Introduction to dynamical systems

In real life, disturbances can interfere with the attainment of your main goals. These disturbances may be small to start with, but might not remain small under some conditions, and may grow with time. The modeling of dynamical systems is an attempt to understand the effect of these disturbances and forms a prelude to the analysis of more complex problems, for instance, systems described by the N–S equation. In this section we introduce some preliminary concepts and notation needed in order to understand the dynamic behavior.

The evolution equation
A dynamical system is represented by an "evolution" equation, which can be expressed in a compact form as

$$\frac{d\boldsymbol{u}}{dt} = \boldsymbol{F}(u; \alpha) \tag{16.1}$$

where $\boldsymbol{u}$ is the solution vector. Here $\boldsymbol{F}$ is a set of non-linear functions (or operators; see below) of u. The vector α represents a set of system parameters. For example, in the context of flow stability $\boldsymbol{u}$ is often a seven-dimensional vector with three components of velocity, three components of vorticity, and pressure. The parameter α is usually the Reynolds number.

In the context of reaction engineering problems, the system equation vector u is usually the concentration (or a set of concentrations for multiple reactions) and the temperature. The system parameter analyzed is usually the Damköhler number for lumped-parameter systems (CSTR) and the Thiele modulus for diffusion–reaction problems (distributed-parameter models), since both of these parameters can be changed easily by changing the operating parameters such as flow rate, particle size, etc.

Classification

Depending on the spatial variation of F, the dynamical systems can be classified into two categories.

1. Lumped-parameter systems (LPSs). The function F here does not depend on the spatial variables. F is a non-linear function of u only. The time derivative in Eq. (16.1) is the ordinary time derivative. An example would be a backmixed reactor where uniform concentration and temperature are assumed.
2. Distributed-parameter systems (DPSs). The function F here depends on the spatial variables and it can be viewed (generally) as a non-linear differential operator. An example would be a plug-flow reactor where the concentration and temperature vary as a function of axial location. A second example would be the transient analysis of flow systems. The time derivative is the local partial derivative with respect to time.

DPSs: the role of spatial discretization

The dynamical behavior of a DPS is usually obtained by a two-step procedure. In the first step, the DPS is converted into an equivalent LPS by suitable discretization of the spatial derivatives. For example, the finite-difference method, finite-element discretization, orthogonal collocation techniques, spectral methods, boundary-integral methods, etc., can be used for the reduction of a DPS to an LPS. The dynamical behavior of the LPS is then analyzed. The procedure is illustrated in Fig. 16.1.

If the system has n variables to start with (i.e., the vector u is of size n) and if n_t discretization points are used to approximate the DPS then we obtain a set of $n \times n_t$ equations in the LPS. Thus the LPS resulting from the original DPS is of much larger vector dimensionality than the size of the original operator. Also, in the overall computational process, the discretization of spatial derivatives (the first step) introduces some errors, which can result in the prediction of spurious results. Hence careful attention has to be paid to the discretization process itself. Higher-order discretization schemes with a larger number of

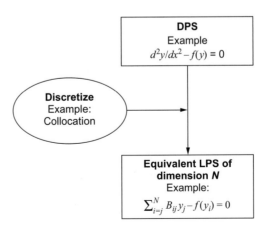

Figure 16.1 Conversion of a DPS into an equivalent LPS by discretization.

discretization nodes are often required in order to obtain a reasonable accuracy in the solution. Such finer approximations of a DPS can increase the dimensionality of the resulting LPS even further. Hence the study of dynamics and the determination of multiple steady states for a DPS is numerically challenging.

We now focus on an LPS, assuming that if the problem was a DPS to start with then it has been reduced to a system of LPSs by discretization of the operator.

Linearized form and the Jacobian

A steady state of the system is described by

$$F(u_s) = 0$$

where u_s is the (vector of) the steady-state solution.

Steady-state solutions can be found by use of a non-linear algebraic equation solver. Most variations use some form of Newton–Raphson method. A simple code is shown later for an illustrative problem. Let us for now assume that we have found a steady state and wish to examine the dynamic system for small perturbations near the steady state. We can then deal with a perturbation variable defined as

$$u_d = u - u_s$$

The function F can then be linearized in the vicinity of the steady state. Thus for the ith component of the vector of F we can write

$$F_i = F_i(u_s) + \sum_j^n \frac{\partial F_i}{\partial u_j}(u_j - u_{s,j})$$

The first term on the RHS is zero since u_s is the vector of steady-state values. The second term is represented as a matrix–vector product, and the equation is compacted by defining the Jacobian matrix $\partial F_i / \partial u_j$ and written in vector notation as

$$F = \tilde{J} \times u_d$$

Hence the dynamic equation in the linearized form is

$$\boxed{\frac{du_d}{dt} = \tilde{J} u_d} \tag{16.2}$$

This equation is easier to analyze than the original problem and is referred to as a linearized form. Note that we have already encountered such equations in transient simulation of LPSs in Chapter 2. The matrix method provides an elegant solution procedure for such problems.

Eigenvalues of the Jacobian

The system behavior can be classified depending on the eigenvalues λ_i of the Jacobian matrix $\tilde{J}$.

The eigenvalues can in general be complex, and can be split into a real part and an imaginary part Also note that the complex part results in sine and cosine function contributions to the time, while the real part contributes to exponential functions of time. The long-time behavior is therefore governed by the real part of the eigenvalues $Re(\lambda_i)$. If the real part is positive even for one of the eigenvalues the perturbations will grow with time, rendering the system unstable. If all the eigenvalues have real negative parts then the perturbations decrease with time, resulting in the system returning to the steady state. However, the system may reach the steady state in oscillatory manner with decreasing amplitude at each cycle, and this behavior is contributed by the imaginary part of the eigenvalues.

Phase-plane plots

If the problem is a two-variable problem then the evolution of the dynamic system can be illustrated by a plot of the first variable $u_d(1)$, vs. the second variable, $u_d(2)$, at each instant of time. Such a plot is known as a **phase-plane plot**. The phase plot starting at any point should approach zero if the system is stable since the disturbance vector should decay to zero. If the plot moves away from the origin then the system is unstable. An illustrative

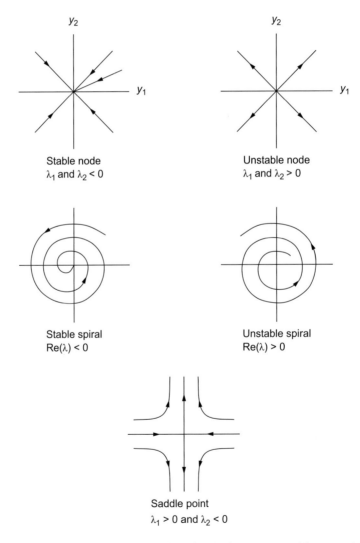

Figure 16.2 Classification of dynamic behavior of LPSs for a system with two variables. The phase-plane plots are of $u_d(1)$ vs. $u_d(2)$ at each instant of time. $Re(\lambda)$ is the real part of the complex eigenvalue λ.

phase plot for various cases is shown in Fig. 16.2, and the nature of the plot depends on values of λ_1 and λ_2. The concepts are illustrated by a simple example case study later in this chapter.

Multiple steady states

Systems can exhibit multiple steady states and hence the bifurcation theory is closely related to the stability analysis. Bifurcation refers to branching. The bifurcation theory examines the steady-state solution as a function of a parameter and examines whether multiple steady states are possible for the same value of the parameter. A simple cubic may serve as an illustration.

Consider the CSTR problem with isothermal conditions. Hence only a mass balance is needed. Also consider the rate function given as the last term on the RHS on the following equation:

$$F(c) = 1 - c - \frac{Dac}{1 + \beta c + \gamma c^2} = 0$$

This type of rate function is often encountered in enzyme-catalyzed reactions that exhibit the phenomenon of substrate inhibition. The (dimensionless) exit concentration c is obtained by solution of above equation. This is a cubic equation and hence can have three real roots for certain ranges of parameters. The goal of bifurcation analysis, in general, is to track down the ranges of parameters for which multiple steady states are possible. A general and quite powerful method is arc-length continuation, which is described below.

16.1.1   Arc-length continuation: a single-equation example

The method of arc-length continuation is best appreciated by considering a single non-linear algebraic equation of the form:

$$F(u; \alpha) = 0 \tag{16.3}$$

The solution dependence may be of the type shown in Fig. 16.3 (the solution curve), where the solution u is plotted as a function of the parameter α. The continuation method requires knowledge of an initial point u_0, α_0 on this curve. Then the dependence on the parameter α can be tracked by moving along the arc length s of the solution curve.

The condition that $dF = 0$ along the arc length leads to

$$\frac{dF}{ds} = \frac{\partial F}{\partial u}\frac{du}{ds} + \frac{\partial F}{\partial \alpha}\frac{d\alpha}{ds} = 0 \tag{16.4}$$

Let F_u denote the partial derivative of F with respect to u and F_α the partial derivative with respect to α. Then the above equation can be represented as

$$F_u\frac{du}{ds} + F_\alpha\frac{d\alpha}{ds} = 0 \tag{16.5}$$

Note that

$$ds^2 = d\alpha^2 + du^2 \tag{16.6}$$

along the arc using the simple Pythagorean rule. Therefore

$$\left(\frac{d\alpha}{ds}\right)^2 + \left(\frac{du}{ds}\right)^2 = 1 \tag{16.7}$$

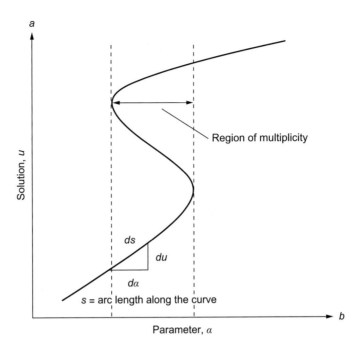

Figure 16.3 The solution plotted as a function of a parameter α. Here s is the distance along the curve

On solving Eqs. (16.5) and (16.7) together, we get a set of two initial-value problems (IVPs) of the type

$$\frac{du}{ds} = \pm \frac{F_\alpha}{\sqrt{F_u^2 + F_\alpha^2}} \tag{16.8}$$

and

$$\frac{d\alpha}{ds} = \pm \frac{F_u}{\sqrt{F_u^2 + F_\alpha^2}} \tag{16.9}$$

These IVPs can be integrated with the initial condition at $s = 0$, $u = u_0$, and $\alpha = \alpha_0$, where (u_0, α_0) is a known point on the solution curve. The initial-value solver then gives the solution u and the corresponding value of p along the arc length at each point on the solution curve. The extension to multiple reactions is straightforward and is given below.

16.1.2   The arc-length method: multiple equations

For a system of n equations, the requirement that the functions F_i do not change along the arc length leads to the following extension of Eq. (16.8):

$$\sum_{j=1}^{n} J_{ij} \frac{du_j}{ds} + F_\alpha \frac{d\alpha}{ds} = 0 \tag{16.10}$$

where instead of F_u, the derivative of F with respect to u, we have used the Jacobian and also summed over the n variables.

This needs to be coupled with the Pythagorean rule in $n + 1$ dimensions:

$$\sum_{j=1}^{n_t} \left(\frac{du_j}{ds}\right)^2 + \left(\frac{d\alpha}{ds}\right)^2 = 1 \tag{16.11}$$

These equations can be rearranged to a set of IVPs in du_j/ds ($j = 1, n$) and $d\alpha/ds$, which are the equivalents of Eqs. (16.8) and (16.9) for the single-equation case. The IVPs can be integrated along the arc length to track the solution curve. The procedure for obtaining the IVPs is slightly different from the single-equation case, and will be described below.

Initial-value representation

The variable set is augmented by introducing an additional variable u_{n+1}, which is set equal to α. The Eq. (16.10) can then be written as

$$\sum_{j=1}^{n+1} \frac{\partial F_i}{\partial u_j} \frac{du_j}{ds} = 0 \tag{16.12}$$

where $\partial F_i/\partial u_{n+1}$ denotes $\partial F_i/\partial \alpha$ and du_{n+1}/ds denotes $d\alpha/ds$.

Equation (16.12) is a linear combination of the derivatives du_j/ds (including α now), and hence these derivatives can all be linearly related to a single quantity, du_k/ds, where k is called the control variable. Hence we can write

$$\frac{du_j}{ds} = \beta_j \frac{du_k}{ds} \tag{16.13}$$

where the β_js are a set of coefficients to be computed and u_k is a chosen control variable.

The equations for calculation of β_j are readily obtained by substituting for du_j/ds from Eq. (16.13) in Eq. (16.12) and rearranging. We obtain the following set of linear equations for the β vector:

$$\sum_{j=1}^{n+1} \frac{\partial F_i}{\partial u_j} \beta_j = -\frac{\partial F_i}{\partial u_k} \tag{16.14}$$

where the summation term on the LHS excludes the variable k. Solution of this set of linear algebriac equation gives us a set of β_j values. Also, using Eq. (16.13) in the Pythagoras expression, we have

$$\frac{du_k}{ds} = \pm \frac{1}{\sqrt{1 + \sum_j \beta_j^2}} \tag{16.15}$$

where the sign is arbitrary and depends on whether we are moving clockwise or counter-clockwise on the solution curve.

Similarly, for all the other variables other than k we have

$$\frac{du_j}{ds} = \pm \frac{\beta_j}{\sqrt{1 + \sum_j \beta_j^2}} \tag{16.16}$$

By this procedure, we obtain the set of IVPs (Eqs. (16.15) and (16.16)) for marching along the solution curve. It may be noted that the integration errors accumulate during the marching and hence the computed values of the vector are used only as predictors. They are used in the solution of the distributed model, and the converged solution is then used as the new solution at this new starting point. Since the predictor of the arc-length solution is close to the correct solution, just two or three iterations will suffice for the distributed model to converge. The new converged solution forms a new starting point on the solution curve. Thus the procedure is similar to a predictor–corrector method. The arc-length integration serves as a predictor while the distributed model solver serves as the corrector. With a proper choice of control variable, k, the algorithm generates a continuous curve of solutions and passes through bifurcation or limit points without any difficulties.

The guideline for the choice of the control variable is as follows. The control variable should change monotonically (either increase or decrease) along the solution curve, and also the linear equations defined by Eq. (16.14) should not be ill-conditioned. If these conditions are not met at any stage of moving along the arc, a new control variable can be used and the procedure continued. The control variable is only a crutch to arrive at an explicit set of IVPs for marching along the arc length. An example application is shown below.

Example 16.1. The dynamic behavior of a CSTR

Examine the behavior of a CSTR for an exothermic chemical reaction around a steady state. The CSTR model is represented as an LPS with two variables. We have a dynamical system that can be expressed in the form of Eq. (16.1).

The concentration is given by

$$\frac{dc}{dt^*} = 1 - c - Daf(c)\exp[\gamma(1 - 1/\theta)]$$

and the temperature by

$$\frac{d\theta}{dt^*} = 1 - \theta + DaBf(c)\exp[\gamma(1 - 1/\theta)] - U^*(\theta - \theta_w)$$

The dimensionless parameters are Da, B, γ, U^*, and θ_w.

What do these parameters represent?

Let us choose a set of parameters for illustration purposes: $Da = 0.02$, $B = 2.0$, $\gamma = 20$, and $\theta_w = 1$, and we set U^* as zero to illustrate the case of an adiabatic reactor, i.e., no wall cooling. Also $f(c)$ is taken as c, i.e., a first-order reaction.

The steady-state solution is obtained by setting the LHS to zero and solving the two algebraic equations for c and θ. This can be done using the following MATLAB code (Box 5). In the Newton–Raphson method we need the evaluation of the Jacobian.

We find that the system has three solutions, and hence there are three steady-state values.

Box 5 Sample MATLAB code for finding the steady-state solution to a CSTR

```
Da  = 0.02; B = 1.0; gamm = 20.0; ustar = 0.0;  Twall = 1.
      y = [ 1; 1.5 ] % initial starting value
nrmdy = 10. ; % inital error; an assumed large error.


% iteration loop
   while nrmdy > 1e-10
```

```
% define the forcing functions.
F(1) =  1 - y(1) - Da *  y(1)  * exp( gamm * ( 1 - 1./y(2) ) )
F(2) = 1.0 - y(2) + B * Da * y(1)  * exp( gamm * ( 1 - 1./y(2) ) )...
          - ustar * ( y(2) - Twall )
% find Jacobian at this point
J(1,1) = -1 - Da * exp( gamm * ( 1 - 1./y(2) ) )
J(1,2) = -Da * y(1) * exp( gamm * ( 1 - 1./y(2) ) ) * gamm/y(2)^2
J(2,1) = B * Da * exp( gamm * ( 1 - 1./y(2) ) )
J(2,2) = -1 + B * Da * y(1) * exp( gamm * ( 1 - 1./y(2) ) ) ...
             * gamm/y(2)^2 - ustar
% Correction vector according to Newton-Raphson
dy = -(J\F');
% new solution and the current residual error
  y = y + dy;  nrmdy = norm(dy)
   end
y % converged solution;
eig(J)  % Jacobian  of the steady state.
```

The system shows three steady states. The values are shown below with variable 1 representing concentration and variable 2 representing the temperature.

(i) The ignition state, with the highest exit temperature, which can be shown to be a stable steady state: $y(1) = 0.0023$ and $y(2) = 1.99770$.

(ii) The extinction state, with the lowest exit temperature, which is also a stable steady state: $y(1) = 0.9541$ and $y(2) = 1.0459$.

(iii) A middle steady state, which is unstable: $y(1) = 0.9235$; and $y(2) = 1.0765$.

The convergence to the steady states when using the Newton–Raphson code shown above is somewhat random and obtained by changing the starting values. A systematic way to achieve this is to do arc-length continuation to track down the entire parameter dependence. Now we examine the stability of the steady state.

The Jacobian is evaluated for each of the above steady states and the eigenvalues of the Jacobian are found. This can be accomplished by the MATLAB command *eig(J)*. For the middle steady state the eigenvalues are $\lambda_1 = -1.0000$ and $\lambda_2 = 0.2369$.

The positive eigenvalue shows that the steady state is unstable. Any disturbance will grow in time, and the system will approach either the lower or the higher steady state.

Since the convergence of the Newton–Raphson code is random and the parametric effects are not captured, it is useful to do a parametric study. Let us do this now, by examining the bifurcation pattern using the arc-length-continuation method. The method shown in Section 16.1.2 was applied to this problem for the values of parameters shown in this example. The parameter Da was used for arc-length continuation. The results are shown in Fig. 16.4, which captures the region of three steady states. It is easy to extend the method for more complex networks of reactions.

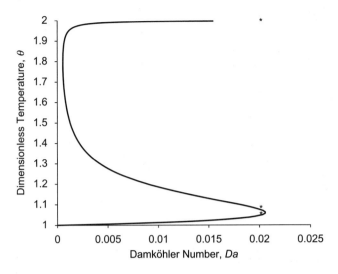

Figure 16.4 A bifurcation diagram for the CSTR.

16.2   Bifurcation and multiplicity of DPSs

The method can be applied to a DVP as well. We will illustrate the key steps for doing this now.

The basic idea is that a distributed system can be discretized and cast as a set of algebraic non-linear equations. If the equations are linearized around a steady state then the set of equations can be put into a form similar to Eq. (16.3). Each differential equation will create n algebraic equations where n is the number of discretization points. The arc-length continuation can then be applied to this expanded LPS. Two examples are shown below.

16.2.1   A bifurcation example: the Frank-Kamenetskii equation

We revisit the Frank-Kamenetskii problem and examine at what point the bifurcation starts. The problem to be solved is restated here:

$$\frac{d^2\theta}{d\xi^2} = -\delta \exp(\theta) \tag{16.17}$$

with the boundary condition of no flux at $\xi = 0$ and zero temperature at $\xi = 1$. The effect of the thermicity parameter δ is examined.

The DPS was discretized by the boundary-integral method. This provides an integral representation of the differential equation and hence is considered to be more accurate than the traditional finite-difference method. The parameter effects on the discretized version were then studied by use of the arc-length algorithm. The details are given elsewhere (Ramachandran and Ramaswamy, 2008). The results are shown in Fig. 16.5, where the center temperature is plotted as a function of the Frank-Kamenetskii parameter δ. It is seen

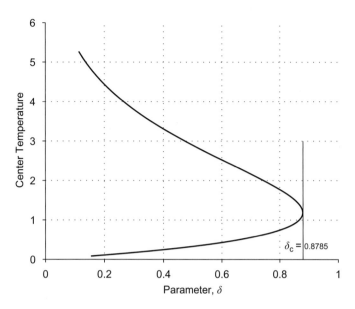

Figure 16.5 The bifurcation diagram for the Frank-Kamenetskii problem, showing the temperature at the center of the slab for the case of exponential heat generation.

that $\delta = 0.8785$ is a branch point. This value is in agreement with earlier work and also shows the usefulness of the arc-length method.

16.2.2   Bifurcation: porous catalyst

The mass balance and energy balance equations describing the diffusion process with irreversible first-order reaction kinetics in a slab geometry can be combined, in the absence of external transport processes, into a single equation. The resulting governing equation is given by

$$\frac{d^2c}{d\xi^2} = \phi^2 c \exp\left(\frac{\gamma\beta(1-c)}{1+\beta(1-c)}\right) \qquad (16.18)$$

Here, c is the concentration of the reactant and the other parameters are defined as follows: ϕ is the Thiele modulus, γ the dimensionless activation energy, and β, dimensionless heat of reaction. The boundary conditions are that there is no flux at $\xi = 0$ and constant concentration $c = 1$ at $\xi = 1$.

For certain ranges of parameters (for example, $\phi = 0.15$, $\gamma = 20$, and $\beta = 0.8$), this system exhibits multiple steady states. It is often difficult to obtain all the steady states, since the results are sensitive to the starting values. Hence it is instructive to examine the arc-length procedure and track the solution with ϕ as the parameter. The spatial discretization is often important, and the boundary-integral method was used since it is an integral method and allows one to avoid the discretization of the operators. The study was done with 20 equispaced elements, and the results obtained match those found with a larger number of nodes. In practice, non-equispaced elements produce more accurate results with fewer elements, but these refinements are not critical to this section, where the focus is on arc-length continuation. The results are discussed below.

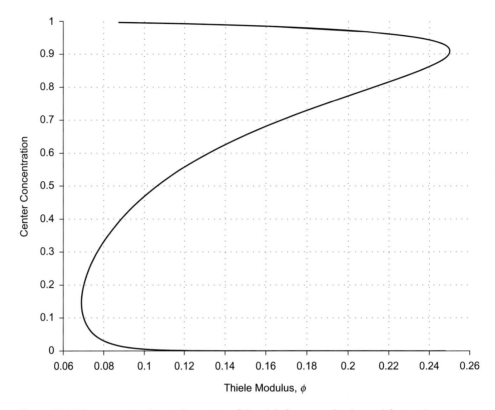

Figure 16.6 The concentration at the center of the slab for a non-isothermal first-order system as a function of the Thiele modulus for the case of $\gamma = 20$ and $\beta = 0.8$.

The center concentration as a function of the Thiele modulus is shown in Fig. 16.6. The continuation analysis indicates that multiple steady states exist in the range of ϕ of 0.069 to 0.25 and these results are in agreement with earlier studies. It is seen that in this range the center concentration can be nearly one or nearly zero. The center temperature can be calculated from the invariance property as $\theta = \beta(1 - c)$.

16.3   Flow-stability analysis

16.3.1   Evolution equations and linearized form

The discussion in this section is adapted from the excellent book by Pozrikidis (1997). Now we move to analysis of flow problems and examine the conditions under which the flow becomes unstable. To prepare the ground for the flow-stability analysis it is useful to recast the N–S equations as the evolution equations. For velocity, this is done simply by writing the equation as

$$\frac{\partial v}{\partial t} = F(v, \alpha) \tag{16.19}$$

where F is a forcing function. This function is obtained from a rearranged version of the N–S equation and reads as

$$F = -(v \cdot \nabla)v + g - \frac{\nabla p}{\rho} + v \nabla^2 v \tag{16.20}$$

The assumptions are that we have (i) a constant viscosity, (ii) Newtonian incompressible flow, and (iii) a constant body force.

The term α is a parameter and is usually the Reynolds number in the dimensionless version of the above equation.

The steady-state solution (indicated by the subscript s) is described by

$$F_s = -(v_s \cdot \nabla)v_s + g - \frac{\nabla p_s}{\rho} + v \nabla^2 v_s = 0 \tag{16.21}$$

where the steady-state pressure field p_s is to be obtained by satisfying the continuity equation, viz., $\nabla \cdot v_s = 0$.

We look for a perturbation velocity v_d defined as

$$v = v_s + v_d$$

The following evolution equation can then be obtained for the velocity perturbation by substituting this into the evolution equation for velocity (16.19):

$$\frac{\partial v_d}{\partial t} = -(v_s \cdot \nabla)v_d - (v_d \cdot \nabla)v_s - \frac{\nabla p_d}{\rho} + v \nabla^2 v_d \tag{16.22}$$

Note that in deriving the above equation the non-linear term in the perturbation velocity has been neglected. An additional equation for the evolution of pressure is needed. The evolution equation for pressure is more complicated, since there is no separate equation for pressure. In an incompressible flow we merely require the instantaneous pressure field to develop in such a manner that the incompressibility condition is satisfied at all times. Hence it is convenient to use the Poisson equation for pressure (as indicated earlier, in Section 5.6.6) for the pressure field:

$$\nabla^2 p = -\rho \nabla \cdot (v \cdot \nabla v)$$

Note that this equation is obtained by taking the dot product of the N–S equation and then using the incompressibility condition.

Using the perturbation term for pressure,

$$p = p_s + p_d$$

we have then the following equation for the "Poisson" equation for the pressure perturbation:

$$\nabla^2 p_d = -\rho \nabla \cdot [(v_s \cdot \nabla v_d) + (v_d \cdot \nabla v_s)]$$

These equations provide us with a set of linear homogeneous partial differential equations for the evolution of the unsteady components of the flow. The disturbances are prescribed as initial conditions. Depending on the structure of the base flow and the form of the disturbance function, the unsteady components of flow may grow with time, and such disturbances are called unstable. If all the disturbances decay with time, the flow is called linearly stable; whereas if certain disturbances grow the flow is called linearly unstable for those perturbations. It may be noted that the procedure of the linear stability analysis is similar to that for heat/mass transfer. However, the computations become more involved due to the vector nature of the equation and the coupling with the continuity equation (which is usually transformed into the pressure Poisson equation).

The above procedure to find the linear stability is rather elegant, but some comments on its limitations should be noted. A flow that is stable according to the linear analysis will not necessarily occur in practice. An example is Poiseuille flow in a circular pipe, which is stable according to the linear stability analysis but is unstable in real life beyond some critical Reynolds-number range. Small non-linear effects, irregular geometry due to wall roughness, etc. (which are not captured by the linear stability analysis) may be responsible for the unstable flow behavior. Nevertheless, the linear stability analysis has proven to be a useful tool for many flow problems.

16.3.2   Normal-mode analysis

An analysis known as normal-mode analysis is often used, since it is practically impossible to study every type of disturbance. Here the disturbance is assumed to be a trigonometric function of time and can be represented as

$$v_d = V(y)\exp(-i\sigma t)$$

Here V is a function of a chosen position coordinate, which is usually the cross-flow direction for analysis of stability of unidirectional flows. The parameter σ is a complex number having the physical meaning of the frequency of oscillation of the disturbances. Note that $2\pi f$ will be σ with f the frequency.

Upon expressing σ in terms of its real and imaginary parts, $\sigma = \sigma_R + i\sigma_I$, the disturbance can be expressed as

$$v_d = V(y)\exp(-i\sigma_R t)\exp(\sigma_I t)$$

Note that a positive σ_I contributes to the increase in the decay and therefore leads to instability. One of the objectives of the normal-mode analysis is to find eigenvalues σ whose imaginary components are zero. This establishes conditions under which the flow is expected to become unstable. The effect of a key dimensionless group such as the Reynolds number is examined, leading to a flow-stability diagram. This diagram is known as the neutral-stability curve.

Illustrative diagrams for some standard cases are presented later in this chapter.

A general computation and analysis for 3D disturbance is a formidable task. In the next section we present a 2D analysis for stability of unidirectional flow leading to a classic equation known as the Orr–Sommerfeld equation. A 2D analysis is sufficient in view of a theorem due to Squire, the details of which are not addressed here. The essence of this theorem is simply that a 2D disturbance is more destabilizing than a 3D disturbance. Note that the parameter σ is often expressed as

$$c = \frac{\sigma}{k}$$

where c is the (complex) phase velocity and k is the (complex) wavenumber.

Two ways of solving the problem are temporal-instability analysis and spatial-instability analysis

Temporal instability

This involves the analysis of a spatially periodic disturbance that evolves in time. Thus we specify a real wavenumber k and solve an eigenvalue problem for c. The complex growth rate σ can then be calculated as kc. This is decomposed into real and imaginary parts. For neutral stability, $\sigma_I = 0$, The solution depends on Re and the wavenumber k. The results are usually presented in a plot of Re vs. k with the region of stability identified as regions on this curve. Illustrative diagrams are shown later.

Spatial instability

Here the disturbance evolves in space. The amplitude at any given point changes with time. Hence k is set as $k_R + ik_I$, the real and imaginary components, while c is set as a real number. The resulting eigenproblem is solved for k. At neutral stability k is real and k_I is zero. The solution depends on Re and σ, the real cycle frequency, which is now equal to Ck_R.

We now illustrate this approach to analyze the stability of a unidirectional flow.

16.4   Stability of shear flows

16.4.1   The Orr–Sommerfeld equation

Here we analyze a base flow that is unidirectional (v_x is a function of y only and $v_y = 0$) and a disturbance flow that has components in just two directions (the main flow, v_{dx}, and in a cross-flow direction, v_{dy}). Thus a 2D problem is analyzed for the perturbation flow. In view of the 2D assumption, the disturbance flow can be analyzed in terms of a streamfunction.

The x-momentum for the disturbance flow is then obtained from Eq. (16.22):

$$\frac{\partial v_{dx}}{\partial t} + v_x \frac{\partial v_{dx}}{\partial x} + v_{dy} \frac{\partial v_x}{\partial y} = -\frac{1}{\rho} \frac{\partial p_d}{\partial x} + v \left[\frac{\partial^2 v_{dx}}{\partial x^2} + \frac{\partial^2 v_{dx}}{\partial y^2} \right] \qquad (16.23)$$

For the y-direction,

$$\frac{\partial v_{dy}}{\partial t} + v_x \frac{\partial v_{dy}}{\partial x} = -\frac{1}{\rho} \frac{\partial p_d}{\partial y} + v \left[\frac{\partial^2 v_{dy}}{\partial x^2} + \frac{\partial^2 v_{dy}}{\partial y^2} \right] \qquad (16.24)$$

The continuity equation should be satisfied for the perturbation velocity. This can be guaranteed if a perturbation streamfunction (denoted simply as ψ here without the subscript d) is introduced. This streamfunction is defined as

$$v_{dx} = \frac{\partial \psi}{\partial y} \qquad (16.25)$$

and

$$v_{dy} = -\frac{\partial \psi}{\partial x} \qquad (16.26)$$

A normal-mode analysis is done. Hence the perturbation streamfunction is assumed to be of the form

$$\psi = \phi(y)\exp[i\alpha(x - ct)]$$

Here the notation α is the wavenumber (the same as k in the previous section) and c is the wave speed.

The x-momentum is now differentiated with y, and the y-momentum is differentiated with x. One of these two equations is subtracted from the other, which eliminates the pressure term. The velocity and their partial derivatives in the resulting equation are now substituted in terms of the streamfunction using Eqs. (16.25) and (16.26) respectively, for the x and y velocities. The algebra is rather lengthy but straightforward. The result is the following fourth-order differential equation for the amplitude of the streamfunction perturbation, ϕ:

$$(v_x - c)[\alpha^2 \phi - \phi''] + v_x'' \phi = \frac{i v}{\alpha}\left[\phi'''' - 2\alpha^2 \phi'' + \alpha^4 \phi\right] \qquad (16.27)$$

This is the famous Orr–Sommerfeld equation. The variables are redefined here once again for clarity: v_x is the base flow velocity in the flow direction which is a specified function of y, c is the wave velocity; α is the wavenumber, ϕ is the magnitude of the perturbation in the streamfunction; and v is the kinematic viscosity. The primes denote differentiation with respect to y.

Note that $1/Re$ is used in place of the parameter v if the equations are expressed in dimensionless form. The boundary conditions for ϕ are as follows: at a solid boundary both ϕ and ϕ' are zero.

This is an eigenvalue problem, and the solution is sought for either c or α. The solution can be obtained by various methods. In particular, the spectral-expansion method proposed in a recent study by McBain et al. (2008) is found to be rather accurate. Numerical computation is discussed in Section 16.4.4.

Illustrative results for some common cases are now presented.

For a Blasius profile (Section 15.6.3) on a flat plate the critical Reynolds number was found to be 520. This is based on the boundary-layer thickness. The corresponding value based on length along the plate is 91 400. An illustrative curve of neutral stability is shown in Fig. 16.7. The shape of the characteristic curve in Fig. 16.7 for the stability envelope is often called the "thumb" curve.

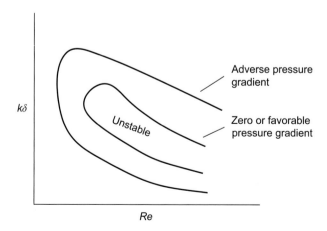

Figure 16.7 The neutral-stability curve for flow over a flat plate; a qualitative plot of $k\delta$ against the Reynolds number. δ is the film thickness.

Figure 16.8 The shear-layer velocity profiles studied by Reynolds in 1883: (a) a velocity profile with no inflexion and (b) a profile with inflexion.

16.4.2 Stability of shear layers: the role of viscosity

We review first qualitatively some aspects of the stability of shear layers studied by Reynolds (1883). The role of viscosity is also explained here. Generally viscosity is assumed to be stabilizing the flow, but we indicate that this is not always true. Reynolds considered the stability of two types of flows, which are shown in Fig. 16.8. The key observations made by Reynolds were that eddies form reluctantly and infrequently in case (a) and readily and regularly near the center for case (b) above. An inviscid flow of type (a) would be stable, whereas that of type (b) would be unstable, which was proven also by the Rayleigh theorem which will be discussed in the next section. The difference between the two cases above is that the velocity has an inflexion point in case (b). Here the inviscid flow is unstable and viscosity will dampen the fluctuations and stabilize the flow. In contrast, for case (a) the inviscid flow is stable and the role of viscosity is a destabilizing. Detailed mathematical analyses performed later by Orr (1907) and Sommerfeld (1908) provided explanations for these effects and remain landmark contributions in this field. For further studies the book by Drazin and Reid (2004) is useful. One qualitative explanation (Kundu and Cohen, 2008) of the role of viscosity is that in simple inviscid parallel flows without inflexion the disturbance field cannot extract energy from the base flow and therefore the disturbance decays. The presence of viscosity, however, changes the relation, and can cause flow instability to be initiated.

A somewhat more quantitative basis is provided by the Rayleigh equation.

16.4.3 The Rayleigh equation

If the viscous terms are small (large-Reynolds-number flows) then the Orr–Sommerfeld equation reduces to the Rayleigh equation,

$$(v_x - c)(\alpha^2 \phi - \phi'') + v_x'' \phi = 0 \tag{16.28}$$

For computational purposes, the equation is usually rearranged to keep the unknown eigenvalue c on the RHS:

$$v_x \phi'' - (v_x \alpha^2 + v_x'')\phi = c(\alpha^2 \phi - \phi'') \tag{16.29}$$

If the above equation is multiplied by the complex conjugate of ϕ it is possible to derive the following condition which was derived by Lord Rayleigh (1883):

$$c_i \int_0^L \frac{v_x'' |\phi|^2}{|v_x - c|^2} \, dy = 0 \tag{16.30}$$

where L is the flow domain under consideration. For the unstable case ($c_i \neq 0$) the integral must be zero. In order for the above integral to be zero, v_x'' must change sign at least once within the interval 0 to L. This means that the integral must have an inflexion point, leading to Rayleigh's first theorem: *a necessary condition for flow instability of an inviscid parallel flow is that the velocity profile must have an inflexion point.* Note that this is a necessary but not sufficient condition for inviscid-flow instability. The theorem agrees qualitatively with the experimental observation of Reynolds. For example, case (a) in Fig. 16.8 has no inflexion point and is considered stable in the absence of viscous forces.

16.4.4   Computational methods

In general, either the Orr–Sommerfeld equation or the Rayleigh equation has to be solved numerically, since analytical solutions are possible only for certain limiting cases or ideal situations. A number of methods can be used, and an extensive discussion of the various numerical methods is given in the book by Pozrikidis (1997). For further details this book is a valuable resource. A recent paper by McBain *et al.* (2008) also provides further information on this topic.

We discuss briefly the differential-equation eigenvalue method wherein the differential equation is solved by spectral methods. This is followed by a computational snippet and an example. You will find the code useful to analyze the stability of flow in other, more complex, problems.

The Orr–Sommerfeld equation is rearranged to the form

$$A\phi = cB\phi$$

where A is an operator comprising the terms in Eq. (16.27), excluding the c term, while B is the operator with the remaining c terms. The eigenvalues of the operator can readily be found using CHEBFUN. Sample code is provided in Box 6. The code was tested for the wavenumber of 1 and $Re = 1 \times 10^4$. The eigenvalue with the largest imaginary part was

$$c = 0.2375 + 0.003\,73i$$

which agrees with the earlier results.

The usual representation of the stability analysis is to plot Re vs. k and show the line where c_I becomes zero. The result for plane Poiseuille flow shows that there is a critical Reynolds number of 5722.0 below which the flow is stable. The corresponding value of c_R is 0.2661 and the wavenumber is 1.021. These results are in agreement with the earlier computational studies and with results generated with the code in Box 6. This code was written by Chris Oxford, modifying the template provided in the CHEBFUN manual.

Box 6 Sample code for the solution of the Orr–Sommerfeld equation using the spectral collocation method with CHEBFUN

```
alpha = 1;
Nre = 1.0e+04;
A = chebop(-1,1);
A.op = @(x,phi) - (diff(phi,4) - 2 * alpha^2 * diff(phi,2)...
        + phi * alpha^4)/Nre/alpha/i.
        + 2 * phi + diag(1 - x.^2) * (diff(phi,2) - alpha^2 * phi);
B = chebop(-1,1);
```

```
B.op = @(x,phi) diff(phi,2) - alpha^2 * phi;
A.lbc = @(phi) [phi , diff(phi)];
A.rbc = @(phi) [phi , diff(phi)];
e = eigs(A,B,5,'LI')  % five eigenvalues are found
```

Now we discuss some classical stability problems studied in the literature

16.5 More examples of flow instability

16.5.1 Kelvin–Helmholtz instability

Problem statement

The Kelvin–Helmholtz problem of instability addresses the behavior of two inviscid fluids with densities of ρ_1 and ρ_2 flowing in the x-direction with velocities (in the x-direction) of v_1 and v_2, respectively. The problem addressed is shown schematically in Fig. 16.9.

The interface is assumed to have been deformed in a sinusoidal manner as shown in Fig. 16.9. The problem is to examine whether this disturbance is stable or grows as a function of time. The book by Drazin and Reid (2004) refers to this as a classic problem that demands little mathematics. I leave the judgment of this statement to the student.

Solution method

The base flow is rather simple, with $v_x = v_1$ for $y < 0$ and $v_x = v_2$ for $y > 0$. The stability with respect to an interface perturbation is then analyzed. Consider the perturbation as

$$y = y(x, t)$$

where y is the perturbed location of the interface, with $y = 0$ being the undisturbed position. Normal-mode perturbation of the interface is used for further analysis. Thus

$$y = A \exp[ik(x - ct)]$$

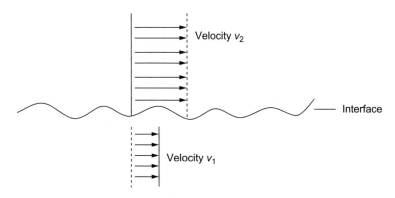

Figure 16.9 The Kelvin–Helmholtz problem of two inviscid fluids in uniform motion in the x-direction.

where A is a complex constant, k is a real wavenumber, and c is the complex wave velocity.

Since the fluid is inviscid and since $v_x'' = 0$, the Rayleigh equation simplifies to the Helmholtz equation $\phi'' - k^2\phi = 0$ and can be solved for both fluids.

The solution to the upper region is therefore $B_1 \exp(-ky)$, while that for the lower region is $B_2 \exp(ky)$. Here B_1 and B_2 are the integration constants. These satisfy the boundary conditions at infinity. In addition the position of the interface needs to be computed.

The kinematic and dynamic conditions are now applied at the interface. The kinematic condition requires that the motion of the fluid particles adjacent to the interface must be consistent with the shape of the interface. This can be shown to yield the following conditions between the constants:

$$A = \frac{B_2}{c - v_2} = \frac{B_1}{c - v_1} \tag{16.31}$$

The dynamic condition will depend on whether the effect of surface tension is included or not. In the absence of surface tension the pressure must be continuous across the interface. The pressure distribution on either side of the interface can be obtained from the Bernoulli equation applied to each side. If the surface tension is included, there is a pressure jump at the interface and the Laplace–Young equation for the normal stress is applied.

After much algebra, which is not presented here, we obtain the following equation for the wave velocity:

$$c = \frac{1}{\rho_1 + \rho_2}\left[\rho_1 v_1 + \rho_2 v_2 \pm \left(\frac{g}{k}(\rho_1^2 - \rho_2^2) + k\sigma(\rho_1 + \rho_2) - \rho_1\rho_2(v_1 - v_2)^2\right)^{1/2}\right]$$

The three terms in the large parentheses on the RHS can be identified as the surface-tension force, inertia force, and gravity force. The surface-tension effect is stabilizing the flow since this term is always positive. Buoyancy can stabilize or destabilize the flow depending on the value of $\Delta\rho$. Inertia is always destabilizing. No viscous effects were included in this model.

If the quantity within the square root is negative, one of the two solutions will have a positive imaginary part and therefore certain disturbances will be unstable. The condition for this to occur can be obtained by rearranging the term within the square root to yield

$$(v_1 - v_2)^2 > \frac{g}{k}\frac{\rho_1^2 - \rho_2^2}{\rho_1\rho_2}\left(1 + \frac{\sigma k^2}{g(\rho_1 - \rho_2)}\right) \tag{16.32}$$

16.5.2   Rayleigh–Taylor instability

This is a special case of the Kelvin–Helmholtz analysis with $v_1 = 0$ and $v_2 = 0$, i.e., the fluids are quiescent. The density of the lower fluid is assumed to be less than that of the upper fluid. The fluids are said to be inversely stratified. It can be shown that there is a critical wavenumber given by

$$k_{\text{crit}} = \left(\frac{g}{\sigma}(\rho_2 - \rho_1)\right)^{1/2} \tag{16.33}$$

Smaller wavenumbers are unstable, while the larger wavenumbers are stabilized by surface tension.

These simple-minded analyses and models find a wide range of applications, for example, finding flow transition in two-phase flow. See, for example, Taitel and Dukler (1976). Other applications are in meteorology. An example from oceanography is shown below.

Example 16.2.
A classical application is a wave on the surface of the sea. Find the relative velocity between water and air for the onset of waves on a surface of a large body of water such as the sea.

Solution.
The following property values are used.
The density of water is $\rho_1 = 1020\,\text{kg/m}^3$.
The density of air is $\rho_2 = 1.25\,\text{kg/m}^3$.
The interfacial tension is $\sigma = 0.074\,\text{N/m}$.
On substituting into Eq. (16.33) we obtain the value for k as 367/m. The cut-off wavelength is therefore $2\pi/k$ or 0.01 m, which is the critical wavelength.
Now, on substituting into Eq. (16.32), we find that $v_2 - v_1 = 6.6$ m/s, which is the minimum air velocity for the onset of waves.

16.5.3   Thermal instability: the Bénard problem

The problem involves fluid contained between two plates, with the bottom plate kept at a higher temperature than the top plate. A critical temperature difference ΔT is needed for flow initiation. Below the critical value, the upward buoyancy force is balanced by a vertical pressure gradient, and there is no flow under these conditions. The heat flux can be calculated using Fourier's law. At and above a critical ΔT, a convective motion sets in in the fluid, as shown in Fig. 16.10. The flow consists of regions of hot fluid rising and regions of cold fluid falling. The resulting flow pattern is in the form of regular cells, called Bénard cells. The transition to this flow condition from the no-flow case is a result of instability caused by perturbation of the buoyancy forces. Consider the no-flow condition. If there is a perturbation in temperature, the buoyancy force is changed. This causes some hot fluid to rise and some to fall. The viscous forces are created due to the relative motion and retard the motion leading back to a stable condition. If the viscous forces are not strong enough to retard such motion, the fluid will continue to accelerate, and a resulting steady-state flow

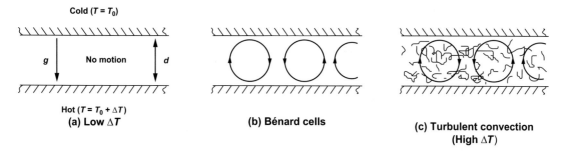

Figure 16.10 A schematic representation of the Bénard problem.

pattern corresponding to the Bénard cells will be set up. Thus we can state the condition for instability as

$$\frac{\text{Buoyancy forces}}{\text{Viscous forces}} \gg 1$$

The dimensionless parameter involving these is seen to be the Rayleigh number. The linear stability analysis shows that flow becomes unstable if the critical Rayleigh number of 1708 is reached:

$$Ra(\text{critical}) = \frac{g\beta\Delta T d^3}{k\nu} = 1708$$

The wavelength at the point of onset of instability was found to be $2d$. The results of laboratory experiments agreed remarkably well with these predictions, and the Bénard problem is considered as a major success story of linear stability analysis.

16.5.4 Marangoni instability

Another type of instability involves gradients in interfacial tension caused by temperature differences. The problem is similar to that for a Bénard problem where a liquid layer is heated from above. The only difference is that the top layer is a gas–liquid interface rather than a solid wall. The mechanism causing the instability is some perturbation that brings slightly warmer liquid to the surface. This creates a local perturbation at the interface temperature; the corresponding change in surface tension acts as if a shear force were being applied at the interface. A motion that brings the slightly warm liquid back to the bottom is set up. A cellular pattern similar to that shown in Fig. 16.10 is set up at the point of onset of instability. Similarly to the Bénard case, we can state the condition for *stability* as

$$\frac{Viscous\ forces}{Surface\text{-}tension\ forces} \gg 1$$

The key dimensionless group is the Marangoni number. The cells formed at the surface are sometimes referred to as roll cells and have a significant effect on the mass transfer rate from the gas to the liquid. An equally important problem is that of the stability of an interface between two liquid layers where the gradient of concentration of a surface-active solute causes the instability. This has important effects in mass transfer in liquid–liquid contactors, since the interfacial mixing caused by the Marangoni effects can increase the mass transfer of the solute. Several studies in this field have been reported, and the papers by Mao and Chen (2004) and Chen *et al.* (2011) are useful starting references.

16.5.5 Non-Newtonian fluids

Stability of non-Newtonian fluid has been examined in several studies. For example, Neofytou and Drikakis (2003) studied the effect of fluid rheological behaviour on flow stability in a channel flow and a flow with a sudden expansion. Nouar and Frigaard (2001) studied the non-linear flow stability of a Bingham fluid. These studies and the references cited therein could be used as a starting point for further study of these types of flows.

Summary

- Non-linear equations often have multiple solutions. A simple example is an algebraic equation of cubic polynomial form, which can have three real roots. Often we have to deal with distributed parameter systems (DPSs) and examine the various solutions. In particular, it may be of interest to track the behavior of the solution as a key parameter is varied. The determination of multiple steady states for DPSs is, however, numerically challenging. A useful method for doing this is arc-length continuation, which is discussed with some illustrative examples in this chapter. Software is also available.

- Bifurcation refers to the branching of a solution as a key parameter is changed. Again the arc-length method can track the various bifurcation patterns.

- The stability of the steady states is analyzed by linearization of the governing equation at the steady-state point by a Taylor series. This leads to a linear dynamic problem typically represented in terms of the Jacobian matrix (see Eq. (16.2)). Stability can be analysed by examining the eigenvalues of the Jacobian matrix. Various standard cases can be itemized using phase-plane plots as shown in the text.

- The method of normal modes consists of introducing a sinusoidal disturbance. A differential equation can be derived for the disturbance for 2D flows, namely the Orr–Sommerfeld equation. Numerical methods for the solution of this equation are available, and the flow can be mapped into a stability diagram where the boundaries between the stable and unstable regions are demarcated.

- Applications of stability analysis have been presented in this chapter. Upon review of these examples, you will develop a basic understanding of the theory of linear stability analysis and how to apply this to study flow problems. You will be able to follow the literature in this field and will be able to understand the use of this theory in many diverse areas such as polymer melt flow, flow in reactors, stability of gas–liquid dispersions, etc.

ADDITIONAL READING

The book by Kubíček and Marek (1983) is a great introduction to bifurcation theory and arc-length continuation.

Flow-stability analysis has been discussed in detail in an edited monograph by Ioss and Joseph (1990). The book by Chandrasekhar (1961) is a classic in this field. The book by Drazin and Reid (2004) is recommended for a detailed study of hydrodynamic stability and includes a chapter on non-linear stability analysis. The book by Pozrikidis (1997) includes a review of all the numerical methods for flow-stability analysis.

Problems

1. The following system of differential equations arises in the modeling of a stirred tank reactor with an autocatalyic reaction:

$$\frac{dx}{dt} = x - xy$$

$$\frac{dy}{dt} = -y + xy + y^2$$

Find the steady states by setting the LHS to zero and solving the set of resulting algebraic equations. Note that the system admits two steady states.

Examine the nature of the solution around these steady states. Show a phase-plane plot of the system (namely a plot of x vs. y at each instant of time).

(This problem has been adapted from Pushpavanam (1999).)

2. The Lorenz equation is widely used in the theory of non-linear equations and was encountered in modeling of natural convection. While working with these equations Lorenz discovered the phenomena of chaos. The system is described by the following set of differential equations:

$$\frac{dx}{dt} = \alpha(y - z)$$

$$\frac{dy}{dt} = -xz + rx - y$$

$$\frac{dz}{dt} = xy - bz$$

where α, r, and b are the parameters.

Fix two parameters $\alpha = 10$ and $b = -8/3$ and examine the stability of the system with r as a parameter. Write MATLAB code to track the trajectories and prepare phase-plane plots.

3. Analyze the Frank-Kamenetskii problem for the three standard geometries of slab, cylinder, and sphere. You will need to discretize the operators suitably for the cylinder and sphere. Plot the bifurcation diagram as a function of the thermicity parameter.

4. **Porous catalyst with external gradients.** Reaction in the catalyst pellet with external transport resistances is another classic problem in chemical reaction engineering. This system needs to be modeled as a set of two second-order differential equations, one for concentration and the other for temperature, with the respective Robin boundary conditions, unlike the previous case shown in the text where the Dirichlet condition was used. The model equations, and the respective boundary conditions used, and the other details can be found in Aris and Hatfield (1969).

It was shown in the original paper that this system exhibits multiple steady states for certain ranges of transport parameters. Up to five steady states are observed for certain values of the Thiele modulus.

Your goal is to reproduce some of these results using arc-length continuation. You should also examine in a qualitative manner the controlling regimes for each of these steady states by comparing the center and surface concentrations as well as the temperature.

5. **Inviscid-flow stability.** Verify the integral obtained by Rayleigh and hence show that the velocity profile needs to have an inflexion point for instability. Show that a simple shear flow is stable. Hence viscosity is needed to cause flow instability for such cases.

6. **Kelvin–Helmholtz analysis.** Derive the kinematic and dynamic conditions needed in the analysis. Set up the equations to find the constants. Now require that the determinant of the coefficient matrix should be zero to obtain a non-trivial solution. Hence verify the algebra leading to Eq. (16.32).

7. **A shear layer between two fluids.** Assume the following velocity distribution between two shear layers:

$$v_x(y) = v_0 \tanh(y/\delta)$$

This is known as the Betchov and Criminale form. Calculate the stability analysis using the Rayleigh equation for the following parameters: $v_0 = 1$ and $\delta = 1$. Set $\alpha = 0.8$. Betchov and Criminale obtained $c_I = 0.1345$ and $c_R = 0$.

8. **Stability of flow in torsional flow.** Taylor determined the critical speed of rotation for flow between concentric cylinders with the inner cylinder rotating. The transition is characterized by a critical Taylor number, Ta, defined as

$$Ta = \frac{R_0^4 \kappa (1 - \kappa)^3 (\omega_1^2 - \omega_2^2)}{v^2}$$

The transition for only the inner cylinder rotating was obtained by Taylor as 1709 by performing a linear stability analysis. Set up a model and analyze this problem.

9. Would the critical Rayleigh number for flow transition for the Bénard problem increase or decrease with the Prandtl number? Explain why in terms of the physics of the problem.

10. **Stability analysis with heat transfer.** Set up the equations for steady state for the Bénard problem.

    Now perturb the temperature and use an energy equation to derive an equation for the temperature perturbation. This in turn needs an equation for the velocity profile. Derive this equation.

    Thus we have two perturbation variables to handle here. How would you proceed further?

11. **The Bénard problem: a sixth-order equation.** Chandrasekhar (1961) has shown that the above (Bénard) problem can be solved in terms of the vorticity, which reduces the problem to a sixth-order eigenvalue problem for the perturbation in temperature. Your task is to work through the algebra and set up this sixth-order problem. Then you can try to extend the code given in the text using the spectral method and finding the critical wavenumber at which the stability sets in. The lowest value of the Rayleigh number was 1707.762, and the corresponding wavenumber was 3.117. Verify these results.

17 Turbulent-flow analysis

Learning objectives

The learning objectives are identified in the following bullets.

- To understand the basic properties of turbulent flow.
- To understand the concept of time averaging and to derive the time-averaged N–S equation, which is also known as the Reynolds-averaged N–S (RANS) equation.
- To introduce the definition of eddy momentum diffusivity and to introduce some closures for this term, especially the Prandtl mixing-length closure.
- To analyze simple problems in turbulent flow using Prandtl's closure, e.g., channel flow and pipe flow.
- To examine and understand properties of turbulent boundary layers.
- To introduce some basic definitions and properties of isotropic turbulence and the energy spectrum, and to study the mode of energy transfer from main flow to fluctuating flow.

Mathematical prerequisites

No additional prerequisites are needed. Some idea of the Fourier transform is, however, useful to understand the last section of this chapter.

Turbulence is the result of flow instability. Any small disturbance could lead to a chaotic type of flow with velocity fluctuating on a small time scale around a mean value. This happens (usually) when the viscous forces are much smaller than the inertial forces. The viscosity effect stabilizes the flow and dampens any disturbance. In the absence of significant viscous forces, any disturbance persists and leads to a continuous fluctuation of velocity, leading to turbulent flow. Thus turbulent flows are characterized by small random fluctuations around a mean value and essentially are chaotic unsteady-state phenomena.

The goal of this chapter is to provide an introductory treatment of turbulent flow. The first goal is look at the time averaging of the Navier–Stokes (N–S) equations.

The main idea is that, although the velocity field may be random in nature, the statistical (average) properties of flow are quite reproducible. This suggests that some time-averaged form of equation of motion can be used for practical applications. The problem associated with time averaging is the closure problem. As we have mentioned many times before, the averaging leads to a loss of information. Hence there is a need for additional closure laws to complete the model. The difficulty in turbulent modeling is due to the fact the closure equations can never be completed. In this chapter, we indicate the closure problem not only for the flow analysis but also for heat and mass transfer under turbulent flow conditions.

Next we introduce the concept of eddy viscosity, which is only a form of closure. Simple models based on the Prandtl mixing length are then suggested for eddy viscosity. We show how these models can be used in conjunction with the momentum balance to obtain velocity profiles in channel flow and pipe flow. A detailed discussion of this approach will be given, and the use of this approach to get engineering information will be studied. Some discussion on turbulent velocity profiles in boundary layers is then provided.

Some additional approaches to close the turbulent models are now available, and many of these have been incorporated into various items of fluid-dynamics-simulation software such as FLUENT, OPENFOAM, etc. Some introduction to these closure models is then presented. You will find this useful if you pursue further studies and research in the area of computational fluid dynamics (CFD).

Finally some properties of turbulent flow are presented, and the nature of energy dissipation in turbulent flow is explained. This discussion is rather brief but provides a basic framework to understand more specialized books and literature in this field.

17.1 Flow transition and properties of turbulent flow

From the discussion in Chapter 16, it appears that the viscous forces have to be large for flow to be laminar. This is more easily characterized by a dimensionless group, the Reynolds number:

$$\text{Reynolds number} = \frac{\text{Inertia force}}{\text{Viscous force}}$$

Thus a low Reynolds number means that viscous forces are larger than inertial forces and the flow will be laminar. At higher Reynolds number the flow becomes turbulent. For pipe flow, the transition Reynolds number is 2100–2300 and the flow is fully turbulent above a Reynolds number of 4000.

For flow in a boundary layer over a flat plate, the transition to turbulence occurs at a Reynolds number of around 2×10^5 and the flow is fully turbulent above 2×10^6. Note that the Reynolds number is defined here as $x v_\infty \rho / \mu$, where x is the distance from the leading edge.

Other examples of flow instability (e.g., in flow over rotating disks, torsional flow, etc.) are common, as indicated in the chapter on flow instability. In this chapter we mainly focus on flow in channels, pipes, and boundary layers.

Properties of turbulent flow

The flow becomes oscillatory at some critical Reynolds number, which is the onset of instability. Beyond this stage the flow eventually becomes fully turbulent. A fully turbulent flow field has the following characteristics (Warsi, 1999).

- It is random. The flow field is a random field. For example, two identical experiments need not leave an identical streamline track of a marker particle released at the same position.
- It is diffusive. The flow disturbance spreads over a larger distance, as if it were a diffusion process. A streak of dye released in a pipe under turbulent conditions spreads over the whole pipe.
- It is dissipative. There is a large energy loss in turbulent flow. For example, the pressure drop has to be increased by a factor of 3.5 times in order to double the volumetric flow rate.
- It is 3D. Even a fully developed unidirectional flow (in a time-averaged sense) has some fluctuating velocity components (superimposed on the main flow) in all three directions.

Stochastic nature

An important property of turbulent flow is that no two identical experiments would give the same track of the velocity profiles **but** the statistical properties remain the same. The flow itself may appear steady if the velocity is measured with devices that are not sensitive to small changes (e.g., a Pitot tube). The applied pressure drop and the resulting flow rate will be constant in the system in a time-averaged sense (if the system is maintained at constant inlet and outlet pressures). Such flows are called "steady-on-average" in order to distinguish them from turbulent flow, which is inherently transient in nature due to random fluctuations in the velocity. Hence it seems logical to look at time-averaged properties of flow rather than the instantaneous velocity vs. time profiles. This can be done by a model-reduction procedure called time averaging.

The N–S equation is a deterministic equation and is clearly defined. But the solution becomes chaotic in turbulent flow, so each experiment or simulation is a different realization of flow and gives a different temporal history of the flow. But the statistical properties are well characterized and reproducible from experiment to experiment. The governing equations for statistical properties can be derived by a procedure known as time averaging. But the averaging leads to more unknowns, so more equations can be derived for these, which leads to even more unknowns (as will be discussed later). Thus there will always be more unknowns than variables for the statistical properties of flow. This is referred to as a closure problem in turbulence modeling. Let us examine the closure problem in more detail by time averaging of the N–S equations.

17.2   Time averaging

Consider the N–S equation written in stress form in a Cartesian coordinate system. Only the 2D case is shown for simplicity, but the 3D case can be simply dealt with by adding the corresponding terms for the z-direction. Only the x-component is shown in order to avoid clutter and to show the idea of time averaging:

$$\rho \left(\frac{\partial v_x}{\partial t} + v_x \frac{\partial v_x}{\partial x} + v_y \frac{\partial v_x}{\partial y} \right) = -\frac{\partial \mathcal{P}}{\partial x} + \frac{\partial \tau_{xx}^{(v)}}{\partial x} + \frac{\partial \tau_{yx}^{(v)}}{\partial y} \tag{17.1}$$

Here the viscous stresses are indicated by a superscript (v) for reasons that will become clear soon.

For averaging purposes it is convenient to rearrange the equation as

$$\rho \left(\frac{\partial v_x}{\partial t} + \frac{\partial (v_x v_x)}{\partial x} + \frac{\partial (v_x v_y)}{\partial y} \right) = -\frac{\partial \mathcal{P}}{\partial x} + \frac{\partial \tau_{xx}^{(v)}}{\partial x} + \frac{\partial \tau_{yx}^{(v)}}{\partial y} \tag{17.2}$$

which is valid in view of the incompressibility condition

$$\frac{\partial v_x}{\partial x} + \frac{\partial v_y}{\partial y} = 0 \tag{17.3}$$

The velocity can be represented as a sum of a mean value and a fluctuating component. For example,

$$v_x = \bar{v}_x + v_x'$$

where the bar indicates a mean or time average of a quantity that is defined, for example, as

$$\bar{v}_x = \frac{1}{T} \int_0^T v_x(t) dt$$

where T is a sufficiently large value of time for the results to be statistically meaningful. The prime symbol is used for fluctuating values.

Time averaged quantities have the following properties.

1. The time average of any fluctuating quantity is zero. Thus, for example, the time average of v_x' is zero.
2. The time average of a product of a (time) averaged quantity and a fluctuating quantity is zero. Thus, for example, the time average of the product of $\bar{v}_x$ and v_x' is zero.
3. The time average of a product of two fluctuating quantities is non-zero. For example, the time average of the product of v_x' and v_y' is non-zero. We say that there is a cross-correlation between v_x' and v_y'. Cross-correlation is an important concept in turbulent-flow analysis. If two random variables, say v_x' and v_y', are completely independent then the cross-correlation is zero, whereas if there is some dependence on each other (as is the case in turbulent flow) then the cross-correlation is non-zero.

Using these rules we can show that

$$\overline{v_x v_x} = (\bar{v}_x)^2 + \overline{v_x' v_x'}$$

and

$$\overline{v_x v_y} = \bar{v}_x \bar{v}_y + \overline{v_x' v_y'}$$

Equation (17.2) can now be time-averaged and the results are as follows. Consider a flow that is steady on average. Then $\bar{v}_x$, etc., do not change with time. Since the time average of the fluctuating quantity is zero we find that the time average of the transient term in Eq. (17.2) is zero for the "steady-on-average" case.

The convection terms on the LHS in Eq. (17.2) time average to

$$\rho \left(\frac{\partial (\bar{v}_x \bar{v}_x)}{\partial x} + \frac{\partial (\bar{v}_x \bar{v}_y)}{\partial y} + \frac{\partial (\overline{v_x' v_x'})}{\partial x} + \frac{\partial (\overline{v_x' v_y'})}{\partial y} \right)$$

and result in two extra terms upon time averaging (the last two terms above involving the fluctuating quantities). Extra terms arise because the products of velocity components are involved in the convective terms and these do not time average to zero. Note that the first two terms are convective terms based on time-average velocity values while the last two terms arise due to cross-correlation of the fluctuations.

The RHS will appear the same except that the time-averaged values would now be involved, since all the terms are linear on the RHS and there are no product terms. The RHS of Eq. (17.2) is therefore

$$-\frac{\partial \bar{\mathcal{P}}}{\partial x} + \frac{\partial \bar{\tau}_{xx}^{(v)}}{\partial x} + \frac{\partial \bar{\tau}_{yx}^{(v)}}{\partial y} \tag{17.4}$$

where the bar means time-averaged values.

The extra quantities which arise on time averaging of the convection terms are usually moved to the RHS, and hence the result can be represented as

$$\rho \left(\frac{\partial (\bar{v}_x \bar{v}_x)}{\partial x} + \frac{\partial (\bar{v}_x \bar{v}_y)}{\partial y} \right) = -\frac{\partial \bar{\mathcal{P}}}{\partial x} + \frac{\partial \bar{\tau}_{xx}^{(v)}}{\partial x} + \frac{\partial \bar{\tau}_{yx}^{(v)}}{\partial y} - \frac{\partial (\rho \overline{v_x' v_x'})}{\partial x} - \frac{\partial (\rho \overline{v_x' v_y'})}{\partial y} \tag{17.5}$$

which is the RANS equation. The last two terms have the look of a stress tensor and are represented by defining a turbulent stress tensor. These turbulent stresses are also known as Reynolds stresses and are related to the time average of the product of the fluctuating component of velocity. These are defined as

$$\tau_{ij}^{(t)} = -\rho \overline{v_i' v_j'}$$

where the superscript (t) denotes the turbulent stress.

The RANS equation can therefore be also written in terms of the turbulent stresses as

$$\rho \left(\frac{\partial (\bar{v}_x \bar{v}_x)}{\partial x} + \frac{\partial (\bar{v}_x \bar{v}_y)}{\partial y} \right) = -\frac{\partial \bar{\mathcal{P}}}{\partial x} + \frac{\partial \bar{\tau}_{xx}^{(v)}}{\partial x} + \frac{\partial \bar{\tau}_{yx}^{(v)}}{\partial y} + \frac{\partial \tau_{xx}^{(t)}}{\partial x} + \frac{\partial \tau_{yx}^{(t)}}{\partial y} \tag{17.6}$$

The terms in the z-direction are not shown to avoid clutter and can readily be added. Also the x-component of the RANS equation is shown above. Similar expressions hold for the y- and z-components.

It should be noted that the turbulent stresses are not stresses in the true sense but arise due to convective momentum transfer from the fluctuating part of the velocity. Hence these "stresses" are flow-dependent and not a fluid property. These have to be modeled in some way and there is no fundamental way of predicting them. This is referred to as the closure problem in turbulence. An equation can be derived from first principles for turbulent stresses, but these will involve triple correlation terms of the type, $\overline{v_i' v_j' v_k'}$ and so on. We end up with more equations for time-averaged properties of flow than unknowns. Classical approaches to close the problem for turbulent stresses will be taken up shortly. Let us move to heat and mass transfer to see what is involved here as well.

17.3 Turbulent heat and mass transfer

The turbulent heat and mass flux vector

The time average of the convective part of the energy equation leads to the following additional terms:

$$\rho c_p \left(\frac{\partial (\overline{v'_x T'})}{\partial x} + \frac{\partial (\overline{v'_y T'})}{\partial y} + \frac{\partial (\overline{v'_z T'})}{\partial z} \right) \tag{17.7}$$

The terms have the appearance of a flux vector and hence a turbulent heat flux vector whose jth component is

$$q_j^{(t)} = \rho c_p \overline{v'_j T'}$$

can be defined. This accounts for the heat transfer due to fluctuating velocities, or the eddies as they are called in the turbulence literature. The terms given in Eq. (17.7) can then be represented as the divergence of the turbulent heat flux vector, $\nabla \cdot q^{(t)}$. The term is usually moved to the RHS of the heat equation and written together with the molecular heat flux (Fourier's law), and the net heat flux is represented as $-\nabla \cdot q^{(m)} - \nabla \cdot q^{(t)}$ in the time-averaged equation for the temperature field.

Similarly, for mass transfer one can define a turbulent molar flux vector whose jth component will be

$$J_{A,j}^{(t)} = \overline{C'_A v'_j}$$

This term accounts for the contribution of velocity fluctuation to mass transfer and is usually written together with the Fick's-law flux in the time-averaged equation.

The reaction contribution

For reacting systems, an additional contribution to the reaction rate arises due to velocity fluctuations if the reaction is a non-linear function of concentration. This can be demonstrated by time averaging the reaction terms as shown in the following exercise.

Consider a second-order reaction with rate R_A defined as $-k_2 C_A^2$. Here C_A is the instantaneous concentration. You should time average this and show that the rate of reaction can be represented as

$$\text{Rate, time-averaged} = -[k_2 (\overline{C_A})^2 + k_2 \overline{C'_A C'_A}]$$

The first term on the RHS is simply the rate based on the mean concentration. The second term is the average of the product of fluctuating concentrations, and leads to an additional contribution to the rate. This term is non-zero and cannot be calculated. Again we have the closure problem of turbulence for reacting systems (whenever the kinetics are non-linear). Some phenomenological models are needed at this stage. Similar effects can be seen when the reaction rate is expressed as an exponential function of the temperature. Terms of the form $\overline{C'_A T'}$ will appear as additional terms in the rate equation.

More details of turbulent heat and mass transfer are addressed in later subsequent chapters. Now we look at some simple closure models for the turbulent stress tensor for flow problems.

17.4   Closure models

One point closure: eddy viscosity

The simplest way to look at the closure is to use some equation similar to Newton's law of viscosity. Here the turbulent stresses are correlated as a linear function of the mean velocity gradient for 1D flows (or the rate of strain based on the mean velocity for 3D flows). We simply use the analogy with Newton's law viscosity and define an eddy diffusivity parameter, v_t, by the following relation:

$$\tau_{yx}^{(t)} = \rho v_t \frac{d\bar{v}_x}{dy} \tag{17.8}$$

for 1D shear-type flows. The parameter introduced, v_t, is called the kinematic eddy viscosity. Also the term ρv_t may be viewed as a "turbulent viscosity" and is often denoted by μ_t. However, these quantities are properties of the flow rather than of the fluid. Thus they will be functions of the local Reynolds number, distance from the wall, etc., and NOT a simple property like viscosity. The closure problem still remains, but is now transferred to v_t or μ_t.

Prandtl's model for eddy viscosity

Prandtl used the analogy with the kinetic theory of gases to obtain a closure for the eddy viscosity. Fundamental to his theory was the concept of a "mixing length", usually denoted by l. This is an average distance an eddy can travel without losing its identity. The eddy can be assumed to retain its momentum up to a distance of l. The (time-averaged) velocity difference between two points separated by a distance l obtained by using Taylor's series is

$$\bar{v}_x(y + l) = \bar{v}_x(l) + l\frac{d\bar{v}_x}{dy}$$

Hence an estimate of the magnitude of the velocity fluctuation is

$$v_x' \approx \pm l\frac{d\bar{v}_x}{dy}$$

The $\pm$ sign is used since the eddy can move both in the $+y$- and in the $-y$-direction.

If one assumes that v_y' is of the same order of magnitude as v_x' (i.e., turbulence is isotropic or uniform in all directions) then an estimate for the product $v_x'v_y'$ can be obtained:

$$\overline{v_x'v_y'} \approx l\frac{d\bar{v}_x}{dy} \cdot l\frac{d\bar{v}_x}{dy}$$

Noting that the turbulent shear stress is equal to $\rho\overline{v_x'v_y'}$ in magnitude, the following equation can be used to represent the turbulent shear stress:

$$\tau_{yx}^{(t)} = \rho l^2 \left|\frac{d\bar{v}_x}{dy}\right| \frac{d\bar{v}_x}{dy} \tag{17.9}$$

The absolute value is used to fit the sign convention that the stress is positive if the velocity gradient is positive and vice versa.

The mixing length is found to depend linearly on the distance from the wall and the relation is expressed as

$$l = \kappa y$$

where κ is a constant (usually taken as 0.4). Hence the turbulent shear stress can also be represented as

$$\tau_{yx}^{(t)} = \rho \kappa^2 y^2 \left| \frac{d\bar{v}_x}{dy} \right| \frac{d\bar{v}_x}{dy} \tag{17.10}$$

The above equation is known as Prandtl closure. Also, on comparing Eqs. (17.8) and (17.10), we have the following closure equation for v_t:

$$v_t = l^2 \left| \frac{d\bar{v}_x}{dy} \right| = \kappa^2 y^2 \left| \frac{d\bar{v}_x}{dy} \right| \tag{17.11}$$

We will now use this closure to evaluate the velocity profiles in unidirectional turbulent flows.

17.5 Flow between two parallel plates

Here we reexamine the problem of flow between two parallel plates. We assume that the conditions are such that a fully developed turbulent flow exists between the two plates. Time-averaged velocities are to be computed in the model. Furthermore, the "usual" assumptions for a fully developed flow are made:

$$\bar{v}_x = f(y \text{ only})$$

with the other (time-averaged) velocity components set as equal to zero.

The RANS equation now simplifies to

$$-\frac{\partial \mathcal{P}}{\partial x} + \frac{\partial \tau_{yx}^{(v)}}{\partial y} + \frac{\partial \tau_{yx}^{(t)}}{\partial y} = 0 \tag{17.12}$$

Here $\mathcal{P}$ is the time-averaged pressure (with the overbar omitted here). The second term is the viscous stress, while the last term is the turbulent stress. The pressure gradient is assumed to be constant, i.e., it is independent of y (which is a good assumption for a fully developed flow):

$$\frac{\partial \mathcal{P}}{\partial x} = \frac{d\mathcal{P}}{dx} = \text{constant}$$

Equation (17.12) can now be integrated once to give

$$\tau_{yx}^{(v)} + \tau_{yx}^{(t)} = \left(\frac{d\mathcal{P}}{dx} \right) y + C \tag{17.13}$$

where C is an integration constant. Let y be the distance measured from one of the walls. In that case the constant C can be interpreted as the wall shear stress, τ_w (the stress acting on the wall, by the sign convention used for the normal). Hence Eq. (17.13) can be written as

$$\tau_{yx}^{(v)} + \tau_{yx}^{(t)} = \left(\frac{d\mathcal{P}}{dx} \right) y + \tau_w \tag{17.14}$$

Also the pressure gradient can be related to the wall shear stress by an overall force balance.

The following relationship, which follows from a macroscopic force balance, holds between the pressure gradient and the wall shear stress:

$$\left(\frac{dP}{dx}\right) = -\tau_w/H$$

where H is half of the width between the plates.

Hence Eq. (17.14) can be written as

$$\tau_{yx}^{(v)} + \tau_{yx}^{(t)} = \tau_w(1 - y/H) \tag{17.15}$$

The viscous stresses can be closed by Newton's law of viscosity:

$$\tau_{yx}^{(v)} = \mu\frac{d\bar{v}_x}{dy}$$

The turbulent stresses are related to the velocity fluctuations and can be represented as

$$\tau_{yx}^{(t)} = -\rho\overline{v_x'v_y'}$$

Hence Eq. (17.15) can be written as

$$\mu\frac{d\bar{v}_x}{dy} - \rho\overline{v_x'v_y'} = \tau_w(1 - y/H) \tag{17.16}$$

Normalization

At this stage some dimensionless representation of the above equation is useful. Neither the reference velocity v_{ref} nor the reference length scale L_{ref} to use is clear. Let us leave these undefined and proceed with the normalization. We define v^+ and y^+ as the dimensionless velocity and distance, respectively:

$$v^+ = \bar{v}_x/v_{ref}$$

and

$$y^+ = y/L_{ref}$$

Then Eq. (17.16) can be written as

$$\left(\frac{\mu}{\rho v_{ref}L_{ref}}\right)\frac{d\bar{v}^+}{dy^+} - \overline{v_x'^+v_y'^+} = \left(\frac{\tau_w}{\rho v_{ref}^2}\right)(1 - y^+/H^+) \tag{17.17}$$

where H^+ is a dimensionless channel height equal to H/L_{ref}. Also $\overline{v_x'^+v_y'^+}$ is a dimensionless turbulent stress.

The reference velocity and length pop out of the above equation. Clearly τ_w/ρ has the dimensions of the square of the velocity. Hence the reference velocity can be chosen as

$$v_{ref} = V_f = \sqrt{\tau_w/\rho}$$

where V_f is called the friction velocity. With this choice the coefficient on the RHS of (17.17) becomes unity.

Similarly, to make the coefficient on the LHS unity, we should choose L_{ref} in the following manner:

$$L_{ref} = \frac{\mu}{\rho v_{ref}} = \frac{\mu}{\rho V_f}$$

Equation (17.17) takes the following simplest form:

$$\frac{d\bar{v}^+}{dy^+} - \overline{v_x'^+ v_y'^+} = (1 - y^+/H^+) \tag{17.18}$$

Note that no free parameter appears in this equation, indicating that our choices of the reference variables are appropriate.

The dimensionless turbulent stress in the above equation is defined as

$$\overline{v_x'^+ v_y'^+} = \overline{v_x' v_y'}/V_f^2$$

The key problem in turbulent-flow modeling is to approximate this term for the turbulent (or Reynolds) stress term, $\overline{v_x'^+ v_y'^+}$. This term varies as a function of distance from the wall, starting with zero at the wall itself. If we have a model for the Reynolds stress as a function of y^+ then we can integrate the equations for the velocity profile. Various expressions are available for this and have been used in the past with varying degrees of success. A semiempirical relation given by Pai (1953) is widely used:

$$\overline{v_x'^+ v_y'^+} = 0.9835 \left(1 - \frac{y^+}{H^+}\right) \left[1 - \left(1 - \frac{y^+}{H^+}\right)^3\right] \tag{17.19}$$

A qualitative sketch of the stress as a function of distance from the wall is shown in Fig. 17.1. The turbulent stress is zero at the wall and the center and peaks somewhere in the middle. The viscous stress is dominant near the wall. The total stress varies linearly with y, which is consistent with the momentum balance.

Pai's expression can be used in Eq. (17.18), and the equation can be integrated numerically. Rather than taking this approach, we discuss the solution based on the simplifications used by Prandtl.

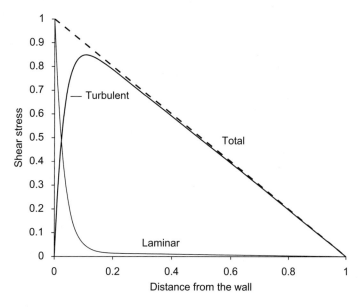

Figure 17.1 An illustration of the variation of laminar and turbulent stress as a function of distance from the wall.

Solution based on Prandtl's approximation

If Prandtl's mixing-length model is used, then the dimensionless stress can be expressed as

$$\overline{v_x'^+ v_y'^+} = -\kappa^2 (y^+)^2 \left(\frac{dv^+}{dy^+}\right)^2 \tag{17.20}$$

Equation (17.18) can then be written as

$$\frac{dv^+}{dy^+} + \kappa^2 (y^+)^2 \left(\frac{dv^+}{dy^+}\right)^2 = \left(1 - \frac{y^+}{H^+}\right) \tag{17.21}$$

The equation can be integrated numerically to get the velocity profile with the no-slip condition $v^+(0) = 0$ at the wall. However, it is better to analyze it for various regions in the channel with appropriate simplifications for each region.

The near-wall region: the viscous sublayer

In this region y^+/H^+ is much smaller than unity. Also the Reynolds or turbulent stresses are not so important here in this "viscous" layer and can be neglected. Hence Eq. (17.21) simplifies to

$$\frac{dv^+}{dy^+} = 1 \tag{17.22}$$

Integration with the boundary condition that v^+ is zero at y^+ of zero (no slip at the walls) gives

$$v^+ = y^+$$

This equation is found to be valid up to $y^+ = 5$, which is called the laminar (viscous) sublayer.

Outside the viscous layer, still near the walls

The first term in Eq. (17.21) is dropped here since the viscous effects are considered not to be important here. Also y^+/H^+ is still considered to be far less than unity and is dropped. Equation (17.21) now simplifies to

$$k^2 (y^+)^2 \left(\frac{dv^+}{dy^+}\right)^2 = 1 \tag{17.23}$$

or

$$ky^+ \left(\frac{dv^+}{dy^+}\right) = 1 \tag{17.24}$$

The positive root is used since dv^+/dy^+ is positive.

Integration of the above equation gives

$$v^+ = \frac{1}{k} \ln y^+ + C$$

Here k and C (an integration constant) are fitted constants; k is usually taken as 0.4. The value of C is found to be near 5.5. Hence the following equation has been found to fit the data. The fit is good even near the center of the channel, although the assumption of y^+/H^+ being far less than one is not applicable near the center. We have

$$v^+ = 2.5 \ln y^+ + 5.5 \tag{17.25}$$

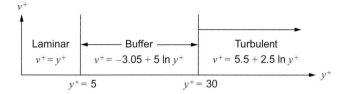

Figure 17.2 An illustration of the velocity profiles in turbulent flow for the three regions near the wall.

This is called the turbulent-core region. The velocity profile is a logarithmic function of distance from the wall in this region.

Experimentally it was found that the laminar region with a linear profile holds up to $y+ = 5$, while the turbulent-core equation is applicable for $y^+ > 30$. The region separating the two is called the buffer zone.

The buffer zone

The region in between the laminar layer and the turbulent core, given by $5 < y^+ < 30$, is called the buffer layer. The velocity here is correlated as

$$v^+ = 5\ln y^+ - 3.05$$

This is more of a curve fit rather than a model-based equation.

The various regions and the expressions for the corresponding profiles are summarized in Fig. 17.2.

Near the center

Equation (17.25) for the fully developed core fits the data well almost up to the center of the channel. At the center the symmetry condition has to be satisfied, i.e., dv^+/dy^+ must be zero at $y^+ = H^+$ but it is not as per Eq. (17.25). This is not surprising, since the term $(1 - y^+/H^+)$ was approximated as one but is actually zero near the center. Corrections to account for this error have been suggested, and these are in general referred to as the velocity-defect law. For further details, interested students should consult advanced books on turbulence or solve for the profiles by numerical integration.

17.6 Pipe flow

The analysis shown above applies with minor modifications to the flow in a circular pipe as well. The starting point is the z-momentum balance which is now augmented by adding the turbulent stress term as

$$-\frac{\partial \mathcal{P}}{\partial z} + \frac{1}{r}\frac{\partial}{\partial r}\left[r\left(\tau_{rz}^{(v)} + \tau_{rz}^{(t)}\right)\right] = 0 \tag{17.26}$$

Integrating from $r = 0$ to any arbitrary location r yields

$$-\frac{r}{2}\frac{\partial \mathcal{P}}{\partial z} + \left[\left(\tau_{rz}^{(v)} + \tau_{rz}^{(t)}\right)\right] = 0 \tag{17.27}$$

Note that here the stress at the center of the pipe, $r = 0$, which is zero, has been used to eliminate the integration constant.

The viscous stress is given by Newton's law of viscosity:

$$\tau_{rz}^{(v)} = \mu \frac{d\bar{v}_z}{dr}$$

The turbulent stresses are related to the velocity fluctuations and can be represented as

$$\tau_{rz}^{(t)} = -\rho \overline{v'_z v'_r}$$

On using these in Eq. (17.27), we obtain

$$\mu \frac{d\bar{v}_z}{dr} - \rho \overline{v'_z v'_r} = \frac{r}{2} \frac{\partial P}{\partial z} \tag{17.28}$$

It is useful to express the pressure gradient in terms of the shear stress on the wall, τ_w,

$$\frac{dP}{dz} = -\frac{2}{R} \tau_w$$

which follows from a macroscopic force balance (see Chapter 1). We now use this in Eq. (17.28) and also change the coordinate to $y = R - r$, the distance measured from the wall. It can be verified that the resulting equation is

$$\mu \frac{d\bar{v}_z}{dy} - \rho \overline{v'_z v'_y} = \tau_w (1 - y/R) \tag{17.29}$$

Equation (17.29) has the same form as (17.16) for flow between the parallel plates. Equation (17.29) can be made dimensionless using a friction velocity V_f defined as before as the scaling velocity. The distance is scaled as before using L_{ref}.

Hence the dimensionless form of Eq. (17.29) is

$$\frac{d\bar{v}^+}{dy^+} - \overline{v'^+_z v'^+_y} = (1 - y^+/R^+) \tag{17.30}$$

Equation (17.30) can be formally integrated to give the velocity profile:

$$v^+ = y^+ - \frac{y^{+2}}{2R^+} + \int_0^{y^+} \overline{v'^+_z v'^+_y} \, dy^+ \tag{17.31}$$

This can also be expressed in terms of the dimensionless radial position $\xi = r/R = (1 - y^+/R^+)$ as

$$v^+ = \frac{R^+}{2}(1 - \xi^2) - \frac{R^+}{2} \int_1^\xi \overline{v'^+_z v'^+_r} \, d\xi \tag{17.32}$$

The first term is the laminar flow profile (parabolic), while the second term is the contribution of the fluctuating velocity. If $\overline{v'^+_z v'^+_r}$ is available as a function of y^+ or ξ then the solution for the velocity profile can be completed.

It is easier to integrate Eq. (17.30) for the three regions with simplifying assumptions as was done for the channel-flow case. The resulting expressions for the velocity profiles are the same as that for channel flow. Thus a logarithmic profile is observed in many of the radial locations. Let us examine some properties that can be derived from these velocity profiles.

Average velocity

The cross-sectional average velocity is an important property and v_b^+, the dimensionless average velocity, is defined as $\langle \bar{v}_x \rangle / V_f$. This is an integral of local value over a cross-section of the pipe and hence can be formulated as

$$v_b^+ = \frac{2}{(R^+)^2} \int_0^{R^+} v^+ (R^+ - y^+) dy^+ \tag{17.33}$$

Let us assume that the logarithmic velocity (Eq. (17.25)) is valid for the entire pipe from $y^+ = 0$ to $y^+ = R^+$. If you use this, the following expression can be derived for the average velocity:

$$v_b^+ = 2.5 \ln R^+ + 1.75 \tag{17.34}$$

It is more convenient to express this in terms of the friction factor and the Reynolds number.

The Reynolds number Re can be shown to be equal to $2R^+ v_b^+$.

Also the friction factor can be represented as

$$\frac{f}{2} = \frac{1}{(v_b^+)^2} \tag{17.35}$$

On using all this in Eq. (17.34), we get a relation between the friction factor and the Reynolds number,

$$\frac{1}{\sqrt{f}} = 4.52 \log_{10}(Re\sqrt{f}) - 2.45 \tag{17.36}$$

This expression does not do a particularly good job of fitting the data. The reason for this is the complete neglect of the buffer region, the laminar layer, and the velocity defect at the center. All these omissions can be corrected with further mathematical manipulations; integration section by section is more accurate but the final solution will not be a simple equation, in contrast to the above method. The key idea was to show how the various parameters are interrelated and how they can be correlated from the basic "theory".

Although Eq. (17.36) does not fit the data, with minor adjustment of the constants it provides a good fit to the data for the friction factor. The fitted equation is

$$\boxed{\frac{1}{\sqrt{f}} = 4.0 \log_{10}(Re\sqrt{f}) - 0.4} \tag{17.37}$$

This equation is sometimes referred to as the Prandtl resistance law and is widely used to correlate the friction factor in turbulent flow in smooth pipes and to relate pressure-drop vs. volumetric-flow data. Note that the equation is implicit in f, and an iterative solution is needed when one wants to find f for a given Reynolds number.

17.6.1 The effect of roughness

The pipe roughness has an effect on the pressure drop if the flow is turbulent. In general, the roughness is characterized by the average height of projections, e, near the wall. The dimensionless measure of this is e/R or e/d_t. If the Reynolds number is large, the effect of the Reynolds number on the friction factor becomes unimportant and the roughness parameter becomes the important parameter which affects the velocity profiles. The condition under which this holds is

$$\frac{1}{Re\sqrt{f}} \frac{d_t}{e} < 0.001 \tag{17.38}$$

corresponding to what is also known as a completely turbulent rough pipe.

Experimental data for these conditions could be correlated as

$$v^+ = 2.5 \ln(y^+/e^+) + 8.5 \tag{17.39}$$

where e^+ is a dimensionless roughness parameter defined as $e\rho V_f/\mu$

The average velocity can be calculated by integration as

$$v_b^+ = 2.5 \ln(R^+/e^+) + 4.75 \tag{17.40}$$

Here the laminar layer, the buffer layer, and the velocity defect are ignored.

Since $v_b^+ = \sqrt{2/f}$ (from Eq. (17.35)) the above expression can be represented in terms of the friction factor as

$$\frac{1}{\sqrt{f}} = 4.00 \log_{10}(R/e) + 3.35 \tag{17.41}$$

or in terms of the pipe diameter as

$$\frac{1}{\sqrt{f}} = 4.00 \log_{10}(d_t/e) + 2.28 \tag{17.42}$$

which is known as Nikuradse's equation for rough pipes. The condition under which this holds is given by Eq. (17.38) above.

17.7 Turbulent boundary layers

A brief description of turbulent boundary layers is presented here for completeness. It is useful to note here that turbulent boundary layers are characterized by a region near the wall where the shear stress is nearly constant, unlike in channel or pipe flow where the shear stress varies linearly as a function of distance from the wall.

Inner and outer regions

The model is simplified by assuming that the shear stress is constant within a large portion of the boundary layer. Again y^+ defined below is used as the dimensionless length: $y^+ = (\sqrt{\tau_w/\rho}/\nu)y$, where τ_w is the wall shear stress.

Typical experimental data on the shear-stress profile show that the shear stress remains constant up to about y^+ of 100 or so. The region where the shear-stress is nearly constant is referred to as the **inner region**. For this region the convective terms can be neglected and τ simply set as a constant equal to τ_w. Also the pressure-drop term is taken as zero for flow over a flat plate. With these modifications, we find that the same expression as that derived for flat plates and circular pipes holds for the turbulent boundary layers as well. Thus the velocity varies as a log function of y^+ for $y^+ > 5$.

For regions near the edge of the boundary layer, the log law does not appear to work well. This is the region where the stress is not equal to the wall stress and starts to fall off, reaching a value of zero at the edge of the boundary layer. This region is referred to as the outer region. The velocity v^+ now depends not only on y^+ but also on the dimensionless distance y/δ. In other words, the data cannot be fitted with y^+ alone as a parameter. There are two length scales, which are equally important now. A correlation for velocity for this region is

$$v^+ = 2.5 \ln y^+ + 5.0 + 2.5 \sin^2\left(\frac{\pi}{2}\frac{y}{\delta}\right) \tag{17.43}$$

which is referred to as the law of the wake or the Coles equation. The equation is valid for $y^+ > 100$. It is quite interesting to note that the equation applies both for the turbulent core and for the outer region since if y/δ is small (for the turbulent core) the sine term is nearly zero. Hence it fits the data over a major portion of the boundary layer. It does not apply to the buffer region or the laminar sublayer.

The integral momentum method

An empirical method for turbulent-boundary-layer analysis is based on the stress correlation of the Blasius equation for the drag and the 1/7th-power law for velocity profiles. The 1/7th power-law provides the following simple relation for the velocity profile within the boundary layer:

$$\frac{\bar{v}_x}{v_\infty} = \left(\frac{y}{\delta}\right)^{1/7}$$

This equation is used in the von Kármán momentum integral to obtain an equation for the variation of δ. The algebra is, however, lengthy and involved. The final result for the thickness of the boundary layer is

$$\frac{\delta}{x} = \frac{0.0376}{Re_x^{1/5}} \tag{17.44}$$

and for the local drag coefficient

$$C_{\mathrm{f}x} = \frac{0.0576}{Re_x^{1/5}} \tag{17.45}$$

These results are in good agreement with the experimental data.

17.8   Other closure models

The one-parameter model: the k model

Now we examine some other closure models for ν_t. The key observation, due to Kolmogorov, is that ν_t is the product of a velocity and a length scale. The expression for ν_t depends on what values are assigned for these scales. Prandtl assumed the velocity scale to be of the order of the time-averaged mean velocity and the length scale to depend on the distance from the wall. This leads to the Prandtl closure, which has the standing of a classic in this field. Let us see some other ways of doing this.

The velocity scale can be related to the mean turbulent kinetic energy per unit mass, k, defined as

$$k = \frac{1}{2}\left[\overline{(v_x')^2} + \overline{(v_y')^2} + \overline{(v_z')^2}\right]$$

The unit of k (J/kg) can also be shown to be equal to m^2/s^2, and the square root of k has the same unit as velocity and hence can be used as a typical length scale to correlate the eddy viscosity. This leads to the one-parameter model for ν_t:

$$\nu_t = \sqrt{k}l$$

where l is a length scale that can be taken to be proportional to the smallest distance from the wall.

An additional differential equation for k is needed in conjunction with this model, as will be discussed later.

17.8.1 The two-equation model: the $k–\epsilon$ model

In this model we also include the time-averaged energy-dissipation contribution due to turbulent fluctuations. This quantity is typically given by the symbol ϵ and is similar to the ϕ_v term for viscous dissipation in laminar flow. The parameter is defined as follows using the index notation:

$$\epsilon = \nu \overline{\epsilon_{ij}\epsilon_{ji}}$$

where ϵ_{ij} is the rate-of-strain tensor based on the fluctuating velocity components. Note that the index notation with both i and j repeating in the above equation implies double summation, and hence the overall quantity ϵ is a scalar. Also the units of ϵ can easily be shown to be $\mathrm{m^2/s^3}$. A dimensionless analysis indicates that a length scale based on this parameter together with the kinetic-energy parameter k is

$$l = \frac{k^{3/2}}{\epsilon}$$

Hence we can relate the velocity and length scale to the eddy viscosity as

$$\nu_t = \sqrt{kl} = C_\mu \frac{k^2}{\epsilon} \tag{17.46}$$

and C_μ is a correlation constant. This model requires two differential equations, one for k and the other for ϵ, in contrast to the one-parameter model discussed earlier. These differential equations are not so difficult to derive, although the algebra is rather lengthy (see Problems 11 and 12, for example). However, the differential equation has many additional closure terms (triple correlations, etc.). Thus the closure problem in turbulence is a never-ending game. Engineering approximations to some of the terms in these equations are therefore used, which leads to the $k–\epsilon$ model which is used widely in engineering analysis. These equations take the form of a transport equation.

The engineering model for k

The following engineering k equation which has a similar form to a transport equation is proposed:

$$\frac{Dk}{Dt} = -\nabla \cdot \mathcal{K} + G - \epsilon \tag{17.47}$$

The term on the LHS is the substantial derivative and shows the change in kinetic energy of a moving control volume. The first term on the RHS is the transport from the bounding surface of the control volume. Here $\mathcal{K}$ is a kinetic-energy flux vector:

$$\mathcal{K} = -(\nu + \nu_t/\sigma_k)\nabla k + \left\langle v_i^2 \right\rangle v_i + pv_i/\rho$$

The second term in Eq. (17.47) is the generation and is modeled as $G = (\tau_{ij}^{(t)})E_{ji}$. Here E is the rate-of-strain tensor based on time-averaged velocity values. The above term represents the interaction of the mean flow with the fluctuating flow. Finally the last term is the loss in kinetic energy by energy dissipation of the eddies. This term ϵ is also modeled as a transport equation.

The engineering model for ϵ

The engineering equation for ϵ is

$$\frac{D\epsilon}{Dt} = \nabla \cdot \left[\left(\nu + \frac{\nu_t}{\sigma_\epsilon} \right) \nabla \epsilon \right] + c_1 \frac{G\epsilon}{k} - c_2 \frac{\epsilon^2}{k} \tag{17.48}$$

There is no rational basis for this equation and, to quote Davidson (2004), the ϵ equation is simply a highly sophisticated exercise in data interpolation.

Nevertheless, the two-parameter model for ν_t using k and ϵ appears to capture some well-documented data, especially for shear flows.

The tunable constants are usually assumed to have the following values: $c_\mu = 0.09$, $\sigma_\epsilon = 1.3$, $\sigma_k = 1.44$, $c_1 = 1.44$, and $c_2 = 1.92$.

The $k–\omega$ model

There are other variations to the two-parameter model, the most common of them being the $k–\omega$ model. Here ω is defined as ϵ/k and is the specific dissipation rate. This parameter should not be confused with the vorticity. The equation for eddy viscosity according to this model is

$$\nu_t = k\omega$$

More details may be found in the book by Wilcox (1998), who has been a developer of, and a key advocate for, the model. The book by Wilcox also contains software that will enable you to compare the various models. Note that for this approach an additional (engineering) equation for ω has been developed and used in conjunction with the k equation.

The need for wall functions

All the turbulence models described above require some additional models in order to capture the region near the wall. Note that the velocity gradients are steep near the wall. Hence the grid resolution and the accuracy of the turbulence models may affect the results. This is overcome by specifying some velocity profile near the wall similar to the $v^+ = y^+$ equation derived by Prandtl. The accuracy of the result often depends on the choice of these functions. Hence caution has to be exercised when using the computational tools for these simulations, and the interpretation of the results requires some intuition into the expected flow behavior. The mass transport calculation from the velocity profiles is especially sensitive to errors in the wall functions.

17.8.2   Reynolds-stress models

Other models used in practice are now discussed briefly to give you an overview. One such model is the Reynolds-stress model. Here we need an additional equation for the Reynolds stresses, which can be derived in a rigorous manner with some lengthy algebra. (You will find the index notation very useful for this.) These equations involve triple correlations in velocity and additional terms such as pressure–rate-of-strain coupling. Hence the closure problem remains and once again some engineering approximations are needed. There appears to be a wide variety of "engineering" Reynolds-stress models. The algebraic model for Reynolds stress seems to be a popular model for CFD simulations. More details can be found in the book by Wilcox (1998).

17.8.3 Large-eddy simulation

Large-eddy simulation (LES) is a class of turbulence model where we abandon the search for only the mean velocity; instead we attempt to compute both the mean flow and the evolution of large-scale eddies. We integrate the N–S equation in time, keeping all turbulent structures up to a certain scale. The unresolved scales need to be modeled (this is again a manifestation of the information-loss principle). They are modeled by some type of sub-grid-scale model. The advantage of LES is that some of the ad hoc aspects of the other models (ν_t expressions) are eliminated in the LES model. The main disadvantage is that large computational times are needed and also the model is not very effective near the boundaries. Some types of approximations (wall functions) are still needed near the boundaries.

17.8.4 Direct numerical simulation

The final type of turbulence modeling is direct numerical solution (DNS), which began with the work of Orszag and Patterson (1972). This approach does not require any turbulence-closure models. Rather, every eddy from the largest to the smallest is computed. However, the need to resolve the eddies of such a wide range of length scales and an equally wide range of time scales make this an impossible approach with which to tackle real-life engineering problems. Only simple geometries with periodic boundary conditions appear to have been studied. Nevertheless, such studies have contributed to our understanding of turbulence in some detail, and the information gained from such computations is very valuable.

17.9 Isotropy, correlation functions, and the energy spectrum

Various definitions and correlation functions are needed in order to quantify turbulent flow and to follow the more advanced literature in this field. The goal of this section is to provide a brief introduction to these definitions and terminologies in turbulent flow.

The velocity correlation function
An important definition in the analysis of turbulence is the two-point correlation function, which is defined as

$$R_{ij}(\boldsymbol{x}_1, \boldsymbol{x}_2, t_1, t_2) = \langle v_i'(\boldsymbol{x}_1, t_1) \times v_j'(\boldsymbol{x}_2, t_2) \rangle$$

This is the most general definition, since we are looking at two spatial points $\boldsymbol{x}_1$ and $\boldsymbol{x}_2$ at two different values of time t_1 and t_2. The tensor R is a measure of the correlation of the fluctuating velocity between two points. The $\langle * \times * \rangle$ notation used above denotes the time average of the product of the quantities indicated within the brackets.

Upon introducing

$$r = x_2 - x_1$$

and

$$\tau = t_2 - t_1$$

the two-point correlation function can be rewritten as

$$R_{ij}(x, r, t, \tau) = v'_i(x, t) \times v'_j(x + r, t + \tau)$$

Similarly, a three-point correlation, pressure–velocity correlation, etc., can be defined, and the latter is a measure of how the pressure fluctuations and the velocity fluctuations are related at a point at some instantaneous time t. These definitions are necessary for following the more advanced books on turbulence but are not needed for following the introductory discussion in this section.

If we look at the data at the same value of time at two points ($\tau = 0$), we have the two-point correlation in space only. This will be a measure of how far a given point affects the flow. When the points are further apart, the function will get smaller and eventually there will be no correlation between the data at these two points. If the two points coincide then the correlation is a measure of the turbulent stresses at that point. Hence the correlation function in space provides us with important information about the properties of the flow.

An equation for the correlation R_{ij} can be derived starting from the N–S equation. Note that this is a tensor field in general. However, this equation will contain the three-point correlation and the pressure–velocity correlation $\langle p'v'_i \rangle$ as unknown terms. Various closure schemes must therefore be constructed to solve for the correlation function. However, if certain assumptions are made regarding the nature of turbulence a simplified equation for the correlation function can be obtained. Some details are taken up in the following discussion.

Homogeneity and isotropy

A random field is statistically steady in time if the correlation function depends only on τ and not the actual values of t_1 and t_2. The time t_1 can then be taken as zero for convenience. Similarly, a random field is stationary in space if the correlation function depends only on r and not on actual values of x_1 and x_2.

Turbulence is classified as *homogeneous* if it is statistically stationary in space. Often the mean velocity is taken to be constant and not time varying. If the mean velocity has only one component, say in the x-direction, and if this varies in the cross-flow direction (y) then the situation is referred to as *homogeneous shear turbulence*.

Turbulence is called isotropic if the time-average values of the square of the fluctuating velocity in each direction are the same, i.e.,

$$\langle (v'_x)^2 \rangle = \langle (v'_y)^2 \rangle = \langle (v'_z)^2 \rangle = \langle (v')^2 \rangle$$

What is the simplified form of the correlation function tensor for these cases?

The mean kinetic energy of turbulence is then defined as

$$k = \frac{1}{2} \langle v' \cdot v' \rangle = \frac{1}{2} \langle v'_i \times v'_i \rangle = \frac{3}{2} \langle v'^2 \rangle \tag{17.49}$$

and is related to a measure of the velocity fluctuation. An equation for this quantity is discussed in Problem 11. This is an important equation in turbulence modeling and also indicates how the energy is drawn from the main flow and dissipated by the eddies in the system.

Other properties for the isotropic case are that (i) the velocity correlation function is now symmetric and (ii) the pressure–velocity correlation is zero. The proofs of these statements are left as an exercise.

The basic analysis of isotropic turbulence was done by Taylor, von Kármán, and Howarth, and the correlation equation for R_{ij} was derived for this case. It may be noted that turbulent pipe flow is neither isotropic nor homogeneous. Experimentally, however, isotropic turbulence can be generated from a grid in a wind tunnel. These experiments, together with accurate measurements of the velocity fluctuations with hot-wire anemometry, have proved to be an important tool in improving our basic understanding of turbulence.

Isotropy: correlation functions

In this section, we look for correlation in space, not in time, and here $\tau = 0$.

A geometric sketch for description of isotropic turbulence is shown in Fig. 17.3. Since the flow field is isotropic, arbitrary translation and rotations as a rigid body should not change the values of the correlation functions. Hence it is sufficient to consider only axes oriented in the r-direction and a direction perpendicular to r, say s. Note that the r-direction is along the line joining the two points under consideration, while the s-direction is perpendicular to it.

Then the two correlation functions can be defined and correlated as

$$\langle v'_x \times v'_x \rangle = \langle v'^2 \rangle f(r)$$

and

$$\langle v'_y \times v'_y \rangle = \langle v'^2 \rangle g(r)$$

since the correlation functions depend only on the distance between the two points.

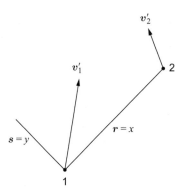

Figure 17.3 A geometric sketch of the local coordinate system for the isotropic turbulence correlation tensor.

Note that x and y are local coordinates as indicated in Fig. 17.3. The function f is called the longitudinal correlation function, while g is referred to as the lateral correlation function. Thus the correlation is described in terms of two scalar functions rather than a tensor-valued function.

Further analysis shows that

$$g = f + \frac{1}{2} r \frac{\partial f}{\partial r} \qquad (17.50)$$

and only the f function is needed in order to reveal the statistical nature of isotropic turbulence. For details of the derivation, more advanced books such as Warsi (1999) should be consulted.

The analysis of Taylor

Taylor showed that the following equation is applicable for small values of r:

$$\frac{d}{dt} \langle v'^2 \rangle = -\frac{10 \nu \langle v'^2 \rangle}{\lambda_T^2}$$

where λ_T is the dissipation length scale.

The rate of dissipation of kinetic energy is therefore given (by multiplying both sides by $3/2$) by

$$\epsilon = -\frac{dK}{dt} = \frac{15 \nu \langle v'^2 \rangle}{\lambda_T^2} \qquad (17.51)$$

The quantity λ_T is the dissipation length scale for turbulence, which is also called the Taylor microlength.

17.10   Kolmogorov's energy cascade

Kolmogorov scales

The concepts behind Kolmogorov's theory of energy transfer can be best understood by reading a little poem by Richardson (1922):

Big whirls have little whirls
which feed on their velocity
And little whirls have lesser whirls
And so on to viscosity

The energy cascade as visualized by Kolmogorov is shown in Fig. 17.4. This cascade involves the transfer of turbulence kinetic energy from large eddies to smaller eddies. At a certain length scale called the Kolmogorov scale, the energy is then lost by viscous dissipation. The velocity of the eddies corresponding to this length is known as the Kolmogorov velocity scale. Both these scales can be correlated by scaling analysis to the kinematic viscosity and the energy-dissipation parameter, E, as shown earlier in Section 14.2.3. The rationale for this theory is that the energy dissipation is effective only at very small scales, which have highly fluctuating strain rates. Larger scales simply transfer energy to smaller scales. In order to appreciate the energies associated with the various length scales, it is

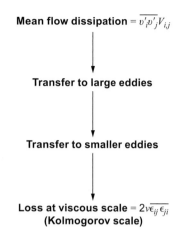

Mean flow dissipation $= \overline{v'_i v'_j} V_{i,j}$

Transfer to large eddies

Transfer to smaller eddies

Loss at viscous scale $= 2v \overline{\epsilon_{ij} \epsilon_{ji}}$
(Kolmogorov scale)

Figure 17.4 The Kolmorogov cascade of energy loss in turbulence.

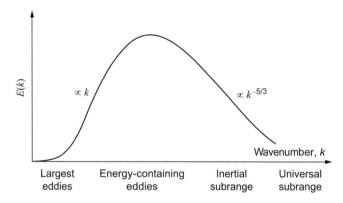

| Largest
eddies | Energy-containing
eddies | Inertial
subrange | Universal
subrange |

Figure 17.5 A figure showing the energy distribution function $E(k)$ for various ranges of wavenumber for isotropic turbulence. Adapted from Warsi (1999).

useful to use the correlation function defined in spectral (Fourier-transformed) space. This is briefly discussed in the following section.

17.10.1   Correlation in the spectral scale

The Fourier transform of R (the two-point correlation tensor discussed above) is useful to find the energy associated with various wavelengths. Consider

$$E(k) = \int_0^\infty R_{11}(x)\cos(kx)dx$$

The parameter k is the wavenumber and $E(k)dk$ can be viewed as the energy associated with eddies in the wavenumber range between k and $k + \delta k$. Small wavenumbers correspond to large eddies and large wavenumbers correspond to small eddies. A qualitative plot of the entire spectrum is shown in Fig. 17.5.

Kolmogorov showed that

$$E(k) = \epsilon^{2/3} k^{-5/3}$$

for the universal subrange which is also known as a 5/3 law. This law has been extended to correlated $E(k)$ for the inner region (the viscous subrange), the outer region (large eddies), and the overlapping region (the universal subrange where the above Kolmogorov equation applies.) The details are reviewed in the books by Wilcox (1998) and by Pope and Bruckner (2010).

Summary

- An important property of turbulent flow is that no two identical experiments would give the same track of the velocity profiles, but the statistical properties remain the same. Hence it seems logical to look at time-averaged properties of flow rather than the instantaneous velocity vs. time profiles. This can be done by a model-reduction procedure called time averaging. The resulting equations are known as Reynolds-averaged Navier–Stokes (RANS) equations

- The RANS equations contain extra unknown terms involving the average of the product of velocity fluctuations. These terms look similar to the viscous stress terms and are referred to as turbulent stress terms. Modeling of these terms is referred to as the closure problem in turbulence.

- One closure is based on the Prandtl mixing-length theory see Eq. (17.10). This model is particularly effective for shear flows but not so applicable for complex 2D flows. Even then it finds wide application in turbulent modeling.

- Equations for the velocity profiles for simple channel flow or pipe flow can be formulated in a similar manner to laminar flow using the Prandtl closure model. The dimensionless form of this equation provides a reference value for the length scale and velocity scale to be used in turbulence modeling. The velocity scale used is also called the friction velocity. Also the scale factors used for length and velocity are different from those used for laminar flow.

- Analytical solutions for the velocity profile for channels and pipe flows applicable to various regions can be obtained using the Prandtl model. The three regions, namely (i) the viscous layer, (ii) the buffer layer, and (iii) the turbulent core should be noted, and the corresponding velocity equations are given in the text. The velocity profile is a logarithmic function of distance in the turbulent core.

- The velocity profiles can be averaged to get the engineering information for flow vs. pressure-drop relations. Some corrections are needed in order to fit the data, but the form of the equation is maintained while fitting the data.

- Pipe roughness has an important impact on the velocity profile in turbulent flow. This parameter is of no importance in laminar flow.

- Turbulent boundary layers are difficult to model. There are two length scales, one near the wall and one representative of the overall thickness of the boundary layer. A modification of the Prandtl logarithmic profile is often used. This equation is called Cole's equation.

- Other closure models include the one-equation (k) and two-equation ($k-\epsilon$) models as well as the $k-\omega$ model. Large-eddy simulation is another tool with which to simulate turbulent flows.

- Direct numerical solution (DNS) is another useful tool to understand turbulent flow behavior. Owing to the excessively large computational costs, this is more of a learning model rather than a design model. DNS is used for simple geometries and provides us with useful information on the basic properties and structure of turbulent flows.

- Turbulence is characterized by eddies or vortices of many length scales. These scales can be represented by defining the various length scales. The Kolmogorov scale is the smallest scale where the turbulent energy is transferred to viscosity and dissipated by viscous dissipation. It is useful to understand the various length scales and the correlation functions that are useful to describe turbulence.

ADDITIONAL READING

Detailed books are available for analysis of turbulent flow. Three commonly used ones are those by Tennekes and Lumley (1972), Pope and Bruckner (2010), and Davidson (2004).

Computational simulations of turbulent shows are illustrated in the books by Tannehill *et al.* (1997), Wilcox (1988), and Farmer *et al.* (2009). These are good starting references for understanding the solution methods, in particular the finite-volume approximations of the flow equations.

Problems

1. Search the literature or web and discuss briefly the principles behind the following flow-measurement devices: a Pitot tube; a hot-wire anemometer; a laser-Doppler velocity meter; and a radioactive-particle tracker. Also comment on the utility of these devices in the study of turbulent flow.

2. Explain the significance of the following cross-correlation terms:

$$\overline{v'_x v'_y}$$

$$\overline{v'_x T'}$$

$$\overline{v'_x C'_A}$$

$$\overline{C'_A C'_A}$$

$$\overline{T' C'_A}$$

Explain what is meant by the closure problem in turbulence modeling.

3. The following data were obtained for the velocity profiles in turbulent flow at a specified point in a channel flow as a function of time. The measurements were taken 0.01 seconds apart. The velocity was reported in m/s.

Find the mean velocity, the turbulent stress tensor (assume $\rho = 1 \, \text{kg/m}^3$), and the kinetic energy of turbulence per unit mass.

4. Verify that the fluctuating component of velocity (2D assumption) satisfies the following equation:

$$\frac{\partial v'_x}{\partial x} + \frac{\partial v'_y}{\partial y} = 0 \tag{17.52}$$

Data for Problem 3

v_x	v_y	v_z
3.84	0.43	0.19
3.50	0.21	0.16
3.80	0.18	0.17
3.60	0.30	0.13
4.20	0.36	0.09
4.00	0.28	0.10
3.00	0.35	0.16
3.20	0.27	0.15
3.40	0.21	0.13
3.00	0.22	0.18
3.50	0.23	0.17
4.30	0.36	0.18
3.80	0.35	0.17

The velocity components can be expanded as a Taylor series in y. A cubic approximation to the velocity is then as follows:

$$v'_x = a_0 + a_1 y + a_2 y^2 + a_3 y^3$$

and

$$v'_y = b_0 + b_1 y + b_2 y^2 + b_3 y^3$$

Use the no-slip condition to show that $a_0 = 0$ and $b_0 = 0$.

Also, since v_x near the wall does not change with x, show that $b_1 = 0$.

Hence suggest an expression for the expansion of turbulent shear stress near the wall. How does it compare with the Prandtl mixing-length theory? Does the linear relation of the turbulent stress with y hold?

5. The velocity profile for turbulent flow in circular pipes is often approximated by the 1/7th-power law:

$$\bar{v}_z = v_{max}(1 - r/R)^{1/7}$$

Find an expression for the cross-sectionally averaged velocity.
A more general expression for the velocity profile is

$$\bar{v}_z = v_{max}(1 - r/R)^{1/n}$$

where n is an index that is fitted as a function of the Reynolds number. The value of n increases with Re, indicating that the profiles becomes steeper at large values of Re.
Find an expression for the average velocity for this general profile.
Answer: $2n^2/[(n+1)(2n+1)]$ for the ratio of average to maximum velocity. Use the above velocity profile and the Prandlt equation for the eddy viscosity and plot the eddy viscosity as a function of the radial location.

6. **The Pai equation for turbulent stress.** Use the equation (17.19) suggested by Pai for the turbulent stress and integrate for the velocity profile. How do the results compare with that of Prandtl?

7. Water is flowing through a long pipe of diameter 15 cm at 300 K. The pressure gradient is 500 Pa/m.

   Using the Blasius equation for the friction factor find the volumetric flow rate and the average velocity. Then find the maximum velocity and the wall shear stress.

   Plot the velocity profile given by the 1/7th-power law and compare your result with the universal velocity profile.

8. Water is flowing in a pipe of diameter 20 cm with a pressure gradient of 3000 Pa/m. $\mu = 0.001\ \text{Pa} \cdot \text{s}$.

   Find the wall shear stress.

   Find the friction velocity.

   Find the thickness of the laminar sublayer.

   Find the velocity at dimensionless radii of 0.6 and 0.8. Find the value of R^+, which is defined as the value of y^+ at the pipe center.

9. Consider a fully developed flow of water in a smooth pipe of diameter 15 cm at a flow rate of 0.006 m^3/s.

   The pressure drop can be calculated by the Blasius relation for this case.

   Verify by an overall force balance that the wall shear stress is equal to $(R/2)(-dp/dz)$. Use the shear-stress value to find the friction velocity and the length scale near the wall.

   Determine the thickness of (a) the laminar sublayer, (b) the buffer layer, and (c) the turbulent core.

   Find the local (time-averaged) velocity at a location 5 cm from the pipe wall.

10. For the above problem find the eddy viscosity as a function of position (based on the universal velocity profile) and plot the eddy viscosity as a function of the distance from the wall.

11. If the RANS equation (17.5) is subtracted from the unaveraged N–S equation (17.1), we obtain the following equation for the perturbation velocity:

$$\frac{\partial v_i'}{\partial t} + V_j(v_i',j) + v_j'(V_i,j) + v_j'(v_i',j) = -(p'\delta_{ij}/\rho)_{,j} + v(v_i')_{,jj} + \left\langle v_i'v_j' \right\rangle_{,j} \qquad (17.53)$$

Verify this equation. Take the dot product of this equation with the perturbation velocity. From this equation derive an equation for the turbulent kinetic energy shown below:

$$\frac{\partial k}{\partial t} + V_j\frac{\partial k}{\partial x_j} = \langle v_i'v_j'\rangle\frac{\partial V_i}{\partial x_j} - \frac{\partial}{\partial x_j}\left\{v_j'[p'/\rho] + k\right\} + v\langle v_i'\nabla^2 v'\rangle \qquad (17.54)$$

The LHS has the standard form for convective transport with a mean velocity and represents the substantial derivative of the kinetic energy. The terms on the RHS are written as the divergence of a flux vector, a generation term, and a loss (dissipation) term. This provides the rationale for the engineering model for k given by Eq. (17.47).

12. Take the product of the perturbation velocity equation given by Eq. (17.53) by any component of the perturbation velocity. This results in an equation for $v_x'v_y'$, which is the turbulent stress. Derive this equation and show that the triple correlation terms remain in this equation, which is a part of the closure problem.

13. Show that, for the isotropic case, (i) the velocity correlation function is now symmetric and (ii) the pressure–velocity correlation is zero.

18 More convective heat transfer

Learning objectives

- To understand the heat transfer with laminar flow in a boundary layer and to study the different methods which can be used for the solution of such problems.
- To study the effect of turbulence on heat transfer; to set up and solve typical problems in turbulent heat transfer both for external and for internal flows. To examine additional complexities in modeling turbulent heat transfer.
- To illustrate how heat transfer in complex geometries can be addressed from a numerical perspective; to briefly examine some issues associated with numerical computing.
- To be able to set up and solve problems in natural convection heat transfer.
- To understand the various flow regimes in boiling heat transfer and to appreciate how these affect the rate of heat transfer in boiling systems.
- To study simple problems in condensation heat transfer.
- To study simple problems with a moving boundary, e.g., a freezing front.

Mathematical prerequisites

No new mathematical prerequisites are needed at this stage.

Convection plays an important role in heat transfer, and analysis of such problems requires simultaneous solution of momentum and heat transport equations. The simplest approach to model such systems is to define and use the heat transfer coefficient as shown earlier. For example, the heat transfer rate from a solid to a fluid is given by

$$q = h \, \Delta T$$

where ΔT is a suitably defined driving force for heat transfer and q is the component of the heat flux vector in the direction normal to the solid surface. The parameter, h, is, however, a function of flow around the solid and also varies locally at each point in the solid surface. Thus q or h is a function of the local flow field, and the central problem in the study of convective heat transfer is that of how to relate q and/or h to the flow field. We have discussed this in connection with heat transfer in laminar flow in channels and pipes. The heat transfer coefficient was related to flow parameters for the case of laminar flow. The key results could be represented in terms of the Nusselt number (a dimensionless heat flux) as a function of the Péclet number (a dimensionless group describing the relative importance of convection and conduction in the fluid). This chapter discusses additional problems in convective heat transfer.

The first class of problems we deal with consists of those which are amenable to a theoretical or semi-theoretical treatment for external flow past a solid. The flow field is known from the boundary-layer theory (forced convection flow) and its effect on heat transfer is examined. This represents a one-way coupled problem. The heat transport in turbulent flow is then examined and additional complexities in modeling are addressed. Both external flow (flat plate) and internal flow (pipe flow) problems are examined as specific examples.

A second class of problem examined in this chapter involves natural convection flow and heat transfer where the density changes caused by heat transfer induce the flow. The flow field now depends on the rate of heat transport itself, and this therefore represents a two-way coupling. An example on the natural convection in a confined enclosure was studied earlier. Also key results for natural convection in unconfined geometries were presented without derivation. Here we take this up in somewhat more detail. The problem of natural convection over a vertical plate is illustrated as a prototype example of natural convection in boundary layers. A somewhat more involved problem in a 2D enclosure requires numerical computations, and we show a simple computational scheme that can be used for a similar class of problems.

A third class of problems in convective heat transport problems involves systems with phase change. These types of problems are very important in engineering practice. These include boiling, condensation, freezing, etc. An introductory treatment of these systems is provided in this chapter. This will provide the student with a basic ability to analyze these processes and use the results for engineering design.

18.1 Heat transport in laminar boundary layers

18.1.1 Problem statement and the differential equation

In Sections 6.9 and 15.5.2 we examined the problem of flow over a flat plate, and a brief introduction to convective heat transfer in boundary layer was provided in Section 8.4.2. Here we consider the problem of heat transfer over a flat plate in further detail. The flow is assumed to take place under laminar conditions. The flow ceases to be laminar if $Re_x > 2 \times 10^5$, where Re_x is the Reynolds number based on the distance from the edge of the plate. We assume that the plate is maintained at a different thermal condition than the

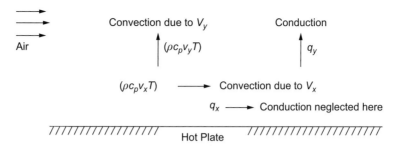

Figure 18.1 Heat transport mechanisms for flow over a plate.

entering fluid. The temperature distribution in the fluid is then given by the convection plus conduction equation.

A schematic diagram of the various transport mechanisms is shown in Fig. 18.1.

The general governing equation is the temperature equation given by Eq. (7.29) in Chapter 7. For the problem under consideration the conduction in the x-direction can be neglected; this assumption can be justified by a scaling analysis. Hence a simplified version of the temperature equation is used:

$$v_x \frac{\partial T}{\partial x} + v_y \frac{\partial T}{\partial y} = \alpha \frac{\partial^2 T}{\partial y^2} \tag{18.1}$$

where α is the thermal diffusivity. The above equation is written in combination with the continuity equation in the following conservative form:

$$\frac{\partial (v_x T)}{\partial x} + \frac{\partial (v_y T)}{\partial y} = \alpha \frac{\partial^2 T}{\partial y^2} \tag{18.2}$$

The boundary conditions are dependent on the manner by which the plate is heated. Two common cases are (i) Dirichlet conditions for the constant-wall-temperature case and (ii) Neumann conditions for a constant-wall-flux case. Let us first do a scaling analysis to find the thickness of the boundary layer.

18.1.2 The thermal boundary layer: scaling analysis

We postulate again a thermal boundary layer of thickness δ_t over which the major temperature change occurs. The commonly accepted definition for this is the point where the temperature is 0.99 of the approach temperature. The integral balance can be used to obtain a relation for this quantity. Once this is known other parameters of design interest (such as the heat transfer coefficient) can be calculated. Before studying the integral heat balance, it may be useful to first do a scaling analysis and see what the results indicate. Note that for a momentum boundary layer the scaling analysis predicted the trends well (see Section 14.2.4 for details). Only the numerical constant could not be evaluated. The integral method or more exact solutions were needed only to fit the constant. The story for heat transfer is the same.

It may be useful to recall the following result from the scaling analysis for the momentum problem: $\delta \propto \sqrt{vx/v_\infty}$, where δ is the thickness of the momentum boundary layer and x is the distance from the leading edge. This result will be needed in this section.

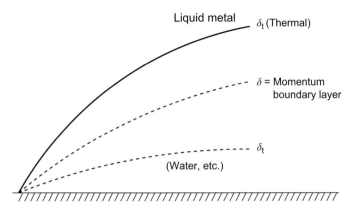

Figure 18.2 Hydrodynamic and thermal boundary layers for transport over a plate. Note the effect of the Prandtl number, which is in the range of 0.02 for mercury and 6.2 for water.

The role of the Prandtl number

The scaling analysis of the thermal boundary layer depends on whether the thermal layer is thicker or thinner than the momentum boundary layer. This depends on the magnitude of the Prandtl number as shown below.

For $Pr \ll 1$, the thermal boundary layer is thicker than the momentum boundary layer. Note that

$$\text{Prandtl number} = \frac{\nu}{\alpha} = \frac{\text{Momentum diffusivity}}{\text{Heat diffusivity}}$$

and therefore gives the relative measure of thermal penetration vs. momentum penetration. Hence if $Pr \ll 1$ then the momentum cannot diffuse far enough and the momentum boundary layer is smaller. Likewise, if $Pr > 1$ the thermal boundary layer is thinner. The role of the Prandtl number is shown schematically in Fig. 18.2.

Scaling analysis for small Pr

The scaling analysis is simple here since the velocity now scales as the free-stream velocity v_∞. The scales are obtained for each term in Eq. (18.2) as follows:

$$\text{Term 1 on LHS} \approx \mathcal{O}(v_\infty \, \Delta T / x)$$

where x is any arbitrary length along the plate. The temperature scale ΔT is $T_w - T_\infty$:

$$\text{Term 2 on LHS} \approx \mathcal{O}(v_y \, \Delta T / \delta_t)$$

where L is any arbitrary length of the plate. In view of the continuity equation, v_y scales as $\delta_t v_\infty / L$. Using this result, we have

$$\text{Term 2 on LHS} \approx \mathcal{O}(v_\infty \, \Delta T / x)$$

which has the same scale as term 1. Hence the two convective terms have similar scales, and both terms should be retained.

Term 3 on the RHS scales as $\alpha \, \Delta T / \delta_t^2$. On comparing the scale of this term with that of term 1 or 2 we find:

$$\alpha \Delta T / \delta_t^2 \approx \mathcal{O}(v_\infty \, \Delta T / x)$$

or

$$\delta_t^2 \approx \mathcal{O}(x\alpha/\nu_\infty) \tag{18.3}$$

On dividing both sides by δ^2 and using an estimate of $\nu x/\nu_\infty$ for δ^2 as shown earlier, the above result can be expressed as

$$\boxed{\Delta^2 \approx \frac{1}{Pr} \text{ or } \Delta \approx \frac{1}{Pr^{1/2}}} \tag{18.4}$$

where Δ is the ratio of the two boundary-layer thicknesses,

$$\Delta = \frac{\delta_t}{\delta}$$

Scaling analysis for large *Pr*

For large *Pr* a modification is needed in the scaling analysis. Only a part of the velocity ν_∞ is seen by the thermal boundary layer, which is now much smaller than the momentum boundary layer. Hence the velocity scales as $\nu_\infty \delta_t/\delta$ or $\nu_\infty \Delta$ rather than as ν_∞. Using this relation in (18.3), we can show that

$$\delta_t^2 \approx \mathcal{O}(x\alpha/[\nu_\infty \Delta])$$

Using the estimate for δ and the definition of *Pr*, the following result holds for large Prandtl number:

$$\boxed{\Delta^3 \approx \frac{1}{Pr} \text{ or } \Delta \approx \frac{1}{Pr^{1/3}}} \tag{18.5}$$

The heat transfer coefficient

Since the heat transfer coefficient scales as k/δ_t, we can suggest a correlating equation for the heat transfer coefficient that is based purely on scaling arguments. The results can be expressed in terms of the key dimensionless groups Nu, Re_x, and Pr, and can be shown to be as follows.

We have $h \propto k/\delta_t$ or $h \propto (k/\delta)(1/\Delta)$. Since Nu_x is equal to hx/k, we obtain

$$Nu_x \propto \frac{x}{\delta}\frac{1}{\Delta}$$

From momentum considerations x/δ is proportional to $Re_x^{1/2}$, and hence

$$Nu_x \propto Re_x^{1/2}\frac{1}{\Delta}$$

Now, using the Prandtl dependence for Δ, the following correlations can be suggested for the Nusselt number, using the scaling analysis:

$$Nu_x = C_1 Re_x^{1/2} Pr^{1/3}$$

for high-Prandtl-number fluids and

$$Nu_x = C_1 Re_x^{1/2} Pr^{1/2}$$

for low-Prandtl-number fluids.

Note that the exponent on the Prandtl number is different for the two cases.

Table 18.1. Values of the Prandtl number for various fluids

Fluid	Pr
Air	0.733
Water	6.75
Alcohol	16.6
Glycerine	7250
Sodium	0.007
Mercury	0.024
Silicon	0.04

It may be noted that the mass transfer case discussed in Section 20.1.2 is analogous, where Sh takes the place of Nu, and Sc takes the place of Pr. The corresponding exponent on the Schmidt number is in most cases 1/3, because Sc is of the order of one or larger, unlike Pr, which varies over a much wider range.

How does the above correlation contrast to that for internal flow in a pipe or channel?

The range of Prandtl values

It may be useful at this stage to note typical values of the Prandtl number for a number of fluids. These are noted in Table 18.1. The values are at 298 K for common fluids, but are at the melting points for mercury, sodium, and silicon.

What is the temperature dependence for Pr? An average value is often used to simplify the problem in convective heat transfer. Is this a good assumption?

Now we proceed to do an integral analysis on this model equation given by Eq. (18.2).

18.1.3   The heat integral equation

The von Kármán heat equation

The integral balance is obtained by integrating Eq. (18.2) with respect to y from $y = 0$ to $y = \delta_t$. The result, after some algebra, can be shown to be

$$\frac{d}{dx}\left[\int_0^{\delta_t} v_x(T - T_\infty)dy\right] = \frac{q_w}{\rho c_p} \tag{18.6}$$

Details of the derivation are left as an exercise. This is called the von Kármán heat equation. Here T_∞ is the temperature outside the boundary layer, which is also equal to the approach temperature of the fluid. q_w is the heat flux from the plate defined as $-k(\partial T/\partial y)$ at $y = 0$ in accordance with Fourier's law. The above integral is analogous to the momentum integral. However, the heat transport problem is coupled to the momentum transfer since the solution of the heat integral requires knowledge of the velocity field v_x in the above equation. We use the value from the von Kármán momentum analysis for this quality, since this is a one-way coupled problem. Also, since information is lost on averaging, the integral balances can be solved only if one assumes a form for the temperature profile within the boundary layer.

Boundary conditions

Note that two types of wall boundary conditions are commonly studied.

1. Type 1, where the plate is maintained at a constant temperature, e.g., by condensing steam on the bottom of the plate. In this case the wall heat flux $q_w = -k(\partial T/\partial y)$ varies as a function of x. The wall temperature T_w remains constant. A variation on the Type 1 problem is where the plate temperature is a prescribed function of x, i.e., $T_w = T_w(x)$, which is a more general case.
2. Type 2, where the plate is maintained at a constant heat flux, e.g., by having an electrical coil on the bottom of the plate or by laser or radiation heating. In this case q_w is a constant and is not a function of x. The wall temperature, T_w, now becomes a function of x. A variation on the Type 2 problem is where the heat flux is a prescribed function of x, i.e., $q_w = q_w(x)$. This is obviously a more general case than is that of constant q_w.

The solution to the heat integral and the corresponding thickness of the thermal boundary layer will depend on the boundary condition. This is not stated explicitly in many books, or only the constant temperature condition is treated. The solutions to the two cases are different, as shown in the following section.

Solution using an assumed profile

The solution of the heat integral will depend also on whether the thermal boundary layer is thicker or thinner than the momentum boundary layer. The case where the thermal boundary layer is thinner is of more interest for common fluids ($Pr > 1$) and is the case analyzed here. The case where $Pr < 1$ is left as an exercise.

For the solution of the heat integral, we need approximations BOTH for the velocity and for the temperature distribution in the boundary layer. The velocity is approximated as a cubic as before:

$$\frac{v_x}{v_\infty} = \frac{3}{2}\eta - \frac{\eta^3}{2} \tag{18.7}$$

where η is defined as y/δ.

The temperature distribution is also assumed to be a cubic:

$$T - T_w = a + by + cy^2 + dy^3$$

where a, b, etc. are constants to be fitted depending on the conditions used. If four (boundary) conditions are assumed for the temperature distribution, as shown in Problem 1, the following approximation is obtained:

$$\frac{T - T_w}{T_\infty - T_w} = \frac{3}{2}\frac{y}{\delta_t} - \frac{1}{2}\left(\frac{y}{\delta_t}\right)^3 \tag{18.8}$$

where δ_t is the thickness of the thermal boundary layer. Using Δ, the ratio of boundary-layer thicknesses, this can be written as

$$\frac{T - T_w}{T_\infty - T_w} = \frac{3}{2\Delta}\eta - \frac{1}{2\Delta}\eta^3 \tag{18.9}$$

The integral term in Eq. (18.6) can then be evaluated as

$$\int_0^{\delta_t} v_x(T - T_\infty)dy = v_\infty(T_w - T_\infty)\delta\left[\frac{3}{20}\Delta^2 - \frac{3}{280}\Delta^4\right] \tag{18.10}$$

You may wish to verify the result using symbolic manipulation in MATLAB or MAPLE. You need to use the command *Simplify ans* to get it in the above form.

The term with Δ^4 in the above equation is usually smaller than the term with Δ^2, since $\Delta < 1$ for the case being analyzed (common fluids) and is neglected for ease of further manipulations. With this approximation, the heat integral can be represented as

$$\frac{3}{20}\frac{d}{dx}\left[(T_w - T_0)\delta\Delta^2\right] = \frac{q_w}{v_\infty\rho c_p} \tag{18.11}$$

Note that the term $(T_w - T_\infty)$ is left inside the differential since it may vary with x.

The heat flux is related to $(T_w - T_\infty)$ by the following relation for a cubic temperature:

$$q_w = \frac{3}{2}\frac{k(T_w - T_\infty)}{\delta_t} = \frac{3}{2}\frac{k(T_w - T_\infty)}{\delta\Delta} \tag{18.12}$$

which follows by applying Fourier's law for the approximate temperature profile given by (18.9).

The above equation can also be expressed in terms of a local Nusselt number defined by

$$Nu_x = \frac{q_w x}{k(T_w - T_0)} = \frac{3}{2}\frac{x/\delta}{\Delta}$$

Recall that δ/x was obtained earlier from the momentum balance,

$$\frac{\delta}{x} = \frac{4.64}{\sqrt{Re_x}}$$

Hence the equation for the Nusselt number can be expressed as

$$Nu_x = 0.323\frac{Re_x^{1/2}}{\Delta} \tag{18.13}$$

Note that a similar relation was indicated by the scaling analysis in the previous section.

Thus, once Δ has been obtained from the solution of Eq. (18.11) the values of the local Nusselt number and other required quantities can be calculated. The cases of constant heat flux and constant wall temperature have to be analyzed separately.

Solution for constant heat flux

Equation (18.11) can be directly integrated for constant wall flux to give

$$\left[(T_w - T_0)\delta\Delta^2\right] = \frac{20q_w x}{3v_\infty\rho c_p} \tag{18.14}$$

Here we assume $\delta_t = 0$ right at $x = 0$, i.e., heat transfer begins at the very start of the plate. Now, using Eq. (18.12) for q_w, we obtain

$$[\Delta^3\delta^2] = 10k/(v_\infty\rho c_p)$$

Using the expression for δ from the momentum balance solution given earlier,

$$\delta^2 = \frac{280}{13}\frac{vx}{v_\infty}$$

we find

$$\Delta^3 = \frac{13}{28}\frac{1}{Pr}$$

where Pr is equal to ν/α.

The result in turn can be used to find the Nusselt number (Eq. (18.13)) or the wall-temperature variation with x using Eq. (18.12). Thus all the quantities of interest can be calculated.

The following result for the Nusselt number for constant wall flux is obtained:

$$Nu_x = 0.4171 Re_x^{1/2} Pr^{1/3}$$

You should verify the details leading to this calculation.

Also note that, since the results are applicable only if $\Delta < 1$, we find that Pr should be greater than $(13/28)$, or 0.464, for these results to hold. Thus the estimate of the scaling analysis of $Pr = 1$ is somewhat higher for the constant-wall flux case.

The variation of the surface (plate) temperature is obtained from Eq. (18.12). Since q_w is a constant here, the plate temperature will vary as δ, which in turn means that the temperature varies as $\sqrt{x}$.

Solution for constant wall temperature
Using Eq. (18.12) in Eq. (18.11), we obtain

$$\frac{1}{10}\frac{d}{dx}[\Delta^2\delta] = \frac{k}{v_\infty \rho c_p \Delta\delta}$$

This is somewhat more difficult to integrate. The integration of this is shown later in the context of mass transfer (which leads to a similar equation). A useful simplification is to assume that Δ is constant with x. This leads to the following result:

$$\Delta^3 = \frac{13}{14}\frac{1}{Pr}$$

This result in turn can be used to find the Nusselt number from Eq. (18.13) or the heat-flux variation with x using Eq. (18.12). Thus all the quantities of interest can be calculated. It can be shown that the surface heat flux will vary as $1/\sqrt{x}$.

The following expression for the Nusselt number is obtained for a constant wall temperature:

$$Nu_x = 0.315 Re_x^{1/2} Pr^{1/3}$$

Although these results are based on some simplifications and are also somewhat sensitive to the assumed velocity and temperature profiles, they are within 2% of the exact solution.

18.1.4 Thermal boundary layers: similarity solution

The similarity approach (similar to the Blasius method for momentum) can also be used for the problem of heat transfer from a flat plate. The starting point is Eq. (18.2). The similarity variable η defined earlier is now introduced, and v_x and v_y can be expressed in terms of the f function defined earlier. Finally, a dimensionless temperature θ defined as

$(T - T_\infty)/(T_w - T_\infty)$ is used. All these manipulations lead to the following equation for the dimensionless temperature profile as a function of η:

$$\theta'' + Prf\theta' = 0 \tag{18.15}$$

The prime indicates differentiation with respect to the similarity variable η. The boundary conditions for an isothermal plate are $\theta(0) = 1$ and $\theta(\infty) = 0$.

The above equation needs to the solved simultaneously with the Blasius equation:

$$f''' + ff'' = 0$$

The integration for the θ equation can be done numerically using tabulated values of f''. An alternative method is to solve for both f and θ simultaneously using a BVP solver, for instance, BVP4C. The final results are left as an exercise.

The slope of the temperature gradient at the surface can be related to the local Nusselt number. The resulting expression is

$$Nu_x = -\frac{\theta'(0)\sqrt{Re_x}}{\sqrt{2}}$$

and hence the value of Nu_x can be calculated for any given Pr number from the value calculated for the slope at the surface. The slope will be a function of the Prandtl number.

18.2 Turbulent heat transfer in channels and pipes

The equation for the temperature profile

The convection–diffusion of heat in a channel of half-width H under turbulent flow conditions can be represented as the temperature equation for the time-averaged value:

$$\rho c_p \bar{v}_x(y)\frac{\partial \bar{T}}{\partial x} = \frac{\partial}{\partial y}(q_y^{(m)} + q_y^{(t)}) \tag{18.16}$$

The only difference between this and the laminar flow is the inclusion of the extra term (the last term) arising from turbulent fluctuations, the so-called eddy thermal conductivity term,

We analyze here the problem under conditions of constant wall heat flux q_w. Since the variation of temperature in the y-direction is more significant, we make a simplifying assumption that the heat flux varies linearly in the y-direction. This means that the convective term (the LHS) is simply assumed to be a constant.

The linear variation of the total heat flux with y can be represented as

$$q_y = q_w(1 - y/H)$$

where q_w is the heat flux at the wall, and the problem can be simplified to

$$(q^{(m)} + q^{(t)}) = q_w$$

The molecular term is written using the Prandtl number, Pr, as

$$q_y^{(m)} = -k\frac{d\bar{T}}{dy} = -[(\rho c_p v)/Pr]\frac{d\bar{T}}{dy} \tag{18.17}$$

The turbulent term is written using a turbulent Prandtl number, Pr_t, as

$$q_y^{(t)} = -\alpha_t \rho c_p \frac{d\bar{T}}{dy} = -\rho c_p [\nu_t / Pt_t] \frac{d\bar{T}}{dy} \tag{18.18}$$

The following notations are to be noted: α_t is the turbulent heat (thermal) diffusivity and ν_t is the eddy diffusivity for momentum as used in the earlier chapter as well. Finally, Pr_t is defined as the ratio of ν_t to α_t. Also we define and use a dimensionless total momentum diffusivity:

$$\nu_T^+ = 1 + \frac{\nu_t}{\nu} \tag{18.19}$$

Note that this is not a physical property and is a turbulence parameter. Thus it varies as a function of y.

Using the above group (ν_T^+) and the definitions for the molecular and turbulent heat fluxes, the total flux can be written as

$$-\rho c_p \nu \left(\frac{1}{Pr} + \frac{\nu_T^+ - 1}{Pr_t} \right) \frac{d\bar{T}}{dy} = q_w (1 - y/H) \tag{18.20}$$

Let us introduce the dimensionless distance y^+ here. Recall that this was defined as $y^+ = y\rho V_f / \mu = yV_f / \nu$.

Hence Eq. (18.20) can be written as

$$-\rho c_p V_f \left(\frac{1}{Pr} + \frac{\nu_T^+ - 1}{Pr_t} \right) \frac{d\bar{T}}{dy^+} = q_w (1 - y^+ / H^+) \tag{18.21}$$

This calls for a dimensionless temperature T^+, which should be defined as

$$T^+ = (T_w - T)\rho c_p V_f / q_w \tag{18.22}$$

Hence Eq. (18.21) is written as

$$\left(\frac{1}{Pr} + \frac{\nu_T^+ - 1}{Pr_t} \right) \frac{dT^+}{dy^+} = 1 - y^+ / H^+ \tag{18.23}$$

The integrated form of the equation is

$$T^+ = \int_0^{y^+} \frac{(1 - y^+ / H^+)}{1/Pr + (1/Pr_t)(\nu_T^+ - 1)} dy^+ \tag{18.24}$$

which is the formal equation to calculate the temperature distribution. If a model for ν_T^+ and for Pr_t as a function of y is proposed, the above equation can be integrated to find the temperature profiles.

Solution for different regions

Again it is convenient to integrate for each region separately. Also the term $(1 - y^+ / H^+)$ is often taken as 1 for simplification. This assumption certainly holds near the walls. Hence Eq. (18.24) is simplified to

$$T^+ = \int_0^{y^+} \frac{dy^+}{1/Pr + (1/Pr_t)(\nu_T^+ - 1)} \tag{18.25}$$

This might not be good near the center, but the effect of this "center defect" on the calculation of the wall heat flux and the Nusselt number is generally small. The solution for T^+ for three regions (laminar, buffer, and turbulent) will now be presented.

The near-wall region: the viscous sublayer

Here ν_t is small and therefore $\nu_T^+ = 1$. Hence Eq. (18.25) can be simplified and it is easy to show that

$$T^+ = Pr\nu^+ \tag{18.26}$$

Since

$$\nu^+ = y^+$$

the temperature profile in the viscous region is

$$T^+ = Pr y^+ \tag{18.27}$$

This is valid for $y^+ < 5$. Note that the (dimensionless) temperature profile is scaled by Pr. Also T^+ is the same as y^+ if $Pr = 1$.

Buffer zone

The region between the laminar layer and the turbulent core, $5 < y^+ < 30$, is called the buffer layer. The velocity here was correlated as

$$\nu^+ = 5 \ln y^+ - 3.05$$

Differentiation of this gives an expression for ν_T^+. It can be shown that

$$\nu_T^+ = 0.2y^+$$

for the above velocity profile. Hence the formal expression for the temperature profile is

$$T^+ - 5Pr = \int_5^{y^+} \frac{dy^+}{1/Pr + (1/Pr_t)(0.2y^+ - 1)} \tag{18.28}$$

The value of Pr_t in the buffer zone is taken as one. Using this value and integrating gives

$$T^+ = 5Pr + 5 \ln \left[Pr([y^+/5] - 1) + 1 \right] \tag{18.29}$$

The value of T^+ at the edge of the buffer zone (at y^+ of 30) is

$$T^+(30) = 5Pr + 5 \ln(5Pr + 1)$$

which is needed as the integration constant for the next step.

The turbulent core

The velocity profile in the turbulent core is given by

$$\nu^+ = 2.5 \ln y^+ + 5.5 \tag{18.30}$$

which is found to be valid for $y^+ > 30$. This is called the turbulent-core region.
    On differentiating this we get ν_T^+:

$$\nu_T^+ = 0.4y^+$$

Also Pr_t is taken as 0.9. Furthermore, $1/Pr - 1/Pr_t$ is much smaller than v_T^+/Pr_t. Hence

$$T^+ - T^+(30) = \int_{30}^{y^+} \frac{dy^+}{(1/0.9)(0.4y^+)} \tag{18.31}$$

Integrating gives

$$T^+ = 5Pr + 5\ln(5Pr + 1) + 2.25\ln(y^+/30) \tag{18.32}$$

These expressions give the temperature profile in the channel as a function of the dimensionless distance from the wall. The temperature profile has a discontinuity in the slope at the edge of the turbulent core. Also dT/dy^+ does not reach zero at $y^+ = H^+$, i.e., at the center there is a temperature "defect." But these issues are not considered to be of great concern in the heat flux predictions.

The equation for the turbulent boundary layer
Owing to the similar nature of the velocity profile near the wall the equations are also applicable to the turbulent boundary layer. The channel thickness H is now replaced by the momentum boundary-layer thickness δ. Since the term $(1 - y^+/H^+)$ is often taken as 1, i.e., $y+ \ll H^+$, the equation for the temperature profile near the wall remains the same for both channel and boundary-layer flow for all three regions.

Stanton number calculation
Using the expression for the temperature and its gradient at the wall, an expression for the heat transfer coefficient can be derived. This is usually represented in terms of the Stanton number defined as $h/[\rho c_p v_\infty]$. We show the equations for turbulent boundary layers. The driving force is defined as $T_w - T_\infty$ and hence $h = q_w/(T_w - T_\infty)$. Now, using the definition of the dimensionless temperature, the local Stanton number can be shown to be related to the drag coefficient by

$$St_x = \left(\frac{C_{fx}}{2}\right)^{1/2} \frac{1}{T_e^+} \tag{18.33}$$

where T_e^+ is the dimensionless temperature at the edge of the boundary layer. This is defined as

$$T_e^+ = (T_w - T_\infty)\rho c_p V_f/q_w \tag{18.34}$$

The expression for T_e^+ can be obtained by applying Eq. (18.32) at $y = \delta$ or correspondingly at $y^+ = \delta^+$. The latter is in turn related to the drag coefficient and hence the resulting expression for T_e^+ is

$$T_e^+ = 5Pr + 5\ln(5Pr + 1) + Pr_t\left[\sqrt{2/C_{fx}} - 14\right]$$

On substituting into the local Stanton number expression we have

$$St_x = \frac{\sqrt{C_{fx}/2}}{5Pr + 5\ln(5Pr + 1) + Pr_t[\sqrt{2/C_{fx}} - 14]} \tag{18.35}$$

The denominator can be viewed as the sum of three resistances in series, namely the viscous sublayer, the buffer region, and the fully developed turbulent core. The relative magnitude of these will depend on the molecular Prandtl number and the Reynolds number. The equation is a good representation of data for $0.5 < Pr < 30$.

Equation (18.35) can be rearranged to

$$St_x = \frac{C_{fx}/2}{Pr_t + \sqrt{C_{fx}/2}\,[5Pr + 5\ln(5Pr + 1) - 14Pr_t]} \tag{18.36}$$

The analogy with momentum transfer

With the above expression for St it is easier to compare it with the various analogies between momentum and heat transfer which have been proposed. The simplest is the Reynolds analogy, which states that

$$St_x = \frac{C_{fx}}{2}$$

Then we have the Prandtl analogy,

$$St_x = \frac{C_{fx}/2}{1 + 5\sqrt{C_{fx}/2}\,(Pr - 1)} \tag{18.37}$$

and the von Kármán analogy,

$$St_x = \frac{C_{fx}/2}{1 + 5\sqrt{C_{fx}/2}[Pr - 1 + \ln[1 + 5(Pr - 1)/6]]} \tag{18.38}$$

All these analogies differ only in the way the denominator term in Eq. (18.36) is handled and what value is assigned to Pr_t.

The role of the molecular Pr

It may be noted that, depending on the value of the molecular Prandtl number Pr, the thermal boundary layer may lie inside the buffer region (large Pr) or may even penetrate deep into the turbulent core (low Pr). This is shown schematically in Fig. 18.3 which has been adapted from the book by Davidson (2004). For the larger-Pr case the region of influence of the conduction effects lies buried within the laminar sub-layer. A thickness of the order of $15/Pr^{1/3}$ is considered to be an indicative magnitude of this region. The velocity profile affects the heat transport rate significantly for this case. Hence, a more accurate velocity profile in the buffer region and also in the viscous region may be needed. It appears that use of the von Kármán profile for the velocity (rather than the Prandtl profile with $v^+ = y^+$) gives a better fit to the experimental data, especially if the Prandtl number is large. However, the following quote from the classic book by Levich (1962) is to be noted (which can be viewed as a more general observation, not restricted to the present case of turbulent heat transport):

The agreement between theory and experiment can not of itself be viewed as corroborating the theory. On one hand theoretical expressions contain a large number of constants, which make it possible to fit the experimental data. On the other hand, experimental data on heat transfer can be relatively inaccurate.

The other case shown in Fig. 18.3 is that of low Prandtl number. The region where conduction effects dominate is here of the order of $5/Pr$ and extends to the buffer zone and to part of the turbulent core. For liquid metals, for which Pr is extremely small, this can extend right up to the channel width and the turbulence has no significant effect on heat transfer in this case.

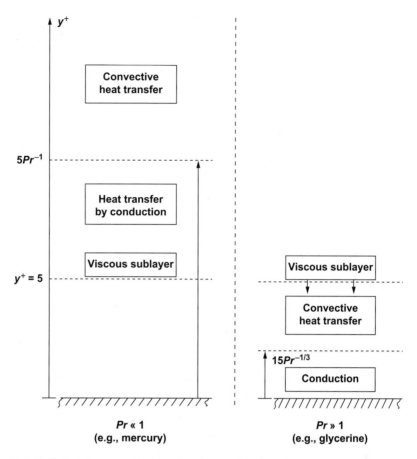

Figure 18.3 Turbulent heat transfer, showing the contribution of molecular Pr in the wall region. For low-Pr fluids the conduction region extends to the turbulent core. Adapted from Davidson (2004).

The role of turbulent Pr

A second point of caution is the value assigned to Pr_t. A constant value of 0.9 is usually assigned, but it appears from the literature that a value depending on Re is more realistic. One explanation is that as Re increases the turbulence energy spectrum moves to larger eddies, which might not be contributing in a proportional manner to heat transfer. A survey of various methods for assigning a proper value to Pr_t has been given in a recent paper by Combest *et al.* (2011). For further study of this topic that paper can be useful.

18.2.1   Pipe flow: the Stanton number

The results for the temperature profile shown for the channel flow and for the boundary layer are also applicable to pipe flow with y defined as the distance from the wall. It is convenient to define the heat transfer coefficient as

$$q_\mathrm{w} = h(T_\mathrm{w} - T_\mathrm{b})$$

where T_b is the average temperature in the pipe (at any given axial location). This is formally defined as

$$T_b = \frac{1}{\langle v \rangle \pi R^2} \int_0^R v 2\pi (R - y) T \, dy$$

or, in terms of dimensionless variables,

$$T_b^+ = \frac{2}{v_b^+ (R^+)^2} \int_0^1 v^+ T^+ (R^+ - y^+) dy^+ \tag{18.39}$$

The heat transfer coefficient data are usually expressed in terms of a dimensionless group, the Stanton number, defined as usual as $St = h/(\rho c_p \langle v \rangle)$. This can be expressed in terms of dimensionless groups as

$$St = \frac{1}{v_b^+ T_b^+}$$

Now, expressing v_b^+ in terms of the friction factor as was done in the previous chapter, we have the following equation for St:

$$St = \sqrt{\frac{f}{2}} \cdot \frac{1}{T_b^+}$$

This is yet another form of the analogy! Note that St can also be expressed as

$$St = \sqrt{\frac{f}{2}} \left(\frac{T_c^+}{T_b^+} \right) \left(\frac{1}{T_c^+} \right)$$

In practice, it is difficult to estimate T_b^+. Note that numerical integration can be done using Eq. (18.39) if detailed v^+ and T^+ profiles are computed. But it is easy to find T_c, the center temperature, by simple substitution in the temperature profile expression for the turbulent-core region. From the expressions derived earlier (Eq. (18.32)), the dimensionless center temperature is easily obtained:

$$T_c^+ = 5Pr + 5 \ln(5Pr + 1) + 2.25 \ln(R^+/30)$$

Since $R^+ = Re/(2v_b^+) = Re\sqrt{f/2}/2$, this can be expressed as

$$T_c^+ = 5Pr + 5 \ln(5Pr + 1) + 2.25 \ln(Re\sqrt{f/2}/60)$$

Hence the Stanton number is given by

$$St = \frac{T_c^+}{T_b^+} \cdot \frac{\sqrt{f/2}}{5Pr + 5 \ln(5Pr + 1) + 2.25 \ln(Re\sqrt{f/2}/60)} \tag{18.40}$$

The expression for St for external flow (Eq. (18.36)) is simpler than that for the pipe-flow case. Why?

The first term, T_c^+/T_b^+, is usually close to unity in turbulent flow due to the fact that the temperature profiles are flatter in the turbulent case than in the laminar case. This term is therefore approximated as one.

Hence the equation is simplified to

$$St = \frac{\sqrt{f/2}}{5Pr + 5 \ln(5Pr + 1) + 2.25 \ln(Re\sqrt{f/2})/60)} \tag{18.41}$$

This expression relates the Stanton number (a measure of heat transfer) to the friction factor (a measure of momentum transfer) and is one of the analogies used in heat transfer.

18.3 Heat transfer in complex geometries

The convection–conduction heat equation can be solved numerically for these cases. An example is for flow over a sphere whose surface is at a different temperature than the bulk gas, and this can be computed, for example, using the OPENFOAM program as the platform. This is an open-source code program in C++. For a sample problem, we investigated the heat transfer from a sphere with a gas flow approaching it at a uniform velocity. The goal was to investigate the temperature profile, the local heat transfer coefficient, and the boundary layer around it. Re is varied as a parameter to cover the flow regimes indicated in Fig. 6.16 in Section 6.9, while we keep $Pr = 1$ for case 1. We set $Pr = 10$ for case 2 to examine how the thermal-boundary-layer thickness changes with the Prandtl number.

The key problem as the Reynolds number increases is the boundary-layer separation. The flow field should accurately portray the wake region if the heat transfer is to be accurately simulated. Significant variation with the angular coordinate is observed both theoretically and experimentally. The local heat transfer coefficient decreases dramatically at the point of flow separation and then increases again in the wake region. The detailed results are not shown here for the sake of brevity.

For practical engineering calculations, the correlations based on dimensionless groups are most convenient. Here we summarize some useful correlations.

Some empirical correlations
For flow past a sphere the average Nusselt number is correlated by the Ranz–Marshall correlation:

$$Nu_m = 2 + 0.6Re^{1/2}Pr^{1/3}$$

Note that the limiting Nu number (for no flow) is 2, as predicted by pure conduction considerations. The Reynolds and Prandtl exponents are consistent with the scaling considerations. For a cylinder the following correlation suggested by Whitaker (1972) is often used:

$$Nu_m = \left(0.4Re^{1/2} + 0.06Re^{2/3}\right)Pr^{0.4}(\mu_\infty/\mu_s)^{1/4}$$

where Nu_m is the average Nusselt number The last term is the viscosity correction factor to account for the effect of viscosity variation on the heat transfer coefficient.

For packed beds the j-factor-based correlations are found to be useful. A correlation proposed by Stewart is

$$j_H = 2.19Re^{-2/3} + 0.79Re^{-0.381}$$

where the j_H factor is defined as

$$j_H = \frac{h}{c_p G}Pr^{2/3} = StPr^{2/3}$$

and Re is the Reynolds number defined as $d_p G/[(1 - \epsilon_B)\mu\psi]$ and St is the Stanton number. Here d_p is the equivalent particle diameter, G is the superficial mass velocity, ϵ_B is the bed porosity, and ψ is a particle-shape factor, which is taken as one for spherical particles and 0.92 for cylindrical pellets. The particle diameter is defined as $6(1 - \epsilon_B)/a_s$ for non-spherical particles, where a_s is the external surface area per unit particle volume.

Conjugated problems

If the sphere has some internal heat generation then the heat transfer problem has to be solved as a conjugated problem. Conjugated problems are generally defined as problems containing two or more subdomains with models described by different differential equations for each subdomain. A detailed review of conjugate problems in convective heat transfer is provided by Dorfman and Renner (2009). A sphere with internal generation is one such a problem and is obviously an important problem in the context of reacting particles. For example, a zeroth-order reaction leads to the Frank-Kamenetskii problem for the sphere and the convection–conduction problem for the external fluid. The temperature and the heat fluxes at the common boundary (the surface of the sphere) are to be matched, leading to a conjugated problem.

Suitably incorporating matching conditions into existing CFD codes can be a challenge, and presents an interesting case-study problem. Many novel computational schemes can be thought of. For example, the dual-reciprocity boundary element (see the book by Ramachandran (1993)) is useful for the solid sphere part while the finite-volume method of Patankar (1980) can be used for the external flow part of the problem. The advantage of the dual-reciprocity method is that it provides direct values for the fluxes at the solid surface, which can then be used iteratively to calculate the fluid-phase temperature distribution. This in turn is to be used for the solid problem for the next cycle of iteration.

The effect of natural convection can also become important under some of these conditions and lead to instability problems. In general natural convection is not dominant under forced-convection conditions. But the angular variation of the temperature due to internal generation as a function of θ (the angle from the stagnation point in this context) leads to a buoyancy-driven flow that can assist (for upflow) or oppose (for downflow) the forced-convection heat transfer. Problems of this type are of importance in the analysis of the stability of reacting systems in packed beds.

18.4 Natural convection on a vertical plate

This represents again a coupled problem involving momentum and heat transfer. Also it is an interesting example where the integral-balance technique finds another useful application. The problem was indicated in Chapter 1 as an example of a coupled transport problem and is restated below for ease of reading.

Consider a vertical heated (or cooled) wall exposed to an ambient at a temperature of T_∞. Since the wall is hotter (or colder) than the ambient, the density of the fluid near the wall is different from that of the ambient. This creates an upward buoyancy force (for a heated wall), which drives a flow near the wall. This flow in turn accelerates the rate of convective heat transfer from the wall to the surroundings. Obviously this is a coupled problem involving the simultaneous solution of the momentum and heat balance. The system is shown schematically in Fig. 1.1 in Chapter 1.

Governing equations

The x-momentum equation can be represented as

$$\rho\left(v_x\frac{\partial v_x}{\partial x} + v_y\frac{\partial v_x}{\partial y}\right) = -\frac{\partial p}{\partial x} - \rho g + \mu\frac{\partial^2 v_x}{\partial y^2} \tag{18.42}$$

Note that here the viscous transport in the x-direction is neglected compared with that in the y-direction, i.e.,

$$\frac{\partial^2 v_x}{\partial x^2} \ll \frac{\partial^2 v_x}{\partial y^2}$$

and the N–S equation has been simplified to some extent. This is a reasonable assumption, which can be justified from the earlier analysis based on the scaling arguments. Also note the sign on the ρg term. The x-direction is taken as vertically upwards and hence this term appears with a negative sign. The y-direction is taken normal to the wall and measured from the wall.

Far away from the plate the equation for hydrostatics holds:

$$-\frac{\partial p}{\partial x} = \rho_\infty g \tag{18.43}$$

where ρ_∞ is the density evaluated at the bulk-fluid temperature. The pressure distribution is the same irrespective of whether it is measured near the plate or far away from it. Hence the term containing the pressure can be eliminated from (18.42), leading to

$$\rho\left(v_x\frac{\partial v_x}{\partial x} + v_y\frac{\partial v_x}{\partial y}\right) = g(\rho_\infty - \rho) + \mu\frac{\partial^2 v_x}{\partial y^2} \tag{18.44}$$

The next step is to express the local density ρ as a function of a local temperature. This can be done by using a Taylor-series expansion for ρ. Let us use a mean temperature $T_f = (T_s + T_\infty)/2$ around which we expand the density:

$$\rho = \rho_f + \left(\frac{\partial \rho}{\partial T}\right)_f (T - T_f) \tag{18.45}$$

Higher-order terms are neglected, which is known as the Boussinesq approximation.

The term $\partial\rho/\partial T$ at constant pressure is related to the coefficient of volume expansion β, which is defined in general as

$$\beta = -\frac{1}{\rho}\frac{\partial\rho}{\partial T} \tag{18.46}$$

Here β will be evaluated at the mean film temperature since the Taylor series is being applied around this point. Hence Eq. (18.45) reduces to

$$\rho = \rho_f - \rho_f\beta_f(T - T_f) \tag{18.47}$$

Similarly ρ_∞ can be expressed as

$$\rho_\infty = \rho_f - \rho_f\beta_f(T_\infty - T_f) \tag{18.48}$$

Hence the density difference term in Eq. (18.44) can be written as

$$g(\rho_\infty - \rho) = g\rho_f\beta_f(T - T_\infty) \tag{18.49}$$

The density term on the LHS of Eq. (18.44) is treated as a constant and also evaluated at the mean film temperature. The same holds for the viscosity value as well. In other words, all properties not associated with the buoyancy effects are evaluated at the mean film temperature and treated as constant. This is also a part of the Boussinesq approximation.

Hence Eq. (18.44) can be expressed as

$$v_x \frac{\partial v_x}{\partial x} + v_y \frac{\partial v_x}{\partial y} = g\beta(T - T_\infty) + v \frac{\partial^2 v_x}{\partial y^2} \tag{18.50}$$

The equation is similar to that considered earlier for a flat plate under forced convection except for the presence of the $g\beta(T - T_\infty)$ term which accounts for the effect of buoyancy due to temperature variation in the boundary layer. This is the term which couples the momentum and heat balance together. The heat balance equation remains the same and is reproduced here for convenience:

$$v_x \frac{\partial T}{\partial x} + v_y \frac{\partial T}{\partial y} = \alpha \frac{\partial^2 T}{\partial y^2}$$

Finally, the continuity equation is also used to complete the set of equations. The boundary conditions that accompany these equations can be easily stated and are not shown here.

Integral balance equations

Since the integral methods have worked well for the forced-convection heat transfer problem, let us apply them to this coupled problem of natural convection. The integral momentum balance can be derived by integrating the momentum balance from 0 to δ. This will have the same form as before except for the additional term due to buoyancy effects. Also the bulk stream velocity is zero. Hence the integral-balance equation takes the following form:

$$\frac{d}{dx}\left(\int_0^\delta v_x^2 \, dy \right) = -\frac{\tau_w}{\rho} + \int_0^\delta g\beta(T - T_\infty) dy \tag{18.51}$$

The heat integral remains the same and is reproduced below for completeness:

$$\frac{d}{dx}\left[\int_0^{\delta_t} v_x(T - T_\infty) dy \right] = \frac{q_w}{\rho c_p} \tag{18.52}$$

Approximations for the velocity and temperature are needed. The temperature approximation is taken to be the same as before:

$$\frac{T - T_s}{T_\infty - T_s} = \frac{3}{2}\frac{y}{\delta_t} - \frac{1}{2}\left(\frac{y}{\delta_t}\right)^3 \tag{18.53}$$

The velocity expression needs to be redone since some of the boundary conditions are different. One assumes a cubic for the velocity as usual:

$$v_x = a + by + cy^2 + dy^3$$

Four conditions are to be imposed to find the four constants a to d. These are as follows:

(i) At $y = 0$, the velocity $v_x = 0$ due to the no-slip condition.
(ii) At $y = \delta$, the velocity $v_x = 0$ due to there being no motion in the bulk.
(iii) At $y = \delta$, the velocity gradient $\partial v_x/\partial y = 0$ due to there being no stress at the edge of the boundary layer.

(iv) At $y = 0$, the second derivative of v_x can be obtained from the governing differential equation after setting v_x and y_y to zero. This gives

$$\frac{\partial^2 v_x}{\partial y^2} = -\frac{g\beta(T_s - T_\infty)}{\nu}$$

at $y = 0$.

Using these conditions, the velocity profile can be expressed as

$$\frac{v_x}{v_c} = \frac{1}{4}\left[\frac{y}{\delta} - 2\left(\frac{y}{\delta}\right)^2 + \left(\frac{y}{\delta}\right)^3\right] \qquad (18.54)$$

where v_c is a reference velocity defined as

$$v_c = g\beta(T_s - T_\infty)\delta^2/\nu \qquad (18.55)$$

Note that v_c is a function of x, since δ is a function of x. The maximum velocity at any x location can be shown to be equal to $0.148v_c$, which occurs at $y = \delta/3$. Also the heat flux at the wall and the Nusselt number are related to the thickness of the thermal boundary layer as before:

$$Nu_x = -\frac{q_w x}{k(T_s - T_\infty)} = \frac{3}{2}\frac{x/\delta}{\Delta}$$

Substitution of the velocity and temperature profile leads to equations for the variation of δ and δ_t with distance. Note that these equations are coupled and have to be solved simultaneously. In the previous case of forced convection, the momentum thickness could be solved separately and used in the heat integral. It is not possible to do so unless some simplification is made.

The one-parameter model

A common simplification is to assume that $\delta = \delta_t$, i.e., the two boundary layers are equal in thickness. This is called a one-parameter model. This seems to be a reasonable assumption since the temperature difference drives the flow. Hence in regions where there are no significant temperature differences there would be no flow as well. Hence the two thicknesses are nearly equal.

The solution proceeds as before by substitution of the profiles, evaluating the integrals, and then solving for δ or δ_t. They are the same here! Once δ is known the velocity and temperature profiles are known, and the heat transport rate can also be calculated.

For the above one-parameter model ($\delta = \delta_t$), the solution is not intimidating. However, the details are left out here. It can be shown that

$$\delta^4 = \frac{4}{3}\frac{315}{2}\frac{\alpha\nu x}{g\beta(T_s - T_\infty)}$$

which can be expressed in dimensionless groups by defining a Grashof number. This is similar to a Reynolds number based on the characteristic velocity for natural convection:

$$Gr_x = v_c x/\nu$$

Using the expression for v_c, the local Grashof number is defined as

$$Gr_x = \frac{g\beta(T_s - T_\infty)x^3}{\nu^2}$$

Also note that $Pr = \nu/\alpha$. Using these results, the thickness of the boundary layer can be expressed in dimensionless form as

$$\frac{\delta}{x} = \frac{3.81}{(Gr_x Pr)^{1/4}} = \frac{3.81}{Ra_x^{1/4}}$$

where the product of the Grashof and Prandtl numbers is known as the Rayleigh number. Hence the Rayleigh number is the primary characteristic parameter for the natural convection process. It may be interpreted as the ratio of the buoyancy forces to the viscous forces.

The expression for the local Nusselt number can then be readily obtained from the equation shown earlier with Δ of one for the the the one-parameter model:

$$Nu_x = \frac{3}{2} \frac{x/\delta}{\Delta} = 0.394 Ra_x^{1/4}$$

The average Nusselt number for a plate of length L can also be calculated.

The accuracy of the integral method has been examined by comparison with the numerical solutions. It is seen that the one-parameter model gives results within 5% of the numerical solution for Pr between 0.6 and 1.4 but degenerates for values of Pr outside this range. The two-parameter model which satisfies both the momentum and the energy is needed for these cases. For the two-parameter model, the Rayleigh number alone is not sufficient to correlate the heat transfer data (Nusselt number). Both the Grashof number and Prandtl number are needed as two separate dimensionless groups. Empirical correlations in terms of these dimensionless groups are also available in the literature.

18.4.1   Natural convection: computations

In this section we illustrate useful computational tools for natural convection heat transfer. For illustration, we take a 2D enclosure that is closed at top and bottom and the two side walls at two different temperatures. This sets up a natural convection-driven flow with the fluid circulating in the enclosure. The temperature and velocity profiles have to be computed together. The vorticity-streamfunction method was used to solve the problem by Chow(1979) together with the equation for the temperature field, and we discuss the rudiments of the method here. The details are left as an as exercise. The advantage of the vorticity-streamfunction method as discussed in Section 5.6.1 is that the pressure field is no longer needed. The continuity is automatically satisfied. This is of special importance in natural convection flows since the system can develop multiple circulation cells that are difficult to solve without fine meshes when solving with the primitive variables (velocity and pressure).

The computational procedure using this method is as follows.

1. Assume a vorticity distribution ω at time zero consistent with the boundary conditions.
2. Solve the Poisson equation $\nabla^2 \psi = -\omega$ to find the streamfunction profiles.
3. Use the values of the streamfunction to find v_x and v_y.
4. Update the temperature profile and vorticity profile for time $t = t + \delta t$.
5. Go back to step 1 and keep marching in time until a steady-state or time-periodic solution is obtained.

Detailed numerical results based on this computation have been given in the book by Glasgow (2010) and illustrative plots have been presented in their book. You may want to set up a code in MATLAB and benchmark your results with the work of Glasgow.

18.5 Boiling systems

18.5.1 Pool boiling

This refers to a case where a pool of liquid is heated from below from a horizontal surface. Consider a case of uniform heat flux applied at the heating surface. The bulk of the liquid is assumed to be at a uniform temperature. We can define ΔT as the temperature difference between the surface and the bulk. The heat flux vs. ΔT curve is called the characteristic curve for pool boiling. A typical curve as shown in Fig. 18.4 has several identifying features and various flow regimes.

18.5.2 Nucleate boiling

This is characterized by the formation of vapor bubbles or nuclei at some points on the heating surface. These break off and condense before reaching the free surface if the temperature difference is small, leading to a very low heat flux. Further increase in ΔT causes the bubbles to reach the surface and make the liquid circulate, causing an increase in the

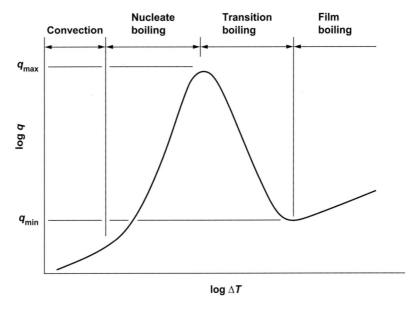

Figure 18.4 A schematic representation of the flow regimes in pool boiling. The characteristic is illustrated by a q vs. ΔT plot

heat flux. Dimensionless analysis indicates that the data in the nucleate boiling regime can be correlated as

$$Nu_b = F(Ja, Pr_l)$$

where Nu_b is the Nusselt number for boiling, Ja is a Jacob number defined as

$$Ja = \frac{c_{pl} \Delta T}{\hat{h}_{lg}}$$

where c_{pl} is the specific heat and $\hat{h}_{lg}$ is the heat needed for vaporization per unit kg mass (with subscripts l for liquid and g for gas). The Jacob number, Ja, is therefore the ratio of sensible heat to latent heat. Finally, Pr_l is the Prandtl number based on liquid properties.

The Nusselt number for boiling indicated above is defined as

$$Nu_b = hL_c/k_l$$

where k_l is the thermal conductivity of the liquid. The characteristic length, L_c, in the definition of the Nusselt number is based on d_b, the maximum bubble diameter as a bubble leaves the surface. The maximum bubble diameter in turn is given by a balance of surface-tension forces and the buoyancy forces:

$$\sigma[\pi d_b(\max)] = (\rho_l - \rho_v)g(\pi d_b^3/6)$$

Hence

$$d_b(\max) = \sqrt{6\frac{\sigma}{g(\rho_l - \rho_v)}} \qquad (18.56)$$

The length scale used to define the dimensionless groups is taken as proportional to the above quantity (without the factor of 6 in the square root for convenience)

$$L_c = \sqrt{\frac{\sigma}{g(\rho_l - \rho_v)}}$$

The correlation suggested by Rohsenow (1952) is commonly used for nucleate boiling:

$$Nu_b = \frac{Ja^2}{c_{nb}^3 Pr^2} \qquad (18.57)$$

where c_{nb} is a constant that depends on the heating surface material and the liquid being boiled. A value of 0.013 is commonly used for copper and stainless steel for boiling of water. Values for various combinations of liquid and surface have been published by Rohsenow and Choi (1961) and are also quoted in many books such as Welty $et\ al.$ (2008). Also the exponent of 2 on the Prandtl number applies for water. For organic liquids the exponent is 4.1 and the constant c_{nb} is of the order of 0.0027 as indicated in Mills (1993). Also the equation provides only order-of-magnitude values, and errors of the order of 100% are common. Nucleate boiling appears to be a difficult problem for theoretical analysis, as pointed out by Dhir (2006). Also see the discussion at the end of the subsection on multiscale efforts to address this problem.

Maximum heat flux

Beyond a certain ΔT, a vapor film collects near the heating surface, forming an insulating layer that now causes the heat flux to decrease with further increases in ΔT as shown in Fig. 18.4. This is known as the transition region. Operation in the transition boiling region is normally avoided in the design of heat transfer equipment and hence the work on developing correlations for engineering analysis in this regime is limited.

The maximum heat flux is an important parameter, since there is a danger of heater burn-out at this point. The correlation developed independently by Kutateladze (1948) in the USSR and Zuber (1958) in the USA is often used for the estimation of this quantity:

$$q_{max} = C_{cr}\hat{h}_{lg}[\sigma g \rho_v^2(\rho_l - \rho_v)]^{1/4} \tag{18.58}$$

where C_{cr} is a constant that depends on the heater geometry. Typical values of this constant are in the range 0.12–0.15.

Minimum flux and the Leidenfrost point

The minimum heat flux can be predicted from a semi-theoretical model developed by Zuber:

$$q_{min} = 0.09\rho_v\hat{h}_{lg}\left[\frac{\sigma g(\rho_l - \rho_v)}{(\rho_l + \rho_v)^2}\right]^{1/4} \tag{18.59}$$

The condition under which the minimum heat flux occurs is referred to as the Leidenfrost point. The Leidenfrost effect is a phenomenon whereby a liquid, in near contact with a surface, which is kept significantly hotter than the boiling point, produces an insulating vapor layer that keeps that liquid from boiling rapidly. This is most commonly seen when cooking; if one sprinkles drops of water in a skillet to gauge its temperature the water may skitter across the metal, and takes longer to evaporate if the temperature of the skillet is at the Leidenfrost point. Mechanistic models are available to predict the onset of this point.

Film boiling

For film boiling over a vertical surface, the thickness of the vapor film can be shown to be

$$\delta(x) = \left[\frac{4k_v \Delta T \mu_v x}{3\rho_v(\rho_l - \rho_v)g\hat{h}_{lg}}\right]^{1/4} \tag{18.60}$$

(See Problem 15) and correspondingly the local heat transfer coefficient is equal to

$$h = \frac{k_v}{\delta(x)} \tag{18.61}$$

The average heat transfer coefficient for a surface of length L can be found by integration of the local value with respect to x. These details are left as exercises.

Radiation effects

Further increase in ΔT after the minimum point causes the radiation effects across the vapor film to play a role, and the heat flux now starts to increase with an increase in ΔT once again.

Boiling as an example of multiscale modeling

Boiling provides an interesting example where a number of important transport effects are occurring significantly: wall heat transfer, nucleation, bubble-growth, bubble detachment, bubble rise, bubble-induced agitation of the liquid, etc. Hence this topic also provides an elegant illustration of multiscale modeling. The models at various hierarchical levels can be developed and coupled as shown in Fig. 18.5. This is adapted from the work by Dhir

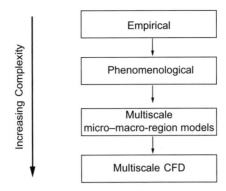

Figure 18.5 Hierarchical levels at which nucleate boiling phenomena can be modeled. The macro-region is the pool of liquid while the micro-region consists of local active cavities on the heating surface, which are responsible for bubble nucleation.

(2006), whose paper provides an interesting study of this topic. A detailed study of this topic has significant educational value since such an approach is not restricted to boiling heat transfer and can be used in many applications.

Example 18.1.

Determine the heat flux when water boils at 1 atm in a polished copper pot kept at 390 K. Also find the maximum heat flux.

Solution.

The physical properties needed are as follows: $c_p = 4211$, $\mu = 278 \times 10^{-6}$, $k = 0.682$, and $Pr = 1.72$ at 373 K, all in SI units.

The surface tension of water as a function of temperature is needed, and for water the following equation is useful:

$$\sigma = 0.1232(1 - 0.00\,146T)$$

The value for water at 373 K is 0.0561 N/m.

The density of steam is calculated from the ideal-gas law as 0.588 kg/m^3.

The Jacob number calculated from the definition given in the text is 0.0317. Now, using the correlation of Rohsenow, we find $Nu = 154$.

The characteristic length scale is proportional to the bubble diameter and is defined by Eq. (18.56). Using this, we find $L_c = 0.0024$ m and the bubbles leaving the surface have diameters in the range of 2.4 mm. Using this value for L_c the heat transfer coefficient is $h = Nu k_l/L_c = 4.38 \times 10^4$ W/m$^2 \cdot$ K. The heat flux is calculated as $h\,\Delta T = 7.34 \times 10^5$ W/m^2.

The maximum heat flux is also of interest and can be calculated from the correlation of Zuber. The value is $q_{max} = 1.24 \times 10^6$ W/m^2 using Eq. (18.58).

The Rohsenow correlation indicates that q is proportional to ΔT to the power of 3. Hence, using the correlation in a backward manner, we find that ΔT is 21 K at this point, or the wall temperature is 394 K at the point of maximum heat flux.

18.6 Condensation problems

Dropwise

Condensation can be of two types: (i) dropwise and (ii) filmwise. Dropwise condensation refers to a process whereby the condensed vapor forms drops on the surface rather than a continuous film. Since a large portion of the surface is exposed to the gas, the condensation rates are rather high. Heat transfer coefficients as high as 250 000 W/m^2 can be achieved. However, two difficulties in using this effectively are as follows: (i) it is difficult to maintain this for a sustained period of time and (ii) the resistance on the coolant side is often the controlling resistance and the heat transfer rate is then determined by how fast the heat can be removed on the coolant side. The high value of the heat transfer coefficient from vapor to drop no longer matters!

Filmwise

In filmwise condensation the surface is blanketed by a layer of liquid film. This liquid film provides a resistance to heat transfer, and therefore the heat transfer coefficients are much lower than for dropwise condensation. The film thickness also increases as a function of the height or the distance along the condensing surface. Hence the heat transfer coefficient decreases with increasing height. In design applications filmwise condensation is usually encountered due to difficulties in maintaining dropwise condensation for prolonged periods of operation. The remainder of the discussion in this section pertains to filmwise condensation, and dropwise condensation is not discussed further.

Flow regimes

It is useful to characterize the various flow patterns at this stage in order to give the student a clearer understanding of the ranges of application of the various correlations and/or models which are commonly used in practice. The flow regimes are (i) laminar, (ii) wavy, and (iii) turbulent. The transition from laminar to wavy occurs around a Reynolds number (defined for film flow as $Re = 4\langle v\rangle\delta/\nu$) of 30, whereas the flow is turbulent beyond Re of 1500.

Laminar film: the Nusselt model

The problem of laminar condensation was analyzed by Nusselt (1916). The schematic basis of the model is shown in Fig. 18.6, where a condensate film of thickness δ is assumed to form on the condensing wall. The film thickness increases as a function of x, the vertical distance along the wall.

The local heat transfer coefficient depends on the local film thickness and can be calculated as $h = k_1/\delta$. The film thickness can be calculated using the mass and momentum balance if the operating conditions are in laminar mode, giving

$$\delta(x) = \left[\frac{4\mu_1 k_1(T_{\text{sat}} - T_{\text{w}})x}{g\rho_1(\rho_1 - \rho_{\text{v}})\hat{h}_{\text{lg}}}\right]^{1/4} \tag{18.62}$$

The model is called the Nusselt model for condensation. The details leading to the above expression are left as an exercise (Problem 15). The vapor velocity is assumed to be small in the Nusselt analysis, and the effect of the vapor velocity is neglected.

What is the effect of the vapor velocity on the rate of condensation? Discuss qualitatively.

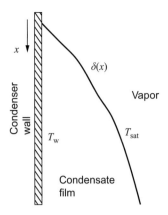

Figure 18.6 A schematic representation for the model used in Nusselt analysis for filmwise condensation in a laminar falling film.

Correction for subcooling and superheating

The Nusselt model described above does not account for the subcooling of the liquid. A correction described below is usually applied. The latent-heat-of-vaporization term, $\hat{h}_{fg}$, in Eq. (18.62) is usually modified to include the liquid subcooling and is replaced by $\hat{h}_{fg}^*$, which is defined as

$$\hat{h}_{fg}^* = h_{fg} + 0.68c_{pl}(T_{sat} - T_w)$$

If the vapor enters as a superheated fluid at a temperature of T_v instead of at T_{sat}, a further correction to the latent heat is applied:

$$\hat{h}_{fg}^* = \hat{h}_{fg} + 0.68c_{pl}(T_{sat} - T_w) + c_{pv}(T_v - T_{Sat})$$

The wavy regime

The effect of waves on the surface, which occurs for $Re > 30$, is to increase the rate of condensation. However, the analysis is complicated, and analytical solutions are not possible, unlike in the case of laminar flow. The increase in heat transfer is of the order of 20%, but can be as high as 50%. The laminar equation can still be used for a conservative design since it will underpredict the heat transfer coefficient. A correction factor of $0.8\,Re^{0.11}$ was recommended by McAdams (1957).

The turbulent regime

Several empirical correlations are proposed for the turbulent regime and can be used in the absence of a detailed CFD analysis. A simple correlation proposed by McAdams (1957) is often used:

$$h = 0.0077 \left[\frac{\rho_l g(\rho_l - \rho_g)k_l^3}{\mu_l^2} \right]^{1/3} Re^{0.4}$$

Phase-change problems

We analyze a simple phase-change problem shown schematically in Fig. 18.7. A liquid at an initial temperature of T_L is brought in contact with a cold surface at a temperature of T_S. The surface temperature is below the freezing point T_f of the liquid. A frozen layer of thickness λ develops, and the frozen front penetrates into the liquid as time progresses. A model for the location of the freezing front as a function of time will be developed.

The governing equations are the transient heat conduction equations for each phase (for the liquid from λ to L and for the solid from 0 to λ), ignoring any convection effects in the liquid. Finally we need an additional equation to locate the freezing front, λ. This changes as a function of time, and hence problems of this type are referred to as moving-boundary problems. A jump heat-balance condition is used here. The heat entering the liquid phase at the interface must be equal to the heat arriving from the solid plus the heat needed to freeze the solid. Thus

$$-k_L \frac{dT_L}{dx} = -k_S \frac{dT_S}{dx} + h_{ls}\dot{m} \text{ at } x = \lambda$$

where h_{ls} is the latent heat for melting per unit mass and $\dot{m}$ is the mass rate of freezing. The mass of solid is $\rho_s\lambda$ per unit surface area. Hence $\dot{m}$ is equal to $\rho_s(d\lambda/dt)$. Thus the freezing condition is

$$\rho_s h_{ls} \frac{d\lambda}{dt} = k_S \frac{dT_S}{dx} - k_L \frac{dT_L}{dx}$$

The pseudo-steady-state model

The solution is often done by assuming a pseudo-steady-state profile for temperature. These profiles are sketched in Fig. 18.7 as dotted lines. The assumption is that the temperature profiles are linear in each of the phases. The equation for the freezing-front location is then

$$\rho_s h_{ls} \frac{d\lambda}{dt} = k_S \frac{T_f - T_S}{\lambda} - k_L \frac{T_L - T_f}{L - \lambda} \tag{18.63}$$

This can be readily integrated by rearranging the variables to find λ as a function of time. The details are left as an exercise (Problem 20).

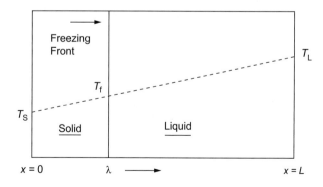

Figure 18.7 A simple example of a phase-change problem, namely solidification of a liquid.

A moving front: liquid at freezing point

A special case occurs when the superheating in the liquid is neglected. The entire liquid is assumed to be at the freezing temperature, and the temperature profile just in the solid is considered. This will be the case, for example, where a freezing front develops over a large body of water that is assumed to be already at its freezing point. Equation (18.63) simplifies for this case and the following expression for the movement of the frozen front is then obtained upon integration:

$$\lambda = \sqrt{\frac{2k_S(T_f - T_S)}{\rho_s h_{ls}}}\sqrt{t}$$

A square-root dependence on time is seen, which is a key feature in many moving-boundary problems. The dimensionless version of this equation is useful, and is given by

$$\frac{\lambda}{\sqrt{\alpha t}} = \sqrt{2Ja}$$

where Ja is the Jacob number defined as $c_{ps}(T_f - T_S)/h_{ls}$ and α is the thermal diffusivity of the solid. This is the model for the moving front based on pseudo-steady-state considerations and assuming that there are no transients in the liquid, i.e., the liquid is already at its freezing point. Let us now examine the conditions for which this pseudo-steady-state model can be used.

The validity of the pseudo-steady-state approximation

Let us assume that the liquid is already near the freezing point. This permits us to concentrate exclusively on the temperature profiles in the solid. The governing equation is the transient conduction equation in the solid. (You should review and restate the equation at this point.) The dimensionless temperature θ defined as $(T - T_S)/(T_f - T_S)$ is also useful. From the earlier analysis for transient problems in Chapter 11 we know that an error-function solution satisfies the transient conduction problem, so let us assume such a solution. Thus let

$$\theta = A\, \mathrm{erf}\left(\frac{x}{2\sqrt{\alpha t}}\right)$$

be used to describe the temperature distribution in the solid, with A a constant to be evaluated. We already know that this satisfies the governing PDE from our earlier experience with transient conduction in a semi-infinite slab. Let us see whether we can fit the boundary conditions with the above equation.

The initial dimensionless temperature is one and the error function for t of zero is $\mathrm{erf}(\infty) = 1$, hence this condition is also satisfied.

At $x = 0$, the $\mathrm{erf}(0) = 0$, and hence $\theta = 0$ and this boundary condition is also satisfied.

Furthermore, at $x = \lambda$ the temperature is equal to one. Hence we should impose the following condition:

$$1 = A\, \mathrm{erf}\left(\frac{\lambda}{2\sqrt{\alpha t}}\right) \tag{18.64}$$

which should hold for any value of time. This is possible only if the argument under the error function is a constant. Hence we require

$$\lambda = 2B\sqrt{\alpha t}$$

where B is a constant. The moving front moves with a square-root dependence on time. Using this condition in (18.64), we find $A = 1/\text{erf}(B)$, and the solution should be of the form

$$\theta = \frac{\text{erf}[x/(2\sqrt{\alpha t})]}{\text{erf}(B)} \tag{18.65}$$

Now, to find B we use the interfacial heat balance. For the case where the liquid is already at its freezing point the interfacial heat balance reduces to

$$\rho_s h_{ls} \frac{\partial \lambda}{\partial t} = k_S \frac{\partial T_S}{\partial x} = k_S (T_f - T_S) \left(\frac{\partial \theta}{\partial x} \right)_{x=\lambda} \tag{18.66}$$

In order to use this we need the derivative of the error function. The standard result $(d/dx)(\text{erf}(x)) = \sqrt{(2/\pi)} \exp(-x^2)$ can be used to evaluate the derivative of θ in Eq. (18.66), and we used this in the interfacial heat balance. The resulting equation, after some algebra, is

$$\sqrt{\pi} B \exp(B^2) \text{erf}(B) = \frac{c_p(T_f - T_S)}{h_{ls}}$$

The dimensionless group on the RHS is seen to be the Jacob number. Hence

$$\boxed{\sqrt{\pi} B \exp(B^2) \text{erf}(B) = Ja} \tag{18.67}$$

which is the required condition to find the constant in the equation for the motion of the freezing front with time. The equation can be solved for B by iteration for a given value of Ja. Once B has been found a relation for the movement of the front is established as a function of time, and the corresponding temperature profiles can be also calculated. (It may be also noted that the Jacob number is known as the Stefan number in phase-change studies.) For small values of Ja the LHS term becomes equal to $2B^2$ and hence $B = \sqrt{Ja/2}$. The solution for the moving front now reduces to

$$\lambda/\sqrt{\alpha t} = \sqrt{2Ja}$$

which is the solution obtained from the pseudo-steady-state model described earlier. Hence the pseudo-steady-state model holds for low values of Ja.

When ice forms over water with a temperature difference of $T_f - T_S$ of 10 K, the value of the Jacob number is 0.058, and the pseudo-steady-state model should hold for such conditions. The corresponding temperature profile is linear. However, the transient analysis is more general and also provides us with the required condition for which a pseudo-steady-state solution can be expected.

Transient analysis of the case where the subcooling of the liquid must also be considered proceeds in a similar line. We have to assume an error-function solution for both sides and match the required conditions. The algebra becomes lengthy, however, and the details are left as an exercise.

Historical vignette. The first work in this area of heat transfer accompanied by a phase change was done by Lamé and Clapeyron in 1831. They found that the thickness of the solid crust formed upon solidification was proportional to the square root of time but could not determine the constant of proportionality.

Equation (18.67) had not been derived then! In 1899 Stefan examined the freezing of ground and introduced the jump heat-balance condition to find the location of the freezing front. Hence this problem is also known as the Stefan problem. Neumann solved the problem and derived Eq. (18.67) and this solution is called the Neumann solution to the classical Stefan problem. Subsequently the problem and its variants have been studied extensively both by engineers and by mathematicians and a notable monograph on this subject is the book by Rubenstein (1971).

Summary

- Heat transfer in boundary layers in laminar flow over a flat plate is a well-studied problem in convective heat transfer and is amenable to a complete theoretical analysis. It can also be approached by both scaling and integral analysis.

- The Prandtl number is the key dimensionless group which determines the relative thickness of the momentum and thermal boundary layers. The numerical value for this number can vary over a wide range, from very low values for liquid metals to very high values for hydrocarbon liquids. The thermal boundary layer is thicker for low-Prandtl-number fluids and *vice versa*.

- The modeling is not so clear cut for turbulent flow. In addition to an adequate model for the velocity profile, one needs an appropriate model for eddy conductivity, or equivalently, the turbulence Prandlt number. A constant value of 0.9 is often used for this parameter as a first estimate. Since the hydrodynamic boundary layer is much thicker than in laminar flow for $Pr > 1$, the thermal boundary layer may be confined to near the laminar sublayer and the buffer zone. A detailed description of velocity profiles in these zones is therefore needed in order to model heat transfer under these conditions (high-Pr fluids). We analyzed the temperature distribution using the Prandtl model for the velocity profiles and also showed the basis for the various analogies for momentum and heat transfer proposed in the literature.

- Natural convection heat transfer over simple geometries, such as a vertical plate, can be analysed by an integral heat balance approach. This is a coupled problem, and momentum and heat transfer boundary layers are solved simultaneously. A simplifying assumption is that the thicknesses of the two boundary layers are the same (the one-parameter model). This leads to simple correlations for the heat transfer coefficients.

- It is seen that the Grashof number is the key parameter in natural convection heat transfer. This replaces the Reynolds number used in forced-convection flow. The natural convection flow becomes unstable as the Grashof number is increased. For a vertical flat plate the flow becomes unstable if $Gr > 10^9$ and turbulent if $Gr > 10^{12}$.

- Phase-change problems, such as boiling and condensation, are modeled by a combination of basic theory and empirical knowledge. Empirical correlations are available for quick calculations, and more complex mechanism-based models are also being developed. These models are illustrative of the multiscale approach used in general for the analysis of transport processes.

- Condensation can be classified as dropwise and filmwise. Heat transfer coefficients are an order of magnitude higher in dropwise condensation. However, industrial applications usually operate in the filmwise region since the dropwise mode is difficult to maintain.

- The liquid flow pattern in filmwise condensation can be classified into laminar, wavy, and turbulent regimes depending on the range of the Reynolds number. For laminar flow the theoretical analysis of Nusselt is useful as a means by which to predict the heat transfer coefficient. For wavy and turbulent flow, empirical correlations or CFD models are useful.

- Freezing of a liquid is an example of a moving-boundary problem where a jump heat-balance condition is needed in order to locate and track the freezing front. The classical problem in 1D was proposed by Stefan, and the analytical solution to this problem is a landmark in this field.

- You should review the learning outcomes stated at the beginning of the chapter and see to what extent you have accomplished them. With this background you will be able to set up and solve a wide range of problems in convective heat transfer. You will also be able to apply these to solve more complex problems, some of which are discussed in the case-study examples.

ADDITIONAL READING

The book by Bejan (2004) is completely dedicated to convective heat transfer, and should provide additional information on natural and forced convection as well as on heat transfer in turbulent flow. The book by Whitaker (1977) is another useful text devoted to fundamentals of heat transfer, including the effects of convection.

Boiling and condensation problems have been discussed extensively in the edited handbook by Kandlikar, Shoji, and Dhir (1999). This book provides detailed models and design correlations and should be consulted for further studies in this field.

The mathematical modeling of phase-change problems has been discussed in detail in the book by Alexiades and Solomon (1993) and also in the book by Crank (1984). These books serve as "single-source" references on this topic which has applications in many fields, including cryosurgery, thermal energy storage, metal processing, etc.

Problems

1. Show all the steps leading to the integral balance equation (18.6) in the text.
   Use the following boundary conditions and verify Eq. (18.7). (i) At $y = 0$, $T = T_S$. (ii) At $y = \delta_t$, $T = T_0$. (iii) At $y = \delta_t$, $\partial T/\partial y = 0$. (iv) At $y = 0$, $\partial^2 T/\partial y^2 = 0$.
   What is the rationale for the fourth condition above? Using the cubic profile, derive Eq. (18.9).

2. For fluids with $Pr < 1$ the velocity profile is assumed to be a cubic for $y < \delta$ and a constant equal to 1 for $\delta < y < \delta_t$. Using this type of profile, show that the following equation can be derived for the ratio of the boundary layers if $\Delta > 1$.

$$\frac{3}{10}\Delta^2 - \frac{3}{10}\Delta + \frac{2}{15} - \frac{3}{140}\frac{1}{\Delta^2} + \frac{1}{180}\frac{1}{\Delta^3} = \frac{37}{315}\frac{1}{Pr}$$

   Evaluate and plot Δ and the corresponding value of Nu for various values of Pr.

3. Consider the flat-plate heat transfer to be solved by the similarity approach. Show all the details leading to Eq. (18.15). What are the boundary conditions on θ? Extend your MATLAB code to solve for both f and θ simultaneously. Test your code for $Pr = 3.0$ (θ' at $y = 0$ should be 0.6860). Calculate the Nusselt number. Compare your answer with the integral solution. Also do further calculations varying Pr as a parameter and fit the slope as a function of the Prandtl number.

4. For the similarity solution, what are the boundary conditions for the constant-wall-flux case? Show that a complete similarity does not exist for this case. Also show the condition for the case where $q_w \propto x^n$. Show that complete similarity exists if $n = -1/2$. To what case does this correspond?

5. Air at 300 K and 1 atm flows along a flat plate at 3 m/s. At a location of 0.3 m from the leading edge, find the thickness of the boundary layer (δ). Calculate the velocity profile at the point at distances of $\delta/2$ and $\delta/5$. Assume a laminar boundary layer. A heat flux of 1 W/m^2 is applied to a wall with air flowing past it. The friction factor is 0.002 and the Prandtl number is 2. Find the temperature at a location 5 mm from the wall if the wall temperature is 300 K.

6. Extend the analysis of Section 18.1.4 to heat transfer over a wedge flow. Derive the following equation for the temperature profile:

$$\theta'' + (m+1)Pr f\theta' = 0 \tag{18.68}$$

where m is related to the wedge angle. Solve for illustrative values of m.

7. Consider heat transfer over a flat plate again but now include an additional term due to viscous heating. Show that the similarity method is applicable to this problem as well, and derive the resulting equation for θ. Solve the problem for various values of Pr and the additional dimensionless group resulting from the viscous heating term. This group is known as the Eckert number,

$$Ec = \frac{v_\infty^2}{c_p T_\infty}$$

What is the physical meaning of this group? Show that $PrEc$ is a dimensionless group similar to the Brinkman number.

8. Integral balances can also be used for heat transfer in a turbulent-flow boundary layer if a form for the velocity profile is assumed. A common form is the 1/7th-power law:

$$\frac{v}{v_\infty} = \left(\frac{y}{\delta}\right)^{1/7}$$

Assume that the temperature is of the same form and also that the thermal and heat boundary layers are of the same thickness. From this derive an expression for Nu_x. How does this compare with the solutions described in the text?

9. A simplified relation for heat transfer in natural convection from a vertical wall is of the form

$$h = A(\Delta T/x)^B$$

where A and B are fitting constants. Determine the values of these constants (a) for air and (b) for water, using the integral balance model given in Section 18.4.

10. A vertical plate 3 m long is at a temperature of 400 K and exposed to air at 300 K. Calculate the thickness of the boundary layer and the value of the local heat transfer coefficient as a function of (the vertical) distance along the plate.
Assume the flow remains laminar all the way.

11. Repeat the calculations for the previous problem if the fluid were water instead of air. In which case do you expect the maximum velocity to be higher?

12. An immersion heater operating at 1000 W is in the form of a rectangular solid with dimensions of 16 cm by 10 cm by 1 cm. Determine the heat transfer coefficient and the surface

temperature of the heater if it is oriented with the 16-cm dimension vertical and immersed in water at 295 K. Assume natural convection heat transfer.

13. Water is boiled in a polished stainless steel pot with a 3 kW heater. The efficiency of the heater is 60%, i.e., only 60% of the heat is transferred to the water. Find the temperature of the inner surface at the bottom of the pot. Also find the temperature of the outer surface if the pot is 6 mm thick.

14. Consider boiling of water under sea-level conditions in a copper vessel. Calculate and plot the heat flux vs. ΔT diagram for water at three different pressures. Show the temperature at the point of q_{max}, the Leidenfrost point, and the burn-out point. Comment on the effect of pressure on the boiling characteristics.

15. **Film boiling: film thickness.** Consider a film of vapor in contact with a liquid. From a heat balance show that the mass flow rate in the vapor, $\dot{m}$, per unit transfer area changes as

$$\hat{h}_{lg}\frac{d\dot{m}}{dx} = \frac{k_v}{\delta}\Delta T$$

where δ is the local film thickness of the vapor film. The film thickness in turn is related to the mass flow rate from momentum-transfer considerations by

$$\dot{m} = \frac{g\delta^3(\rho_l - \rho_v)\rho_v}{3\mu_v}$$

Combine these expressions and derive the formula given in the text for film thickness for the film boiling.

16. Find the temperature at which the surface tension of water becomes zero. What is the physical significance of this temperature?

17. **The Nusselt model for condensation heat transfer.** Consider a vapor condensing on a wall and forming a liquid film. Assume that locally the film thickness is related to the flow rate per unit transfer area by momentum-transfer considerations.

Then show that the change in mass flow rate is related to the film thickness as per the following equation:

$$\frac{d\dot{m}}{dx} = \frac{\rho^2 g\delta^2}{\mu}\frac{d\delta}{dx}$$

From a heat balance show that the change in mass flow rate is given by

$$\hat{h}_{lg}\frac{d\dot{m}}{dx} = \frac{k_l}{\delta}[T_{sat} - T_w]$$

Equate the two expressions for the change in $\dot{m}$ and and derive an expression for $d\delta/dx$. Integrate this expression and derive an expression for δ as a function of x. Verify that the resulting expression is the same as Eq. (18.62). Find an expression for the local heat transfer coefficient and verify that it is proportional to $x^{-1/4}$.

Find an expression for the average heat transfer coefficient for a wall of height L by integration of the local value.

18. Saturated steam at 356 K condenses on a vertical tube of diameter 5 cm whose surface is maintained at 340 K.

Find the height at which the flow becomes wavy. Calculate and plot the film thickness and the heat transfer coefficient up to this height, and also find the average heat transfer coefficient.

At what height would the film become turbulent?

19. Saturated steam at 55 °C is to be condensed at a rate of 10 kg/h on the outside of a vertical tube of diameter 3 cm by maintaining the surface at 45 °C. Find the tube length required for this purpose.

20. Rearrange Eq. (18.63) in dimensionless form. What are the key dimensionless parameters? Integrate the equation to find an expression for λ/L as a function of dimensionless time. Is a steady state reached in this system?

21. A deep pool of water is initially at 4 °C and suddenly the surface is exposed to a freezing front at −4 °C and remains so exposed for a long time.
    Find the velocity of the freezing front as a function of time.
    Determine the temperature profiles in both phases.

22. A common procedure in metallurgy is solidification from a melt in a cylindrical mold. If the Jacob number is small we can use the pseudo-steady-state model, which should now include the conduction in the cylindrical shell of the solidified melt as well as in the mold wall. Develop such a model, and derive an expression for how to completely solidify the melt in terms of the system's physical properties, the radius of the initial melt, and the thickness of the mold.

23. A stainless-steel component is exposed to laser heating at an initial temperature of 300 K. After a short transient the surface reaches its melting point, and the surface recedes at a rate of 230 μm/s thereafter. The molten melt is assumed to be drained away and the solid surface is always exposed to the laser. The melting temperature is 1670 K. Estimate the heat flux at the surface delivered by the laser. Find the initial time needed to reach the melting point. Also find the temperature profile in the solid. The density of stainless steel is 7000 kg/m^3, $c_p = 600$ W/kg/K and $\alpha = 4 \times 10^{-6}$ m^2/s.

24. **Crystal growth by directional solidification: a case-study problem.** A schematic diagram of crystal growth by directional solidification is shown in Fig. 18.8. Analyze the various flow mechanisms and indicate how they affect the heat transfer, the movement of the frozen front, and the shape of the melt–solid interface. Problems of this type are important in production of solar cells.

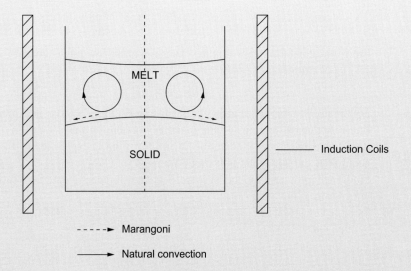

Figure 18.8 A schematic representation of directional solidification of silicon and the important convective transport mechanisms.

25. **Design of a heat pipe: a case-study problem.** A heat pipe is an evaporator–condenser system in which the liquid is returned to the evaporator by capillary action. In the simplest form it consists of a wire-mesh region that serves to act as a wick. Heat pipes can be used in many applications, although the original application was for space vehicles, since in space gravity effects are weak, and hence surface tension was used as the motive force for liquid transport. The applications include solar thermal devices, pipelines over permafrost, etc. A heat pipe also provides an interesting multi-faceted example of application of models for convective heat transfer. A phenomenologically based modeling approach for heat pipes is discussed in the book by Mills (1993). The book by Faghri (1995) is another important reference on this field. Analyze the models for prediction of the performance of heat pipe.

19 Radiation heat transfer

Learning objectives

The main learning objectives of this chapter are listed below.

- To learn the basic definitions and properties of radiative heat transfer.
- To understand the concept of view factors and to be able to calculate radiation exchange between black (non-reflecting) surfaces.
- To study the calculations for radiation exchange between absorbing and diffusively reflecting surfaces.
- To study the modeling of radiation in participating media and to derive the RTE.

Radiation can be viewed as the transfer of energy due to electromagnetic wave propagation. Heat transfer at high temperatures is always accompanied by radiation. This chapter provides the theory needed to analyze these processes and illustrates the theory with sample problems. First we review some basics of electromagnetic waves and introduce several laws to describe the energy distribution in radiating systems. In particular, the Stefan–Boltzmann law is widely used in radiation heat transfer calculations.

The media involved in radiation heat transfer can be either non-participating or participating. The distinction and how this affects the type of modeling effort is described. An important simplification in non-participating media is that only the heat exchange between surfaces present in the system needs to be considered. These types of models are often referred to as surface-to-surface models. How the surfaces are oriented towards each other and how they see each other becomes an important consideration in the calculation of the rate of heat exchange between the surfaces. This is characterized by a view-factor parameter. Illustrative examples of calculation of this parameter are presented. If the surfaces absorb the radiation completely, they are called black bodies. The calculation of the radiation heat exchange in such cases is rather simple. Knowledge of the view factor and the Stefan–Boltzmann law is sufficient to calculate the heat transfer rate in such systems.

If the surface is only partly absorbing then a part of the radiation is reflected, and this needs to be accounted for. If the reflection is diffuse (i.e., not mirror-like) then the additional concept of radiosity is needed in order to model the heat exchange. Radiation exchange involving a number of surfaces can be then expressed as a set of linear algebraic equations, with the radiosities of the surfaces as the unknowns. Once the radiosity has been calculated the heat transfer rate can be calculated. Illustrative examples and some simple MATLAB code are presented as an aid to understanding the method.

Participating media absorb and emit radiation within the volume. Hence the modeling has to be done on a volume basis, and surface-to-surface models are not sufficient. A further complexity arises if the radiation is scattered, in which case the directionality has to be tracked as well. The resulting model is an integro-differential equation known as the radiation transport equation (RTE). Obviously the solution of such models is complex. Some computational methods for solving such problems are indicated as a case-study problem.

19.1   Properties of radiation

The basic properties of radiation can be summarized as follows.

- It is the energy emitted by an object as a consequence of its temperature.
- It does not need a participating medium or an intervening material for transmission. Thus direct radiation from the Sun to the Earth through a vacuum occurs.
- Radiation is basically an electromagnetic wave-propagation phenomenon. Energy is transported at the speed of light.
- For an electromagnetic wave we can define the wave length λ and the frequency ν. The two are related by

$$\lambda \nu = c$$

where c is the velocity of light. Note $c = 3 \times 10^8$ m/s in vacuum.
- The visible spectrum has a wavelength in the range of 0.38–0.76 μm.
- The region of interest in heat transfer is the infrared region with wavelengths in the range 0.8–100 μm. This is known as thermal radiation.
- The wavelength at which the maximum radiation is emitted is a function of temperature and is given by

$$\lambda_{max}(\mu m) = 2897/T$$

This is called Wien's displacement law. Thus the radiation shifts to shorter wavelength as the temperature increases.

19.2   Absorption, emission, and the black body

The thermal radiation incident on a body may be reflected, absorbed, or transmitted. If ρ, α, and τ are the fractions of the incident radiation which are reflected, absorbed, and transmitted, respectively, then

$$\rho + \alpha + \tau = 1 \tag{19.1}$$

For most solids the transmissivity, τ, is zero and hence

$$\rho + \alpha = 1 \tag{19.2}$$

An ideally absorbing body for which the absorptivity is $\alpha = 1$ is called a black body. The reflectivity of this body is zero, $\rho = 0$, as per the above equation.

Two concepts needed in the study of emission of radiation are the intensity of radiation and emissive power. A measure of the amount of energy traveling in a certain direction is called the intensity, denoted by I. The value of I may be different for different wavelengths in which case we define I_λ as the intensity in the interval near λ. We ignore the wavelengths dependence here, and assume that I refers to an intensity averaged over the entire wavelength.

Solid angle

The concept of solid angle (Fig. 19.1) is needed here in order to define the intensity. This is a 3D version of the angle and is best understood by revisiting the definition of an angle and its relation to the arc length, ds, namely

$$ds = R\,d\theta$$

Hence $d\theta$ is the angle corresponding to an arc length of ds located at a distance R from the center, $d\theta = ds/R$. This may be viewed as the definition of an angle.

The solid angle is likewise defined as an extension from 2D to 3D. The surface area, dA, is given by

$$dA = R^2\,d\omega$$

The solid angle, $d\omega$, has the units of steradians (sr), and can be calculated by using the spherical coordinate system and integrating over a projected surface area on a sphere,

$$d\omega = \sin\theta\,d\theta\,d\phi \tag{19.3}$$

where θ and ϕ are the directions in the spherical coordinate system.

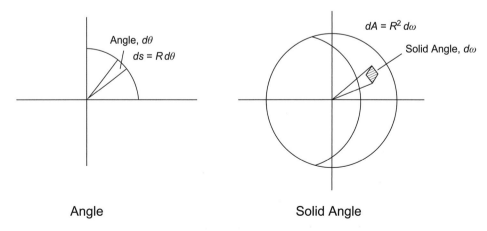

Angle   Solid Angle

Figure 19.1 The concept of solid angle and intensity of radiation.

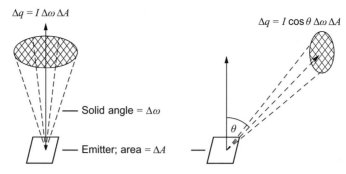

Figure 19.2 A figure showing the geometrical parameters needed for the definition of the intensity. Δq is the rate at which energy is propagating in the given direction.

Intensity definition

The formal definition of intensity (Fig. 19.2) can be expressed using the solid angle. It is the energy emitted in the direction θ per unit time per unit area of the emitting surface (ΔA) per unit solid angle:

$$\Delta q = I \cos \theta \, d\omega \, \Delta A$$

Note the $\cos \theta$ dependence since $\Delta A \cos \theta$ is the projected area of the emitter in the direction θ. Recall here that dA is the area of the emitting surface.

The intensity defined as above is independent of the direction θ for a diffuse surface. Then the total radiation emitted by a surface of area dA is obtained by integrating the above equation over the hemisphere. This is defined as the emissive power of the surface:

$$E = \frac{q_T}{\Delta A} = \int_0^{2\pi} I \cos \theta \, d\omega$$

where q_T is the total heat flux leaving a differential area ΔA in all directions over the hemisphere, or, using the definition of the solid angle,

$$E = \frac{q_T}{\Delta A} = \int_{\theta=0}^{\pi/2} \int_{\phi=0}^{2\pi} I \cos \theta \, \sin \theta \, d\theta \, d\phi$$

On performing the integration we find

$$\boxed{E = \pi I} \tag{19.4}$$

Hence the energy due to radiation leaving through a solid angle $d\omega$ is then equal to

$$\Delta q = \frac{E}{\pi} \cos \theta \, d\omega \, \Delta A \tag{19.5}$$

The radiation emitted by a black body is designated by an emissive power denoted as E_b. This in turn is given by the Stefan–Boltzmann law,

$$E_b = \sigma T^4 \tag{19.6}$$

where σ is called the Stefan–Boltzmann constant and has a value of 5.676×10^{-8} W/m^2 K^4 in SI units and T is the absolute temperature of the emitted surface in K. Note that E has the same units as a heat flux. The Stefan–Boltzmann law is in turn derived from the basic quantum theory of Planck.

Spectral dependence

Actually there is a spectral dependence and we have to deal with $E_{b,\lambda}$, a distribution function, rather than E_b for a more general analysis.

Planck's law gives the intensity of radiation as

$$I(\lambda) = \frac{C_1 \lambda^{-5}}{\exp[C_2/(\lambda T)] - 1}$$

where $C_1 = 2hc^2$ and $C_2 = hc/k_B$.

This is shown schematically in Fig. 19.3, a plot of the emissive power as a function of wavelength in accordance with Planck's law. The maximum peaks occur in accordance with Wien's displacement law, which can be obtained from basic calculus.

The gray body

A black body emits the maximum radiation. The black body is only an ideal concept. The ratio of the radiation emitted by a real body to that of a black body is called the emissivity, denoted, by ϵ:

$$\epsilon = \frac{E}{E_b}$$

Again this is a value averaged over all wavelengths. In reality, we have different values for ϵ for each wavelength, denoted by ϵ_λ. For a simplified treatment, as in this chapter, we ignore these subtleties.

A relation between the absorptivity and the emissivity is given by Kirchhoff's law. This states that they are equal:

$$\alpha = \epsilon \tag{19.7}$$

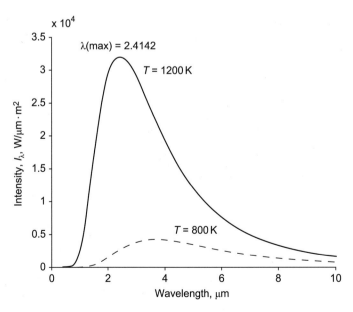

Figure 19.3 The distribution of the spectrum according to Planck's law for the intensity of radiation as a function of the wavelength.

which applies to systems at thermodynamic equilibrium. The law is used, however, as an approximation for systems where the departures from equilibrium are small, such as practical furnaces, high-temperature reactors, etc. It cannot be used, for instance, for studying radiation exchange between the Sun and the Earth. The temperature differences are too large!

With this background, we are ready to study the radiation exchange between different surfaces, for example, the walls of a furnace. (We assume first that the intervening medium is non-participating.) For modeling black surfaces we need an additional geometric concept, *viz.*, the view factor.

19.3   Interaction between black surfaces

We wish to find the radiation emitted from surface 1 which is intercepted by surface 2. We assume both surfaces are black.

View-factor definition

We define a view factor F_{12} (or F_{ij} in general) as follows:

$F_{12} =$ Fraction of the radiation emitted from surface 1 that is intercepted by surface 2

The heat transferred from surface 1 to surface 2 is therefore equal to

$$Q_{1\to 2} = F_{12}A_1 E_{b1} \tag{19.8}$$

The following properties of the view factor should be noted. The reciprocity rule states that

$$F_{12}A_1 = F_{21}A_2$$

The summation rule states that

$$\sum_{I=1}^{N} F_{1I} = 1$$

The view-factor equation

An expression for F_{12} is now derived from geometric considerations (Fig. 19.4).

Consider an emitter of area dA_1 and let us focus on a direction θ_1 measured from the normal to the emitter. Then the radiation rate in this cone formed by the solid angle ω is

$$\delta q = (E/\pi)dA_1 \cos\theta_1\, d\omega \tag{19.9}$$

Now consider a receiving surface with area dA_2. If this is in the direct line of sight from 1, then all the radiation in the cone $d\omega$ will reach this surface. If the surface is oriented at an angle $\cos\theta_2$ then the radiation in the cone from 1 seen by 2 will be reduced by this inclination. The projected area of the receiver is

$$dA_2 \cos\theta_2$$

The corresponding solid angle is

$$d\omega = dA_2 \cos\theta_2/R^2$$

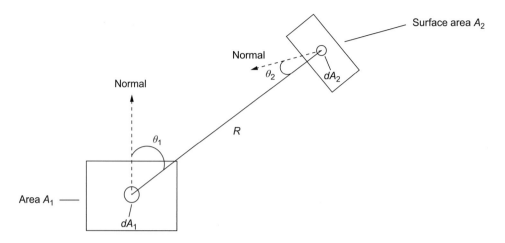

Figure 19.4 Geometric parameters involved in the calculation of the view factor.

Hence the radiation coming from 1 that is intercepted by 2 is given by

$$\delta q = (E/\pi)(dA_1 \cos \theta_1)(dA_2 \cos \theta_2 / R^2) \tag{19.10}$$

Note the inverse square of the distance dependence, which is often referred to as Lambert's law. This arises automatically due to the definition of the solid angle.

$Q_{1\to 2}$ is the integral of the above value and hence the view factor is defined as

$$\boxed{F_{12} = \frac{1}{A_1} \int_{A_1} \int_{A_2} \frac{\cos \theta_1 \cos \theta_2}{\pi R^2} \, dA_1 \, dA_2} \tag{19.11}$$

Note that the view factor is based on unit area of surface 1 and hence the term A_1 appears in the denominator above. If one switches 1 and 2 in Eq. (19.11), the corresponding equation for F_{21} is obtained and the reciprocity rule is also evident.

View-factor algebra

View factors for various geometries are available in the literature. For other cases numerical integration using Eq. (19.11) is needed. However, for some (simple) cases, the view factors can be determined without integration. Consider a hemispherical dome (surface 1) and a circular field which forms the base (surface 2). See Fig. 19.5.

We have

$$A_1 F_{12} = A_2 F_{21}$$

Substituting for the areas gives

$$2\pi R^2 F_{12} = \pi R^2 F_{21}$$

Hence $F_{12} = F_{21}/2$.

Also $F_{21} + F_{22} = 1$. But $F_{22} = 0$, since surface 2 cannot see itself. Hence $F_{21} = 1$ and therefore $F_{12} = 1/2$.

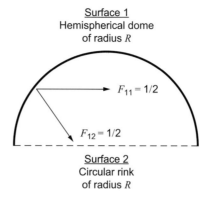

Figure 19.5 An example of view-factor algebra.

Now, since $F_{11} + F_{12} = 1$, we have $F_{11} = 1/2$.

Thus all the view factors have been calculated.

Now, if surface 1 is at 1000 K and surface 2 is at 900 K, the net radiation exchanged between them can be calculated as

$$Q_1 = Q_{1\rightarrow 2} - Q_{2\rightarrow 1} = (E_{b1} - E_{b2})A_1 F_{12}$$

The result is 15 323 W.

Interaction among black surfaces

It is easy to extend this concept to N black surfaces. Consider an enclosure formed by various surfaces, numbered 1, 2, 3, etc., which are at different temperatures T_1, T_2, etc. The heat lost by a particular surface Q_i can be calculated as follows:

$$Q_i = A_i \sigma T_i^4 - \sum_{k=1}^{N} F_{ki} A_k \sigma T_k^4 \tag{19.12}$$

The first term is the black-body emission from surface i. The second term is the radiation from all the surfaces which is intercepted by surface i. All of it is absorbed by surface i for a black body. The surface may receive radiation from itself if F_{ii} is non-zero, i.e., if it can see itself.

It is convenient to rearrange the above equation using view-factor algebra. Using the reciprocity rule,

$$A_i F_{ik} = A_k F_{ki}$$

Eq. (19.12) can be written as

$$Q_i = A_i \left[\sigma T_i^4 - \sum_{k=1}^{N} F_{ik} \sigma T_k^4 \right]$$

On noting that

$$\sum_{k=1}^{N} F_{ik} = 1$$

the equation for Q_i can be written as

$$Q_i = A_i \left[\sum_{k=1}^{N} F_{ik}\sigma T_i^4 - \sum_{k=1}^{N} F_{ik}\sigma T_k^4 \right]$$

or

$$Q_i = \sigma A_i \sum_{k=1}^{N} F_{ik}(T_i^4 - T_k^4) \qquad (19.13)$$

This equation can also be written in terms of the emissive power as

$$Q_i = A_i \sum_{k=1}^{N} F_{ik}(E_{bi} - E_{bk}) \qquad (19.14)$$

19.4   Gray surfaces: radiosity

The problem becomes complicated for radiation exchange among gray enclosures. Consider the radiation arriving at surface i. Only part of it is absorbed. The remaining portion is reflected back, assuming that there is no transmission (an opaque surface). This reflected radiation from surface i will reach other surfaces, and part of it will be absorbed and reflected again. The problem obviously becomes more complicated than that of black surfaces. A number of methods can be used, and a general method is illustrated in the following sections. Here we assume that the reflected radiation is diffuse, i.e., it is emitted in all directions, and that the reflection is not mirror-like (which is referred to as specular reflection). If the surfaces are not highly polished, such as furnace walls, the assumption of diffuse reflection is a good one.

The concept of radiosity is introduced to account for reflected radiation. The radiation reflected from surface i is considered as though it originated from the surface itself. The total radiation from surface i (per unit area of surface i) is called the radiosity of the surface and has two components, (i) its own emission and (ii) the unabsorbed (or reflected) radiation coming from all the surfaces. See Fig. 19.6.

A related concept is the irradiation, G, on a surface i. This is defined as the total radiation (per unit area) impinging on the surface i. An expression for G can be obtained from the radiosity of all the surfaces forming the enclosure and their view factors as follows. The radiation arriving from surface k to surface i is equal to $A_k F_{ki} J_k$. The total radiation arriving from all the surfaces impinging on i is therefore equal to

$$\sum_{k=1}^{N} A_k F_{ki} J_k$$

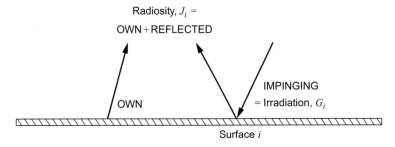

Radiosity, $J_i =$
OWN + REFLECTED

OWN

IMPINGING
= Irradiation, G_i

Surface i

Figure 19.6 An illustration of the concepts of radiosity and irradiation.

Hence the irradiation impinging on i is obtained by dividing this by A_i, since it is based on unit area of the receiving surface:

$$G_i = \frac{1}{A_i} \sum_{k=1}^{N} A_k F_{ki} J_k$$

Using the view-factor algebra,

$$A_k F_{ki} = A_i F_{ik}$$

the equation for G_i above reduces to

$$G_i = \sum_{k=1}^{N} F_{ik} J_k \qquad (19.15)$$

A fraction ϵ_i of this is absorbed by surface i and the remaining portion is reflected back. (Note Kirchhoff's law that the absorptivity is equal to the emissivity is used here.) Hence the radiosity from surface i is equal to

$$J_i = \epsilon_i E_{bi} + (1 - \epsilon_i) G_i \qquad (19.16)$$

Note that this consists of two terms. The first term on the RHS is the emissive power of surface i (its own radiation) and the second term is the reflected radiation from surface i. The emissive power in turn is equal to the black-body emission E_{bi} multiplied by the emissivity of the surface.

Using the expression for G from Eq. (19.15), the above equation can be written more explicitly as

$$J_i = \epsilon_i E_{bi} + (1 - \epsilon_i) \sum_{k=1}^{N} F_{ik} J_k \qquad (19.17)$$

Note that the radiosity of surface i depends on the radiosities of all the other surfaces and cannot be solved directly. The solution, however, becomes easy if all the surface temperatures are known. The matrix algebra can be used directly here. The surface temperatures of some of the surfaces forming the enclosure, e.g. heat shields, reradiating surfaces, etc., might not be known, and a different procedure or iterative solution assuming a surface temperature is needed for such cases. Let us illustrate the calculations assuming that all the temperatures are known. In this case Eq. (19.17) can be rearranged to a vector–matrix form:

$$\tilde{A} J = r \qquad (19.18)$$

where J is the vector of unknown radiosities (the solution vector).

$\tilde{A}$ contains the matrix coefficients in (19.17) of the form

$$A_{ik} = -\delta_{ik} + (1 - \epsilon_i)F_{ik}$$

where δ_{ik} is the Kronecker delta. This is defined as equal to one if $i = k$ and zero otherwise. This is needed in order to put the LHS of Eq. (19.17) into matrix form.

Likewise, r is the first term on the RHS of the set of linear equations defined by (19.17). The coefficients of this vector are

$$r_i = -\epsilon_i E_{bi} = -\epsilon_i \sigma T_i^4$$

The linear equation (19.18) can be solved directly in MATLAB using the backslash operator:

```
J = A \ r
```

Note that the solution can be written as

$$J = [\tilde{A}]^{-1} r \tag{19.19}$$

but a numerically more stable and faster way is to use the backslash operator shown above. This operator does the same thing and solves for the vector J but does not explicitly calculate the inverse of the coefficient matrix $\tilde{A}$. It uses the factorization approach instead. In general, inversion of matrices should be avoided from numerical-calculation perspectives, especially for large matrices.

19.5 Calculations of heat loss from gray surfaces

Once the radiosity values have been calculated by the above procedure, the net loss from a surface can be calculated. The relevant formula is first stated and a derivation of the formula then follows:

$$Q_i = \frac{A_i \epsilon_i}{1 - \epsilon_i}(E_{bi} - J_i) \tag{19.20}$$

The derivation of this equation is as follows. Consider the radiation flux impinging on a surface i. A fraction ϵ_i of this is absorbed by the surface. The surface itself emits an amount of radiation equal to $\epsilon_i E_{bi}$. Hence the net flux from the surface is

$$Q_i = A_i \epsilon_i (E_{bi} - G_i) \tag{19.21}$$

Equation (19.16) for the radiosity can be rearranged to give an expression for G_i as follows:

$$G_i = \frac{1}{1 - \epsilon_i}(J_i - \epsilon_i E_{bi})$$

On substituting this into Eq. (19.21) and simplifying, one obtains Eq. (19.20), which is the working or design equation needed to do the heat balance on enclosures. The calculation of heat exchange by use of Eq. (19.22) is also known as the Gebhart method. Three examples are provided.

Example 19.1.

Consider two parallel plates facing each other (Fig. 19.7). Plate 1 is at a temperature of T_1 and has an emissivity of ϵ_1, while plate 2 is at a temperature of T_2 and has an emissivity of ϵ_2. Assume a non-participating medium. Calculate the heat exchanged between these surfaces.

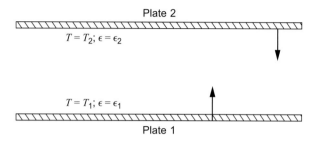

Plate 2

$T = T_2$; $\epsilon = \epsilon_2$

$T = T_1$; $\epsilon = \epsilon_1$

Plate 1

Figure 19.7 Two plates facing each other.

Solution.

This represents a problem where an analytic solution can be obtained since there are only two surfaces. Using Eq. (19.20), the heat loss from 1 is given by

$$Q_1 = \frac{\epsilon_1}{1 - \epsilon_1}(E_{b1} - J_1)$$

Two equations can be set up for J_1 and J_2, and can be solved analytically using Kramer's rule. On substituting for J_1, the value of Q_1 can be obtained. The result is

$$Q_1 = \frac{1}{1/\epsilon_1 + 1/\epsilon_2 - 1}(E_{b1} - E_{b2}) \tag{19.22}$$

An expression for Q_2 can be derived in a similar manner. But, since there are only two surfaces, we deduce that $Q_2 = -Q_1$.

Example 19.2.

Consider a square enclosure of dimensions 3 m by 3 m with the bottom (surface 1) at 600 K, the sides (surface 2) at 1700 K, and the top at 1400 K (Fig. 19.8). Find the radiosity and the heat exchanged between the various surfaces.

Use an emissivity of 0.5 for all the surfaces.

Solution.

Since there are three surfaces, it is easier to solve the set of algebraic equations numerically. We will illustrate the procedure using MATLAB to implement the Gebhart method. A small piece of code can be written for this. It is shown below.

```
% Radiation calculations by Gebhart method.
% Data section; to be changed as per problem specification
sigma = 5.67E-08  ;
n =  3 ;       %   Number of surfaces
```

```
emiss = [ 0.5  0.5  0.5 ]; % emissivity of surfaces
                           % put as a row vector.
area = [ 3  6  3 ] ;  % Surface area of each surface.
temp = [ 600 1700 1400 ] % temperature of each surface in degrees K
% VIEW FACTOR DATA. Provide an n by n matrix of view factors.
VIEW = [ 0     0.586   0.414
         0.293  0.414  0.293
         0.414  0.586  0.0 ] ;
   % set up the matrices and the right-hand side.
eblack = sigma * temp.^4  % Black-body fluxes, a vector.
   for i = 1:n
      rhs (i) = -emiss(i) * eblack (i)

      for k = 1:n
         a(i,k) = ( 1 - emiss(i) ) * VIEW(i,k)
      end
      a(i,i) = a(i,i) - 1 % correct for diagonal term.
   end
% solve the linear equation
Rad = a\rhs'
% calculate net transfer from each surface
for i = 1:n
   q(i) = emiss(i) * area(i) * (eblack(i) - Rad(i) ) / (1 - emiss(i) )
end
```

The solution gives the radiosities of the three surfaces. The values are

$$J_1 = 1.6638 \times 10^5 \ \text{W/m}^2$$

$$J_2 = 3.7617 \times 10^5 \ \text{W/m}^2$$

$$J_3 = 2.5357 \times 10^5 \ \text{W/m}^2$$

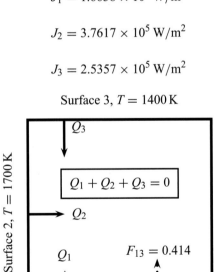

Figure 19.8 The system of Example 19.2.

The corresponding values of Q_i computed from Eq. (19.20) are as follows:

$$Q_1 = 4.7710 \times 10^5 \text{ W/m-width}$$

$$Q_2 = -5.8435 \times 10^5 \text{ W/m-width}$$

$$Q_3 = 1.0725 \times 10^5 \text{ W/m-width}$$

Surface 1 and surface 3 receive heat, while surface 2 loses heat. Note that

$$\sum Q_i = 0$$

as required by the first law of thermodynamics.

Example 19.3.

Consider the same problem as Example 19.2 but now surface 3 is well insulated. Modify the analysis and solve for the heat transfer rate from the various surfaces, and also estimate the temperature of surface 3.

Solution.

In this case there is no loss from surface 3, the surface is adiabatic, and Q_3 must be zero. The simplest way is to assume different temperatures for surface 3 and see which value leads to Q_3 of zero. Using the above MATLAB program, the sign of Q_3 changes for T_3 of between 1518 and 1519 K, and the temperature of surface 3 is in this range. Further iterations can narrow down this value, but that is not needed for practical applications.

A more direct way is to reformulate the equation for the radiosity of surface 3 and solve by using the matrix method with modified coefficients and RHS for surface 3. Note that the following equation holds for surface 3:

$$J_3 = \epsilon_3 E_{b3} + (1 - \epsilon_3) \sum_{k=1}^{N} F_{3k} J_k \qquad (19.23)$$

But, since Q_3 is zero, we have $E_{b3} = J_3$ from Eq. (19.20). Using this, E_{b3} can be eliminated and the equation for the radiosity of 3 can be rearranged to

$$0 = (\epsilon_3 - 1)J_3 + (1 - \epsilon_3) \sum_{k=1}^{N} F_{3k} J_k$$

Hence the RHS vector coefficient r_3 is to be set to zero for surface 3. Also for the matrix $\tilde{A}$ the Kronecker term is to be replaced by $(\epsilon_3 - 1)\delta_{3k}$. With these changes the MATLAB code can be rerun to solve for the radiosity and the heat flux values. The results for the heat flux are

$$Q_1 = -5.1671 \times 10^5 \text{ W/m-width}$$

$$Q_2 = 5.1671 \times 10^5 \text{ W/m-width}$$

$$Q_3 = -0.0096 \times 10^5 \text{ W/m-width}$$

Also the temperature of surface 3 is calculated as $(E_{b3}/\sigma)^{1/4}$, which is also equal to $(J_3/\sigma)^{1/4}$. The value is 1518.4 K, which agrees with the answer obtained by the direct iteration method.

19.6 Radiation in absorbing media

So far we have considered radiation emitted and absorbed by surfaces. The media separating the surfaces were considered to neither emit nor absorb the radiation. Such media are appropriately called non-participating media. Examples are systems under vacuum, monatomic inert gases such as He, and diatomic molecules such as oxygen and nitrogen, as in air, etc. In this section, we consider the case where the medium is emitting and absorbing radiation. Gases with molecules of relatively asymmetric structure such as CO_2 and H_2O are examples of participating media. Combustion gases, for instance, contain sufficient amounts of CO_2 and water that the analysis of radiative heat transfer in such systems should include the absorption and radiation of such modules. The absorption of solar radiation by such gases is also responsible for global warming. The problem is more complex than the case of non-participating media for several reasons, some of which are enumerated below. See Fig. 19.9.

1. The radiation is now a volumetric phenomena, and the analysis cannot be restricted exclusively to the boundaries.
2. Radiation is usually emitted (or absorbed) over a narrow range wavelength range, unlike for a solid surface, which can emit or absorb over the whole spectrum. Hence the assumption of gray-body absorption might not be valid. A spectral emissivity and absorptivity is needed.
3. The emission properties are dependent on the concentration of the gases.
4. The propagation of radiation can be complicated by the presence of small particles such as soot, aerosols, etc., which scatter the radiation. The scattering spreads the radiation in all angular directions in spherical coordinates, and hence the angular intensity of radiation should be included in the analysis.

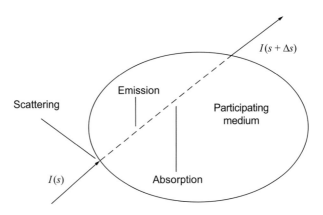

Figure 19.9 The diagram for derivation of the RTE for an absorbing (participating) medium.

The intensity of radiation is again a convenient quantity and is defined in the same manner as the radiation energy leaving a direction per unit time per unit area placed at that point. For a participating medium the intensity is defined as if the radiation originated at the point under consideration. The change in intensity along a path is the key parameter for absorbing media, and this can be stated in words as follows:

Change in intensity along a path = Emission − Absorption − Scattering

First we consider just the absorption term. The basic modeling of absorption starts with the analysis of the attenuation of the radiation passing through an absorbing medium, which is given by Beer's law. This law states that the decrease in intensity of radiation is in direct proportion to the intensity of the radiation,

$$\frac{dI}{ds} = -\kappa I \tag{19.24}$$

The proportionality constant κ is known as the absorption coefficient or the extinction coefficient.

A note on the wavelength dependence

Here we assume that the intensity is averaged over all wavelengths. More generally we defined I_λ as the intensity in the wavelength interval $d\lambda$ near λ and an extinction coefficient κ_λ corresponding to that wavelength. In general κ_λ is a function of λ, temperature, pressure, and the concentration of the absorbing species. We now continue, ignoring the wavelength dependence, and this assumption is often referred to as the gray-gas assumption. Also let us assume that there is no emission (which is a good assumption at low temperatures) and no scattering, and see how the intensity changes along a path.

Bouguer's law for attenuation of intensity

The integrated form of Eq. (19.24) is known as the Beer–Lambert equation,

$$I(s) = I_O \exp(-\kappa s) \tag{19.25}$$

where I_O is the radiation at location $s = 0$, a point on the emitting source. This simplistic expression is often used in modeling of photochemical reactors,

More generally κ varies along the radiation path, and the integrated form can be expressed as

$$I(s) = I_O \exp\left[-\int_0^s \kappa(S)dS\right] \tag{19.26}$$

where S is the dummy variable for position. Using a dimensionless parameter τ defined as

$$\tau = \int_0^s \kappa(S)dS$$

the integrated form is represented as

$$I(s) = I_O \exp(-\tau) \tag{19.27}$$

This expression is called Bouguer's law, after Bouguer, who in 1730 was the first to show on a quantitative basis how light intensities could be compared, but is also known as the Beer–Lambert law. Here τ has the status of a dimensionless distance but includes the variation of κ along the path. This parameter is called the optical thickness parameter.

If this parameter is much less than one, the domain is called optically thin. If it is much larger than unity, the domain is optically thick. These represent the two limiting cases of a non-participating gas, and a heavily absorbing gas, respectively. If κ does not vary along a path then τ is equal to κs.

Emission

The intensity variation is now modeled using an additional source to include the contribution of the emission:

$$\frac{dI}{ds} = -\kappa I + s_v$$

where s_v represents the emission contribution to the intensity change along the direction s. The emission term is modeled on the basis of a local thermodynamic equilibrium and is equal to the intensity of a black-body emission I_b times the absorption coefficient of the medium. Hence

$$s_v = \kappa I_b$$

The intensity I_b for a black body is given by

$$I_b = \frac{E_b}{\pi} = \frac{n^2 \sigma T^4}{\pi}$$

where n is the refractive index of the medium. The value of n is one for gases but this parameter is thrown for generalities because the internal radiation is also important in many cases for heat transport in molten materials such as glasses. For further discussion we use $n = 1$. Hence the change in intensity due to both absorption and scattering is given by

$$\frac{dI}{ds} = -\kappa I + \kappa \sigma T^4 / \pi \tag{19.28}$$

The equation has to be solved together with the heat balance since the black-body source term appearing here is a function with the temperature in accordance with the Stefan–Boltzmann law, which is discussed briefly now.

Coupling with the heat balance

The intensity needs the temperature equation which is given by the energy balance in the system. For this a control volume has to be considered, and the energy arriving in the control volume from all incident directions has to be added as a flux term. The energy produced by emission is added as a source term. It is useful to define a radiant heat flux vector q_R as follows:

$$\nabla \cdot q_R = \kappa \left[4\sigma T^4 - \int_{4\pi} I(\omega) d\omega \right] \tag{19.29}$$

where the first term accounts for the emission per unit control volume and the second term is the energy absorbed by the control volume from the radiation intensity at the surface of the control volume. The integration is over all the solid angle from 0 to 4π.

The q_R term is added to the heat-balance term. The temperature equation and the intensity equation are solved together. An example from the literature is shown here.

Radiation between two gray plates

Gray gas between two plates is one of the simple and well-studied problems in radiation. This is essentially an extension of the problem of radiation exchange between two gray plates studied in Example 19.1. Heaslet and Warming (1965) provided complete solutions to the problem, and indicated that a simple correction factor to account for internal radiation can be used. A plot of the correction factor is shown in Fig. 19.10.

The radiation exchanged is then calculated as

$$q_{1 \to 2} = \sigma(T_1^4 - T_2^4) \frac{\psi_b}{1 + \psi_b(1/\epsilon_1 + 1/\epsilon_2 - 2)}$$

where ψ_b is the correction factor shown in Fig. 19.10.

Other approximations are also useful, and a more detailed account of these approximations can be found in the book by Siegel and Howell (1992). Here we discuss the diffusion approximation which is often used in CFD codes since it provides a simple way of modeling the system in analogy to Fourier's law.

The diffusion approximation

The basis of this is that the two terms on the RHS of Eq. (19.28) are nearly equal, so the intensity contribution remains constant along a path. The intensity term does not contribute to the heat balance, and the second term on the RHS of Eq. (19.29) may be neglected, resulting in

$$\nabla \cdot q_R = 4\kappa\sigma T^4$$

An equivalent heat-conduction vector can be extracted from the above equation:

$$q_R = -k_R \nabla T$$

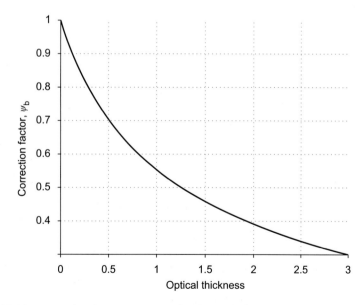

Figure 19.10 The correction factor to account for the participating medium. The solution applies to radiation between two plates facing each other.

where k_R is the radiation contribution to heat conductivity,

$$k_R = \frac{16\sigma T^3}{3\kappa}$$

and this conductivity term is simply added to the molecular (Fourier) term. The parameter k is often slightly modified to give a parameter κ_R referred to as an effective ("Rosseland") absorption coefficient. No correction is needed for gray gases, and κ is equal to κ_R here. Note that the radiation conductivity defined above is a function of temperature, and the problem is to be solved as a variable-conductivity problem. The diffusion assumption does not hold near the solid boundary. A slip parameter is often used to match the jump.

The effect of scattering

Scattering is another beast altogether; the resulting equations are integro-differential equations as shown in the following discussion:

$$\frac{dI}{ds} = -\kappa I - \sigma_s I + \kappa n^2 \sigma T^4 / \pi + \frac{\sigma_s}{4\pi} \int_0^{4\pi} I(r, s)\phi(s, s')d\Omega \qquad (19.30)$$

Here $\phi(s, s')$ is a scattering coefficient, also called the scattering albedo, and represents the fraction of the radiation arriving along the path s which gets scattered in the direction s'. The parameter σ_s is the scattering coefficient.

The terms on the RHS of Eq. (19.30) can be interpreted as follows. The first term is the loss by absorption, the second term is the loss by scattering, the third term is the gain by emission, and the last term is the gain by scattering from all other directions towards the direction s. This type of model is used in photo-catalytic reactions where the presence of fine catalyst particles can cause considerable scattering effects.

Summary

- The physics of radiation is different from that of conduction and convection. In particular, the geometric consideration represented by the view factor plays an important role. The (diffusively) reflected radiation needs to be accounted for in some cases. Thus the radiation analysis often involves the solution of a set of algebraic equations rather than a set of differential equations.

- For non-participating media the concept of radiosity is useful in analyzing the radiation interaction between gray surfaces. A set of linear algebraic equations for the radiosity can be set up and solved (after specifying the boundary conditions (temperature or flux) for each surface). Once the surface radiosity has been calculated for each surface, the net loss from any given surface can be calculated. The view factors for all the surface pairs are needed. These view factors depend on the geometry and orientation of the surfaces.

- For participating media, the radiation is a volumetric process. A shell-balance approach for intensity variation (an approach very familiar to us) is used to derive the radiative transport equation (RTE). This equation can be solved by various methods and coupled with the energy equation for the temperature field.

ADDITIONAL READING

The standard textbook for radiation heat transfer is the book by Siegel and Howell (1992). This book should be consulted for a further study of this topic. Another useful book is that by Modest (2013).

Problems

1. Using the basic physical constants, show that $C_1 = 3.742 \times 10^8 \, \text{W} \, \mu\text{m}^4/\text{m}^2$ and $C_2 = 1.4389 \times 10^4 \, \mu\text{m}/\text{K}$ in the Planck model.

   Take the derivative of the Planck equation with temperature. Set this equal to zero, and verify Wien's displacement law.

2. The solar constant is defined as the average flux of solar energy incident on the outer fringes of the Earth's atmosphere and a commonly used value is $1.353 \, \text{kW}/\text{m}^2$. If the Sun has a diameter of $1.39 \times 10^6 \, \text{km}$ and is assumed to radiate like a black body, what is its effective temperature?

   If the Sun is treated as a black body with a temperature of 5765 K, show that 99% of the radiation emitted by the Sun is in the wavelength range 0.25–3.96 μm.

3. Verify the algebra leading to Eq. (19.22) for radiation exchange between two plates facing each other.

4. **The effect of a radiation shield.** Consider a shield with an emissivity of ϵ_s separating two parallel plates as shown in Fig. 19.11. Assume that the shield is adiabatic and that there is no net loss of heat. Find the temperature of the shield and the factor by which radiation between the plates is reduced due to the presence of the shield.

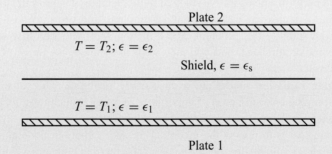

Plate 2

$T = T_2; \epsilon = \epsilon_2$

Shield, $\epsilon = \epsilon_s$

$T = T_1; \epsilon = \epsilon_1$

Plate 1

Figure 19.11 Two plates separated by a radiation shield.

5. Consider radiation exchange between two annular cylinders of radii R_i, and R_o facing each other. This is similar to the radiation between flat plates considered in the text except for the fact that the geometry is different.

   What are the view factors? How would you calculate them? Use them to set up the equation for radiosity and develop an expression for the radiation interchange between these surfaces.

6. Two parallel disks each of radius R are facing each other and separated by a distance H. Derive an expression for the view factor between these disks.

7. Two rectangular plates of sizes L and W are facing each other and separated by a distance H. Derive an expression for the view factor between these disks. Find the limit if W is large, which represents two line elements facing each other.

8. Consider two plates at temperatures of 600 K and 500 K with emissivities of 0.8 and 0.4, respectively. The plates are separated by a gray gas that has an absorption coefficient of 0.2 cm^{-1} and the plates are 10 cm apart. Calculate the heat flux transferred from the hot plate to the cold plate.

9. **Cooling of a solid with radiation.** A solid at an initial temperature of T_0 is exposed to colder surroundings at T_a and is cooling down by radiation heat loss. Develop a model based on the lumping approximation for the temperature of a solid as a function of time.

10. A gray gas with an absorption coefficient of κ is contained between two plates separated by a distance L. Using the diffusion approximation for radiation, derive the relation

$$Q_1 = \frac{1}{3\kappa L/4 + 1/\epsilon_1 + 1/\epsilon_2 - 1}(E_{b1} - E_{b2}) \tag{19.31}$$

for the radiation exchanged from plate 1 to plate 2.

11. Two plates are at temperatures of T_1 and T_2 and a chemical reaction is producing heat at a constant rate within the system. Derive a model to predict the temperature distribution within the system.

12. **Computational aspects of RTE: a case-study problem.** The common methods for the computational solution of radiative transport equations are summarized briefly below:

   1. The discrete ordinate method
   2. The Monte–Carlo method
   3. Hottel's zonal method
   4. The integral-equation methods

   Your task in this case study is to review these methods, apply them to a simple problem, and perform a comparative study of these methods. The case of two gray plates facing each other is a classical and well-studied problem, and you may wish to apply your computational skills to this problem. The diffusion approximation discussed in the text provides a simpler approach, and you may wish to examine the range of validity of this method as well.

13. **Solar radiation modeling: a case-study problem.** The calculation of solar radiation impinging on a surface is of importance in many applications, for example, design of solar collectors, temperature control of buildings, etc.

   Your goal is to review the modeling of solar radiation modeling and discuss the models which are suitable for the design of a solar thermal collector.

14. **Radiation interactions in a crystal puller: a case-study problem.** Radiation interactions are of importance in crystal pullers for growing crystals from melts since these operate at very high temperatures. The radiation view factors change at various stages of crystal growth, and accurate modeling requires tracking of these at various stages of cooling. For short crystals it may turn out that the (colder) crystal surface is actually gaining heat (from radiation from the hot melt surface) instead of losing it, thereby affecting the rate of solidification. You will need to be able to compute the various view factors and the radiation interactions at various stages of the growth with the background material presented in this chapter and suggest proper cooling strategies at various stages of the crystal growth. The paper of Srivastava *et al.* (1986) is useful as an additional source here.

15. **Photochemical reactor modeling: a case-study problem.** Although radiation is important in heat transfer, an analogous model can be used in the design of photochemical reactors. The modeling of these reactors requires that the radiation intensity be tracked in the reactor as a function of position and coupled to the kinetics of chemical reaction. The RTE becomes an important sub-model for such reactors.

The book by Cassano and Alfabo (1991) is the most valuable source for photochemical reactor modeling. Study this paper or related papers, and set up and solve a case-study example of the light-intensity distribution in a photochemical reactor.

20 More convective mass transfer

Learning objectives

The learning objectives of this chapter are as follows.

- To understand and to be able to use the common tools for analysis of convective mass transport processes.
- To understand the importance of length scale for mass transfer compared with flow and how the near-wall region has to be modeled more accurately for turbulent mass transport problems.
- To understand the dynamics of gas–liquid interfaces and how mass transfer rates are affected by these interactions.
- To understand the concept of dispersion and the usefulness of this concept in engineering applications.

This chapter continues the theme of convective transfer from Chapter 12, which discussed mainly problems with internal laminar flow. This chapter extends the study to other flow problems (e.g., external flow) and parallels Chapter 18 on convective heat transfer. The main goal is to examine how mass transfer rates and transfer coefficients can be calculated for illustrative problems for a prescribed or computed velocity profile. First we start with a well-defined flow problem, viz., laminar external flow over a flat plate with a low-flux model for mass transport taking place from (or to) the plate surface. This problem of mass transfer in boundary-layer flow is analyzed by four techniques, namely (i) dimensionless analysis, (ii) scaling analysis, (iii) integral analysis, and (iv) the similarity-transformation method. As discussed earlier, these are some of the widely used tools for solution of transport problems. Applying them to the same problem will give the student a good understanding of these tools. The similarity to heat transfer is also pointed out. One important difference is that mass transfer is often accompanied by chemical reaction. Hence we also examine how the chemical reaction modifies the rate of mass transfer.

The next problem examined in this chapter is the high-flux case where convection induced by mass transfer cannot be neglected. Various engineering models useful for such systems are presented. This is followed by sections on mass transfer in turbulent boundary layers and at a gas–liquid interface, mass transfer from a spherical bubble, and bubble swarms.

Finally, on a slightly different note, the model for axial dispersion in flow systems introduced in Chapter 12 is revisited. There the concept of axial dispersion was introduced as a consequence of model reduction in the context of a laminar-flow reactor. Here we provide a detailed derivation of the Taylor model for axial dispersion in laminar flow by considering a transient version of the problem. It is shown that the axial dispersion represents a combined effect of axial convection and radial diffusion. This concept, which was developed first by Taylor in the context of laminar flow, finds wide application in many fields, for example, in dispersion of pollutants in the atmosphere, separation of solutes by chromatography, etc.

20.1 Mass transfer in laminar boundary layers

The classical problem in mass transfer in external flows is the flow over a plate with a species, A, being transferred to or from the plate. Problems of this type have many applications, including that of chemical vapor deposition processes for electronic and solar materials. The problem is shown schematically in Fig. 20.1. Various general tools to study transport problems introduced earlier are illustrated for this problem. In particular the dimensionless analysis and scaling analysis provide a simple way of analyzing data without rigorous mathematical calculations.

20.1.1 The low-flux assumption

The problem of mass transfer in boundary layers may be classified into two types depending on whether the rate of mass transfer from (to) the surface to (from) the flowing gas is large enough to affect the flow. Note that the mass transfer creates a net mass flux n_t (in the y-direction) at the surface as noted earlier. A flow in the y-direction is introduced as the result of this flow:

$$v_y(y = 0) = n_t/\rho = N_t/C \tag{20.1}$$

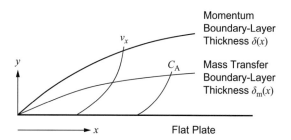

Figure 20.1 A Schematic representation of velocity and concentration profiles in a boundary layer for laminar flow over a flat plate. Δ is defined as the ratio δ_m/δ and is usually less than one.

where n_t is the component of total mass flux in the y-direction at the surface. Hence the no-slip boundary condition for momentum transfer at $y = 0$ has to be modified accordingly. The x-component of velocity at the surface, v_x, is still zero at $y = 0$, but the y-component is equal to n_t/ρ. Hence the momentum-balance boundary conditions are altered by the rate of mass transfer. This is referred to as the high-flux case or detailed model. The problem becomes a two-way coupled problem. In contrast, if n_t is small (or even zero as a consequence of equimolar counterdiffusion) then the momentum transfer can be solved independently of mass transfer. This is referred to as a low-flux case and is the focus of this section. The problem is now one-way coupled and the flow analysis for a boundary layer from Chapter 14 can then be directly used. The high-flux case is somewhat complicated in terms of the mathematics, and some aspects of it are discussed in the next section.

20.1.2   Dimensional analysis

From the discussion and the linear-algebra-based calculations presented in Chapter 13 we find that the dimensionless mass transfer coefficient is the Sherwood number Sh, which should be correlated in terms of Re and Sc. The local distance is used in the definition of the Reynolds number. Hence we denote this as Re_x, which is defined as $Re_x = v_\infty x/\nu$, where x is the length from the entrance. The Schmidt number is $Sc = \nu/D$, which is the ratio of the momentum diffusivity to the mass diffusivity. The dimensionless rate of mass transfer is the Sherwood number, $Sh_x = k_m x/D$.

A power-law type of correlation is used. Thus the equation

$$Sh_x = A Re_x^\alpha Sc^\beta$$

appears to be a good form for the correlation based on dimensionless analysis.

Dimensional analysis does not resolve what the exponents (α and β above) should be, but merely gives us a form for the correlation of the experimental data. However, this problem has been subjected to a complete theoretical analysis, according to which, the following relation appears to do the job well:

$$Sh_x = 0.332 Re_x^{1/2} Sc^{1/3} \tag{20.2}$$

for laminar flow. Note that the exponent on Sc is 1/3. This is characteristic of problems for cases where mass transfer is taking place from a no-slip boundary,

The above expression gives the local mass transfer coefficient. An integrated average value is often used in engineering-design calculations. By integrating over a length L of plate, an average Sherwood number can be defined and shown to be given by the following expression:

$$Sh_L = 0.664 Re_L^{1/2} Sc^{1/3}$$

Note that the equation should not be used if $Re_L > 2 \times 10^5$. The flow ceases to be laminar under these conditions. The boundary layer becomes fully turbulent if $Re_x > 2 \times 10^6$. Turbulent analysis is presented later in this chapter, but for completeness we present the commonly used empirical correlation for this case.

Turbulent flow

The form of the correlation suggested by dimensionless analysis is also applicable to turbulent flow with the following correlation for the local value being the most commonly used form:

$$Sh_x = 0.0292 Re_x^{0.8} Sc^{1/3} \tag{20.3}$$

Note the change in the dependence on Re, which is now much stronger in turbulent flow, namely to a power of 0.8. The exponent on Sc remains the same.

Reacting systems

Often mass transfer data in a reacting system are needed. The Sherwood number, Sh, now depends also on the local Damköhler number, Da_x, which is defined as $k_1 x/D$, where k_1 is the rate constant for a first-order reaction. A power-law correlation of the type

$$Sh_x = [A + BDa_x^c] Re_x^\alpha Sc^\beta$$

may be useful. The factor in the square brackets can then be thought of as an enhancement in transport due to reaction. However, this approach is purely empirical and can be in error, as will be shown in the next section. This also shows that a purely power-law model using the various dimensionless groups might not be useful as the correlating form of the equation. A more precise form of the expression can be obtained on the basis of the scaling analysis shown in the next subsection.

20.1.3   Scaling analysis

The scaling analysis is demonstrated here with reference to mass transfer on a flat plate.

We assume that there is a mass transport boundary layer of thickness δ_m near the surface, as shown in Fig. 20.1. Then an estimate of the mass transfer coefficient is

$$k_m \approx \frac{D}{\delta_m} = B\frac{D}{\delta_m} \tag{20.4}$$

where B is a numerical constant. The value of this constant will depend on the form of the concentration profile assumed within the boundary layer. For the moment we will leave it as a constant since the scaling analysis provides only relative values, rather than absolute numbers.

The mass transfer boundary layer is usually buried within the hydrodynamic boundary layers (since Sc is one or larger) and the ratio of the two boundary layers is denoted by Δ,

$$\Delta = \frac{\delta_m}{\delta}$$

where Δ is usually less than one.

Equation (20.4) can be arranged as

$$k_m = B\frac{D}{\delta}\frac{\delta}{\delta_c} = B\frac{D}{\delta}\frac{1}{\Delta}$$

where δ is the hydrodynamic boundary layer. The expression in terms of the Sherwood number is

$$Sh_x = \frac{k_m x}{D} = B\frac{x}{\delta}\frac{1}{\Delta} \tag{20.5}$$

The hydrodynamic thickness is known from the momentum balance. The equation based on the scaling is (for laminar flow)

$$\frac{\delta}{x} = B_1 \frac{1}{\sqrt{Re_x}} \tag{20.6}$$

where B_1 is a constant that depends on where the boundary layer is marked. The usual value of B_1 is 5 from the Blasius model and 4.64 from the cubic approximation of the von Kármán model. Using this in (20.5) and rearranging gives

$$Sh_x = B'\sqrt{Re_x}\frac{1}{\Delta}$$

where the constants have been combined to B'. Thus, if an estimate of Δ can be provided as a function of Sc (and Da_x for reacting systems) we are in a position to correlate the mass transfer parameters in terms of these dimensionless groups. This estimate for a non-reacting system based on the scaling analysis is

$$\boxed{\Delta = \frac{\delta_\mathrm{m}}{\delta} = \frac{1}{Sc^{1/3}}} \qquad (20.7)$$

the verification of which is left as Problem 5. Hence

$$Sh_x = B'\sqrt{Re_x}Sc^{1/3}$$

which is the form that should be used to correlate the data based on scaling considerations.

Note that the exponents on Re_x and Sc are provided by the scaling analysis, in contrast to dimensionless analysis. Only the fitting constant cannot be provided by the scaling model, and is found to have a value of 0.332 from detailed models that will be shown later.

Reacting systems: fast reactions

The scaling analysis for a reacting system is left as Problem 5. But one key result that is useful for a fast (first-order) reaction is

$$\Delta = \frac{1}{(Da_x Sc)^{1/2}} \qquad (20.8)$$

On substituting into (20.5) we have

$$Sh_x = B'\sqrt{Re_x Da_x Sc}$$

Further simplification results, since

$$\sqrt{Re_x Da_x Sc} = x\sqrt{k_1/D}$$

which is a type of Hatta number. Using this for Sh_x and rewriting in dimensional terms gives

$$\boxed{k_\mathrm{m} = \sqrt{Dk_1}} \qquad (20.9)$$

The model for fast pseudo-first-order reaction is recovered now and we get an equation that we have seen before in Section 10.5.1. Note that the hydrodynamic parameter (the thickness of the boundary layer) does not affect the mass transfer rate under this condition. The transport rate is simply a balance of diffusion and reaction, and the whole process is confined to a thin (reacting) layer near the solid boundary.

20.1.4   The low-flux case: integral analysis

Here we can directly proceed with the species mass balance, which is written for a binary system in terms of the concentration of A denoted as usual by C_A. A first-order reaction and validity low-flux model are assumed. The governing equation is repeated here for convenience:

$$v_x \frac{\partial C_A}{\partial x} + v_y \frac{\partial C_A}{\partial y} = D \frac{\partial^2 C_A}{\partial y^2} - k_1 C_A \tag{20.10}$$

Here the diffusion in the x-direction is neglected. A chemical-reaction term is added, assuming that the species can react in the fluid phase. This term can be set as zero if there is no reaction. The integral version of this equation is obtained by integration with respect to y from 0 to δ_m:

$$\frac{d}{dx}\left[\int_0^{\delta_m} v_x(C_A - C_{A\infty})dy\right] = N_{A0} - \int_0^{\delta_m} k_1 C_A \, dy \tag{20.11}$$

The solution can be obtained if some form for the velocity and the concentration profile is assumed as a function of y. A cubic profile is normally used and the constants in this cubic equation are fitted using four "boundary" conditions as shown before. The velocity approximation remains the same for the low-flux case since v_y is taken as zero for this approximation.

$$\frac{v_x}{v_\infty} = \frac{3}{2}\eta - \frac{\eta^3}{2} \tag{20.12}$$

where η is defined as y/δ.

The momentum thickness remains the same as before:

$$\delta(x) = \sqrt{\frac{280}{13}}\sqrt{\frac{\nu x}{v_\infty}} = 4.64\sqrt{\frac{\nu x}{v_\infty}} \tag{20.13}$$

The concentration profile is approximated by a cubic equation as

$$c_A = \frac{C_A - C_{As}}{C_{A\infty} - C_{As}} = \frac{3}{2}\frac{y}{\delta_m} - \frac{1}{2}\left(\frac{y}{\delta_m}\right)^3 \tag{20.14}$$

What boundary conditions are used to fit this cubic profile for concentration?

Note the manner in which the concentration is normalized. It takes a value of zero at the surface and one in the bulk gas. This puts the problem into conformity with momentum balance, where the dimensionless velocity also takes the same values. Also note that this expression is actually true only for the case of no reaction since $d^2 C_A/dy^2$ is zero at $y = 0$. For a reacting case, this term is equal to $k_1 C_{As}/D$ and a minor correction is needed. We ignore this correction here since the assumed concentration profile is also an approximation.

It is convenient to write Eq. (20.14) with the dimensionless ratio of the boundary-layer thickness defined by Δ as was done for heat transfer,

$$c_A = \frac{3\eta}{2\Delta} - \frac{\eta^3}{2\Delta^3} \tag{20.15}$$

On substituting the profiles for v_x and c_A into the integral mass balance Eq. (20.11) and integrating, we get the following equation for the variation of Δ with the x.

$$\frac{1}{10}\frac{d}{dx}[\delta\Delta^2] = \frac{D}{v_\infty\delta\Delta} - \frac{1}{4}\frac{k}{v_\infty}\delta\Delta$$

which has the same form as that for heat transfer if there is no reaction term. It may be noted that for the reacting case we take $C_{A\infty}$ to be zero to arrive at the above expression. For the non-reacting case the expression is general.

Now, on substituting for δ (the momentum thickness) as a function of x using (20.13), the following differential equation is obtained:

$$\frac{4}{3}x\frac{d}{dx}\Delta^3 + \Delta^3 = \frac{13}{14}\frac{1}{Sc} - 5\frac{k}{v_\infty}x\Delta^2 \tag{20.16}$$

which is a non-linear equation for Δ^3, solutions for which are now presented.

The no-reaction case

If there is no homogeneous chemical reaction, this reduces to a linear differential equation for Δ^3:

$$\frac{4}{3}x\frac{d}{dx}\Delta^3 + \Delta^3 = \frac{13}{14}\frac{1}{Sc} \tag{20.17}$$

The solution of the differential equation is not simple except for the students in Green Hall. They will immediately recognize that the equation is a (linear) differential equation of the Cauchy–Euler type with a non-homogeneous constant term and will solve it quickly to obtain

$$\Delta^3 = c_1 x^{-3/4} + \frac{13}{14}\frac{1}{Sc} \tag{20.18}$$

where the integration constant c_1 is obtained by applying the proper initial condition for Δ.

Note that the mass-transfer boundary layer need not start at the same point as where the momentum boundary layer starts. If the mass transfer starts at some point $x = X$ then $\Delta = 0$ at this point. The above equation now becomes

$$\Delta = 0.9076 Sc^{-1/3}[1 - (X/x)^{3/4}]^{1/3} \tag{20.19}$$

If the mass transfer starts right at $x = 0$ then

$$\Delta = \frac{13}{14}\frac{1}{Sc^{1/3}} \tag{20.20}$$

and the resulting expression for the Sherwood number is

$$Sh_x = 0.36 Re_x^{1/2} Sc^{1/3}$$

which is close to the exact solution shown in the next section. The error is 8%.

The reaction case

The reaction case is more complicated since the differential equation is non-linear due to the presence of the kx/v_∞ term in Eq. (20.16). Note that this term is the local Damköhler number, denoted as Da_x. Two cases can be analyzed as limiting cases.

1. **Slow reaction: small Da_x.** The solution can be represented as a power series:

$$\Delta = \frac{1}{Sc^{1/3}}\left[1 + a_1 Da_x + a_2 Da_x^2 + \cdots\right]$$

The coefficients can be determined by substitution into the differential equation. The method is a variation of the regular perturbation method discussed in Chapter 14. The leading term $a_1 = -5/7$ and is negative. Hence we conclude that the chemical reaction thins the boundary layer.

2. **Fast reaction: large** Da_x. The leading term is obtained by setting the entire LHS of (20.16) to zero:

$$\Delta = \sqrt{5}(ScDa_x)^{1/2}$$

Note that the exponent for the dependence on Sc is now 1/2. Thus the exponent is dependent on the reaction rate constant k.

20.1.5 The low-flux case: exact analysis

The governing equation is (20.10), and for the flat plate a similarity solution can be obtained for the low-flux case. The similarity solution closely follows the Blasius method discussed for thermal boundary layer in Section 15.6.3. Note that, if we make the low-flux approximation, then the momentum problem is uncoupled from the mass transfer problem. The solution for the momentum problem is the same as the Blasius equation discussed in Section 15.6.3. This equation is repeated here for convenience:

$$f''' + ff'' = 0 \tag{20.21}$$

Here the derivatives are with respect to a similarity variable η defined as

$$\eta = \frac{y}{2}\sqrt{\frac{v_\infty}{vx}}$$

and the velocity v_x and v_y can be related to the f function as shown in Section 15.6.3.

What are the boundary conditions which are needed for the above third-order ordinary differential equation?

For the mass transfer problem defined by Eq. (20.10) we introduce a dimensionless concentration c_A defined as

$$c_A = \frac{C_A - C_{A\infty}}{C_{As} - C_{A\infty}} \tag{20.22}$$

Note that this takes the values of 1 at the surface and 0 in the bulk, and is one minus the value used in the previous section. This definition is more useful here since we wish to focus on the effect of chemical reaction of a diffusing species near the wall.

The convection–diffusion equation is now changed in terms of the f function and the dimensionless concentrations. After some algebra the following equation is obtained:

$$\boxed{c_A'' + Scfc_A' - 4Da_xScc_A = 0} \tag{20.23}$$

which can be solved numerically together with the equation for f. The boundary conditions to be used are as follows: $\eta = 0$, $c_A = 1$; and $\eta = \infty$, $c_A = 0$.

The infinite domain has to be truncated to a finite domain for numerical integration. Usually $\eta = 5$ is sufficient.

The concentration gradient at the surface evaluated at $\eta = 0$ can then be related to the Sherwood number by

$$Sh_x = -\sqrt{\frac{Re_x}{2}}\left(\frac{dc_A}{d\eta}\right)_{\eta=0} \tag{20.24}$$

Table 20.1. Results for concentration gradients at the plate evaluated using the BVP4C program

Da_x	$Sc = 1$	$Sc = 10$	$Sc = 100$
0.0	0.6641	1.4563	3.1438
1.0	2.0407	6.3361	20.0416
10	6.3287	20.004	63.24

Illustrative results obtained by simultaneous solution of the f and c_A equation using the BVP4C code are shown in Table 20.1.

The no-reaction case is shown in the first line of Table 20.1. The data may be fitted with $Sc^{1/3}$ and the exponent is in conformity with the earlier analysis. The last line is for a fast reaction. It can be seen that the data may be fitted with $Sc^{1/2}$ as also indicated from the scaling analysis.

20.2 Mass transfer: the high-flux case

The high-flux case is distinguished from the low-flux case by assuming that $v_y(y = 0)$ is non-zero and dependent on the mass transfer rate. The mass transport and momentum transport now become a two-way coupled problem. The mass transfer rate and the **direction** of mass transfer (to or away from the plate) affect both the hydrodynamic and the mass-transfer boundary-layer thickness.

The classical film theory is a useful crutch for modeling systems with high flux and provides a simple framework to analyze the essential features of the two-way coupling. The advantages of the film model are the following. The effect of the cross-velocity term v_0 on the enhancement of transport can then be examined, at least to a first level of modeling. Another advantage is that the situation of simultaneous heat, mass, and momentum transport can be examined. It will be shown that the momentum (and heat) transport is affected by mass transport. We therefore review the applications of the film model to the high-mass-flux case. This is followed by a similarity solution method for the high-flux case.

20.2.1 The film model revisited

In the film model, the velocity profile in the entire mass (or heat)-transfer boundary layer is approximated as a constant, i.e., $v_y = v_0$. The continuity equation then makes $v_x = 0$ in the boundary layer, leading to the film being called the "stagnant" film. Also a thickness for the mass boundary layer (now called the film thickness) is assumed or estimated from the average mass transfer coefficient for the plate based on the low-flux model as $\delta_f = D/k_m^o$.

The equation for mass balance of A (written in terms of mass concentration here) then simplifies to

$$v_0 \frac{d\rho_A}{dy} = D\frac{d^2\rho_A}{dy^2} - k\rho_A \tag{20.25}$$

The no-reaction case will now be examined, since the goal is to evaluate the role of high flux, i.e., the contribution of the v_0 term on the LHS of the above equation. The velocity v_0

is caused by mass transfer and is often called the induced velocity. Its effects on heat and momentum transfer can be studied in a simple model framework.

First consider the solution for mass transport and the case of no reaction. The governing equation given by (20.25) with $k = 0$ can be integrated once to obtain the following equation:

$$\frac{\rho_{Ab} - (n_{A0}/v_0)}{\rho_{As} - (n_{A0}/v_0)} = \exp(v_0 \delta_f/D) \tag{20.26}$$

For a binary system with non-diffusing B, n_{A0}/v_0 is equal to ρ, the total density. Hence the above equation can be written in terms of the mass fraction as

$$\frac{\omega_{Ab} - 1}{\omega_{As} - 1} = \exp(v_0 \delta_m/D) \tag{20.27}$$

On rearranging, the induced velocity is calculated as

$$v_0 = \frac{D}{\delta_f} \ln\left(\frac{\omega_{Ab} - 1}{\omega_{As} - 1}\right) \tag{20.28}$$

and the corresponding flux is calculated as

$$n_{A0} = \frac{D\rho}{\delta_f} \ln\left(\frac{\omega_{Ab} - 1}{\omega_{As} - 1}\right) \tag{20.29}$$

Note that the flux is not linearly proportional to the concentration gradient. The linearity will hold only for the low-mass-flux case.

The above expression, Eq. (20.29), can also be written as the low-flux value multiplied by a correction factor as follows:

$$n_{A0} = k_m^0 \rho(\omega_{As} - \omega_{Ab}) \times \mathcal{F}_m \tag{20.30}$$

where $\mathcal{F}$ is a correction factor to the linear law. Note that k_m^0 is defined in the usual manner as D/δ_f. The correction factor is often referred to as the drift factor in the chemical engineering literature. On comparing Eq. (20.29) with Eq. (20.30) we find that the correction factor is given by

$$\mathcal{F}_m = \frac{\ln[(\omega_{Ab} - 1)/(\omega_{As} - 1)]}{\omega_{As} - \omega_{Ab}} \tag{20.31}$$

Noting that $(\omega_{As} - \omega_{Ab})$ can be written as $(1 - \omega_{Ab}) - (1 - \omega_{As})$, the correction factor is often written in terms of the mass fraction of the inert (non-diffusing) species:

$$\mathcal{F}_m = \frac{1}{\text{log mean mass fraction inert species}} \tag{20.32}$$

The results are identical to that obtained for evaporation in a stagnant gas in Section 10.1. The model is therefore often referred to as the stagnant-film model or simply as the film model in the chemical engineering literature. Note that the film is not really stagnant. Only the x-component of velocity was set as zero, and the model permitted finite velocity in the y-direction.

The effect on momentum transfer

The momentum equation based on the film model can be represented as

$$v_0 \frac{dv_x}{dy} = \nu \frac{d^2 v_x}{dy^2} \tag{20.33}$$

which has the same form now as the species mass balance equation. Hence the solution for the velocity profile is similar to the concentration profile,

$$\frac{v_x - v_0}{v_\infty - v_0} = \frac{\exp(v_0 y/v) - 1}{\exp(v_0 \delta_f/v) - 1} \tag{20.34}$$

The wall shear stress can now be calculated as $\mu\, dv_x/dy$ evaluated at $y = 0$, and the result is

$$\tau_f = \frac{\mu v_0/v}{\exp(v_0 \delta_f/v) - 1}$$

The limiting case of this is when v_0 is zero and the corresponding wall flux denoted with a $^\circ$ is given by

$$\tau_f^\circ = \frac{\mu v_\infty}{\delta_f}$$

which is just Newton's law of viscosity. The ratio of the above two values of the wall flux can be viewed as a measure of the effect of mass transfer on momentum transfer:

$$\frac{\tau_f}{\tau_f^\circ} = \frac{v_0 \delta_f/v}{\exp(v_0 \delta_f/v) - 1} \tag{20.35}$$

The parameter $v_0 \delta_f/v$ can be considered as the augmentation factor for momentum in the presence of high-flux mass transfer.

Effect on heat transfer

Similarly, the heat transport rate is augmented by mass transfer and can be accounted for by the blowing factor discussed in Sections 8.4.1 and 13.6.1.

20.2.2 The high-flux case: the integral-balance model

The integral analysis can also be used for the high-flux model and the equations are derived in the same manner as for the low-flux model, i.e., by integrating the transport equations with respect to y from zero to the corresponding boundary-layer thickness. The only difference is that v_y at the plate surface is not set as zero and is retained in the integral balance. Here we present the complete set of integral equations for the high-flux case:

$$\frac{d}{dx}\left(\int_0^\delta v_x(v_x - V)dy\right) - v_0 V = -\frac{\tau_w}{\rho} \tag{20.36}$$

$$\frac{d}{dx}\left[\int_0^{\delta_t} v_x(T - T_\infty)dy\right] - v_0 V(T_s - T_\infty) = \frac{q_w}{\rho c_p} \tag{20.37}$$

$$\frac{d}{dx}\left[\int_0^{\delta_m} v_x(\rho_A - \rho_{A,\infty})dy\right] - v_0(\rho_A - \rho_{A,\infty}) = \frac{n_w}{D} - \int_0^{\delta_m} k\rho_A\, dy \tag{20.38}$$

The solution proceeds by assuming suitable profiles for velocity, temperature, and concentration. Differential equations for the three boundary-layer thicknesses as a function of x can be derived. Often, to simplify the problem, a quadratic profile is assumed, rather than a cubic profile.

In general, these are coupled ordinary first-order equations, and can be numerically integrated and solved. Further details are not presented here.

20.2.3   The high-flux case: the similarity-solution method

The case of high mass flux can be analyzed using the similarity variable. This is similar to the Blasius approach used for momentum transfer, which is Eq. (15.78) for f. The only difference is in the boundary conditions.

The true similarity exists only if $v_y(y = 0)$ denoted as v_0 varies inversely as the square root of x. This can be ascertained by looking at the definition of v_y in terms of the f function. Recall from Section 15.6 that

$$v_y = \frac{1}{2}\left(\frac{\nu v_\infty}{x}\right)^{1/2}[\eta f' - f]$$

At $y = 0$ we have $\eta = 0$, and hence

$$v_0 = -\frac{1}{2}\left(\frac{\nu v_\infty}{x}\right)^{1/2} f$$

which is one of the boundary conditions at $\eta = 0$. The variable x remains in the boundary condition unless $v_0 = 0$ (the low-flux case) and hence a complete similarity is not achieved for the high-flux case. Let us assume that v_0 varies inversely as the square root of x. Let

$$v_0 = K\frac{1}{2}\left(\frac{\nu v_\infty}{x}\right)^{1/2}$$

The boundary condition (at $\eta = 0$) now is $f = -K$, where K is a constant called the blowing or sucking factor. The case of K positive refers to positive v_0, i.e., mass transfer from the surface (blowing), whereas K negative would be the case of mass transfer to the surface (sucking). Thus the effect of mass transfer on the momentum transfer can be examined using the similarity-solution method. The temperature and concentration equations remain the same, and all three equations (for f, θ, and c_A) can be solved simultaneously using BVP4C. The numerical implementation is left as an exercise. The key result that you may want to verify is that the boundary layer gets thicker if K is positive (blowing) and thinner if K is negative (sucking).

20.3   Mass transfer in turbulent boundary layers

The treatment of mass transfer in turbulent flow follows the same lines as that of heat transfer. The equations are very similar. However, a complication arises for mass transfer in liquids, where Sc can be as high as 1000! Here the concentration gradients are confined mostly to the viscous layer and to some extent to the buffer zone, and any error in the velocity profile in this region will affect the predictions considerably. We illustrate the difference by looking at alternative models for the mixing length near the wall.

Near the wall, the eddy viscosity varies as y^3 rather than as y (Problem 4, Chapter 17). The Prandtl-mixing-length model with $l = ky$ would predict a square dependence. A more precise form of l is important in modeling mass transfer processes where the concentration boundary layer is expected to be thin (for higher Sc). One such model was suggested by van Driest (1956), who proposed a simple exponential damping factor for l:

$$l = \kappa y\left[1 - \exp\left(-\frac{y^+}{26}\right)\right] \tag{20.39}$$

This permits a smooth transition from the viscous layer to the turbulent core without the need to introduce an artificial buffer zone. This is one of many equations suggested for the near-wall region, and many improvements have been put into place. For example, Cebeci and Smith (1974) have proposed that the constant 26.0 should be replaced by a function of dimensionless quantities involving pressure gradients, the mass transfer rate, the fluid compressibility, and other factors.

A modified van Driest equation has also been proposed by Hanna *et al.* (1981). According to their model a denominator term is added to the original van Driest model.

$$l = \kappa y \frac{1 - \exp(-y^+/26)}{\sqrt{1 - \exp(-0.26 y^+)}} \tag{20.40}$$

This also predicts correctly that the turbulent viscosity will be proportional to y^3 near the wall. Taking the limit as $y \to 0$, the eddy viscosity near the wall can be represented as

$$\nu_t = \nu \left(\frac{y^+}{C} \right)^3 \tag{20.41}$$

where C includes all the constant terms in the modified model. An adjusted value of 14.5 that is based on experimental results is assigned to this. The relation for the mass transfer coefficient based on the modified van Driest model is now examined.

The mass flux is given by

$$N_A = -(D + \nu_t/Sc_t)\frac{dC_A}{dy} \tag{20.42}$$

where Sc_t is the turbulence Schmidt number. Note that ν_t/Sc_t is equal to D_t, the turbulent diffusivity.

The flux N_A is assumed to be a constant equal to the wall mass flux N_{Aw} in the boundary layer.

Also we introduce a dimensionless total momentum diffusivity as was done earlier:

$$\nu_T^+ = 1 + \frac{\nu_t}{\nu} \tag{20.43}$$

Equation (20.42) can now be expressed as

$$\frac{N_{Aw}}{\nu} = -\left[\frac{1}{Sc} + \frac{1}{Sc_t}(\nu_T^+ - 1) \right] \frac{dC_A}{dy} \tag{20.44}$$

Also a dimensionless length $y^+ = y(\tau_w/\rho)^{1/2}/\nu$ is introduced.

Equation (20.44) can be formally integrated across the boundary layer to obtain an expression for the wall mass flux:

$$(\sqrt{\tau_w/\rho})\frac{C_{A\infty} - C_{As}}{N_{Aw}} = \int_0^{\delta^+} \frac{1}{1/Sc + (1/Sc_t)(\nu_T^+ - 1)}\, dy^+ \tag{20.45}$$

Using the definition $\tau_w = C_{fx}\rho V^2/2$ and k_m as the transfer coefficient,

$$\frac{N_{Aw}}{V(C_{A\infty} - C_{As})} = \frac{k_m}{V} = St_m = Sh/(ReSc)$$

Equation (20.45) can be represented as

$$\frac{\sqrt{C_{fx}/2}}{St_m} = \int_0^{\delta^+} \frac{1}{1/Sc + (1/Sc_t)(v_T^+ - 1)} \, dy^+$$

Integration can be done if the variation of v_T^+ with y^+ is specified. For the modified van Driest model, the following equation for v_T^+ can be derived:

$$v_T^+ = 1 + \left(\frac{y^+}{C}\right)^3$$

Also it is convenient to use ∞ as the upper limit of integration since the concentration gradients are confined to only small values of y^+ compared with δ^+. The results after integration and some rearrangement can be expressed as

$$\frac{Sh}{ReSc} = St_m = \frac{1}{17.5} \frac{\sqrt{C_{fx}/2}}{Sc^{2/3}}$$

which fits the data with some adjustment in the numerical constant.

20.4   Mass transfer at gas–liquid interfaces

20.4.1   Turbulent films

We mainly focus on turbulent mass transfer here since the transport in laminar film was examined earlier in Section 12.3.2. A useful review of mass transfer in turbulent-film flow is provided in the paper by Bin (1983).

The convection–diffusion equation is now augmented by incorporating an eddy diffusion coefficient D_t for mass transfer,

$$v_z(y) \frac{\partial C_A}{\partial z} = (D + D_t) \frac{\partial^2 C_A}{\partial y^2} \tag{20.46}$$

The velocity profiles are steeper in turbulent flow and hence a constant value of v equal to v_{max} can be used in the interface region. The main problem emerges in the formulation of the expression for the turbulent eddy diffusion coefficient in Eq. (20.46). Levich (1962) suggested the following correlation:

$$D_t = ay^n$$

where y is the distance normal to the interface. The value of $n = 2$ seems to fit the data best (which can be justified by a scaling argument) while the parameter a was fitted by the following equation:

$$a = 1.17 \times 10^{-6} (g^2/v)^{1/3} Re_L^{1.448}$$

where Re_L is the Reynolds number for film flow. Other correlations are available; see for example Grossman and Heath (1984) and Won and Mills (1982). The role of surface tension in the dampening of the turbulence at the interface does not appear to have been clearly established.

One of the first analytical studies based on the above model was that by King (1966), who obtained two asymptotic solutions: (i) for a short contact time and (ii) for a long contact time. For a very short contact time the eddy diffusivity has no effect on the rate of mass transfer since the penetration distance is small and the limit of penetration is within the region where the turbulence is damped. The classical penetration model holds for this case.

For a long contact time the concentration profiles become fully established and the steady-state mass-transfer model holds. The solution then gives

$$k_L = \frac{n}{\pi} \sin(\pi/n) a^{1/n} D^{1-1/n}$$

The analysis of simultaneous heat and mass transfer was done in a paper by Grossman and Heath (1984), and can be a useful case-study problem.

20.4.2   Single bubbles

Mass transfer from a rigid bubble to a fluid can be approached from a theoretical angle. The velocity profile is computed and substituted into the convection–diffusion equation. For axisymmetric flows around a sphere the governing equation is

$$v_r \frac{\partial C_A}{\partial r} + \frac{v_\theta}{r} \frac{\partial C_A}{\partial \theta} = \frac{D}{r^2} \left[\frac{\partial}{\partial r} \left(r^2 \frac{\partial C_A}{\partial r} \right) + \frac{1}{\sin \theta} \frac{\partial}{\partial \theta} \left(\sin \theta \frac{\partial C_A}{\partial \theta} \right) \right]$$

The equation must be solved with the velocity profile computed from momentum transfer considerations. The numerical solution up to Re of 48 was done by Conner and Elghobashi (1987). Note that the mass transfer boundary layer is usually thin, and hence an accurate description of the velocity near the bubble is needed. Also the flow field for the wake region has to be accurately captured. Further complications arise due to the presence of surface contaminants that affect the interfacial boundary conditions due to surface-tension gradients. A computational study by Dani et al. (2006) is useful in this context.

The results from computational studies indicate that considerable variance in the local mass transfer coefficient is seen as a function of θ. Illustrative results are shown in the book by Clift et al. (1978) and other sources. For example, for $Pe = 1000$ and $Re = 1$ the local Sherwood number is about 15 at the forward stagnation point ($\theta = \pi$) and decreases to a value close to 2 near the rear stagnation point ($\theta = 0$). Note that the value of the Sherwood number is two for no-flow conditions and the results show that the solute is swept away near the rear stagnation point. The average value is about 11 for this case. The results of Conner and Elghobashi (1987) are similar.

For practical applications an average mass transfer coefficient is used. The results for the average mass transfer coefficient can be represented in terms of a Sherwood number expressed as a function of Pe. For rigid spheres the average mass transfer coefficient can be fitted to within 2% accuracy by the following correlation:

$$Sh = 1 + (1 + Pe)^{1/3}$$

in the Reynolds number range of one. It may be noted that, for a stagnant fluid ($Pe = 0$), Sh has a value of two, which is the limiting value.

For the Hadamard–Rybczynski profile the following correlation was shown to be suitable:

$$Sh = 1 + (1 + 0.564 Pe^{2/3})^{3/4}$$

20.4.3   Bubble swarms

Mass transfer from a swarm of bubbles to a liquid finds many important applications in industrial processes. For example, bubble-column reactors are used in a wide variety of processes, including oxidation, hydroformylation, Fisher–Tropsch synthesis, etc.; see Ramachandran and Chaudhari (1983). In this contactor we often have a swarm of bubbles creating the circulation in the liquid, and hence the prediction of mass transfer coefficient is a difficult task since the velocity field cannot be accurately calculated. Furthermore, one needs to classify the flow regime since the hydrodynamics is different for small spherical bubbles, large ellipsoidal bubbles and large spherical cap bubbles. Also the results depend on whether there is an internal circulation within the bubbles or not. We can only provide a brief (empirical) discussion on these topics. The book by Clift *et al.* (1978) provides additional details.

The correlation of Calderbank and Moo-Young (1961) is widely used in the literature for bubbles of diameter greater than 2.5 mm:

$$Sh = 0.42 Gr^{1/3} Sc^{1/2}$$

where Gr is defined as

$$Gr = \frac{d_B^3 \rho_L g \, \Delta \rho}{\mu_L^2}$$

The penetration model is used as an alternative, with the contact time taken as d_B/v_B, where v_B is the effective rise velocity of the swarm of bubbles in the liquid.

Now we move to a different theme and revisit the concept of Taylor dispersion introduced in Section 12.5.

20.5   Taylor dispersion

Here we show the classical transient analysis of Taylor and how the value of D_E, the axial dispersion coefficient, introduced in Section 12.5.1 can be derived. Consider a pulse of dye introduced into a liquid laminar flow as shown in Fig. 20.2. This pulse is sheared due to convection. This creates a concentration gradient in the radial direction and causes the spread of the dye by diffusion. After some entry region, the dye is smeared into a slug, and further on it appears that the dye is spreading on either side of the slug by a diffusion-type process. The process is sketched in Fig. 20.2. The spreading of the pulse is called the dispersion, and is a combined effect of convection and radial diffusion. The goal of

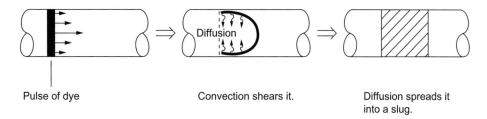

Pulse of dye   Convection shears it.   Diffusion spreads it
                                                               into a slug.

Figure 20.2 An illustration of the concept of Taylor diffusion.

the Taylor dispersion analysis is to derive an expression for the spread of the dye using a Fick's-law type of model, for the dispersion process.

Consider the convection–diffusion model, now with the time derivative but no reaction. This is reproduced below:

$$\frac{\partial C_A}{\partial t} + 2\langle v\rangle(1-\xi^2)\frac{\partial C_A}{\partial z} = \frac{D}{R^2}\left[\frac{1}{\xi}\frac{\partial}{\partial \xi}\left(\xi\frac{\partial C_A}{\partial \xi}\right)\right] \tag{20.47}$$

Dimensional variables are used here, except for the radial coordinate, which is represented by ξ equal to r/R.

The coordinate system can be transformed to moving coordinates as

$$z^* = z - \langle v\rangle t$$

and

$$t^* = t$$

Here z^* is the axial position in a coordinate system moving with the average velocity. The variable t^* is the same time variable t, but a different symbol is used to distinguish the two sets of independent variables in the two systems (stationary vs. moving). Note that t^* and z^* here have the units of time and distance, respectively, and are not dimensionless in this section.

The partial derivatives in (20.47) have to be transformed using the chain rule. The following results apply:

$$\frac{\partial C_A}{\partial t} = \frac{\partial C_A}{\partial t^*}\frac{\partial t^*}{\partial t} + \frac{\partial C_A}{\partial z^*}\frac{\partial z^*}{\partial t}$$

Note that

$$\frac{\partial z^*}{\partial t} = -\langle v\rangle$$

and

$$\frac{\partial t^*}{\partial t} = 1$$

and hence

$$\frac{\partial C_A}{\partial t} = \frac{\partial C_A}{\partial t^*} - \langle v\rangle\frac{\partial C_A}{\partial z^*}$$

By similar mathematical manipulations, the student should verify that

$$\frac{\partial C_A}{\partial z} = \frac{\partial C_A}{\partial t^*}\frac{\partial t^*}{\partial z^*} + \frac{\partial C_A}{\partial z^*}\frac{\partial z^*}{\partial z} = \frac{\partial C_A}{\partial z^*}$$

since the term $\partial t^*/\partial z^*$ is zero and the term $\partial z^*/\partial z$ is one.

Hence the model in the moving reference frame is

$$\frac{R^2}{D}\frac{\partial C_A}{\partial t^*} + \frac{\langle v\rangle R^2}{D}[2(1-\xi^2)-1]\frac{\partial C_A}{\partial z^*} = \frac{1}{\xi}\frac{\partial}{\partial \xi}\left(\xi\frac{\partial C_A}{\partial \xi}\right) \tag{20.48}$$

where the term D/R^2 in the radial-diffusion term has been multiplied out. The time-derivative term can be dropped if $t > R^2/D$. The interpretation is that after time greater than the radial-diffusion time in a moving coordinate the system reaches a pseudo-steady-state case with the convection term balancing the diffusion term. Referring to Fig. 20.2, this means that the smearing of the tracer due to diffusion has already taken place (the third stage of the diagram). The corresponding distance will be $\langle v \rangle R^2/D$, and the analysis will be valid if the observation point is larger than this length.

Also, in a moving reference frame $\partial C_A/\partial z^*$ is not expected to vary significantly and may be assumed to be a constant. Let this constant be called A:

$$A = \frac{\partial C_A}{\partial z^*} = \frac{\partial \langle C_A \rangle}{\partial z^*} \tag{20.49}$$

where $\langle C_A \rangle$ is the cross-sectional average concentration. Hence Eq. (20.48) reduces to

$$(1 - 2\xi^2)\frac{A\langle v \rangle R^2}{D} = \frac{1}{\xi}\frac{\partial}{\partial \xi}\left(\xi \frac{\partial C_A}{\partial \xi}\right) \tag{20.50}$$

This can be integrated twice with respect to ξ to obtain an approximate expression for the variation of the concentration as a function of the dimensionless radial position. The result after applying the two boundary conditions of no flux both at the center and at the wall is

$$C_A = \left(\frac{\xi^2}{4} - \frac{\xi^4}{8}\right)\frac{A\langle v \rangle R^2}{D} + C_0 \tag{20.51}$$

where C_0 is the unknown center-line concentration.

On taking the cross-sectional average we have

$$\langle C_A \rangle = \frac{1}{12}\frac{A\langle v \rangle R^2}{D} + C_0 \tag{20.52}$$

Taking the flow average

$$C_b = \frac{1}{16}\frac{A\langle v \rangle R^2}{D} + C_0 \tag{20.53}$$

The difference may expressed as

$$C_b = \langle C_A \rangle - \frac{A\langle v \rangle R^2}{48D}$$

On substituting the value of A from Eq. (20.49) into the above equation and then multiplying throughout by $\langle v \rangle$ we have

$$\langle v \rangle C_b = \langle v \rangle \langle C_A \rangle - \frac{1}{48}\frac{\langle v \rangle^2 R^2}{D}\frac{\partial \langle C_A \rangle}{\partial z^*}$$

The first term can be viewed as the mass of A crossing any axial position (per unit cross-sectional area). The second term is the mass of A crossing computed using the mean (area-averaged) concentration. Hence the last term is an extra term that can be interpreted as a type of diffusion or a dispersion term, and the expression can be written as

$$\langle v \rangle C_b = \langle v \rangle \langle C_A \rangle - D_E \frac{\partial \langle C_A \rangle}{\partial z^*} \tag{20.54}$$

By comparing the last two equations, the following expression for D_E can be obtained:

$$\boxed{D_E = \frac{1}{48}\frac{\langle v \rangle^2 R^2}{D}} \tag{20.55}$$

which is the classical result obtained by Taylor for the dispersion coefficient in laminar flow.

Although it wad derived originally for laminar flow, the concept has been used in many applications, including chemical-reactor analysis, where the deviation from plug flow is accounted for by including an extra dispersion term. The spread of a solute in a chromatography column is another example where the concept of dispersion is useful.

Summary

- Convective mass transfer in boundary layers in laminar flow over a flat plate is a well-studied problem and is amenable to a complete theoretical analysis. The problem is similar to that for heat transfer with a constant wall temperature, and similar tools can be used for this problem as well. The Schmidt number is the key dimensionless group which determines the relative thickness of the momentum and mass boundary layers and is similar to the Prandtl number for heat transfer problems.

- An additional consideration unique to mass transfer is that the velocity v_y at the surface can be non-zero. The effect of this becomes important in high-mass-flux cases. In many cases this effect may dominate. In such cases it is reasonable to assume a constant boundary-layer thickness and use the simplistic film model.

- Mass transfer processes are often accompanied by chemical reactions. We studied the effect of chemical reactions on the boundary-layer thickness. An important application is in chemical vapor deposition processes.

- The modeling is not so clear cut for turbulent flow. In addition to dealing with the complexities of the flow, one needs an appropriate model for the eddy diffusivity, or, equivalently, the turbulent Schmidt number. A constant value of 0.7 is often used for this parameter as a first estimate. Also the mass-transfer boundary layer is often very thin and contained within the laminar sublayer or buffer zone. A detailed description of velocity profiles in these zones is therefore needed in order to model mass transfer under these conditions. We examined some approaches suggested in the literature.

- For turbulent reactive systems, the local micro-mixing ($\overline{C'_A C'_A}$ parameter) is important for fast non-linear reactions. This is a field of considerable research at this moment and is important in many applications, e.g., combustion processes.

- The concept of dispersion arises from averaging of the model equation; the dispersion coefficient turns out to be the key parameter and the dispersion Péclet number is the important dimensionless group. This can be predicted from theory for simple flows such as laminar flow. In general this is a fitted parameter based on experimental data. Dispersion effects are important in pollutant transport in atmospheres, chromatography, etc.

ADDITIONAL READING

The classic book by Levich (1962) is an important source on convective mass transfer. In addition, this book has detailed information on the effect of surface-tension-driven flows on mass transfer. Another useful source for convective mass transfer is the book by Ghiaasiaan (2011).

Mass transfer from drops and bubbles has been examined in detail in the monograph by Clift, Grace, and Weber (1978).

Problems

1. Develop an expression for the average mass transfer coefficient for a plate of length L, where part of the flow is turbulent for part of the plate. Assume that there is no transition and that the flow is turbulent right at the point where $Re_x = 2 \times 10^5$ and onwards. Note that integration in two parts is needed.

   Use the expression to find the average mass transfer coefficient for flow of air at velocity 7.5 m/s blowing over a pond of water 300 m long. At what point does the flow start to become turbulent?

2. The factor j_D is also used in the correlation. This is defined as

   $$j_D = St(Sc)^{2/3} = \frac{k_m}{v_{ref}}(Sc)^{2/3}$$

   where St is the Stanton number for mass transfer. Show that the following correlations apply for a flat plate:

   $$j_D = 0.664 Re^{-1/2}$$

   for laminar flow, and

   $$j_D = 0.037 Re^{-0.2}$$

   for turbulent flow.

3. Verify that the j-factor is related to the drag coefficient by the relation

   $$j_D = \frac{c_D}{2}$$

   for mass transfer for flow over a flat plate.

4. A silicon substrate 10 cm long is exposed to a gas stream containing an arsenic precursor so that a GaAs film can be deposited on the surface. Estimate the mass-transfer coefficient, and the average rate of mass transfer. If the deposition is mass-transfer-controlled, how would the deposit thickness vary along the length of the plate? In practice, uniform deposition is needed and the flat plate may therefore not be a good arrangement under these conditions.

5. The scaling analysis for the mass-transfer boundary-layer thickness is parallel to that for heat transfer analysis discussed in Section 18.1.2. Use the same method to show that Δ is proportional to $Sc^{-1/3}$ for a non-reacting system.

   Extend the analysis for a reacting system and suggest a suitable expression for Δ.

   Then show that, in the limit of fast reaction, $Da_x \gg 1$, the expression simplifies to Eq. (20.8).

6. **Secondary-emission measurement: a case-study problem.** An indirect way of measuring of secondary emission from ponds or large bodies of water used in waste treatment is to measure the concentration and velocity over the surface. The data can then be fitted to a model of the type presented in Section 20.1.4.

   In a typical experiment benzene concentration and velocity were measured at various locations above the water surface and the data are as follows.

Location (cm)	Velocity (cm/s)	Concentration (ng/l)
10	230	470
100	700	390

Fit the data to the boundary-layer model and evaluate the rate of emission from the surface. D for benzene is 0.077 cm^2/s in air.

Usually the data are measured at about six points above the surface, but we use only two points in order to simplify the calculations.

Write a critique on this technique of secondary-emission measurement.

7. Use BVP4C to simulate the Blasius flow with non-zero velocity in the y-direction at the plate. Use the assumption in Section 20.2.3 so that a true similarity exists. Examine the effect of the parameter K (the blowing or suction parameter) on the wall shear stress and the local mass transfer coefficient.

8. **Chemical vapor deposition (CVD) on an inclined susceptor: a case-study problem.** An important application of convective mass transfer theory is in CVD processes employed to coat surfaces with thin films of metals or semi-conductors. In fact, this turns out to be an example of simultaneous heat and mass transfer with chemical reaction.

A flat plate is used as a susceptor to deposit a material for a semi-conductor application. The inclination of the plate is θ to the horizontal. The flow will then be governed by the Falkner–Skan equation described in Section 15.6.

Your goal is to set up a mass transfer model and evaluate the variation of the local mass transfer coefficient at various locations in the plate. Usually a constant or nearly constant boundary-layer thickness is preferable. Investigate the conditions leading to this requirement.

9. **Mass transfer from a bubble.** Calculate the mass transfer coefficient for the air–water system for bubbles rising at a gas velocity of 5 cm/s in a pool of stagnant liquid. Use the penetration model for k_L.

10. **Axial dispersion in channel flow.** Consider the pressure-driven laminar flow in a channel of height $2h$. Derive the following formula for the axial dispersion coefficient:

$$D_E = \frac{2h^2 \langle v \rangle^2}{105D}$$

Here $\langle v \rangle$ is the averaged velocity and D is the diffusivity of the species under consideration.

11. **Axial dispersion in turbulent flow.** Taylor showed that the following expression is suitable for the calculation of the axial dispersion coefficient for turbulent flow in a pipe:

$$D_E = 10RV_f$$

Here R is the pipe radius and V_f is the friction velocity. Show the basis for this equation, and write a review of the many correlations proposed for this and the improvements suggested for the original Taylor model.

12. **A model for a hemodialyser with simulation of the patient–artificial-kidney system: a case-study problem.** A useful case study is the paper by Ramachandran and Mashelkar (1980), where a mesoscopic model with axial dispersion was used for the blood side and plug flow was used for the dialysate side. The solution was analytic and was then combined with a compartmental model to simulate the patient–artificial-kidney system. Your goal in this case study is to review the paper and write a comprehensive MATLAB simulation code.

The second task is to do a parametric sensitivity analysis and then suggest how the model can be used to optimize conditions to get the desired end results. This case study combines several key modeling concepts discussed in this text and may be viewed as a capstone example.

13. **A model for chromatographic separation: a case-study problem.** An important application of Taylor dispersion is in chromatography. Here pulses of a mixture of solutes are introduced into one end of a packed-bed reactor containing an adsorbent and washed through the bed with a solvent. Since different species adsorb and diffuse at different rates, we obtain a separation. The axial dispersion will often interfere in the separation (it causes a pulse broadening).

Your goal is to set up a model for chromatographic processes wherein you should include the adsorption on the solids as an additional term. You can examine how the dispersion coefficient changes with the adsorption equilibrium constant.

21 | Mass transfer: multicomponent systems

Learning objectives

This chapter has several learning objectives which are listed below.

- To understand the Stefan–Maxwell constitutive model for multicomponent mass transfer.
- To be able to set and solve common problems in multicomponent mass transfer in non-reacting as well as reacting systems.
- To learn the concept of the generalized version of Fick's law and the Fick matrix for multicomponent diffusivities.
- To understand the similarity in the relations for binary diffusion and multicomponent diffusion using the Fick matrix.
- To appreciate that mass transfer can be caused by other driving forces such as gradients in pressure and even in temperature.
- To be able to set and solve simple problems in pressure diffusion and thermal diffusion.

Mathematical prerequisites

- Matrix representation for the solution of a set of first-order differential equations (see Chapter 2, Section 2.2.2).
- Solution methods for boundary-value problems (see Section 10.9).

The main focus of this chapter is on models for multicomponent (systems with more than two components) diffusion and applications of these to common problems in mass transfer. The basic constitutive equation for modeling mass transport is usually based on Fick's law. The key postulate of Fick's law is that the flux of each component is proportional to its own concentration gradient. In other words, the diffusion rate for each component is treated independently by invoking its own diffusivity, and interaction effects are ignored. Strictly speaking, this assumption is valid only for binary systems for ideal gases. However, in view of its simplicity it is also used as an approximation for multicomponent systems. Here some value of a pseudo-binary diffusion coefficient for each of the species in the mixture is assumed, and Fick's law is used as the transport law for each of the species. This is a good approximation under several conditions, e.g., (i) if the components of interest are present in small quantities in a mixture with a large excess of one component (usually referred to as the solvent), (ii) for mixtures where the components are of similar size and chemical type. If these conditions are not satisfied, more advanced constitutive models for multicomponent diffusion are needed. Also such systems exhibit many complexities, as pointed out in Section 9.6 (e.g., reverse diffusion). The goal of this chapter is to introduce transport laws for multicomponent systems and to demonstrate tools by the use of which multicomponent models are solved. In particular, the Stefan–Maxwell model is introduced and applied to many illustrative problems.

The framework for modeling multicomponent systems is rooted in the concepts from irreversible thermodynamics. It is seen from this formalism that diffusion can also be caused by pressure and temperature gradients, and this chapter includes a brief analysis of such systems as well.

21.1 A constitutive model for multicomponent transport

21.1.1 Stefan–Maxwell models

In this section we show the Stefan–Maxwell (S–M) model as a constitutive model for multicomponent mass transfer. An alternative representation based on a matrix of diffusivities, which is actually an inverted form of the S–M model, is presented in Section 21.4. We start with binary diffusion first.

Binary revisited

It is useful to recall some relations for binary diffusion that follow directly from Fick's law. In particular, the difference in velocity between the species A and B can be shown to be equal to

$$v_A - v_B = -D_{AB} \, \nabla \ln(y_A/y_B) \tag{21.1}$$

The velocity can be with reference to either a stationary frame or a moving frame, since the difference of two velocity values rather than an individual species velocity appears on the RHS of the above equation. Here y is the mole fraction of the subscripted species.

The flux of A (in a stationary frame of reference) is related to the velocity of A by the relation $N_A = Cy_A v_A$; a similar relation for N_B holds as well. Hence

$$\frac{N_A}{Cy_A} - \frac{N_B}{Cy_B} = -D_{AB} \, \nabla \ln(y_A/y_B)$$

With minor rearrangement, this reduces to the following equation:

$$\nabla y_A = \frac{y_A N_B - y_B N_A}{C D_{AB}} \tag{21.2}$$

This is the same as Fick's law, but the form is inverted; i.e., the mole fraction of a species (A or B) is given as a combination of the fluxes of the two species.

The following equation holds as well since the velocity difference can be taken also with respect to a moving frame in Eq. (21.1):

$$\nabla y_A = \frac{y_A J_B - y_B J_A}{C D_{AB}} \tag{21.3}$$

which is again an inverted form with the mole-fraction term appearing explicitly.

21.1.2   Generalization

Generalization of this way of writing the diffusion model for a binary leads to the S–M model for multicomponent systems:

$$-\nabla y_A = \sum_{j=1}^{n_s} \frac{y_j N_i - y_i N_j}{C D_{ij}} \tag{21.4}$$

where n_s is the number of components and D_{ij} is the binary pair diffusivity for the $i-j$ pair. Note that the equation can also be written in terms of the diffusive fluxes as

$$-\nabla y_A = \sum_{j=1}^{n_s} \frac{y_j J_i - y_i J_j}{C D_{ij}} \tag{21.5}$$

You should verify that the two forms are equivalent. For problem-solving purposes, the form in terms of the combined flux (Eq. (21.4)) is more useful. The constitutive equations are supplemented with the species mass-balance equations, which have the same form as in Chapter 9. Thus, for a steady-state case we have

$$\nabla \cdot (N_A) = R_A \tag{21.6}$$

The simultaneous solution to Eqs. (21.4) and (21.6) will be our focus now, and we shall present the application to illustrative multicomponent mass transfer problems. We show first a number of examples that can be solved analytically. This is followed by examples where numerical solutions are more convenient. We consider the problem in three categories in the same spirit as the binary problems considered in Chapter 10, *viz.* (i) non-reacting systems, (ii) heterogeneous reactions, and (iii) homogeneous reactions. The principal difficulty, compared with binary problems, is that the equations are flux-implicit and hence the species-balance equation (21.6) cannot be directly expressed in terms of species mole fractions (or species concentrations). Many methods have been developed for this, mainly based on matrix manipulations, and the book by Taylor and Krishna (1993) provides considerable details on this. Here we use simple numerically based tools and focus on applications to practical problems, which is also a somewhat novel feature of this chapter.

21.2 Non-reacting systems and heterogeneous reactions

21.2.1 Evaporation in a ternary mixture

This is the multicomponent analogue of the Arnold problem. Here a species A is evaporating into a mixture of inert gases B and C in a tube. The flux of A (evaporation rate) is to be calculated.

Since B and C are inert species, we can set

$$N_B = 0 \text{ and } N_C = 0$$

which is the determinacy condition. Also

$$y_A = 1 - y_B - y_C$$

With these substitutions the Stefan–Maxwell expressions for B and C are as follows:

$$\frac{dy_B}{dz} = \frac{N_A}{CD_{AB}} y_B$$

and

$$\frac{dy_C}{dz} = \frac{N_A}{CD_{AC}} y_C$$

Each equation can be integrated separately here. Integration with the boundary condition at $z = L$ (the bulk gas) gives

$$y_B = y_{BL} \exp[-N^*(1 - z/L)]$$

and

$$y_C = y_{CL} \exp[-N^* \beta(1 - z/L)]$$

where N^* is a dimensionless flux defined as

$$N^* = \frac{N_A}{CD_{AB}/L}$$

and β is the diffusivity ratio defined as D_{AB}/D_{AC}.

Finally, on applying the boundary condition at the surface (where the mole fraction of A is given by equilibrium considerations) we have

$$y_{As} = 1 - y_{BL} \exp(-N^*) - y_{CL} \exp(-N^* \beta) \tag{21.7}$$

which provides a transcendental equation for finding N^*.

We also note that the binary expression is recovered if $\beta = 1$, i.e., when the binary diffusivity of A in B is the same as that for A in C, $D_{AB} = D_{AC}$. Note that the (dimensionless) flux of A, N^*, which is the required quantity, is given by an implicit equation now rather than in an explicit manner.

Evaporation of a binary liquid mixture

Here we consider a similar problem with a liquid containing now a mixture of A and B. This mixture is evaporating into an inert gas C. The problem is now that of how to compute both N_A and N_B.

Since C is an inert species we can set $N_C = 0$, which is the required deteminacy condition. Also, since the mole fractions add up to unity, we have $y_C = 1 - y_A - y_B$.

When these simplifications are introduced into the S–M equations we obtain the following equations for the mole-fraction profiles of A and B:

$$\frac{dy_A}{dz} = y_A \left(\frac{N_B}{CD_{AB}} + \frac{N_A}{CD_{AC}} \right) + y_B \left(\frac{N_A}{CD_{AC}} - \frac{N_A}{CD_{AB}} \right) - \frac{N_A}{CD_{AC}} \tag{21.8}$$

and

$$\frac{dy_B}{dz} = y_A \left(\frac{N_B}{CD_{BC}} - \frac{N_B}{CD_{AB}} \right) + y_B \left(\frac{N_A}{CD_{AB}} + \frac{N_B}{CD_{BC}} \right) - \frac{N_B}{CD_{BC}} \tag{21.9}$$

The boundary conditions needed are as follows. At the evaporating surface, $z = 0$, the mole fractions are given by the vapor-pressure values (assuming an ideal liquid mixture). In the bulk gas, $z = L$, the mole fractions are specified from the bulk-gas values. Often a value of zero is used as a simplification.

The above differential equations are of first order and look like initial-value problems (IVPs) and candidates for explicit marching in z or solvers such as ODE45. However, the fluxes are unknown and have to be computed by imposing the boundary condition at $x = L$. Hence this is a boundary-value problem! We discuss two methods for solution of this problem.

The matrix method

Equations (21.8) and (21.9) can be expressed in matrix form as

$$\frac{d\mathbf{y}}{dz} = \tilde{A}\mathbf{y} + \mathbf{R}$$

where y is the solution vector; in the present case the solution vector is of dimension two, representing y_A and y_B.

The term A above is the coefficient matrix and R contains the non-homogeneous terms. Since both the coefficient matrix and the non-homogeneous terms are constants, the solution can be represented in matrix form:

$$\mathbf{y} = (\mathbf{y}_0 + \tilde{A}^{-1}\mathbf{R})\mathrm{expm}(Az) - \tilde{A}^{-1}\mathbf{R}$$

where $\mathbf{y}_0$ is the mole fraction at the evaporating surface.

Now, using the conditions in the bulk gas, the following algebraic relation is obtained:

$$\mathbf{y}_L = (\mathbf{y}_0 + \tilde{A}^{-1}\mathbf{R})\mathrm{expm}(AL) - \tilde{A}^{-1}\mathbf{R}$$

This equation contains the unknowns N_A and N_B, which are buried in the $\tilde{A}$ matrix. The solution of this algebraic equation gives the fluxes. This can be accomplished, for instance, by using a non-linear algebraic solver such as FSOLVE in MATLAB.

The BVP method

A second method is to solve the equations directly as a BVP (boundary-value problem). Two additional differential equations for the mole fluxes,

$$\frac{dN_A}{dz} = 0$$

and

$$\frac{dN_B}{dz} = 0$$

are added to close the problem. These equations are now supplemented with Eqs. (21.8) and (21.9). Thus we have four first-order differential equations to solve. We need therefore four boundary conditions in total. The boundary conditions for y_A and y_B are specified at both $z = 0$ and $z = L$, providing us with the four conditions needed to close the problem. The conditions for N_A and N_B are not needed. The solver will compute these values at each step of integration and use them in the S–M model. Note that the concentration profiles are also calculated simultaneously, while in the matrix method the values of fluxes are computed first, and then, using these values, the concentration profiles can be calculated at any desired point. The BVP method has the advantage that the reacting systems can be handled by the same procedure as that shown in Section 21.3.

The two methods are illustrated by the following example.

Example 21.1.

The problem analyzed here is an Arnold cell containing a binary mixture of acetone (1) and methanol (2). The mixture is evaporating into an inert gas, air (3). The pressure and temperature are maintained at 99.4 kPa and 328.5 K, respectively. The gas composition at the liquid surface is $y_1 = 0.319$ and $y_2 = 0.528$. The height of the Arnold cell was 0.238 m and the composition of both vapor species (1) and (2) at this end was zero. Find the evaporation rate and sketch the mole-fraction profiles in the cell. This problem is adapted from the book of Taylor and Krishna (1993).

The binary diffusion coefficients are

$$D_{12} = 8.48 \times 10^{-6} \, \text{m}^2/\text{s}$$
$$D_{13} = 13.72 \times 10^{-6} \, \text{m}^2/\text{s}$$
$$D_{23} = 19.91 \times 10^{-6} \, \text{m}^2/\text{s}$$

Solution.

The problem was solved both by the matrix method using the MATLAB *FSOLVE* function for the solution of the algebraic equation and by the boundary-value method using the BVP4C solver in MATLAB. The results for these two cases agree and are as follows:

$$N_1 = 1.78 \times 10^{-3} \, \text{mol}/(\text{m}^2 \cdot \text{s})$$
$$N_2 = 3.13 \times 10^{-3} \, \text{mol}/(\text{m}^2 \cdot \text{s})$$

The concentration profiles are sketched in Fig. 21.1. These profiles are in good agreement with the experimental data shown in Taylor and Krishna (1993) as well. The student should verify these results using the sample codes posted on the web.

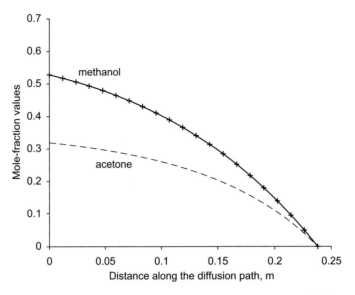

Figure 21.1 Composition profiles in an Arnold-type cell for evaporation of a binary liquid mixture.

The next example uses the equimolar counterdiffusion condition as the determinacy condition to relate the fluxes.

Example 21.2.

Estimate the mass transfer rates during distillation of ethanol (1), t-butyl alcohol (2), and water (3) at a point in the column where the following conditions are specified.

The bulk vapor composition is $y_{10} = 0.6$; $y_{20} = 0.13$.

The interface vapor composition is $y_{1\delta} = 0.5$; $y_{2\delta} = 0.14$.

The film thickness $\delta = 1$ mm. The following values may be used for the three binary pair diffusion coefficients:

$$D_{12} = 8.0 \times 10^{-6} \, \text{m}^2/\text{s}$$
$$D_{13} = 21.0 \times 10^{-6} \, \text{m}^2/\text{s}$$
$$D_{23} = 17.0 \times 10^{-6} \, \text{m}^2/\text{s}$$

Solution.

Since distillation is usually done under adiabatic conditions, we assume equimolar diffusion. This assumes that the molar heats of vaporization of the three species are nearly equal. Hence the total flux is zero or

$$N_C = -N_A - N_B.$$

Of course, the mole fractions have to add up to unity, and hence

$$y_C = 1 - y_A - y_B.$$

The mole fraction y_C and flux N_C are then eliminated from the S–M equations for y_A and y_B, respectively. The resulting equations can be solved by a matrix method (as is followed by FSOLVE) or by use of the BVP4C routine. The problem was set up by the two methods as shown above, and the following values were obtained for the fluxes.

$$N_1 = 8.09 \times 10^{-2}\,\text{mol}/(\text{m}^2 \cdot \text{s})$$
$$N_2 = 0.434 \times 10^{-2}\,\text{mol}/(\text{m}^2 \cdot \text{s})$$
$$N_3 = -8.52 \times 10^{-2}\,\text{mol}/(\text{m}^2 \cdot \text{s})$$

It is interesting to note that component 2 diffuses against its concentration gradient, an example of reverse diffusion.

We now change the problem specification slightly to indicate the phenomena of osmotic diffusion. We set $y_{20} = 0.13$, adjusting the mole fraction of C. The following values can now be computed, and the student should verify these by running the code.

$$N_1 = 8.30 \times 10^{-2}\,\text{mol}/(\text{m}^2 \cdot \text{s})$$
$$N_2 = 0.79 \times 10^{-2}\,\text{mol}/(\text{m}^2 \cdot \text{s})$$
$$N_3 = -9.10 \times 10^{-2}\,\text{mol}/(\text{m}^2 \cdot \text{s})$$

Species 2 diffuses to the liquid, even though its own mole fraction is the same both in the vapor and at the interface. The effect is due to the multicomponent diffusion coupling of the transport of three species.

21.2.3 Ternary systems with heterogeneous reactions

Consider a case where the reaction

$$A \rightarrow B + C$$

is taking place over a catalyst surface.

Then, by stoichiometry, the following conditions apply for the fluxes:

$$N_B = -N_A \text{ and } N_C = -N_A$$

Thus N_B and N_C can be eliminated from the S–M equations. In addition the mole fraction of one of the species, say, C, can also be eliminated using

$$y_C = 1 - y_A - y_B$$

The S–M equation for A, which reads

$$\frac{dy_A}{dz} = \frac{y_A N_B - y_B N_A}{CD_{AB}} + \frac{y_A N_C - y_C N_A}{CD_{AC}}$$

then reduces to

$$\frac{dy_A}{dz} = -\frac{y_A N_A}{CD_{AB}} + y_B N_A \left(\frac{1}{CD_{AC}} - \frac{1}{CD_{AB}} \right) - \frac{N_A}{CD_{AC}} \tag{21.10}$$

and

$$\frac{dy_B}{dz} = y_A N_A \left(\frac{1}{CD_{AC}} - \frac{1}{CD_{BC}} \right) + y_B N_A \left(\frac{1}{CD_{BA}} - \frac{2}{CD_{BC}} \right) + \frac{N_A}{CD_{AC}} \tag{21.11}$$

The condition that no homogeneous reaction takes place provides the third equation:

$$\frac{dN_A}{dz} = 0 \tag{21.12}$$

Three boundary conditions are needed to solve the problem. These equations need to be integrated with an additional condition to be imposed at the heterogeneous surface as discussed below.

For fast reactions, the mole fraction of A at the surface is now set to zero, which provides the third boundary condition. These equations can be used to solve for N_A and also for $y_{B,L}$, the mole fraction of B at the reacting surface.

For a finite reaction rate, the balance of flux of A at the reacting surface leads to

$$N_A(z = L) = -R_A = k_s C y_{As}$$

where the last term is specific for a first-order reaction at the catalyst surface. Here $z = L$ represents the reacting surface. Thus a Robin-type boundary condition is now specified at the reaction surface.

Example 21.3.

Vapor-phase dehydrogenation of ethanol (designated as A or 1) to produce acetaldehyde (2) and hydrogen (3) was studied by Froment, Bischoff, and de Wilde (2011), and is an example of the above type of system where a heterogeneous catalytic reaction takes place on a surface,

$$\text{ethanol} \rightarrow \text{acetaldehyde} + \text{hydrogen}$$

The kinetics of the heterogeneous reaction was modeled as follows.

The rate is $-k_s y_A$, with k_s equal to $10.0 \, \text{mol}/(\text{m}^2 \cdot \text{s})$.

The bulk gas-phase concentrations were $y_1 = 0.6$, $y_2 = 0.2$, and $y_3 = 0.2$.

Find the rate of reaction, assuming a mass-transfer film thickness of 1 mm. The temperature is 548 K and the pressure 101.3 kPa.

The binary diffusion coefficients are

$$D_{12} = 7.2 \times 10^{-5} \, \text{m}^2/\text{s}$$
$$D_{13} = 23.0 \times 10^{-5} \, \text{m}^2/\text{s}$$
$$D_{23} = 23.0 \times 10^{-5} \, \text{m}^2/\text{s}$$

Solution.

The S–M model is written for each of the species, resulting in three differential equations. These are simplified using the determinacy condition and the constraint on the mole fraction to two equations, similarly to other examples shown in the text. The equations can be set up in matrix form so that the solution can be written, *in principle*, in terms of the *expm* function. Finally the finite-reaction-rate boundary conditions at the surface are applied to close the problem and the resulting non-linear algebraic equations are solved by FSOLVE or similar routines. The unknown variables are the mole fractions of species 1 and 2 at the catalytic surface.

An alternative is to integrate the system of equations by using the BVP4C solver in MATLAB. The following values of the mole fraction at the catalytic surface should be obtained upon using either of these procedures: $y_1 = 0.0592$, $y_2 = 0.6055$, and $y_3 = 0.3353$.

Note that the low value of y_1 (compared with the bulk) indicates significant mass-transport resistance. The flux of A is computed as $0.9473 \, \text{mol}/(\text{m}^2 \cdot \text{s})$.

21.3   Application to homogeneous reactions

Examples are found in gas–liquid reactions and gas–solid catalytic reactions. The species mass balance should now include the reaction term. For the case where there are multiple reactions, the rate of reaction can be calculated by using the stoichiometric matrix and a rate function for each reaction. Thus the rate of production for species i is given by the summation of its rate of production from all the reactions (subscript r) as

$$R_i = \sum_{r}^{n_r} \nu_{ri} \mathcal{R}_r$$

where n_r is the number of reactions, ν_{ri} is the stoichiometry of the ith species in the rth reaction and $\mathcal{R}_r$ is the rate term for the rth reaction (which represents the molar rate of production of any species with unit stoichiometry). The rate term is now introduced into the species-balance equations:

$$\frac{dN_i}{dx} = \sum_{r}^{n_r} \nu_{ri} \mathcal{R}_r \tag{21.13}$$

Equations (21.4) and (21.13) have to be integrated together now. There are two equations for each species and hence there are $2 \times n_s$ first-order differential equations to be solved simultaneously. Note that the number of equations can be simplified using the rank of the stoichiometric matrix and this is left as an exercise. Here we keep all the $2 \times n_s$ equations for simplicity and illustrate the computational algorithm.

The problem is of the boundary-value type, since the composition is usually specified at one end ($z = 0$, the bulk or the pore mouth), while some condition (no flux, for example) is specified at the other end ($z = L$). Since the source term appears directly in the differential equation, the matrix method (which is applicable for non-reacting systems) is no longer suitable. It is now best to use a numerical method such as that involving the BVP4C solver. An example relating to diffusion in a porous catalyst is now presented.

21.3.1   Multicomponent diffusion in a porous catalyst

Mass transport in isothermal multicomponent systems in the gas phase is described by the S–M equation as shown earlier. For porous catalysts the equation is modified by including both Knudsen and bulk diffusion:

$$-\frac{dC_i}{dx} = \frac{N_i}{D_{\mathrm{Ke},i}} + \sum_{j=1}^{n_s} \frac{y_j N_i - y_i N_j}{D_{ij,\mathrm{e}}} \tag{21.14}$$

Here $D_{\mathrm{Ke},i}$ is the effective Knudsen diffusivity of i, while $D_{ij,\mathrm{e}}$ is the effective pair gas phase diffusivity (D corrected with the porosity and tortuosity factor). We use the concentration form rather than the mole-fraction form which is more convenient for reacting systems. Strictly speaking, the concentration form is valid for constant-pressure and constant-total-concentration conditions.

The relative magnitude of the two terms in (21.14) determines the nature of the solution method. If Knudsen diffusion dominates, the fluxes are uncoupled, whereas if the bulk gas diffusion dominates we could expect multicomponent coupling effects, especially if the molecular weights of the diffusing species are vastly different. The uncoupled problem can be reduced to a set of second-order differential equations in concentrations and solved directly. Coupled problems are best solved by simultaneous computations of fluxes and concentrations.

The constitutive model for flux Eq. (21.14) is augmented by a set of mass balance equations:

$$\frac{dN_i}{dx} = \sum_r^{n_r} \nu_{ri} \mathcal{R}_r \tag{21.15}$$

where the RHS is the total rate of production of species i from the various reactions. The term $\mathcal{R}_r$ is the rate function for the rth reaction, which is the function of the local concentrations of the various species. Equations (21.14) and (21.15) have to be integrated together now.

Note that Eq. (21.14) can also be written in terms of the component partial pressures as

$$-\frac{1}{R_G T}\frac{dp_i}{dx} = \frac{N_i}{D_{\mathrm{Ke},i}} + \sum_{j=1}^{n} \frac{p_j N_i - p_i N_j}{p_T D_{ij,e}} \tag{21.16}$$

where p_T is the total pressure and R_G is the gas constant. Since the rates of reactions are usually correlated as functions of partial pressures, the above form is often more convenient. The viscous contribution to flow is neglected in these equations. That does not mean that the catalyst pellets are isobaric; pressure gradients can exist in the pellets and they are implicitly accounted for by the use of the constitutive equations in terms of the partial pressures rather than in terms of component mole fractions.

Computer implementation of this problem in MATLAB is discussed below.

21.3.2   MATLAB implementation

The program shown here calculates the concentration profiles and the effectiveness factor based on the S–M model for a porous catalyst assuming a slab model. The test reaction is a single reaction, the water-gas shift reaction, shown below, and it is easy to extend this to multiple reactions. These test reactions are based on a paper by Haynes (1984), who did a detailed computational study of the problem based on orthogonal collocation. Here we show the computational implementation based on MATLAB BV4PC code.

The user actions to run the code are to specify parameters, a guess function for trial solution, a function *odes* to define the set of first-order differential equations, and a function *bcs* to specify the boundary conditions. You also need to define rate in the function *ratemodel* which defines $\mathcal{R}$ for each reaction.

The code is shown for the water-gas shift reaction taking place in a porous catalyst:

$$CO + H_2O \rightleftharpoons CO_2 + H_2$$

$$R_A = -\rho_p k p(1)^{0.9} p(2)^{0.3} p(3)^{-0.6}[1 - p(3)p(4)/(K_{eq}p(1)p(2))]$$

where the $p(i)$ are the partial pressures with $i = 1, 2, 3, 4$ denoting CO, H_2O, CO_2, and H_2. Other parameters are defined in the program. K_{eq} is the thermodynamic equilibrium constant and k is the rate constant based on the mass of catalyst. The main driver was written using CHEBFUN, which is used as a wrapper for BVP4C. Note that the initial guess is provided as a CHEBFUN and the call to BVPINIT is not required in this case. The specified values of the surface mole fractions and the rate based on surface values were used as initial guesses. There are eight variables here: four mole fractions and four fluxes. It may be noted that only one flux is independent, since there is only one reaction, but we keep four fluxes in order to make the program somewhat general so that we are able to handle multiple reactions.

The main driver is as follows:

```
global temp pressure
global ctot
temp = 400 + 273; pressure = 25.0;   % atm
Rgas = 82.06;                        % gas constant in cm^3 atm/g-mole K
ctot = pressure/Rgas/temp; % total concentration mole/c.c

% surface (pore mouth) mole fraction
global ys
ys(1) = 0.1; ys(2) = 0.6; ys(3) = 0.08; ys(4) = 0.22;
ys = ys * ctot   % surface molar concentrations.

% rate and other parameters
global k rhop Keq
k = 2.05e-04; rhop = 1.84; Keq = 12.0

% diffusion parameters; dk = Knudsen; db = binary pair (a matrix)
global dk db
dk(1) = 0.00649; dk(2) = 0.00494;
dk(3) = 0.0098; dk(4) = 0.0231;
db = zeros (4,4) % bulk is not used. Knudsen-dominated case

length = 0.1588; % catalyst properties

d = [0,length];
one = chebfun(1,d);
  eta = chebfun('eta', d);
    rate = reactionrate(ys); % initial guess of rate

initguess = [ys(1)*one ys(2)*one ys(3)*one ys(4)*one ...
    rate*one rate*one -rate*one -rate*one];

%%%%% call bvp4c now
sol = bvp4c (@odes, @bcs, initguess) % BVP solved,
y1 = sol(:,1); % mole fration of CO
plot(y1,'--' )

eta = y(5,1)/rate_s /0.1588 % effectiveness factor
```

The differential equations are coded as a set of dy_i/dz and dN_i/dz values. Those with $i = 1-4$ are mole fractions, while those with $i = 5-8$ are the fluxes.

```
function dydx = odes(x,y)
global dk db
rate = reactionrate(y);
% bulk diffusion terms can be  added as extra contributions
 ns = 4; bulkterm = zeros (ns,1);
dydx = [-y(5)/dk(1) + bulkterm(1)
           -y(6)/dk(2)
           -y(7)/dk(3)
           -y(8)/dk(4)
           -rate
           -rate
           rate
           rate];
```

The boundary conditions are set as the mole-fraction values at the surface and zero flux at the pore end:

```
%---------------------------------------------------------------
function res = bcs(ya,yb)
global a0 b0
global ys
res = [ya(1) - ys(1)
         ya(2) - ys(2)
         ya(3) - ys(3)
         ya(4) - ys(4)
         yb(5)
         yb(6)
         yb(7)
         yb(8)];
```

The kinetic model is set as a separate subroutine to provide code flexibility:

```
function ratemodel = reactionrate(y)
global k rhop Keq
global temp pressure
global ctot
p = y * 82.06 * temp    % species partial pressures
ratemodel = rhop *  k * p(1)^0.9 * p(2)^0.3 * p(3)^(-0.6)...
 * (1 - p(3) * p(4)/Keq/p(1)/p(2));
```

An illustrative profile simulated by this code is shown as a plot of CO mole fraction as a function of distance into the catalyst in Fig. 21.2. The parameters are representative of severe diffusional gradients and the CO concentration drops sharply in the pellet. Note that CO does not become zero at the center of the slab but approaches an equilibrium value since the reaction is reversible.

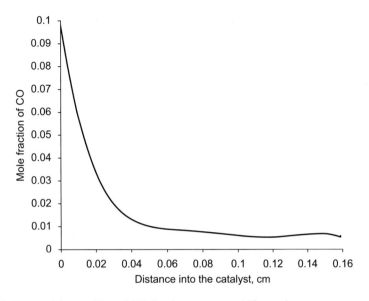

Figure 21.2 Composition profiles of CO for the water-gas shift reaction.

21.4 Diffusion-matrix-based methods

In this formulation we deal with diffusion fluxes rather than total fluxes. The two fluxes are related as usual by

$$N_A = J_A + y_A N_t$$

with similar expressions for other components. Here N_t is the total flux as usual. It turns out that the S–M equation has the same form as that shown earlier also when it is expressed in terms of diffusion fluxes. We repeat the equation in terms of diffusion fluxes for convenience of easy reference. Thus we have

$$-\frac{dy_i}{dz} = \sum_{j=1}^{n_s} \frac{y_j J_i - y_i J_j}{CD_{ij}} \quad \text{for } i = 1, n_s \tag{21.17}$$

The diffusion fluxes, however, add up to zero. Hence we can write for the n_sth species (which is chosen arbitrarily and usually called the solvent)

$$J_{n_s} = -\sum_{j}^{n_s-1} J_j$$

The mole fraction of the n_sth species is equal to

$$y_{n_s} = 1 - \sum_{j}^{n_s-1} y_j$$

These two equations permit us to eliminate both y_{n_s} and J_{n_s} from Eq. (21.17) and reduce the set of equations to $(n_s - 1)$ equations. The resulting equation after some simple algebra is

$$-\frac{dy_i}{dz} = -J_i \sum_{j=1}^{n_s} \left(\frac{y_i}{CD_{ij}} \right) + J_j \sum_{j=1}^{n_s} \left(\frac{y_i}{CD_{ij}} - \frac{y_i}{CD_{ij}} \right) \quad \text{for } i = 1, (n_s - 1) \qquad (21.18)$$

The equation can be represented in a compact vector–matrix form as

$$C \nabla y = -\tilde{B} J \qquad (21.19)$$

The matrix representation can be inverted to obtain J in terms of the mole-fraction gradients and the resulting equation can be represented as

$$\boxed{J = -C\tilde{K} \nabla y} \qquad (21.20)$$

where K is equal to

$$K = [\tilde{B}]^{-1}$$

Equation (21.20) has a form similar to Fick's law, except that it is now in a vector–matrix form. The matrix K is the tensor equivalent to the binary diffusivity. The form above is often referred to as the generalized version of Fick's law. The matrix K is referred to as the diffusivity matrix. The symbol K is used here for the diffusivity matrix to avoid confusion with D, since D was used as the diffusivity for the binary pair values. The coefficients of the matrix K should not be confused with the coefficients of the matrix D. The matrix D is symmetric, whereas the matrix K is not, in general, symmetric. Thus, for $n = 3$, there are three values for the D matrix ($D_{12}, D_{13},$ and D_{23}) while there are four values for the K matrix.

The nature of the matrix becomes clearer if the generalized version of Fick's law is written out explicitly for a three-component system ($n_s = 3$). In this case the diffusivity matrix is a 2×2 matrix and the fluxes are given by:

$$J_1 = -CK_{11} \nabla y_1 - CK_{12} \nabla y_2$$

and

$$J_2 = -CK_{21} \nabla y_1 - CK_{22} \nabla y_2$$

The flux of species 1 (or 2) depends on the mole-fraction gradient of both species 1 and 2. This coupling is responsible for some of the complexities observed in multicomponent-transport. The complexities are significant if the off-diagonal terms are large, whereas they are negligible if these terms are small.

The explicit forms for the K matrix can be obtained for a ternary system analytically and are as follows:

$$K_{11} = \left[\frac{y_1}{D_{12}} + \frac{y_2 + y_3}{D_{23}} \right] \bigg/ \text{DENO}$$

$$K_{12} = \left[y_1 \left(\frac{1}{D_{12}} - \frac{1}{D_{13}} \right) \right] \bigg/ \text{DENO}$$

$$K_{21} = \left[y_2 \left(\frac{1}{D_{12}} - \frac{1}{D_{23}} \right) \right] \Big/ \text{DENO}$$

$$K_{22} = \left[\frac{y_2}{D_{12}} + \frac{y_1 + y_2}{D_{13}} \right] \Big/ \text{DENO}$$

where

$$\text{DENO} = \frac{y_1}{D_{12}D_{13}} + \frac{y_2}{D_{12}D_{23}} + \frac{y_3}{D_{13}D_{23}}$$

Example 21.4.

An often-quoted and well-studied problem is that of the hydrogen (1)–methane (2)–argon (3) system and the values of K for this case are calculated from the above equations for a composition of $y_1 = 0.2$, $y_2 = 0.2$, and $y_3 = 0.6$ as follows:

$K_{11} = 0.76 \times 10^{-4} \, \text{m}^2/\text{s}$
$K_{21} = -0.01 \times 10^{-4} \, \text{m}^2/\text{s}$
$K_{21} = -0.12 \times 10^{-4} \, \text{m}^2/\text{s}$
$K_{22} = 0.25 \times 10^{-4} \, \text{m}^2/\text{s}$

Note the relatively large value for the cross-diagonal term K_{21}. Complex diffusion effects are signalled by this term.

Also the coefficients are concentration-dependent, unlike the binary pair values. We have calculated for one composition, but if you do it for another composition you will get a different set of values. An average composition along the diffusion path is often used as an approximation.

Some applications of the generalized formulation of Fick's law are now presented. The form of this is similar to the binary system (except for the interaction terms). Hence extension of binary solution methods can be used for multicomponent systems. A key point to appreciate is that the matrix K is composition-dependent. Hence it is not possible to get exact solutions without recourse to numerical methods. A number of solution methods use the assumption that K is constant along with the total concentration C. If one takes up this assumption, the solution is simplified and analytical solutions are possible. Such a model is referred to as the linearized model for multicomponent diffusion and was first analyzed by Toor (1957) and also independently by Stewart and Prober (1964). We provide one example of the use of the method rather than going into extensive mathematical details of the solutions using matrix computations.

The simplest case to analyze is that of equimolar counterdiffusion. Here $N_t = 0$ and hence $J_i = N_i$ for each species.

Example 21.5. Transient diffusion in a two-bulb apparatus.

Two bulbs containing gas mixtures with different compositions at time zero are connected by a porous plug, and transient diffusion is allowed to occur. First we analyze the binary diffusion and then show how the results may be generalized for a multicomponent system based on the linear model.

The binary solution is

$$y_i - y_{i\infty} = (y_{i0} - y_{i\infty})\exp(-\beta Dt)$$

where β is the cell constant and D is the binary diffusivity. This was indicated in Problem 6 of Chapter 10, using the pseudo-steady-state hypothesis.

The multicomponent solution is simply the vector extension of this result:

$$\boldsymbol{y} - \boldsymbol{y}_\infty = (\boldsymbol{y}_0 - \boldsymbol{y}_\infty) = \text{expm}(-\beta \tilde{K} t) \tag{21.21}$$

where the expm is the matrix exponential of the matrix term within the brackets.

We show the results for a gas mixture of hydrogen, nitrogen, and carbon dioxide. The initial composition in bulb 1 was chosen to be 0.5 for nitrogen and 0.5 for carbon dioxide. The initial composition in bulb 2 was chosen to be 0.5 for nitrogen and 0.5 for hydrogen. The K matrix was computed for a pressure of 100 kPA and a temperature of 298 K. The composition profile can then be computed directly with Eq. (21.21).

The calculated results for the composition profile of nitrogen are shown in Fig. 21.3. This shows some interesting behavior. During the early stage of the transient process nitrogen accumulates in the right bulb (bulb 2). This is an example of reverse diffusion. After some accumulation, the nitrogen diffuses back from bulb 2 to bulb 1. In the absence of diffusional interactions, there would not be any concentration profile for nitrogen, and the mole fraction in both bulbs would remain at the initial value of 0.5. The experimental data from Duncan and Toor (1962) (not shown here) are in complete conformity with these findings.

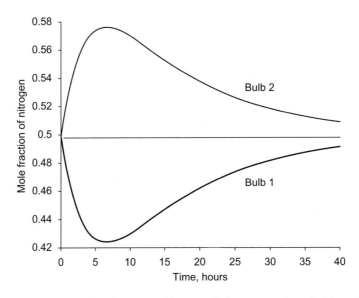

Figure 21.3 Composition profiles in an Arnold-type cell for evaporation of a binary liquid mixture.

21.5   An example of pressure diffusion

A gradient in chemical potential can be considered as the driving force for mass transfer, as discussed in the earlier section. The pressure gradients can therefore drive the diffusion process since the chemical potential varies as a function of pressure. The phenomenon is known as pressure diffusion. In order to model and develop a constitutive equation for pressure diffusion, it is therefore necessary to examine how the chemical potential changes with pressure. The relevant relation is obtained from thermodynamics as

$$\delta\mu = \left(\frac{\partial\mu_A}{\partial P}\right)_{T,y} \delta P = \tilde{V}_A \, \delta P \tag{21.22}$$

since

$$\frac{\partial\mu_A}{\partial P} = \tilde{V}_A$$

where $\tilde{V}_A$ is the partial molar volume of species A.

Upon introducing these concepts and after some algebraic manipulations of the generalized driving force, the flux expression for an ideal binary system can be shown to be

$$J_A^* (\text{Fick's law} + \text{pressure diffusion}) = -CD_{AB} \, \nabla x_A - \frac{D_{AB}}{R_G T}(\phi_A - \omega_A)\nabla P \tag{21.23}$$

where ϕ_A is the volume fraction of A, while ω_A is the mass fraction of A. The first term on the RHS is the contribution due to the concentration gradient, while the second term is the contribution due to the pressure gradient.

The term $\phi_A - \omega_A$ is positive for a lighter species and hence it diffuses in a direction opposite to the pressure gradient. Similarly, the heavier species diffuses in the direction of the pressure gradient. Students well-versed in thermodynamics will appreciate the similarity with Le Chatelier's principle.

For ideal gases, the volume fraction, ϕ_A, is equal to the mole fraction, x_A, and is related to the mass fraction by

$$\omega_A = x_A M_A / M = \phi_A M_A / M$$

where M is the average molecular weight of the components of the mixture.

Hence the difference between the volume fraction and the mass fraction is given by

$$\phi_A - \omega_A = x_A - \omega_A = x_A(1 - M_A/M)$$

This term is positive if $M_A < M$, i.e., species A is lighter than species B. Note that the direction of diffusion of this species is therefore opposite to that of the pressure gradient in accordance with Eq. (21.23), as also discussed earlier.

The contribution of pressure to diffusion is therefore written as

$$J_A^* (\text{pressure diffusion}) = -D_{AB}\frac{x_A}{R_G T}(1 - M_A/M)\nabla P \tag{21.24}$$

Since

$$M = x_A M_A + x_B M_B$$

the above equation can be rearranged to

$$J_A^* \text{(pressure diffusion)} = -D_{AB}\frac{x_A x_B}{MR_G T}(M_B - M_A)\nabla P \qquad (21.25)$$

The student should verify the simple algebra needed here for clarity. Since $C = P/(R_G T)$, the above expression can be also be rearranged to

$$J_A^* \text{(pressure diffusion)} = -CD_{AB}\frac{x_A x_B}{M}(M_B - M_A)\nabla \ln P \qquad (21.26)$$

It is convenient to define a pressure diffusion coefficient by the following relation (Haase, 1968):

$$D_P = \alpha_P D_{AB} x_A x_B$$

where α_P is a pressure diffusion factor given by

$$\alpha_P = \frac{M_B - M_A}{M} = \frac{M_B - M_A}{x_A M_A + x_B M_B}$$

Hence the pressure diffusion can be expressed as

$$J_A^* \text{(pressure diffusion)} = -CD_P \nabla \ln P \qquad (21.27)$$

It may be noted here that the thermal diffusion treated in the next section has an analogous representation in terms of a thermal diffusion factor. Note that $\nabla \ln P$ is the effective driving force for pressure diffusion. Likewise $\nabla \ln T$ is the driving force for thermal diffusion.

The various forms indicated above are useful depending on the problem specifications.

A steady-state concentration profile in the presence of a pressure field is readily computed. For a steady state $J_A^* = 0$. Using Eq. (21.23) for the pressure contribution to diffusion and, as usual, Fick's law for concentration diffusion leads to

$$\frac{dx_A}{dz} = -\frac{x_A x_B}{M}(M_B - M_A)\nabla \ln P \qquad (21.28)$$

The equation of hydrostatics requires

$$\nabla P = \rho g$$

Finally, the equation of state for a gaseous system requires

$$\rho = \frac{MP}{R_G T}$$

and hence

$$\nabla \ln P = \frac{Mg}{R_G T}$$

On using this in the expression for the combined flux we have

$$\frac{dx_A}{dz} = -\frac{g x_A x_B}{R_G T}(M_B - M_A) \qquad (21.29)$$

This is readily integrated to obtain the mole-fraction profile.

The integrated expression can also be expressed as

$$\frac{x_A}{x_{A0}}\frac{1 - x_{A0}}{1 - x_A} = \exp\left(\frac{g[M_B - M_A]z}{R_G T}\right) \qquad (21.30)$$

which provides an expression for the segregation caused by the pressure gradient.

Some examples where pressure diffusion effects are significant are (i) species distribution in a natural gas well and (ii) protein separation in an ultracentrifuge.

The concentration profile induced by pressure diffusion is illustrated in the following example.

Example 21.6.

As an example consider methane and carbon dioxide at a mole fraction of 0.5 each at the surface. What is the mole fraction of methane in a deep natural gas well 1 km deep? What is the pressure-diffusion factor?

Solution.

We use $M_A = 16 \times 10^{-3}$ kg/mole and $M_A = 44 \times 10^{-3}$ kg/mole to get consistency for SI units.

Using Eq. (21.30), we find $x_A = 0.47$.

The average molecular weight at the top is 30. Hence the pressure diffusion factor is

$$\alpha_P = \frac{M_B - M_A}{M} = 0.9333.$$

The value computed at the bottom of the well is 0.9079.

21.6   An example of thermal diffusion

The additional mass flux resulting from thermal diffusion is modeled as

$$J_A^* = C D_{AB} k_T \nabla \ln T \tag{21.31}$$

where k_T is the thermal diffusion factor.

For a liquid a Soret coefficient σ_T defined as

$$k_T = \sigma_T x_A x_B T \tag{21.32}$$

is more commonly used.

In general, for binary systems, species with larger molecular weight move to the colder region and vice versa.

Example 21.7. The Clusius–Dickel cell.

This consists of two bulbs maintained at different temperatures similar to a diffusion cell. It is named after Clusius and Dickel, who published the first paper on thermal diffusion in 1938. See Clusius and Dickel (1939a, 19939b). At steady state a concentration profile would be established in the system, and this can be calculated by setting J_A^* as zero. This leads to

$$\frac{dx_A}{dz} + \frac{k_T}{T}\frac{dT}{dz} = 0$$

Ignoring the effect of composition on k_T (since the composition variation is small), the equation can be integrated across the system to obtain the composition difference between the two ends:

$$x_{AL} - x_{A0} = -k_T \int_{T_1}^{T_2} \frac{dT}{T} = -k_T \ln(T_2/T_1)$$

Isotope separation by thermal diffusion and natural convection

The degree of separation that can be achieved by thermal diffusion is usually small. However, the combination of thermal diffusion with natural convection can be used to enhance the degree of separation. Here we have two vertical walls. The hot wall causes the fluid to flow upward, while the cold wall causes it to flow down. This was used to separate uranium hexafluoride gas during World War II.

In this system, the transfer of heat across a thin film of liquid or gas is used to accomplish the temperature difference needed for isotope separation by thermal diffusion. This can be used, for instance, for separation of ^{235}U and ^{238}U. On cooling a vertical film on one side and heating it on the other side, the resultant convection currents will produce an upward flow along the hot surface and a downward flow along the cold surface. Under these conditions, the lighter ^{235}U gas molecules will diffuse toward the hot surface, and the heavier ^{238}U molecules will diffuse toward the cold surface. These two diffusive motions combined with the convection currents will cause the lighter ^{235}U molecules to concentrate at the top of the film and the heavier ^{238}U molecules to concentrate at the bottom of the film.

Summary

- The Stefan–Maxwell (S–M) equations provide a convenient constitutive model for multi-component systems. The model is shown to be accurate for ideal gases at low pressures. The model equation is given by Eq. (21.4) and is worth memorizing. The values of only binary pair diffusion coefficients, one for each pair in the mixture, are needed. Thus for a ternary mixture we need three diffusion coefficients: D_{12}, D_{13}, and D_{23}.

- The S–M equations need to be supplemented with some determinacy condition relating the fluxes of various species, similarly to what was done for a binary mixture. Common conditions useful in many problems are (i) the assumption that there is only one non-diffusing gas, i.e., as unimolecular diffusion (UMD), (ii) equimolar counterdiffusion (EMD), and (iii) reacting systems where the fluxes are related by the reaction stoichiometry.

- The main difficulty in computations with the S–M model is that the equations are implicit in fluxes. We often wish to compute the fluxes given the mole-fraction gradients, but S–M gives the mole-fraction gradients in terms of all the component fluxes.

- One solution method, which is applicable when there are no homogeneous reactions, is the matrix method. A matrix–vector solution can be written formally in terms of the exponential matrix. The fluxes are implicit in the coefficients of this exponential matrix and can be solved by a non-linear algebraic-equation solver. Once the fluxes have been computed, the mole-fraction profiles can be calculated using the formal matrix–vector solution.

- The second method is to augment the S–M equations with species mass-balance equations. The resulting problem is of a boundary-value type and can be solved using any standard package for the solution of boundary-value problems. You should endeavor to understand both methods of solution by undertaking a careful study of the number of examples presented in the text and be able to apply them to new problems.

- An inverted formulation of the S–M model leads to the generalized version of Fick's law for diffusion fluxes. The form of the equation is similar to that for a binary mixture and is the vector analogue of the law. The matrix appearing in this expression, the K matrix, can be calculated from the binary pair values at an average concentration in the system. Once this matrix is known, the solution method for binary systems has a direct vector–matrix analogue and can be useful for solution of some problems. The assumption is that the concentration dependence of K is ignored. This is often referred to as a linear model for multicomponent diffusion.

- Complexities associated with multicomponent systems can be associated with the off-diagonal terms of the K matrix. If these terms are small or zero then each species diffuses in proportion to its own gradient. Complexities such as reverse diffusion, osmotic diffusion, etc. are absent in such cases.

- The K matrix can also be viewed as a fitted parameter relating the fluxes to the mole-fraction gradient and used as such in an empirical setting. Such an approach, for instance, is useful for non-ideal systems.

- The irreversible thermodynamics provides a platform for development of the constitutive models for diffusion. The S–M equations have a basis rooted in this theory. The theory also shows that diffusion can be caused by pressure gradients and thermal gradients. Problem 3 illustrates this further.

ADDITIONAL READING

The book by Taylor and Krishna (1993) is a classic source for multicomponent mass transfer. Many examples in the current chapter have been suitably adapted for this book. You may wish to note that Taylor and Krishna look at the solution from a matrix point of view (the famous bootstrap-matrix method), whereas we view this here more from a numerical point of view with direct use of various MATLAB functions.

Another useful book with many example is that by Wesselingh and Krishna (2000).

Irreversible thermodynamics and its application to develop constitutive models for transport are developed in the classic book by Haase (1968). The book by De Groot and Mazur (1962) is another classic in this field. These books provide the connection between classical thermodynamics and the phenomenological theory of irreversible processes.

Problems

1. Show that the two representations of the Stefan–Maxwell model given by Eqs. (21.4) and (21.5) are equivalent.

2. Show that the Stefan–Maxwell model can be rearranged to define a pseudo-binary diffusivity, D_{im}, for species i in the mixture:

$$\frac{1}{D_{im}} = \frac{\sum_{k=1}^{N} \left(y_k - y_j N_k/N_j \right)/D_{jk}}{1 - y_j \sum_k [N_k/N_j]}$$

The pseudo-binary diffusivity is seen to be a function of the local mole fraction. A mean average value is often used, and this is treated as a constant. The flux ratios appearing in the above equation are often related by stoichiometry and hence appropriate values can be assigned for heterogeneous reactions occurring at a catalyst surface. Show that for a limiting case in which species 1 diffuses through a "stagnant" species 2, species 3, etc. (unimolecular diffusion) the above expression reduces to the Wilke equation, Eq. (9.34). Hence show that the Wilke equation is useful only for a dilute solution of species 1 and also when the fluxes of other species are zero.

3. Support for the form of the generalized version of Fick's law introduced in this chapter can be found in the thermodynamics of irreversible processes. Here we introduce a brief description of this field. More detailed descriptions can be found in the books by Haase (1968) and De Groot and Mazur (1962). The starting point is the rate of entropy production described in the following subsection.

Entropy production due to heat transport was discussed in Chapter 8. We recall the expression here:

$$g_s = -\frac{1}{T^2} q \cdot \nabla T$$

The starting point was the definition of the change in internal energy,

$$d\hat{u} = T \, d\hat{S} - p \, d\hat{V}$$

For a multicomponent system the change in internal energy is given by

$$d\hat{u} = T \, d\hat{S} - p \, d\hat{V} + \sum_i \mu_i \, d\mathcal{M}_i$$

where $\mathcal{M}_i$ is the number of moles of component i in the mixture and μ_i is its chemical potential.

Show that, in the presence of a diffusion flux, the entropy production due to diffusion contributes the following additional term:

$$g_s(\text{diffusion}) = \sum_i^n J_i \cdot \nabla(\mu_i/T)$$

If the term is split into a flux and a driving force, we obtain a phenomenological model for diffusion. Follow up this line of thought and show that the irreversible thermodynamics provides a platform for development of the constitutive models for diffusion. Note that there is an additional contribution to entropy production due to chemical reaction, which is not included in the transport models.

4. Acetone is evaporating in a mixture of nitrogen and helium. Find the rate of evaporation and compare it with the rates in pure nitrogen and pure helium. Also compare it with the model using a pseudo-binary diffusivity value for acetone. The pseudo-binary diffusivity can be calculated from the correlation given by Wilke, Eq. (9.34).

5. Catalytic oxidation of CO is an important reaction in pollution prevention. The reaction scheme is

$$O_2(A) + 2CO(B) \rightarrow 2CO_2(C)$$

Set up the Stefan–Maxwell model for this problem. State the determinacy condition and simplify the equations into two differential equations for the mole-fractions of any two of these species. Calculate the solution for the mole-fraction profiles and find the values of the fluxes using the matrix method and the BVP method. Assume fast reaction at the catalyst surface.

6. Examine how the rank of the stoichiometric matrix can be used to reduce the number of equations to be solved. For example, if there are n_s components and n_r equations we can show that only $n_s - n_r$ independent mass-balance equations are needed. The remaining variables form an invariant. Write MATLAB code to find the invariants.

7. Extend the analysis in Section 21.2.3 to two reactions, making it applicable for steam reforming of methane. The first reaction is

$$CH_4 + H_2O \rightleftharpoons CO + 3H_2$$

and this is followed by the water-gas shift reaction shown in the text. Kinetic and diffusion parameters can be found in the paper by Haynes (1984) or other sources.

8. **Composition dependence of the Fick matrix $\tilde{K}$.** Consider the three-component system consisting of acetaldehyde (1), hydrogen (2), and ethanol (3). The binary diffusivity values at 548 K and 101.3 kPa are given in Example 21.3.

   Find the K matrix at the bulk gas composition and at the catalyst surface concentration. Use ethanol as the component eliminated (the solvent). Comment on the values and the extent of multicomponent interactions.

9. **Effect of composition numbering on the Fick matrix $\tilde{K}$.** In the matrix representation the species n chosen for elimination is usually referred to as the solvent. The choice of "solvent" species is arbitrary, but it can have an effect on the coefficient and the structure of the resulting form as shown below.

   A system with a very large variation in binary diffusion values was studied by Wesselingh and Krishna (2000). The three components are hydrogen (1), nitrogen (2), and dichlorodifluoromethane (3).

   Estimate the binary pair values from Eq. (1.53) given in Chapter 1.

   Find the K matrix if species (3) is the solvent. Also show that the K matrix has different values if species (2) is the solvent.

   Although the three matrices look different, show that the eigenvalues are the same. Also show that the component fluxes calculated are the same.

10. Derive the solutions for transient concentration profiles in the two-bulb apparatus (Example 21.5 in the text) for the binary case, and show that the multicomponent case can be derived as an extension of this. What assumption is implicit in extending the binary case to the multicomponent case?

11. **Diffusion across a porous plug: the effect of a third component (adapted from BSL).** When two gases A and B are forced to diffuse through a third gas C, there is a tendency of A and B to separate because of the difference in their diffusivities in gas C. This phenomenon could possibly be used for isotope separation. Consider a "diffusion tube" of diameter d and length L packed with some non-reacting material such as glass wool. One end of the tube, $z = 0$, the feed side, is maintained at mole fractions of x_{Af} and x_{Bf}. The other end,

the product end, $z = L$, is maintained at x_{Ap} and x_{Bp}. Your task is to model the degree of separation that can be achieved in this system and to find the fluxes of the various species across the tube.

(a) Set up the model to describe the system.
(b) Express the model in terms of dimensionless quantities.
(c) Show or discuss a procedure to solve the model. Set up computations in MATLAB.

12. Show that another way of writing the flux due to pressure diffusion is in terms of partial molar volume and the total mixture density is

$$J_A^* \text{(pressure diffusion)} = -CD_{AB}\frac{1}{R_G T}(\tilde{V}_A - \tilde{V})\nabla P$$

13. Estimate the steady-state concentration profile when a typical albumin solution is subjected to a centrifugal field of 30 000 times the force of gravity under the following conditions: the cell length is 1.0 cm, the molecular weight of albumin is 45 000, the apparent density of albumin in solution is 1.34 g/cm³, and the mole fraction of albumin at $z = 0$ (one end of the cell) is 5×10^{-6}.

14. A mixture of H_2 and D_2 is contained in two bulbs connected by a porous plug. The bulbs are maintained at different temperatures of 200 and 600 K. The thermal diffusion parameter $k_T = 0.0166$. The mole fraction of deuterium is initially equal to 0.1. Find the differences in the mole fractions in the two bulbs when a steady state is established due to the combined effect of Fick and thermal diffusion. In which bulb does hydrogen accumulate?

Learning objectives

The learning objectives of this chapter are as follows.

• To study the role of electric field in the transport of ions and to introduce the important Nernst–Planck equation for electrochemical transport.

• To understand how the electric field modifies the transport rate and the diffusivity of ions; to study how the field in turn is itself modified by diffusion.

• The concept of charge neutrality and how this concept can be used to simplify the models for electrochemical transport.

• To identify the regions where charge neutrality fails (the electrical double layer). To understand how the charge distribution in the double layer affects the flow in micro-channels and leads to various electrokinetic effects.

• To study briefly transport effects and electron distribution in ionized gases (plasma processes).

Transport of ions or charged species is encountered in a wide variety of processes. For example, sodium chloride is ionized in water and exists as sodium and chloride ion-pairs. Thus the diffusion of sodium chloride in reality involves the diffusion of positively charged sodium ions and negatively charged chloride ions. Such systems are of importance in electrochemical reaction engineering. Examples can be found, for example, in electro-winning of metals, fuel cells, batteries, etc. Equally important is the transport of charged species in charged membranes or solids carrying a net surface charge. An example of such a system is the ion-exchange membrane, as is widely used in water purification, metal recovery, and selectivity modulation of chemical reactors. Transport of charged species in biological membranes is another example of transport across charged membranes. In addition, a number of techniques to separate large molecules such as DNA and proteins involve application of an electric field. Electrophoresis is such an example. The ionic-transport effects are also important in the study of the surface properties of colloids.

The key feature to be included in the model for transport in such systems is the charge migration due to the electric field. The constitutive equation for diffusion in such systems is the classical Nernst–Planck equation, which is discussed in Section 22.1.2. Further complexity in models for these systems arises due to the requirement that the electric field may have to be computed simultaneously with the concentration field. We explore these situations in detail in this chapter.

Many materials possess a surface charge, which can affect the velocity profiles in narrow channels made of such materials. These electrical forces are confined to a thin layer near the channel walls and cause many interesting phenomena such as electro-osmosis, the streaming potential, the sedimentation potential, etc. These effects are collectively known as electrokinetic phenomena. This chapter provides an introductory analysis for such problems. Finally, transport of ions in gaseous systems is encountered in chemical vapor processing, thin-film growth, and a number of other important applications, and we review some models for such systems.

22.1   Transport of charged species: preliminaries

22.1.1   Mobility and diffusivity

A positively charged species with a unit charge of 1 Coulomb moves in an electric field with a velocity proportional to the electric field:

$$v_i = \mu_i E$$

where the proportionality constant, μ_i, is defined as the mobility of species i. Note that the velocity v_i is defined for a system with zero total velocity, i.e., in a frame of reference moving with a mixture velocity. Also the above equation applies to a positive charge. A negative charge moves in the opposite direction to the field. The two cases can be reconciled by introducing z_i, the valency of the ion. Hence the above equation can be written as

$$v_i = z_i \mu_i E \tag{22.1}$$

where z_i is positive for cations and negative for anions.

The unit for electric field is N/C since it is force on a coulomb of charge and hence the unit of mobility is m · C /(N · s).

The electric field can be represented as the negative of the gradient of a scalar potential,

$$E = -\nabla \phi \tag{22.2}$$

ϕ is then the electric potential with units of J/C.

The mobility is related to the diffusivity by the Einstein equation,

$$\boxed{\mu_i = D_i \frac{F}{R_G T}} \tag{22.3}$$

where D_i is the diffusion coefficient for the ion. Here F is the Faraday constant, which has a numerical value of 96485 C/mole and represents the charge equal to one mole of electrons. Note that

$$F = e\mathcal{N}$$

where e is the unit charge on an electron (equal to 1.602×10^{-19} C) and $\mathcal{N}$ is Avogadro's number, 6.23×10^{23}.

It may also be noted that $R_G T/F$ has the units of volts and has a value of 25.69 mV at 298 K. Also it may be noted that $R_G T/F$ is equal to $k_B T/e$, where k_B is the Boltzmann constant (1.38×10^{-23} J/K).

22.1.2 The Nernst–Planck equation

The flux of a charged species caused by the electric field is therefore equal to $C_i v_i$. This is in turn equal to $C_i z_i \mu_i E$ using Eq. (22.1) or equal to $C_i z_i D_i (F/(R_G T)) E$ using the Einstein relation for mobility and diffusivity.

Finally, expressing the electric field in terms of the potential gradient, $E = -\nabla\phi$, we have

$$\text{Migration flux due to electric field} = -D_i C_i z_i (F/(R_G T))\nabla\phi.$$

The total flux of a charged species is then obtained by combining Fick's law, which gives the diffusion due to the concentration gradient, and the flux due to mobility under the electric field. The x-component of this flux, for example, is therefore

$$J_{i,x} = -D_i \frac{dC_i}{dx} - D_i z_i C_i \frac{F}{R_G T} \frac{d\phi}{dx} \tag{22.4}$$

Here z_i is positive for cations, whereas it is negative for anions. Thus the flux of anions (z_i negative) is in the same direction as the electric-potential gradient. The cations move in the direction opposite to the potential gradient.

If there is a net system velocity, the contribution of the convective term is added to obtain the combined flux:

$$\boxed{N_{i,x} = -D_i \frac{dC_i}{dx} + v_x C_i - D_i z_i C_i \frac{F}{R_G T} \frac{d\phi}{dx}} \tag{22.5}$$

This equation is known as the Nernst–Planck equation. The vector representation is

$$N_i = -D_i \nabla C_i - D_i z_i C_i \frac{F}{R_G T} \nabla\phi + v C_i \tag{22.6}$$

The three terms on the RHS can be identified as diffusion, migration, and convection, respectively. The migration term is the key to transport in charged systems and does not appear in the flux model for uncharged species. Also it may be noted that the above equation is valid for dilute systems, where we assume that the diffusion of each species can be described independently of other species. For concentrated systems, the multicomponent interactions need to be accounted for and some type of Stefan–Maxwell model is needed to describe the transport. These details are not presented in this chapter, since the goal of this chapter is to provide a general overview. As noted by Taylor and Krishna (1993), the multicomponent interactions can be safely ignored if the concentrations of the ions are less than 10 mol/m^3.

It is common to express the fluxes in terms of electric current in electrochemistry. A flux of a charged species produces a current. The total current density in an electrolyte solution is related to the fluxes by the following equation:

$$i = F \sum_i z_i N_i \tag{22.7}$$

Also, due to conservation requirements, the divergence of the current density is zero,

$$\nabla \cdot i = 0$$

It may be noted that both the concentration field and the electric-potential field have to be computed simultaneously in order to calculate the flux of each ionic species. The additional equations needed to compute the potential field are now presented.

22.1.3   Potential field and charge neutrality

The general equation for the potential field is Poisson's equation, which is given by

$$\nabla^2 \phi = -\frac{\rho_c}{\epsilon} \tag{22.8}$$

which follows from the Maxwell equation.

Here ρ_c is the electrical charge density given by

$$\rho_c = F \sum_i^N z_i C_i \tag{22.9}$$

and ϵ is the dielectric permittivity of the medium. This is usually defined as the permittivity of free space ϵ_0 multiplied by the dielectric constant of the medium. Using this in Eq. (22.8), we have

$$\nabla^2 \phi = -\frac{F}{\epsilon} \left(\sum_i^N z_i C_i \right) \tag{22.10}$$

What are the units of various quantities in the above Poisson equation, (22.10)?

Typical values of ϵ and the value of the coefficient term on the RHS are examined now. The value of ϵ_0 (the free-space permittivity) is 8.8×10^{-12} C/(V · m) and the dielectric constant for water is 80.2. Hence the value of the parameter F/ϵ is equal to 1.392×10^{16} V · cm/gmol for water.

This means that the charge density will be nearly zero in most cases. This condition is called the electroneutrality condition:

$$\sum_i^N z_i C_i = 0 \tag{22.11}$$

This condition is valid everywhere except near the solid (electrode) surface where there is a thin layer (of thickness of the order of 1–10 nm) over which the charge density is non-zero. This thin region is known as the electrical double layer. Electroneutrality does not hold here, but holds everywhere else to a first approximation. However, often electroneutrality is assumed to hold all the way to the electrode surface and the effects due to the double layer are incorporated indirectly into the boundary condition at the electrode. More details of the electrical double layer and some effects arising due to it are presented in a later section. The assumption of electroneutrality simplifies the calculation of the potential field as shown in Example 22.1.

A note of caution is in order here. If the condition that the charge density is zero is substituted into the Poisson equation, Eq. (22.10), we find that the Laplace equation holds for the potential field: $\nabla^2 \phi = 0$. But this equation for the electrical potential is valid only under certain conditions such as constant electrical conductivity of the medium and there being no concentration gradients in the system (e.g., bulk liquid outside the diffusion boundary layers). In such cases, it is appropriate to use the Laplace equation to calculate the potential field, but such calculations may be inaccurate near the concentration boundary layer (diffusion film) near the electrodes. In general, electroneutrality is assumed to hold only as an approximation, and this does not imply that the Laplace equation is automatically applicable for the field calculation. This is discussed in more detail later in this chapter.

A simple application to an electrolysis process where the charge-neutrality condition is used to find the electric field is now presented. Once the field is known, the mass flux can be calculated from the Nernst–Planck equation. This example also illustrates how the transport rate is augmented by the electric field.

Example 22.1.

Consider a binary salt MX under the action of an electric field. The cation M is deposited on the cathode. The concentration of the salt is M_b in the bulk liquid and M_S at the cathode.

Find the concentration distribution and the flux of the ions to the cathode. Assume quasi-steady-state conditions and use the concept of a diffusion film near the cathode.

Also neglect convection effects and use the low-mass-flux approximation: $J_A = N_A$.

Solution.
The Nernst–Planck equation is applied for both of the species X^- and M^+. Since there is no flux of species X^- (denoted by subscript X) we can equate J_X to zero:

$$J_X = -D_X \frac{dC_X}{dx} + D_X \frac{F}{R_G T} C_X \frac{d\phi}{dx} = 0 \tag{22.12}$$

Note that z_i is taken as -1 in the Nernst–Planck equation. The electric field can be solved from the above equation:

$$\frac{d\phi}{dx} = \frac{R_G T}{F} \frac{1}{C_X} \frac{dC_X}{dx} \tag{22.13}$$

Charge equality means that $C_M = C_X$. (M means here the species M^+.) Hence the electric field can also be expressed as

$$\frac{d\phi}{dx} = \frac{R_G T}{F} \frac{1}{C_M} \frac{dC_M}{dx}$$

For M^+ we have, from the Nernst–Planck equation,

$$J_M = -D_M \frac{dC_M}{dx} - \frac{D_M F}{R_G T} C_M \frac{d\phi}{dx}$$

Note the minus sign on the second term of the RHS since $z_i = +1$ now.

On using the expression for the electric field given by Eq. (22.13) and simplifying, we find

$$J_M = -2D_M \frac{dC_M}{dx}$$

A simple application of Fick's law would have produced the above result but without the factor of two. Hence the effect of the electric field is to enhance the transport by a factor of two over that given by the simple version of Ficks' law.

Since J_M is constant in the diffusion film, we find that the concentration profile is linear. Also the concentration profiles for both the species M and X are the same. Thus there is a concentration gradient for species X in the film as well, but this does not imply that X is diffusing towards the cathode. The diffusive flux of X caused by the favorable concentration gradient is balanced by the migration flux of X in the opposite direction. The net flux of X is therefore zero! This is another avatar of the concentration-polarization phenomena.

The transport processes of diffusion and migration both for cations and for anions near a cathode are shown schematically in Fig. 22.1.

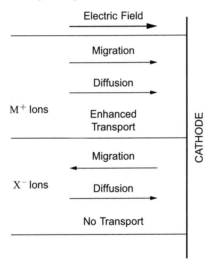

Figure 22.1 Transport processes near a cathode for cations and anions. Cations are deposited at the cathode in this example and have a non-zero net flux while anions have no net flux.

The electric-potential field can be computed by integration of Eq. (22.13) keeping the concentration gradient as a constant. The result is

$$\phi^*(0) - \phi^*(\delta) = \ln\left(\frac{C_X(0)}{C_X(\delta)}\right)$$

where ϕ^* is the dimensionless electric potential $\phi/(R_G T/F)$. This is a very common way of writing the potential in electrochemistry, i.e., all potentials are scaled by $R_G T/F$. The parameter δ is the thickness of the diffusion film as usual.

What is the direction of the electric field? Why?

Further notes and comments on the above example are shown below since they are important in modeling of electrochemical reactors.

The effect of a second electrolyte

The enhancement by a factor of two (over the Fick's-law value) in transport rate is valid if there are no additional ions in the systems. If there is another salt of the composition NX then the effect of the migration term will be smaller than one, and in the limiting case there will be no enhancement and Fick's law simply holds. This happens because electroneutrality now requires $X = M + N$ and, since N is not reacting, the overall contribution of migration is reduced.

Conductivity and transference numbers

If there is no concentration gradient in the systems, the current is simply related to the electric-potential gradient and Ohm's law is applicable. For example, in the bulk region between the two electrodes, the liquid is often assumed to be mixed and to have a uniform concentration. Here Ohm's law can be used:

$$i = -\kappa \, \nabla \phi \tag{22.14}$$

where κ is the conductivity of the solution. Here i is the current density. Using the expression for current given by Eq. (22.7) and the Nernst–Planck equation for the flux, the following expression for the conductivity can be derived:

$$\kappa = F \sum_i z_i^2 \mu_i c_i$$

This expression is written in terms of the mobilities of the ions μ_i. We can also write it in terms of the diffusivity using the Einstein relation as

$$\kappa = \frac{F^2}{R_G T} \left(\sum_i z_i^2 D_i c_i \right)$$

The two expressions are completely equivalent.

The fraction of the current carried by species i is called the transference number,

$$t_i = \frac{z_i^2 \mu_i c_i}{\sum_i z_i^2 \mu_i c_i}$$

It should be emphasized here that these expressions are valid only if there are no concentration gradients. If there are concentration gradients, the approach described in the following section should be used.

A general expression for the electric field

The following expression can be derived starting from Eq. (22.7) for the current and the Nernst–Planck equation for the fluxes. The student should verify this before proceeding further, since it is an important equation in electrochemical transport:

$$\nabla \phi = -\frac{i}{\kappa} - \frac{F}{\kappa} \sum_i z_i D_i \, \nabla c_i$$

which is an inverted form, similar to Ohm's law, relating the current to the potential. Note the concentration-gradient term in the above expression. This term is often referred to as the diffusion potential. The pure version of Ohm's law is often used even in cases where there are concentration gradients, but such an approach is technically incorrect, as has been pointed out nicely in the book by Newman and Thomas-Alyea (2004).

If there is a net velocity in the system, the expression for the potential is given by adding the contribution of the convection term. The resulting contribution is referred to as the streaming potential. The following is then the general expression for the potential:

$$\nabla \phi = -\frac{i}{\kappa} - \frac{F}{\kappa}\left(\sum_i z_i D_i \nabla c_i\right) + \frac{Fv}{\kappa}\left(\sum_i z_i c_i\right) \tag{22.15}$$

which is a general expression. The three terms on the RHS can be interpreted as (i) the Ohmic term with the actual current, (ii) the potential caused by diffusion, and (iii) the streaming potential caused by the bulk flow of the charges. The last term is zero whenever electroneutrality holds, but need not be zero in some cases, for example, transport in charged membranes. Here electroneutrality applies only if the immobile charge of the membrane is also included. Thus the term $\sum z_i c_i$ is not zero for the mobile charges, and this term contributes to the streaming potential.

The mass balance for reacting systems
The flux expression N_i is coupled with the species mass balance in the usual manner as described earlier. Thus

$$\nabla \cdot (N_i) + R_i = 0$$

where R_i is the rate of production of species i by homogeneous reaction in the liquid. Note that the reaction on the electrode surface is not included here and will appear as the wall boundary condition.

The reaction boundary condition
The reaction at the electrode surface is modeled as a Robin boundary condition at the wall. Simply stated, this is the balance of the flux to the surface to the reaction rate for each species. The surface concentration of a reacting species at the electrode is determined by the kinetics of the reaction. The kinetics of the reaction is modeled by the Butler–Volmer kinetics or by a simple Tafel model. In general, the rate is a function of the overpotential (the potential at the surface minus the equilibrium value). The rate is also a function of the concentration and temperature. A detailed expression for the rate may be found in the electrochemical literature and books. The reaction rates are often expressed in terms of the current rather than in terms of the number of moles reacted.

If the surface reaction is rapid, the concentration of the reacting species is zero at the electrode. This condition determines the maximum current in the system, which is also known as the limiting current.

22.2   Electrolyte transport across uncharged membranes

Consider a binary electrolyte diffusing across a membrane due to a concentration gradient on either side of the membrane. Here we derive an expression for the flux and also for the potential gradient developed across the membrane.

We assume that the electrolyte is fully ionized and diffuses as M^+ and X^- ions. The fluxes are given by the Nernst–Planck equation,

$$J(M^+) = -D_+ \frac{dC}{dx} - z_+ D_+ \frac{F}{R_G T} C \frac{d\phi}{dx}$$

A similar equation holds for X^-

$$J(X^-) = -D_- \frac{dC}{dx} - z_- D_- \frac{F}{R_G T} C \frac{d\phi}{dx}$$

Here C is either $C(M^+)$ or $C(X^-)$, which are assumed to be equal due to electroneutrality. Also the two fluxes are equal, which helps us to eliminate the potential term. Hence the following expression for $d\phi/dx$ can be obtained:

$$\frac{d\phi}{dx} = -\frac{R_G T}{F} \left(\frac{D_+ - D_-}{z_+ D_+ - z_- D_-} \right) \frac{1}{C} \frac{dC}{dx}$$

Integrating across the system gives an expression for the potential:

$$\phi(L) - \phi(0) = \left(\frac{D_+ - D_-}{z_+ D_+ - z_- D_-} \right) \frac{R_G T}{F} \ln \left(\frac{C_0}{C_L} \right) \tag{22.16}$$

This potential is known as the diffusion potential and arises when the diffusion coefficients of the cation and anion are not equal.

Using this, we can now calculate the flux of either of the species. The result can be expressed in a Fick's-law type of equation as

$$J(M^+) = J(X^-) = J(\text{salt}) = -D_{\text{eff}} \frac{dC}{dx}$$

where the effective diffusion coefficient turns out to be

$$D_{\text{eff}} = \left(\frac{(z_+ - z_-)D_+ D_-}{z_+ D_+ - z_- D_-} \right) \tag{22.17}$$

which is known as the ambipolar diffusion coefficient.

The expression can be written in a simple form for the case of equal charges $z_+ = -z_-$ as

$$D_{\text{eff}} = \frac{2D_+ D_-}{D_+ + D_-}$$

or as

$$\frac{2}{D_{\text{eff}}} = \frac{1}{D_+} + \frac{1}{D_-}$$

Thus the effective diffusivity is the harmonic mean of the diffusivities of the positive and negative ions. The concept of ambipolar diffusion is shown in Fig. 22.2.

Typical values of diffusion coefficients of the various common ions are shown in Table 22.1. Note the relatively high value for H^+ ions (protons), which is somewhat inconsistent with their size. The higher value is due to molecular interaction with the water molecules, the so-called Grotthuss effect.

Example 22.2.

A membrane separates two bulk solutions of 0.5 M NaCl and 0.1 M NaCl. Find the potential difference developed across the membrane and the flux of NaCl across the membrane.

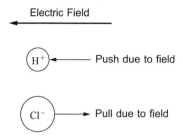

Figure 22.2 A schematic diagram of the diffusion of ions in an uncharged membrane or barrier. Note that the faster-moving H^+ ions set up an electric field in the opposite direction to that of their motion. This in turn slows it down. The slower Cl^- ions are accelerated by the field.

Table 22.1. Diffusion coefficients of ions in water at 25 °C.

	H^+	Na^+	K^+	Ca^{2+}	OH^-	Cl^-	$SO_4{}^{2-}$
$D\ (10^9\ m^2/s)$	9.313	1.334	1.957	0.7920	5.860	2.032	1.065

Solution.
The potential gradient can be calculated using Eq. (22.16). We find $\Delta\phi = -0.0086$ V or -8.6 mV.

Using expression (22.17), the effective diffusion coefficient can be calculated to be

$$D_{\text{eff}} = 1.61 \times 10^{-9}\ m^2/s$$

which is an intermediate value between the diffusion coefficient of sodium ions and that of chloride ions. Thus the faster-diffusing chloride ions are dragged down by the slower-moving sodium ions.

The flux can then be calculated using Fick's law, using the above value of the ambipolar diffusion coefficient. The flux is then $4.64 \times 10^{-7}\ \text{mol}/(m^2 \cdot s)$.

22.3 Electrolyte transport in charged membranes

The analysis of transport in a charged membrane is complicated due to the fact that both the concentration and the potential distribution in the membrane are affected by the charge distribution. Various methods can be used. The simplest is to assume that the potential can be assumed to be a linear function, i.e., the electric field is assumed to be constant. A model based on this assumption is discussed in this section.

The Nernst–Plank equation now reads

$$J_i = -D_i \frac{dC_i}{dx} + D_i \frac{F}{R_G T} z_i C_i E_x \tag{22.18}$$

where E_x is equal to $-d\phi/dx$. Since the potential field, ϕ, is assumed to be linear in x, the electric field E_x can be treated as a constant.

The flux is constant across the membrane due to the species-conservation requirement. Hence we can set dJ_i/dx as zero. Hence

$$\frac{d^2 C_i}{dx^2} - \frac{F}{R_G T} z_i E_x \frac{dC_i}{dx} = 0 \tag{22.19}$$

This is now a convection–diffusion type of equation. The solution can be obtained with the prescribed boundary conditions at $x = 0$ and $x = L$.

It is easier to compute in terms of a dimensionless distance $\zeta = x/L$, where L is the membrane thickness, and a dimensionless group reflective of the trans-membrane electric field:

$$E^* = \frac{LFE_x}{R_G T}$$

The governing equation is

$$\frac{d^2 c_i}{d\zeta^2} - z_i E^* \frac{dc_i}{d\zeta} = 0 \tag{22.20}$$

where c_i is a suitably defined dimensionless concentration. The solution can be represented as

$$c_i(\zeta) = c_i(0) + [c_i(1) - c_i(0)] \frac{1 - \exp(z_i E^* \zeta)}{1 - \exp(z_i E^*)} \tag{22.21}$$

where the concentrations at $\zeta = 0$ and $\zeta = 1$ are specified as $c_i(0)$ and $c_i(1)$ and used as the boundary conditions. It may be noted that the concentration profile in the membrane is not linear in this case.

The corresponding flux in dimensionless units is given by

$$J_i^* = -z_i E^* \frac{c_i(0)\exp(z_i E^*) - c_i(1)}{\exp(z_i E^*) - 1} \tag{22.22}$$

where J_i^* is defined as J_i divided by $D_i C_{\text{ref}}/L$.

The current across the membrane for a binary pair of ions is defined as

$$i = F(z_1 J_1 + z_2 J_2)$$

The dimensionless current can then be plotted as a function of the dimensionless transmembrane potential as shown in Fig. 22.3.

If the channel size is small (which is often the case in biological membranes), then electroneutrality cannot be assumed. Also, if the number of ions transported is small, a continuum approximation cannot be used. The problem is amenable to solution by molecular dynamics, although it appears that only the simple cases can be solved by this procedure due to computational limitations, the complex geometry of the pores, and the additional gate-closing/opening mechanisms inherent in biological membranes *in vivo*. Biological channels exhibit a number of other complexities, and hence the basic model presented above is more of academic learning value. Some of the complexities are discussed in the book by Hille (2001). For more details, that book and the paper by Corry *et al.* (1999) may be consulted.

22.4 Transport effects in electrodialysis

What is electrodialysis? Electrodialysis refers to transport across membranes that are selective to ions with only one type of charge. Thus cation-selective membranes permit only the

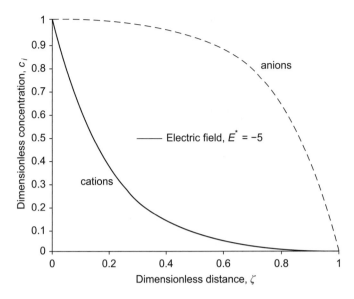

Figure 22.3 The species distribution in a charged membrane as a function of the transmembrane potential.

transport of positive ions and the negative ions are rejected. Conversely, anion-selective membranes permit transport only of negative ions. By stacking these in an alternating manner it is possible to remove salt from water and produce a solution with a lower concentration of salt.

The analysis of transport in such systems is similar to that for transport in a charged membrane considered in the previous section. The main difference is that the electric field is set up for the transport of the selected ion and the type of membrane charge and cannot be assumed to be a simple function *a priori*. The basic model and analysis of electrodialysis are now presented.

The flux of a charged species in the x-direction is given by

$$N_i = -D_i \frac{dc_i}{dx} + v_x c_i - D_i z_i c_i F^* \frac{d\phi}{dx} \tag{22.23}$$

where F^* is defined as $F/(R_G T)$ for compactness. The velocity term is often neglected (the low-flux assumption), but is retained here for completeness.

Using the continuity equation

$$\frac{dN_i}{dx} = 0 \tag{22.24}$$

leads to the convection–diffusion–migration equation in the membrane region:

$$D_i \frac{d^2 c_i}{dx^2} - v_x \frac{dc_i}{dx} + D_i z_i c_i F \frac{d^2\phi}{dx^2} + D_i z_i F^* \left(\frac{d\phi}{dx}\right)\left(\frac{dc_i}{dx}\right) = 0 \tag{22.25}$$

This equation also applies to the mass-transfer boundary layer (diffusion film) adjacent to the membrane surface. The only difference is that the diffusivity is that for the solution rather than for the membrane. As a first approximation, the external mass transfer is assumed to be rapid and this resistance is not included in the model.

An additional equation to be imposed to calculate the electric field is the electroneutrality condition:

$$\sum_i^N z_i c_i = -\omega \Theta(x) \tag{22.26}$$

Note that the RHS is not zero now. Here the $\Theta(x)$ term on the RHS is the fixed charge concentration in the system, and ω is the sign of these charges (positive or negative) depending on the type of membrane.

Finally, the current density across the system can be represented as

$$i = F \sum_i^N z_i N_i \tag{22.27}$$

On multiplying Eq. (22.23) by Fz_i and summing over all the species, we obtain

$$i = F \sum_i^N \left[-z_i D_i \frac{dc_i}{dx} + v_x z_i c_i - D_i z_i^2 c_i F^* \frac{d\phi}{dx} \right] \tag{22.28}$$

The second term on the RHS is equal to $-v_x \omega \Theta(x)$ in view of the electroneutrality condition given by Eq. (22.26). An equation for the potential gradient is therefore obtained by rearranging the above equation as

$$\frac{d\phi}{dx} = \frac{1}{\kappa} \left[-i + F v_x \omega \Theta(x) + \sum_i^N \left(-z_i D_i F \frac{dc_i}{dx} \right) \right] \tag{22.29}$$

where κ is the local value of the conductivity defined as

$$\kappa = FF^* \sum_i^N D_i z_i^2 c_i \tag{22.30}$$

The second derivative of the potential is given by

$$\frac{d^2\phi}{dx^2} = \frac{1}{\kappa} \left[F v_x \omega \frac{d\Theta(x)}{dx} + \sum_i^N \left(-z_i D_i F \frac{d^2 c_i}{dx^2} \right) \right] \tag{22.31}$$

In the above derivation κ is not differentiated, and is kept constant for simplicity of numerical implementation. The value of κ is, however, adjusted to its local value at each iteration in the numerical computation.

Simultaneous solution of (22.25) and (22.31) gives the concentration profiles and potential profiles in the system. The problem requires a numerical solution since both the concentration field and the electric field must be solved simultaneously now. MATLAB routines such as BVP4C are useful, since the problem is in a form similar to a non-linear second-order differential equation. An elegant solution method for these problems based on the boundary-element method was provided by Barbero *et al.* (1994). In this method the need for discretization of the differential operators is avoided by representing the problem in an integral manner, thereby leading to a greater accuracy. For further details this paper may be consulted.

Boundary conditions

Usually the concentration values in bulk solutions are known. Since the charge of the membrane is fixed and does not penetrate into the solution there is a concentration jump at the solution–membrane interface. Hence the boundary conditions at the membrane surface are not equal to the bulk concentrations but must be calculated by the following procedure. The discussion is for a univalent electrolyte of the type MX with $z_1 = 1$ and $z_2 = -1$ but can be extended to other types of electrolytes.

The ionic product of the cations and anions is fixed. Thus

$$C^+(b) \times C^-(b) = C^+(s) \times C^-(s)$$

where (b) represents the bulk values and (s) represents the surface values. The value of C^- at the surface is equal to

$$C^-(s) = C^+(s) + \omega\Theta(s)$$

Also in the bulk $C^+(b)$ is equal to $C^-(b)$ and denoted as $C(b)$, which is the known bulk concentration of the electrolyte. Hence the surface concentration of the positive ions is calculated by solving the following quadratic equation:

$$C^+(s)^2 + C^+(s)\omega\Theta(s) - C(b)^2 = 0$$

This equation is often called the Donnan equilibrium condition. A similar equation can be derived for $C^-(s)$ as well. These equations applied at each end provide the required boundary conditions and account for the concentration jump at the solution–membrane interface.

22.5 Departure from electroneutrality

Now we move to discuss the electric-field distribution near a solid surface. The assumption of electroneutrality does not hold here, and there is a thin region where the potential varies sharply. This region is known as the electrical double layer. Some phenomena and transport processes involving the effect of this double layer are addressed later. In this section, we derive the expression for the charge distribution in the double layer.

Consider a positively charged surface in contact with an electrolyte solution as shown in Fig. 22.4. The anions are attracted to this, and we expect a higher concentration of anions near the surface. Such a redistribution of charges leads to what is called a double layer. The intuitive term double layer arises due to there being two layers of charges: (i) one on the surface and (ii) the second due to the accumulated layer of counterions at the surface. This concept was postulated by Helmholtz (1879) (Fig. 22.4(a)), but in practice random thermal motion of the molecules causes a diffuse double layer as shown in Fig. 22.4(b). The potential distribution in the diffuse layer was studied by Guoy (1910) and Chapman (1913) and is presented in the following section.

The Poisson equation for the charge distribution in one dimension is

$$\frac{d^2\phi}{dx^2} = -\frac{\rho_c}{\epsilon} \tag{22.32}$$

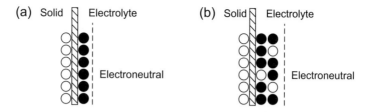

(a) Solid Electrolyte Electroneutral (b) Solid Electrolyte Electroneutral

Figure 22.4 Illustration of an electric double layer; (a) a sharp double layer according to Helmholtz and (b) the diffuse layer due to thermal motions according to the Guoy–Chapman model.

where ρ_c is the charge density and ϵ is the permittivity of the medium. The charge density is given by

$$\rho_c = F \sum_i z_i C_i$$

where F is the Faraday constant (equal to 96 485 C/mol), and z_i is the charge of the i th ion, and C_i is its concentration in mol/m^3.

Near the electrode the concentration is assumed to be related to the local potential by a Boltzmann distribution,

$$\frac{C_i}{C_{i\infty}} = \exp\left[-E_a/(R_G T)\right]$$

where C_∞ is the concentration far away from the electrode and E_a is the activation energy. Since E_a for a charged species is given by

$$E_a = z_i F \phi$$

the following relation can be used for the distribution of concentration:

$$\frac{C_i}{C_{i\infty}} = \exp\left[-F z_i \phi / (R_G T)\right] \tag{22.33}$$

This relation is in conformity with the fact that, as the potential increases, the concentration of positive ions will decrease and that of the negative ions will increase.

Hence the charge density is given by

$$\rho_c = F \sum_i z_i C_\infty \exp\left[-F z_i \phi / (R_G T)\right]$$

For further discussion consider a single electrolyte MX ionizing into M$^+$ and X$^-$, with M and X having equal and opposite charges. Thus we consider symmetric electrolytes, e.g. NaCl, CuSO$_4$ etc. The extension to a case with ion pairs of different valences is mathematically lengthy but involves no new concepts. Hence it is not considered here. Let $\nu = z_i$ be the valency of the charge. Hence $z_i = \nu$ for positively charged ions and $z_i = -\nu$ for negatively charged ions. The charge density is therefore

$$\rho_c = F \nu C_{i\infty}[\exp[-F\nu\phi/(R_G T)] - \exp[F\nu\phi/(R_G T)]]$$

or

$$\rho_c = -2F\nu C_{i\infty} \sinh[F\nu\phi/(R_G T)]$$

Hence the potential distribution is given by the solution of the Poisson equation in the following form:

$$\frac{d^2\phi}{dx^2} = -\frac{2vFC_\infty}{\epsilon} \sinh[Fv\phi/(R_GT)] \tag{22.34}$$

This is the Guoy–Chapman model for the diffuse double layer. The above differential equation is a singular perturbation problem, since ϵ is usually a small quantity. Hence there is a thin region over which a major change in potential can be expected. This confirms the fact that the double-layer thickness is small, and that it is confined to a region near the solid boundary. Exact analytical solutions can be obtained by the p-substitution method discussed earlier (see Example 8.3 and Section 10.4.6). However, simplified solutions are used in practical applications.

For further simplifications, the hyperbolic sinh term is replaced by a one-term Taylor approximation,

$$\sinh[vF\phi/(R_GT)] \approx \frac{Fv\phi}{R_GT}$$

leading to the following linear differential equation for the potential:

$$\frac{d^2\phi}{dx^2} = -\left(\frac{2v^2F^2C_\infty}{\epsilon R_GT}\right)\phi \tag{22.35}$$

This linear equation is known as the Debye–Hückel model. A "Debye" length parameter is introduced in order to make the equation dimensionless:

$$\lambda = \sqrt{\frac{\epsilon R_GT}{2v^2F^2C_\infty}} \tag{22.36}$$

The governing equation for one dimension can then be represented as

$$\frac{d^2\phi}{dx^{*2}} = -\phi$$

where the length has been scaled by the Debye length ($x^* = x/\lambda$) and the potential has been left in dimensional units.

The solution to the above differential equation is simply

$$\phi = \phi_0 \exp(-x^*) = \phi_0 \exp(-x/\lambda) \tag{22.37}$$

where ϕ_0 is the potential at the electrode (also known as the zeta potential). Here we assume that the potential decays to zero (a base value) as x^* tends to infinity.

The thickness of the double layer is of the order of the Debye length. The satisfaction of electroneutrality can therefore be assumed beyond this length.

Once the potential distribution is known, the charge density can be calculated using the rearranged form of the Poisson equation:

$$\rho_c = -\epsilon\left(\frac{d^2\phi}{dx^2}\right)$$

The final result for ρ_c can readily be derived as

$$\rho_c = -\frac{\epsilon\phi_0}{\lambda^2} \exp(-x/\lambda)$$

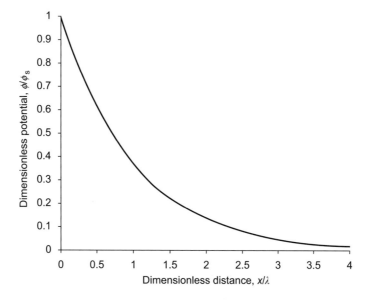

Figure 22.5 A plot of the potential distribution near a charged solid surface.

This equation is needed in the next section on electro-osmosis. We find that the charge decays to 63% of its initial value at a Debye length of one. It decays to 5% at a Debye length of three (Fig. 22.5). Hence the Debye length is the correct scale to view what is happening near the solid surface.

Once the potential near the surface is known, the concentrations of the co-ions and counterions can be calculated using Eq. (22.33). The equations can be expressed in terms of the dimensionless surface potential ϕ_s^* defined as $\phi_0/(R_G T/F)$:

$$\frac{C_i}{C_{i\infty}} = \exp[-z_i \phi_s^* \exp(-x/\lambda)] \tag{22.38}$$

An illustrative plot of the co-ions and counterions is shown in Fig. 22.6. Note the increase in the counterion concentration near the surface and the decrease in that of co-ions, as expected from electrostatic considerations.

How does the Debye length change with the electrolyte concentration?

The effect of the double layer on the velocity profile in a micro-channel tube is illustrated in the next section.

22.6   Electro-osmosis

Flow can be caused by an imposed electric field in tubes of small diameter, provided that the channel walls carry a surface charge and the flowing fluid contains ions (e.g., salt solution). Such flows are called electro-osmosis. Here we develop a model for the the velocity distribution in the system. Assume that the surface of the pipe has a charge density that causes a surface potential equal to ϕ_0. No external pressure gradient is imposed on the system. The problem analyzed is shown in Fig. 22.7.

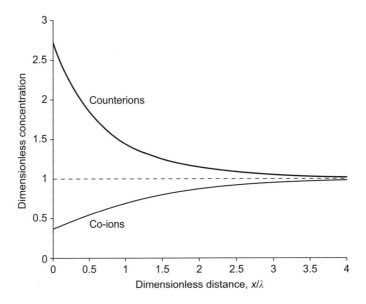

Figure 22.6 A plot of co-ion and counterion concentrations near a charged solid surface.

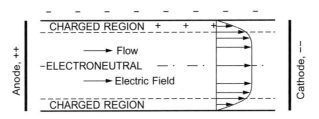

Figure 22.7 A diagram for analysis of electro-osmotic flow in a micro-channel.

Momentum balance in the x-direction leads to

$$\mu \frac{d^2 v_x}{dy^2} + \rho_c E_x = 0 \tag{22.39}$$

where the additional term due to the force on a charged particle in an electric field has been added. Here ρ_c is the charge density and E_x is the component of the electric field in the x-direction. The value of ρ_c can be calculated by taking the second derivative of (22.37) and using the Poisson equation. Note that ρ_c varies as a function of y now:

$$\rho_c = -\frac{\epsilon \phi_0}{\lambda^2} \exp(-y'/\lambda)$$

On substituting into Eq. (22.39) momentum balance leads to

$$\mu \frac{d^2 v_x}{dy^2} - E_x \epsilon \frac{\phi_0}{\lambda^2} \exp(-y/\lambda) = 0 \tag{22.40}$$

The boundary conditions are as follows: (a) no slip at the wall, $v_x = 0$ at $y = 0$; and (b) symmetry at the center of the channel, $dv_x/dy = 0$ at $y = H$, where H is half the channel width. Often the channel width is much larger than the Debye length and a simpler condition, namely

$$\text{as } y \to \infty, \quad v_x \to \text{constant}$$

is used. This is tantamount to using a semi-infinite domain as the solution region. Integration and the use of the modified boundary conditions gives the velocity profile as

$$v_x = \frac{\epsilon \phi_0 E_x}{\mu}[1 - \exp(-y/\lambda)] \tag{22.41}$$

A flat velocity profile is observed for much of the region if the plate spacing is much greater than the Debye length. The velocity profile is then plug flow for most of the tube, given by the following equation:

$$v_x = \frac{\epsilon \phi_0 E_x}{\mu} \tag{22.42}$$

This result is also applied to circular channels under similar assumptions and is known as the Smoluchowski equation. One of the interesting properties of this flow is that the flow is irrotational outside the double layer, and the velocity potential is therefore given by the Laplace equation.

The model can be modified for flow in a channel with a finite-gap as well as pipe flow. The finite-gap boundary condition should be used if the gap width H is comparable to the Debye length, λ. Details of the finite-gap channel are not provided and are left as exercise problems, but the relevant equations are shown below.

The finite-channel case

First the Debye–Hückel model is solved with the no-flux potential condition at $y = H$. Here H is half the channel width. The result is

$$\phi = \phi_0 \frac{\cosh[(H - y)/\lambda]}{\cosh(H/\lambda)}$$

Where ϕ_0 is the surface potential, λ is the Debye thickness defined in the usual manner, and y is the distance measured from the wall.

The corresponding charge density is calculated as

$$\rho_c = -\frac{\epsilon \phi_0}{\lambda^2} \frac{\cosh[(H - y)/\lambda]}{\cosh(H/\lambda)}$$

This is now used in the x-momentum balance and integrated with the no-slip condition at the wall and the no-shear condition at the center. The final answer is

$$v_x = \frac{\epsilon \phi_0 E_x}{\mu} \left[\frac{\cosh[(H - y)/\lambda]}{\cosh(H/\lambda)} - 1 \right] \tag{22.43}$$

A schematic diagram of the dimensionless velocity profile is shown in Fig. 22.8. The corresponding volumetric flow rate is

$$Q/W = \frac{2H\epsilon \phi_0 \lambda}{\mu} \left[1 - \frac{\tanh(H/\lambda)}{H/\lambda} \right]$$

In the limit of large value of H/λ, say 50, the profile of plug flow is nearly achieved as illustrated in Fig. 22.8. For large values of H/λ, the volumetric flow rate approaches a value of $2H\epsilon \phi_0 \lambda / \mu$, which is also the value obtained using the semi-infinite-domain approximation.

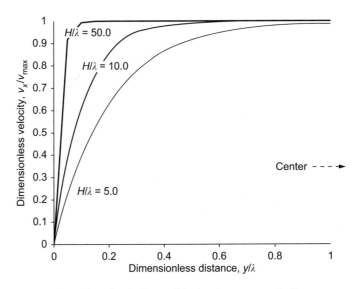

Figure 22.8 An illustrative plot of velocity profiles in electro-osmotic flow.

Circular pipes

The analysis can be extended to circular pipes, but the resulting solutions involve the Bessel functions as expected for cylindrical coordinates. The details are left as a set of exercise problems. It is interesting to note that the plug-flow model for the semi-infinite geometry holds for this case as well, provided that R/λ is greater than 20.

How would you model the system if there were also an imposed pressure gradient in the system?

22.7   The streaming potential

Consider a pressure-driven (Poiseuille) flow in a capillary with residual charge on the wall (Fig. 22.9). This charge created a Debye layer near the wall where the charge density is non-zero. These charges are now carried by the flow, and this generates a current in the system. This current leads to a potential gradient in the system. This potential generated by the Poiseuille flow is called the streaming potential. The magnitude of this potential is given by the following equation:

$$\Delta\phi = \frac{\epsilon\phi_0}{\sigma\mu}\Delta p \tag{22.44}$$

where σ is the conductivity of the solution.

The derivation of the above formula is presented below. Recall that the velocity profile is parabolic and given by

$$v_z = \frac{\Delta p}{4\mu L}(R^2 - r^2)$$

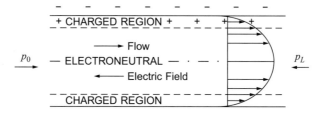

Figure 22.9 Pressure-driven flow in a micro-channel with surface charges, illustrating the streaming potential.

It is more convenient to measure distance from the wall for this problem. Hence we let $y = R - r$ or $r = R - y$. The velocity profile expressed as a function of y is

$$v_z = \frac{\Delta P}{4\mu L}(2Ry - y^2)$$

The charge distribution is given by

$$\rho_c = -\frac{\epsilon\phi_0}{\lambda^2}\exp(-y/\lambda)$$

Hence the current generated is equal to the charge density times the volumetric flow rate. Hence

$$I = \int_0^R 2\pi r v_z \rho_c \, dr$$

This can be replaced in terms of y by

$$I = \frac{\Delta p}{4\mu L}\int_0^R 2\pi(R - y)(2Ry - R^2)\rho_c \, dy$$

This can be simplified to

$$I = \frac{\Delta p}{4\mu L}(4\pi R^2)\int_0^\infty y\rho_c \, dy$$

which is another example of the "art of approximation" whereby the y^2 term is neglected. Since we are dealing only with the wall region, $R - y$ is taken as R. Also the limit of integration is changed to ∞ rather than R.

On substituting for ρ_c as a function of y and integrating we have

$$I = \frac{\Delta p\,\pi R^2\epsilon\phi_0}{\mu L}$$

The conductance of the system is $\pi R^2\sigma/L$, where σ is the specific conductance. By Ohm's law the current is related to the potential gradient by

$$I = (\pi R^2/L)\sigma\,\Delta\phi$$

Upon equating the two expressions for the current, the expression for the streaming potential given by Eq. (22.44) is obtained.

Example 22.3.

Find the streaming potential when a pressure drop of 0.1 MPa is applied to a liquid with a specific conductivity of 0.0014 (Siemens/m). The viscosity of the fluid is 0.001 Pa · s. The surface has a charge density causing a surface potential of 110 mV.

Solution.

On substituting into the expression for the streaming potential we find $\Delta\phi = 976 \, \text{mV}$.

Measurement of the streaming potential is a useful technique to find the surface charge density or the zeta potential. The values are not too small, so accurate values can be measured and related to surface properties. Effects of surface modifications can then be examined and related to molecular-level models.

22.8   The sedimentation potential

When a charged particle is released in a liquid it will attain a relative velocity equal to the terminal settling velocity. Movement of the particle in an electrolyte solution causes the distortion of the double layer adjoining the particle. The liquid behind the particle carries an excess of counterions compared with the liquid ahead of the particle. Since the net current is zero, an induced electric field is set up such that the net current is made zero. Thus, if electrodes are placed at the top and bottom of the settling tube, a potential known as the sedimentation potential can be measured. Detailed analysis of the phenomena and transport models are given in the book by Masliyah and Bhattacharjee (2006) and not presented here. However, two key equations based on a simple model are shown below for completeness.

The sedimentation electric field

The magnitude of the electric field generated due to sedimentation of a charged particle is given by

$$E(\text{sedimentation}) = \frac{\epsilon\zeta(\rho_p - \rho_L)\epsilon_s g}{\mu\sigma_\infty} \tag{22.45}$$

where ρ_P is the particle density, ρ_L is the liquid density, σ_∞ is the conductivity of the solution, and ϵ_s is the solid holdup (fraction of solid volume per total volume). Here ϵ is the permittivity of the medium and ζ is the zeta potential of the particle. ζ is the same as ϕ_0, the potential of the electrode, in the notation used earlier. The term zeta potential is more commonly used by physical chemists.

Terminal velocity

The terminal settling velocity of the particle, v_t, is also reduced due to the electric force and can be calculated by application of the Smoluchowski equation:

$$v_t = v_s \left[1 + \frac{\epsilon^2 \zeta^2}{R^2 \sigma_\infty \mu}\right]^{-1}$$

Here v_s is the terminal velocity given by Stokes' law.

22.9 Electrophoresis

Electrophoresis refers to movement of charged particles in solution under the action of an electric field. This area is becoming important in what is now called the subject of proteomics, which requires separation of proteins, DNA molecules, etc. They can be separated effectively by electrophoretic methods since these "particles" carry appreciable surface charges. The motion of charged particles is similar to that for ions, except that the mobility is rather low compared with that for ions. However, the number of unit charges carried by the particles, z_i, is often large compared with values for ions, and hence the overall quantity $z_i \mu_i$ is comparable to that for ions. Thus separation of such particles by electrophoresis is a viable technique.

How the molecule moves in a fluid depends on the nature of the colloidal system. Colloids are suspensions of fine particles in fluid media with the particle size generally less than in the micrometer range and can be classified into hydrophobic or hydrophilic types. Typical examples of hydrophobic colloids are suspensions of metals, colloidal sulfur, etc., and these move in a fixed direction in an electric field depending on the type of the charge carried by these particles. Sols of metallic particles are positively charged while those of metal oxides, etc., are negatively charged. However, for hydrophilic (lyophilic, to be more general) sols, the direction of the motion is not unique and is very sensitive to the hydrogen-ion concentration. Proteins are examples of such systems. In fact, one can define an iso-electric pH value at which the protein will not move in an electric field. Thus the motion of these particles can be controlled by simply changing the pH of the solution, which provides an additional operational flexibility (iso-electric focusing).

A simple model for a flat-channel system for electrophoresis is shown here in order to illustrate the application of transport-phenomena concepts. Consider a channel with a prescribed velocity field and an electric field applied across the system. The protein solution is introduced as a point source at the inlet. This is convected by flow in the x-direction and moves to the electrode at $y = H$ by electrophoretic migration. (Here we assume that $y = H$ is of the opposite polarity to the charge on the protein.)

The model shown below is for plug flow of the liquid. A coordinate system in the direction of the velocity vector and perpendicular to the velocity is set up as shown in Fig. 22.10. For this system the angle between the direction of convection and the x-axis can be computed as

$$\tan \theta = \frac{\mu_p E_y}{v_x}$$

where μ_p is the electrophoretic mobility, which is defined as $z_i \mu_i$. The convection–diffusion equation holds in this system:

$$v_p \frac{\partial C}{\partial x_c} = D_p \frac{\partial^2 C}{\partial y_c^2}$$

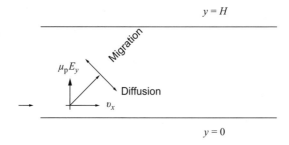

Figure 22.10 Electrophoretic transport of a charged species.

The subscript c on the coordinates x and y indicates that these are the coordinates in the direction of and perpendicular to the velocity, respectively. In the above equation v_p is the convective velocity of the protein, defined as

$$v_p = \sqrt{v_x^2 + (\mu_p E_y)^2}$$

The differential equation above is solved subject to the assumption of a point source at the origin. The resulting solution is

$$C = m_p \sqrt{\frac{v_p}{2\pi D_p x_c}} \exp\left(-\frac{v_p y_c^2}{4 D_p x_c}\right)$$

where m_p is the source strength equal to the mass or number of moles of protein being injected at the source point per unit width of the slab. This is a Gaussian type of distribution and the standard deviation is

$$\sigma = \sqrt{4 D_p x_c / v_p}$$

The separability of two protein species by electrophoresis can be assessed by this simple model. In general a standard deviation of $\sigma = 4$ will maintain a concentration difference of 0.1%.

The above discussion provides a basic model for the process, and is useful to provide some guidelines for the design and choice of suitable equipment. The factors to consider are as follows. (i) The length of the device must be reasonable. (ii) Multiple species must be separable in a continuous manner. (iii) The Joule heating due to the imposed field causes a temperature rise, which can cause protein denaturation. Hence the design should limit the maximum temperature rise in the system, and heat-transfer models also become important. (iv) Other effects such as Taylor dispersion, and double-layer distortion affecting the mobility of the particle become important in many cases.

22.10   Transport in ionized gases

Plasma is defined as a partly ionized gas. The gas now consists of positively charged and negatively charged particles in addition to neutral molecules. The ionized species, especially the electrons, have a high energy level to activate species that can then take part in chemical reactions. Thus the kinetic limitations of reactions can be overcome by application of plasmas. The reactions can be carried out at much lower temperatures (assuming that the thermodynamics favors it) than would be possible if the reactions were carried out in the absence of plasma. Hence plasma systems are of importance in many fields, including micro-electronic fabrications, energy systems, deposition of amorphous silicon for solar cells, methane activation or pyrolysis, etc.

Modeling a plasma system is a classic example of multiscale analysis. Thus the models can be build at three levels, with the most complex model shown at the top.

1. Detailed models involving the solution of Boltzmann equation.
2. Simpler Monte–Carlo simulation models.
3. Approximate representation using a continuum approach.

Owing to the low pressure of operation, the continuum approximation might not be valid, but it is still useful as a means to gain an understanding of the effect of transport processes in such systems. The goal of this section is to introduce the basic properties of a plasma system based on transport concepts and on a continuum approximation.

Consider argon gas contained between two electrodes subject to a high electric potential. The system is maintained at high enough vacuum that the frequency of molecular collisions is reduced. This permits part of the gas to be ionized. Thus we have free electrons, positive Ar^+ ions, and, in addition, neutral Ar atoms. This represents the simplest example of a d.c. plasma. The problem considered here is how to model the electron distribution and the current–voltage relations in the system.

A plasma can be divided into a region near the cathode (the cathode sheath) and the bulk region (also known as the glow region). (Note that there is also a sheath near the anode, which is ignored in the first level of analysis.) Charge neutrality holds in the glow region, but not in the cathode sheath. Hence the modeling can be done on the basis of a two-region model, as was done in the paper by Pirooz et al. (1991) and other groups. In this section we present some simple models for the two regions. The model provides us with a basic understanding of the system and also demonstrates how the transport methodology can be effectively applied.

The bulk or glow region

In the bulk region we have $n_+ = n_- = n$ due to electroneutrality, where n is the number concentration of electrons. Also the diffusion of both ions can be characterized by defining a single diffusion coefficient, the ambipolar diffusivity. This is the harmonic mean of the two diffusivity values as indicated in Section 22.2,

$$\frac{2}{D_a} = \frac{1}{D_+} + \frac{1}{D^-}$$

Hence the transport equation is

$$D_a \frac{d^2 n}{dx^2} = -\mathcal{R}$$

Again we find the ubiquitous presence of the diffusion–reaction equation. The RHS term $\mathcal{R}$ is the rate of production of electrons by ionization. Assuming a first-order process $\mathcal{R} = kn$, The solution is readily obtained as

$$n = n_0 \frac{\sin(x\sqrt{k/D_a})}{\sin[(L - \delta_c)\sqrt{k/D_a}]}$$

where L is the width of the plasma and δ_c is the cathode-sheath thickness. The boundary conditions used to derive the above equation are zero concentration at the anode ($x = 0$) and a value of n_0 at the border of the glow and sheath regions. The value of n_0 can be established by matching the solutions for the electron distributions in the two regions.

Very often δ_c is neglected and the equation is assumed to hold for the entire gap ($L - \delta_c \approx L$). Note that no acceptable solutions are possible if $\sin[(L - \delta_c)\sqrt{kD_a}] \geq \pi$. This sets the gap width which should be used in practice. If the system parameters are such that the above criterion is exceeded, the plasma gets extinguished. We find that even such simple models provide useful insights and are also useful to set operability conditions.

The cathode sheath

In this region electroneutrality does not hold. The model is formulated by writing species-continuity equations for the electrons and ions together with a generation term. In addition an additional equation for the energy of the electrons is added, since the generation depends on the energy level of the electrons. The models are then patched at the point where the glow region ends (where the ion and electron concentrations become nearly equal) and the cathode sheath begins. Detailed models can be found in the paper of Pirooz *et al.* (1991)

Summary

• Mass transport of charged species is important in many applications. The transport rate is now also affected by the electric field. The Nernst–Planck equation (22.6) is commonly used as the transport law for such systems.

• The electric field (or equivalently the negative gradient of the electrical potential) has to be either specified or simultaneously calculated to find the transport rates of charged species. In general the electrical potential is given by the Poisson equation (22.10). However, the electroneutrality condition (positive and negative charges are equal) holds for most of the region and is used as an implicit condition to find the electric field. Some applications are shown in the text, and a careful study is important in order to understand these complex interactions.

• The current in an electrochemical system cannot be computed simply by application of Ohm's law based on the applied potential. The potential gradient is composed of three components as indicated by Eq. (22.15): (i) the Ohmic term with the actual current, (ii) a diffusion potential as needed for electroneutrality, and (iii) a streaming potential due to bulk flow of the liquid.

• Electrodialysis refers to transport in a charged membrane that permits selective transport of only positive or only negative ions.

• Electroneutrality does not hold in a thin region near a charged surface. This layer is called the Debye layer, and equations to calculate the thickness of the Debye layer were derived. Ions with opposite charge to that of the surface (counterions) accumulate near the surface. Some important electrokinetic effects are dependent on the charge distribution in the Debye region.

- Electro-osmosis represents the flow, due to an applied electric field, of an electrolyte solution relative to a charged surface (e.g., a capillary tube with surface charge). The momentum balance coupled with the electric force due to the charge distribution near the surface provides a model for the system. The velocity profile is very nearly plug flow in such systems.

- The streaming potential refers to a potential generated by flow created by a pressure gradient in a capillary or tube with surface charge. Similarly, the sedimentation potential is the potential generated when a charged particle moves relative to a liquid. Both phenomena are of importance in surface sciences and for colloids.

- Electrophoresis refers to the migration of a charged species relative to a liquid caused by an applied electric field. It is a useful technique to separate proteins, DNA, etc., and finds a wide range of applications in bio-separation. A simple diffusion–migration–convection model was presented to track the particle trajectory in such systems. The model can be used to specify conditions for separation of particles.

- Ionized gases, also known as plasmas, are of importance in many applications and represent another example of mass transport in charged systems. A simple model can be used to understand electron distributions in such systems.

ADDITIONAL READING

An excellent treatise on this subject is the book by Newman and Thomas-Alyea (2004).

For a more detailed treatment of electrokinetic and colloidal transport phenomena, the book by Masliyah and Bhattacharjee (2006) is a great source.

Glow discharge plasma systems and examples of applications are covered in the book by Chapman (1980).

Problems

1. Verify by inputting the various constants that $R_G T/F$ is equal to $k_B T/e$.
2. Find the mobility of H^+, OH^-, and other ions from the diffusivity data given in Table 22.1.
3. Show from combination of the equation for current and the flux expression that the conductivity of an electrolyte solution is given by

$$\kappa = \frac{F^2}{R_G T} \sum_i^N D_i z_i^2 C_i$$

when there are no concentration gradients in the solution.

Calculate the value for 0.1-M and 1-M solutions of NaCl.

Also show that the divergence of the current is equal to zero. Hence show that the Laplace equation holds for the potential field in such cases with no concentration gradients.

Estimate the electrical conductivity of pure water.

4. When there is a concentration gradient in the system, show that the potential gradient is composed of two terms, (i) an Ohm's-law contribution and (ii) a diffusional contribution. State the equation for the current. Now take the divergence of the current and show that the

following Poisson equation holds for the potential field:

$$\nabla \cdot (\kappa \nabla \phi) = -F \sum_i z_i \nabla \cdot (D_i \nabla c_i)$$

Show the similarity to heat transport with generation with a variable thermal conductivity. What assumption is implicit in the above equation?

5. Copper is deposited at a cathode from solution with a bulk concentration of 0.5 M at the rate of 3.0 g/m$^2 \cdot$ s. Find the surface concentration of Cu^{2+} at the cathode if the mass transfer coefficient from the bulk to the surface is 1×10^{-4} m/s.
   Find the current density.
   What will the maximum rate of deposition be?

6. Consider a system with a "supporting" electrolyte, for example CuSO$_4$, and a second salt such as a mixture of CuSO$_4$ and Na$_2$SO$_4$. The system now consists of Cu^{2+}, SO$_4^{2-}$, and Na$^+$. Assume that only Cu^{2+} can react at the cathode. Hence the flux of Cu^{2+} should be assumed to be non-zero, while those of the other species are zero.
   Develop an equation to find the diffusion potential developed in this system and the factor by which the Cu^{2+} flux is enhanced.

7. Calculate the diffusion potential for an uncharged membrane with a concentration of 0.5 M on one side and 0.1 M on the other side for solutions of (a) CuSO$_4$, (b) MgCl$_2$, and (c) KCl. Also calculate the effective diffusion coefficients for these systems.

8. Consider an uncharged membrane of thickness 100 μm with concentrations of 1 M HCl on one side and 0.1 M HCl on the other side. Assume that HCl is completely ionized.

   (a) Find the flux of HCl in the system.
   (b) Find the diffusion potential generated in the system.
   (c) Which side of the membrane is at a higher potential? Explain your answer in terms of the physics of the system.
   (d) Find the diffusion flux and migration flux of Cl$^-$ and H$^+$ ions and tabulate them. Comment on the results.

9. Two solutions are separated by a porous sintered disk (1 mm thick) that permits diffusion across the disk. On one side we have a mixture of 1 M HCl and 1 M BaCl$_2$ while on the other side we have pure water. It is required to find the flux across the system. Both salts are completely ionized and diffuse as H$^+$, Cl$^-$, and Ba^{2+} across the disk. Set up the model to compute the fluxes. From the fluxes, find the effective diffusivities of these ions across the disk.
   Use for the ionic diffusivity for Ba ions 0.85×10^{-9} m^2/s. For the other ions, use the values in Table 22.1.
   Use the condition of no net current to solve for the potential.
   Answers: H, 6.14; Cl, 2.2268; and Ba, 0.271.

10. Diffusion of weakly ionized acids such as acetic acid is an interesting and complex problem in diffusion. The diffusion involves the ionized species CH$_3$COO$^-$ as well as the un-ionized acid CH$_3$COOH. The diffusion of acetate ions is affected by H ions due to electroneutrality while that for un-ionized molecules of acetic acid is a constant. Set up a model to compute the flux based on this model. Express the results in terms of the total concentration of acetic acid on one side of a region. The concentration on the other side is zero. What is the dependence of the flux on the concentration?

11. Consider a **charged** membrane of thickness 100 µm with concentrations of 1 M HCl on one side and 0.1 M HCl on the other side. The membrane has a negative charge with a constant electric field of 1 N/C.

    Find the flux of HCl in the system using the model for transport across a charged membrane given in Section 22.3.

12. Extend the analysis of a system for transport in a charged membrane to the case of a mixture of two electrolyte salts with a common anion (a mixture of NaCl and KCl, for example).

    Derive a formula for the membrane potential when one side of the membrane is exposed to concentrations of $C_1(0)$ for KCl, and $C_2(0)$ for NaCl while the other side at L is exposed to $C_1(L)$ and $C_2(L)$.

    Calculate the potential for a membrane that is 100 µm thick with concentrations of 0.1 M and 0.01 M on either side.

13. Calculate the Debye length for a 0.1-M solution of a univalent (1, 1) electrolyte, i.e., one with single-charged cations and single-charged anions, in water.

    If the surface has a negative charge with a zeta potential of 100 mV, plot the concentrations of the co-ions and counterions near the electrode surface. The co-ion is defined as the ion with the same charge as the electrode. Repeat for a divalent (2,2) electrolyte. Which case has the larger Debye length?

14. A solid is in the form of a long cylinder and has a surface charge of q_S. Derive an expression for the potential in the external region. Use the Debye–Hückel approximation and solve as a linear differential equation.

15. A solid is in the form of a sphere and has a surface charge of q_S. Derive an expression for the potential in the external region.

16. Solve the non-linear equation for the double-layer potential given by

$$\nabla^2\phi = -\frac{2\nu F C_\infty}{\epsilon}\sinh[F\nu\phi/(R_G T)] \tag{22.46}$$

    Use the dimensionless variables

$$\phi^* = \phi\frac{F\nu}{R_G T}$$

    and scale the distance by the Debye length. Show that the following dimensionless equation holds for the potential:

$$\frac{d^2\phi^*}{dx^{*2}} = \sinh(\phi^*)$$

    Solve the equation subject to the boundary condition that $\phi^* = \phi_s^*$ at $x^* = 0$ and is zero at $x^* \to \infty$. Derive the following solution to the potential field:

$$\phi^* = 2\ln\left[\frac{1 + \exp(-x^*)\tanh(\phi_s^*/4)}{1 - \exp(-x^*)\tanh(\phi_s^*/4)}\right]$$

    How does this compare with the linear model represented by Eq. (22.37)? What is the limit of the above expression for small values of the surface potential ϕ_s^*?

17. Solve the Debye–Hückel equation for circular pipes whose walls carry a surface potential of ϕ_0. Show that the potential is given by

$$\phi = \phi_s\frac{I_0(\kappa r)}{I_0(\kappa R)}$$

    where I_0 is the modified Bessel function of the first kind.

From this derive an expression for the charge distribution function ρ_c in a circular channel. This expression is useful to analyze electro-osmotic flow in a circular channel.

18. Verify that the current carried in electro-osmotic flow can be expressed as

$$I = \int_0^R 2\pi r v_z \rho_c \, dr$$

Using this relation, derive an expression for the current in electro-osmotic flow.

19. Analyze the problem of electro-osmotic flow with an additional imposed pressure gradient. Show that the volumetric flow rate can be expressed as

$$Q = L_{11} \, \Delta\phi + L_{12} \, \Delta P$$

where $\Delta\phi$ is an applied potential gradient in the x-direction equal to E_x/L.

Also show that the current can be expressed as

$$I = L_{21} \, \Delta\phi + L_{22} \, \Delta P$$

Derive expressions for the coefficients L_{11} etc. Verify that $L_{12} = L_{21}$, a type of Onsager reciprocity relation (Onsager, 1931).

20. Find the flow rate for an electro-osmotic flow in a tube of diameter 5 μm and length 10 cm filled with an aqueous solution. The applied potential is 1 kV. The zeta potential is given as −100 mV.
Answer: $Q = 1.39 \times 10^{-14} \, \text{m}^3/\text{s}$.

21. An effectiveness factor of electro-osmotic flow is often used. This factor is defined as the volumetric flow rate due to electo-osmosis divided by that due to an applied pressure gradient.

Derive an expression for this.

22. Derive expressions for the velocity profile and the volumetric flow rate for a circular channel. Examine the limiting cases of this equation for small R and large R (in comparison with the Debye length).

23. A circular capillary has a diameter of 10 μm and a length of 3 cm. The surface carries a zeta potential of −0.5 V. The flowing fluid is an aqueous solution of 0.01 M $CaSO_4$, and the viscosity is 0.001 Pa · s at the specified temperature.

Find the volumetric flow rate. Also compute the pressure gradient that would need to be applied to create the same flow if no electric field were applied.

24. Find the streaming potential when a pressure drop of 0.1 MPa is applied to a liquid with a specific conductivity of 0.0014 S/m. The surface charge is 100 mV.

25. Consider the settling of charged spherical particles with a zeta potential of 50 mV in a solution of 0.0 5 M NaCl. The particle density is 2060 kg/m^3 and the particle holdup is 0.1.

Calculate and plot the sedimentation electric field as a function of the particle diameter. Also calculate and plot the particle settling velocity and show that the correction to Stokes' law is negligible for this case.

26. **Separation distance for proteins.** It is required to separate two proteins with the mobilities of $\mu_1 = 8 \times 10^{-5}$ m · C/(N · s) and $\mu_2 = 6 \times 10^{-5}$ m · C/(N · s), The diffusion coefficient is 6×10^{-11} m^2/s. The flow velocity is 0.2 mm/s. The applied field is 2000 V/m.

Find the trajectory of the "plumes" of the two proteins and the distance at which the two proteins can be separated.

27. **Design arrangements in electrophoresis: a case-study problem.** Various methods have been developed in order to increase the throughput in electrophoresis. Most of these designs vary in the flow arrangement and the changes in the direction of the electric field. These include the Philpot design, Hanning design, annular design, and rotating annular column, among many others. Discuss these arrangements. A comparison of the performance analysis has been done by Yoshisato *et al.* (1986), who used the well-studied glycine–glutamic-acid solute pair as a model system. Study this and related papers and do a case study on performance analysis of various designs.

28. **Other electrophoretic methods: a case-study problem.** In the text we considered only the zone electrophoresis where a zone of a solute-rich layer is created by the action of an electric field. Other techniques are isotochorphoresis and iso-electric focusing. Review what these techniques are. In particular, emphasize the transport modeling issues involved in each of these techniques. The book by Westermeier (2005) would be a good starting point.

29. **The proton-exchange membrane (PEM) fuel cell: a case-study problem.** A hydrogen fuel cell using a PEM consists of a gas-diffusion backing layer with a Pt on C supported catalyst as anode and cathode. The two electrodes are separated by a membrane, which permits selective transport of H^+ ions. The protons are released by reaction of hydrogen gas at the anode:

$$H_2 \rightarrow 2H^+ + 2e^-$$

These protons diffuse across the membrane and react at the cathode with the oxygen (air) gas:

$$1/2\,O_2 + 2H^+ + 2e^- \rightarrow H_2O$$

The overall reaction is simply oxidation of hydrogen to produce water. This reaction is spontaneous with a negative free energy, $\Delta G^\circ = 24\,000$ kJ/mol at 298 K. An equivalent voltage is generated in the system under conditions of nearly zero current and is the maximum potential that can be generated in the fuel cell. Transport effects reduce the voltage, and an analysis of the various transport effects together with the rates of the anodic and cathodic reactions produces the current versus voltage distribution in the system.

Develop a model to design a fuel cell and to calculate the current–voltage relations.

30. **Diffusion through ordered force fields: a case-study problem.** The classical Fick or Einstein model for diffusion is based on the assumption that the diffusing molecule is exposed to a random force field arising from molecular motion. The assumption is reasonable and predicts phenomena such as Knudsen diffusion for small-sized pores but might not be true for nanopores, where there may be a directional force field. This variation is due to local concentration differences, the presence of surface charge, etc. Such systems can be simulated by using the Fokker–Planck equation and the Smoluchowski equation, and an illustrative study has been presented by Wang *et al.* (2009). Your goal is to review this paper, write a computational model that permits incorporation of a prescribed force field, and calculate the additional contribution to diffusion arising from such ordered force fields. Diffusion in zeolites is an example in chemical reaction engineering where such effects can become important.

31. **Soil dewatering by electro-osmosis: a case-study problem.** The phenomenon of electro-osmosis can be used for dewatering and consolidation of soils and mine tailing and waste

sludges. The idea is that by appropriate placement of electrodes the flow can be diverted away from a contaminated site in a controlled fashion. In a reverse manner, a flow of a suitable chemical sealant can also be directed towards the spill site. Porous soils such as clays have a negative surface charge and hence the salt-bearing ground water flows from the anode to the cathode, thereby lowering the water table near the contaminated site. Simple models can be based on the equations presented in this chapter and models of increasing levels of complexity may be developed using a 2D or 3D geometry and a computational method such as finite-element analysis. The paper by Shapiro and Probstein (1993) is a starting source for this problem. Develop models based on the material in this chapter and other literature sources for simulation of such systems.

CLOSURE

The following quote (*in italics*) at the end of the book by BSL is an appropriate closing remark for this book.

No engineering project can be conceived, let alone completed, purely through the use of descriptive disciplines, such as transport phenomena. Transport phenomena can, however, prove immensely helpful by providing useful approximations, starting with order-of-magnitude estimates and going on to successively more accurate approximations.

I have tried to provide the knowledge base for this with scaling concepts, dimensionless arguments, and modeling at three scales, together with some computational snippets.

Much remains to be done, but the utility of transport phenomena can be expected to increase rather than diminish. Many challenges remain to be met. The quantitative undertones provided by transport phenomena will prove to be an immense help. I hope that you, the students and the readers, will able to participate in this exciting field.

REFERENCES

Abramowitz, M. and Stegun, J. A. (1964) *Handbook of Mathematical Functions*. Washington, DC: National Bureau of Standards.

Acrivos, A. and Taylor, T. E. (1962) Heat and mass transfer from single sphere in Stokes flow. *Phys. Fluids*, **5**, 387–394.

Alexiades, V. and Solomon, A. D. (1993) *Mathematical Modeling of Melting and Freezing Processes*. Washington, DC: Hemisphere Publishing Corporation.

Arfken, G. B., Weber, H. J., and Harris, F. E. (2013) *Mathematical Methods for Physicists*, 7th edition. Waltham, MA: Academic Press.

Aris, R. (1962) *Vectors, Tensors and Basic Equations of Fluid Mechanics*. Englewood Cliffs, NJ: Prentice-Hall.

Aris, R. (1975) *Mathematical Theory of Diffusion and Reaction in Porous Catalysts*. Oxford: Clarendon Press.

Aris, R. and Hatfield, B. (1969) Combined effect of external and internal diffusion for nonisothermal case. *Chem. Eng Sci.*, **24**, 1213–1220.

Astarita, G. (1967) *Mass Transfer with Chemical Reaction*. Amsterdam: Elsevier.

Barbero, A. J., Mafé, S. and Ramírez, P. (1994) Application of the boundary element method to convective electrodiffusion problems in charged membranes. *Electrochim. Acta*, **39**, 2031–2035.

Barrer, R. M. (1951) *Diffusion in and through Solids*. Cambridge: Cambridge University Press.

Battles, Z. and Trefethen, L. N. (2004) An extension of Matlab to continuous functions and operators. *SIAM J. Scient. Computing*, **25**, 1743–1770.

Becker, S. M. and Kuznetsov, A. V. (editors) (2013) *Transport in Biological Media*. Amsterdam: Elsevier.

Bejan, A. (2004) *Convective Heat Transfer*. New York: John Wiley & Sons.

Bender, C. M. and Orszag, S. A. (1978) *Advanced Mathematical Methods for Engineers and Scientists*. New York: McGraw-Hill Publishers.

Bhagwat, S. S. and Sharma, M. M. (1988) Intensification of solid–liquid reactions with microemulsions. *Chem. Eng. Sci.*, **43**, 195–205.

Bin, A. K. (1983) Mass transfer into a turbulent liquid film. *Int. J. Heat Mass Transfer*, **26**, 981–991.

Bird, R. B. (2008) Five decades of transport phenomena. *AIChE J.*, **1**, 1–17.

Bird, R. B. (2010) Chemical engineering education: A gallimaufry of thoughts. *Ann. Rev. Chem. Biomol. Eng.*, **1**, 1–17.

Bird, R. B., Armstrong, R. C., and Hassager, O. (1987) *Dynamics of Polymeric Fluids*. New York: Wiley-Interscience,

Bird, R. B., Stewart, W. E., and Lightfoot, W. E. (2007) *Transport Phenomena*. New York: John Wiley and Sons.

Blasius, H. (1910) Funktionentheoretische Methoden in der Hydrodynamik. *Z. Math. Phys.* **58**, 90–110.

Bothe, D. and Prüss, J. (2010) Two phase Navier–Stokes equation with Boussinesq–Scriven model. *J. Math. Fluid Mech.*, **12**, 133–150.

Boussinesq, J. (1903) *Théorie analytique de la chaleur*, vol. 2. Paris: Gauthier-Villars.

Brennen, E. B. (2005) *Fundamentals of Multiphase Flow*. Cambridge: Cambridge University Press.

Buckingham, E. (1914) On physically similar systems: illustrations of the use of dimensional equations. *Phys. Rev.*, **4**, 345–376.

Calderbank, P. H. and Moo-Young, M. B. (1961) Mass transfer in bubble swarms. *Chem. Eng. Sci.*, **16**, 39–44.

Carslaw, H. S. and Jaeger, J. C. (1959) *Conduction of Heat in Solids*. Oxford: Clarendon Press.

Cassano, A. C. and Alfabo, O. M (1991) *Photochemical Reactor Design: A Rigorous Approach*. New York: Kluwer Academic Publishers.

Cebeci, T. and Bradshaw, P. (1977) *Momentum Transport in Boundary Layers*. New York: Hemisphere.

Cebeci, T. and Smith, A. M. O. (1974) *Analysis of Turbulent Boundary Layers*. New York: Academic Press.

Chabra, R. P. and Richardson, J. F. (2004) *Non-Newtonian Flow: Fundamentals and Engineering Applications*. Oxford: Butterworh-Heinemann.

Chandrasekhar, S. (1961) *Hydrodynamic and Hydromagnetic Stability*. Oxford: Clarendon Press.

Chapman, B. (1980) *Glow Discharge Processes: Sputtering and Plasma Etching*. New York: Wiley.

Chapman, D. L. (1913) Contribution to the theory of electroencapillarity. *Phil. Mag.*, **25**, 475–481.

Chen, S., Yuan, X., Fu, B., and Yu, K. (2011) Simulation of interfacial Marangoni convection in gas–liquid mass transfer. *Front. Chem. Sci. Eng.*, **5**(4), 448–454.

Chow, C. Y. (1979) *An Introduction to Computational Fluid Mechanics*. Altamonte Springs, FL: Seminole Publishing, p. 700.

Churchill, S. W. and Chu, H. H. S. (1975) Correlating equations for laminar and turbulent free convection from a vertical plate. *Int. J. Heat Mass Transfer*, **18**, 1323–1329.

Clark, M. M. (1996) *Transport Modeling for Environmental Engineers and Scientists*. New York: John Wiley & Sons.

Clift, R., Grace, J. R., and Weber, M. E. (1978) *Bubbles, Drops and Particles*. New York: Dover Publications.

Clusius, K. and Dickel, G. (1939a) Das Trennrohr. I. Grundlagen eines neuen Verfahrens zur Gasentmischung und Isotopentrennung durch Thermodiffusion. *Z. Phys. Chem.* B, **44**, 397–450.

Clusius, K. and Dickel, G. (1939b) Das Trennrohr. II. Trennung der Chlorisotope. *Z. Phys. Chem.* B, **44**, 451–473.

Cochran, W. G. (1934) The flow due to a rotating disk. *Proc. Camb. Phil. Soc.*, **39**, 365–375.

Colburn, A. P. and Drew, T. N. (1937) Condensation of mixed vapours. *Trans. Am. Inst. Chem. Eng.*, **33**, 197–215.

Cole, J. D. (1968) *Perturbation Methods in Applied Mathematics*. Waltham, MA: Blaisdell Publishers.

Combest, D. P., Ramachandran, P. A., and Duduković, M. P. (2011) *Ind. Eng. Chem. Res.*, **50**, 8817–8823.

Conner, J. M. and Elghobashi, S. E. (1987) Numerical solution of laminar flow past a sphere with surface mass transfer. *Numerical Heat Transfer*, **12**, 57–68.

Corry, B., Kuyucak, S., and Chung, S. H. (1999) Test of the Poisson–Nernst–Planck theory in ion channels. *J. Gen. Physiol.*, **114**, 597–599.

Crank, J. (1975) *Mathematics of Diffusion*. Oxford: Clarendon Press.

Crank, J. (1984) *Free and Moving Boundary Problems*. Oxford: Clarendon Press.

Crochet, M. J., Davies, A. R., and Walters, K. (1984) *Numerical Simulation of Non-Newtonian Flow*. Amsterdam: Elsever Publishing Company.

Cussler, E. L. (2009) *Diffusion: Mass Transport in Fluid Systems*. Cambridge: Cambridge University Press.

Cussler, E. L. and Moggridge, G. D. (2001) *Chemical Product Design*. Cambridge: Cambridge University Press.

Danckwerts, P. V. (1970) *Gas–Liquid Reactions*. New York: McGraw-Hill Publishing Co.

Dani, A., Cockx, A., and Guiraud, P. (2006) Direct numerical simulation of mass transfer from spherical bubbles: The effect of interface contamination at low Reynolds numbers. *Int. J. Chem. Reactor Eng.*, **4**, 18–26.

Davidson, P. A. (2001) *An Introduction to Magnetohydrodynamics*. Cambridge: Cambridge University Press.

Davidson, P. A. (2004) *Turbulence: An Introduction for Scientists and Engineers*. Oxford: Oxford University Press.

Deen, W. M. (2011) *Analysis of Transport Phenomena*. New York: Oxford University Press.

De Bleecker, K., Bogaerts, A., and Goedheer, W. (2005) Role of thermophoretic force on transport of nanoparticles. *Phys. Rev.*, **71**, 66405–66412.

De Groot, S. R. and Mazur, P. (1962) *Irreversible Thermodynamics*. New York: Dover Publications.

Dhir, V. K. (2006) Mechanistic prediction of nucleate boiling heat transfer – Achievable or a hopeless task? *J. Heat Transfer*, **128**(1), 1–12.

Dorfman, A. and Renner, Z. (2009) Conjugate Problems in Convective Heat Transfer: A Review in Mathematical Problems in Engineering, Hindawi Publishing Corporation, Article ID 927350.

Drazin, P. G. and Reid, W. H. (2004) *Hydrodynamic Stability*. Cambridge: Cambridge University Press.

Duncan, J. B. and Toor, H. L. (1962) Experimental study of three-component gas diffusion. *AIChE J.*, **8**, 38–41.

Edwards, D. A., Brenner, H., and Wasan, D. T. (1991) *Interfacial Transport Processes and Rheology*. Stoneham: Butterworth-Heinemann.

Faghri, A. (1995) *Heat Pipe: Science and Technology*. Basingstoke: Taylor & Francis.

Faghri, A. and Zhang, Y. (2006) *Transport Phenomena in Multiphase Systems*. Amsterdam: Elsevier.

Farmer, R. C., Cheng, G. C., Chen, Y. S., and Pike, R. W. (2009) *Computational Transport Phenomena for Engineering Analysis*. Boca Raton, FL: CRC Press.

Finlayson, B. A. (2006) Using Comsol Multiphysics to model viscoelastic fluid flow. Excerpt from the proceedings of the COMSOL Users Conference 2006 Boston (available online).

Fournier, R. L. (2011) *Basic Transport Phenomena in Biomedical Engineering*. Boca Raton, FL: CRC Press.

Froment, G. F., Bischoff, K. B., and de Wilde, J. (2011) *Chemical Reactor Analysis and Design*, 3rd edition. New York: John Wiley & Sons Inc.

Gebhart, B., Jaluria, Y., Mahajan, R. L., and Sammakia, B. (1988) *Buoyancy-Induced Flows and Transport*. New York: Hemisphere Publishing Company.

Geiger G. H. and Poirier D. R. (1998) *Transport Phenomena in Metallurgy*. New York: John Wiley & Sons.

Ghiaasiaan, S. M. (2011) *Convective Heat and Mass Transfer*. Cambridge: Cambridge University Press.

Glasgow, L. A. (2010) *Transport Phenomena: An Introduction to Advanced Topics*. Hokoken, NJ: John Wiley & Sons.

Gmehling, J. and Onken, U. (1984) *Vapor–Liquid Equilibrium Data Collection*. Frankfurt: DECHEMA.

Green, H. L. and Lane, W. R. (1957) *Particulate Clouds, Dusts, Smokes and Mists*. London: E. and F. N. Spon Ltd.

Grossman, G. and Heath, M. T. (1984) Simultaneous heat and mass transfer in absorption of gases in turbulent liquid films. *Int. J. Heat Mass Transfer*, **27**, 2365–2376.

Guoy, G. (1910) Sur la constitution de la charge électrique à la surface d'un électrolyte. *J. Phys. Radium*, **9**, 457–468.

Haase, R. (1968) *Thermodynamics of Irreversible Processes*. New York: Dover Publishing Company.

Hadamard, J. S. (1911) Mouvement permanent lent d'une sphère liquide et visqueuse dans un liquide visqueux. *Compt. Rend. Acad. Sci.*, **152**, 1735–1738.

Hanna, O. T., Sandall, O. C., and Mazet, P. R. (1981) Mixing length model for mass transfer. *AIChE. J.*, **27**, 693–697.

Happel, J. and Brenner, H. (1983) *Low Reynolds Number Hydrodynamics*. Den Haag: Martinus Nijhoff Publishers.

Harriott, P. (2003) *Chemical Reactor Design*. New York: Marcel-Dekker.

Haynes, H. W. (1984) Multicomponent diffusion and reaction in porous catalysts. In *Chemical and Catalytic Reactor Modeling*, ed. P. L. Mills and M. P. Duduković. Washington, DC: American Chemical Society, pp. 217–238.

Heaslet, M. A. and Warming, R. R. (1965) Radiative transport and wall temperature slip in an absorbing planar medium. *Int. J. Heat Mass Transfer*, **102**, 979–999.

Helmholtz, H. V. (1879) Studien über elektrische Grenzschichten. *Ann. Phys. Chem.*, **7**, 337–387.

Hiemenz, K. (1911) Die Grenzschicht an einem in den gleichförmigen Flüssigkeitsstrom eingetauchten geraden Kreiszylinder. *Dinglers Polytechn.*, **326**(21), 321–340.

Higbie, R. (1935) The rate of absorption of a pure gas into a still liquid during short periods of exposure. *Trans. AICHE*, **31**, 368–389.

Hikita, H. and Asai, S. (1963) *Kagaku Kogaku*, **11**, 823–830 [article in Japanese]; English translation: (1964) Gas absorption with (*m*, *n*)th order irreversible chemical reactions. *Int. Chem. Eng.*, **4**, 332–340.

Hille, B. (2001) *Ion Channels of Excitable Membranes*, 3rd edn. Sunderland, MA: Sinauer Associates, Inc.

Hirshfelder, J. O., Curtiss, C. F., and Bird, R. B. (1954) *Molecular Theory of Gases and Liquids*. New York: Wiley & Sons.

Ioss, G. and Joseph, D. D. (1990) *Elementary Stability and Bifurcation Theory*. New York: Springer-Verlag.

Johnson, R. S. (2005) *Singular Perturbation Theory with Applications to Engineering*. New York: Springer.

Kandlikar, S. G., Shoji, M., and Dhir, V. K. (1999) *Handbook of Phase Change: Boiling and Condensation*. Philadelphia, PA: Taylor & Francis Publishers.

Kedem, O. and Katchalsky, A. (1958) Thermodynamic analysis of permeability of biological membranes to non-electrolytes. *Biochim. Biophys. Acta*, **27**, 229–246.

Kim, S. and Karrila, S. J. (1991) *Microhydrodynamics*. Boston, MA: Butterworth-Heinemann.

King, C. J. (1966) Turbulent phase mass transfer at a free gas–liquid interface. *Ind. Eng. Chem. Fundamentals*, **5**, 1–8.

Krantz, W. B. (2007) *Scaling Analysis in Modeling Transport and Reaction*. New York: Wiley-AIChE.

Kreyszig, E. (2011) *Advanced Engineering Mathematics*. New York: Wiley.

Krogh, A. (1919) The number and distribution of capillaries in muscles with calculations of the oxygen pressure head necessary for supplying the tissue. *J. Physiol.*, **52**, 409–415.

Kubíček, M. and Marek, M. (1983) *Computational Methods in Bifurcation Theory and Dissipative Structures*. New York: Springer-Verlag.

Kundu, P. K. and Cohen, I. M. (2008) *Fluid Mechanics*. Amsterdam: Elsevier.

Kutateladze, S. S. (1948) On the transition to film boiling under natural convection. *Kotloturbostroenie*, **3**, 10–12.

Leal, L. G. (2007) *Advanced Transport Phenomena: Fluid Mechanics and Convective Transport Processes*. Cambridge: Cambridge University Press.

Levich, V. G. (1962) *Physicochemical Hydrodynamics*. Englewood Cliffs, NJ: Prentice-Hall.

Lewis, W. K. and Whitman, W. G. (1924) Principles of gas adsorption. *Indust. Eng. Chem.*, **16**, 1215–1220.

Lightfoot, E. N. (1974) *Transport Phenomena in Living Systems*. New York: Wiley and Sons Publishers.

Maginn, E. J. (2009) From discovery to data: what must happen for molecular simulation to become a mainstream chemical engineering tool. *AIChE J.*, **55**, 1304–1310.

Majda, A. J. and Bertozzi, A. L. (2002) *Vorticity and Incompressible Flow*. Cambridge: Cambridge University Press.

Mao, Z. and Chen, J. (2004) Numerical simulation of Marangoni effect on mass transfer to single slowly moving drop in liquid–liquid systems. *Chem. Eng. Sci.*, **59**, 1815–1828.

Masliyah, J. H. and Bhattacharjee, S. (2006) Electrokinetic and colloid transport phenomena. Hoboken, NJ: Wiley-Interscience.

Matsuhisa, S. and Bird, R. B. (1965) Analysis of Ellis fluid in various geometries. *AIChE. J.*, **11**, 588–595.

McAdams, W. H. (1957) *Heat Transmission*, 3rd edition. New York: McGraw-Hill Book Company.

McBain, G. D., Chubb, T. H., and Armfield, S. W. (2008) Numerical solution to Orr–Sommerfeld equation using viscous Green's function. *J. Comput. Phys.*, **224**, 397–404.

McLaughlin, H. S., Mallikarjun, R., and Naumann, E. B. (1986) Effect of radial velocities in laminar flow. *AIChE. J.*, **32**, 419–425.

Middleman, S. (1988a) *Introduction to Mass and Heat Transfer*. New York: John Wiley and Sons.

Middleman, S. (1988b) *Fluid Dynamics: Analysis and Design*. New York: John Wiley and Sons.

Middleman, S. and Hochberg, A. K. (1993) *Process Engineering Analysis in Semiconductor Device Fabrication*. New York: McGraw-Hill.

Mills, A. F. (1993) *Heat and Mass Transfer*. Toronto: Irwin Publishing.

Modest, M. F. (2013) *Radiative Heat Transfer*, 3rd edn. New York: Academic Press.

Morbidelli, M. and Varma, A. (1997) *Mathematical Methods in Chemical Engineering*. New York: Oxford University Press.

Mukherjee, B., Wrenn B. A., and Ramachandran, P. A. (2012) Relationship between size of oil droplet and energy dissipation during chemical dispersion of crude oil. *Chem. Eng. Sci.*, **68**(1), 432–442.

Nayfeh, A. H. (1981) *Introduction to Perturbation Methods*. New York: Wiley.

Neofytou, P. and Drikakis, D. (2003) Non-Newtonian flow instability in a channel with sudden expansion. *J. Non-Newtonian Fluid Mech.*, **111**, 127–150.

Neumann, E. B. (1987) *Chemical Reactor Design*. New York: John Wiley & Sons.

Newman, J. and Thomas-Alyea, K. E. (2004) *Electrochemical Systems*. New York: John-Wiley Publishers.

Noble, R. D. and Stern, S. A. (editors) (1995) *Membrane Separations Technology: Principles and Applications*. Amsterdam: Elsevier Science Publishing Company.

Nouar, C. and Frigaard, I. A. (2001) Nonlinear stability of Poiseuille flow of a Bingham fluid. *J. Non-Newtonian Fluid Mech.*, **100**, 127–149.

Nusselt, W. (1916) Die Oberflachenkondensation des Wasserdampfes. *Z. Vereines Deutsch. Ing.*, **60**, 541–546.

Oksendal, B. (2010) *Stochastic Differential Equations*. New York: Springer-Verlag.

Onsager, L. (1931) Reciprocal relations in irreversible thermodynamics. *Phys. Rev.*, **37**, 405–426.

Orr, W. (1907) The stability or instability of the steady state motion of a perfect and of a viscous liquid. *Proc. R. Irish Acad. Ser. A*, **27**, 9–138.

Orszag, S. A. and Patterson, G. S. (1972) Numerical simulation of three dimensional isotropic turbulence. *Phys. Fluids*, **8**, 3128–3148.

Oseen, C. W. (1910) Über die Stokessche Formel und über eine verwandte Aufgabe in der Hydrodynamik. *Ark. Mat. Astron. Fys.*, **6**(29), 1–20.

Ostrach, S. (1953) An analysis of laminar free-convection flow and heat transfer about a flat plate parallel to the direction of the generating body force. NACA report 1111 (http://naca.central.cranfield.ac.uk/reports/1953/naca-report-1111.pdf).

de Pablo, J. J. (2005) Molecular and multiscale modeling in chemical engineering; current view and future perspectives. *AIChE J.*, **51**, 2372–2376.

Pai, S. I. (1953) On turbulent flow between parallel plates. *J. Appl. Mech.*, **20**, 109–114.

Patankar, S. V. (1980) *Numerical Heat Transfer and Fluid Flow.* Washington DC: Hemisphere.

Peyret, R. and Taylor, T. D. (1983) *Computational Methods for Fluid Flow.* New York: Springer-Verlag.

Pirooz, S., Ramachandran, P. A., and Abraham-Shrauner, B. (1991) Two region computational model for DC glow discharge plasma. *IEEE Trans. Plasma Sci.*, **19**, 408–418.

Poling, R. C., Prausnitz, J. M., and O'Connnell, J. P. (2000) *The Properties of Gases and Liquids.* New York: McGraw-Hill Publishers,

Pope, S. B. and Bruckner, D. (2010) *Turbulent Flows.* Cambridge: Cambridge University Press.

Pozrikidis, C. (1997) *Introduction to Theoretical and Computational Fluid Dynamics.* New York: Oxford University Press.

Proudman, I. and Pearson, J. R. A. (1957) Expansions at small Reynolds numbers for the flow past a sphere and a circular cylinder. *J. Fluid Mech.*, **2**, 237–262.

Pushpavanam, S. (1999) *Mathematical Methods in Chemical Engineering.* New Delhi: Prentice-Hall India Publications.

Ramachandran, P. A. (1992) A numerical-solution method for boundary-value-problems containing an undetermined parameter. *J. Comput. Phys.*, **102**(1), 63–71.

Ramachandran, P. A. (1993) *Boundary Element Methods in Transport Phenomena*, Southampton: Computational Mechanics Publications.

Ramachandran, P. A. and Chaudhari, R. V. (1983) *Three Phase Catalytic Reactors.* New York: Gordon and Breach Science Publications.

Ramachandran, P. A. and Mashelkar, R. A. (1980) Lumped parameter model for haemodialyser with application to simulation of patient–artificial kidney system. *Med. Biol. Eng. Computing*, **18**, 179–188.

Ramachandran, P. A. and Ramaswamy, R. C. (2008) Multiple steady states in distributed parameter systems using boundary elements and arc length continuation. *J. Res. Eng. Technol.*, **5**(3), 255–275.

Ramachandran, P. A. and Sharma, M. M. (1971) Absorption of two gases. *Trans. Inst. Chem. Eng.*, **28**, 253–286.

Ramkrishna, D. and Amundson, N. R. (1985) *Linear Operator Methods in Chemical Engineering.* Englewood Cliffs, NJ: Prentice-Hall.

Ranade, V. V. (2002) *Computational Flow Modeling for Chemical Reactor Engineering.* San Diego, CA: Academic Press.

Rayleigh, Lord (1883) Investigation of the character of equilibuium of a heavy fluid of variable density. *Proc. London Math. Soc.*, **14**, 170–177.

Reid, R. C., Prausnitz, J. M., and Poling B. E. (1987) *Properties of Gases and Liquids.* New York: McGraw Hill-Publishing Co.

Renken, E. M. (1954) Filtration, diffusion and molecular motion in porous cellulosic membranes. *J. Gen. Physiol.*, **38**, 225.

Reynolds, O. (1883) An experimental investigation of the circumstances which determine whether the motion of water shall be direct or sinuous, and of the law of resistance in parallel channels. *Phil. Trans. Roy. Soc.*, **174**, Part III, 935–982.

Richardson, L. F. (1992) *Weather Prediction by Numerical Process.* Cambridge: Cambridge University Press.

Rohlf, K. and Tenti, G. (2001) The role of Womersley number in pulsatile blood flow. *J. Biomech.*, **34**, 141–148.

Rohsenow, W. M. (1952) Method of correlating heat transfer data for surface of boiling liquids. *Trans. ASME*, **74**, 969–976.

Rohsenow, W. M. and Choi, H. Y. (1961) *Heat, Mass and Momentum Transfer*. Englewood Cliffs, NJ: Prentice Hall Inc.

Rubenstein, L. I. (1971) *The Stefan Problem*. Providence, RI: American Mathematical Society.

Ruckenstein, E. (1987) Scaling and physical models in transport phenomena. *Adv. Chem. Eng.*, **13**, 11–32.

Rybczyński, W. (1911) Über die fortschreitende Bewegung einer flüssigen Kugel in einem zähen Medium. *Bull. Acad. Sci. Cracovie* A, 40–46.

Saffman, P. G. (1993) *Vortex Dynamics*. Cambridge: Cambridge University Press.

Sandler, S. I. (2006) *Chemical, Biochemical, and Engineering Thermodynamics*, 4th edn. New York: Wiley.

Schiesser, W. E. (1994) *Computational Mathematics in Engineering and Applied Science: ODEs, DAEs and PDEs*. Boca Raton, FL: CRC Press.

Schlichting, H. and Gersten, K. (2000) *Boundary Layer Theory*. New York: Springer Publishing Co.

Scriven, L. E. (1960) Dynamics of a fluid interface. *Chem. Eng. Sci.*, **12**, 98–108.

Seader, J. D., Henley, E. J., and Roper D. K. (2011) *Separation Process Principles*. Hoboken, NJ: John Wiley.

Shapiro, A. P. and Probstein, R. F. (1993) Electroosmosis for in-situ remediation. *Environ. Sci. Technol.*, **27**, 283–292.

Sharma, M. M. and Danckwerts, P. V. (1963) Fast reactions of CO_2 in alkaline solutions – (a) Carbonate buffers with arsenite, formaldehyde and hypochlorite as catalysts (b) Aqueous monoisopropanolamine (1-amino-2-propanol) solutions. *Chem. Eng. Sci.* **18**, 729–806.

Siegel, R. and Howell, J. R. (1992) *Thermal Radiative Heat Transfer*, 3rd edn. New York: Hemisphere.

Sindo, K. (1996) *Transport Phenomena and Materials Processing*. New York: Wiley-Interscience.

Slattery, J. C., Sagis, L., and Oh, E. S. (2007) *Interfacial Transport Phenomena*, 2nd edn. New York: Springer Publishers.

Smith, J. M., van Ness, H. C. and Abbott, M. M. (2005) *Introduction to Chemical Engineering Thermodynamics*, 7th edn. New York: McGraw-Hill.

Sommerfeld, A. (1908) Ein Betrag zur hydrodynamischen Erklärung der turbulenten Flüssigkeitsbewegungen. *Proceedings of the 4th International Congress of Mathematicians*, Rome, Vol. III, pp. 116–124.

Sparrow, E. M. and Gregg, J. L. (1956) Laminar free convection from a vertical plate with uniform heat flux. *Trans. ASME*, **78**, 435–440.

Srivastava, R., Ramachandran, P. A. and Duduković, M. P. (1986) Radiation view factors in Czochralski crystal-growth apparatus for short crystals. *J. Cryst. Growth*, **74**(2), 281–291.

Staverman, A. J. (1951) The theory of measurement of osmotic pressure. *Rec. Trav. Chim. Pays-Bas*, **70**, 344–352.

Sten-Knudsen, O. (2002) *Biological Membranes: Theory of Transport*. Cambridge: Cambridge University Press.

Stern, S. A. (1994) Polymers for gas separations: The next decade. *J. Membrane Sci.*, **94**, 1–65.

Sternling, C. V. and Scriven, L. E. (1959) Interfacial turbulence: Hydrodynamic instability and Marangoni effect. *AIChE J.*, **5**, 514–523.

Stewart, W. E. and Prober, R. (1964) Linearized model for multicomponent diffusion. *Ind. Eng. Chem. Fundamentals*, **3**, 224–230.

Sutera, S. and Skalak, R. (1993) History of Poiseuille's flow. *Ann. Rev. Fluid Mech.*, **25**, 1–19.

Szekely, J. and Themelis, N. J. (1991) *Rate Phenomena in Process Metallurgy*. New York: Wiley-Interscience.

Taitel, Y. and Dukler, A. E. (1976) A model for predicting the flow regime transition in horizontal and near-horizontal gas–liquid flows. *AIChE J.*, **22**, 47–55.

Tannehill, J. C., Anderson, D. A., and Pletcher, R. H. (1997) *Computational Fluid Mechanics and Heat Transfer*. Philadelphia, PA: Taylor & Francis Publishers.

Taylor, R. and Krishna, R. (1993) *Multicomponent Mass Transfer*. New York, NY: Wiley.

Tennekes, H. and Lumley, J. L. (1972) *First Course in Turbulence*. Cambridge, MA: MIT Press.

Toor, H. L. (1957) Diffusion in three-component gas mixtures. *AIChE J.*, **3**, 198–207.

Trefethen, L. N. (2007) Computing numerically with functions instead of numbers. *Math. Computer Sci.*, **1**, 9–19.

Truskey, G. A., Yuan, F., and Katz, D. F. (2004) *Transport Phenomena Biological Systems*. Upper Saddle River, NJ: Prentice Hall.

van Driest, E. R. (1956) On turbulent flow near a wall. *J. Aero. Sci*, **23**, 1007–1011.

van Dyke, M. (1975) *Pertubation Methods in Fluid Dynamics*. Stanford, CA: Parabolic Press.

Varma, A., Morbidelli, M., and Hua Wu (1999) *Parametric Sensitivity in Chemical Systems*. Cambridge: Cambridge University Press:

Vrentas, J. S. and Vrentas, C. M. (2012) *Diffusion and Mass Transfer*. Boca Raton, FL: CRC Press.

Wang, F. Y., Zhu, Z. H., and Rudolph, V. (2009) Diffusion through ordered force fields in nanopores represented by Smoluchowski equation. *AIChE J.*, **55**, 1225–1337.

Warsi, Z. U. A. (1999), *Fluid Dynamics: Theoretical and Computational Approaches*. Boca Raton, FL: CRC Press.

Welty, J. R., Wicks, C. E., Wilson, R. E., and Rorrer, G. L. (2008). *Fundamentals of Momentum, Heat, and Mass Transport*. New York: John Wiley & Sons.

Wesselingh, J. A. and Krishna, R. (2000) *Mass Transport in Multicomponent Mixtures*. Amsterdam: VSDD Publishers.

Westermeier, R. (2005) *Electrophoresis in Practice*. New York: Wiley-Blackwell.

Whitaker, S. (1972) Forced convection heat transfer correlations for flow in pipes, past flat plates, single cylinders, single spheres, and for flow in packed beds and tube bundles. *AIChE J.*, **18**, 361–371.

Whitaker, S. (1977) *Fundamental Principles of Heat Transfer*. Malabar, FL: Krieger Publishing Co.

White, R. E. and Subramanian, V. (2010) *Computational Methods in Chemical Engineering with MAPLE*. Berlin: Springer-Verlag.

Wilcox, D. C. (1998) *Turbulence Modeling for CFD*. La Canada, CA: DCW Industries.

Wilkes, J. O. (2006) *Fluid Mechanics for Chemical Engineers*. Upper Saddle River, NJ: Prentice Hall.

Womersley, J. R. (1955) Method of calculation of flow and viscous drag in arteries when the pressure gradient is known. *J. Physiol.*, **127**, 553–563.

Won, Y. S. and Mills, A. F. (1982) Correlation of the the effects of viscosity and surface tension on gas absorption in freely falling turbulent films. *Int. J. Heat Mass Transfer*, **25**, 223–229.

Yoshisato, R. A., Korndorf, L. M., Carmichael, G. R., and Datta, R. (1986) Performance analysis of a continuous rotating annular electrophoresis column. *Separation Sci. Technol.*, **21**(8), 727–753.

Zlokarnik, M. (1991) *Dimensionless Analysis and Scaleup in Chemical Engineering*. New York: Springer-Verlag.

Zuber, N. (1958) On the stability of boiling heat transfer. *Trans. ASME*, **80**, 711–720.

INDEX